Ravi P. Gupta

Remote Sensing Geology

Springer

*Berlin
Heidelberg
New York
Hong Kong
London
Milan
Paris
Tokyo*

Ravi P. Gupta

Remote Sensing Geology

Second Edition

With 453 Figures and 53 Tables

 Springer

Dr. Ravi Prakash Gupta

Professor of Earth Resources Technology
Department of Earth Sciences
Indian Institute of Technology Roorkee
(Formerly, University of Roorkee)
Roorkee – 247 667, India

E-mail: rpgesfes@iitr.ernet.in

Cover page:

Doubly plunging folds in rocks of the Delhi Super Group (Landsat MSS4 infrared band image).

Figure 16.24

ISBN 3-540-43185-3 Springer-Verlag Berlin Heidelberg NewYork

Cataloging-in-Publication-Data applied for

Bibliographic information published by Die Deutsche Bibliothek. Die Deutsche Bibliothek lists this publication in the Deutsche Nationalbibliografie; detailed bibliographic data is available in the Internet at <http://dnb.ddb.de>

This work is subject to copyright. All rights are reserved, whether the whole or part of the material is concerned, specifically the rights of translation, reprinting, reuse of illustrations, recitation, broadcasting, reproduction on microfilms or in any other way, and storage in data banks. Duplication of this publication or parts thereof is permitted only under the provisions of the German Copyright Law of September 9, 1965, in its current version, and permission for use must always be obtained from Springer-Verlag. Violations are liable for prosecution under the German Copyright Law.

Springer-Verlag a member of BertelsmannSpringer
Science + Business Media GmbH

http://www.springer.de

© Springer-Verlag Berlin Heidelberg 2003
Printed in Germany

The use of general descriptive names, registered names, trademarks, etc. in this publication does not imply, even in the absence of a specific statement, that such names are exempt from the relevant protective laws and regulations and therefore free for general use.

Product liability: The publishers cannot guarantee the accuracy of any information about dosage and application contained in this book. In every individual case the user must check such information by consulting the relevant literature.

Production: PRO EDIT GmbH, 69126 Heidelberg, Germany
Cover design: Erich Kirchner, Heidelberg
Typesetting: Camera-ready by the author
Printed on acid-free paper   32/3141Re – 5 4 3 2 1 0

To,

M. S. R.

for inspiration, faith and persistence

Preface to the Second Edition

The first edition of this book appeared in 1991, and since then there have been many developments in the field of remote sensing, both in the direction of technology of data acquisition as well as in data processing and applications. This has necessitated a new edition of the book.

The revised edition includes new and updated material on a number of topics – SAR interferometry, hyperspectral sensing, digital imaging cameras, GPS principle, new optical and microwave satellite sensors, and some of the emerging techniques in digital image processing and GIS. Besides, a host of new geological applications of remote sensing are also included.

The book has been thoroughly revised; nevertheless, it retains the original long axis and style, i.e. discuss the basic remote sensing principles, systems of data acquisition, data processing and present the wide ranging geological applications.

The following individuals reviewed parts of the manuscript, suggested improvements and furnished missing links: R. P. Agarwal, M. K. Arora, R. Gens, U. K. Haritashya, K. Hiller, H. Kaufmann, D. King, J. Mathew, F. vander Meer, R. R. Navalgund, S. Nayak, A. Prakash, S. K. Rath, A. K. Saha, A. K. Sen, and A. N. Singh. I am greatly obliged to them for their valuable inputs and suggestions in arriving at the final presentation.

I deeply appreciate the infinite patience and endurance of Sarvesh Kumar Sharma in typing and computer-finishing the manuscript.

Finally, I am indebted to my wife Renu, for her encouragement and support, particularly in times when no end appeared in sight.

Roorkee Ravi P. Gupta
November 2002

Preface to the First Edition

There has been phenomenal growth in the field of remote sensing over the last two to three decades. It has been applied in the fields of geology, mineral exploration, forestry, agriculture, hydrology, soils, land use etc. – that is, in all pursuits of sciences dealing with the features, processes, and phenomena operating at the Earth's surface. The status of geological remote sensing has rapidly advanced and the scientific literature is scattered. The aim of the present book is to systematically discuss the specific requirements of geological remote sensing, to summarize the techniques of remote sensing data collection and interpretation, and to integrate the technique into geo-exploration.

The main conceptual features of the book are:

- To combine various aspects of geological remote sensing, ranging from the laboratory spectra of minerals and rocks to aerial and space-borne remote sensing.
- To integrate photo-geology into remote sensing.
- To promote remote sensing as a tool in integrated geo-exploration.
- To elucidate the wide-spectrum geoscientific applications of remote sensing, ranging from meso to global scale.

The book has been written to satisfy the needs of mainly graduate students and active research workers interested in applied Earth sciences. It is primarily concept oriented rather than system or module oriented.

The organization of the book is detailed in Chapter 1 (Table 1.1). The book has three chief segments: (1) techniques, sensors and interpretation of data in the optical region; (2) techniques, sensors and interpretation of data in the microwave region; and (3) data processing, integration and applications.

The idea for the book germinated as I prepared a course in remote sensing at the University of Roorkee for graduate students, during which extensive lecture notes were made. The book is an outcome of my teaching and research at the University of Roorkee, and partly also at the University of Munich.

A wide-spectrum book in a field like remote sensing, where advancements are taking place at such a fast pace, can hardly be exhaustive and up-to-date. Although every effort has been made to incorporate recent developments, the priority has been on concepts rather than on compilation of data alone (SPOT data examples could not be included because of copyright limitations).

Sincere thanks are due to many individuals and organizations who have contributed in various ways to the book. Particularly, I am grateful to Dr. Rupert Haydn, Managing Director, Gesellschaft fur Angewandte Fernerkundung mbH, Munich, Germany, and formerly at the University of Munich, for supplying numerous illustrations. He kindly provided many images for the book, and offered blanket permission to select illustrations and examples from his wide and precious collection. Dr. Haydn also spent valuable time reviewing parts of the text, offered fruitful criticism and is responsible for many improvements.

Dr. Konrad Hiller, DLR Germany and formerly at the University of Munich, provided what was needed most – inspiration and warm friendly support. Many stimulating discussions with him promoted my understanding of the subject matter and led to numerous reforms. Without Konrad's encouragement, this book may not have seen the light of the day.

I am grateful to a number of people, particularly the following, for going through parts of the manuscript of their interest, suggesting amendments and furnishing several missing links: K. Arnason, R. Chander, R.P.S. Chhonkar, F. Jaskolla, H. Kaufmann, F. Lehmann, G. Philip, A.K. Saraf, K.P. Sharma, V.N. Singh, B.B.S. Singhal, R. Sinha, D.C. Srivastava, U. Terhalle, R.S. Tiwari, L.C. Venkatadhri and P. Volk.

Thanks are also due to Prof. Dr. J. Bodechtel, Institut für Allgemeine und Angewandie Geologie (Institute for General and Applied Geology), University of Munich, for his advice, suggestions and free access to the facilities at Munich. The Alexander von Humboldt Foundation, Bonn, and the Gesellschaft für Angewandte Fernerkundung mbH, Munich (Dr. R. Haydn) kindly provided financial support for my visits and stay in Germany, during which parts of the book were written.

A book on remote sensing has to present many pictures and illustrations. A large number of these were borrowed from colleagues, organizations, instrument manufacturers, commercial firms and publications. These are acknowledged in the captions.

For the excellent production of the book, the credit goes to Dr. W. Engel, Ms. I. Scherich, Ms. G. Hess, Ms. Jean von dem Bussche and Ms. Theodora Krammer of Springer-Verlag, Heidelberg.

Although a number of people have directly and indirectly contributed to the book, I alone am responsible for the statements made herein. It is possible that some oversimplifications appear as erroneous statements. Suggestions from readers will be gratefully accepted.

Finally, I am indebted to my wife Renu for not only patiently enduring 4 years of my preoccupation with the book, but also extending positive support and encouragement.

If this book is able to generate interest in readers for this newly emerging technology, I shall consider my efforts to be amply rewarded.

Roorkee, June 1991 R. P. Gupta

Contents

Chapter 1: Introduction

1.1	Definition and Scope	1
1.2	Development of Remote Sensing	1
1.3	Fundamental Principle	3
1.4	Advantages and Limitations	4
1.5	A Typical Remote Sensing Programme	6
1.6	Field Data (Ground Truth)	9
1.6.1	Timing of Field Data Collection	9
1.6.2	Sampling	10
1.6.3	Types of Field Data	11
1.6.4	GPS Survey	14
1.7	Scope and Organization of this Book	16

Chapter 2: Physical Principles

2.1	The Nature of EM Radiation	19
2.2	Radiation Principles and Sources	20
2.2.1	Radiation Terminology	20
2.2.2	Blackbody Radiation Principles	20
2.2.3	Electromagnetic Spectrum	23
2.2.4	Energy Available for Sensing	24
2.3	Atmospheric Effects	24
2.3.1	Atmospheric Scattering	25
2.3.2	Atmospheric Absorption	26
2.3.3	Atmospheric Emission	28
2.4	Energy Interaction Mechanisms on the Ground	28
2.4.1	Reflection Mechanism	28
2.4.2	Transmission Mechanism	30
2.4.3	Absorption Mechanism	32
2.4.4	Earth's Emission	32

Chapter 3: Spectra of Minerals and Rocks

3.1	Introduction	33
3.2	Basic Arrangements for Laboratory Spectroscopy	34
3.3	Energy States and Transitions – Basic Concepts	36
3.3.1	Electronic Processes	36
3.3.2	Vibrational Processes	39
3.4	Spectral Features of Mineralogical Constituents	39
3.4.1	Visible and Near-Infrared Region (VNIR) (0.4–1.0 µm)	39
3.4.2	SWIR Region (1–3 µm)	39
3.4.3	Thermal-IR Region	42
3.5	Spectra of Minerals	44
3.6	Spectra of Rocks	45
3.6.1	Solar Reflection Region (VNIR + SWIR)	45
3.6.2	Thermal-Infrared Region	48
3.7	Laboratory vs. Field Spectra	49
3.8	Spectra of Other Common Objects	50
3.9	Future	52

Chapter 4: Photography

4.1	Introduction	53
4.1.1	Relative Merits and Limitations	53
4.1.2	Working Principle	54
4.2	Cameras	55
4.2.1	Single-Lens Frame Cameras	56
4.2.2	Panoramic Cameras	58
4.2.3	Strip Cameras	58
4.2.4	Multiband Cameras	58
4.3	Films	59
4.3.1	Black-and-White Films	59
4.3.2	Colour Films	64
4.4	Filters	68
4.5	Film–Filter Combinations for Spectrozonal Photography	69
4.6	Vertical and Oblique Photography	70
4.7	Ground Resolution Distance	71
4.8	Photographic Missions	72
4.8.1	Aerial Photographic Missions	72
4.8.2	Space-borne Photographic Missions	72
4.8.3	Product Media	74

Chapter 5: Multispectral Imaging Systems

5.1	Introduction	75
5.1.1	Working Principle	75
5.2	Factors Affecting Sensor Performance	78
5.2.1	Sensor Resolution	80
5.3	Non-Imaging Radiometers	81
5.3.1	Terminology	81
5.3.2	Working Principle	82
5.4	Imaging Sensors (Scanning Systems)	83
5.4.1	What is an Image?	83
5.4.2	Imaging Tube (Vidicon)	84
5.4.3	Optical-Mechanical Line Scanner (Whiskbroom Scanner)	86
5.4.4	CCD Linear Array Scanner (Pushbroom scanner)	88
5.4.5	Digital cameras (CCD-Area-Arrays)	92
5.5	Space-borne Imaging Sensors	97
5.5.1	Landsat Programme	97
5.5.2	IRS Series	105
5.5.3	SPOT Series	108
5.5.4	MOMS Series	110
5.5.5	JERS-1 (Fuyo-1) OPS	111
5.5.6	CBERS Series	112
5.5.7	RESURS-1 Series	112
5.5.8	ASTER Sensor	113
5.5.9	MTI	114
5.5.10	Space Imaging/Eosat – Ikonos	115
5.5.11	DigitalGlobe – Quickbird	115
5.5.12	Other Programmes (Past)	116
5.5.13	Planned Programmes	119
5.6	Products from Scanner Data	121

Chapter 6: Geometric Aspects of Photographs and Images

6.1	Geometric Distortions	123
6.1.1	Distortions Related to Sensor System	125
6.1.2	Distortions Related to Sensocraft Altitude and Perturbations	128
6.1.3	Distortions Related to the Earth's Shape and Spin	131
6.1.4	Relief Displacement	132
6.2	Stereoscopy	136
6.2.1	Principle	136
6.2.2	Vertical Exaggeration	137
6.2.3	Aerial and Space-borne Configurations for Stereo Coverage	138
6.2.4	Photography vis-à-vis Line-Scanner Imagery for Stereoscopy	140
6.2.5	Instrumentation for Stereo Viewing	140
6.3	Photogrammetry	142

6.3.1	Measurements on Photographs	142
6.3.2	Measurements on Line-Scanner Images	144
6.3.3	Aerial vis-à-vis Satellite Photogrammetry	145
6.4	Transfer of Planimetric Details and Mapping	146

Chapter 7: Image Quality and Principles of Interpretation

7.1	Image Quality	147
7.1.1	Factors Affecting Image Quality	148
7.2	Handling of Photographs and Images	151
7.2.1	Indexing	151
7.2.2	Mosaic	152
7.2.3	Scale Manipulation	153
7.2.4	Stereo Viewing	153
7.2.5	Combining Multispectral Products	153
7.3	Fundamentals of Interpretation	154
7.3.1	Elements of Photo Interpretation	155
7.3.2	Geotechnical Elements	157

Chapter 8: Interpretation of Data in the Solar Reflection Region

8.1	Introduction	161
8.2	Energy Budget Considerations for Sensing in the SOR Region	162
8.2.1	Effect of Attitude of the Sun	162
8.2.2	Effect of Atmospheric Meteorological Conditions	165
8.2.3	Effect of Topographic Slope and Aspect	165
8.2.4	Effect of Sensor Look Angle	167
8.2.5	Effect of Target Reflectance	168
8.3	Acquisition and Processing of Solar Reflection Image Data	168
8.4	Interpretation	169
8.4.1	Interpretation of Panchromatic Black-and-White Products	169
8.4.2	Interpretation of Multispectral Products	174
8.4.3	Interpretation of Colour Products	177
8.5	Luminex Method	180
8.6	Scope for Geological Applications	180

Chapter 9: Interpretation of Data in the Thermal-Infrared Region

9.1	Introduction	183
9.2	Earth's Radiant Energy – Basic Considerations	184
9.2.1	Surface (Kinetic) Temperature	185
9.2.2	Emissivity	190
9.3	Broad-Band Thermal-IR Sensing	190

9.3.1	Radiant Temperature and Kinetic Temperature	191
9.3.2	Acquisition of Broad-Band Thermal-IR Data	192
9.3.3	Processing of Broad-Band TIR Images	194
9.3.4	Interpretation of Thermal-IR Imagery	195
9.3.5	Thermal Inertia mapping	198
9.3.6	Scope for Geological Applications – Broad-Band Thermal Sensing	201
9.4	Temperature Estimation	206
9.4.1	Use of Landsat TM Data for Temperature Estimation	206
9.4.2	Use of Landsat-7 ETM+ Data for Temperature Estimation	209
9.5	Thermal-IR Multispectral Sensing	210
9.5.1	Multispectral Sensors in the TIR	211
9.5.2	Data Correction and Enhancement	213
9.5.3	Applications	214
9.6	LIDAR Sensing	215
9.6.1	Working Principle	215
9.6.2	Scope for Geological Applications	216
9.7	Future	216

Chapter 10: Digital Image Processing of Multispectral Data

10.1	Introduction	217
10.1.1	What is Digital Imagery?	217
10.1.2	Sources of Multispectral Image Data	219
10.1.3	Storage and Supply of Digital Image Data	220
10.1.4	Image Processing Systems	220
10.1.5	Techniques of Digital Image Processing	222
10.2	Radiometric Image Correction	224
10.2.1	Correction for Atmospheric Contribution	224
10.2.2	Correction for Solar Illumination Variation	226
10.2.3	Correction for Topographic Effects	226
10.2.4	Sensor Calibration	228
10.2.5	De-striping	229
10.2.6	Correction for Periodic and Spike Noise	231
10.3	Geometric Corrections	232
10.3.1	Correction for Panoramic Distortion	232
10.3.2	Correction for Skewing Due to Earth's Rotation	232
10.3.3	Correction for Aspect Ratio Distortion	233
10.4	Registration	233
10.4.1	Definition and Importance	233
10.4.2	Principle	234
10.4.3	Procedure	235
10.5	Image Enhancement	238
10.6	Image Filtering	242
10.6.1	High-Pass Filtering (Edge Enhancement)	243

10.6.2	Image Smoothing	248
10.6.3	Fourier Filtering	248
10.7	Image Transformation	250
10.7.1	Addition and Subtraction	253
10.7.2	Principal Component Transformation	255
10.7.3	Decorrelation Stretching	258
10.7.4	Ratioing	258
10.8	Colour Enhancement	262
10.8.1	Advantages	262
10.8.2	Pseudocolour Display	263
10.8.3	Colour Display of Multiple Images – Guidelines for Image Selection	263
10.8.4	Colour Models	264
10.9	Image Fusion	267
10.9.1	Introduction	267
10.9.2	Techniques of Image Fusion	267
10.10	2.5-Dimensional Visualization	270
10.10.1	Shaded Relief Model (SRM)	271
10.10.2	Synthetic Stereo	271
10.10.3	Perspective View	272
10.11	Image Segmentation	274
10.12	Digital Image Classification	274
10.12.1	Supervised Classification	276
10.12.2	Unsupervised Classification	281
10.12.3	Fuzzy Classification	282
10.12.4	Linear Mixture Modelling (LMM)	283
10.12.5	Artificial Neural Network Classification	283
10.12.6	Classification Accuracy Assessment	284

Chapter 11: Hyperspectral Sensing

11.1	Introduction	287
11.2	Spectral Considerations	289
11.2.1	Processes Leading to Spectral Features	289
11.2.2	Continuum and Absorption Depth – Terminology	290
11.2.3	High-Resolution Spectral Features – Laboratory Data	291
11.2.4	Mixtures	294
11.2.5	Spectral Libraries	296
11.3	Hyperspectral Sensors	296
11.3.1	Working Principle of Imaging Spectrometers	297
11.3.2	Sensor Specification Characteristics	299
11.3.3	Airborne Hyperspectral Sensors	300
11.3.4	Space-borne Hyperspectral Sensors	300
11.4	Processing of Hyperspectral Data	302
11.4.1	Pre-processing	302

11.4.2	Radiance-to-Reflectance Transformation	304
11.4.3	Data Analysis for Feature Mapping	307
11.5	Applications	311

Chapter 12: Microwave Sensors

12.1	Introduction	317
12.2	Passive Microwave Sensors and Radiometry	317
12.2.1	Principle	317
12.2.2	Measurement and Interpretation	318
12.3	Active Microwave Sensors – Imaging Radars	320
12.3.1	What is a Radar?	320
12.3.2	Side-Looking Airborne Radar (SLAR) Configuration	321
12.3.3	Spatial Positioning and Ground Resolution from SLAR	325
12.3.4	SLAR System Specifications	328
12.3.5	Aerial and Space-borne SLAR Sensors	329

Chapter 13: Interpretation of SLAR Imagery

13.1	Introduction	337
13.2	SLAR Image Characteristics	337
13.2.1	Radiometric Characteristics	337
13.2.2	Geometric Characteristics	342
13.3	SLAR Stereoscopy and Radargrammetry	345
13.4	Radar Return	346
13.4.1	Radar Equation	346
13.4.2	Radar System Factors	347
13.4.3	Terrain Factors	350
13.5	Processing of SLAR Image Data	355
13.6	Polarimetry	357
13.7	Field Data (Ground Truth)	358
13.7.1	Corner Reflectors (CRs)	359
13.7.2	Scatterometers	359
13.8	Interpretation and Scope for Geological Applications	359

Chapter 14: SAR Interferometry

14.1	Introduction	367
14.2	Principle of SAR Interferometry	367
14.3	Configurations of Data Acquisition for InSAR	369
14.4	Baseline	372
14.5	Airborne and Space-borne InSAR Systems	373
14.5.1	Airborne Systems	373

14.5.2	Space-borne Systems	374
14.5.3	Ground Truth and Corner Reflectors	376
14.6	Methodology of Data Processing	377
14.7	Differential SAR Interferometry (DInSAR)	381
14.8	Factors Affecting SAR Interferometry	382
14.9	Applications	383
14.10	Future	392

Chapter 15: Integrating Remote Sensing Data with Other Geodata (GIS Approach)

15.1	Integrated Multidisciplinary Geo-investigations	393
15.1.1	Introduction	393
15.1.2	Scope of the Present Discussion	395
15.2	Geographic Information System (GIS) – Basics	395
15.2.1	What is GIS?	395
15.2.2	GIS Data Base	397
15.2.3	Continuous vs. Categorical Data	398
15.2.4	Basic Data Structures in GIS	399
15.2.5	Main Segments of GIS	400
15.3	Data Acquisition (Sources of Geo-data in a GIS)	400
15.3.1	Remote Sensing Data	400
15.3.2	Geophysical Data	400
15.3.3	Gamma Radiation Data	403
15.3.4	Geochemical Data	404
15.3.5	Geological Data	404
15.3.6	Topographical Data	404
15.3.7	Other Thematic Data	405
15.4	Pre-processing	405
15.5	Data Management	413
15.6	Data Manipulation and Analysis	413
15.6.1	Image Processing Operations	413
15.6.2	Classification	416
15.6.3	GIS Analysis	420
15.7	Applications	424

Chapter 16: Geological Applications

16.1	Introduction	429
16.2	Geomorphology	431
16.2.1	Tectonic Landforms	433
16.2.2	Volcanic Landforms	434
16.2.3	Fluvial Landforms	435
16.2.4	Coastal and Deltaic Landforms	441

16.2.5	Aeolian Landforms	442
16.2.6	Glacial Landforms	444
16.3	Structure	445
16.3.1	Bedding and Simple-Dipping Strata	448
16.3.2	Folds	450
16.3.3	Faults	456
16.3.4	Neovolcanic Rift Zone	460
16.3.5	Lineaments	460
16.3.6	Circular Features	477
16.3.7	Intrusives	480
16.3.8	Unconformity	480
16.4	Lithology	481
16.4.1	Mapping of Broad-Scale Lithologic Units – General	481
16.4.2	Sedimentary Rocks	482
16.4.3	Igneous Rocks	486
16.4.4	Metamorphic Rocks	490
16.4.5	Identification of Mineral Assemblages	493
16.5	Stratigraphy	497
16.6	Mineral Exploration	498
16.6.1	Remote Sensing in Mineral Exploration	498
16.6.2	Main Types of Mineral Deposits and their Surface Indications	501
16.6.3	Stratigraphical–Lithological Guides	502
16.6.4	Geomorphological Guides	502
16.6.5	Structural Guides	503
16.6.6	Guides Formed by Rock Alteration	505
16.6.7	Geobotanical Guides	515
16.7	Hydrocarbon Exploration	519
16.8	Groundwater Investigations	523
16.8.1	Factors Affecting Groundwater Occurrence	524
16.8.2	Indicators for Groundwater on Remote Sensing Images	526
16.8.3	Application Examples	526
16.9	Engineering Geological Investigations	536
16.9.1	River Valley Projects – Dams and Reservoirs	536
16.9.2	Landslides	539
16.9.3	Route Location (Highways and Railroads) and Canal and Pipeline Alignments	542
16.10	Neotectonism, Seismic Hazard and Damage Assessment	542
16.10.1	Neotectonism	543
16.10.2	Local Ground Conditions	551
16.10.3	Disaster Assessment	555
16.11	Volcanic and Geothermal Energy Applications	555
16.11.1	Volcano Mapping and Monitoring	555
16.11.2	Geothermal Energy	560
16.12	Coal Fires	563

16.13	Environmental Applications	571
16.13.1	Vegetation	572
16.13.2	Land Use	573
16.13.3	Soil Erosion	575
16.13.4	Oil Spills	576
16.13.5	Smoke from Oil Well Fires	581
16.13.6	Atmospheric Pollution	583
16.14	Future	583

Appendices 585

References 593

Illustrations – Location Index 623

Subject Index 627

Chapter 1: Introduction

1.1 Definition and Scope

Remote sensing, in the simplest words, means obtaining information about an object without touching the object itself. It has two facets: the technology of acquiring data through a device which is located at a distance from the object, and analysis of the data for interpreting the physical attributes of the object, both these aspects being intimately linked with each other.

Taking the above definition literally, various techniques of data collection where sensor and object are not in contact with each other could be classed as remote sensing, e.g. looking across a window or reading a wall-poster, as also many standard geophysical exploration techniques (aeromagnetic, electromagnetic induction, etc.), and a host of other methods. Conventionally, however, the term remote sensing has come to indicate that the sensor and the sensed object are located quite remotely apart, the distance between the two being of the order of several kilometres or hundreds of kilometres. In such a situation, the intervening space is filled with air (aerial platform) or even partly vacuum (space platform), and only the electromagnetic waves are able to serve as an efficient link between the sensor and object.

Remote sensing has, therefore, practically come to imply data acquisition of electromagnetic radiation (commonly between the 0.4 µm and 30 cm wavelength range) from sensors flying on aerial or space platforms, and its interpretation for deciphering ground object characteristics.

1.2 Development of Remote Sensing

Remote sensing has evolved primarily from the techniques of aerial photography and photo interpretation. It is a relatively young scientific discipline, and is an area of emerging technology that has undergone phenomenal growth during the last nearly three decades. It has dramatically enhanced man's capability for resource exploration and mapping, and monitoring of the Earth's environment on local and global scales (Fischer 1975; Williams and Southworth 1984).

A major landmark in the history of remote sensing was the decision to land man on the moon. As a sequel to this, the space race between the US and the erstwhile USSR began, which led to rapid development of space systems. The US National Aeronautics and Space Administration (NASA) has led the development of many aerial and spaceborne programmes, which have provided remote sensing data world-wide. In addition, the European Space Agency and national space agencies of a number of countries, viz. Canada, Japan, India, China, Brazil and Russia have also developed plans for remote sensing systems. All these missions have provided a stimulus to the technology and yielded valuable data and pictures of the Earth from space.

The first space photography of the Earth was transmitted by Explorer-6 in 1959. This was followed by the Mercury Program (1960), which provided orbital photography (70-mm colour) from an unmanned automatic camera. The Gemini mission (1965) provided a number of good-quality stereo vertical and oblique photographs, which formally demonstrated the potential of remote sensing techniques in Earth resources exploration (Lowman 1969). Later, the experiments in the Apollo Program included Earth coverage by stereo vertical photography and multispectral 70-mm photography. This series of photographic experiments finally paved the way for unmanned space orbital sensors.

Meanwhile, sensors for Earth observations had already been developed for meteorological purposes (TIROS-I, ITOS and NOAA series) and were in orbit in the early 1960s. The payload of the weather satellite NOAA was modified for inclusion in the first Earth Resources Technology Satellite (ERTS-1).

With the launching of ERTS-1, in 1972 (later renamed Landsat-1), began a new era in the history of remote sensing of the Earth. ERTS-1 carried a four-channel multispectral scanning system (MSS) and a tape recorder on-board, which provided extremely valuable data on world-wide distribution. In fact, the period from the 1970s to the early 1980s, with the availability of MSS image data in multispectral bands with good geometric and radiometric quality, world-wide coverage and low costs can certainly be called a milestone in the history of remote sensing.

At the same time, valuable data were accumulating on the spectral behaviour of the atmosphere, as well as on spectral signatures of minerals, rock materials and vegetation. Based on this knowledge, a new sensor called the Thematic Mapper (TM), was developed and launched in 1982 (Landsat-4). The Landsat-ETM+ (Enhanced Thematic Mapper Plus) presently in orbit is a modified version of Landsat-TM. The TM/ ETM+ type satellite sensor data with good spatial resolution and appropriately selected spectral channels has been extensively used in remote sensing world-wide since 1982.

Concurrently, developments in the space transportation system took place, and reusable space shuttles came on the scene. Many short-term lower-altitude space experiments have been conducted from the space shuttle. The space shuttle has provided an easy launching, on-board modular approach, and facilitated the trial of sensors from an orbital platform. The most important of these experiments have been the Metric Camera, Large Format Camera, electronic scanner MOMS, and the Shuttle Imaging Radar series (SIR-A, -B, -C).

Developments in electronic technology led to the design of solid-state scanners, the first of these being the German space mission MOMS-1, flown on the space shuttle. Subsequently, many sensors utilizing this technology have been placed in orbit on free-flying platforms. Examples include: the French SPOT series, the Indian IRS series, the Japanese MOS and Fuyo, and the China–Brazil series.

The development of CCD-area array technology (at the beginning of the 1980s) has been another important step. Subsequent developments led to the design and fabrication of area-array chips with a large number of detector cells. This technology forms the heart of most modern high-spatial-resolution remote sensing systems (e.g. IKONOS-2, QuickBird-2 in orbit and several planned missions).

Another important development has been the hyperspectral imaging sensor which provide images in several hundred channels. These image data have utility in lithologic identification and possibly quantification of mineral content.

The use of Side-Looking Airborne Radar (SLAR) imaging techniques in the 1960s and early 1970's from aerial platforms indicated their great potential for natural resources mapping and micro relief discrimination. Seasat (1978) was the first free-flying space sensor which provided SLAR imagery. Subsequently, a series of shuttle imaging radar experiments (SIR-A, -B, -C) were flown to understand the radar response in varying modes like multi-frequency, multi-polarization and multi-look configurations. Further, the Japanese Fuyo, Canada's Radarsat and ESA's ERS-1/2 programmes provided SAR data of the Earth from space.

The interferometric synthetic aperture radar (SAR) data processing has been another great step in remote sensing application, allowing monitoring of ground terrain from space, with accuracy in the range of centimetres.

In addition, during the past few years, micro-electronics and computer technology have been revolutionized. Image-processing facilities, which were earlier restricted to selected major research establishments, have now become widely available, and form almost a house-to-house facility. This has also been responsible for greater dissemination of remote sensing/image processing knowledge and interest in such studies.

1.3 Fundamental Principle

The basic principle involved in remote sensing methods is that in different wavelength ranges of the electromagnetic spectrum, each type of object reflects or emits a certain intensity of light, which is dependent upon the physical or compositional attributes of the object (Fig. 1.1). Figure 1.2 shows a set of multispectral images in green, red and near-infrared bands of the same area, indicating that various features may appear differently in different spectral bands. Thus, using information from one or more wavelength ranges, it may be possible to differentiate between different types of objects (e.g. dry soil, wet soil, vegetation, etc.), and map their distribution on the ground.

The curves showing the intensity of light emitted or reflected by the objects at different wavelengths, called *spectral response curves,* constitute the basic infor-

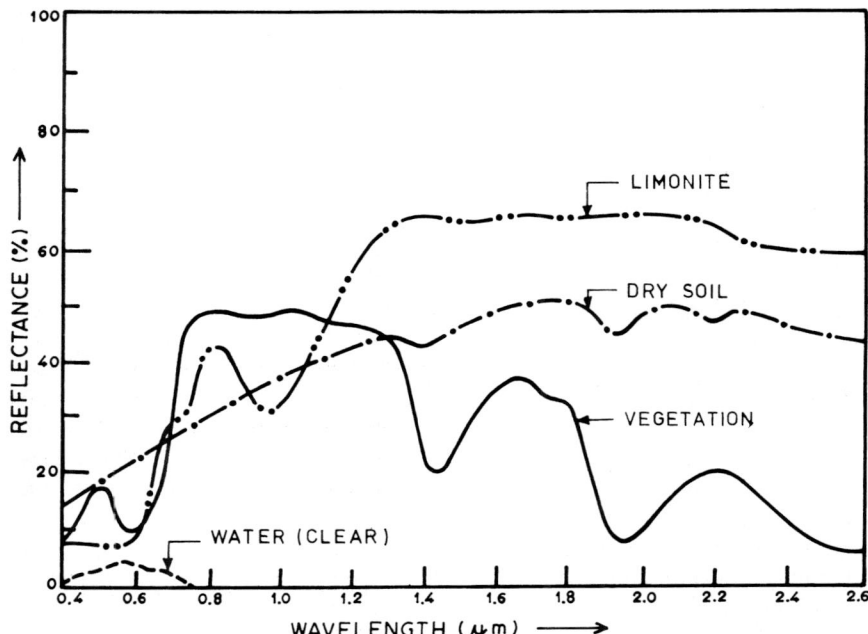

Fig. 1.1. Typical spectral reflectance curves for selected common natural objects – water, vegetation, soil and limonite

mation required for successful planning of a remote sensing mission. The remote sensing data acquired from aerial/spaceborne sensors are processed, enhanced and interpreted for applications.

1.4 Advantages and Limitations

The major advantages of remote sensing techniques over methods of ground investigations are due to the following:

1. Synoptic overview: Remote sensing permits the study of various spatial features in relation to each other, and delineation of regional features/trends/phenomena (Fig. 1.3).

2. Feasibility aspect: As some areas may not be accessible to ground survey, the only feasible way to obtain information about such areas may be from remote sensing platforms.

3. Time saving: The technique saves time and manpower, as information about a large area is gathered quickly.

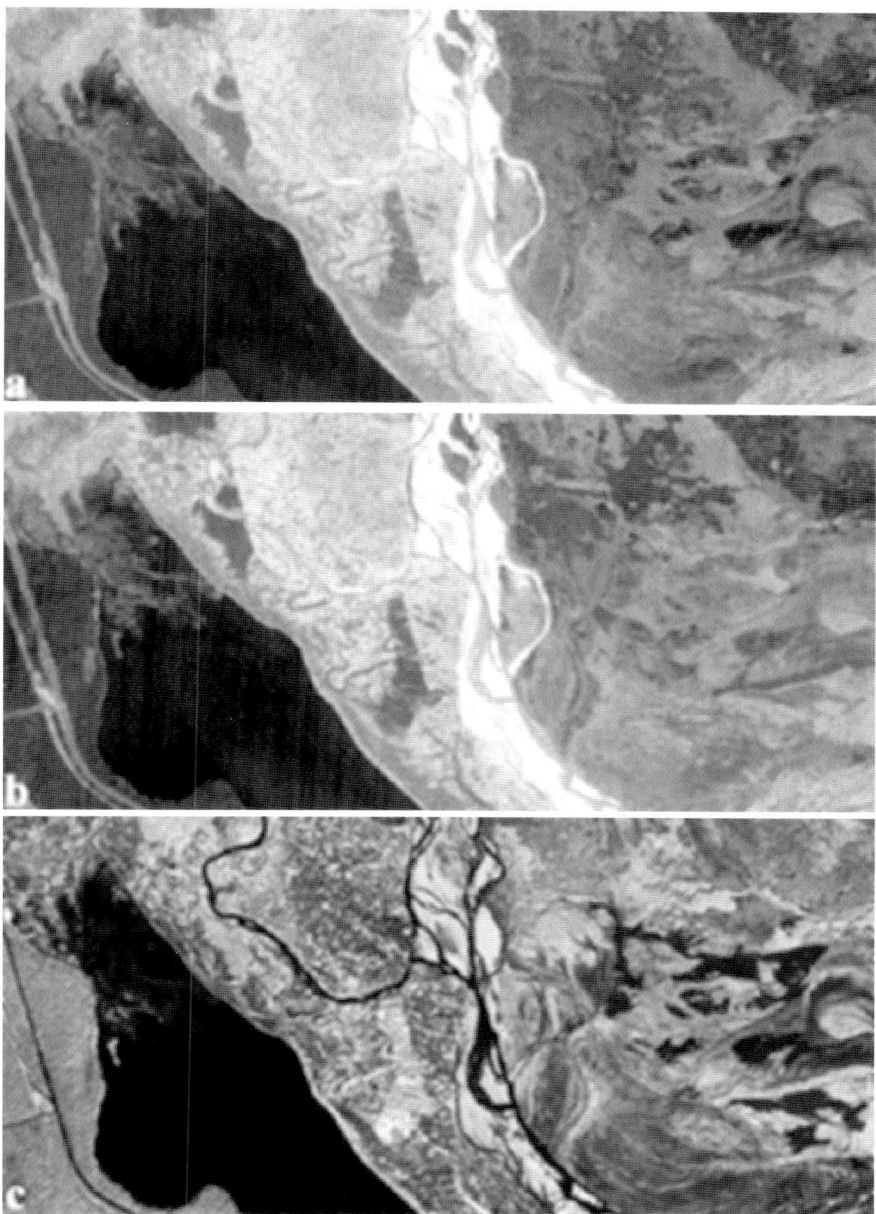

Fig. 1.2a–c. Multispectral images in **a** green, **b** red and **c** near-infrared bands of the same area. Note differences in spectral characters of various objects in the three spectral bands. (IRS-LISS-III sensor images of a part of the Gangetic plains)

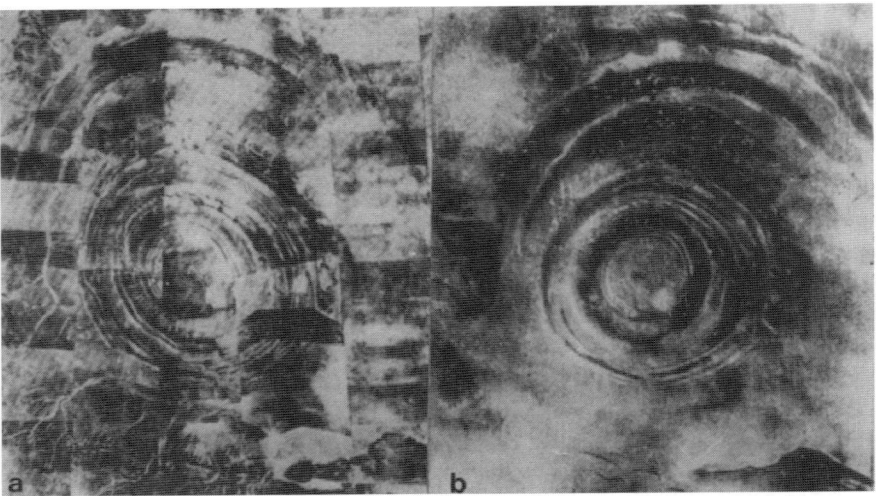

Fig. 1.3a,b. One of the chief advantages of remote sensing lies in providing a synoptic overview – an altogether different scale of observation, which may give new insights into the problem. This illustration shows the Richat structure in Mauritania. **a** Air-photo mosaic, and **b** satellite image. (Beregovoi et al. in Kats et al. 1976)

4. Multidisciplinary applications: The same remote sensing data can be used by researchers/workers in different disciplines, such as geology, forestry, land use, agriculture, hydrology etc., and therefore the overall benefit-to-cost ratio is higher.

There are additional specific advantages associated with individual sensors, namely: the photographic systems are marked by analogy to the eye system and have high geometric fidelity; scanners provide remote sensing data such that the digital information is directly telemetered from space to ground; and imaging radars possess the unique advantages of all-weather and all-time capability. Such specific advantages are highlighted for various sensor types, at appropriate places.

1.5 A Typical Remote Sensing Programme

A generalized schematic of energy/data flow in a typical remote sensing system is shown in figure 1.4. Most remote sensing programmes utilize the sun's energy, which is the predominant source of energy at the Earth's surface. In addition, some remote sensors also utilize the blackbody radiation emitted by the Earth. Also, active sensors such as radars and lasers illuminate the Earth from artificially generated energy. The electromagnetic radiation travelling through the atmosphere is selectively scattered and absorbed, depending upon the composition of the atmosphere and the wavelength involved.

A Typical Remote Sensing Programme

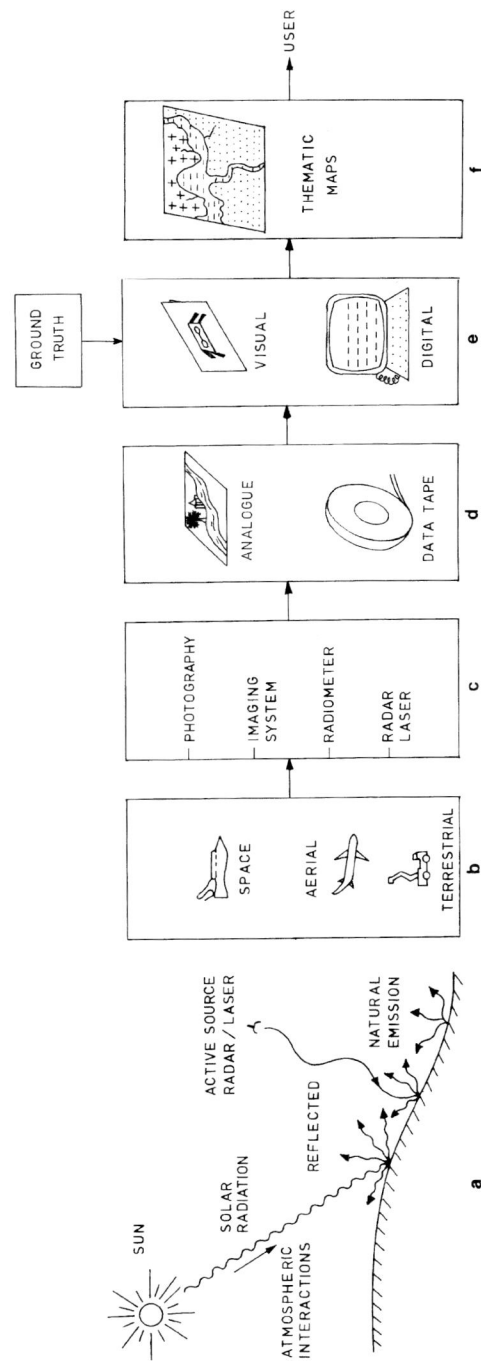

Fig. 1.4a–f. Scheme of a typical remote sensing programme. **a** Sources of radiation and interaction. **b** Platforms. **c** Sensors. **d** Data products. **e** Interpretation and analysis. **f** Output. (After Lillesand and Kiefer 1987)

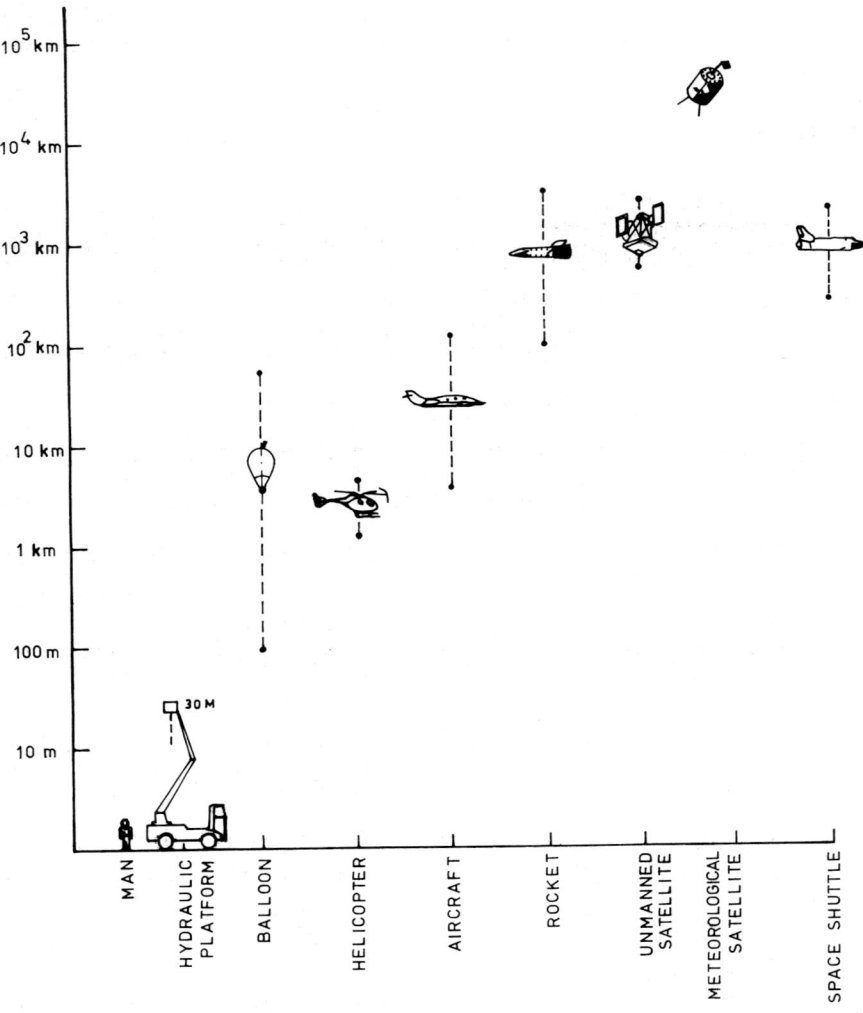

Fig.1.5. Remote sensing platforms for earth resources investigations. (After Barzegar 1983)

Sensors such as photographic cameras, scanners or radiometers mounted on suitable platforms record the radiation intensities in various spectral channels. The platforms for remote sensing data acquisition could be of various types: aerial (balloons, helicopters and aircraft) and space-borne (rockets, manned and unmanned satellites) (Fig. 1.5). Terrestrial platforms are used to generate ground truth data. The remotely sensed data are digitally processed for rectification and enhancement, and integrated with 'ground truth' and other reference data. The processed products are interpreted for identification/discrimination of ground objects. Thematic maps may be integrated with other multidisciplinary spatial data and ground truth data and used for decision making by scientists and managers.

1.6 Field Data (Ground Truth)

Ground truth implies reference field data collected to control and help remote sensing image interpretation. In the early days of remote sensing research, ground investigations were used to verify the results of remote sensing interpretation, e.g. the soil type, condition of agricultural crops, distribution of diseased trees, water ponds etc. Hence the term *ground truth* came into vogue. The same term (ground truth) is still widely applied in remote sensing literature, although now somewhat erroneously, as the reference data may now be obtained from diverse sources, not necessarily involving ground investigations.

The main purposes of field data collection are the following:
a. To calibrate a remote sensor
b. To help in remote sensing data correction, analysis and interpretation
c. To verify the thematic maps generated from remote sensing.

The parameters/physical properties of interest are different in various parts of the electromagnetic (EM) spectrum, from visible, near-IR, thermal-IR to the microwave region (for terminology, see Sect. 2.2.3). Table 1.1 gives a brief overview.

There are four main considerations while planning the ground truth part of a remote sensing project:
1. Timing of ground truth data collection
2. Sampling
3. Types of field data
4. GPS survey.

1.6.1 Timing of Field Data Collection

Ground data can be collected before, during or after the acquisition of remote sensing data. The field data may comprise two types of parameters: (a) intrinsic and (b) time variant. An *intrinsic parameter* is a time-stable parameter that could be measured any time, e.g. albedo, spectral emissivity, rock type, structure etc. A *time-variant* (or time-critical) parameter varies with time and must be measured during the remote sensing overpass, e.g. temperature, rain, condition of crop etc. Generally, data on meteorological conditions are collected for about one week before and during the remote sensing overpass; this is particularly important for thermal-IR surveys.

Table 1.1. Main physical properties for study during field data collection

A. General
Topography, slope and aspect
Atmospheric–meteorological conditions – cloud, wind, rain etc.
Solar illumination – sun azimuth, elevation

B. Solar Reflection Region (Visible-Near Infrared)
Spectral reflectance
Sun–object–sensor angle
Bidirectional reflectance distribution function
Surface coatings, leachings, encrustations etc.
Soil - texture, moisture, humus, mineralogy etc.
Rock type, structure
Vegetation characteristics, land use/land cover types, distribution etc.

C. Thermal-Infrared Region
Ground temperature
Emissivity
Soil–texture, moisture, humus, mineralogy etc.
Vegetation characteristics, land use/land cover types, distribution etc.
Rock type, mineralogy, structure

D. Microwave Region
Microwave roughness (surface and sub-surface)
Volume scattering and complex dielectric constant
Rainfall pattern

1.6.2 Sampling

Different methods of sampling ground truth may be adopted depending upon the time and resources available (Townshend 1981). The most commonly used method of sampling is *purposive sampling*. In this method, observations are made in linear traverses in such frequency and intensity as seems appropriate to the field worker. It utilizes the skills and local knowledge of the field worker. The method is time and cost effective. It is well suited to making point observations and interpretations, and interpreting any anomalous features observed on remote sensing images. However, the drawback is the difficulty in statistically extrapolating results and deducing quantitative results for the whole study area.

Other methods include probability sampling, random sampling, systematic sampling etc., which are more time and cost consuming (for details, see Townshend 1981).

Field Data (Ground Truth) 11

Fig. 1.6. Field spectroradiometer. (Courtesy of Geophysical & Environmental Research)

6.3 Types of Field Data

The ground data may be considered to be of two main types: (1) thematic maps and (2) spectral data. These may be derived from a variety of sources, such as: (a) dedicated field measurements/surveys; (b) aerial photographic interpretation; and (c) library records/reports.

1. Thematic Maps: show distribution of features which may be of interest for a particular remote sensing project, e.g. landforms, drainage, distribution of agricultural crops, water bodies, lithological boundaries, structure. In addition, field data may also involve maps exhibiting special features, such as landslides, or suspended silt distribution in an estuary, or isotherms on a water body etc. Such thematic maps may be derived from aerial photographic interpretation, existing records/reports, or generated through dedicated field surveys.

2. Spectral data: are generally not available in the existing reports/records and have almost invariably to be specifically collected. The instruments could be field-portable, or may be mounted on a hydraulic platform or used on an aerial plat-

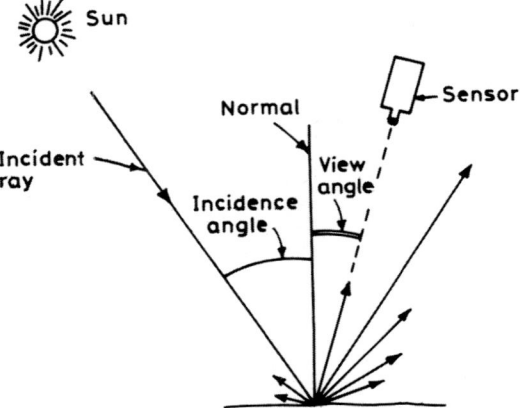

Fig. 1.7. Geometry of bidirectional arrangement in solar reflection region

form. Instrumentation for generating field spectral data is different in various parts of the EM spectrum. However, they all have in common two components: an optical system to collect the radiation, and a detector system to convert radiation intensity into electrical signal. Usually, a PC notebook is integrated to record the data, which provides flexibility in data storage and display.

In the solar reflection region (visible, near-IR, SWIR), two types of instrument are used: (a) multichannel radiometers, and (b) spectro radiometers. *Multichannel radiometers* generate spectral data in selected bands of wavelength ranges, which commonly correspond to satellite sensor bands (such as the Landsat TM, IRS-LISS, SPOT-HRV etc.). Some instruments permit calculation of band ratios and other computed parameters in the field, in real time. Data from multichannel field radiometers help in interpreting the satellite sensor data.

Spectroradiometers are used to generate spectral response curves, commonly in the wavelength range of 0.4–2.5 μm (Fig. 1.6). The system acquires a continuous spectrum by recording data in more than 1000 narrow spectral bands. Further, matching spectra to a library of previously recorded/stored spectra is also possible in some instruments.

In the solar reflection region, we have a typical bidirectional arrangement in which illumination is from one direction and observation from another (Fig. 1.7). Taking a general case, sunlight incident on an object is scattered in various directions, depending upon the reflection characteristics of the material. The intensity of light in a particular viewing direction depends upon the angle of incidence, angle of view and the reflection characteristics. Therefore, the sun–target–sensor goniometric geometry has a profound influence on the radiance reaching the sensor. For the same ground surface, the radiance reaching the sensor may be different depending upon the angular relations. This property is given in terms of the *bidirectional reflectance distribution function* (BRDF) of the surface, which mathematically relates reflectance for all combinations of illumination and viewing an

Field Data (Ground Truth)

Fig. 1.8. Goniometer for measuring the BRDF; the device is integrated with the GER3700 field spectroradiometer shown in figure 1.6. (Courtesy of NASA)

gles at a particular wavelength (e.g. Silva 1978). For field measurements of the BRDF, a goniometer is used (Fig. 1.8). This consists of a semi-circular arch on which a spectroradiometer is mounted. The spectroradiometer is moved over the arch from one side to the other to view the same ground area from different angles. In this way, reflectance data can be gathered to describe the BRDF for the surface.

In the thermal-IR region, field surveys are carried out to measure two ground parameters: (a) ground temperature, and (b) spectral emissivity. Suitable *thermometers* (0.1°C least count) are commonly used for measuring ground temperatures. Temperature measurements are made for several days at a point, or along traverse lines at the same time, as per the requirement. Measurement of *spectral emissivity* in the field is done through a reflection arrangement. Spectral reflectance (R_λ) over narrow spectral ranges is measured in the field and spectral emissivity (ε_λ) is computed using Kirchoff's Law ($\varepsilon_\lambda = 1-R_\lambda$) (for details, see Sect.

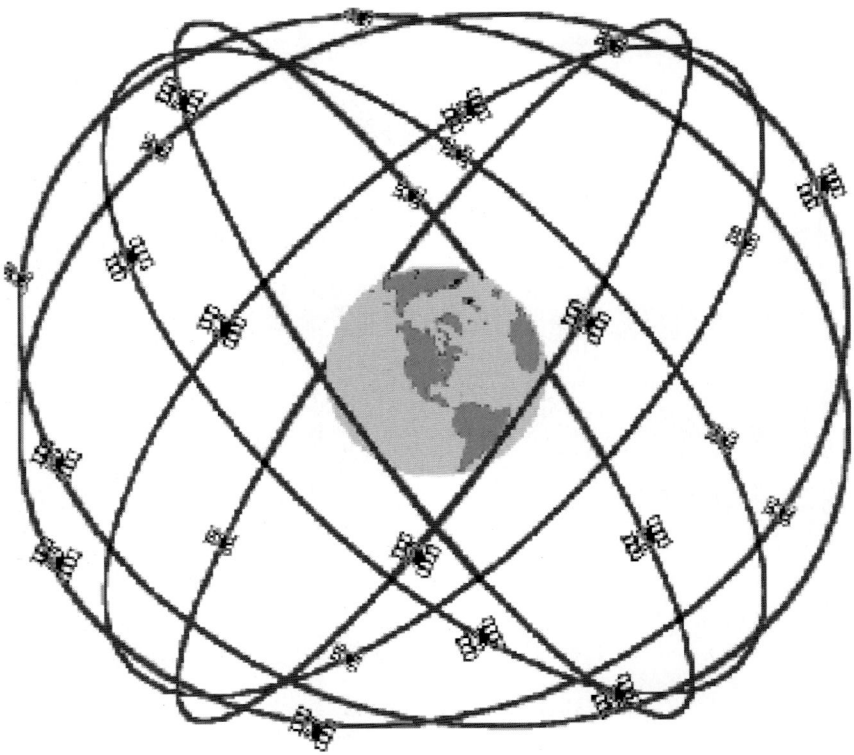

Fig. 1.9. Satellites orbitting around the Earth forming the Global Positioning System. (Courtesy of P.H. Dana)

2.2.2). In addition, it is also important to monitor the heat energy budget, for which a host of parameters are measured (see Sect. 9.3.2.3).

In the microwave region (SAR sensors) the main parameter of interest is the back-scattering coefficient. *Scatterometers* which can be operated at variable wavelengths and incidence angles are used to gather the requisite field data (also see Sect. 13.7.2).

1.6.4 GPS Survey

Global Positioning System (GPS) devices are used to exactly locate the position of field observations. The GPS is a satellite-based navigation system, originally developed by the US Department of Defense. It includes a group of nominally 24 satellites (space vehicles, SV), that orbit the Earth at an altitude of about 20,200 km, with an orbital period of 12 hours. There are six orbital planes (with four satellites in each), equally spaced 60° apart along the equator and inclined at 55°

Fig. 1.10. A GPS receiver. (Courtesy of Garmin Corp.)

from the equatorial plane. The orbits are nearly circular, highly stable and precisely known. In all, there are often more than 24 operational satellites at any point in time (as new ones are launched to replace the older satellites) (Fig. 1.9).

The satellites transmit time-coded radio signals that are recorded by ground-based GPS receivers (Fig. 1.10). At any given time, at least four and generally five to eight SVs are visible from any point on the Earth (except in deep mountain gorges) (Fig. 1.11). The method of satellite ranging is used to compute distances and locate the position. Four GPS satellite signals are used to compute the position of any point in three dimensions on the Earth's surface. Basically, this works on the principle of measuring the time for a signal to travel between the SV and the GPS receiver. With the speed of light known (3×10^8 m s^{-1}), the distance can be calculated.

There may be errors due to the clock (errors in clock synchronization, called clock bias), uncertainties in the satellite orbit, errors due to atmospheric conditions (influencing the travel of EM radiation through the atmosphere), GPS receiver errors etc. The exact ephemeris (orbital) data and SV clock corrections are monitored for each SV and also recorded by the ground GPS receiver. This corrects for the corresponding errors (for more details on GPS, refer to Kaplon 1996; Leick 1995).

Differential GPS method aims at eliminating the above errors and providing refined estimates of differential or relative distances on the ground. It involves one stationary or base GPS receiver and one or more roving (moving) receivers. Simultaneous signals from SVs are recorded at the base and rover GPS receivers. As positional errors are similar in base and rover GPS receivers, it is possible to obtain highly refined differential estimates of distances.

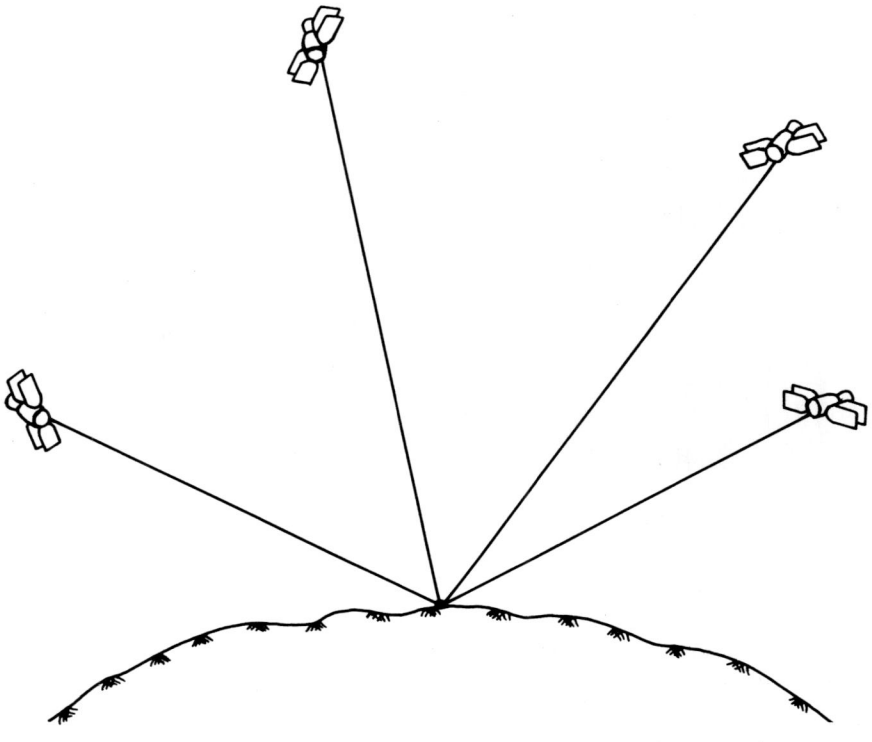

Fig. 1.11. Signals from four GPS satellites being received at a field site

Accuracies using GPS depend upon the GPS receiver and data processing, both of which are governed by project costs. Some estimates are as follows:
- Low cost, single receiver: 10–30m
- Medium cost, differential receiver: 50 cm–5 m
- High cost, differential GPS: 1 mm to 1 cm.

1.7 Scope and Organization of this Book

Remote sensing techniques have proved to be of immense value in mapping and monitoring various Earth's surface features and resources such as minerals, water, snow, agriculture, vegetation etc., and have attained an operational status

Scope and Organization of this Book

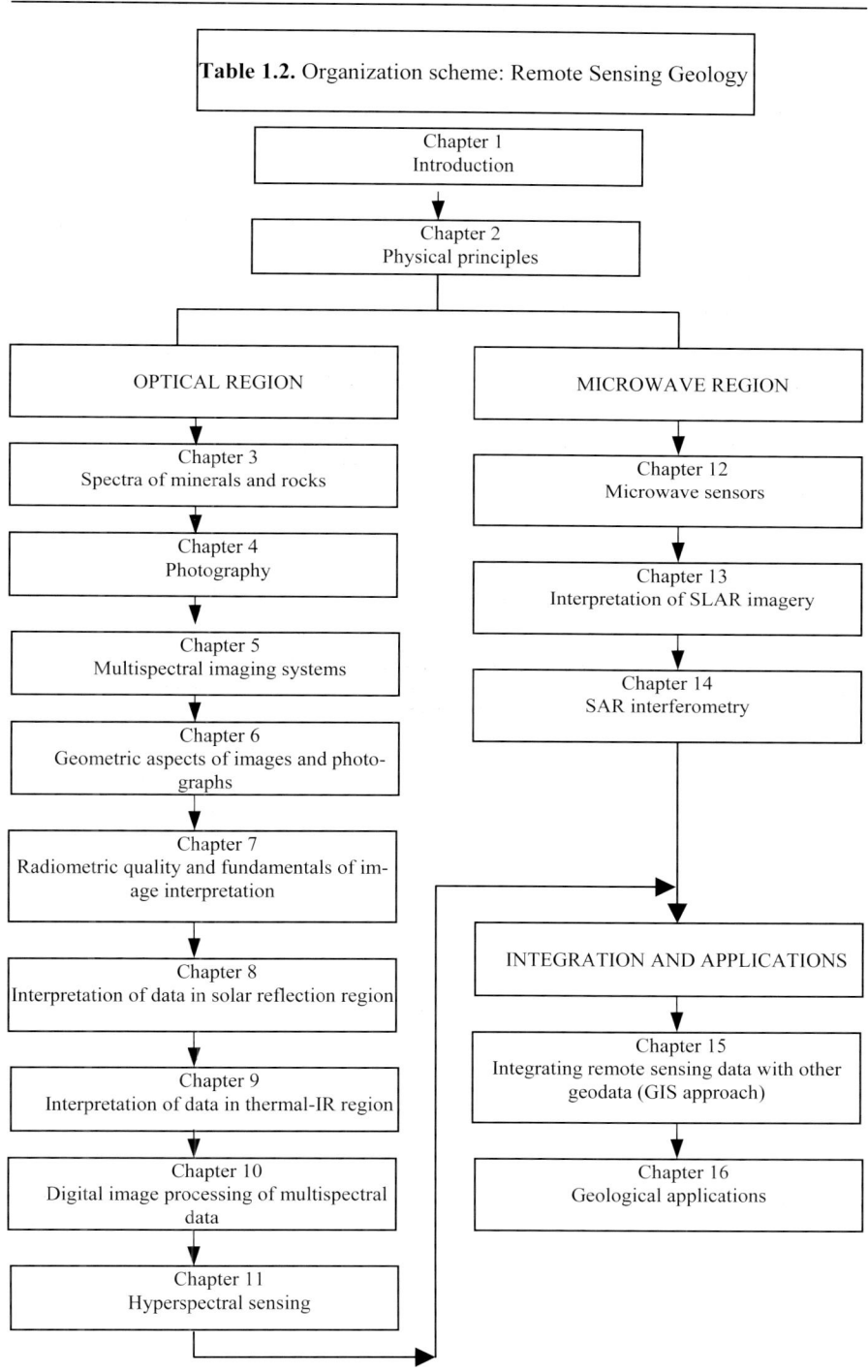

Table 1.2. Organization scheme: Remote Sensing Geology

in many of these disciplines. Details of broad-spectrum applications for these resources can be found elsewhere (e.g. Colwell 1983). Here, in this work, we concentrate specifically on geological aspects including sensors, investigations and applications.

The organization of this book is schematically shown in Table 1.2. In Chapter 1, we have introduced the basic principles involved in the remote sensing programme. Chapter 2 discusses the physical principles, including the nature of EM radiation and the interaction of radiation with matter. Chapters 3 to 11 present various aspects of remote sensing in the optical region of the EM spectrum. Chapters 12 to 14 discuss microwave and radar remote sensing. Chapter 15 deals with the GIS approach of image-based data integration. Finally, Chapter 16 gives examples of thematic geological applications.

Chapter 2: Physical Principles

2.1 The Nature of EM Radiation

As discussed in Chapter 1, in remote sensing, the electromagnetic (EM) radiation serves as the main communication link between the sensor and the object. Fraser and Curran (1976), Silva (1978) and Suits (1983), provide valuable reviews on the nature of EM radiation and physical principles. The properties of EM radiation can be classified into two main groups: (1) those showing a wave nature and (2) those showing particle characteristics.

Maxwell gave a set of four differential equations, which forms the basis of the electromagnetic wave theory. It considers EM energy as propagating in harmonic sinusoidal wave motion (Fig. 2.1), consisting of inseparable oscillating electric and magnetic fields that are always perpendicular to each other and to the direction of propagation. The wave characteristics of EM radiation are exhibited in space and during interaction with matter on a macroscopic scale. From basic physics, we have

$$c = \nu \lambda \tag{2.1}$$

where c is the speed of light, ν is the frequency and λ is the wavelength. All EM radiation travels with the same speed in a particular medium. The speed varies from one medium to another, the variation being caused due to the change in wavelength of the radiation from medium to medium. The speed of EM radiation in a vacuum is 299793 km s^{-1} (approx. 3×10^8 m s^{-1}). The frequency (ν), given in hertz (cycles per

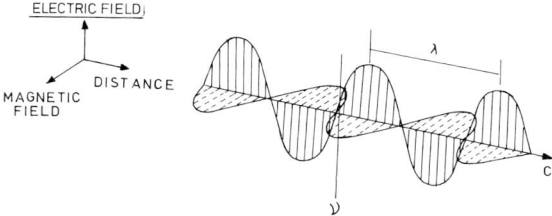

Fig. 2.1. Electromagnetic wave – the electric and magnetic components are perpendicular to each other and to the direction of wave propagation; λ = wavelength, c = velocity of light and ν = frequency

second), is an inherent property of the radiation that does not change with the medium. The wavelength (λ) is given in µm (10^{-6} m) or nm (10^{-9} m).

The particle or quantum nature of the EM radiation, first logically explained by Max Planck, postulates that the EM radiation is composed of numerous tiny indivisible discrete packets of energy called photons or quanta. The energy of a photon can be written as:

$$E = h\nu = h \cdot c/\lambda \qquad (2.2)$$

where E is the energy of a photon (Joules), h is a constant, called Planck's constant (6.6.2 × 10^{-34} J s) and ν is the frequency. This means that the photons of shorter wavelength (or higher frequency) radiation carry greater energy than those of larger wavelength (or lower frequency). The quantum characteristics are exhibited by EM radiation when it interacts with matter on an atomic–molecular scale and explain strikingly well the phenomena of blackbody radiation, selective absorption and photoelectric effect.

2.2 Radiation Principles and Sources

2.2.1 Radiation Terminology

Several terms are used while discussing EM radiation. *Radiant energy* is given in joules. *Radiant flux* or power is the radiant energy per second and is given in watts. *Irradiance* implies the amount of radiant energy that is incident on a horizontal surface of unit area per unit time. It is called *spectral irradiance* when considered at a specific wavelength. *Radiance* describes the radiation field as dependent on the angle of view. If we consider the radiation passing through only a small solid angle of view, then the irradiance passing through the small solid angle and incident on the surface is called radiance for the corresponding solid angle.

2.2.2 Blackbody Radiation Principles

Blackbody radiation was studied in depth in the 19th century and is now a well-known physical principle. All matter at temperatures above absolute zero (0 K or −273.1°C) emits EM radiation continuously. The intensity and spectral composition of the emitted radiation depend upon the composition and temperature of the body. A blackbody is an ideal body and is defined as one that absorbs all radiation incident on it, without any reflection. It has a continuous spectral emission curve, in contrast to natural bodies that emit only at discrete spectral bands, depending upon the composition of the body.

Temperature has a great influence on the intensity of blackbody emitted radiation (Fig. 2.2). Experimentally, it was found initially that the wavelength at which

Radiation Principles and Sources

most of the radiation is emitted depends on the temperature of the blackbody. The relationship, called Wien's Displacement Law, is expressed as

$$\lambda_{max} = A/T \qquad (2.3)$$

where λ_{max} is the wavelength (cm) at which peak of the radiation occurs, A is a constant (= 0.29 cm K) and T is the temperature (K) of the object. This relationship is found to be valid for shorter wavelengths, and gives the shift in λ_{max} with temperature of the radiating object. Using this law, we can estimate the temperature of objects by measuring the wavelength of peak radiation. For example, for the Sun, λ_{max} occurs at 0.48 μm, which gives the temperature of the Sun as 6000 K (approx.); similarly for the Earth, the ambient temperature is 300 K and λ_{max} occurs at 9.7 μm (Fig.2.2).

The Stefan–Boltzmann Law gives the total radiation emitted by a blackbody over the entire EM range

$$W = \int_0^\infty w_\lambda \, d\lambda = \sigma T^4 \text{ watts m}^{-2} \qquad (2.4)$$

where w_λ is the spectral radiance, i.e. the energy radiated per unit wavelength per second per unit area of the blackbody, T is the temperature of the blackbody and σ

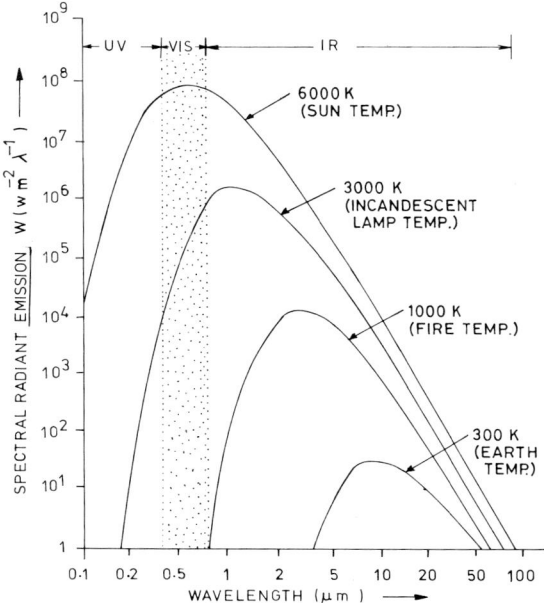

Fig. 2.2. Spectral distribution of energy radiated from blackbodies of various temperatures, such as that of Sun, incandescent lamp, fire and Earth. The spectral radiant power w_λ is the energy emitted ($m^{-2}\lambda^{-1}$). Total energy radiated, W, is given by the area under the respective curves

is the Stefan–Boltzmann constant. It implies that the total energy emitted is a function of the fourth power of temperature of the blackbody. This relation applies to all wavelengths of the spectrum shorter than microwaves.

Another important radiation relationship is the Rayleigh–Jeans Law, valid for microwaves and written as

$$w_\lambda \cong \frac{2\pi ck}{\lambda^4} \cdot T \tag{2.5}$$

where k is called Boltzmann's constant. This implies that spectral radiance is directly proportional to temperature.

Max Planck, using his quantum theory, developed a radiation law to inter-relate spectral radiance (w_λ in watts) and wavelength (λ in m) of the emitted radiation to the temperature (T in K) of the blackbody

$$w_\lambda = \frac{2\pi hc^2}{\lambda^5} \left(\frac{1}{e^{hc/\lambda kT} - 1} \right) \tag{2.6}$$

where h is Planck's constant (= 6.62×10^{-34} J s), c is the speed of light in m s^{-1} and k is Boltzmann's constant (1.38×10^{-23} J deg^{-1}).

Planck's Law was able to explain all the empirical relations observed earlier. Integrating Planck's radiation equation over the entire EM spectrum, we can derive the Stefan–Boltzmann Law. Wien's Displacement Law is found to be a corollary of Planck's radiation equation when λ is small. The Rayleigh–Jean's Law is also found to be an approximation of Planck's radiation equation when λ is large.

As mentioned earlier, a blackbody is one which absorbs all radiation incident on it, without any reflection. It is observed that the fraction of the radiation absorbed exactly equals the fraction that is emitted by any body. Good absorbers are good emitters of radiation. This was stated by Kirchoff, in what is now called Kirchoff's Law

$$a_\lambda = \varepsilon_\lambda \tag{2.7}$$

where a_λ is the spectral absorptivity and ε_λ is the spectral emissivity. Both a_λ and ε_λ are dimensionless and less than 1 for natural bodies. A blackbody has

$$a_\lambda = \varepsilon_\lambda = 1 \tag{2.8}$$

A blackbody radiates a continuous spectrum. It is an idealization, and since $a_\lambda = \varepsilon_\lambda = 1$, radiations are emitted at all possible wavelengths (Fig. 2.2). Real materials do not behave as a blackbody. A natural body radiates at only selected wavelengths as permitted by the atomic (shell) configuration. Therefore, the spectrum of a natural body will be discontinuous, as typically happens in the case of gases. However, if the solid consists of a variety of densely packed atoms (e.g. as in the case of the Sun and the Earth), the various wavelengths overlap, and the resulting spectrum has *in toto* a near-continuous appearance.

The emitting ability of a real material compared to that of the blackbody is referred to as the material's emissivity (ε). It varies with wavelength and geometric configuration of the surface and has a value ranging between 0 and 1:

$$0 \leq \varepsilon_\lambda \leq 1 \tag{2.9}$$

A graybody has an emissivity less than 1, but constant at all wavelengths. Natural materials are also not graybodies.

To account for non-blackbodiness of the natural materials, the relevant parts of the various equations described above are multiplied by the factor of spectral emissivity, i.e.

$$(w_\lambda)_{object} = (\varepsilon_\lambda)_{object} \cdot (w_\lambda)_{blackbody} \tag{2.10}$$

$$(W)_{object} = \int_0^\infty (\varepsilon_\lambda)_{object} \cdot (w_\lambda)_{blackbody} \cdot d\lambda \tag{2.11}$$

$$= (\varepsilon_\lambda)_{object} \cdot \sigma \cdot T^4 \tag{2.12}$$

Similarly

$$(w_\lambda)_{object} = (\varepsilon_\lambda)_{object} \frac{2\pi hc^2}{\lambda^5} \cdot \left(\frac{1}{e^{hc/\lambda KT} - 1}\right) \tag{2.13}$$

2.2.3 Electromagnetic Spectrum

The electromagnetic spectrum is the ordering of EM radiation according to wavelength, or in other words, frequency or energy. The EM spectrum is most commonly presented between cosmic rays and radiowaves, the intervening parts being gamma rays, X-rays, ultra-violet, visible, near-infrared, thermal-infrared, far-infrared and microwave (Fig. 2.3). The EM spectrum from 0.02-μm to -m wavelength can be divided into two main parts, the optical range and the microwave range. The optical range refers to that part of the EM spectrum in which optical phenomena of reflection and refraction can be used to focus the radiation. It extends from X-rays (0.02-μm wavelength) through visible and includes far-infrared (1 mm wavelength). The microwave range is from 1-mm to 1-m wavelength.

For remote sensing purposes, as treated later, the most important spectral regions are 0.4–14 μm (lying in the optical range) and 2 mm–0.8 m (lying in the microwave range). There is a lack of unanimity amount scientists with regard to the nomenclature of some of the parts of the EM spectrum. For example, the wavelength at 1.5 μm is considered as near-IR (Fraser and Curran 1976; Hunt 1980), middle-IR (Silva 1978), and short-wave-IR (Goetz et al. 1983). The nomenclature followed throughout the present work is shown in Fig. 2.3b.

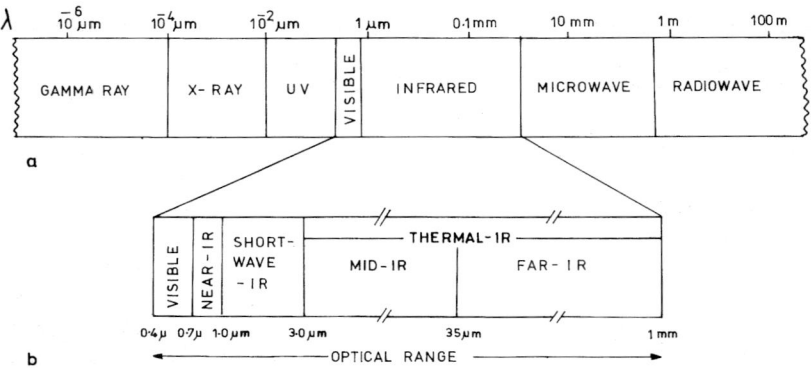

Fig. 2.3. a Electromagnetic spectrum between 10^{-8} μm and 10^2 m. **b** Terminology used in the 0.4-μm–1-mm region in this work, involving VIS, NIR, SWIR, MIR and FIR

2.2.4 Energy Available for Sensing

Most commonly, in remote sensing, we measure the intensity of naturally available radiation – such sensing is called *passive sensing*, sensors being accordingly called *passive sensors*. The Sun, due to its high temperature (≈ 6000 K), is the most dominant source of EM energy. The radiation emitted by the sun is incident on the Earth and is back scattered. Assuming an average value of diffuse reflectance of 10%, the spectral radiance due to solar reflection is as shown in Fig. 2.4a. Additionally, the Earth itself emits radiation due to its thermal state (Fig. 2.4a). All these radiation – the Sun's radiation reflected by the Earth and those emitted by the Earth - carry information about ground materials and can be used for terrestrial remote sensing. On the other hand, in some cases, the radiation is artificially generated (*active sensor*!), and the back-scattered signal is used for remote sensing.

2.3 Atmospheric Effects

The radiation reflected and emitted by the Earth passes through the atmosphere. In this process, it interacts with atmospheric constituents such as gases (CO_2, H_2O vapour, O_3 etc.), and suspended materials such as aerosols, dust particles etc. During interaction, it gets partly scattered, absorbed and transmitted. The degree of atmospheric interaction depends on the pathlength and wavelength.

Pathlength means the distance travelled by the radiation through the atmosphere, and depends on the location of the energy source and the altitude of the sensor platform. Sensing in the solar reflection region implies that the radiation travels through the atmosphere twice – in the first instance from the Sun towards the Earth, and then from the Earth towards the sensor, before being sensed. On the other hand, the radiation emitted by the Earth traverses the atmosphere only once. Further, pathlength also depends upon the altitude of the platform – whether it is at low aerial altitude, high aerial altitude or space altitude.

Attenuation of the radiation due to atmosphere interaction also depends on the wavelength. Some of the wavelengths are transmitted with higher efficiency, whereas others are more susceptible to atmospheric scattering and absorption. The *transmissivity* of the atmosphere at a particular wavelength is a measure of the fraction of the radiance that emanates from the ground (due to solar reflection or self-emission) and passes through the atmosphere without interacting with it. It varies from 0 to 1. The transmissivity is inversely related to another attribute called the *optical thickness* of the atmosphere, which describes the efficiency of the atmosphere in blocking the ground EM radiation by absorption or scattering.

Thus the atmosphere acts as scatterer and absorber of the radiation emanating from the ground. In addition, the atmosphere also acts as a source of EM radiation due to its thermal state. Therefore, the atmosphere–radiation interactions can be grouped into three physical processes: scattering, absorption and emission.

A remote sensor collects the total radiation reaching the sensor – that emanating from the ground as well as that due to the atmospheric effects. The part of the signal emanating from the atmosphere is called *path radiance*, and that coming from the ground is called *ground radiance*. The path radiance tends to mask the ground signal and acts as a background noise.

2.3.1 Atmospheric Scattering

Atmospheric scattering is the result of diffuse multiple reflections of EM radiation by gas molecules and suspended particles in the atmosphere. These interactions do not bring any change in the wavelength of the radiation and are considered as elastic scattering. Several models have been proposed to explain the scattering phenomena.

There are two basic types of scattering: (a) nonselective scattering and (b) selective scattering. *Nonselective scattering* occurs when all wavelengths are equally scattered. It is caused by dust, cloud and fog, such that the scatterer particles are much larger than the wavelengths involved. As all visible wavelengths are equally scattered, clouds and fog appear white.

Amongst *selective scattering,* the most common is *Raleigh scattering*, also called molecular scattering, which occurs due to interaction of the radiation with mainly gas molecules and tiny particles (much smaller than the wavelength involved). Raleigh scattering is inversely proportional to the fourth power of the wavelength. This implies that shorter wavelengths are scattered more than longer wavelengths. This type of scattering is most severe in the ultraviolet and blue end of the spectrum and is negligible at wavelengths beyond 1 µm. This is responsible for the blue colour of the sky. If there were no atmosphere, the sky would appear just as a dark space.

In the context of remote sensing, Raleigh scattering is the most important type of scattering and causes high path radiance at the blue-end of the spectrum. It leads to haze on images and photographs, which results in reduced contrast and unsharp

Table 2.1 Major atmospheric windows (clearer windows shown in boldface)

Name	Wavelength range	Region
Ultraviolet–visible	**0.30–0.75 μm**	Optical
Near-IR	**0.77–0.91 μm**	Optical
Short-wave-IR	1.00–1.12 μm	Optical
	1.19–1.34 μm	
	1.55–1.75 μm	
	2.05–2.4 μm	
Mid-IR (Thermal-IR)	3.50–4.16 μm	Optical
	4.50–5.0 μm	
	8.00–9.2 μm	
	10.20–12.4 μm	
	(**8–14 μm** for aerial sensing)	
	17.00–22.0 μm	
Microwave	2.06–2.22 mm	Microwave
	7.50–11.5 mm	
	20.0 + mm	

pictures. The effect of this type of scattering can be reduced by using appropriate filters to eliminate shorter wavelength radiation.

Another type of scattering is the large-particle scattering, also called *Mie scattering*, which occurs when the particles are spherical. It is caused by coarse suspended particles of a size larger than the wavelength involved. The main scatterers of this type are suspended dust particles and water vapour molecules, which are more important in lower altitudes of the atmosphere, close to the Earth's surface. Mie scattering influences the entire spectral region from near-UV up to and including the near-IR, and has a greater effect on the larger wavelengths than Raleigh scattering. Mie scattering depends on various factors such as the ratio of the size of scatterer particle to the wavelength incident, the refractive index of the object and the angle of incidence. As it is influenced by water vapour, the Mie effect is more manifest in overcast atmospheric conditions.

2.3.2 Atmospheric Absorption

The atmospheric gases selectively absorb EM radiation. The atoms and molecules of the gases possess certain specific energy states (rotational, vibrational and electronic energy levels; see Chap.3). Photon energies of some of the EM radiation may be just sufficient to cause permissible energy level changes in the gas molecule leading to selective absorption of EM radiation (Fig. 2.4b). The most important atmospheric constituents in this regard are H_2O vapour, CO_2 and O_3. The spectral regions of least absorption are called *atmospheric windows*, as they can be used for looking at ground surface phenomena from aerial or space platforms across the atmosphere. Important atmospheric windows available for space-borne sensing are listed in Table 2.1. The visible part of the spectrum is marked by the presence of an excellent atmospheric window. Prominent windows occur throughout the EM

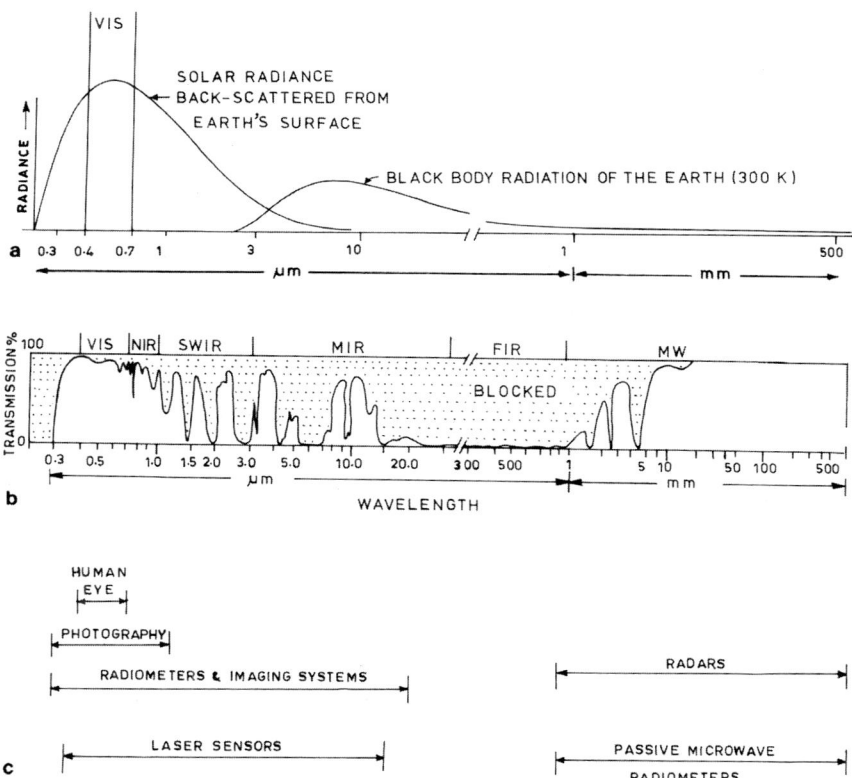

Fig. 2.4. a Energy available for remote sensing. The solar radiation curve corresponds to the back-scattered radiation from the Earth's surface, assuming the surface to be Lambertian and having an albedo of 0.1. The Earth's blackbody radiation curve is for 300 K temperature; **b** Transmission of the radiation through the atmosphere; note the presence of atmospheric absorption bands and atmospheric windows. **c** Major sensor types used in different parts of the spectrum

spectrum at intervals. In the thermal-IR region, two important windows occur at 8.0-9.2 μm and 10.2-12.4 μm, which are separated by an absorption band due to ozone, present in the upper atmosphere. For sensing from aerial platforms, the thermal channel can be used as 8–14 μm. The atmosphere is essentially opaque in the region of 22 μm to 1mm wavelength. Microwaves of wavelength greater than 20 mm are propagated through the atmosphere with least attenuation.

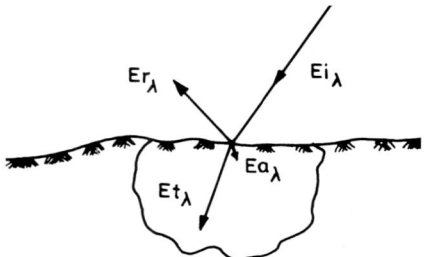

Fig. 2.5. Energy interaction mechanism on ground; Ei_λ is the incident EM energy and Er_λ, Ea_λ and Et_λ are the energy components reflected, absorbed and transmitted respectively

2.3.3 Atmospheric Emission

The atmosphere also emits EM radiation due to its thermal state. Owing to its gaseous nature, only discrete bands of radiation (not forming a continuous spectrum) are emitted by the atmosphere. The atmospheric emission would tend to increase the path radiance, which would act as a background noise, superimposed over the ground signal. However, as spectral emissivity equals spectral absorptivity, the atmospheric windows are marked by low atmospheric emission. Therefore, for terrestrial sensing, the effect of self-emission by the atmosphere can be significantly reduced by restricting remote sensing observations to good atmospheric windows.

2.4 Energy Interaction Mechanisms on the Ground

The EM energy incident on the Earth surface may be reflected, absorbed and or transmitted (Fig. 2.5). Following the Law of Conservation of Energy, the energy balance can be written as

$$Ei_\lambda \equiv Er_\lambda + Ea_\lambda + Et_\lambda \qquad (2.14)$$

where Ei_λ is the spectral incident energy; Er_λ, Ea_λ and Et_λ are the energy components reflected, absorbed and transmitted respectively. The components Er_λ, Ea_λ and Et_λ differ for different objects at different wavelengths. These inherent differences build up the avenues for discrimination of objects by remote sensing measurements.

2.4.1 Reflection Mechanism

The reflection mechanism is of relevance to techniques in the solar reflection region and in active microwave sensing, where sensors record the intensity of EM radiation reflected from the ground. In the reflectance domain, the reflected energy can be written as

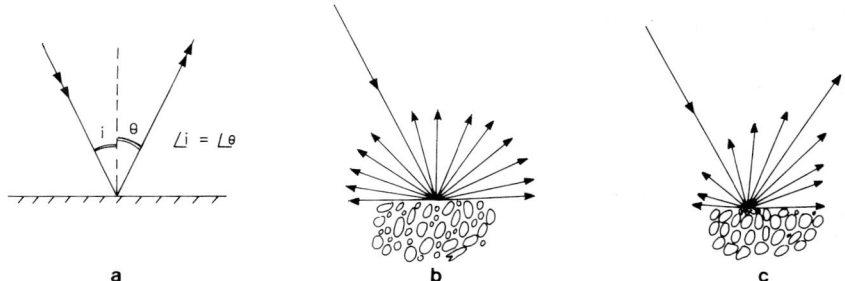

Fig. 2.6a–c. Reflection mechanisms. **a** Specular reflection from a plane surface; **b** Lambertian reflection from a rough surface (diffuse reflection); **c** semi-diffused reflection (natural bodies)

$$Er_\lambda = Ei_\lambda - \left(Ea_\lambda + Et_\lambda \right) \tag{2.15}$$

Therefore, the amount of reflected energy depends upon the incident energy, and mechanisms of reflection, absorption and transmission. The reflectance is defined as the proportion of the incident energy which is reflected

$$R_\lambda = Er_\lambda / Ei_\lambda \tag{2.16}$$

When considered over a broader wavelength range, it is also called *albedo*. Further, the interactions between EM radiation and ground objects may result in reflection, polarization and diffraction of the wave, which are governed mainly by composite physical factors like shape, size, surface features and environment. These phenomena occur at boundaries, and are best explained by the wave nature of light.

If the surface of the object is an ideal mirror-like plane, specular reflection occurs following Snell's Law (Fig. 2.6a). The angle of reflection equals the angle of incidence, and the incident ray, the normal and the reflected ray are in the same plane. Rough surfaces reflect in multitudes of directions, and such reflection is said to be scattering or non-specular reflection. This is basically an elastic or coherent type of phenomenon in which no change in the wavelength of the radiation occurs. The uneven surfaces can be considered as being composed of numerous small non-parallel plane surfaces and fine edges and irregularities, the dimensions of which are of the order of the wavelength of the incident radiation. This results in multitudes of reflections in numerous directions and diffraction at fine edges and small irregularities, leading to a sum total of scattered radiation from the surface (Fig. 2.7). An extreme ideal case is the Lambertian surface in which the radiation is reflected equally in all directions, irrespective of the angle of incidence (Fig. 2.6b). Most natural bodies are in between the two extremes of specular reflection and Lambertian reflection, and show a semi-diffuse reflection pattern. The radiation is scattered in various directions, but is maximum in one direction, which corresponds to Snell's Law (Fig. 2.6c).

Further, whether a particular surface behaves as a specular or a rough surface depends on the dimension of the wavelength involved and the local relief. For exam-

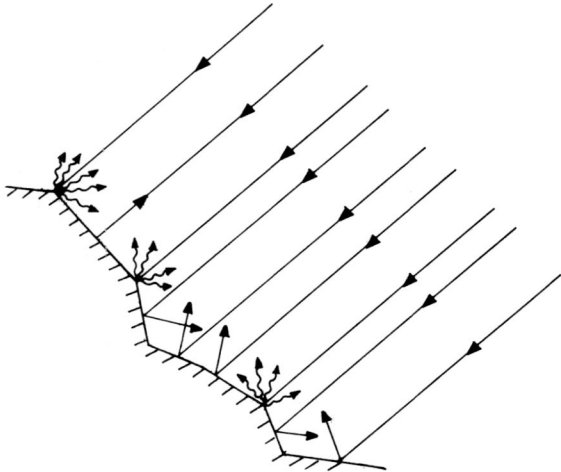

Fig. 2.7. Mechanism of scattering in multiple directions from natural uneven surfaces

ple, a level bed composed of coarse sand (grain size e.g. 1 mm) would behave as a rough surface for VIS–NIR wavelengths and as a smooth surface for microwaves.

Therefore, the intensity of reflected EM radiation received at the remote sensor depends on, beside other factors, geometry – both viewing and illuminating (see Fig. 1.7). In practice, a number of variations occur: (a) In solar reflection sensing, commonly the Sun is obliquely illuminating the ground, and the remote sensor is viewing the terrain near-vertically from above (Fig. 2.8a). (b) The SLAR imaging involves illumination and sensing from an oblique direction (Fig. 2.8b). (c) In LIDAR the sensors operate in near-vertical mode and record back-scattered radiation (Fig. 2.8c). It is important that the goniometric aspects are properly taken into account while interpreting the remote sensing data.

Some special phenomena may occur in specific circumstances during reflection, the most important of which is polarization. The reflected wave train may become polarized or depolarized in a certain direction depending upon the ground attributes. The potential for utilizing the polarization effects of waves in remote sensing appears to be quite distinct in the case of microwaves (see Sect. 13.6).

A remote sensor measures the total intensity of EM radiation received the sensor, which depends not only on the reflection mechanism but also on factors influencing absorption and transmission processes.

2.4.2 Transmission Mechanism

When a beam of EM energy is incident on a boundary, for example on the Earth's surface, part of the energy gets scattered from the surface (called surface scattering) and part may get transmitted into the medium. If the material is homo-

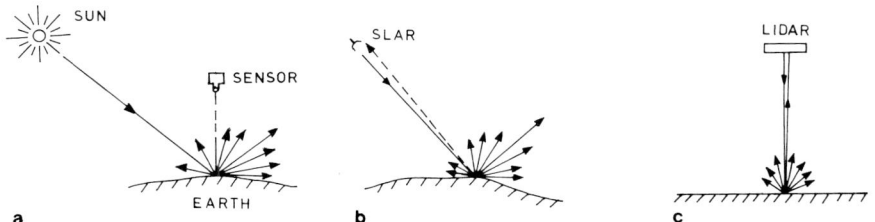

Fig. 2.8a–c. Common geometric configurations in reflection sensing. **a** Solar reflection sensing. **b** SLAR sensing. **c** LIDAR sensing

geneous, then this wave is simply transmitted. If, on the other hand, the material is inhomogeneous, the transmitted ray gets further scattered, leading to volume scattering in the medium. In nature, both surface and volume scattering happen, side by side, and both processes contribute to the total signal received at the sensor.

As defined, the depth of penetration is considered as that depth below the surface at which the magnitude of the power of the transmitted wave is equal to 36.8% (1/e) of the power transmitted, at a point just beneath the surface (Ulaby and Goetz 1987).

The transmission mechanism of EM energy is still not fully understood. It is considered to depend mainly on an electrical property of matter, called the complex dielectric constant (δ). This varies spectrally and is different for different materials. When the dielectric constant is low, the radiation penetrates to a greater depth and the energy travels through a larger volume of the material (therefore there is less surface scattering and greater volume scattering). Conversely, when the object has a higher δ, the energy gets confined to the top surficial layer with little penetration (resulting in dominantly surface scattering). As the complex dielectric constant of materials varies with wavelength, the depth penetration also varies accordingly. For example, water bodies exhibit penetration at visible wavelengths but mainly surface scattering at microwave frequencies, whereas the reverse happens for dry rock/soil (Table 2.2).

It is implicit that the transmission characteristics also influence the amount of energy received at the sensor, for the simple reason that transmission characteristics

Table 2.2. Bearing of the spectral complex dielectric constant (δ_λ) of matter on depth penetration (transmission)

Wavelength range	Water/sea body	Dry rock/soil
Visible	Low δ_λ; transmission of radiation and volume scattering	High δ_λ; surface scattering
Microwave	High δ_λ; surface scattering	Low δ_λ; transmission of radiation and volume scattering

govern surface vis-à-vis volume scattering, as also the component of the energy which is transmitted and does not reach the remote sensor.

2.4.3 Absorption Mechanism

Interaction of incident energy with matter on the atomic–molecular scale leads to selective absorption of the EM radiation. An atomic–molecular system is characterized by a set of inherent energy states (i.e. rotational, vibrational and electronic). A different amount of energy is required for transition from one energy level to another. An object absorbs radiation of a particular wavelength if the corresponding photon energy is just sufficient to cause a set of permissible transitions in the atomic–molecular energy levels of the object. The wavelengths absorbed are related to many factors, such as dominant cations and anions present, solid solutions, impurities, trace elements, crystal lattice etc. (for further details see Chapter.3).

2.4.4 Earth's Emission

The Earth, owing to its ambient temperature, is a source of blackbody radiation, which constitutes the predominant energy available for terrestrial sensing at wavelengths > 3.5 µm (Fig. 2.4a). The emitted radiation depends upon temperature and emissivity of the materials. These aspects are presented in greater detail in Chapter 9.

In the above paragraphs, we have discussed the sources of radiation, atmospheric effects and the mechanism of ground interactions. The sensors used in the various spectral regions are shown in Fig. 2.4c. They include the human eye, radiometer, scanner, radar, lidar and microwave passive sensors.

Further discussion is divided into two main parts: optical range (Chapters 3–11) and microwave range (Chapters. 12–14).

Chapter 3: Spectra of Minerals and Rocks

3.1 Introduction

Interactions of the EM radiation with matter at atomic–molecular scale result in selective absorption, emission and reflection, which are best explained by the particle nature of light. The relationship between the intensity of EM radiation and wavelength is called the *spectral response curve*, or broadly, *spectral signature* (Fig. 1.1). A single feature or a group of features (pattern) in the curve could be diagnostic in identifying the object. In the context of remote sensing, objects can be marked by the following types of spectral behaviour.

1. Selective absorption. Some of the wavebands are absorbed selectively and the spectral character is marked by a relatively weak reflection; this phenomenon is widely observed in the solar reflection region.

2. Selective reflection. At times, a particular wavelength is strongly reflected, leading to a 'resonance-like' phenomenon; the selective reflection may be so intense that it may lead to separation of a monochromatic beam (called residual rays or *Reststrahlen*). There occur numerous Reststrahlen bands in the thermal-infrared region.

3. Selective higher or lower emission. Some of the objects may exhibit selective higher or lower emission at certain wavelengths.

Obviously, spectral signatures constitute the basic information needed for designing any remote sensing experiment or interpreting the data. During the last three decades or so, this subject matter has received a good deal of attention. In the following pages, we first give a brief introduction to laboratory spectroscopic methods of spectral data collection and then discuss the atomic–molecular processes which lead to spectral features. After this, spectra of selected mineral–ionic constituents, minerals and rocks are summarized and field and laboratory aspects are discussed. Finally, some aspects of the spectral response of other common objects such as vegetation are presented.

3.2 Basic Arrangements for Laboratory Spectroscopy

A number of methods have been developed for laboratory spectral data measurements. These methods differ with regard to ranges of wavelength (visible, infrared, or thermal-IR) and also the physical phenomenon (reflection, absorption or emission) utilized for the investigation. Here, we will discuss only some of the basic concepts, so that the variation in spectral response curves due to differing spectroscopic arrangements is understandable.

Laboratory spectroscopic arrangements can be conceived as being of three types: reflection, emission and absorption (also called transmission) (Fig. 3.1). Precise inter-relationships between emission, reflection and absorption spectra are still not well understood.

1. Reflection arrangement. This has been the most extensively used arrangement in the optical region. The EM radiation from an external source (the Sun or an artificial light source) impinges upon the object-sample and is reflected onto a detector (Fig. 3.1a). It is customary to express the spectral reflectance as a percentage of the reflectance of MgO, in order to provide a calibration.

2. Emission arrangement. The basic phenomenon that all objects above zero Kelvin (K) temperature emit radiation can form the basis for spectroscopy (Fig. 3.1b). In order to measure radiation emitted by the object-sample at room temperatures, devices have to be cooled to very low temperatures (so that the measuring instrument itself does not constitute a source of emission). Another possibility is to heat the sample, in order to measure the emitted radiation; however,

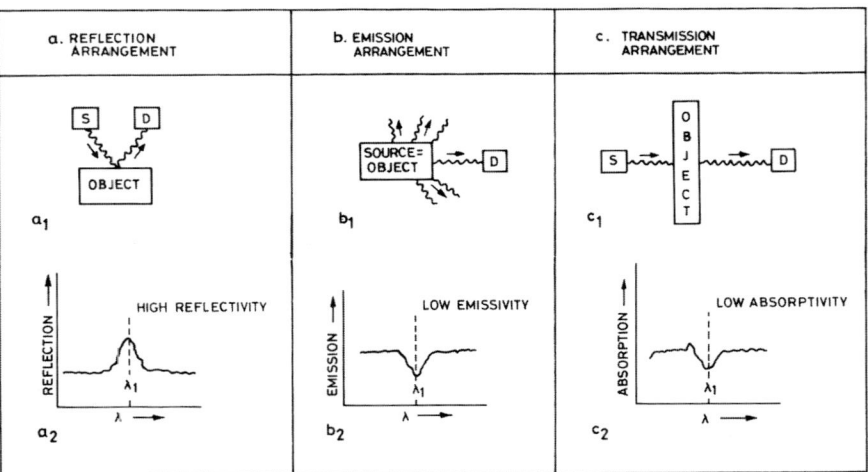

Fig. 3.1. Laboratory schemes of spectroscopic study and resulting spectral curves. **a** Reflection. **b** Emission. **c** Absorption/transmission (S = source of radiation; D = detector)

Basic Arrangements for Laboratory Spectroscopy

there are practical problems of non-uniform heating. Owing to the above difficulties, in general, the emission spectra are computed from reflection or transmission spectral measurements.

3. Transmission arrangement. This is most suited for gases and liquids; fine particulate solids suspended in air or embedded in suitable transparent pellets can also be studied under this arrangement. The sample is placed before a source of radiation; the radiation intensity transmitted through the sample is measured by a detector (Fig. 3.1c). As the physical phenomenon involved is spectral absorption, it is also called an *absorption arrangement*.

The type of laboratory arrangement employed for spectral studies also depends upon the EM wavelength range under investigation. In the visible–near-infrared (VNIR)–SWIR region, reflection is the most widely used arrangement, although transmission spectra are also reported. An emission arrangement in this region would require the sample to be heated to several thousand degrees, which is impractical. In the thermal-IR range, all three, viz. reflection, emission and transmission arrangement have been used by different workers.

If an object exhibits selective high reflection at a particular wavelength, the same feature will appear as selective low emission and selective low absorption (or high transmission) at the same spectral band in other arrangements. Hence, the shape and pattern of any spectral curve depend on the spectroscopic arrangement used. In this treatment, all the spectra in the VNIR–SWIR range are given for reflection, and that in the TIR range for emission (unless otherwise specified).

A spectral curve appears as a waveform comprising positive and negative peaks and slopes. The negative peaks in all types of spectral curves are commonly called *absorption bands*, irrespective of whether they are related to reflection, absorption, emission or transmission or may imply high or low spectral absorptivity.

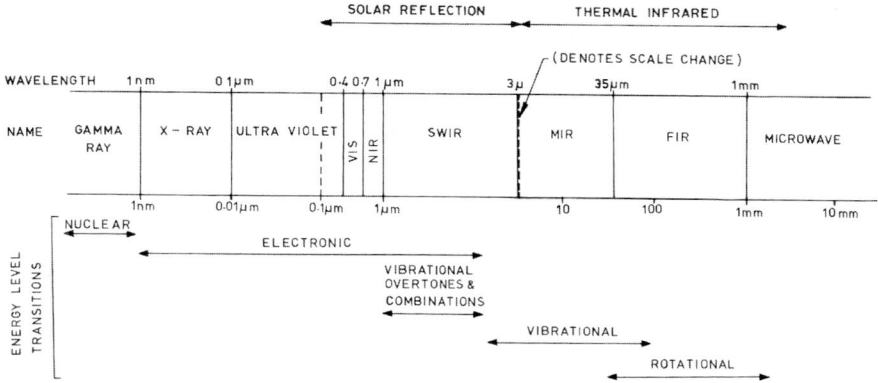

Fig. 3.2. Types of energy level changes associated with different parts of the EM spectrum

3.3 Energy States and Transitions – Basic Concepts

The energy state of an object is a function of the relative position and state of the constituent particles at a given time. The sum-total energy of an atomic–molecular system can be expressed as the sum of four different types of energy states: translational, rotational, vibrational and electronic. A different amount of energy is required for each of these types of transitions to occur. Therefore, different energy transitions appear in different parts of the EM spectrum (Fig. 3.2).

Translational energy, because of its unquantized nature, is not considered here. *Rotational energy*, which is the kinetic energy of rotation of a molecule as a whole in space, is also not considered here because of the physical property of solid substances. The *vibrational energy* state is involved with the movement of atoms relative to each other, about a fixed position. Such energy level changes are caused by radiation of the thermal-IR and SWIR regions. The overtones and combinations of vibrational energy level changes are caused by SWIR radiation. The *Electronic energy* state is related to the configuration of electrons surrounding the nucleus or to the bonds; their transitions require an even greater amount of energy, and are caused by photons of the near-IR, visible, UV and X-ray regions. (Photons of the gamma ray are related to nuclear transitions, i.e. radioactivity.)

3.3.1 Electronic Processes

Electronic processes (transitions) occur predominantly in the UV–VIS–near-IR region. Several models and phenomena have been conceived to explain the electronic processes which lead to selective absorption.

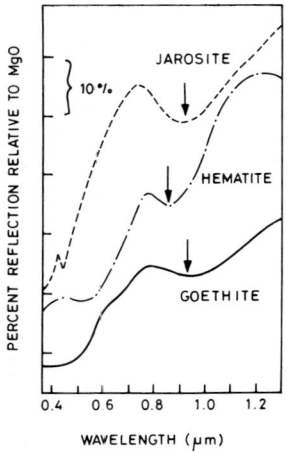

Fig. 3.3. Spectral reflectance curves for jarosite, hematite and goethite showing sharp fall-off in reflectance in the UV–blue region due to the charge-transfer effect. Also note a ferric-ion feature at 0.87 μm present in all three. (Curves offset for clarity) (Segal 1983)

Energy States and Transitions – Basic Concepts

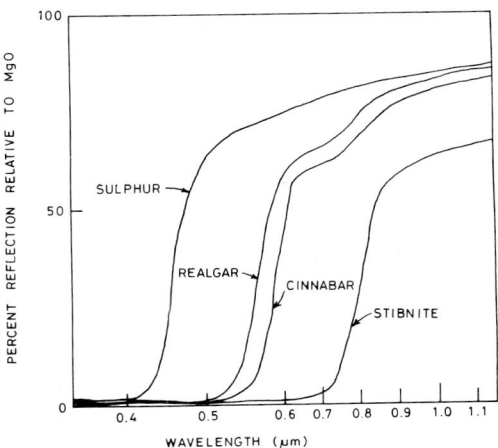

Fig. 3.4. Reflection spectra of particulate samples of stibnite, cinnabar, realgar and sulphur, displaying sharp conduction-band absorption edge effect. (After Hunt 1980)

1. Charge-transfer effect. In some materials, the incident energy may be absorbed, raising the energy level of electrons so that they migrate between adjacent ions, but do not become completely mobile. This is called the charge-transfer effect and may be caused by photons of the UV–visible region. It is typically exhibited by iron oxide, a widespread constituent of rocks. Fe-O absorbs radiation of shorter wavelength energy, such that here there is a steep fall-off in reflectance towards blue (Fig. 3.3); it therefore has a red colour in the visible range. Another example of charge-transfer effect is the uranyl ion (UO_2^{2+}) in carnotite, which absorbs all energy less than 0.5 μm, resulting in the yellow colour of the mineral.

2. Conduction-band absorption effect. In some semi-conductors such as sulphides, nearly all the photons of energy greater than a certain threshold value are absorbed, raising the energy of the electrons to conduction-band level. Thus, the spectra show a sharp edge effect (Fig. 3.4).

3. Electronic transition in transition metals. Electronic processes frequently occur in transition metals, for example:
Ferrous ion: 1.0 μm, 1.8–2.0 μm and 0.55–0.57 μm
Ferric ion: 0.87 μm and 0.35 μm; sub-ordinate bands around 0.5 μm
Manganese: 0.34 μm, 0.37 μm, 0.41 μm, 0.45 μm and 0.55 μm
Copper: 0.8 μm
Nickel: 0.4 μm, 0.74 μm and 1.25 μm
Chromium: 0.35 μm, 0.45 μm and 0.55 μm
The presence of these elements leads to absorption bands at the appropriate wavelengths.

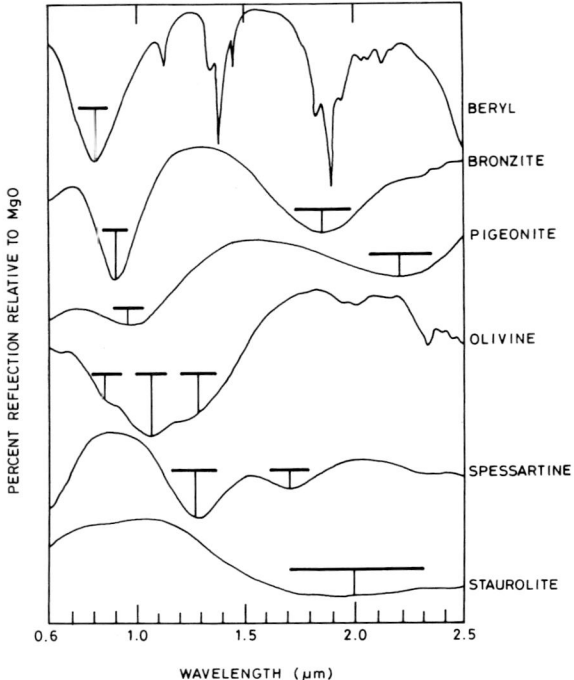

Fig. 3.5 Crystal field effect. Reflection spectra showing ferrous-ion bands in selected minerals. The ferrous ion is located in an aluminium octahedral six-fold co-ordinated site in beryl, in a distorted octahedral six-fold co-ordinated site in olivine, in an octahedral eight-fold co-ordinated site in spessartine, and in a tetrahedral site in staurolite. (Spectra separated for clarity) (Hunt 1980)

4. Crystal field effect. The energy level for the same ion may be different in different crystal fields. This is called the crystal field effect. In the case of transition elements, such as Ni, Gr, Cu, Co, Mn etc., it is the 3d shell electrons that primarily determine the energy levels. These electrons are not shielded (outer shell orbital). Consequently, their energy levels are greatly influenced by the external field of the crystal, and the electrons may assume new energy values depending upon the crystal fields. In such cases, the new energy levels, the transition between them, and consequently their absorption spectra, are determined primarily by the valence state of the ion (e.g. Fe^{2+} or Fe^{3+}) and by its co-ordination number, site symmetry and, to a limited extent, the type of ligand formed (e.g. metal–oxygen) and the degree of lattice distortion. Hunt (1980) has given a set of examples of ferrous ions which, when located in different crystal fields, produce absorption peaks at different wavelengths (Fig. 3.5).

In some cases, the influence of the crystal field on spectral features may not be significant, for example in the case of rare earth atoms, where the unfilled shells are the 4f electrons, which are well shielded from outside influence, and show no significant crystal field effect.

3.3.2 Vibrational Processes

Most of the rock-forming minerals (including silicates, oxides, hydroxyls, carbonates, phosphates, sulphates, nitrates etc.) are marked by atomic–molecular vibrational processes occurring in the SWIR and TIR parts of the EM spectrum. These are the result of bending and stretching molecular motions and can be distinguished as fundamentals, overtones and combinations. The fundamental tones occur mainly in the thermal-infrared region (> 3.5 µm) and their combinations and overtones in the SWIR region (1-3 µm).

3.4 Spectral Features of Mineralogical Constituents

Hunt and his co-workers have published considerable data on this aspect, which has been summarized in Hunt (1977, 1979, 1980) and Salisbury and Hunt (1974). In addition, Lyon (1962, 1965), Farmer (1974), Karr (1975) and Kahle et al. (1986) have also contributed significantly to the understanding of these aspects. The following summary is based mainly on the above investigations and reviews. The discussion is divided into three parts: (1) the VNIR region, (2) the SWIR region and (3) the TIR region.

3.4.1 Visible and Near-Infrared Region (VNIR) (0.4–1.0 µm)

Spectral features in this part of the EM spectrum are dominated by electronic processes in transition metals (i.e. Fe, Mn, Cu, Ni, Cr etc.). Elements such as Si, Al, and various anion groups such as silicates, oxides, hydroxides, carbonates, phosphates etc., which form the bulk of the Earth's surface rocks, lack spectral features in this region. Iron is the most important constituent having spectral properties in this region. The ferric-ion in Fe-O, a ubiquitous constituent in rocks, exhibits strong absorption of UV–blue wavelengths, due to the charge-transfer effect. This results in a steep fall-off in reflectance towards blue, and a general rise towards infrared (Fig. 3.3), with the peak occurring in the 1.3–1.6 µm region. The absorption features due to iron, manganese, copper, nickel and chromium have been mentioned above (Sect. 3.3.1).

3.4.2 SWIR Region (1–3 µm)

The SWIR region is important as it is marked by spectral features of hydroxyls and carbonates, which commonly occur in the Earth's crust. The exact location of

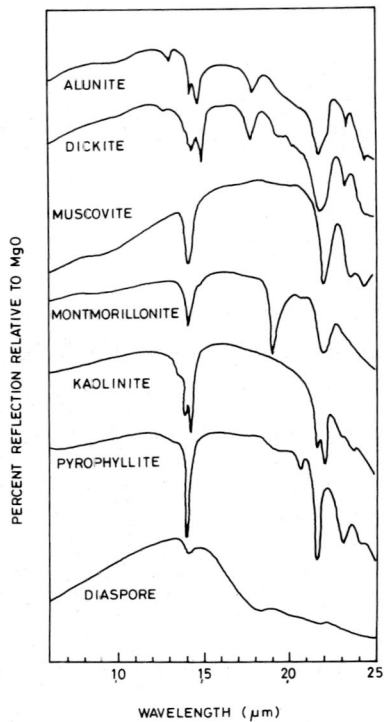

Fig. 3.6. Reflectance spectra of some common clay minerals. (Rowan et al. 1983)

peaks may shift due to crystal field effects. Further, these absorption bands can be seen on solar reflection images.

1. Hydroxyl ion. The hydroxyl ion is a widespread constituent occurring in rock-forming minerals such as clays, micas, chlorite etc. It shows a vibrational fundamental absorption band at about 2.74–2.77 μm, and an overtone at 1.44 μm. Both the fundamental (2.77 μm) and overtone (1.44 μm) features interfere with similar features observed in the water molecule. However, when hydroxyl ions occur in combination with aluminium and magnesium, i.e. as Al-OH and Mg-OH, which happens to be very common in clays and hydrated silicates, several sharp absorption features are seen in the 2.1–2.4 μm region (Fig. 3.6).

The Al-OH vibrational band typically occurs at 2.2 μm and that of Mg-OH at 2.3 μm. If both Mg-OH and Al-OH combinations are present, then the absorption peak generally occurs at 2.3 μm and a weaker band at 2.2 μm, leading to a doublet, e.g. in kaolinite (Fig. 3.6). In comparison to this, montmorillonite and muscovite typically exhibit only one band at 2.3 μm, due to Mg-OH. Iron may substitute for either aluminium or magnesium, and this incremental substitution in clays

Spectral Features of Mineralogical Constituents

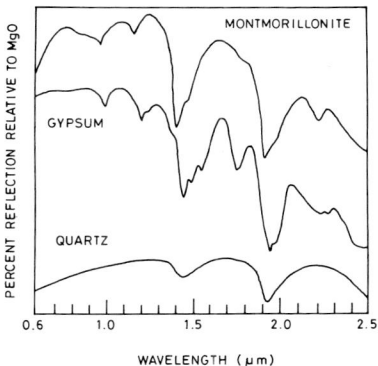

Fig. 3.7. Reflectance spectra of selected water-bearing minerals. Note the absorption bands at 1.4 μm. (Hunt 1980)

structurally reduces the intensity of the Al-OH band (2.2 μm) or the Mg-OH band (2.3 μm) (and increases the electronic bands for iron in the 0.4–1.0 μm region).

The strong absorption phenomenon within the 2.1–2.4 μm range due to Al-OH and Mg-OH leads to sharp decreasing spectral reflectance beyond 1.6 μm, if clays are present. This broad-band absorption feature at 2.1–2.4 μm is used to diagnose clay-rich areas. As the peak of the reflectance occurs at about 1.6 μm (Fig. 3.6), the ratio of broad bands 1.55–1.75 μm / 2.2–2.24 μm, is a very powerful parameter in identifying clay mineral assemblages (Abrams et al. 1977). Further, it is also useful in identifying hydrothermal alteration zones as many clay minerals, e.g. kaolinite, muscovite, pyrophyllite, alunite, dickite, montmorillonite and diaspore, are associated with such zones.

2. *Water molecules.* Important absorption bands due to water molecules occur at 1.4 μm and 1.9 μm (Fig. 3.7). Sharp peaks imply that water molecules occur in well-defined sites, and when the peaks are broad it means that they occur in unordered sites.

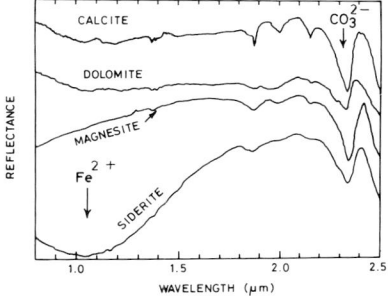

Fig. 3.8. Reflectance spectra of carbonates. Note the carbonate absorption band at 2.35 μm. (Whitney et al. 1983)

3. Carbonates. Carbonates occur quite commonly in the Earth's crust, in the form of calcite ($CaCO_3$), magnesite ($MgCO_3$), dolomite [$(Ca-Mg)\,CO_3$] and siderite ($FeCO_3$). Important carbonate absorption bands in the SWIR occur at 1.9 μm, 2.35 μm and 2.55 μm, produced due to combinations and overtones (Fig. 3.8). The peak at 1.9 μm may interfere with that due to the water molecule and that at 2.35 μm with a similar feature in clays at around 2.3 μm. However, a combination of 1.9 μm and 2.35 μm and also an extra feature at 2.5 μm is considered diagnostic of carbonates. Further, presence of siderite is accompanied by an electronic absorption band due to iron, occurring near 1.1 μm (Fig. 3.8) (Whitney et al. 1983).

3.4.3 Thermal-IR Region

This part of the EM spectrum is characterized by spectral features exhibited by many rock-forming mineral groups, e.g. silicates, carbonates, oxides, phosphates, sulphates, nitrates, nitrites, hydroxyls. The fundamental vibration features of the above anionic groups occur in the thermal-IR region. Physical properties such as particle size and packing can produce changes in emission spectra in terms of relative depth of the absorption, but not in the position of the spectral band.

Typical spectra of representative anionic groups, plotted as percent emission, are shown in Fig. 3.9. The negative peaks (popularly called absorption bands) in these curves indicate low spectral emissivity, which implies low spectral absorp-

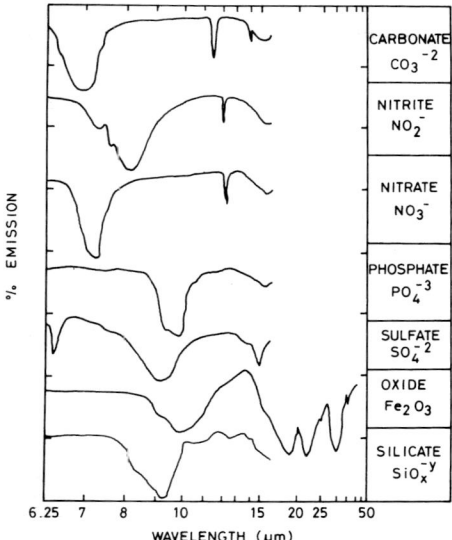

Fig. 3.9. Thermal-infrared spectra of the major anionic mineral groups (generalized; data in all figures are reflectance spectra converted to emission spectra using Kirchoff's law). (Christensen et al. 1986)

Spectral Features of Mineralogical Constituents

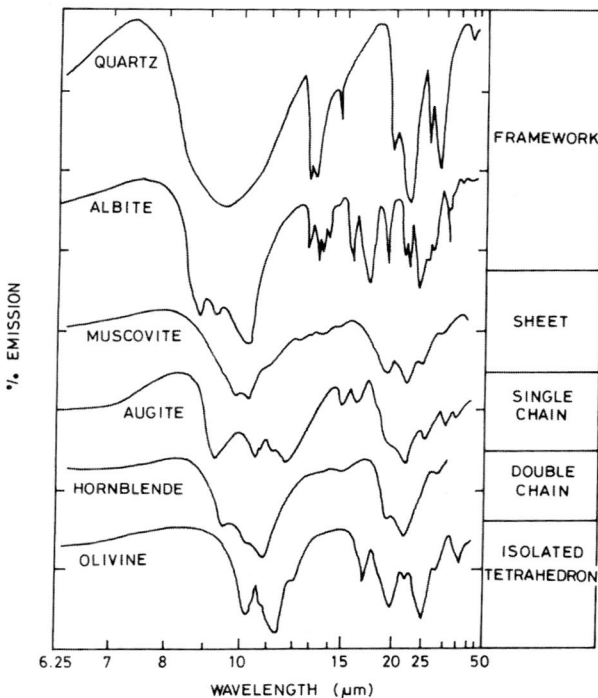

Fig. 3.10. Emission spectra of selected silicate minerals showing the correlation between band location (vibrational energy) and mineral structure. (Christensen et al. 1986)

tivity or, in other words, high spectral reflectivity (Reststrahlen bands – physically, simply the inverse of the reflection spectra).

The carbonates show a prominent absorption feature at 7 μm. However, as this is outside the atmospheric window (8–14 μm), it cannot be used for remote sensing; instead, the weak feature around 11.3 μm can possibly be detected. The sulphates display bands near 9 μm and 16 μm. The phosphates also have fundamental features near 9.25 μm and 10.3 μm. The features in oxides usually occupy the same range as that of bands in Si-O, i.e. 8 to 12 μm (discussed below). The nitrates have spectral features at 7.2 μm and the nitrites at 8 μm and 11.8 μm. The hydroxyl ions display fundamental vibration bands at 11 μm, e.g. in the H-O-Al bending mode in the aluminium-bearing clays.

The silicates, which form the most abundant group of minerals in the Earth's crust, display vibrational spectral features in the TIR region due to the presence of SiO_4-tetrahedron. Considering the spectra of common silicates such as quartz, feldspars, muscovite, augite, hornblende and olivine, the following general trends can be outlined (Hunt 1980; Christensen 1986) (Fig. 3.10).

a) In the region 7–9 μm there occurs a maximum, called the Christiansen peak; its location migrates systematically with the composition, being near 7 μm for felsic minerals and near 9 μm for ultramafic minerals.

b) In the region 8.5–12 μm an intense silicate absorption band occurs, overall centered around 10 μm; therefore, 10 μm is generally designated as the Si-O vibration absorption (Reststrahlen) region; however, its exact position is sensitive to the silicate structure and shifts from nearly 9 μm (framework silicates or felsic minerals) to 11.5 μm (chain silicates or mafic minerals).

c) The 12–15 μm region is sensitive to silicate and aluminium-silicate structure of tectosilicate type; other silicates having structures of sheet, chain or ortho types do not show absorption features in this region; the absorption patterns in the form of numbers and location of peaks are different for different feldspars, thus permitting possible identification.

From a geological point of view, therefore, the thermal-IR is the most important spectral region for remote sensing aiming at compositional investigations of terrestrial materials.

3.5 Spectra of Minerals

Minerals are naturally formed inorganic substances and consist of combinations of cations and anions. They may be chemically simple or highly complex. Some of the ions may occur as major, minor or trace constituents. The spectrum of a mineral is governed by the *in toto* effect of the following factors:

– Spectra of dominant anions
– Spectra of dominant cations
– Spectra of ions occurring as trace constituents
– Crystal field effect.

A few examples are given below to clarify the above.

1. Limonite (iron oxide) exhibits a wide absorption band in the UV–blue region, due to the Fe-O charge-transfer effect; in addition, a ferric-ion absorption feature occurs in the near-IR (0.87 μm) region (Fig. 3.3). If the mineral is hydrated, the water molecule bands occur at 1.44 μm and 1.9 μm.
2. Quartz consists of simple silicate tetrahedra and its absorption bands occur only in the thermal-IR. However, the occurrence of impurities in quartz leads to absorption bands in the visible region and hence coloration, e.g. iron impurities give brown and green colours to the mineral.
3. In pyroxenes, absorption bands due to ferrous ion occur in the VNIR region and those due to silicate ion in the thermal-IR. The position and intensity of the absorption bands are governed by crystal field effect.

4. In amphiboles, micas and clays, absorption bands due to iron occur in the VNIR region; those due to hydroxyl ion in the SWIR region and those due to silicates in the thermal-IR region.
5. In all carbonate minerals, the carbonate absorption bands occur in the SWIR and thermal-IR regions. Additionally, absorption features due to iron in siderite, and due to manganese in rhodocrosite are seen in the VNIR region.
6. Clays exhibit the characteristic SWIR bands for Al-OH/ Mg-OH (bands at 2.2–2.3 µm) and water molecule bands at 1.4 µm. In the thermal-IR region, the Al-OH feature appears at 11 µm. The presence of iron leads to electronic transition bands in the near-IR region.

Therefore, it can be summarily concluded that the spectrum of a mineral is a result of the combination of the spectra of its constituents and crystal field effects.

3.6 Spectra of Rocks

Rocks are aggregates of minerals and are compositionally more complex and variable than minerals. Defining diagnostic spectral curves of rocks is difficult. However, it is possible to broadly describe the spectral characters of rocks, based on the spectral characters of the constituent minerals. The discussion here is divided into two parts: (a) the solar reflection region and (b) the thermal-IR region.

3.6.1 Solar Reflection Region (VNIR + SWIR)

Spectra of rocks depend on the spectra of the constituent minerals and textural properties such as grain size, packing, and mixing etc. Several models, semi-empirical methods and analytical procedures have been proposed to understand the spectra of polymineral mixtures (e.g. Johnson et al. 1983; Clark and Roush 1984; Smith et al. 1985; Adams et al. 1986; Huguenin and Jones 1986). This is a potent research field.

Four types of mixtures are distinguished: areal, intimate, coatings and molecular mixtures. These are discussed in greater detail in Section 11.2.4, as they are closely related to the concepts of spectral unmixing used in hyperspectral sensing. Briefly it may be mentioned here that areal mixtures exhibit linear additive spectral behaviour, whereas intimate mixtures show a non-linear spectral mixing pattern. Further, surface coatings and encrustations have immense influence on the reflection spectra, as the depth of penetration of EM radiation is of the order of barely 50 µm.

1. Igneous rocks. The representative laboratory spectra of igneous rocks in the visible, near-infrared and SWIR regions are shown in Fig. 3.11a. The graphic granites display absorption bands at 1.4 µm, 1.9 µm and 2.2 µm, corresponding to absorption bands of OH and H_2O. Biotite granites and granites have less water, and therefore the OH absorption bands are weaker. Mafic rocks contain iron, py-

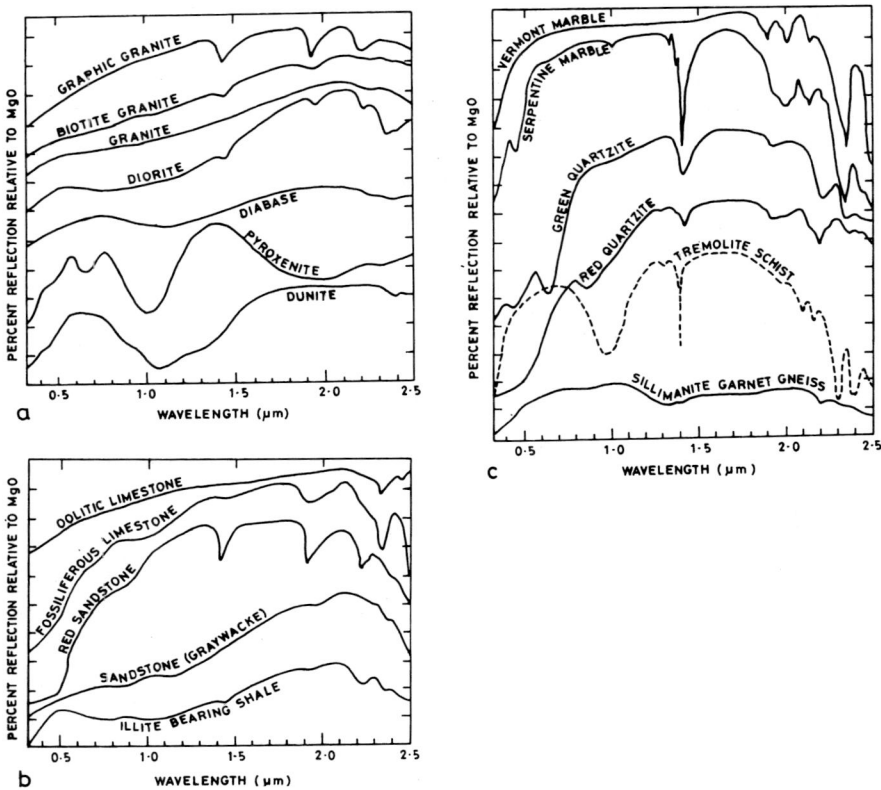

Fig. 3.11. Laboratory reflectance spectra of selected common rocks. **a** Igneous rocks, **b** sedimentary rocks and **c** metamorphic rocks. (Reflectance divisions are 10%) (Salisbury and Hunt 1974)

roxenes, amphiboles and magnetite, and therefore absorption bands corresponding to ferrous and ferric ion appear at 0.7 μm and 1.0 μm. Ultramafic rocks contain still larger amounts of opaque mineral and Fe^{2+}-bearing minerals, and therefore the ferrous bands become still more prominent, e.g. in pyroxenite. Dunite is almost all olivine and hence there is a single broad absorption band at 1.0 μm.

2. Sedimentary rocks. The laboratory spectral response of important sedimentary rock types in the VNIR + SWIR region is shown in Fig. 3.11b. All sedimentary rocks generally have water absorption bands at 1.4 μm and 1.9 μm. Clay-shales have an additional absorption feature at 2.1–2.3 μm. Ferrous and ferric ions produce absorption features in the VNIR. Carbonaceous shales are featureless. Pure siliceous sandstone is also featureless. However, sandstones usually have some iron-oxide stains, which produce spectral features (0.87 μm). Limestones and calcareous rocks are characterized by absorption bands of carbonates (at 1.9 μm and

2.35 μm, the latter being more intense); the ferrous ion bands at 1.0 μm are more common in dolomites, due to the substitution of Mg^{2+} by Fe^{2+}.

3. Metamorphic rocks. Typical laboratory spectra of common metamorphic rock types are shown in Fig. 3.11c. The broad absorption due to ferrous ion is prominent in rocks such as tremolite schists. Water and hydroxyl bands (at 1.4 μm and 1.9 μm) are found in schists, marbles and quartzites. Carbonate bands (at 1.9 μm and 2.35 μm) mark the marbles.

4. Alteration zones. Alteration zones, which form important guides for mineral exploration, are usually characterized by the abundance of such minerals as kaolinite, montmorillonite, sericite, muscovite, biotite, chlorite, epidote, pyrophyllite, alunite, zeolite, quartz, albite, goethite, hematite, jarosite, metal hydroxyls, calcite and other carbonates, actinolite–tremolite, serpentine, talc etc. These alteration minerals can be broadly organized into five groups:

(a) *quartz + feldspar (framework silicates)*, which exhibit no spectral feature in the VNIR–SWIR range and lead to general increased reflectance;

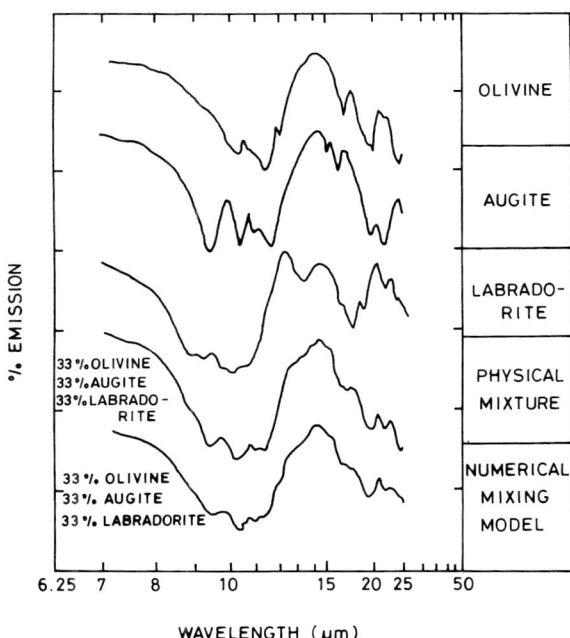

Fig. 3.12. Emission spectra modelling of mineral mixtures. The emission spectra of three minerals, olivine, augite and laboradorite, are shown. When these minerals are physically mixed to form an artificial rock composed of 1/3 of each of these components, the observed emission spectrum is shown as a physical mixture. Further, if the spectra of the individual minerals are combined (weighted by the above relative amounts) a virtually identical spectral curve is obtained. This shows that mineral spectra are additive in the TIR. (Christensen et al. 1986)

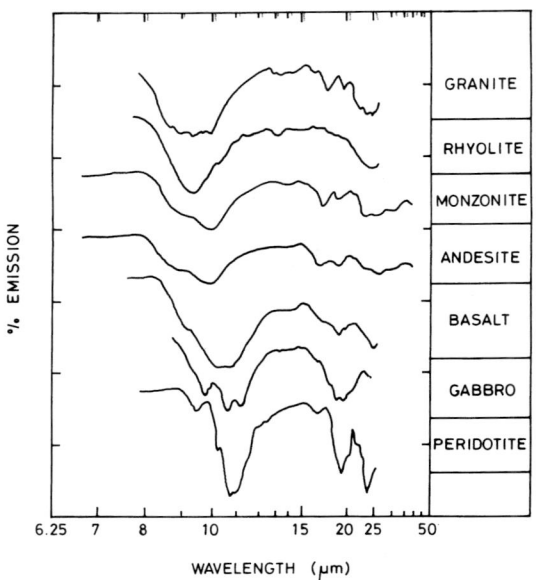

Fig. 3.13. Thermal-infrared spectra of common rocks varying from high SiO_2 (granite) to low SiO_2 (peridotite). Note the systematic shift in the absorption band with varying SiO_2 content. (After Vickers and Lyon 1967; Christensen et al. 1986)

  (b) *clays (sheet silicates)*, marked by absorption bands at 2.1–2.3 µm;
  (c) *carbonates*, which possess spectral features at 1.9 µm and 2.35 µm;
  (d) *hydroxyls, water and metal hydroxides,* which produce absorption features at 1.4 µm and 1.9 µm;
  (e) *iron oxides*, due to which spectral features in the VNIR region may occur.

The relative amounts of these assemblages may vary, which may result in corresponding variations in spectral response.

3.6.2 Thermal-Infrared Region

For the thermal-IR region, the mineral spectra are basically additive in nature (see e.g. Fig. 3.12). Therefore, rock spectra in the TIR region are readily interpretable in terms of relative mineral abundances.

The thermal-infrared spectra of selected igneous rocks, arranged in decreasing silica content from top to bottom, are shown in Fig. 3.13. It can be seen that the centre of the minimum emission band gradually shifts from about 9 µm in granite to about 11 µm in olivine-peridotite. This is due to the corresponding shift in the Si-O absorption band (Fig. 3.10) in mineral groups which form the dominant silicates in the above igneous rocks.

Additional diagnostic bands in the TIR region are associated with carbonates, hydroxyls, phosphates, sulphates, nitrites, nitrates and sulphides.

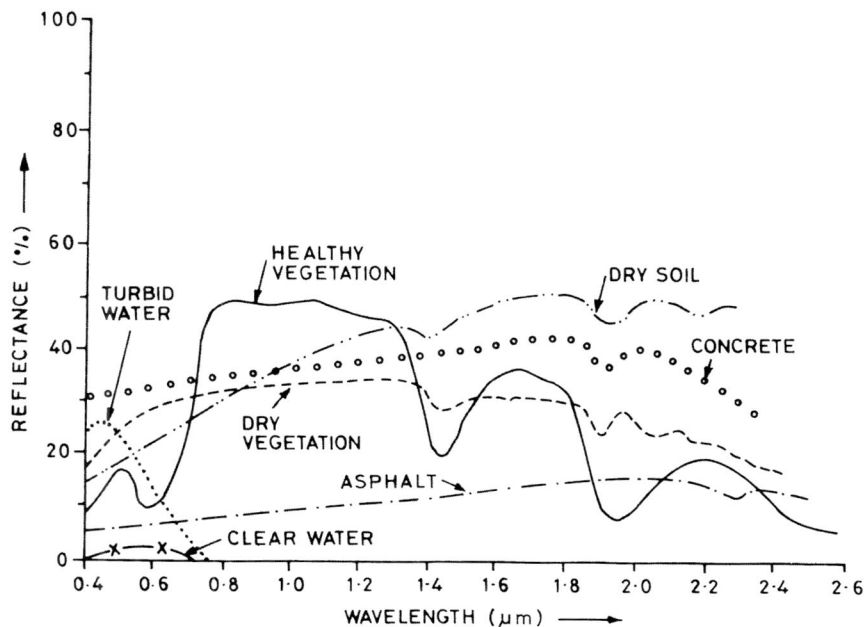

Fig. 3.14. Generalized spectra of selected common objects

3.7 Laboratory vs. Field Spectra

Laboratory data are generally free from complexities and interference caused by factors such as weathering, soil cover, water, vegetation, organic matter and man-made features, which affect the in-situ spectra (Siegal and Abrams 1976; Siegal and Goetz 1977). The extent to which *in-situ* spectra are identical to laboratory spectra may be quite variable and therefore field spectra should be interpreted with care. In general, freshly cut surfaces show higher reflectance than weathered surfaces. Scanty dry grass cover, thin soil and poor organic content in the soil tend to increase the similarity between field and laboratory spectra.

In the solar reflection region, the information comes from about the top 50 μm surface layer zone, and so the spectra are affected by surface features; therefore the correspondence between laboratory and field (geological) reflectance data may be only limited and has to be carefully ascertained.

In the thermal-IR region, the information is related to about the top 10-cm-thick surface zone. Therefore, the spectral features of the bedrock are more readily observable on TIR remote sensing data, even if surficial coatings, encrustation, varnish etc. are present.

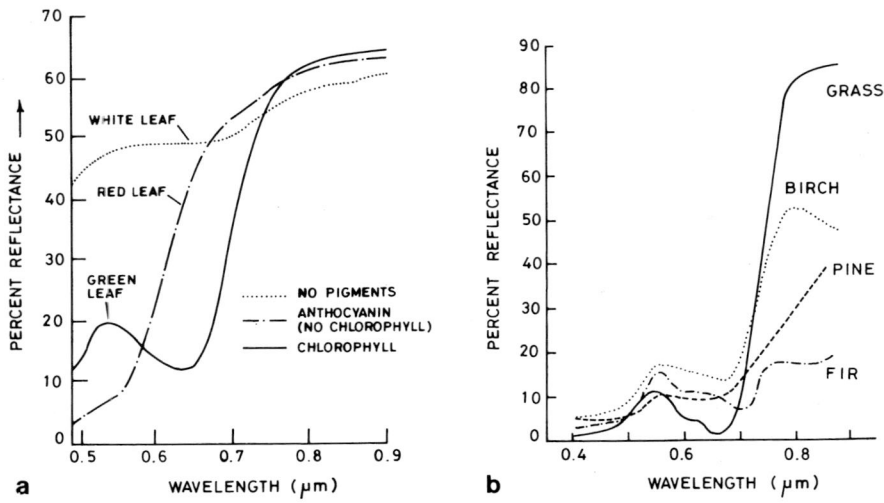

Fig. 3.15. a Spectral response of leaves with different types of pigmentation. (Hoffer and Johannsen in Schanda 1986) **b** Spectral reflectance curves for vegetation differing in foliage and cell structure. (Goetz et al. 1983)

3.8 Spectra of Other Common Objects

From the point of view of object discrimination and data interpretation, it is necessary to have an idea of the spectra of other common objects.

The spectra of selected common natural objects in the VNIR–SWIR region are shown in Fig. 3.14. The deep clear water body exhibits low reflectance overall. The turbid shallow water has higher reflectance at the blue end, due to multiple scattering of radiation by suspended silt, and due to bottom reflectance. Fresh snow generally has a high reflectance. Melting snow develops a water film on its surface and therefore has lower reflectance in the near-IR. Soil reflectance is governed by a number of factors such as the parent rock, type and degree of weathering, moisture content and biomass. Common sandy soil exhibits even-tenor reflectance in the visible region and generally increasing reflectance towards the near-IR, which may be greatly modified by the presence of other ingredients such as iron oxide and water. Concrete and asphalt exhibit medium and low reflectances respectively, which are nearly uniform throughout the VNIR–SWIR region.

A large part of Earth's surface is covered with vegetation and therefore vegetation spectra have drawn greater attention, especially recently from the geobotanical point of view. Valuable contributions have been made by numerous workers (Gates 1970; Gausmann et al. 1977, 1978; Horler et al. 1980, 1983; Collins et al, 1981, 1983; Milton et al. 1983; Sellers 1985; Peterson et al. 1988; Rock et al. 1988).

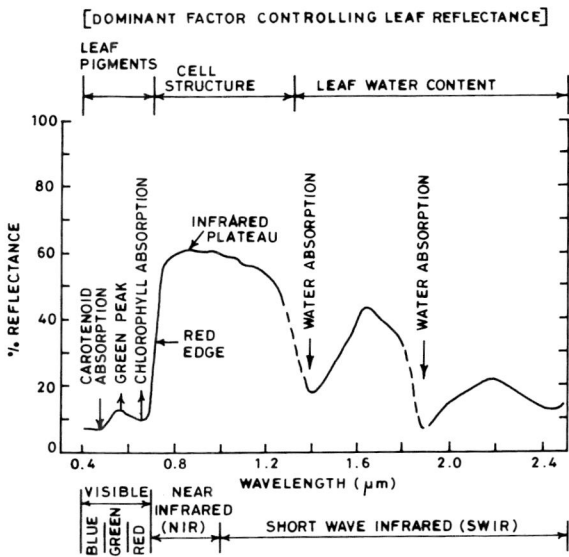

Fig. 3.16. A typical reflectance curve of green vegetation in the visible, near-infrared and short-wave-infrared region. (After Goetz et al. 1983)

In general, in the visible region, leaf pigments govern the leaf spectrum (Fig. 3.15). The normal chlorophyll-pigmented leaf has a minor but characteristic green reflection peak. In the anthocyanin-pigmented leaf, the green reflection is absent and there is greater reflection in the red wavelength, leading to a red colour. The spectrum of the white leaf (no pigments) has a nearly constant level in the visible region. In the near-IR region, in general, the spectral reflectance depends on the type of foliage and cell structure. Some leaves, such as fir and pine, reflect weakly in the near-IR, whereas grass reflects very strongly (Fig. 3.15b). This region can therefore be used for identifying vegetation types.

The characteristic spectrum of a healthy green leaf is shown in Fig. 3.16 in the VNIR–SWIR region. Leaf pigments absorb most of the light in the visible region. There is a minor peak at 0.55 µm leading to a green colour. The absorption feature at 0.48 µm is due to electronic transition in carotenoid pigments, which work as accessory pigments to the chlorophyll in the photosynthetic process, and the 0.68 µm absorption is due to electronic transition in the chlorophyll molecule centred around the magnesium component of the photoactive site. The region 0.8–1.3 µm shows a general high reflectance and is called the near-IR plateau. The reflectance in this region is governed by leaf tissue and cellular structure. The sharp rise near 0.8 µm, which borders the near-IR plateau, is called the red edge of the chlorophyll absorption band. The near-IR plateau also contains smaller and potentially diagnostic bands, which could be related to cellular structure and water content in the leaf. The ratio of the near-IR to visible reflectance is, in general, an indication of the photosynthetic capacity of the canopy, and is used as a type of *vegetation index* (also see Sect. 16.13.1). The region 1.0–2.5 μm (SWIR) contains prominent water absorption bands at 1.4 μm, 1.9 μm and 2.45 μm (Fig. 3.16). The reflectance in the SWIR is related to biochemical content in the canopy, such as proteins (nitrogen concentration), lignin and other leaf constituents.

The spectra of vegetation over mineral deposits has drawn considerable attention recently, mainly because of the growing awareness among researchers that vegetation spectra undergo fine modification due to geochemical stresses. These aspects related to hyperspectral sensing are discussed in Chapter 11 in more detail.

3.9 Future

Major directions of future development in this field are anticipated to be the following (Kahle et al. 1986): (1) compiling a spectral library for representative and well-characterized rocks, minerals, soils etc., and understanding the effects of coatings and differing particle sizes etc.; (2) understanding the effects of chemical changes including elemental substitution, solid-solution, lattice distortion etc. on spectral characters; (3) determination of 'real-world' spectral properties, incorporating mixing models of mineral abundance from field data; (4) investigation of directional effects on spectra obtained in the field and under natural atmospheric conditions; and (5) determination of the exact relationship between emission, reflection and transmission spectra.

Chapter 4: Photography

4.1 Introduction

Photography was invented in 1839 by N. Niepce, W.H.F. Talbot and L.J.M. Daguerre, and since then photographic techniques have been used in applied sciences for various applications. Photographic pictures of the Earth were first acquired from balloons and kites, until the aeroplane was invented in 1903. World Wars I and II provided a new opportunities and challenges to apply photographic techniques from aerial platforms. Soon afterwards, man started exploring space and observing the Earth from space platforms, using improved photographic techniques. Photographic systems for remote sensing have been discussed by Smith and Anson (1968), Colwell (1976), Slater (1980, 1983), Eastman Kodak Company (1981, 1990, 1992), Curran (1985), Teng et al. (1997) and Lillesand and Kiefer (2000), and in numerous other publications.

4.1.1 Relative Merits and Limitations

The main advantages of using photographic systems for remote sensing of the Earth are the following.

1. Analogy to the eye system. The photographic camera works in a manner analogous to the human eye – both having a lens system. Photographic products are, therefore, easy to study and interpret.

2. Geometric fidelity is another great advantage associated with photographic systems, as intraframe distortions do not occur.

3. Stereo capability. The photographic technique still remains the most widely used one for stereoscopic studies.

Besides the above, these systems also have advantages associated with all remote sensing techniques viz. synoptic overview, permanent recording, feasibility aspects, time saving capability, and multidisciplinary applications (see Sect. 1.4).

Limitations or disadvantages in using photographic systems arise from the following.

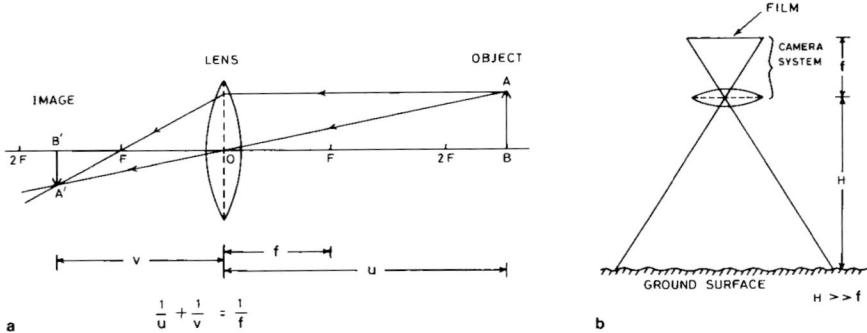

Fig. 4.1a,b. Working principle of photographic system. **a** Image formed by a converging lens. **b** Configuration for remote sensing photography

1. Limited spectral sensitivity. Photographic films are sensitive only in the 0.3–0.9 µm wavelength range and the rest of the longer wavelengths cannot be sensed by photographic techniques.

2. Retrieval of films. The exposed film containing the information has to be retrieved, i.e. a transportation system of some sort (aircraft, parachute, space shuttle etc.) is necessary to retrieve the film. This aspect of the photographic system is in contrast to that of scanners, where data is telemetered on microwave links to ground receiving stations, making the latter a highly versatile technique. For this reason, the use of photographic techniques for Earth-observation purposes has generally remained confined to aerial platforms and some selected space missions.

3. Deterioration of film quality with age is a common problem.

4. Deterioration in photographic duplication. The duplication of photographic products for distribution etc. is often accompanied by some loss of information content.

4.1.2 Working Principle

The working principle of an aerial/space camera system is simple. For an object being focused by a convex lens the well-known relation giving distances is (Fig. 4.1):

$$\frac{1}{u} + \frac{1}{v} = \frac{1}{f} \qquad (4.1)$$

where, u = distance of object from lens centre, v = distance of image from lens centre, and f = focal length of the lens.

If the object is far away (i.e. u ≈ ∞), the image is formed at the focal plane (v = f). Cameras for remote sensing purposes utilize this principle and are of fixed-

Cameras

Fig. 4.2. A typical single-lens frame camera. (Courtesy of Carl Zeiss, Oberkochen)

focus type. They carry a photosensitive film placed at the focal plane and the objects falling in the field-of-view of the lens are imaged on the film.

A photographic system consists of three main components: camera, filters and film.

4.2 Cameras

Commonly, the cameras used in remote sensing are precision equipment. Their main element is a highly sophisticated lens assembly, which images the ground scene on the film. The cameras are placed on stable mounts, as even very slight shaking would seriously affect the quality of photographs and their resolution. A variety of cameras have been used for photographic remote sensing.

Depending upon the film format (i.e. size), four types of cameras are distinguished: (a) small-format (35 mm), (b) medium-format (70 mm), (c) medium to large-format (126/140 mm) and (d) large-format (240 mm). Large-format cameras have been used quite extensively in aerial photography. The use of smaller-format cameras (35 mm, 70 mm and 126 mm) has been mainly for non-metric photography, especially when ease of use in terms of camera size and mobility is an important consideration. For example, such cameras have been used in a hand-held manner by astronauts on the space shuttle. Another important application of the small-format camera is in low-altitude aerial photography with high temporal resolution (weekly or even more frequently); this could be done over an area, for example to check spatial variation in flowering time of plants in a geobotanical survey.

Four types of cameras can be distinguished, depending upon their construction, objective and working (for details, refer to e.g. Slater 1975, 1983; Colwell 1976): (1) single-lens frame cameras, (2) panoramic cameras, (3) strip cameras, and (4) multiband cameras.

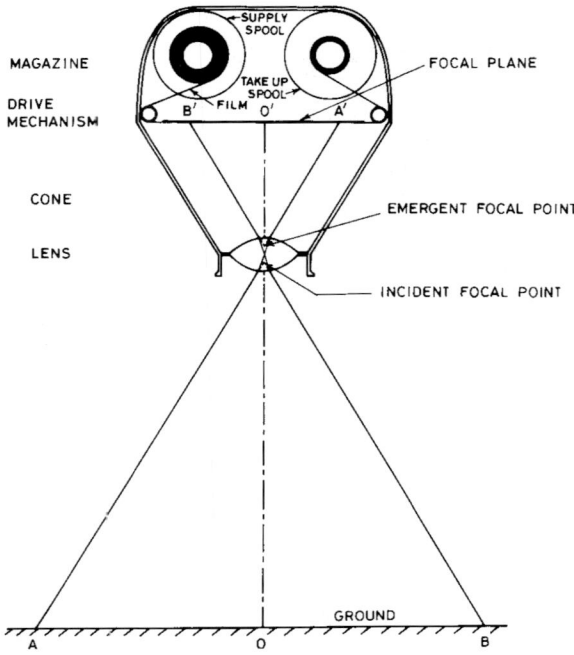

Fig. 4.3. Basic components of a single-lens frame camera. (After Colwell 1976)

4.2.1 Single-Lens Frame Cameras

The single-lens frame camera, also known as a single-frame conventional camera, (Fig. 4.2) has been the most widely used type in aerial photography. Its basic components are the magazine, drive-mechanism, cone and lens (Fig. 4.3). The magazine holds the film (common width 240 mm) and includes a supply spool and a take-up spool. The drive mechanism is a series of mechanical devices for forward motion of the film after exposure and for image motion compensation. The film for exposure is held in a plane perpendicular to the optical axis of the lens system, and after exposure is successively rolled up. The cone is a light-tight component, which holds the lens at a distance 'f' from the film plane. The lens system is a high-quality chromatically corrected assembly to focus the ground objects on the film plane. It also comprises filters, diaphragm and shutter etc. Attached to the camera are a view-finder (to sight the camera), an exposure meter (to measure the light intensity) and an intervalometer (to set the speed of the motor drive and obtain the desired percentage of overlap for stereoscopic purposes). On-board GPS could be employed to locate the precise position of the photo-frame.

In a camera, the angle subtended at the lens centre from one image corner to the diagonally opposite image corner is called the angular field-of-view (FOV) or

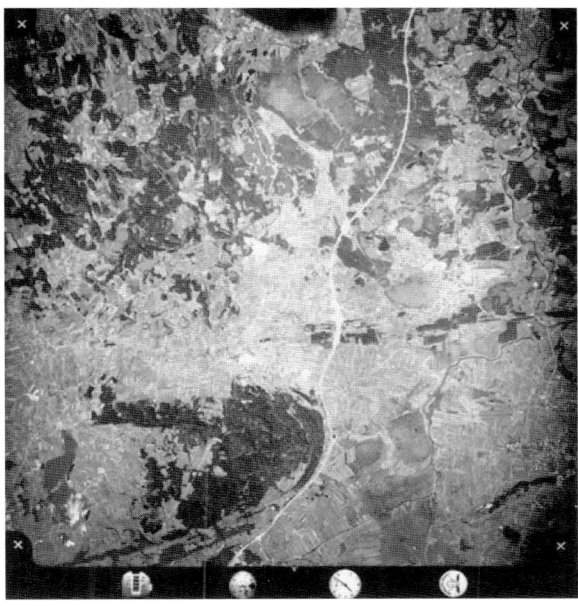

Fig. 4.4. A typical aerial photograph showing fiducial marks and flight data. The terrain is Alpine foreland, Germany; note the Murnau syncline in the folded molasse sediments to the south. (Courtesy of Hansa Luftbild GmbH)

angle of the lens. In aerial photography, the lens angles are called normal (50°–75°), wide (75°–100°) and super wide (100°–125°), the lens of 90° being the most commonly used type. The focal length of the lens system directly controls the scale of photography as scale S = f/H (Fig. 4.1b). Larger focal length implies smaller angular field-of-view, less areal coverage, and larger scale of photography, other factors remaining the same. On the basis of focal length, the aerial cameras are called short focal length (f $\approx$ 88 mm), medium focal length (f $\approx$ 150 mm) and long focal length (f > 210 mm) cameras. In space photography, the altitude is very high and, in general, a camera lens of a smaller FOV (about 15°–20°) and large focal length (about 300–450 mm) is used.

Single-lens frame cameras are often distinguished as two types: (a) frame reconnaissance cameras and (b) mapping cameras. *Reconnaissance cameras* are less expensive than mapping cameras, both to buy and to operate, and come in a variety of configurations. Commonly, they have a long focal length (ranging from a few cm to more than 1 m, the 15 cm, 30 cm and 45 cm being more common) and a narrow angular field-of-view (10°–40°). Further, their lens design may sometimes also preclude the use of colour films. A number of frame reconnaissance cameras have been flown on space missions, such as Gemini, Apollo etc., the most worthy of mention being the Earth Terrain Camera (ETC) flown on SKYLAB during 1973–74.

Mapping cameras are used to obtain high-quality vertical photographs. These are also variously named metric cameras, photogrammetric cameras or cartographic cameras. A distinctive feature of such cameras is very high geometric accuracy, which enables photogrammetric measurements. Reseau marks (consisting of several fine cross-marks across the photographs) are exposed on the film to enable determination of any possible dimensional change in the photographic product. Principal point, fiducial marks and reseau marks are exposed simultaneously with the exposure of the ground scene, and extensive flight and camera data are also shown alongside each frame (Fig. 4.4). The main application of mapping cameras is to acquire vertical photogrammetric photography.

4.2.2 Panoramic Cameras

Panoramic cameras have been deployed in the past for photo-reconnaissance surveys, particularly for mapping dynamic resources. A panoramic camera typically carries a long focal length lens (about 600 mm) and scans the ground from horizon to horizon, providing a large total angular field-of-view (typically > 100°). The resulting photographic products are geometrically highly distorted images. These cannot be used for geometric mapping, even after many corrections. This type of camera was used in the past for acquiring information on the temporal status of dynamic features (e.g. floods). However, the erstwhile unique advantage of panoramic cameras, that they cover large tracts in a single flight, has now been overshadowed by the satellite data, which provide regional information in a geometrically correct format.

4.2.3 Strip Cameras

Strip cameras were developed for detailed photography in selected strips. They were used initially for military intelligence and later for civilian tasks such as route alignment, where details along an alignment are required. The strip camera has a slit-shaped FOV, through which the film is exposed. The film is moved in the focal plane of the camera at exactly the same velocity as that with which the image is moving past the slit, which automatically renders image motion compensation. However, as the total angular FOV of strip cameras is small, their use has declined, especially as a result of the significant improvements in satellite sensor products.

4.2.4 Multiband Cameras

A multiband camera (also called a multispectral camera) is used to photograph the same ground scene in different wavelength ranges. The cameras are co-sighted (i.e. view the same area on the ground) and are triggered simultaneously. The number of cameras has varied from two to nine in different versions, although four or six cameras have been most frequently used. Commonly, 70-mm format cameras are used for this purpose. Photography in different wavelength ranges is made possible by using different film/filter combinations, the most commonly used configuration being the following:

Camera I   Green band
Camera II   Red band
Camera III Infrared band (black-and-white)
Camera IV   Colour infrared film

Many aerial and space missions yielding interesting multiband photographic products have been flown in the past. However, the film used is commonly of 70-mm width, which provides a relatively low ground distance resolution. As multispectral imagery from space sensors, which has high spatial resolution, is readily available, interest in this technique has also lately declined.

4.3 Films

The films used for photography are of two main types: black-and-white and colour films.

4.3.1 Black-and-White Films

Spectral sensitivity. A black-and-white film portrays only brightness variations across a scene. Depending upon the range of EM spectrum to which a film is sensitive, it can be called panchromatic (sensitivity 0.3–0.7 μm) or infrared (sensitivity 0.3–0.9 μm or sometimes up to 1.2 μm). The sensitivity of panchromatic film is quite analogous to that of the human eye. It is also called *standard film* and is extensively used in aerial photography. *Infrared films* are sensitive in the UV–visible and a part of the infrared region. Spectral sensitivity curves of black-

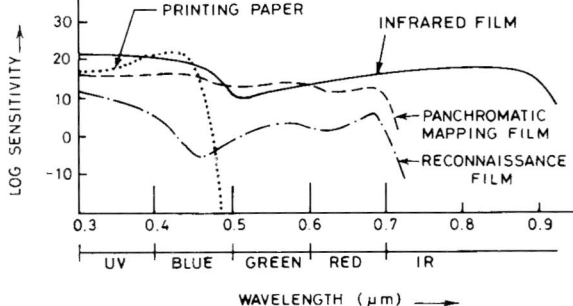

Fig. 4.5. Typical spectral sensitivity curves for different types of films and printing paper. (Adapted after Wolf 1983; Eastman Kodak 1992)

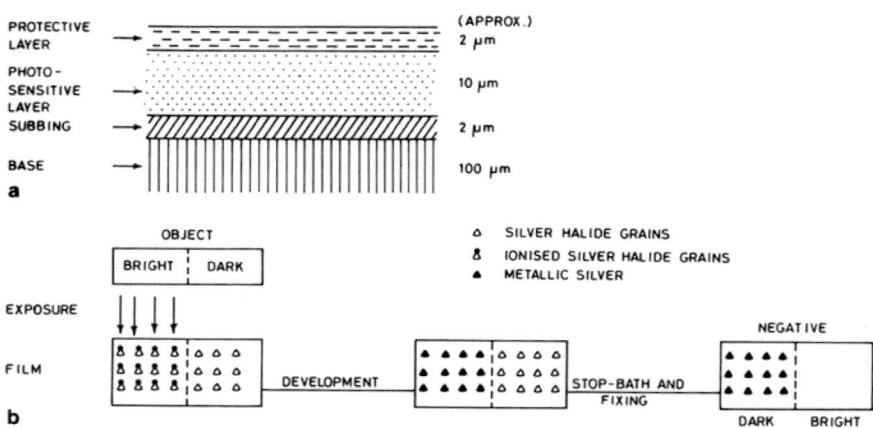

Fig. 4.6. a Structure of a black-and-white photographic film. **b** Working principle of a black-and-white film

and-white aerial films are shown in Fig. 4.5. The films generally show higher sensitivity to red wavelength. Most of the black-and-white films are quite sensitive to blue light and need a filter to cut off these wavelengths for optimum results. Further, if an infrared black-and-white film is to be used for only infrared photography, a filter to cut off the visible radiation is necessary.

Design. In construction, a black-and-white film consists of several layers (Fig. 4.6a). The main layer is a flexible transparent polyester base about 100 μm thick, the purpose of which is to impart strength and geometric consistency, vital for photogrammetric purposes. The top surface of the polyester layer is generally somewhat uneven and this is made smooth by a thin layer called subbing. Over this lies the photo-sensitive emulsion layer – the heart of the film. The emulsion layer contains highly irregularly shaped silver halide grains, a few microns or less in diameter, uniformly dispersed in a solidified gelatin. Over this is a thin protective layer. Below the polyester base, a fine anti-static layer is provided to facilitate rolling/movement of the film.

Exposure and processing. As the film is exposed to light, photons impinge on the silver halide grains and knock out electrons, causing their ionization (Fig. 4.6b). At this stage, the film carries an invisible impression of the scene called a latent image. The exposed film is immersed in an alkaline solution (reducing agent) for development, which does not react with unexposed silver halide grains, but reduces the exposed grains to metallic silver. After this, the film is put in a stop-bath containing acidic solution to stop further development, and then in a chemical fixing solution, which dissolves and takes out the unexposed silver halide grains. In this manner a negative is formed. On this negative, the originally bright objects appear dark, as they correspond to areas of metallic silver deposition. A negative can then be used to produce positive prints and enlargements. The theory of the photographic process has been discussed in detail by Mees and James (1966) and Jones (1968).

Film sensitometry (D–log E curve). Films vary in their sensitivity to light – some are fast and others slow. Film sensitometry deals with the evaluation of sensitivity of films to light, i.e. to describe a relation between exposure (E) and resulting density (D).

Exposure (E) is a measure of light energy received at any point on the film plane as the film is exposed. It is given by

$$E = \frac{sd^2 t}{4f^2} \qquad (4.2)$$

where E = film exposure, s = brightness of the object (W mm^{-2} s^{-1}), d = diameter of the lens opening (mm), t = time of exposure (s) and f = lens focal length (mm).

It can be seen from the above equation that various combinations of d and t can be made to give the same value of E. Density (D) is used to describe "darkness" at a point on the film, resulting from the deposition of metallic silver consequent upon exposure and development. It is equal to the log of the opacity at a point. If a beam of light is incident on a film at a point, opacity at that point (O_p) is the ratio of the intensity of the incident beam to that of the transmitted beam. The reverse of this, i.e. the ratio of the transmitted intensity to the incident intensity, is called transmittance (T_p) at that point. As opacity varies by several orders of magnitude, log of opacity ($\log_{10} O_p$) is used to describe density at a point.

A direct correlation exists between exposure and density. Larger exposure leads to deposition of more metallic silver on the film and hence higher opacity and higher density. However, the rate at which density increases with exposure is im-

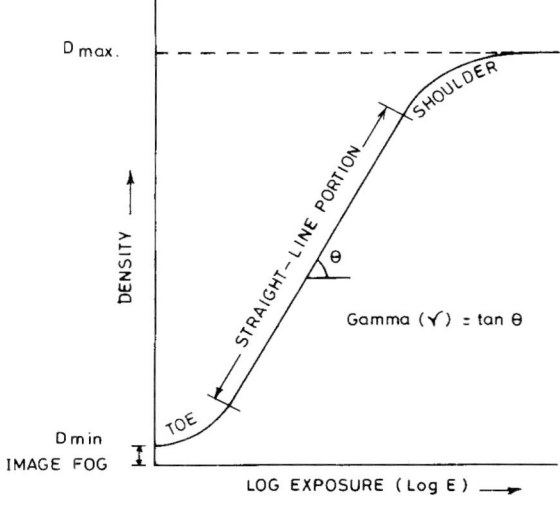

Fig. 4.7. A typical D–log E curve of a black-and-white film

portant and governs the speed and contrast parameters of the film. To evaluate this, the relation between density (D) and log of exposure ($\log_{10} E$) is experimentally plotted for every batch of films (Fig. 4.7). The curve is called the D–log E curve, or film characteristic curve, or also the H and D curve (after Hurter and Driffeld, who first conceived these curves in the 1890s). The film is exposed over different timings and the resulting respective densities are measured using a densitometer.

A D–log E curve typically exhibits three main segments – toe, straight line and shoulder. Even at the time of no-exposure, the film has a certain opacity, although quite small, and this is referred to as minimum density (D_{min}). This D_{min} is partly due to fog in the gelatin layer. As the exposure is increased, the density rises at an ever-increasing rate (Fig. 4.7), this portion of the curve being called the toe. After this, the relation between D and log E is rectilinear and this is the straight-line portion. Beyond this, any further increase in exposure leads to increase in density, but at a decreasing rate, until a maximum density (D_{max}) is reached. This part of the curve is called the shoulder. The straight-line portion of the curve is of greatest interest for photographic applications as through this the observed density variations in the photograph can be directly related to brightness variations in the scene.

The D–log E character may differ from batch to batch or may also be affected by storage etc. For optimum results, the D–log E plot is determined just before the actual flight.

Film contrast. The slope of the straight-line portion of the D-log E curve is called gamma (γ)

$$\gamma = \frac{D}{\log E} \tan \theta \tag{4.3}$$

where γ is a measure of film contrast. Consider two different cases, where θ [= $\tan^{-1}(\gamma)$] is steep (film F_1), and gentle (film F_2) (Fig. 4.8). If these two films are exposed through the same range of exposure values, the resulting variation in density is higher in film F_1 and lower in film F_2. Film F_1 is called high-contrast film and film F_2 is called low-contrast film. Films having moderate slopes (about 45°) are called normal films.

Speed of aerial film. The speed of an aerial film (or aerial film speed, AFS) is a measure of the rapidity (in terms of exposure) with which a certain density is attained. In general, high-contrast films gain density faster and hence are faster, and low-contrast films are accordingly slower.

Film resolution denotes the spatial resolution capability of a film and is given for high-contrast objects. It is expressed in line-pairs/mm (e.g. 100 line-pairs/mm or 400-line pairs/mm). The capability to resolve objects depends on the contrast ratio of the objects (i.e. ratio of maximum brightness to minimum brightness in the objects). The capability is high for a high-contrast object and low for a low-contrast object. Film resolution is conventionally given for high-contrast objects.

Films

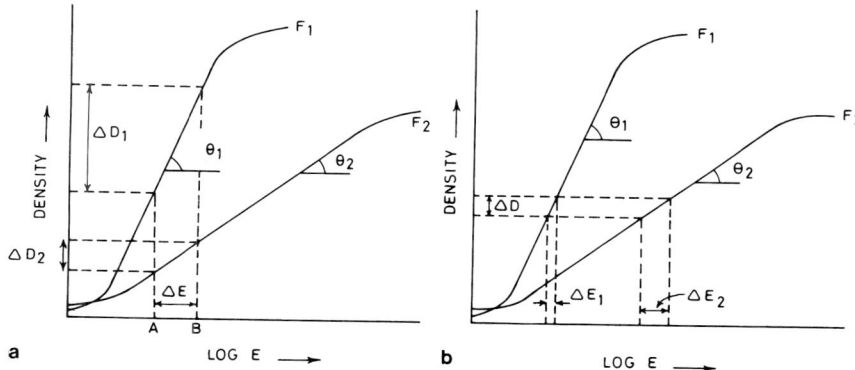

Fig. 4.8. a Film contrast. Consider two films F_1 and F_2; for the same exposure difference, ΔE, film F_1 gives a greater contrast than film F_2 ($\Delta D_1 > \Delta D_2$); therefore F_1 is a higher contrast film than F_2. **b** Radiometric resolution. If ΔD is the resolution of a densitometer used for measuring density of films, then F_1 provides detection of finer exposure differences than F_2 ($\Delta E_1 > \Delta E_2$); hence, a faster film has a higher radiometric resolution

If an object consisting of many equally spaced dark and bright lines or bars is imaged on a film, then the closest spacing that the film can resolve is the film resolution. It is given in line-pairs mm^{-1}.

Film resolution and speed of the film are inversely related, both being dependent on the size of the silver halide grains. A film having finer grains is slower and provides higher *spatial resolution* than a film with coarser grains (Table 4.1).

Radiometric resolution is the smallest detectable exposure change possible in a film. The high-contrast or faster films provide finer radiometric resolution than low-contrast or slower films (Fig. 4.8b). It is not possible to have a film with high spatial resolution as well as high radiometric resolution; a trade-off between speed and resolution has to be made.

Table 4.1. Speed and resolution of black-and-white films. (After Curran 1985)

Kodak film number	Film type	Film speed	Granularity (1=low, 4=high)	Film spatial resolution in lines/mm at a contrast of 1.6:1	Likely ground resolution of film in cm at an image scale of 1:15,000
3414	Reconnaissance	8(very slow)	1	250	6
3410	Reconnaissance	40	2	80	19
2402	Mapping	200	3	50	30
2403	Mapping	640(very fast)	4	25	60

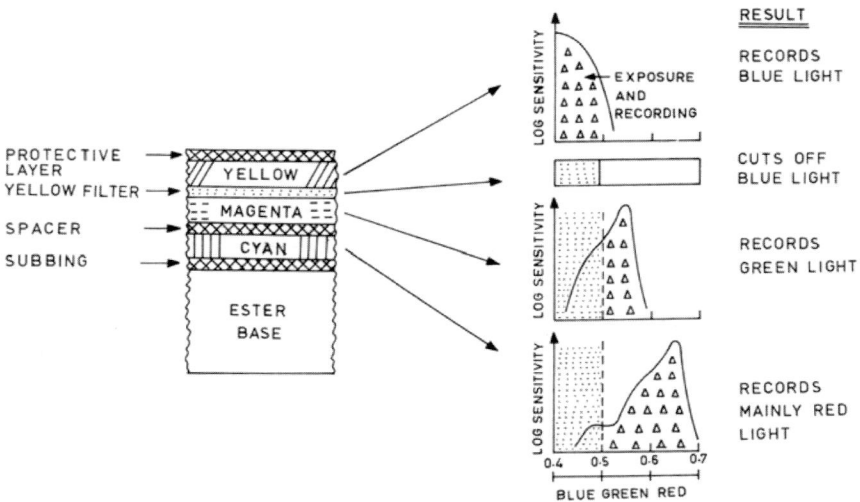

Fig. 4.9. Design of a tripack colour negative film. (After Curran 1985; Eastman Kodak 1992)

4.3.2 Colour Films

There are three primary additive colours (blue, green and red), and complementary to these there are three primary subtractive colours (yellow, magenta and cyan) (Appendix 4.1). Mixing any one set of the primary colours in different proportions can produce the entire gamut of colours. The colour films utilize this principle. All colour films consist of three layers of emulsions, superimposed over each other (Figs. 4.9 and 4.10). Each of the three layers is sensitive to a particular wavelength range of EM radiation and is colour-dyed in a certain primary subtractive colour (one in yellow, second in magenta and third in cyan). The amount of colour dye that is introduced in a layer is related to the degree of exposure that particular layer has undergone. Thus, depending upon the cumulative effect of dye-colouring in the three layers, different colours show up. The pack of three emulsion layers is mounted on a flexible polyester base in order to impart physical strength to the film, as in the case of black-and-white films.

Colour films are basically of two types: colour negative films, which result in a colour negative, and reversal films, which directly yield a colour positive transparency.

4.3.2.1 Colour Negative Film

In a colour negative film, the three photo-sensitive emulsion layers are sensitive to radiation in the visible range of the spectrum, one to blue, one to green, and the

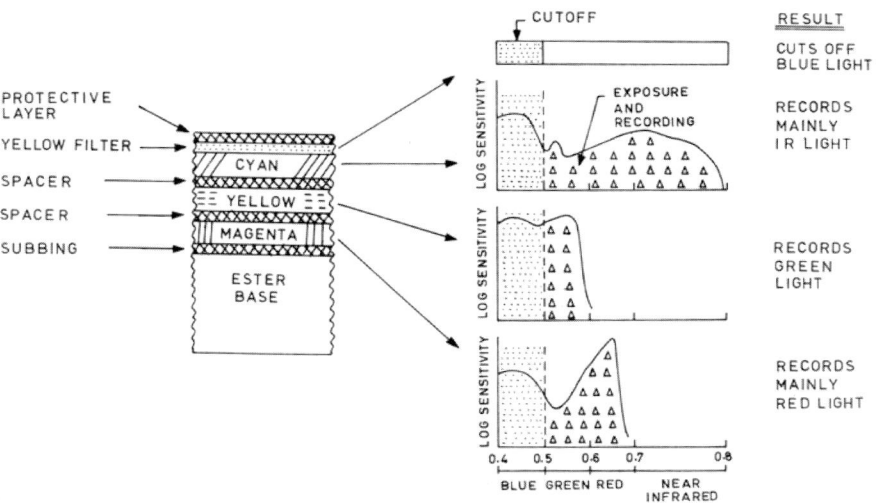

Fig. 4.10. Design of a tripack colour infrared film. (After Curran 1985; Eastman Kodak 1992)

third to red light (Fig. 4.9). A yellow filter is used to cut off blue light from reaching and sensitizing the green- and red-sensitive layers. The emulsion layer sensitive to radiation of a particular colour is colour-dyed in its complementary primary subtractive colour. This yields a negative, i.e. the objects finally appear in their complementary colours (Fig. 4.11 and Table 4.2). In remote sensing investigations, this film is useful as it can be used to generate paper positive prints, which are convenient to handle, particularly in the field.

Table 4.2. Manifestation of selected colour – objects on different types of colour films

Object colour	Colour negative film	Colour positive film	Colour infrared film
White	Black	White	White
Blue	Yellow	Blue	Black
Green	Magenta	Green	Blue
Red	Cyan	Red	Green
Infrared-reflecting	Black	Black	Red
Black	White	Black	Black

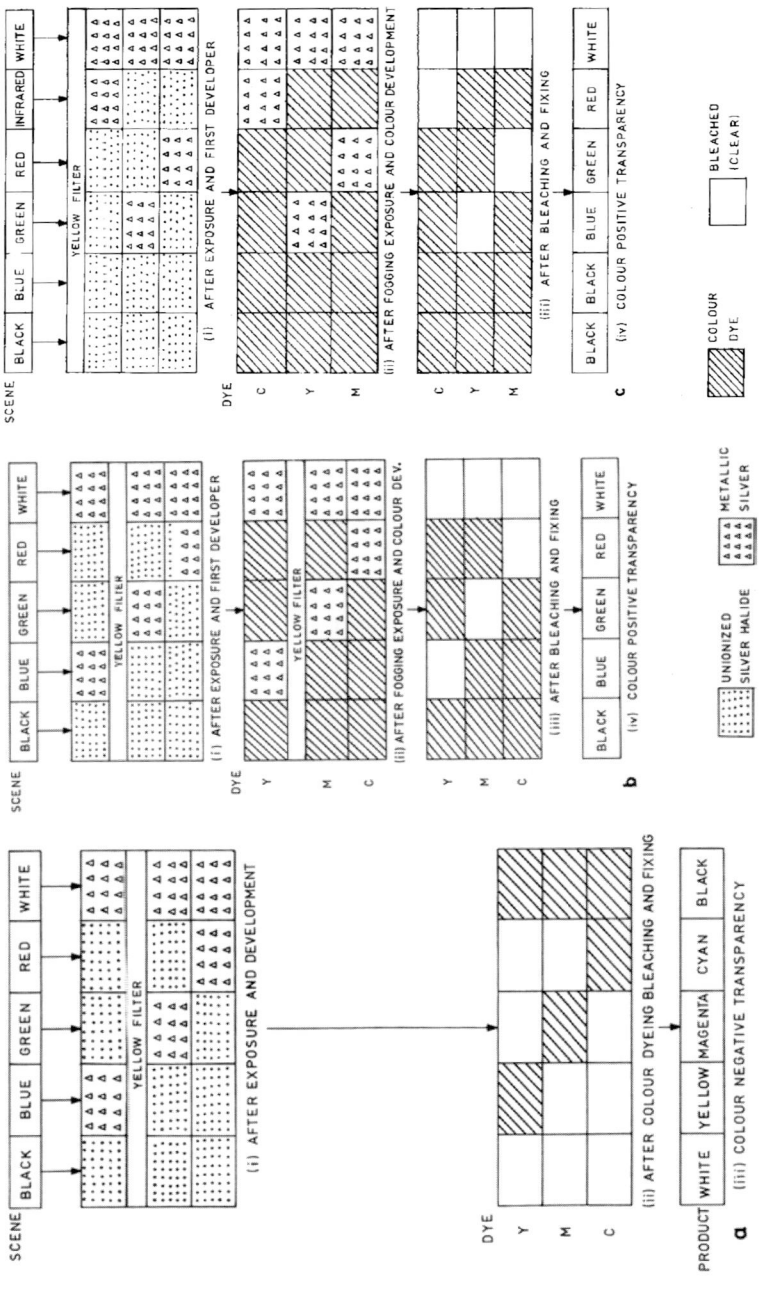

Fig. 4.11a–c. Working principle of colour films. **a** Colour negative film. **b** Colour positive film. **c** Colour infrared film. (Adapted after Slater 1980)

4.3.2.2 Colour Reversal Films

Reversal films directly yield positive transparencies. It has been mentioned above that the photographic negatives are produced by cultivating the ionized silver halide grains, but how about the unionized silver halide grains? They carry a latent positive image, and have the potential to yield a positive image directly. Reversal films utilize this principle. As a sort of inversion or reversal process is involved, the films are called reversal films. This type of film has found very wide usage in remote sensing (and also in popular photography). Depending upon their sensitivity to the range of the EM spectrum, two types of reversal films are distinguished: colour positive film and colour infrared film.

1. Colour positive film. The range of sensitivity of the colour positive film is similar to that of a colour negative film. However, the working principle is slightly different (Fig. 4.11). As the film is first exposed, a latent image is formed, as in a colour negative film. At this stage the film is put into a first developer, which develops the ionized silver halide grains and turns them into metallic silver. After this, the film is thoroughly washed and is re-exposed to light, called flashing. This activates the earlier-unionized silver halide grains, which carry a latent positive image. Then, the film is placed into a second developer, which is a dye-coupler. It replaces the newly ionized silver halide grains by colour dyes (yellow, magenta and cyan) in the three respective layers. Finally, the film is bleached (to remove the metallic silver which was formed during the first development), washed and fixed. This gives a colour positive film product. If this product is held against the light, normal colours of objects are seen (Table 4.2).

2. Colour infrared film. In a colour infrared (CIR) film, the three emulsion layers are made sensitive to green, red and near-infrared regions of the EM spectrum respectively (Fig. 4.10). A minus-blue or yellow filter is mounted on the top, to cut off blue radiation from reaching any of the layers, and this also enhances image contrast. The colour-dyeing scheme of the tripack is: green-sensitive layer is dyed in yellow, red-sensitive layer in magenta and infrared-sensitive layer in cyan. The working principle of a CIR film is analogous to that of the colour positive film, involving exposure, formation of a latent image, first development, washing, flashing, second development with a dye-coupler, bleaching, fixing and washing (Fig. 4.11). If this film is held against the light, the various objects appear in false colours (Table 4.2). This type of film is therefore also called a *false colour film*. Further, in view of its specific ability to discriminate green colour from vegetation, it is also called a *camouflage detection film*.

The characteristic curves for colour films have the same significance as for black-and-white films. A colour film comprises three photosensitive emulsion layers and for each layer one D–log E curve exists, depicting its variation in density with exposure. The speed of a colour aerial film is a relative film speed, which is assessed by subjective comparison of photographs, taken the colour or black-and-white films of known speed, under similar conditions.

In general, the normal colour positive and negative films have a somewhat limited application in remote sensing, as they have a blue-sensitive layer which re-

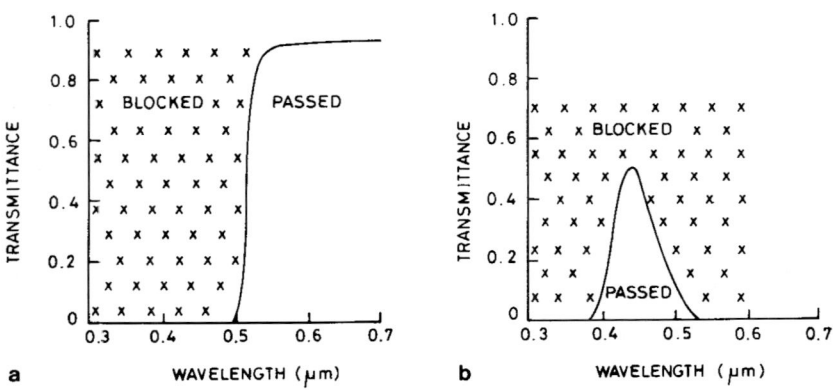

Fig. 4.12a,b. Typical transmittance curves for **a** absorption filter and **b** interference filter

duces image contrast. The CIR film provides higher image contrast and is particularly helpful when objects have distinct differences in infrared reflectance.

4.4 Filters

Filters form a very important component of modern camera systems. They permit transmission of selected wavelengths of light by absorbing or reflecting the unwanted radiation. They are applied in order to improve image quality and for spectrozonal photography. On the basis of the physical phenomenon involved, filters can be classified as (a) spectral filters, (b) neutral density filters and (c) polarization filters.

1. Spectral Filters. These filters lead to spectral effects, i.e. transmitting some selected wavelengths and blocking the rest. These are of two types: absorption filters and interference filters.

Absorption filters work on the principle of selective absorption and transmission. They are composed of coloured glass or dyed gelatine. A filter of this type typically absorbs all the shorter wavelengths than a certain threshold value and passes the longer wavelengths (Fig. 4.12a). Therefore, it is also called a 'long-wavelength pass' or 'short-wavelength blocking' filter (short-wavelength pass filters of the absorption type do not exist). A number of long-wavelength pass filters are available (Fig. 4.13). For example, Wratten 400 cuts off all radiation shorter than 0.4 μm wavelength and effectively acts as a haze cutter. Wratten 12 (yellow filter, also called minus-blue filter) cuts off all radiation shorter than 0.5 μm, and use of this filter with a black-and-white film or colour infrared film enhances image contrast. Wratten 25A cuts off all radiation shorter than red, and Wratten 89B eliminates all radiation of visible range and passes only near-IR radiation.

Interference filters work on the principle of interference of light. An interference filter consists of a pack of alternating high and low refractive index layers, such that the required wavelengths pass through, and other shorter and longer wavelengths are blocked either by destructive interference or by reflection. This effectively acts as a band-pass filter, i.e. an optical window, on either side of which the view is blocked (Fig. 4.12b). The band-pass width is susceptible to the angle of incidence of incoming rays. Band-pass filters are used in multiband cameras and also in vidicon cameras and CCD-pushbroom scanner cameras, where information in only a specified wavelength range is required. Examples are as follows: Wratten 47 transmits blue radiation; Wratten 58 transmits only green radiation; Wratten 18A passes radiation in the UV (0.3–0.4 µm) region; similarly, Wratten 39 passes radiation in the ultraviolet–blue region (Fig. 4.13).

2. Neutral Density Filters. Neutral density (including graded neutral density) filters have no spectral character. Their most frequent use is to provide uniform illumination intensity over the entire photograph. An *anti-vignetting filter* is a typical example. In a lens system, the intensity of light transmitted is greater in the central part and less in the peripheral region, leading to non-uniform illumination. An anti-vignetting filter is darker in the central part and becomes gradually lighter in the peripheral region, and compensates for the above geometric fall-off in light intensity from the centre outwards. It is also called a graded neutral density filter, and lacks spectral features. Sometimes the anti-vignetting effect is incorporated in the other types of filters, in order to reduce the number of filters to be physically handled.

3. Polarization Filters. Polarization filters use the principle of polarization of light. Such a filter permits passage of only those rays that vibrate in a particular plane, and blocks the rest. However, the potential of this type of filter has yet to be adequately demonstrated in remote sensing.

In addition to the above, there are colour compensating filters which are used to compensate for the change in spectral sensitivity of film emulsion layers, a deterioration which occurs as a result of storage over long periods of time.

4.5 Film–Filter Combinations for Spectrozonal Photography

The spectral characters of filters and films have been discussed above. The two can be combined for spectrozonal photography in selected wavelength bands, e.g. in blue, green, red, near-IR and ultraviolet bands (Table 4.3, Fig. 4.13). Applications of multiband products are discussed in detail elsewhere (see Chapters 8, 10 and 16). In brief, the spectral characters in the visible and near-IR region are used to study type and density of vegetation, diseased plants, broad rock type, geological structure, landform, soil and soil moisture etc. Ultraviolet photography has a particular application in detecting the presence of oil film on water surfaces.

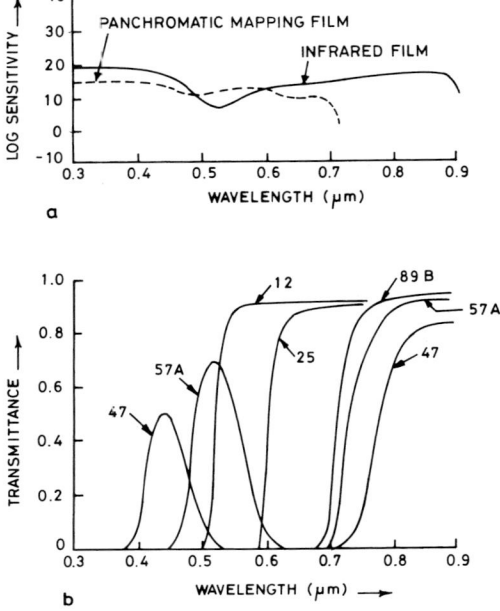

Fig. 4.13. Film–filter combinations for spectrozonal photography. **a** Spectral sensitivity curves of films and **b** filter transmittance curves (see also Table 4.3)

4.6 Vertical and Oblique Photography

The geometric fidelity of photographs is controlled by the orientation of the optic axis of the lens system. If the optic axis is vertical, geometric fidelity is high, and accurate geometric measurements such as heights, slopes and distances (or X, Y, Z co-ordinates of objects in a co-ordinate system) from photographs are possible (Sect. 6.3). Vertical stereoscopic photographs from photogrammetric and frame reconnaissance cameras are common remote sensing data products from aerial platforms. The utility of this technique as a practical tool still remains beyond question.

Table 4.3. Film–filter combinations for spectrozonal photography

	Film and filter combination	Photography in
1	Panchromatic film + Wratten 47	Blue band
2	Panchromatic film + Wratten 57A	Green band
3	Panchromatic film + Wratten 25A	Red band
4	Infrared film + Wratten 89 B	Near-infrared band
5	Panchromatic film + Wratten 18A	Ultraviolet band
6	Panchromatic film + Wratten 39	Ultraviolet + blue band

The photographs are said to be oblique or tilted when the optic axis is inclined. Oblique photography may be done for some specific purpose, for example: (1) to view a region from a distance for logistic or intelligence purposes, (2) to study vertical faces, e.g. escarpments, details of which would not show up in vertical photography, (3) to read vertical snow-stacks in snow surveys, and (4) to cover large areas in only a limited number of flights. Such photographs, however, may have limited photogrammetric applications due to higher geometric distortions.

4.7 Ground Resolution Distance

Broadly speaking, resolution is the ability to distinguish between closely spaced objects. In relation to photographic data products, it is used to denote the closest discernible spacing of bright and dark lines of equal width (Fig .4.14). It is given as line pairs mm^{-1} (e.g. 100 or 300 line-pairs mm^{-1} etc). The photographic resolution depends on several factors.

1. Lens resolution. The characteristics of a camera lens are important, as it forms the heart of the photographic system. The resolving power of the lens depends on several factors, namely wavelength used, f-number of the lens, relative aperture and angular separation from the optic axis (for details, see Slater 1980). The modulation transfer function (MTF) (Appendix 4.2) of lenses used in remote sensing missions is quite high, and constraints put by other factors are usually more stringent in limiting the overall system resolution.

2. Film resolution. This is the inherent resolution of the film, as discussed earlier, given in line-pairs mm^{-1}.

3. Object contrast ratio. This is the ratio of the intensity of radiation emanating from two adjacent objects that are being imaged (Sect.7.1). For objects with a high contrast ratio, resolution is greater.

In addition to the above, there are several other factors such as navigational stability, image motion, and atmospheric conditions which also affect the photographic image quality and resolution.

Thus, photographic resolution is dependent on several factors. It is usually found to be in the range of 20–100 lines mm^{-1}. This gives the ground resolution distance (GRD) as

$$\text{GRD} = \frac{H}{f} \times R_s \qquad (4.4)$$

where R_s = resolution of the system, f = focal length and H = flying height. For example, a photograph at a scale of 1:25,000, taken with a system with overall resolution of 50 lines mm^{-1}, would have a GRD of (25,000/50 mm) = 0.5 m.

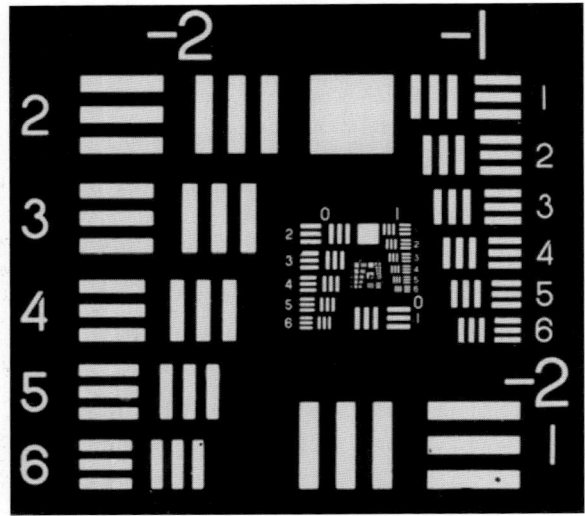

Fig. 4.14. Resolving power test chart. (Courtesy of Teledyne Gurley Co.)

4.8 Photographic Missions

4.8.1 Aerial Photographic Missions

Planning an aerial photographic mission involves a number of considerations related to technological feasibility and mission-specific needs, such as (1) type of camera, (2) vertical or oblique photography, (3) type of film–filter combination, (4) scale of photography, including focal lengths available and flying heights permissible, (5) area to be covered, and (6) degree of overlap required.

4.8.2 Space-borne Photographic Missions

As far as space photography is concerned, as mentioned earlier, the difficulties in film retrieval from orbital platforms have made photography a secondary tool for civilian applications. Some of the more important space-borne photographic missions have been the following.

Earth Terrain Camera (ETC). This was the most noteworthy space camera experiment in early 1970s. The ETC was flown on SKYLAB and was a high-performance frame reconnaissance camera, which yielded photographs at a scale of nearly 1:950,000, providing a ground distance resolution of nearly 30 m.

Photographs from Handheld Cameras. The Space Transportation System (space shuttle) flights have provided opportunities for Earth photography from space. These photographs have been taken from smaller-format cameras (35-, 70- and 140-mm), using natural colour and in black-and-white films, and some in CIR film, and provide ground resolution of the order of 30–80 m.

Corona Photographs. The photographs from Corona cameras, acquired for intelligence purposes by the United States, were declassified in 1995 (McDonald 1995). These photographs focusing mainly on the Sino-Soviet bloc, provide spatial resolution of about 2–8 m.

Metric Camera and Large-Format Camera. The space shuttle provided opportunities for two dedicated cartographic experiments, the Metric Camera (Konecny 1984) and the Large-Format Camera (Doyle 1985). These sensors provided limited Earth coverage and rekindled the interest of scientists and engineers in stereo-space photography for mapping applications from space (Table 4.4).

Photographs from Russian Cameras. The Russian unmanned spacecraft KOSMOS, orbiting the Earth at about 220 km, has carried two sophisticated cameras: KVR-1000 and TK-350 (Table 4.4). These cameras operate in conjunction with each other and enable generation of rectified imagery even without ground control. The TK-350 is a 10 m resolution topographic camera with 350 mm focal length and provides overlapping photographs for stereoscopic analysis. The KVR-1000 camera employs a 1000 mm lens, and provides 2 m ground-resolution photographs covering a large area (160 km × 40 km) in a single frame with minimal

Table 4.4. Characteristics of space-borne photographic systems

	Characteristic	Metric Camera (Europe)	Large Format Camera (USA)	KVR-1000 (Russia)	TK-350 (Russia)
1.	Satellite altitude (km)	≈ 250	≈ 250	220	220
2.	Flight vehicle	Space shuttle	Space shuttle	KOSMOS	KOSMOS
3.	Scene coverage (km)	Variable	Variable	34 × 57	175 × 175
4.	Spatial resolution (m)	20–30	≈ 20	2–3	5–10
5.	Film type	Panchromatic, and CIR film	Panchromatic, normal colour and CIR film	Panchromatic film	Panchromatic film
6.	Stereo-overlap	60–80%	60–80%	Minimal	60–80%

overlap. Both these systems use panchromatic film. The film is scanned to produce digital images. Their image-products, called SPIN-2, are now available for general
distribution in digital and analog forms (web-sites: www.Terra-Server.com; www.spin-2.com).

4.8.3 Product Media

The products of aerial/space photography have conventionally been stored and distributed on photographic media (films of various types and/or paper prints). Recently, a trend has evolved of scanning the photographic products and store/distribute the data on CD-ROMs in digital format. In this way, photographic products, particularly from the old archives, can be integrated and processed with other remote sensing/ancillary GIS data. This also helps minimize degradation in quality with age and duplication for distribution.

The geometric aspects of photographs are discussed in Chapter 6, the radiometric quality and interpretation in Chapters 7 and 8, and applications in Chapter 16.

Chapter 5: Multispectral Imaging Systems

5.1 Introduction

In this chapter we discuss the non-photographic multispectral imaging systems operating in the optical range of the electromagnetic spectrum. The optical range has been defined as that range in which optical phenomena of reflection and refraction can be used to focus the radiation. It extends from X-rays (0.02-µm wavelength) through visible and infrared, reaching up to microwaves (<1-mm wavelength) (Fig. 2.3). However, as the useful region for remote sensing of the Earth lies between 0.35 µm and 14 µm, we largely focus our attention on this specific region. For valuable reviews on non-photographic sensors, see Lowe (1976), Silva (1978), Slater (1980, 1985), Joseph (1996) and Ryerson et al. (1997).

5.1.1 Working Principle

A non-photographic sensor consists of two main parts: an optical part and a detector part.
1. Optical part: this includes radiation-collecting optics and radiation-sorting optics. The optics for radiation collection primarily comprise mirrors, lenses and a

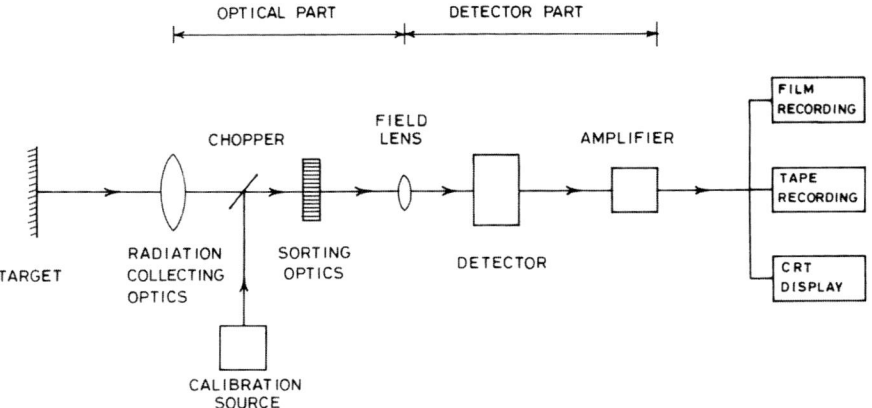

Fig. 5.1. Main components of a non-photographic remote sensor

telescopic set-up to collect the radiation from the ground and focus it onto radiation-sorting optics (Fig. 5.1). A calibration source is often provided on board, and a chopper enables radiation from the calibration source to be viewed by the detector at regular intervals. Radiation-sorting optics use optical devices such as gratings, prisms and interferometers to separate radiation of different wavelength ranges. In some cases, such as vidicons and pushbroom scanners, spectral separation may be carried out by appropriate band-pass optical filters, which cover the lens or detectors. After sorting, the radiation of selected wavelength ranges is directed to detectors.

2. Detector part: this primarily includes devices which transform optical energy into electrical energy. The heart of the device is a quantum or photo-detector unit. The incident photons interact with electronic energy level of the detector material, and electrons or charge carriers are released (photoelectric effect). The response in photo-detectors is very quick, and the intensity of the electrical signal output is proportional to the intensity of photons incident in a specified energy range. Major limitations of photo-detectors are due to the fact that: (a) their response varies quickly with wavelength (Fig. 5.2a) and (b) photoconductors operating at longer wavelengths have to be operated at very low temperature (195 K, 77 K or sometimes 5 K; Fig. 5.2b) to avoid noise. This is done by placing the detector within a double-walled vessel called a Dewar, filled with liquid helium or nitrogen for cryogenic cooling.

There are the two basic types of quantum detectors: photo-emissive detectors and photo-conductive detectors. In photo-emissive detectors (e.g. photo-multiplier tubes or PMT), the absorption of photons leads to emission of electrons, which are accelerated to an anode. PMTs are used only up to about 1 μm or shorter wave-

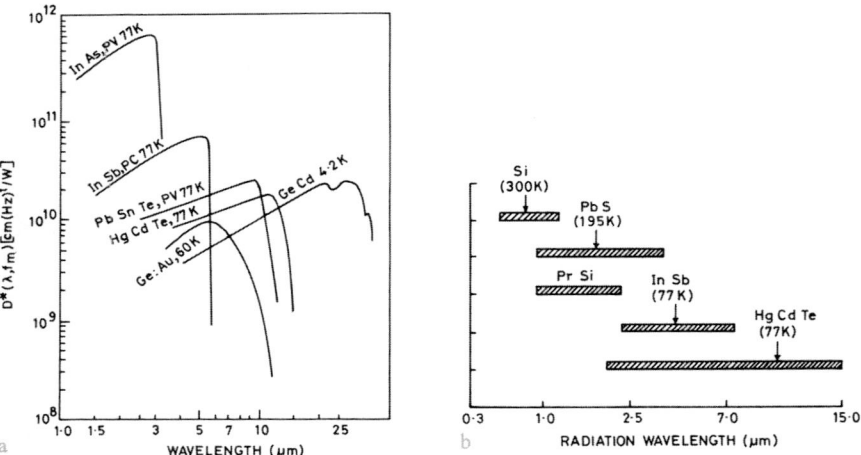

Fig. 5.2ab. a Spectral detectivity curves of some selected photo-detectors (2π steradians FOV, 295 K background temperature) (Hughes Aircraft Company, Santa Barbara Research Centre).; **b** Detector materials commonly used in different wavelength ranges in the optical region (operating temperature shown in parentheses above)

Introduction

Table 5.1. Relative merits of imaging systems over photographic sensors

Advantages	Disadvantages
1. Provide digital information which can be telemetered to ground from space	1. Limitations are imposed by: (a) high data rate and (b) technical sophistication required in generating images and data processing
2. Problem of film retrieval associated with photographic sensors is absent	
3. Remote sensing in extended wavelength range of 0.3 μm–1 mm possible, in contrast to photographic range of 0.3–0.9 μm	
4. Higher spectral resolution	
5. Higher radiometric resolution	
6. Good spatial resolution from modern sensors	
7. The information can be stored and is reproducible	
8. Amenability of data to digital processing for enhancement and classification	
9. Flexibility in handling of data	
10. Repeatability of results	

lengths, e.g. in Landsat MSS 1,2,3 bands. As they do not require deep cooling, their operation is more convenient.

In photo-conductive detectors, the absorption of the incident photon is accompanied by the raising of the energy levels of the electrons from valence levels (where they are bound) to conduction level (where they become mobile). Thus, the bulk conductivity of the detector is increased in proportion to the photon flux, and this can be measured. As there is no emission of electrons (solid-state technology), the energy requirements in photo-conductors are lower than for photo-emission devices, and therefore lower-energy radiation (SWIR and TIR) can also be detected by such devices. The development of appropriate photo-conductors has been an area of intensive and priority research. In many cases, dopants (impurities) are used to make alloys so that photons of a certain wavelength range can be detected. Some of the photo-conductors commonly used in VNIR–SWIR–TIR ranges are silicon, lead sulphide, indium antimonide, and merecury–cadmium-telluride (Fig. 5.2b). Photodiodes use the same material as the photo-conductors, and differ in operation only in the way that noise is reduced. An important evolution of the photodiode array is the charge-coupled device (CCD) which forms the heart of the most modern remote sensing devices.

The electrical signal from quantum detectors is amplified and quantized, i.e. given one of several possible integer numbers depending upon the intensity. It is recorded on film (analogue recorder) or more commonly on tape (digital recorder), relayed down to the Earth receiving station, and may be used for real-time display and/or subsequent processing and applications (Fig. 5.1, see also Fig. 5.28).

Relative Merits. The main advantages and disadvantages of non-photographic devices as compared to photographic ones are listed in Table 5.1. As such, the advantages far outweigh the limitations and make the non-photographic sensors ideally suited for use on free-flying space platforms.

In the following pages, first, the various factors affecting sensor performance are considered, after which non-imaging and imaging instruments, particularly from space platforms, are described.

5.2 Factors Affecting Sensor Performance

Physical processes governing the energy emitted and reflected from the ground have been discussed in Chapter 2. Consider the case of a remote sensor viewing a uniform object on the ground (Fig. 5.3). The radiant power (I_λ) illuminating the detector is (after Lowe 1976).

$$P_\lambda = \frac{I_\lambda \cdot T_{a(\lambda)} \cdot T_{0(\lambda)} \cdot A_s \cdot A_0}{H^2} \tag{5.1}$$

where
I_λ = spectral radiance of the source (object) in W cm^{-2} sr^{-1} μm^{-1}
$T_{a(\lambda)}$ = spectral transmittance of the atmosphere, to take into account the atmospheric losses
$T_{0(\lambda)}$ = spectral transmittance of the optical system, to take into account losses within the optical system
A_s = area of the source under view
A_0 = effective area of the collector optic system collecting the radiation
H = distance of the sensor from the object.

The noise equivalent spectral power (NEP$_\lambda$) of any detector is given by

$$NEP_\lambda = \frac{\sqrt{Ad \cdot \Delta f}}{D_\lambda^*} \tag{5.2}$$

where
Ad = area of the detector
Δf = electronic bandwidth (being physically inversely proportional to the observation time)
D_λ^* = spectral detectivity of the material used (a measure of sensitivity of the material).

Therefore, signal-to-noise ratio (S/N), over a certain wavelength range, can be written as

$$\frac{S}{N} = \int_{\lambda_1}^{\lambda_2} \frac{P_\lambda}{NEP_\lambda} \, d\lambda \qquad (5.3)$$

$$= \int_{\lambda_1}^{\lambda_2} \frac{I_\lambda \cdot T_{a(\lambda)} \cdot T_{0(\lambda)} \cdot A_s \cdot A_0 \cdot D_\lambda^*}{H^2 \sqrt{A_d \cdot \Delta f}} \cdot d\lambda. \qquad (5.4)$$

Further, if β is the angle of instantaneous field-of-view (IFOV), then

$$\frac{A_s}{H^2} = \frac{A_d}{f^2} = \beta^2 \qquad (5.5)$$

where A_d is detector area and f is focal length of the system. This yields:

$$\frac{S}{N} = \int_{\lambda_1}^{\lambda_2} \frac{I_\lambda \cdot T_{a(\lambda)} \cdot T_{0(\lambda)} \cdot \beta \cdot A_0 \cdot D_\lambda^*}{f \sqrt{\Delta f}} \cdot d\lambda \qquad (5.6)$$

This is a fundamental equation of great importance in understanding and evaluating performance of a sensor. A good sensor is one which provides a high S/N ratio. Several factors affect the S/N ratio: (1) I_λ (brightness of the ground object) is directly related, therefore conditions of higher scene brightness (reflection or emission) are better suited; (2) $T_{a(\lambda)}$ (transmittance of the wavelength through the atmosphere) is directly related, therefore the wavelength used should have minimum attenuation through the atmosphere; (3) $T_{0(\lambda)}$ (transmittance of the wave-

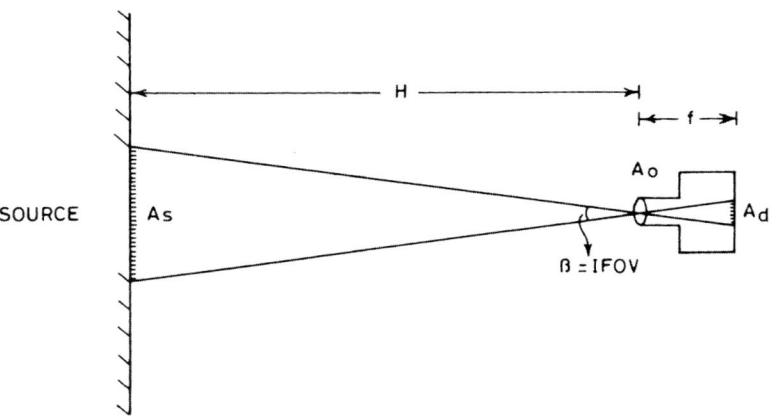

Fig. 5.3. Schematic of geometric relations involved in radiant power reaching the sensor. (After Lowe 1976)

length through the optical system) is directly related – an efficient optic system permits through-put with minimum losses; (4) A_0 (area of the collecting optics) is directly related, however collecting optics with very large areas are not permitted owing to size and weight constraints; (5) β (IFOV) is directly related, however increase in β reduces spatial ground resolution and therefore a trade-off between S/N ratio and β is necessary; (6) D_λ^* (spectral detectivity) is directly related – several materials are available for use as detectors in different wavelength ranges with varying spectral detectivity values (Fig. 5.2a) and a detector with higher spectral detectivity should be used; (7) f (focal length) is inversely related, i.e. short focal length systems give higher signal-to-noise ratio, but on the other hand they lead to a decreased scale and lower spatial resolution, and hence again a trade-off is required; (8) Δf (electronic bandwidth) is inversely related, i.e. the dwell time is directly related; (9) S/N ratio is also directly proportional to the width of the wavelength band ($\lambda_1-\lambda_2$), but on the other hand increasing the spectral range results in reduced spectral resolution, and therefore again a trade-off has to be made.

In a spectral region like the visible, the spectral radiance (I_λ) is very high (Fig. 2.4a), and hence the sensor S/N ratio is also generally high. In such a situation, both spectral and spatial resolutions can be made much finer, in comparison to a situation where scene brightness is relatively lower, as in the case of the thermal IR region. Furthermore, several trade-offs are possible between spectral and spatial resolutions. A judicious decision must be based on a very careful analysis of the available technology and understanding of the requirements.

The *Modulation Transfer Function* (MTF) is a useful parameter for evaluating the performance of sensors. It has the same connotation as in photographic sensors. The MTF evaluates how faithfully and finely the spatial variation of radiance in the scene is emulated in the image (see Appendix 4.2). The MTF is used to evaluate the performance of the sensor or its components, such as optics, detector system etc., on an individual basis.

5.2.1 Sensor Resolution

It will be appropriate here to become conversant with the various terms related to sensor resolution, viz. spatial resolution, spectral resolution, radiometric resolution and temporal resolution.

1. Spatial resolution of a sensor implies the area on the ground, which fills the IFOV of the sensor. It is also called the ground element or ground resolution cell.

2. Spectral resolution means the span of wavelength range ($\lambda_1 \sim \lambda_2$) over which a spectral channel operates. The band-pass response of a channel is most commonly of a gaussian type (Fig. 5.4). The width of the band pass (spectral bandwidth) is defined as the width in wavelength at 50% response level of the function. For example, in figure 5.4, the 50% response level of the detector is from 0.55 μm to 0.65 μm, providing a spectral bandwidth of 0.1μm. {It is also called 'full-width at half maximum' (FWHM) for the band.}

3. Radiometric resolution of a sensor means the degree of sensitivity of a sensor to radiation intensity variation. Basically, it corresponds to *noise equivalent power*

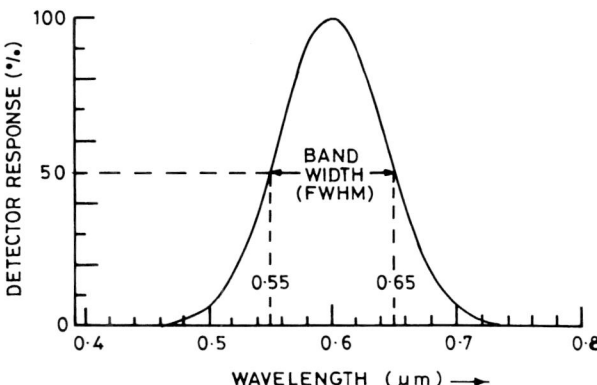

Fig. 5.4. Spectral bandwidth (resolution) of a detector. The spectral range of this detector is from 0.55 μm to 0.65 μm, giving a bandwidth of 0.1 μm

difference (NEΔP$_\lambda$), or *noise equivalent temperature difference* (NEΔT) in the case of thermal sensors. It is the brightness difference between two adjacent ground resolution elements that produces a signal-to-noise ratio of unity at the sensor output. A good scanner has a smaller noise-equivalent power difference. However, it is at times loosely used to imply the total number of quantization levels used by the sensor (e.g. 7-bit, 8-bit, 10-bit etc.).

4. *Temporal resolution* refers to the repetitiveness of observation over an area, and is equal to the time interval between successive observations. It depends on orbital parameters and swath-width of the sensor.

5.3 Non-Imaging Radiometers

5.3.1 Terminology

A non-imaging sensor measures the intensity of EM radiation falling in its field-of-view (FOV) and provides a profile-like record of intensity with distance in the direction of line of flight (or time). It does not sample the scene in the across-track direction, and therefore does not produce an image.

By definition, the term radiometer means an instrument used for measuring radiation intensity. Generally, it is used to imply a non-imaging sensor, although there is no universal acceptance of terminology, and the term radiometer is also used by a few workers for imaging sensors (e.g. in MESSR, ASTER). We use the term radiometer here only for a non-imaging sensor working in profiling mode. It is a passive sensor. Photometer is a term used for a similar device operating at shorter wavelengths ($\lambda < 1$ μm). Multi-band radiometer and spectro-radiometer are

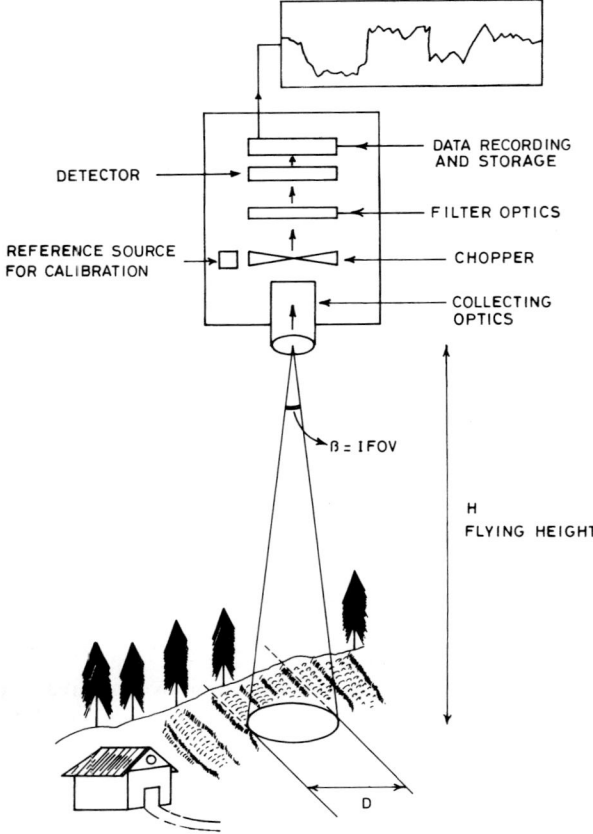

Fig. 5.5. Design and working principle of a radiometer in profiling mode

terms applied to radiometers which measure radiation intensity in more than one wavelength band. Another term sometimes seen in the literature is scanning radiometer, e.g. in VISSR (in SMS-GOES geostationary satellite); we group such sensors under scanners.

5.3.2 Working Principle

The working principle of a radiometer is quite simple (Fig. 5.5). An optic system (usually a refractive lens system) collects and directs radiation onto an optical filter (such as a grating, prism, interferometer or filter wheel etc.), where the radiation is sorted out wavelength-wise. The selected radiation is directed to the detector system, which quantizes the radiation intensity.

In many radiometers, especially in those operating at longer wavelengths ($\lambda > 1$ μm), reference brightness sources are kept within the housing for calibration, since the various objects including parts of the radiometer also form a source of radiation at these wavelengths. The detector is made to receive radiation alternately from the ground and the reference sources by a moving chopper.

The ground resolution element of the radiometer is the area corresponding to IFOV on the ground (Fig. 5.5). It is usually a circle, the diameter D of which is given by $D = H \times \beta$ (where H is the flying height, β is the IFOV). The ground resolution cell is given as $\pi (D/2)^2$. As the remote sensing platform keeps moving ahead, measurements are made at successive locations, which result in profile data.

The design of a radiometer, i.e. the spectral bandwidth, angular field-of-view etc., depends on numerous factors such as the purpose of the investigation, scene brightness, detector technology etc., all of which affect the S/N ratio. The wavelength range of operation extends from visible to microwaves.

5.4 Imaging Sensors (Scanning Systems)

5.4.1 What is an Image?

Imaging sensors or scanning devices build up a two-dimensional data array of spatial variation in brightness over an area. The entire scene to be imaged is considered as comprising a large number of smaller, equal-sized unit areas, which form ground resolution cells (Fig. 5.6). Starting from one corner, line-by-line and cell-by-cell, the radiation from each unit area is collected and integrated by the sensor to yield a brightness value which is ascribed to that unit area. In this manner, spatial information is converted into a time-dependent signal. This process is called scanning, and the data are called scanner data.

The photo-radiation emanating from a unit area is collected, filtered, and quantized to yield an electrical signal. The signal, depending upon its intensity, is classified into one of the various levels, called quantization levels (since measuring actual photo-current at each unit area would be a too time-consuming and laborious exercise and, for most of our investigations, it is sufficient to have relative brightness values). In this manner, a scanner provides a stream of Digital Numbers (DN's). These data are stored on tape recorders on-board the remote sensing platform and/or relayed down to the ground receiving station using a microwave communication link. On the ground, the scanner data can be rearranged as a two dimensional array and be presented as an optical analogue by choosing a suitable gray scale. The various brightness values measured over ground unit areas are depicted as gray tones at the corresponding positions on the optical analogue. This is called an image. Therefore, *an image is an optical analogue of a two-dimensional data array*. The unit area on the ground is variously termed ground resolution/ spatial resolution/ ground resolution cell/spot size/ground IFOV. The same unit area on the image is called the picture element or *pixel*. A critical review of

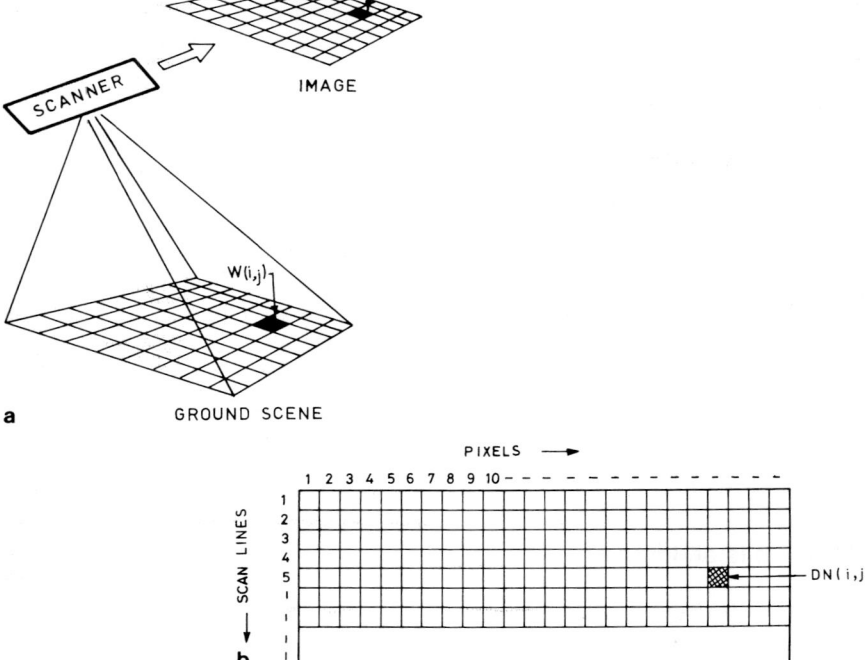

Fig. 5.6a,b. The scanning process. **a** The entire scene is considered to be comprised of smaller, equal-sized unit areas, and radiation from each unit area ($W_{i,j}$) is collected by the sensor and integrated to yield a brightness value ($DN_{i,j}$). **b** The scanner data has a structure of rows (scan lines) and columns (pixels)

'what's in a pixel' – in terms of both a geometrical point of view and a more physical point of view – is given by Cracknell (1998). Obviously, an image consists of a large number of pixels. Table 5.2 gives the salient differences between a photograph and an image.

Broadly, four types of imaging sensors can be identified: (1) imaging tubes, (2) optical mechanical line scanners, (3) CCD linear arrays and (4) digital cameras (CCD area arrays). As far as basic principle is concerned, there is a lot of similarity between photographic systems and imaging systems. Table 5.3 gives a comparison of the photographic systems and imaging systems, in terms of the sensor operation technique.

5.4.2 Imaging Tube (Vidicon)

Several types of imaging tubes were used in earlier remote sensing programmes, starting in 1960s. However, they are now of rather historical interest, in

Table 5.2. Salient differences between a photograph and an image

Photograph	Image
1. Originally produced in analogue form	1. Originally produced in digital form
2. Generated by photographic film system	2. Generated from line scanners and digital cameras
3. Does not have any pixels	3. Basic element is a pixel
4. Lacks row and column structure	4. Possesses rows and columns
5. Scan lines absent	5. Scan lines may be observed
6. Zero indicates no data	6. Zero is a value, does not indicate absence of data
7. No numbering at any point	7. Every point has a certain digital number
8. Photography is restricted to photographic range of EM spectrum	8. Image can be generated for any part of the EM spectrum, or any field
9. Once a photograph is acquired, colour is specific and cannot be (inter-)changed	9. Colour has no specific role and can be changed during processing

Table 5.3. Similarity in basic techniques of photography and imaging

Basic technique	Photographic system	Imaging system
1. Collection of video data for the entire frame concurrently	Single-frame camera	Digital imaging camera
2. Collection of video data as the optical system scans the ground, across the flight path	Panoramic camera	Opto-mechanical line scanner (panoramic scanner)
3. Collection of video data in strips/lines, orthogonal to the flight-path	Strip camera	Pushbroom line scanner

view of the availability of other better devices. Only the Return Beam Vidicon (RBV) type of sensor used in the initial Landsat (1,2,3) program, will be briefly mentioned here. The RBV instrument utilized a photo-sensitive, photo-conductive target in the form of a thin film to act as the imaging surface. The film was supported on a glass or quartz face plate. The whole assembly (including electronics) was placed in a camera housing (Fig. 5.7). When the camera shutter opened, the photo-sensitive layer got charged due to exposure to the incident radiation, and the scene was first imaged on the tube. This image was scanned by a rapidly moving electronic beam and the signal was derived from the depleted electronic beam reflected from the target (hence the name 'return beam vidicon').

However, there have been many drawbacks associated with vidicon sensors, such as: (a) occurrence of gross geometric errors, especially near the perimeter of the tube; (b) variation in radiometric response; (c) migration of electrons from local concentration, reducing image sharpness; and (d) often, the earlier picture leaves a remnant at the tube called the ghost, which is subtly present and has to be subtracted from the new picture. Due to the above geometric and radiometric limitations, vidicon devices no longer find favour in remote sensing systems.

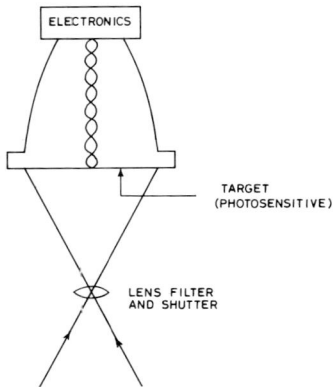

Fig. 5.7. RBV sensor – working principle. (After NASA 1976)

5.4.3 Optical-Mechanical Line Scanner (Whiskbroom Scanner)

This has been a very widely used scanning technology. The major advantages of the optical-mechanical (OM) line scanners have been that they can easily operate in multispectral mode, at a wide range of wavelengths from 0.3 μm to 14 μm, and generate digital image data. Pioneering work in this field was carried out at the University of Michigan (Braithwaite 1966; Braithwaite and Lowe 1966; Lowe 1969).

The collector optics of an aerial/spaceborne OM scanner includes a plane mirror, which revolves or oscillates along an axis parallel to the nadir line (flight line) (Fig. 5.8). This permits radiation from different parts of the ground, lying in the across-track direction, to be reflected onto the filter and detector assembly. The radiation is separated according to wavelength by grating, prism etc. and directed onto photo-detectors, where it is converted into an electrical signal. The signal is amplified, integrated and chopped at regular time intervals and quantized. The chopping provides subdivision in the scan line (pixels!). The forward motion of the sensorcraft allows advancing of the scene of view. The signal is recorded, transmitted, stored, or displayed as per the available output mode. In this fashion, across-track scanning is facilitated by the moving mirror, and along-track by the forward motion of the sensorcraft, and the entire scene is converted into a two-dimensional data array. Owing to its similarity in operation, it is also known as a whiskbroom scanner.

An important consideration in OM scanners is the V/H (velocity/height) ratio. The scene in the along-track direction is advanced by the movement of the sensorcraft. In order for the ground to be contiguously scanned (without any over- or underscanning), it is necessary for the V/H ratio to be commensurate with the rate of scanning (V/H = ß × n, where n = number of cycles s^{-1} of active scan mirror movement and ß = IFOV).

Imaging Sensors (Scanning Systems)

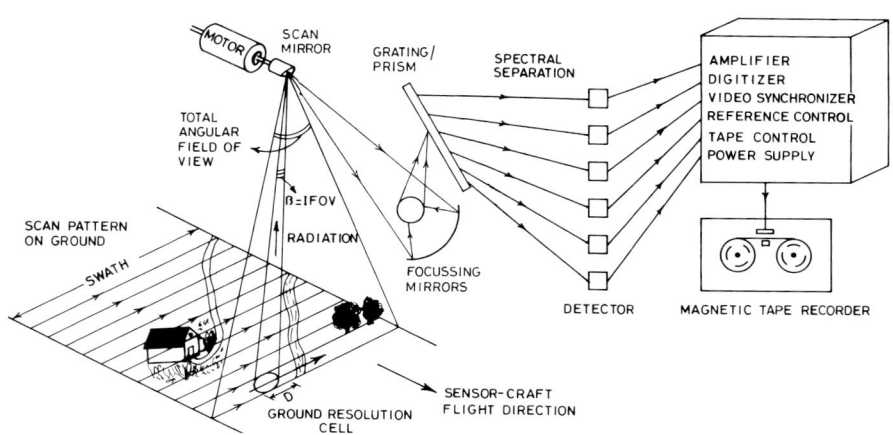

Fig. 5.8. Working principle of an opto-mechanical line scanner

The spectral resolution (i.e. bandwidth), radiometric resolution (NEΔP_λ) and spatial resolution (IFOV) are all interdependent and critical parameters, and suitable trade-offs have to be judiciously arrived at [see Eq. (5.6)]. Although the working principle is similar in both aerial and spaceborne OM scanners, the actual resolution specifications differ in the two cases.

The instantaneous field-of-view (IFOV = ß) (Fig. 5.8) is the angle of field-of-view at a particular instant. It governs the dimension of the ground resolution cell or spatial resolution. For most aerial scanners, it is about 1–3 millirad and corresponds to about 5–10 m on the ground (Table 5.4). The total angle of view, through which the radiation is collected as the mirror moves, is called the total angular field-of-view or simply angular field-of-view (FOV).

In airborne OM scanners, the mirror is rotated by a constant-speed motor and the total angular field-of-view is about 90° to 120°. During the rest of the rotation of 270–240°, the mirror either directs radiation from standard lamps onto the detector for calibration (which is a must at wavelengths > 1 μm) or performs no useful function. In spaceborne OM scanners, the total FOV is much smaller (10°–15°) and oscillating mirrors are commonly deployed in radiation-collecting optics.

The total angular FOV together with the altitude of the sensorcraft control the swath width (length in across-track direction sensed in one pass or flight). Ground swath can be given as

Ground swath = 2H . tan (FOV/2) (5.7)

Further, if successive pixels are contiguous, without any overlap/underlap, then

Ground swath = number of pixels × pixel dimension along the swath (5.8)

Table 5.4. Common characteristics of aerial OM line scanners

FOV	: Commonly 90°–120°
Altitude	: Variable
Swath width	: Variable
IFOV	: ≈ 2–3 mrad
Ground resolution	: Variable (commonly 5–10 m)
Spectral bands	: Varying, for example:
	Bendix 11-channel }
	Bendix 24-channel } (operating in VIS–NIR–SWIR–TIR)
	Daedulus 11-channel }
	Daedulus TIMS 6-channel scanner (operating in TIR)

For evaluating the performance of a scanner, several parameters are considered, the more important of which are: spatial, spectral and radiometric resolution, calibration, D.C. drift, roll stabilization and occurrence, rectification of various types of geometric distortions, and finally output mode.

In the VNIR region, the energy available is relatively high and good detector materials are known, and therefore scanners with finer spectral, spatial and radiometric resolution have been in operation for many years. In the SWIR and TIR regions, the energy available for sensing is relatively less and it is only with developments in sensor design that scanners with higher resolution levels are now coming onto the scene. The Bendix 24-channel scanner and Daedulus 11-channel scanner, both operating in the VNIR to TIR region, were flown from aerial platforms for many experiments in the 1970s and 1980s. The TIMS aerial scanner is an improved 6-channel OM scanner operating in the TIR region which became became operational in 1980s. On space platforms, the important OM scanners include MSS and TM on Landsats, MTI and ASTER (TIR part).

5.4.4 CCD Linear Array Scanner (Pushbroom Scanner)

Charge-Coupled Device (CCD) linear arrays were first conceived and tested for operation in the early 1970s and the tremendous interest in this new technology has given it a big boost (Melen and Buss 1977; Thompson 1979; Beynon and Lamb 1980). We will consider here some of the fundamental concepts relevant to imaging systems. The CCDs utilize semi-conductors, i.e. photo-conductors, as the basic detector material, the most important so far having been silicon. An array of a large number of detector elements, individually similar to a photodiode, is used. The detectors are endowed with leads to act as electrodes. The entire fabrication is carried out using micro-electronics, so that a chip about 1 cm long carries several thousand detector elements (Fig. 5.9a).

The basic technique of pushbroom line scanning is illustrated in figure 5.9b. A linear CCD array, comprising several thousand detectors, is placed at the focal plane of a camera lens system. The lens system focuses radiation emanating

Imaging Sensors (Scanning Systems) 89

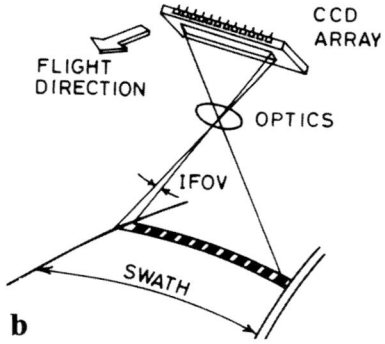

Fig. 5.9. a Photograph of a linear array with 1024 elements; in the background is an enlarged view of a silicon chip. (Courtesy of Reticon Co., USA) **b** Principle of line-imaging by CCD linear array; the array placed at the focal plane of an optic system, records radiation, the pattern of signal being analogous to that of scene illumination along the scan line

from the ground on to the CCD line array. At each detector element, the radiation is integrated for a short period and recorded. As the sensorcraft keeps moving, the data are collected line by line.

In a simplified manner, the CCD line imager may be considered to have three main segments: (a) a photo-gate, (b) a transfer gate and (c) a shift register (Fig. 5.10). The photo-gate comprises a series of radiation-sensing photo-conductor elements or cells. They are separated from each other by a channel-stop diffusion to avoid cross-talk, and all cells are placed at a high positive potential of equal value. As the photo-gate is exposed to radiation, the incident photons raise the energy level of the electrons in the semi-conductor to conduction band level, and the electrons become concentrated at the positively charged electrodes. The number of electrons collected at a given detector electrode within a given time period, called integration period, or dwell time, is proportional to the local photon flux. Thus, the pattern of charge collected at the detector array becomes analogous to the pattern of scene illumination along the scan line.

At the end of the integration period, the charge collected is quickly transferred through the transfer gate to the shift register located at the back. The purpose of the shift register is to store the charge temporarily and to permit read-out at convenience. To understand how the charge is read out in the shift register, we have to go a little deeper into the matter. In the shift register, there are three electrodes corresponding to each element in the photo-gate (Fig. 5.10), which are interconnected in a cycle of three. The charge is laterally moved from one element to another, by cyclic variation of voltage, and read out sequentially at one output point (hence the name 'charge-coupled device').

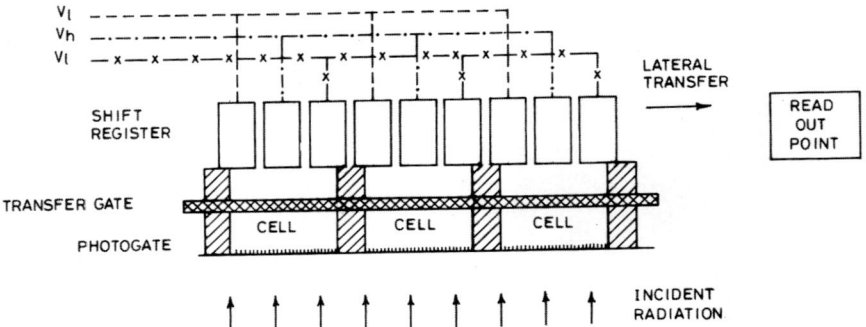

Fig. 5.10. Main segments of CCD linear imaging device: photo-gate transfer gate and shift register

The mechanism of lateral charge transfer is illustrated in figure 5.11. To understand how the charge is moved from one potential well to another, consider a set of four closely spaced electrodes, A, B, C and D such that B is at a high potential (V_H). The other three electrodes are at rest or low potential (V_L). At the end of the integration period, the charge collected at the photo-gate gets transferred to the B-electrode (with the high positive potential, V_H). Now, if the potential of C-electrode is raised to V_H level, then the charge will flow from B to C (Fig. 5.11), provided that the inter-electrode spacing is small enough. At the same time, if the bias (voltage) of B-electrode is reduced to V_L, all the charge will be transferred to C. In this manner, the charge can be transferred from one element to another by cyclic variation of bias. The read-out is carried out at one output point (Fig. 5.10). The procedure is repeated until the charge transfer and read-out are completed in the entire array in the shift register.

Thus, as the sensorcraft keeps moving, the charge is read out in the shift register by transfer technique, and at the same time the photo-gate keeps collecting video information for the next line. The large number of detectors in the CCD line array provides scanning in the cross-track direction, and along-track scanning is made possible by the forward motion of the sensorcraft. This generates an image. Owing to its similarity, this is also referred to as the pushbroom system.

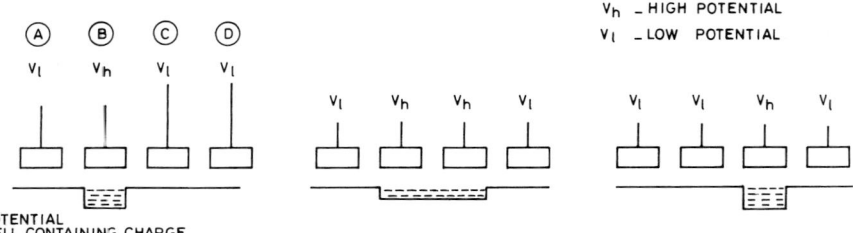

Fig. 5.11. Mechanism of lateral charge transfer by cyclic variation of bias (voltage) in the shift register of a CCD line imaging device (for details, see text)

Table 5.5. Relative advantages and disadvantages of pushbroom scanners over OM-type scanners[a]

Advantages	Disadvantages
1. Light weight	1. Owing to the presence of large number of cells striping effects occur in the CCD line scanners which necessitate radiometric calibration and correction
2. Low voltage operation	
3. Long life	
4. Smaller size and compactness	
5. No moving parts	
6. Higher camera stability and geometric accuracy owing to absence of moving parts	2. The CCD technology has operational status only in the VNIR (0.4–1.1 µm) region
7. Higher geometric accuracy along each scan line	3. Low temperature operation is necessary to minimize noise
8. Dwell time significantly higher, providing significant increase in S/N ratio 9. Higher radiometric resolution (NEΔP_λ)	4. Number of detector elements fabricated on a single chip is limited, with the present technology; use of several chips in precision alignment is necessary
10. Higher spectral resolution	
11. Higher spatial resolution	
12. Lower cost	5. The chip junctions act as dead elements; to counter this, either the chips have to be staggered or data interpolation is required
13. Off-nadir viewing configuration facilitates in-orbit stereo viewing and increased temporal resolution	

[a] Summarized after Colvocoresses (1979); Thompson (1979); Tracy and Noll (1979); Slater (1980).

The 'opening' at each of the detector elements is the IFOV. It is given in terms of centre-to-centre spacing of the detector cells, and is of the order of a few micrometers. It is a measure of the along-track aperture. The dimension of the ground resolution cell is governed by IFOV, flying height, velocity vector and integration period. The parameters are usually designed to generate square-shaped ground resolution cells.

The CCD-based pushbroom line scanner can easily operate in multispectral mode. The spectral separation is provided by discrete optical filters, which may be mounted on the camera lens or alternatively may be used as coatings on the CCD chips. Typically, one detector array is required for each spectral channel.

The relative advantages and disadvantages of the pushbroom scanners over the opto-mechanical scanners are given in Table 5.5. The main advantages of CCD line scanner arise from two aspects: (a) the solid-state nature and (b) the fact that the dwell time of the sensor is appreciably increased. In OM line scanners, the various cells along a scan line are viewed successively, one after the other. On the other hand, in CCD linear array scanners all the cells along a scan line are viewed concurrently; the sensor integrates photon flux at each cell for the entire period during which the IFOV advances by one resolution element. This leads to increment in dwell time by a factor of about 500–1000. The increased S/N ratio can be traded off with reduction in size of collecting optics (size and weight) and superior spectral and/or spatial resolution [c.f. Eq. (5.6)].

Fig. 5.12. Digital Camera. (Courtesy of Eastman Kodak Company)

CCD linear array scanners have been fabricated and flown on airborne platforms primarily as experimental sensors from the hardware point of view (Wharton et al. 1981). Since the mid 1980s, a number of CCD linear imaging systems have been successfully flown for remote sensing from space (see Sect. 5.5).

5.4.5 Digital Cameras (CCD Area Arrays)

The digital imaging camera constitutes a recent development in remote sensing technology as it provides high-definition images at reduced costs. An example of a digital camera is given in figure 5.12. The camera employs a solid-state area array, which produces a digital two-dimensional frame image. Other terms used synonymously for the same type of sensor/technology include: digital electronic imaging, digital photographic camera, digital snapshot system, staring arrays, CCD area array, CCD matrix sensor, and FPA system.

The basic difference between a film-based conventional photographic camera and a digital imaging camera is that, in the latter, film is replaced by solid-state electronics such as 'charge coupled devices' (CCDs) or the newer CMOS (complementary metal-oxide semi-conductor). The electronic detectors, with millions of energy-sensitive sites, form a two-dimensional array, called an area array or matrix, which is placed at the focal plane of the camera system (Fig. 5.13) (hence the name 'focal plane array', FPA). The incident light is focused by the optical

Imaging Sensors (Scanning Systems) 93

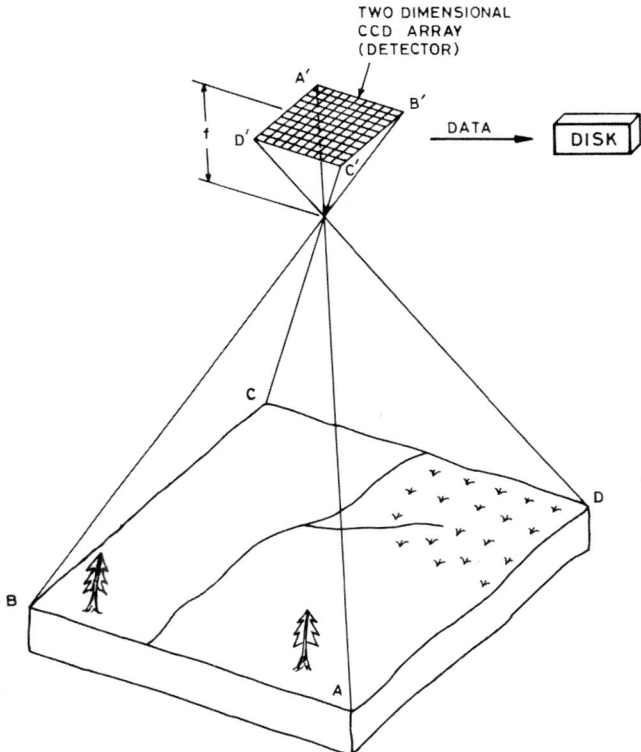

Fig. 5.13. Working principle of a digital imaging camera

device onto the CCD area array, which generates an analogue signal. The signal is quantized to yield an image. Therefore, in a digital imaging camera, data storage takes place directly in digital form (using magnetic, optical and solid-state media).

Table 5.6 gives the relative advantages and disadvantages of digital imaging vis-à-vis film photographic systems. The main advantages of digital imaging cameras arise from their direct digital output, fast processing, higher sensitivity, better image radiometry, higher geometric fidelity and lower cost. The main limitations are imposed by present-day detector technology, i.e. the availability of chips with only a limited number of photo sites. The advantages of digital imaging cameras far outweigh the disadvantages. A valuable review on digital cameras is given by King (1995).

5.4.5.1 Design and Working Principle

The working principle of the digital imaging camera is quite similar to that of the CCD line array pushbroom scanner, except that here an area array is used instead of a linear array to form the detector.

Table 5.6. Relative advantages and disadvantages of digital camera vis-à-vis film-photographic systems

---

Advantages

1. A great advantage of digital imaging cameras is their direct computer-compatible output, which enables direct digital processing of data. The various advantages of digital techniques automatically follow (Table 5.1)
2. Another prime advantage is in-flight viewing of imagery to ensure image quality (exposure, contrast etc.) and target coverage
3. Ability of rapid change detection (following less turn-around time) is of high tactical importance in military reconnaissance and also of significance in various civilian applications, such as for disaster missions (monitoring floods, storm damage, large fires, volcanic eruptions etc.)
4. The dynamic range of digital cameras is about 12-bit (4096 shades of gray) which is much superior to the 6- or 7-bit (about 64–128 shades of gray) typical of films.
5. Silicon chips (CCDs) possess much higher sensitivity in the near-IR as compared to standard films
6. The high sensitivity of the CCD sensor, along with its wide dynamic range, permits markedly higher contrast to be recorded in the image
7. Exposure time in digital cameras may be shorter than in photography, reducing the problem of image motion
8. Another important advantage of the matrix sensor is the rigid two-dimensional geometry of the CCD pixels, which can provide an even higher stability and reproducibility than a film-based camera. The polyester substrate used in films is not perfectly rigid in its geometry; effects of stretching or differences in humidity may produce non-linear geometric errors in the range of 5–10 μm, which put a limit on photogrammetric applications of film products
9. The photogrammetric solutions developed for aerial photography can be applied to solid-state two-dimensional array images
10. Silicon sensor response is linear in all spectral bands, i.e. the digital gray level is directly proportional to the target radiance. This permits fairly simplified radiometric calibration of digital cameras (although the two-dimensional format of the sensor presents some difficulties)
11. Lower cost of sensor components
12. Additional advantages are: no film development, no scanning, no chemical waste

Disadvantages

1. A major limitation arises due to the smaller size of CCD area arrays available presently. This implies reduced areal coverage (at the same photo scales) and greater expenses for data collection
2. Technical sophistication required for two-dimensional radiometric corrections/calibrations is another constraint
3. As the number of cells on each chip (rectangular array) presently available is limited, several chips may have to be used in conjunction, and their optical butting etc. for sensor design needs special care

Imaging Sensors (Scanning Systems)

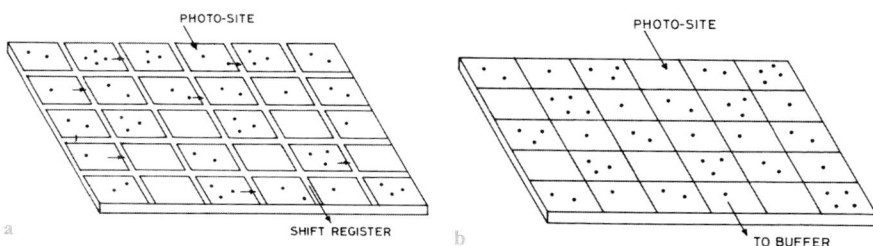

Fig. 5.14a,b. Techniques of charge transfer in a CCD area array. **a** Interline transfer, and **b** frame transfer

The shutter of the camera opens for a fraction of a second, allowing light energy to fall on the solid-state imaging sensor (CCD matrix chip). The incident light energy (photons) is absorbed, raising the energy level of electrons in the chip material, which jump to outer shells. These electrons become mobile and are then held in electrodes (which are embedded in the substrate) by applications of voltage. By varying the voltage across the electrodes, the electrons are laterally transferred/brought to the chip boundary. The entire sequence is timed, so that the sensor chip can be exposed to incident energy for a given time interval, after which the generated charge carriers are drawn off to form the signal.

There are several methods of transferring charge out of the photo-sites, the two common techniques being: (a) interline transfer technique and (b) frame transfer technique (Fig. 5.14). In the interline transfer technique, the photo-sites do not touch each other but are separated from each other by column-and-row spaces. The charge from each photo-site is read off to the adjacent space ('shift register') and then down the spaces (usually columns) to the edge of the chip. Most cameras use this technique. In the frame transfer technique, charge is moved through all photo-sites across the chip, and is transferred to a buffer of the same size. In the interline transfer architecture, the rows and columns of shift register are interlaced with photo-gate cells (and hence there is some loss of video information, although only minimal). In the frame transfer configuration the photo-gate cells are quite close to each other.

5.4.5.2 Performance Characteristics

Spatial Resolution in digital cameras is variable, depending on the number and size of the photo-sites on the sensor chip. In terms of signal-to-noise ratio, it is also related to target contrast, illumination, lens aperture, and spectral band [see Eq. (5.6)].

There is an inter-relation between field-of-view (FOV), focal length (f), sensor size, spatial resolution, scale and flying height. Photo-scale is given as f/H. For a given f and H, the sensor size (area) affects FOV, or in other words areal coverage. A bigger sensor size implies larger FOV and larger coverage of ground area. This also means that with f, H and sensor size constant, spatial resolution increases with increase in the number of photo-sites in the sensor. The primary factor governing spatial resolution is photo-site size: 6 μm is better than 9 μm, 9 μm is better than 12 μm.

With the number of photo-sites in a sensor matrix constant, as the sensor size increases, the photo-site size tends to be larger, which leads to a reduced spatial resolution but increased radiometric sensitivity. This can be traded-off with a number of parameters: decrease in spectral bandwidth (higher spectral resolution!), increase in the shutter speed (decreased image motion effects!), decrease in the aperture size (reduced vignetting effects!), or decrease in the gain setting (decreased electronic noise!).

Radiometric Effects. There are two main radiometric view angle effects observed in digital imaging camera products: (a) shading and (b) vignetting (also see Sect. 7.1.1). As view angle increases, radiation intensity at the sensor surface decreases, theoretically as a function of $\cos^4$ or $\cos^3$ of the view angle; this variation involves often termed 'shading' or '$\cos^4$ fall-off' (Slater, 1980). Further, vignetting is absorption and blocking of more radiation by the lens wall, thus decreasing the radiation reaching the sensor. In normal medium- to wide-angle imaging, both shading and vignetting, in combination, produce significantly decreased image brightness towards the corners of images.

Image Motion. Image motion is an important consideration in high-resolution imaging. For a given ground pixel size requirement, a commonly accepted upper limit on image motion is the shutter speed which will generate ½-pixel image motion for a specific sensorcraft velocity. For instance, for imaging at 1–5 cm resolution, in a medium-speed aircraft (40–60 ms^{-1}), image motion of ½-pixel or more may be present, even if shutter speed in the range of 1/1000s is used. Therefore, image motion compensation is necessary, which is carried out either by purely electronic methods (TDI, time-delay integration), or by photo-mechanical devices.

Calibration. For radiometric accuracy, pre-flight and in-flight calibration is generally followed. Cooling of the detector may also be used for noise reduction.

Multispectral Digital Cameras. In view of the limited range of spectral sensitivity of silicon chips (0.4–1.0 μm), multispectral sensors using digital imaging are limited to this spectral range. Several designs for multispectral digital imaging from aerial platforms have been employed which include beam-splitting, hyperspectral filtering etc. From space platforms, digital cameras have been developed to acquire images in frequently four spectral bands: blue, green, red and near-IR, quite similar to the multispectral spaceborne sensors in the VNIR range. These systems employ multiple cameras, typically one area array for each spectral band.

5.4.5.3 Development, Present Status and Future

The development and growth of digital imaging technology has been very fast. The earliest solid-state sensor developed for airborne remote sensing incorporated a 100 × 100 CCD imager (Hodgson et al. 1981) and the first aerial multispectral digital camera included a 4-camera assembly (Vlcek and King 1985). Gradually, larger and larger array sizes were developed and sophistication in imaging technique evolved. With the developments in detector technology, it has been possible to manufacture larger chips with greater numbers of photo-sites; for example, today's rectangular array sizes are about 4k × 4k, 4k × 7k, 7k × 9k and 9k × 9k. This has raised visions of large-format digital imaging cameras employing CCD matrix-sensor chips of 20k × 20k (Hintz 1999). This target may not be far in the future and would eventually allow fuller utilization of this technology for remote sensing.

With the technological advances in digital imaging cameras, and resolutions approaching photography, it is expected that most topographic mapping missions will prefer to use CCD matrix cameras as the basic data source. This is due to the simple fact that developing digital elevation models from digital imaging camera data is easier than from photography, because of the direct digital interface and extra flexibility and accuracy in data acquisition.

Space sensors using this technology include modern and future satellite sensors such as Ikonos, QuickBird, OrbView and ISI-EROS series (Sect. 5.5). Examples of images from digital cameras are given in figures 16.16 and 16.103.

5.5 Space-borne Imaging Sensors

5.5.1 Landsat Programme

Following the successful photographic experiments on the initial manned space missions (Mercury, Gemini and Apollo), NASA developed plans for unmanned orbital imaging sensors for Earth resources. The first in the series, called the Earth Resources Technology Satellite (ERTS), later renamed the Landsat Program, has been a phenomenal success. It has given tremendous impetus to remote sensing programmes and technology world-wide. In 1970s-80s, the main space sensor was MSS on Landsats. The chief factors leading to the ascendancy of Landsat MSS have been (Curran 1985): (1) world-wide data coverage due to a tape recorder on board; (2) unrestricted availability of data without any political or security constraints; (3) ready availability of multispectral and multi-temporal data, without the necessity of planning dedicated flights; (4) low cost; (5) fairly good geometric accuracy; and (6) the digital nature of the data, amenable to digital processing and enhancement.

The Landsat programme was initiated by NASA as an experimental programme, wherein the data was largely distributed by the Eros Data Centre (EDC), US Geological Survey. Subsequently, the commercial activities of the Landsat programme were taken over by the NOAA and later (since January 1985) by a private company, EOSAT Inc.

Landsat web-sites: http://www.spaceimage.com; http://edcwww.cr.usgs.gov/; http://landsat.gsfc.nasa.gov/

Table 5.7. Landsat series of satellites – general characters (after NASA)

Space-craft	Launch	Altitude (km)	Orbit inclination	Orbital period (min)	Repeat cycle (days)	Sensors
Landsat-1	July 1972	918	99°	103	18	MSS: 4 bands RBV: 3 bands
Landsat-2	Jan. 1975	918	99°	103	18	MSS: 4 bands RBV: 3 bands
Landsat-3	March 1978	918	99°	103	18	MSS: 5 bands RBV: 1 band
Landsat-4	July 1982	705	98.2°	98.9	16	MSS: 4 bands TM: 7 bands
Landsat-5	March 1984	705	98.2°	98.9	16	MSS: 4 bands TM: 7 bands
Landsat-6	Oct. 1993	Failed				ETM: 8 bands
Landsat-7	April 1999	707	98.2°	98.9	16	ETM+: 8 bands

1. Landsat Orbit

Landsat-1 was launched in July 1972 and since then six more satellites in this series (Landsat-2, -3, -4, -5, -6 and -7) have been launched (Table 5.7). All these satellites have a near-polar, near-circular, Sun-synchronous orbit (Sun-synchronous means that the orbit is synchronized with the Sun, which implies that adjacent areas are seen with almost no change in solar illumination angle or azimuth, permitting good comparison and mosaicking). Landsat-1, -2 and -3 had similar appearance (Fig. 5.15a) and identical orbital parameters (Table 5.7), each completing the Earth's coverage every 18 days. Landsat-4, -5, -6 were redesigned (Fig. 5.15b) for higher stability and placed in a relatively lower orbit to permit higher spatial resolution of the TM sensor, and each of these have provided complete coverage of the Earth every 16 days.

For a Sun-synchronous orbit like that of Landsat, a typical ground trace for one day is as shown in figure 5.16. Landsat sensors record data in the descending node (i.e. as the satellite moves from north to south). The data are pre-processed, sometimes stored on magnetic tapes (only MSS data) and relayed down to the Earth, either directly or through TDRS (tracking and data relay satellite – a communication satellite). A number of ground receiving stations have been built to receive the Landsat data around the globe (Fig. 5.15c). The satellite sensor data are reformatted and indexed in terms of path-and-row numbers for easy referencing and distribution world-wide.

2. MSS Sensor

It was the Multispectral Scanning System (MSS) which made the Landsat Program a tremendous success. In 1970s–1980s, it gathered a large amount of remote

Space-borne Imaging Sensors

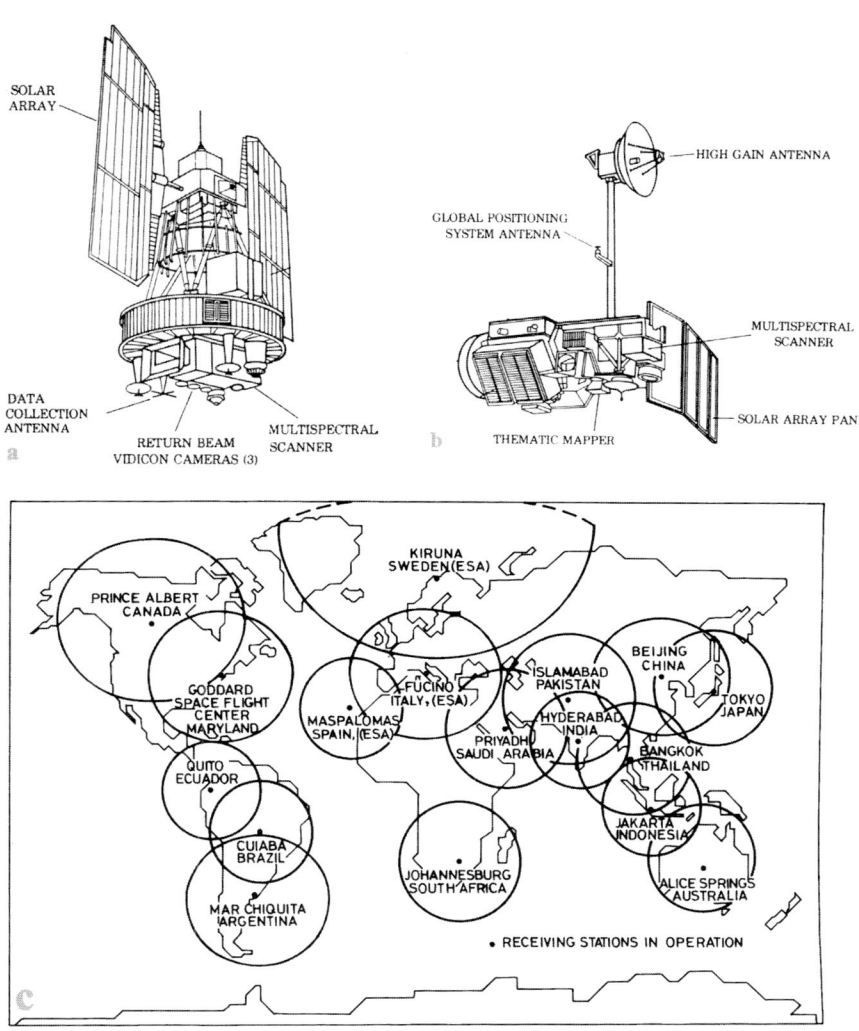

Fig. 5.15. a Design of Landsat-1, -2 and -3. **b** Design of Landsat-4 and -5. **c** Location and areal coverage of various receiving stations of Landsat data. (All figures after NASA)

sensing data world-wide. In some areas it may constitute the only available remote sensing data in archives for 1970-80s, for example as required for comparative studies and/or environmental investigations etc.; therefore, some discussion on this early sensor is not out of place.

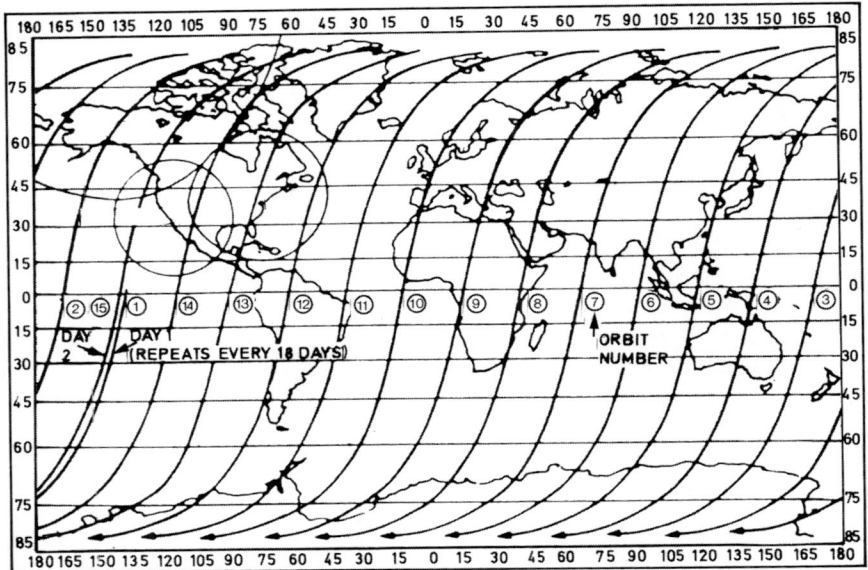

Fig. 5.16. Typical ground trace of Sun-synchronous orbit for 1 day (only southbound passes shown); the example corresponds to Landsat-1, but all Sun-synchronous satellites such as IRS, SPOT, JERS, Radarsat etc. have a similar ground trace pattern, differing only in number of passes per day and repeat cycle

The MSS used a plane mirror, oscillating along an axis parallel to the flight direction, to scan the ground in an across-track direction (Fig. 5.17). On Landsat-1, -2 and -3, which orbited at an altitude of 918 km, mirror oscillation through ±2.9° provided an FOV of 11.56°, which corresponds to a ground swath of 185 km. The active scan was during the west-to east phase of the mirror oscillation. Owing to the high ground track velocity (6.47 km s^{-1}) and a mirror oscillation rate of 13.6 Hz, the MSS needed to generate six scan lines (or use six detectors adjacent to one another in a row) per spectral band to provide along-track ground resolution of 79 m.

The MSS recorded data in four spectral channels (Table 5.8) and therefore a total of 24 detectors were used concurrently. The MSS channels 1, 2 and 3 used photo-multiplier tubes and the MSS channel 4 used silicon photodiodes. For radiometric calibration, MSS used on-board calibration lamps, which were scanned by the mirror during each oscillation cycle.

The MSS ground resolution cell is 79 × 79 m; however, sampling of the signal was carried out in an overlapping fashion, such that sampling interval corresponds to only 56 m on the ground in the scan direction (Table 5.6). A Landsat MSS scene is nominally 185 × 185 km in dimension.

For use on Landsat-4 and -5, which orbited at a slightly lower altitude of 705 km, minor changes in the MSS instrumentation were incorporated; for example, the FOV was increased to 14.93° to achieve ground resolution of 82 × 82 m.

Space-borne Imaging Sensors

Table 5.8. Landsat MSS, TM and ETM+ sensor specifications – Landsat-1, -2, -3, -4, -5 and -7 (summarized after NASA)

(a) MSS – Landsat-1, -2, -3, -4, -5

Band no.[a]	Spectral range (μm)	Name	Ground resolution	Number of scan lines run concurrently	Quantization	Image size (km)
MSS-1	0.5–0.6	Green	79×79m	6	6-bit	185 × 185
MSS-2	0.6–0.7	Red	spaced	6	6-bit	
MSS-3	0.7–0.8	Near-IR	at 56 m	6	6-bit	
MSS-4	0.8–1.1	Near-IR	interval	6	6-bit	

[a]Note: Initially, in Landsat–1, -2, and -3, the MSS bands -1, -2, -3, and -4 were numbered as MSS-4, -5, -6 and -7 respectively.

(b) TM – Landsat-4, -5

Band no.	Spectral range (μm)	Name	Ground resolution (m)	Number of scan lines run concurrently	Quantization	Image size (km)
TM-1	0.45–0.52	Blue-green	30	16	8-bit	185 × 185
TM-2	0.52–0.60	Green	30	16	8-bit	
TM-3	0.63–0.69	Red	30	16	8-bit	
TM-4	0.76–0.90	Near-IR	30	16	8-bit	
TM-5	1.55–1.75	SWIR-I	30	16	8-bit	
TM-7	2.08–2.35	SWIR-II	30	16	8-bit	
TM-6	10.4–12.5	Thermal-IR	120	4	8-bit	

(c) ETM+ – Landsat-7

The bands are similar to Landsat-TM as above with two changes:
- Thermal-IR band (Band-6) has an improved ground resolution of 60 m (in place of 120 m)
- There is an additional 'eighth panchromatic' band with 15 m ground resolution

The geometric and radiometric characteristics of the data are discussed in Chapters 6 and 7. Performance characteristics of the MSS sensor have been described by Slater (1979), Markham and Barker (1983) and Rice and Malila (1983).

3. RBV Sensor

The working principle of the Return Beam Vidicon (RBV) sensor has been briefly discussed in Sect. 5.4.2. This sensor was used in the Landsat-1, -2 and -3 missions. Landsat-1 and -2 carried a set of three RBV cameras in multispectral mode (spectral ranges: 0.47–0.57 μm = green, 0.58–0.68 μm = red, and 0.69–0.83

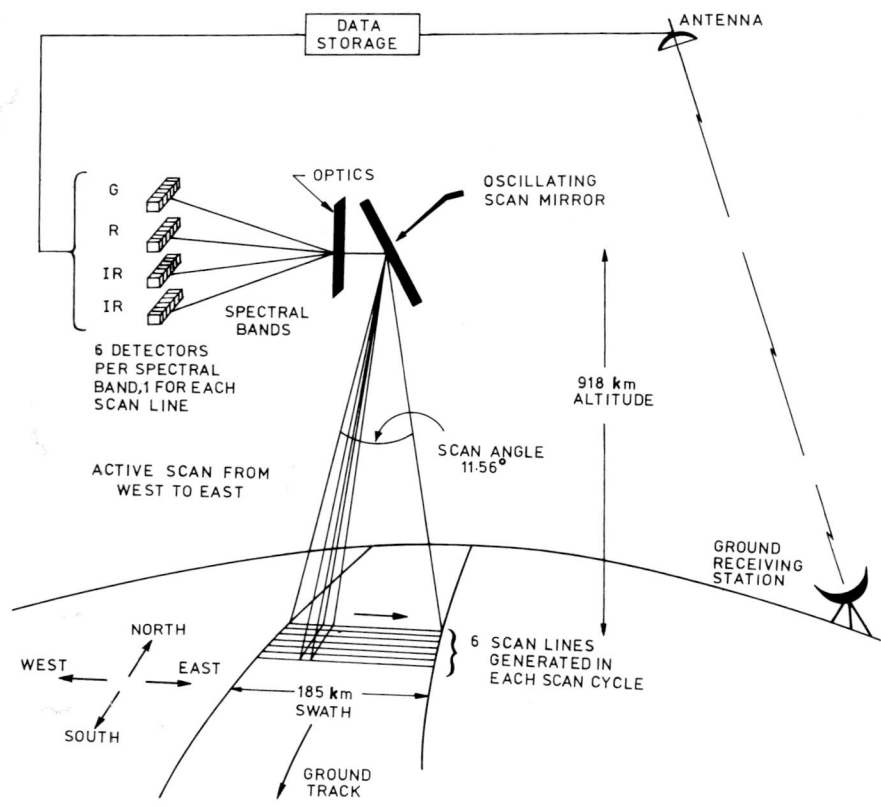

Fig. 5.17. Working principle of Landsat MSS (Landsat-1, -2 and -3). Each oscillation cycle of the mirror generates six scan lines. The data are recorded in four spectral bands by a total of 24 detectors. On Landsat -4 and -5, the altitude is 705 km and the scan angle is 14.9°

μm = infrared), each RBV image covering a ground area of about 185×185 km^2. Landsat-3 carried a set of two broad-band panchromatic RBV cameras (spectral range 0.5–0.75 μm), each of which imaged adjacent areas of nominally 98×98 km on the ground, concurrently. This produced higher spatial resolution images, as each area of 185×185 km was now imaged in four RBV frames. However, the RBV sensor had only limited success and was therefore discontinued in subsequent Landsat missions.

4. TM Sensor

The Thematic Mapper (TM) sensor is a refined multispectral scanner used on Landsat-4 and -5 missions, so much so that it is also considered as a second-generation sensor. It operates in seven wavelength bands, lying in the range of 0.45–12.5 μm (Table 5.8). It is an OM-type scanner and uses an oscillating mirror

Space-borne Imaging Sensors 103

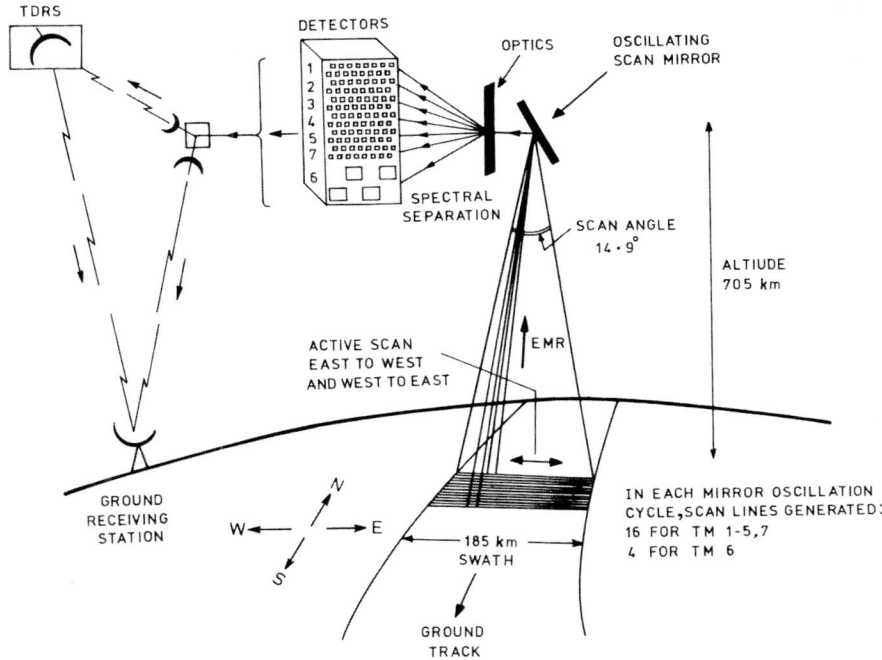

Fig. 5.18. Working principle of Landsat TM. The sensor records data in seven bands. The detectors in the VNIR region are placed at the primary focal plane and those in the SWIR–TIR region at the cooled focal plane. A total of 16 scan lines are concurrently generated for bands 1 to 5 and 7 and four lines for band 6

to scan the ground (Fig. 5.18). The main differences of the TM in comparison to the MSS sensor are as follows.

1. The TM sensor uses seven wavebands lying in the VIS, NIR, SWIR and TIR regions, in comparison to the four wavebands of MSS lying in the VIS and NIR.

2. In TM, spectral limits of the wavebands have been based on the knowledge of spectral characteristics of common natural objects and transmission characteristics of the atmosphere. This is in contrast to the spectral limits of the MSS sensor, which were quite arbitrarily chosen, as it was an initial experimental programme and the information on spectral characteristics of objects and transmission characteristics of the atmosphere was limited.

3. The MSS used active scan only during the west-to-east phase of the oscillation mirror; the retrace-cycle phase could not be used for active scanning due to scan time shift (Fig. 5.19a). In the TM, a scan-line corrector mechanism compensates for this angular distortion and provides scanning perpendicular to the

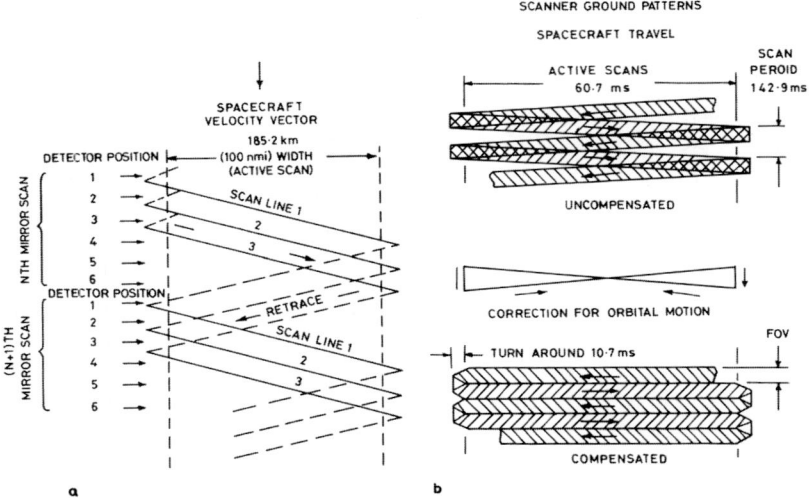

Fig. 5.19. a Scan-time shift due to forward motion of the sensor-craft, as in the case of a typical simple OM scanner such as the MSS. **b** Scan-time shift correction in the TM leading to scan lines perpendicular to the flight direction

flight line (Fig. 5.19b). This enables both back and forth oscillation phases of the mirror to be used for active scanning.

4. In TM, the detector assemblies for bands 1–4 are located behind the primary mirror as a focal plane array and employ 16 detectors for each waveband. The detector assemblies for bands 5, 6 and 7 are placed at the cooled focal plane and comprise sets of 16 detectors for TM-5 and TM-7 and a set of 4 detectors for TM-6. In this way, in each sweep of the mirror, 16 scan lines are generated for bands 1–5 and 7, and 4 scan lines for band 6 (Table 5.8). In MSS, 6 scan lines were generated for each of the four spectral channels.

5. The TM sensor has relatively higher resolution specifications – spatial, spectral and radiometric. The quantization level is 8-bit (256 levels) in all bands.

In TM, silicon detectors are used for VNIR bands (1–4); indium antimonide (In Sb) detectors are employed for SWIR bands (5 and 7); mercury cadmium telluride (Hg Cd Te) detectors are used in the thermal IR band (band 6). At any instant of time, all 100 detector elements view a different area on the ground, due to spatial separation of individual detector elements, within the TM focal planes. However, as it is a systematic geometric variation per design, accurate band-to-band registration is accomplished using data on the scan mirror's angular position as a function of time. For radiometric calibration, the TM employs three tungsten filament lamps and a blackbody for the TIR band. Each time the scan mirror changes direction, calibration sources are viewed. This enables monitoring of the radiometric response of various detectors over the sensor's life.

The data from the TM is relayed either directly or over the TDRS (Tracking and Data Relay Satellites) to the Earth receiving stations, where it is pre-processed and made available to users. There are no on-board tape recorders. The TM sensor provides high-quality data, which have been invaluable in numerous types of applications for Earth resources investigations, particularly for geological exploration (for applications, see Chapter 16).

5. ETM Sensor

Landsat-6, carrying Enhanced Thematic Mapper (ETM), was launched in October 1993, but failed. Its main aim was to provide continuity of Landsat-4 and -5 TM data. Therefore, it had a similar planned orbit as Landsat-4/-5 and sensor configuration (seven bands) as in TM. In addition, it carried an eighth "panchromatic" band (0.5–0.9 μm) with spatial resolution of 15 m, hence the name Enhanced Thematic Mapper.

6. ETM+ Sensor

Landsat-7 was launched in April 1999, carrying the Enhanced Thematic Mapper Plus (ETM+). Its main aim is again to provide continuity of Landsat-4 and -5 TM data. Therefore it has a similar orbit and repeat pattern as Landsat-4/-5. ETM+ has eight spectral bands in all. Bands 1 to 6 are in blue, green, red, near-IR, SWIR-I and SWIR-II, and are exactly the same as those in Landsat TM with 30 m ground resolution. The thermal-IR band has an improved ground resolution of 60 m (in comparison to the 120 m of TM and ETM). The panchromatic eighth band (15 m resolution) is the same as in ETM (Table 5.8).

In addition, in the ETM+, the detector system has been redesigned to allow automatic registration of all bands as the data are acquired. This capability is called 'monolithic' detector design. Another improvement is the ability to set the gains for individual bands from the ground. The high gain can be used in areas of low reflectance (e.g. a water body), and the low gain can be employed over bright regions (e.g. snow, desert etc.). The ETM+ also has an improved calibration system (absolute calibration within 5%).

Landsat-7 also carries an on-board solid-state recorder for temporary storage of remote sensing data. This allows acquisition of remote sensing data over areas outside the reach of ground receiving stations.

5.5.2 IRS Series

Under the Indian Remote Sensing Satellite (IRS) programme, the Indian Space Research Organization (ISRO) has launched a number of satellites. Figure 5.20a shows a schematic of the IRS platform. The first to be launched was IRS-1A (1988), followed by IRS-1B (1991), both from the Soviet Cosmodrome, Baik-

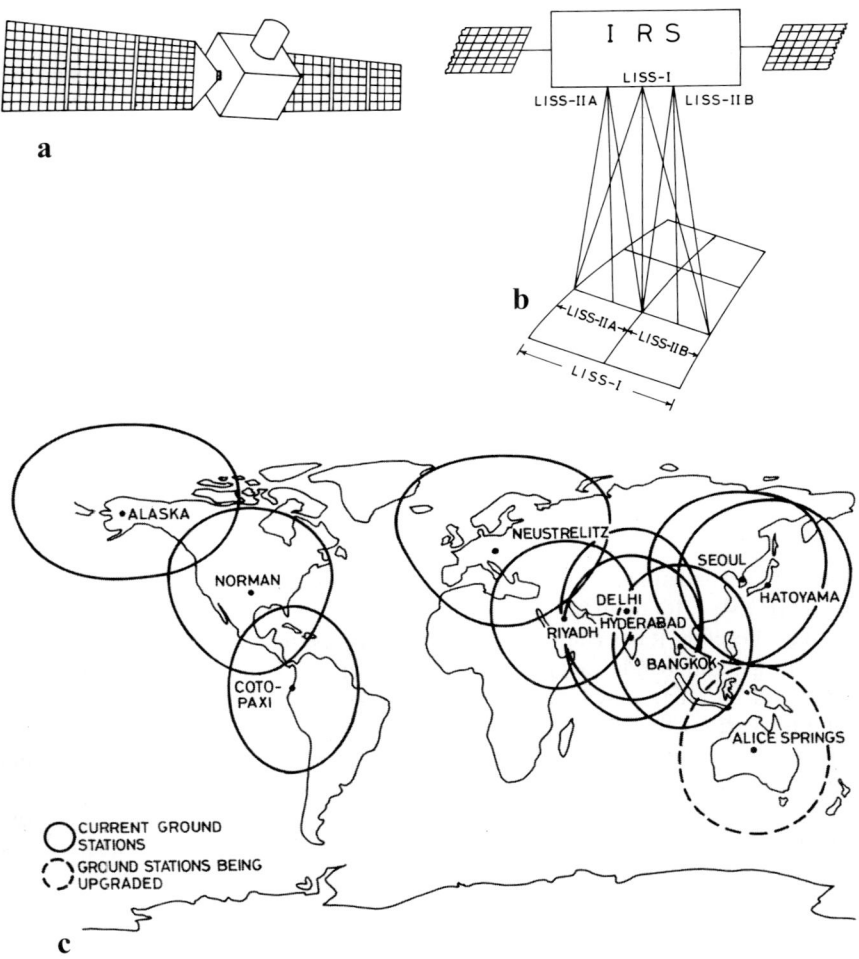

Fig. 5.20. a Schematic of IRS satellite. **b** Relative ground coverages of LISS-I and LISS-II cameras. **c** Location and area coverage of various receiving stations of IRS data. (All figures courtesy of ISRO)

noure. IRS-1A and -1B were exactly similar in orbit and sensors. They were placed in an orbit very similar to that of the early Landsats (Sun-synchronous orbit, 904 km altitude, equatorial crossing at 10.00 h, descending node). The second-generation satellites in this series are IRS-1C (1995) and -1D (1997) and these two are similar in orbit and sensors.

1. LISS-I and LISS-II Sensors:

The Linear Imaging Self Scanning (LISS-I and LISS-II) sensors, carried onboard IRS-1A and -1B, were CCD linear array pushbroom scanners. Both LISS-I

Space-borne Imaging Sensors

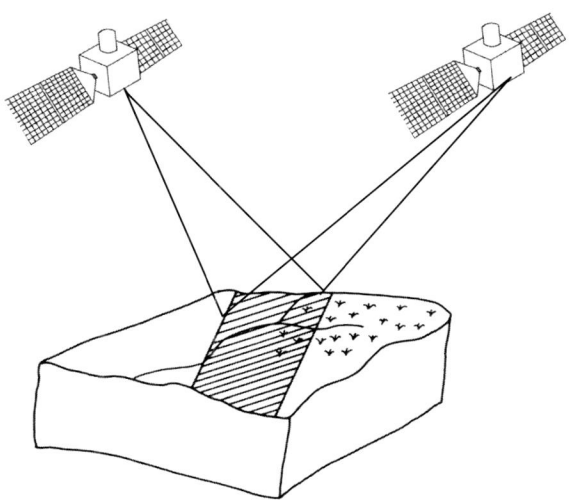

Fig. 5.21. Stereo capability by across-track pointing (e.g. IRS-PAN, SPOT-HRV sensors)

and -II provided data in four spectral bands: 0.45–0.52 µm (blue), 0.52–0.59 µm (green), 0.62–0.68 µm (red), and 0.77–0.86 µm (near-IR). Light-emitting diodes (LED) provided on-board calibration of CCD detector arrays. The LISS-I used a 152-mm focal length optics and provided data with 72.5 × 72.5 m resolution in a swath width of 148 km, quite comparable to Landsat MSS. The LISS-II used a set of two cameras, A and B, each of 304 mm focal length optics, and provided data with 36.25 × 36.25 m ground resolution, quite comparable to Landsat TM sensor data. The combined swath of LISS-II A and B cameras was 145 km (Fig. 5.20b; Table 5.9) (Thiruvengadachari and Kalpana 1986).

2. IRS-PAN Sensor:

The panchromatic sensor (PAN) is carried on-board IRS-1C and -1D. It is a high spatial resolution (5.8 m) broad-band (0.50–0.75 µm) sensor, with a swath width of 70 km. It can be steered across-track for stereo capability (Fig. 5.21).

3. LISS-III Sensor:

The LISS-II has been modified to the LISS-III sensor, which now includes bands equivalent to TM-2, -3, -4 and -5 (Table 5.9). The resolution of green, red and near-IR bands is 23.5 m and that of the SWIR band is 70.5 m, the swath width being about 140 km.

4. IRS-WiFS Sensor:

The IRS-1C and -1D also carry an WiFS (Wide Image Field Sensor), which is a coarse-resolution (188 m) two-band (red and near-IR) large swath (800 km) sensor designed mainly for regional vegetation studies. This is also particularly useful for

Table 5.9. Salient specifications of IRS sensors (summarized after ISRO)

Sensor Space- craft	LISS-I IRS-1A, -1B		LISS-II IRS-1A, -1B	LISS-III IRS-1C, -1D		PAN IRS-1C, -1D
Spectral bands	4[a]		4[a]	4		1
Name Blue Green Red Near-IR SWIR-I	Code B1 B2 B3 B4	Spectral range 0.45–0.57 µm 0.52–0.59 µm 0.62–0.68 µm 0.77–0.86 µm Absent		Code B1 B2 B3 B4 B5	Spectral range Absent 0.52–0.59 µm 0.62–0.68 µm 0.77–0.86 µm 1.55–1.70 µm	Broad spectral range 0.50–0.75 µm, with off-nadir viewing capability
Ground resolution	72.5 m		36.25 m	For B2–4: 23.5 m For B5: 70.5 m		5.8 m
Swath (km)	148.5		74.2 × 2	141 (B2–4); 148 (B5)		70

[a] LISS-I and LISS-II have identical spectral bands, differing only in ground resolution.

imaging with a greater repeat cycle (three-day interval), needed for example in natural disaster studies.

Data from IRS satellites are received by a large number of ground stations world-wide (Fig. 5.20c). Several examples of IRS-LISS and -PAN data are given later (Chapters 10 and 16).

IRS web-sites: http://www.isro.org; http://www.nrsa.gov.in

5.5.3 SPOT Series

The French Satellite System SPOT (Syste'me Probatoire de l'Observation de la Terre, which literally means Experimental System for Observation of the Earth) is one of the typical second-generation Earth resources satellites. It is also the first free-flying European Earth resources satellite, and is operated under the French Centre National d'Etudes Spatiales (CNES) and SPOT-IMAGE Inc. SPOT-1 (Fig. 5.22a) was placed in orbit in February 1986 (near-polar, Sun-synchronous, 830 km high orbit, repeat cycle 26 days, inclination 98.7°, crossing the equator in descending node at 10.30 h local time). SPOT-2 (1990) was identical to SPOT-1. SPOT-3 was designed as identical to SPOT-1/-2, and launched in 1993, but failed.

SPOT-1and -2 carry a set of two identical sensors called HRV (High Resolution Visible) sensors, each of which is a CCD linear array pushbroom scanner. The HRVs can acquire data in several interesting configurations:

1. In nadir-looking panchromatic (PAN) mode, each HRV, using an array of 6000 CCD detectors, provides data with 10-m ground resolution, in a swath width of

Space-borne Imaging Sensors

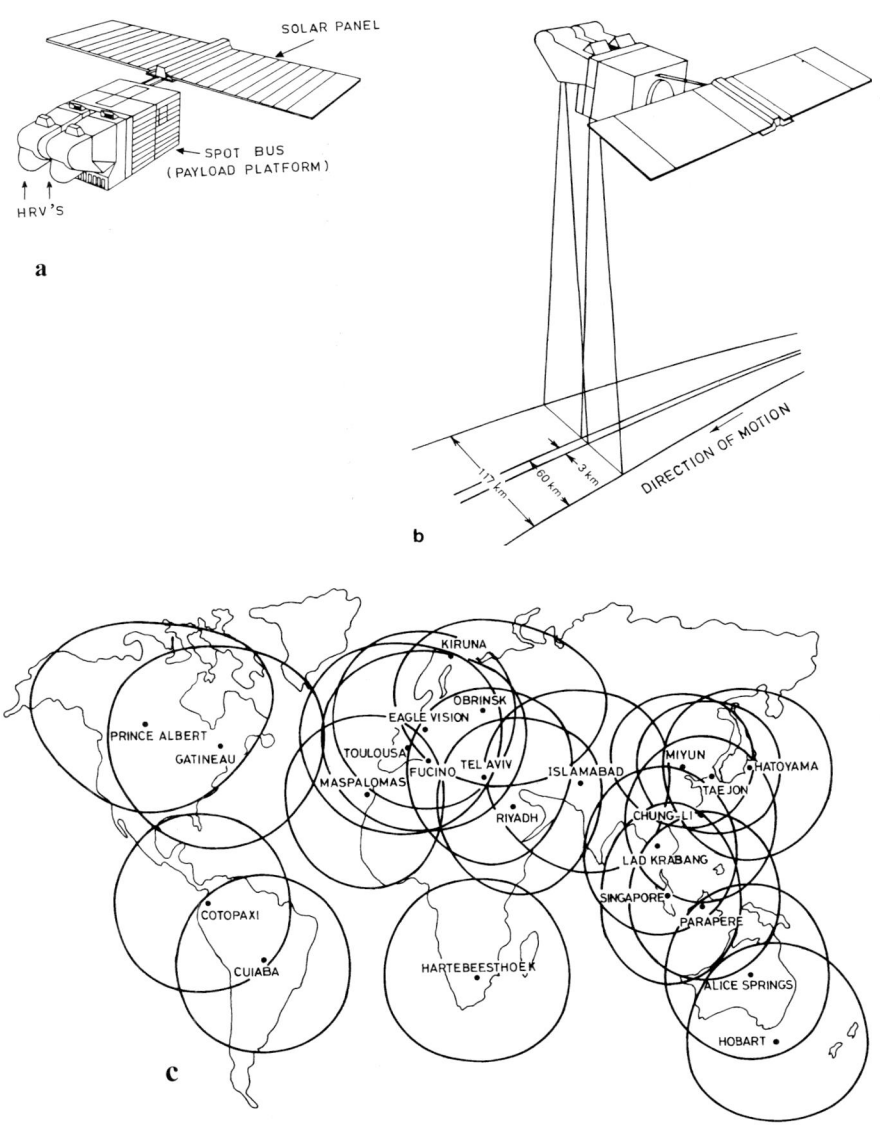

Fig. 5.22. a SPOT platform. **b** Nadir-viewing HRV sensor. **c** World-wide ground receiving stations of SPOT data. (All figures after CNES/SPOT Image)

60 km. The two HRVs together cover a 117 km ground swath with 3 km overlap (Fig. 5.22b).
2. In multispectral (XS) mode, the HRVs acquire data in three spectral channels with a ground resolution of 20 m, and swath of 60 km.

3. The HRVs can be directed to view off-nadir, across-track; this is employed for stereoscopy, and also for higher repeat cycle.

The SPOT-HRV data have been widely used for various Earth resources investigations. Figure 5.22c shows the global distribution of receiving stations of SPOT sensor data.

SPOT-4 (1998) incorporates some changes. The sensor is called HRVIR (High Resolution Visible and Infrared). It is again a CCD-pushbroom line scanner and it also comprises two HRVIRs – as in the earlier HRVs. The major changes in SPOT-4 are as follows.

(a) The higher-resolution 10-m panchromatic band is changed to red band (0.61–0.68 µm).
(b) In the multispectral mode, there is an additional band in the SWIR (1.5–1.75 µm), bringing the total number of bands to four, each with 20 m resolution.
(c) Further, SPOT-4 also carries a 'Vegetation Instrument' in which all the five bands in HRVIR are present, but with a coarse resolution of 1 km and a swath of 2250 km.

SPOT web-sites: http://www.spotimage.fr; http://www.spot.com

5.5.4 MOMS Series

The German Modular Optoelectronic Multispectral Scanner (MOMS-1) was the first ever pushbroom-type scanner used for remote sensing from space (STS-7), in the year 1983. It operated in two spectral bands (Band1: 0.57–0.62µm; and Band2: 0.82–0.92 µm). Each spectral band had a set of two cameras, thus four cameras in all were used. The sensor provided spatial resolution of 20 × 20 m, and covered selected parts of the Earth in a swath width of 140 km from the shuttle (Hiller 1984; Bodechtel et al. 1985; Bodechtel and Zilger 1996).

MOMS-2 was an upgraded version, emphasizing upon the in-orbit stereo capability and high-resolution panchromatic band sensor. It was flown on ESA-Spacelab Mission (altitude 300 km) in 1993. In fact, this was the first system to acquire in-orbit stereo images of the Earth from space. The MOMS-2 generated

Table 5.10. MOMS-02P: sensor specifications

Channel	Mode	Orientation	Wavelength range (µm)	Ground element size (m^2)	Swath width (km)
1	MS	Nadir	0.440–0.505	18 × 18	105 / 50
2	MS	Nadir	0.530–0.575	18 × 18	105 / 50
3	MS	Nadir	0.645–0.680	18 × 18	105 / 50
4	MS	Nadir	0.770–0.810	18 × 18	105 / 50
5	HR	Nadir	0.520–0.760	6 × 6	105 / 50
6	Stereo	+21.4°	0.520–0.760	18 × 18	105 / 50
7	Stereo	−21.4°	0.520–0.760	18 × 18	105 / 50

Space-borne Imaging Sensors 111

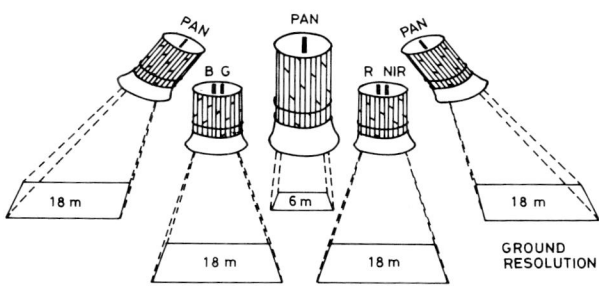

Fig. 5.23. MOMS-02P system: there is a high-resolution nadir-looking PAN camera, two nadir-looking multispectral cameras for imaging in blue, green, red and near-infrared bands, and a pair of inclined (fore and aft) PAN cameras for stereo

images of selected strips of the Earth with a ground resolution of 4.5 m in panchromatic band and 13.5 m in the multispectral band.

After the ESA-Spacelab mission, the sensor MOMS-2 camera was refurbished to fly on the Russian space-station MIR module PRIRODA, and the sensor was called MOMS-02P, during the period 1996–98. Due to its longer space life, MOMS-02P covered several parts of the Earth providing interesting and useful image data. The data were recorded in a magnetic tape recorder on-board.

MOMS-02P employed CCD linear array detectors with the optical concept shown in figure 5.23. There are three main sensor components: (a) a multispectral camera with four bands (channels 1, 2, 3, 4 = B, G, R, and NIR respectively) in nadir-looking mode; (b) a high-resolution panchromatic nadir-looking camera (channel 5); and (c) a set of two identical panchromatic cameras (channels 6 and 7) inclined (+21.4° and –21.4°) in fore and aft directions for in-orbit stereo. The sensor has been operated in a variety of modes making combinations of high-resolution, stereo and multispectral channels, covering different parts of the Earth in different modes. Table 5.10 summarizes the sensor specifications and modes of operation. Examples of MOMS-02P images are given in later chapters in figures 8.4, 10.41 and 16.4.

MOMS web-sites: http://www.nz.dlr.de/moms2p; www.scanex.ru/data/MOMS

5.5.5 JERS-1 (Fuyo-1) OPS

Japan's Earth Resources Satellite-1 (renamed Fuyo-1) (1992) employs both an optical sensor (OPS) and a radar sensor (SAR) on the same spacecraft. Fuyo-1 has been placed in a near-circular, Sun-synchronous, near-polar orbit at a nominal altitude of 568 km with a 44-day repeat cycle. The OPS sensor system uses CCD linear arrays, and can be broadly divided into two sub-systems: VNIR and SWIR. Its salient specifications are summarized in Table 5.11. Bands 1 to 3 are similar to Landsat TM bands 2 to 4, operating in green, red and near-IR. OPS bands 3 and 4

Table 5.11. Salient specification of JERS-1 OPS sensor system (after Nishidai 1993)

Band	Wavelength (μm)
VNIR	1. 0.52–0.6 (green)
	2. 0.63–0.69 (red)
	3. 0.76–0.86 (NIR)
	4.*0.76–0.86 (NIR)
	* Forward viewing for stereo imaging
SWIR	5. 1.60–1.71
	6. 2.01–2.12
	7. 2.13–2.25
	8. 2.27–2.40
Detector	: CCD linear array
Ground resolution	: 18.3 × 24.2 m
Swath width	: 75 km

provide stereoscopic capabilities (with B/H = 0.3). Bands 5 to 8 cover critical regions in the SWIR, useful in mineralogical discrimination.

JERS web-site: http://hdsn.eoc.nasda.go.jp/

5.5.6 CBERS Series

China and Brazil have jointly formed a remote sensing programme called CBERS. The CBERS-1 was launched in October 1999 in a Sun-synchronous orbit (778 km altitude, 26 day repeat cycle). It carries three main sensors.

1. *CCD Camera*: with five channels, each with 20 m ground resolution. The spectral ranges cover panchromatic, blue, green, red and near-IR bands. Swath width is 113 km.
2. *IR-MSS:* the Infrared Multispectral Scanning System operates in four spectral bands – one in panchromatic, two in SWIR and one in thermal-IR bands. Its main purpose is to extend the remote sensing capability in the SWIR–TIR bands. The sensor has a ground resolution of 80–160 m.
3. *WFI:* the Wide Field Imager has a large swath of 890 km and a revisit interval of 3 to 5 days. It is basically intended for vegetation monitoring.

CBERS web-site: http://satellites.satellus.se/cbers.asp; http://www.dgi.inpe.br

5.5.7 RESURS-1 Series

A series of Russian remote sensing satellites (RESURS-01) is now represented by two vehicles: RESURS-01-3 (launched in October 1994) and RESURS-01-4 (July 1998). Both satellites operate in polar sun-synchronous orbits with a mean altitude of 678 km (RESURS-01-3) and 835 km (RESURS-01-4). The remote sensing systems consist of CCD linear arrays. There are two sensors: MSU-E (high-resolution) and MSU-SK (medium-resolution). The MSU-E sensor provides data in three spectral bands (green, red and near-IR) with pixel dimensions of about 35 m × 45 m. The MSU-SK sensor has four spectral bands (green, red, near-IR and thermal-IR) with pixel dimensions of about 140 m × 185 m in VNIR bands, and about 600 m × 800 m in TIR band.

RESURS web-site: http://www.scanex.ru

5.5.8 ASTER Sensor

ASTER is the 'Advanced Spaceborne Thermal Emission and Reflection' radiometer, launched by NASA as a part of the first 'Earth Observation Satellite' (EOS-AM-1) programme on 18 December 1999. ASTER is one of the five imaging sensors carried on-board the EOS-AM-1, which aims at comprehensively understanding global changes, especially climatic changes. ASTER carries moderate resolution imaging sensors and aims at contributing to the understanding of local and regional phenomena on the Earth's surface and in the atmosphere.

The EOS-AM-1 is launched in a near polar, near circular orbit, with nominal altitude 705 km, orbit inclination 98.2°, Sun-synchronous, descending node, equatorial crossing 10.30 h, and a repeat cycle of 16 days.

In the ASTER instrument, several improvements have been incorporated in order to exceed the performance of existing optical sensors such as Landsat-TM, SPOT-HRV, JERS-OPS, and IRS-LISS/PAN. ASTER has a total of 14 spectral channels spread over the range of 0.53–11.65 µm (VNIR–SWIR–TIR) and an increased B/H ratio of 0.6 (as compared to 0.3 of earlier sensors) for better stereo imaging (Kahle et al. 1991).

In order to provide wide spectral coverage, the ASTER instrument has three radiometer subsystems: (a) a visible and near-infrared (VNIR) radiometer subsystem, (b) a short-wave infrared (SWIR) radiometer subsystem, and (c) a thermal-infrared (TIR) radiometer subsystem. Table 5.12 gives the design performance specification of the ASTER radiometers.

(a) VNIR radiometer: this is a pushbroom (CCD line) scanner and acquires data in the VNIR range (0.52–0.86 µm). Its main features are as follows. (a) For multispectral imaging, there are three bands – green, red and near-IR in nadir-looking mode, with 15 m ground resolution and swath width of 60 km. (b) The instrument has cross-track pointing capability (± 24°) which leads to a 232 km swath. (c) There is a back (or aft)-looking CCD linear array in the near-IR, called 3B. This enables generation of image pairs (bands 3N and 3B) for stereoscopic viewing with a B/H ratio of 0.6. The image data are automatically corrected for positional deviation due to the Earth's rotation around its axis. (d) The stereo image pair are generated in the same orbit, in real time (time lag of barely 55 s).

(b) SWIR radiometer: this is also a pushbroom (CCD line) scanner with 30 m spatial resolution. It acquires multispectral image data in the SWIR range (1.60–2.43 µm) in six bands. These data should be highly useful for the study of minerals, rocks, volcanoes, snow, vegetation etc. The SWIR scanner has an off-nadir cross-track pointing capability of ± 8.55°, by rotation of the pointing mirror.

Table 5.12. ASTER sensor specifications (summarized after NASA)

	Spectral band	Spectral range (μm)	Scanner type	Detector	Spatial resolution (m)	Signal quantization
VNIR			5000 cells	Si-CCD	15	8-bit
Green	1	0.52–0.60				
Red	2	0.63–0.69				
NIR	3	0.76–0.86				
NIR	3B[a]	0.76–0.86				
SWIR	4	1.60–1.70	2048 cells	Pr-Si-CCD	30	8-bit
	5	2.145–2.185				
	6	2.225–2.245				
	7	2.235–2.285				
	8	2.295–2.365				
	9	2.360–2.430				
TIR	10	8.13–8.48	OM	Hg-Cd-Te	90	12-bit
	11	8.48–8.83				
	12	8.90–9.25				
	13	10.25–10.95				
	14	10.95–11.65				

– Swath width: 60 km
– Coverage in cross-track direction by pointing function: 232 km
– Base to height (B/H) ratio of stereo capability: 0.6 (along-track)
– Cross-track pointing VNIR: ± 24°, SWIR: ± 8.55°, TIR: ± 8.55
3B[a] Band 3B is backward-looking for stereo

(c) TIR radiometer: this is an opto-mechanical line scanner and collects multispectral image data in five bands in the thermal-IR range (8–12 μm), with a spatial resolution of 90 m and swath of 60 km.

ASTER can acquire data in various modes, making combinations of various radiometers in day and night. For geological studies, ASTER is expected to generate the most interesting type of remote sensing data from sensors currently in orbit.

ASTER web-sites: http://asterweb.jpl.nasa.gov/; http://eos-am.gsfc.nasa.gov/

5.5.9 MTI

The Multispectral Thermal Imager (MTI) was launched in March 2000 in a near-polar, Sun-synchronous orbit of 555 km altitude, at 97° inclination to the equator. It is an advanced multispectral thermal imaging sensor and carries 15 spectral bands ranging from visible to thermal-IR. The sensor has a highly accurate radiometry. Resolution is 5 m in the three visible bands and 20 m in the remaining 12 bands lying in the NIR, SWIR and TIR region.

MTI web-site: http://nis-www.lanl.gov/nis-projects/mti/

Fig. 5.24. The IKONOS platform (Courtesy: Space Imaging)

5.5.10 Space Imaging / Eosat – IKONOS

IKONOS is the first commercial high-resolution satellite (Fig. 5.24). Launched by Space Imaging/Eosat Inc. in September 1999, it orbits in a Sun-synchronous, near-polar orbit (681 km altitude, 98.1° inclination). The satellite provides remote sensing image data of high geometric accuracy. It has two sensors: (a) a panchromatic sensor (0.45–0.90 µm) with 1 m spatial resolution and (b) a multispectral sensor with 4 m spatial resolution, and working in four (blue, green, red and near-IR, spectral bandwidth similar to Landsat TM) bands. Repeat cycle is approximately 11 days. An interesting feature is that both fore–aft and cross-track stereoscopic capability exists by ± 45° tilt capabilities, leading to a high B/H ratio (0.6) and VE of about 4. Further, the revisit interval is about 1.5–3 days. Individual scenes are about 11 × 11 km, with strips of up to 100 km in length. Quantization is 11 bit. An example of the IKONOS image is given in figure 16.103.

IKONOS web-site: http://www.spaceimage.com

5.5.11 DigitalGlobe – QuickBird

DigitalGlobe (earlier called EarthWatch) Inc. developed early plans to deploy its own satellites for commercially utilizing space-borne remote sensing. Its first satellite, EarlyBird, was launched in December 1997, but lost contact with the Earth receiving station several days after launch. It utilized digital imaging camera technology, aiming at imaging in two modes with ground resolutions of 3 m (panchromatic band) and 15 m (multispectral sensor in G, R and NIR bands).

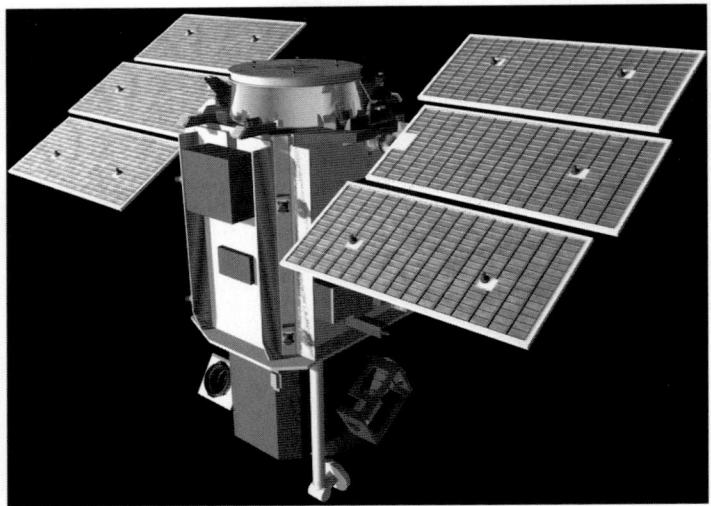

Fig. 5.25. The QuickBird platform. (Courtesy of DigitalGlobe)

QuickBird-2, the next mission from DigitalGlobe Inc., was successfully launched in October 2001 (Fig. 5.25). The selected orbit is Sun-synchronous (450 km altitude and 98° inclination). The sensor provides ground resolution of 61cm in the panchromatic band and 2.5 m in four multispectral bands (B, G, R and NIR). It uses digital camera technology. The panchromatic band image is approx. 28,000 pixels in swath width, each frame being 16.5 × 16.5 km. In multispectral mode, the image is 7000 pixels wide. Image data can be acquired with ± 25° inclination, both along-track and across-track.

DigitalGlobe web-site: http://www.digitalglobe.com/

5.5.12 Other Programmes (Past)

A number of other space-borne sensors have been developed and deployed for remote sensing observations (see e.g. Ryerson et al. 1997). Only some of these will be mentioned here, with the view that data from the earlier satellites may be available in archives for some applications.

SKYLAB was a manned space station and was launched by NASA during 1973–74 into an oblique orbit with 430 km altitude. It became news in 1979 when it re-entered the Earth's atmosphere and disintegrated, scattering itself near the Australian continent. One of the experiments carried by SKYLAB involved a 13-channel opto-mechanical line scanner. It used conical scanning to allow uniform atmospheric path length (Fig. 5.26). However, the data from conical scan lines are difficult to manipulate and rearrange for generating the image, and hence could not be much used by scientists.

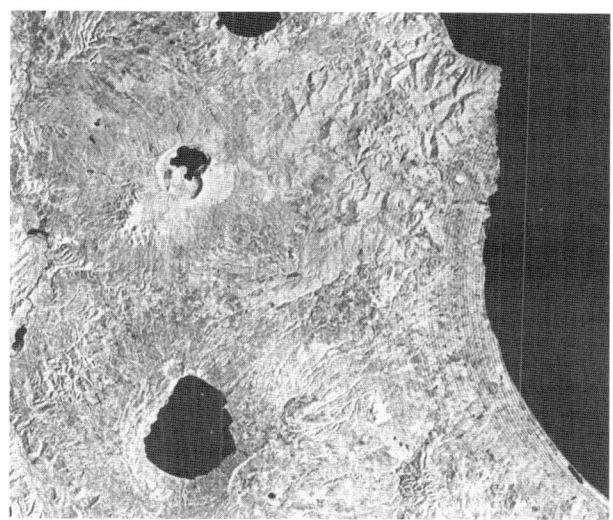

Fig. 5.26. Example of imagery from the SKYLAB scanner. Note the conical scan lines

The Heat Capacity Mapping Mission (HCMM) was another important mission from NASA. It was dedicated to thermal mapping of the Earth's surface from orbit (Short and Stuart 1982). The spacecraft HCMM was launched in April 1978 and provided data until September 1980. It circled the Earth at 620 km altitude in a near-polar orbit. The sensor, named Heat Capacity Mapping Radiometer (HCMR), consisted of a visible channel (0.55–1.1 µm; 500 m resolution) and a thermal channel (10.5–12.5 µm; NE ΔT = 0.4 K at 280 K, 600 m resolution). In order to provide repetitive coverage, the radiometer had a very wide FOV (scan angle of ± 60°), resulting in a swath width of 720 km. The ground resolution varied from 0.6 km at the centre of the image to about 1 km at the edge. The satellite had no on-board tape recorder and so did not provide world-wide data, the HCMR data having been limited to North America, Europe and Australia. The repetitive coverage was used for detecting thermal differences in rock/soil properties (see Chapter 9).

The Japanese MOS-1 (Marine Observation Satellite) was launched in early 1987, in an orbit very similar to that of the Landsat. An important sensor on MOS-1 was the MESSR (Multispectral Electronic Self-Scanning Radiometer). This sensor used a pushbroom scanner to yield video data in four wavelength bands (VNIR) with a ground resolution of 50 × 50 m.

In addition, a number of other spacecraft have been launched for Earth observations, such as the TIROS–NOAA series, Nimbus series etc. in polar orbits, and the GOES series, Meteosats, Himawari, INSATS etc. in geostationary orbits. These satellites are primarily for meteorological–oceanographical purposes. Their

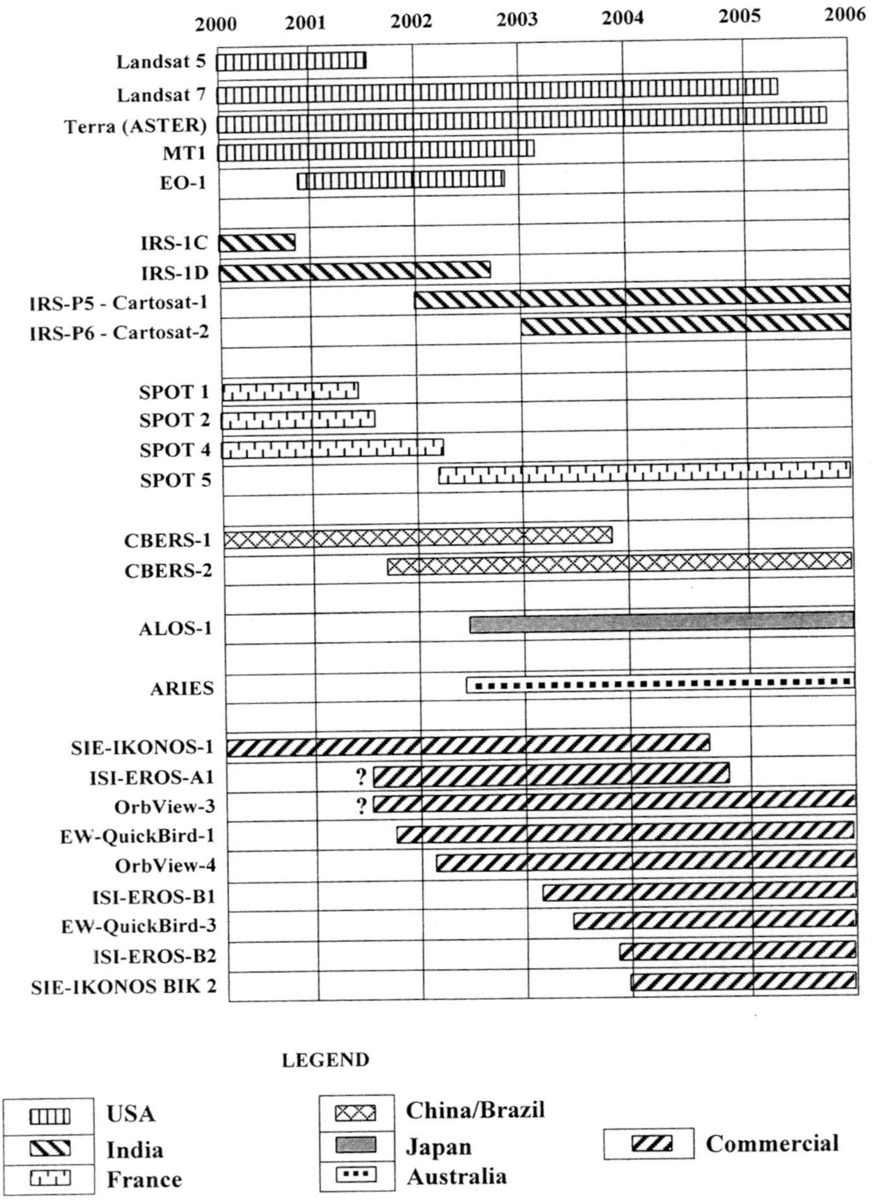

118   Chapter 5: Multispectral Imaging Systems

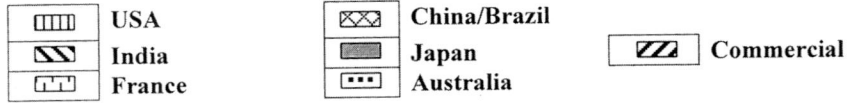

Fig. 5.27. Space-borne optical sensors (2000–2006)

large FOV, substantial geometric distortions and coarse spatial resolution, and therefore their data are of limited utility for geological applications, particularly as better-quality data for this purpose are available from other sensors.

5.5.13 Planned Programmes

Several space-borne remote sensing systems are proposed for launch in the near future, including some from government organizations and some from commercial ones (Fig. 5.27).

1.ORBIMAGE–OrbView: the Orbital Imaging Corporation (ORBIMAGE) has proposed a series of low-cost, high spatial resolution remote sensing satellites, called OrbView. OrbView-1and -2 have been launched and OrbView-3 and -4 are proposed. OrbView-1 is a weather satellite. OrbView-2 carries the SeaWiFs sensor comprising eight bands – six visible and two NIR. OrbView-3 is proposed to consist of a 1-m resolution panchromatic (0.45–0.90 μm) band and a 4-m resolution multispectral sensor in four (blue, green, red and near-IR) bands, both sensors having 8-km swath width. OrbView-4 will have the same panchromatic and multispectral sensors as OrbView-3, and will additionally carry a 280-channel hyperspectral imaging instrument (8-m ground resolution, 5-km swath).

ORBIMAGE web-site: http://www.orbimage.com/

2. ALOS: the Japanese Advanced Land Observation Satellite (ALOS) is a follow-up programme to JERS (Fuyo)-1, and is proposed for launch in 2001. Its main application will be in land observation – resources surveying, cartography, disaster monitoring etc. It will have three main sensors: (a) PRISM – a Panchromatic Remote sensing Instrument for Stereo Mapping; (b) AVNIR-2 – an Advanced Visible and Near Infrared Radiometer-2; and (c) PALSAR – Phased Array-type L-band Synthetic Aperture Radar.

ALOS web-sites: http://www.nasda.go.jp/; http://www.eoc.nasda.go.jp/

3. CBERS-2: is scheduled for launch in October 2001 as a follow-up to CBERS-1, and the two are identical. CBERS-3 and -4 are also proposed (2002–03) with higher resolution sensors (panchromatic 5 m, multispectral 10 m, infrared 40–80 m). (web-site as for CBERS-1.)

4.CARTOSAT: proposed for launch in 2002 as a follow-up to the IRS-programme, in a near-polar Sun-synchronous orbit (altitude 618 km). It is meant as a mapping mission and will carry two panchromatic cameras, one looking aft (−5°) and the other fore (+26°) to provide a stereo pair. The resolution contemplated is 2.5 m for both cameras and a height resolution of better than 5 m is expected in stereo products. The swath will be 30 km and data will have 10-bit quantization. It will also carry a solid-state data recorder on board for storing payload data. (web-site as for the IRS series.)

5. *SPOT-5*: proposed for launch in 2002–2003. It will carry a panchromatic sensor with a higher resolution (5 m/2.5 m), and a multispectral sensor with bands in green (10 m), red (10 m), infrared (10 m) and SWIR-I (20 m). (web-sites as for the SPOT series.)

6. *ISI-EROS Series*: Image Sat International has made plans for a constellation of high-resolution low-cost, agile, light weight, low-orbit remote sensing satellites with frequent daily re-visit. It has been named the EROS series. There are two classes of EROS satellites: A and B. EROS-A1 and -A2 will orbit at an altitude of 480 km and carry CCD area array (FPA) detectors, to produce images of 1.8 m ground resolution and 12.5 km swath width. EROS-B1 and -B2 will orbit in polar orbit at an altitude of 600 km. They will use a CCD/TDI (charge-coupled device/time delay integration) focal plane. This will allow imaging even under poor lighting conditions. It is expected to produce images with 0.8-m ground resolution and 16 km swath width. Further, the EROS satellites can turn up to 45° in any direction for in-orbit stereo imaging.

EROS-A satellites would scan asynchronously, i.e. they could slew backwards to allow longer dwell time over ground area. In this way, they would be able to receive more radiation and improve contrast and signal-to-noise ratio for optimal imaging.

ISI-EROS web-site: http://www.imagesatintl.com/

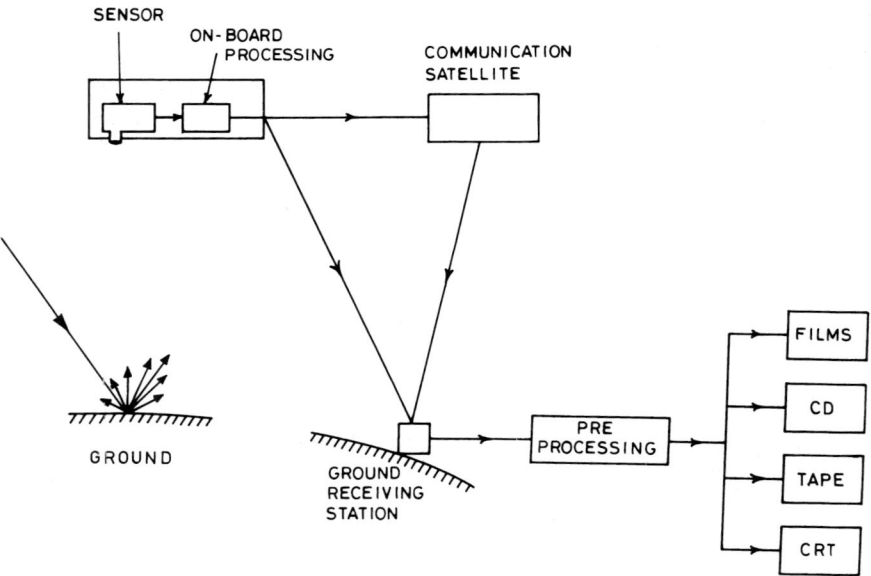

Fig. 5.28. Schematic of data flow in a non-photographic remote sensing mission

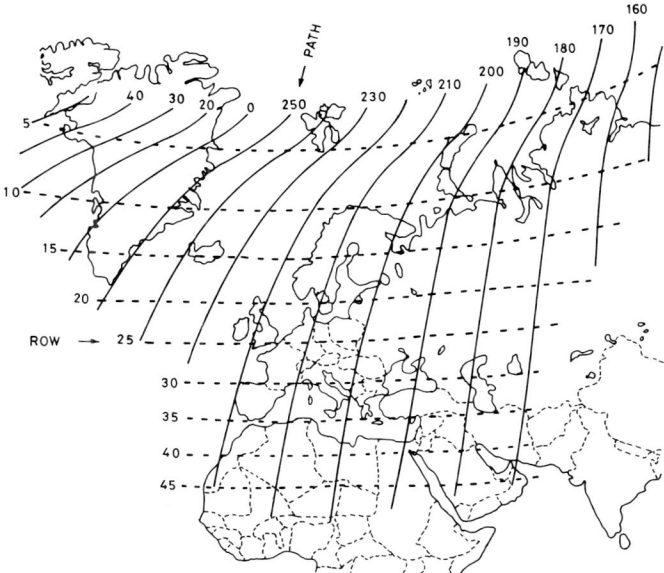

Fig. 5.29. Example of system of path and row numbers for global indexing and referencing

5.6 Products from Scanner Data

The satellite/aerial scanner data are pre-processed on board, and relayed down to the Earth receiving stations (Fig. 5.28). On the ground, the data are pre-processed, partly rectified and formatted in scenes per path-and-row numbers. They may be fed to a monitor for real-time display or could be readied for distribution as photographic products and digital data products. For distribution, commonly all satellite-sensor remote sensing data are reformatted and organized in scenes, which are indexed in terms of path and row (Fig. 5.29).

Photographic products have been the most widely used and convenient products for visual display and interpretation. The basic technique is that a film writer exposes a film using a fine laser beam or electronic beam (Fig. 5.30a). The photographic film, mounted on a rotating drum, is exposed so that the exposure at each unit area corresponds to the radiance level (Digital Number, DN) at the pixel. In this way, the digital numbers are converted into shades of grey (Fig. 5.30b). In order to generate a good-quality photographic image from remote sensing data, it is necessary to know the statistical distribution of DN values in the scene and the D–log E character of the film. The pixel exposures, which correspond to the pixel DN values, are made on the film in such a way that a greater part of the rectilinear

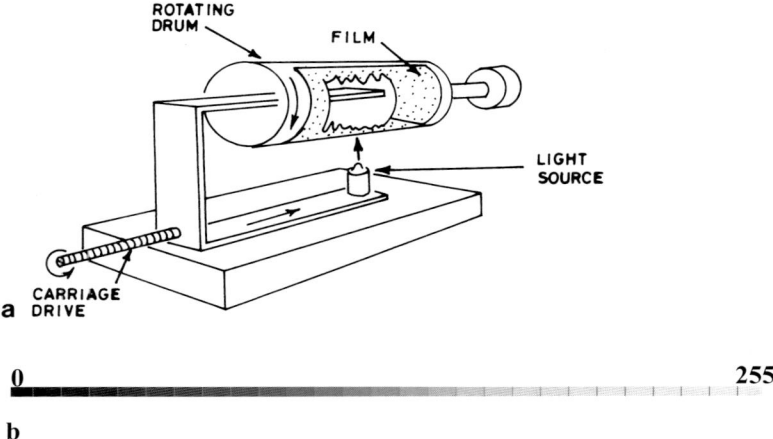

Fig. 5.30. Generation of film products from scanner data. **a** Working principle of film writers. **b** A typical gray wedge and the corresponding digital number from scanner data

segment of the D–log E curve is utilized and the points are well distributed on the D–log E curve. This provides a good contrast on the image.

The photographic products are of several types. *Browse products* are used for referencing and indexing and as quick-look products, to help examine and select the data. *Standard photographic products* are the normal products available and are the most extensively used. Sometimes *precision products* and special products such as map-oriented digital products are also available, but may have to be specially ordered. *Digital data products* are usually made available in media such as compact disks (CDs), cartridges, cassettes etc.

A number of factors which affect the geometric and radiometric quality of the image data are discussed in Chapters 6 and 7. The principles of interpretation of the remote sensing data are presented in Chapters 7, 8 and 9, digital image processing in Chapter 10 and applications in Chapter 16.

Chapter 6: Geometric Aspects of Photographs and Images

It is often pertinent to know not only what the object is, but also where it is; therefore, a universal task of remote sensing scientists is to deliver maps displaying spatial distribution of objects. Geometrically distorted image data provide incorrect spatial information. Geometric accuracy requirements in some applications may be quite high, so much so that the entire purpose of the investigation may be defeated if the remote sensing data are geometrically incorrect beyond a certain level. In brief, geometric fidelity of the remote sensing data is essential for producing scaled maps and for higher application prospects.

6.1 Geometric Distortions

The geometric distortions occurring in remote sensing data products can be considered in many ways. For planning and developing rectification procedures, it is important to know whether a certain geometric distortion occurs regularly or irregularly. On the basis of regularity and randomness in occurrence, the various geometric distortions can be grouped into two categories: systematic and non-systematic. *Systematic distortions* result from planned mechanism and regular relative motions during data acquisition. Their effects are predictable and therefore easy to rectify. Many of the systematic distortions are removed during preprocessing of the raw data. *Non-systematic distortions* arise due to uncontrolled variations and perturbations. They are unpredictable and require more sophisticated processing (e.g. rubber sheet stretching) for removal, and are generally ignored in routine investigations. Digital processing for geometric rectification is discussed in Chapter 10.

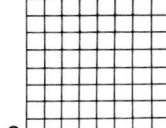 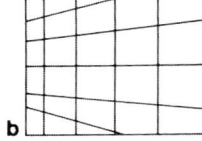 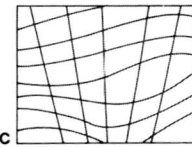

Fig. 6.1a–c. Interframe and intraframe distortions. **a** Nominal ground, **b** interframe distortion, **c** intraframe distortion

Table 6.1. Factors affecting geometry of images and photographs

Factors	Whether systematic (S)/ non-systematic (NS)
1. Sensor system factors	
– Instrument error	NS, S
– Panoramic distortion	NS, S
– Aspect ratio	NS, S
– Scan-time shift	S
2. Sensorcraft attitude and perturbations	
– Variation in velocity and altitude of the sensorcraft	NS
– Pitch, roll and yaw distortions due to platform instability	NS
3. Earth's shape and spin	
– Effects of the Earth's curvature	NS
– Skewing due to the Earth's rotation	S
4. Relief displacement	
– Local terrain relief	NS
– Sensor look angle	S

There is another way to classify geometric distortions. If several sets of photographs/images are available, two different levels of distortions can be distinguished depending upon whether the variations occur frame-to-frame, or within a frame, line-to-line (see e.g. Welch and Marko 1981). These are called interframe type and intraframe type respectively (Fig. 6.1). In the *interframe* type, distortion is uniform over one full frame and varies from frame to frame. This typically occurs in products of photographic systems and digital cameras, where the entire scene is covered simultaneously; each individual frame has uniform distortion parameters, although the distortion may vary from frame to frame. In the *intraframe* type, the distortion varies within a frame. Intraframe distortion can be further subclassified into two types: (a) *interline*, i.e. distortion is uniform over one scan line but varies from line to line; and (b) *intraline*, in which distortion varies within the line, from pixel to pixel. In a linear array CCD device, one line is imaged concurrently and therefore the distortion is uniform in each line, although it may differ from line to line. In an opto-mechanical line scanner, scanning is carried out pixel by pixel and therefore the geometric distortion may vary within one line, from one pixel to another. However, as the speed of OM scanning is very high, within-line distortions can usually be ignored.

The headings under which geometric effects are described in this chapter are guided by genetic considerations, and can be classified into four broad groups as follows (Table 6.1):

1. Distortions related to sensor system factors.
2. Distortions related to sensor craft altitude and perturbations.
3. Distortions arising due to the Earth's curvature and rotation below.
4. Effects of relief displacement.

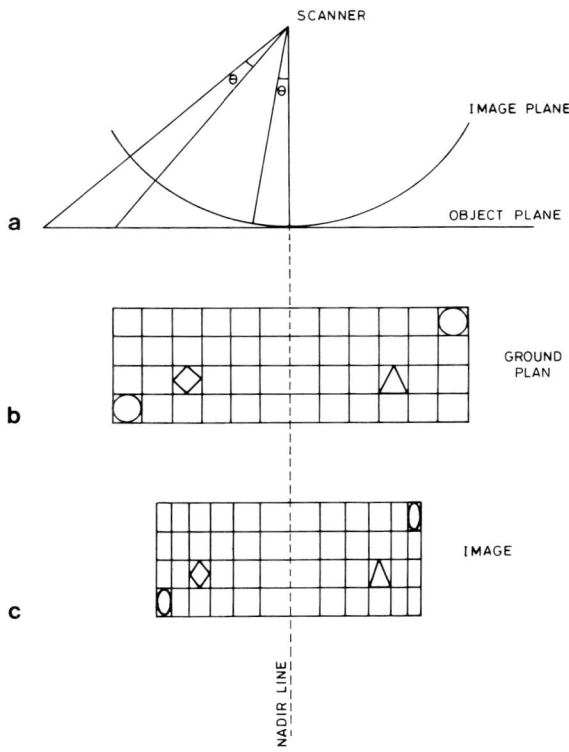

Fig. 6.2a–c. Panoramic distortion resulting from non-verticality of optic axis causing compression of scale at margins. **a** Mechanism, **b** ground plan, **c** image. Note the decrease in ground element size with increase in inclination of optic axis

The resulting geometry of a photograph/image is governed by a complex interplay between various types of distortions occurring concurrently.

6.1.1 Distortions Related to Sensor System

6.1.1.1 Instrument Error

Sensor instruments may not function uniformly or perfectly, especially if a moving mechanics is involved in data collection, and this may affect the image geometry. For example, in OM line scanners, the mirror motion may have non-linear characteristics, especially in those scanning devices which use an oscillating mirror (e.g. Landsat TM). Other possible instrument errors could occur due to distortion in optical system or alignment, non-uniform sampling rate, etc. The instrument errors may be systematic or non-systematic.

6.1.1.2 Panoramic Distortion

Panoramic distortion results from non-verticality of the optical axis in the optical imaging device. Typical examples have been products from OM line scanners and panoramic cameras. As the sensor sweeps over the area, across the flight line, data are collected with line of sight inclined at varying angles to the vertical (Fig. 6.2). As a result, the ground element size varies as a function of angle of inclination of the optical axis such that the scale at margins is squeezed. This results in scale distortion. A correction for this distortion is necessary (see Sect. 10.3.1), particularly in aerial surveys, where inclination is typically about 50° (total angular field-of-view, FOV, about 100°); however, in satellite surveys, as the inclination is very small (total FOV about 10–12°), the error is often ignored.

More recently, a number of space-borne CCD line and area array sensors have been launched which acquire video data in oblique viewing modes, i.e. with the optical axis inclined in the fore, aft and/or across-track direction. Such image data also carry panoramic distortions.

6.1.1.3 Aspect Ratio

The ratio of linear geometric scales along the two rectangular arms of an image is called aspect ratio, and a distortion arising out of this not being unity is called

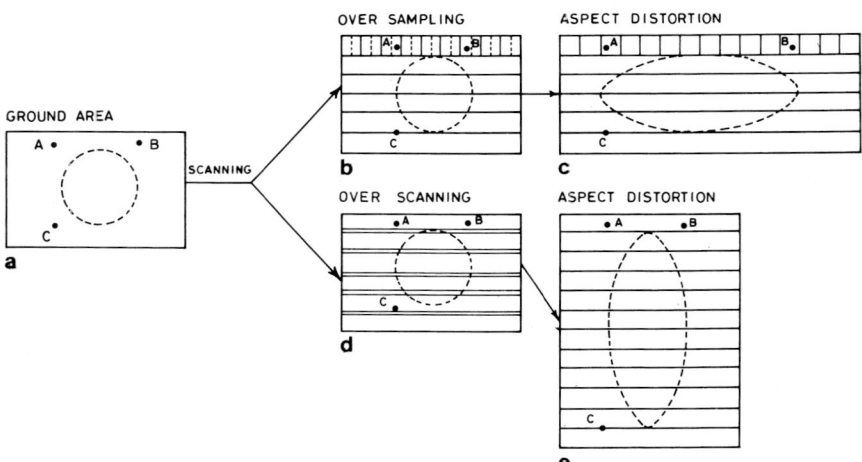

Fig. 6.3a–e. Aspect ratio distortion. **a** Nominal ground. **b** Oversampling along scan line direction and **c** corresponding aspect distortion – image is elongated along the scan line direction. **d** Overscanning in along-track direction and **e** corresponding aspect distortion – image is elongated in the along-track direction

Geometric Distortions

aspect distortion. Aspect distortion may be associated with scanner image data, such as those from OM and CCD line scanners. A line scanner will produce a geometrically correct image if the velocity and height of the sensorcraft, i.e. V/H factor, is commensurate with the rate of scan cycle, IFOV, and rate of sampling in the scan direction. Any variation in this relationship leads to aspect distortion, which could be in the form of oversampling/undersampling or overscanning/ underscanning (Fig. 6.3). This could be systematic (e.g. as in the case of Landsat MSS where oversampling was incorporated in the design) or non-systematic (viz. generated due to uncontrolled V/H variations).

6.1.1.4 Scan-Time Shift

In OM line scanners, the scanner sweeps across the path collecting video data from one end of the swath to the other. During this short interval of time, the sensorcraft keeps moving ahead. This motion results in scan lines being inclined with respect to the nadir line, and is called scan-time shift (Fig. 5.19a). The erstwhile aerial OM line scanners had prominent distortions due to this factor; however, the new aerial scanners have a very high speed of scanning, reducing this type of distortion substantially.

The space-borne OM scanner uses an oscillating mirror, as the FOV is quite small. In such a case, the active scan of mirror motion is restricted to only one direction of the mirror oscillation cycle, to avoid a criss-cross effect on the scan lines arising due to scan-time shift (e.g. in Landsat MSS). On the other hand, if both back and forth phases of mirror oscillation are to be used for active scan (as in Landsat TM), an on-board corrector mechanism is required for parallel alignment of scan lines (see Fig. 5.19b).

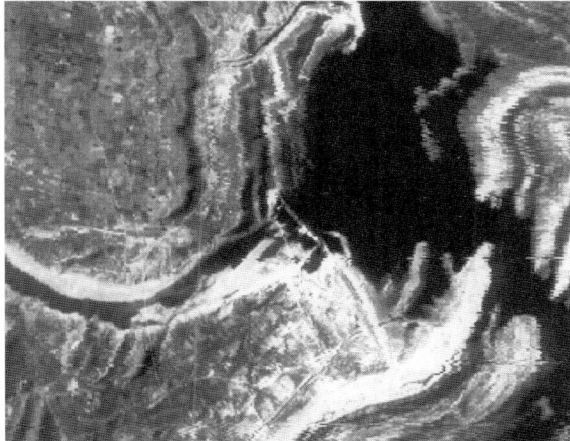

Fig. 6.4. Degradation due to sampling and quantization; aerial scanner imagery (red band) of part of the Mahi river area, India. (Courtesy of Space Applications Centre, ISRO)

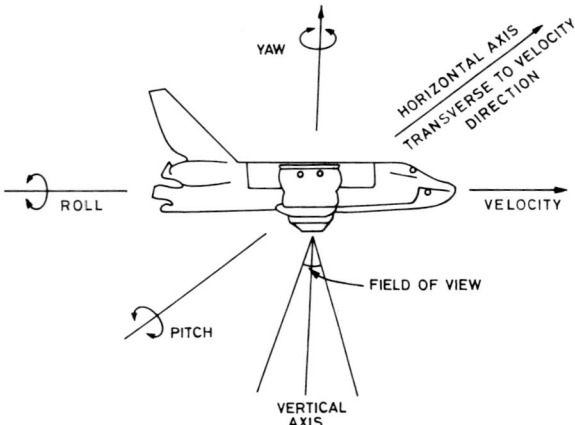

Fig. 6.5. Pitch, roll and yaw distortions – terminology

6.1.1.5 Degradation Due to Sampling and Quantization

To form a digital image, a continuous scene is broken into discrete units, and an average digital number (brightness value) is assigned to each discrete unit area. This artificial segmentation, i.e. sampling and quantization, is a type of degradation inherent in all data where analog-to-digital (A/D) conversion is involved (Fig. 6.4).

6.1.2 Distortions Related to Sensorcraft Altitude and Perturbations

6.1.2.1 Pitch, Roll and Yaw Distortions Due to Sensor-Platform Instability

One of the most important parameters governing the geometric quality of remote sensing images is the orientation of the optical axis. When the optic axis is vertical, image data has high geometric fidelity. Many of the sensors are designed to operate in this mode. However, sensor platform instability may lead to angular distortions. Any angular distortion can be resolved into three components: pitch, roll and yaw (Fig. 6.5).

Yaw is the rotation of the sensorcraft about the vertical. This rotation leads to a skewed image such that the area covered is changed; however, no shear or scale deformation occurs. *Roll* is the rotation of the sensorcraft about the axis of flight or velocity vector; it leads to scale changes. *Pitch* is the rotation along the across-track axis; it also leads to scale changes. Figure 6.6 shows schematically the above types of distortions in photographs and scanner images. The angular distortions may occur in combination.

Geometric Distortions

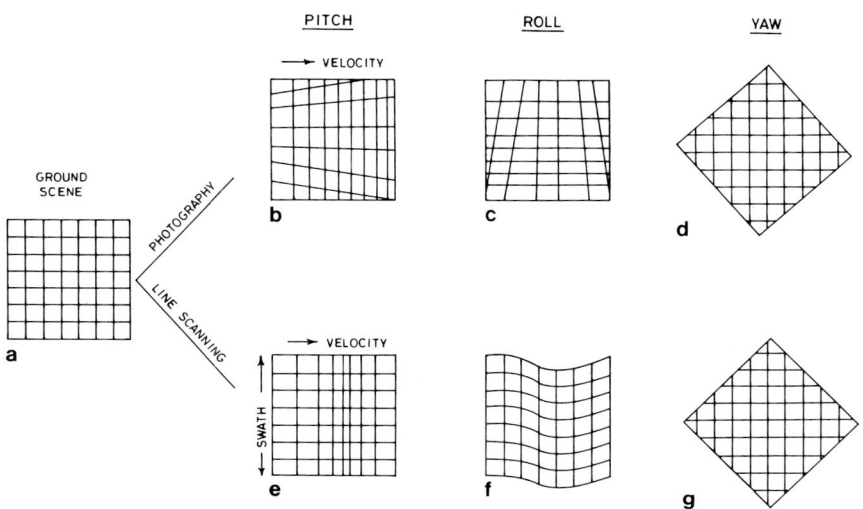

Fig. 6.6a–g. Schematic of pitch, roll and yaw distortions. **a** is the nominal ground; **b, c, d** show the pitch, roll and yaw distortions in photographs (interframe type); **e, f, g** show the same in scanner image data (intraframe type)

The aerial platforms are subject to greater turbulence due to wind etc., and their data may carry larger amounts of pitch, roll and yaw errors as compared to those from orbital platforms, which are more stable and steady. In most aerial scanners, the effect of roll is rectified by an electronic device, which adjusts the start of line scan in the imagery. The effects of yaw and pitch are not usually rectified, but the effect of pitch is eliminated by over-scanning, and yaw leads only to skewed images, which can be later accounted for by appropriate alignment. The errors due to sensor platform instability are typically non-systematic (Table 6.1).

6.1.2.2 Variations in Velocity and Altitude of the Sensorcraft

A remote sensor is designed to operate at a certain altitude and velocity combination. Variations in these parameters produce geometric distortion in the images or over-/under-coverage. They are typically non-systematic. The geometric distortions depend on the type of sensor. Examples include the following
(a) In the case of photographic systems and digital cameras, an increase in altitude leads to larger areal coverage and decreased photographic/image scale. On the other hand, an increase in platform velocity leads to decreased area of overlap in stereo coverage, i.e. under-coverage (Fig. 6.7).
(b) In OM line scanners, the governing factor is the V/H ratio. Increase in the V/H ratio leads to under-scanning (skipping of some areas) and decrease in the V/H ratio leads to over-scanning (repetition of some areas).

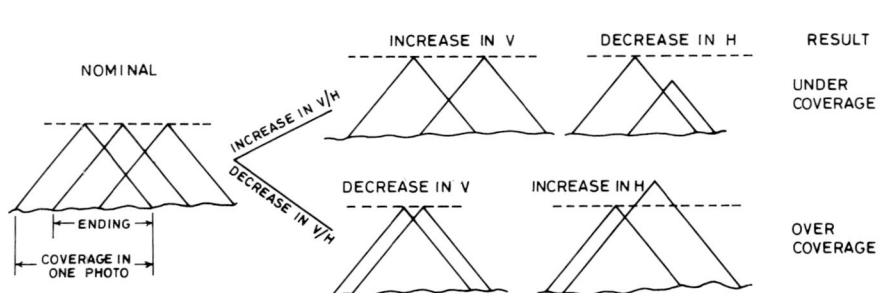

Fig. 6.7. Schematic variation in velocity and altitude (V/H factor) causing over- and under-coverage in photographic systems

(c) In CCD line scanners, the forward motion of the pushbroom is accompanied by integration of radiation over a particular time (dwell time), after which the process of accumulating radiation for the next line starts. In this way, there is no omission/repetition of ground area. The length of the along-track arm of the ground element is controlled by the satellite velocity, altitude and integration time. The length of the across-track arm of the ground element is governed by the altitude and IFOV. The V/H ratio is thus very critical and any variation in V/H ratio produces aspect distortion (Fig. 6.8). A relative increase in H increases the across-track arm of the pixel, and a relative decrease in H produces a shortening of the across-track arm. A change in velocity affects the along-track arm of the ground element; higher velocity leads to longer along-track arm, and vice versa. Variations in the V/H factor thus produce aspect distortions (Fig. 6.8).

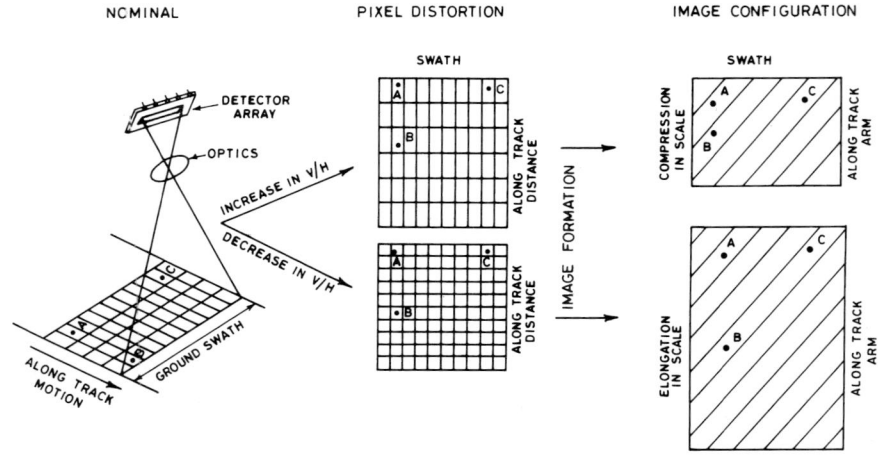

Fig. 6.8. Schematic variation in V/H factor causing aspect and scale distortion in CCD line scanner images

Geometric Distortions

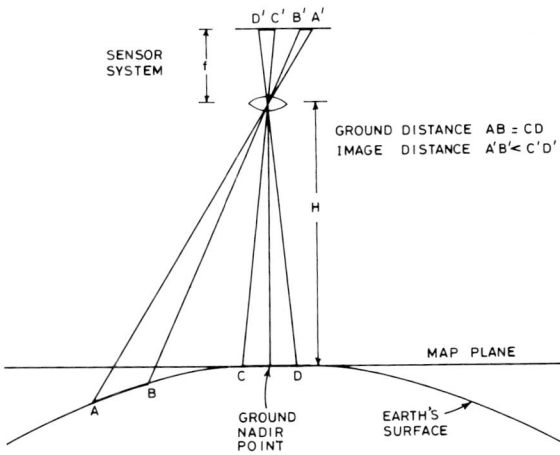

Fig. 6.9. Effects of the Earth's curvature on scale. Ground distance AB = CD, but image distance A' B' < C' D'

6.1.3 Distortions Related to the Earth's Shape and Spin

6.1.3.1 Effect of the Earth's Curvature

If an image or photo covers extensive areas on the Earth's surface, the Earth's curvature affects geometric scale and exerts a type of panoramic effect (Fig. 6.9). This phenomenon is commonly observed in scenes acquired from high altitudes. The Earth's curvature varies with latitude, and therefore the effects of the Earth's curvature are more pronounced at higher latitudes. Although the shape of the Earth is well known, for remote sensing image construction purposes it is considered as a non-systematic type of distortion.

6.1.3.2 Skewing Due to the Earth's Rotation

This geometric distortion typically occurs in scanner (OM and CCD) images obtained from high altitudes and space-borne sensors. The effect of the Earth's rotation is quite negligible in low-altitude (aerial) scanner surveys. In photographs, as the entire scene is pictured simultaneously, the distortion is absent. Figure 6.10 shows a typical situation with the Earth rotating around its axis from west to east and the satellite making an orbital pass around the Earth. As the satellite sensor completes one scan operation and positions itself for the next scan operation, the Earth below rotates around its axis from west to east over a certain distance. This brings relatively west-located parts of the scene before the sensor system every time a fresh scan cycle starts, thus causing image skew. The skew is maximum in

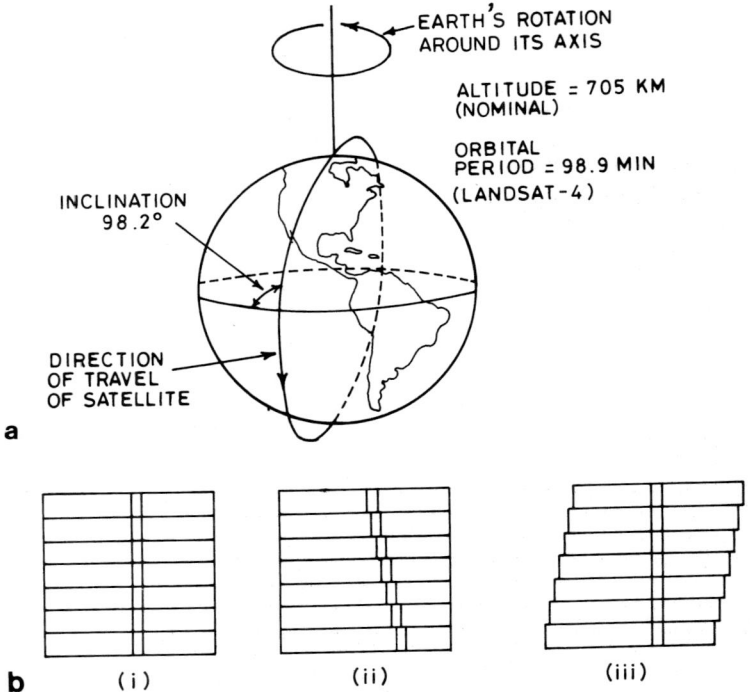

Fig. 6.10a,b. Earth's rotation and image skewing. **a** Relative motion of a satellite in polar orbit around the Earth and the Earth's rotation around its axis. **b** Scheme showing genesis of image skew. (i) Nominal ground scene; a number of scan lines and a straight road segment are shown. (ii) Effect due to the Earth's rotation; note the disrupted road in the uncorrected image data. (iii) To relocate the image data in proper geometry, the scan lines have to be shifted successively westwards, leading to image skew

polar orbits and zero in the equatorial one. The method for skew correction is given in Sect. 10.3.2.

The geometric quality of the photographs and images used for interpretation is also influenced by the susceptibility of photographic material to expansion. For this reason, sometimes regularly placed reseau marks are printed on the film for high-precision requirements. Furthermore, in photogrammetric investigations, glass-plate diapositives are used for photography instead of flexible base films in order to eliminate shrinkage.

6.1.4 Relief Displacement

Displacement in the position of the image of a ground object due to topographic variation (relief) is called *relief displacement*. It is a common phenomenon on all remote sensing data products, particularly those of high-relief terrain. The magnitude of relief displacement is given as

Geometric Distortions

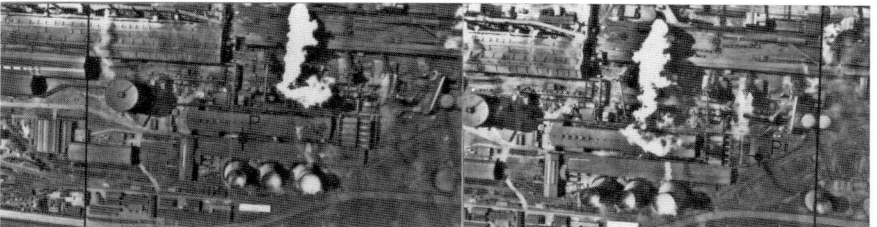

Fig. 6.11. Relief displacement seen on a photographic stereo pair. (Courtesy of Aerofilms, London)

$$\text{Relief displacement} \approx r.h/H \tag{6.1}$$

where r is the distance of the object from the principal point, h is the object height and H is the flying height. Therefore, relief displacement is dependent upon local terrain relief and look angle at the satellite (which in turn depends upon sensor-craft altitude and distance of the ground feature from the nadir point).

The pattern of relief displacement depends upon the perspective geometry of the sensor. Two main types can be distinguished: (1) sensors with central perspective geometry; and (2) imaging systems with line-scanning devices.

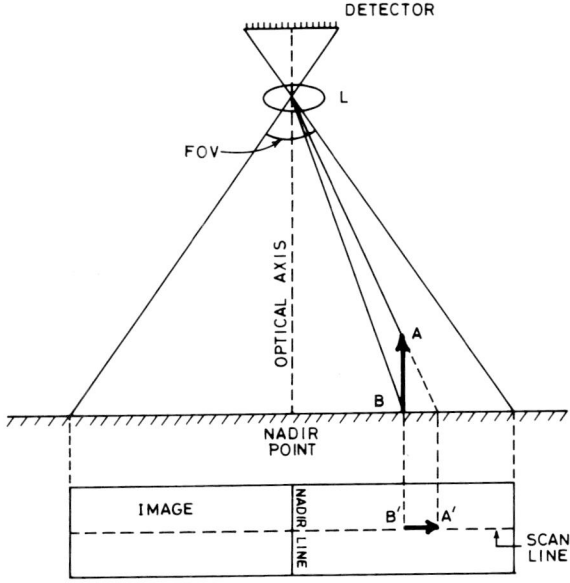

Fig. 6.12. Schematic relief displacement on line-scanning images. The relief displacement occurs outward and away from the nadir line, in the scan direction

1. Relief displacement on central perspective geometry products: This includes photographic-film camera systems and digital cameras. Here, the key optical–geometrical feature is that for each frame (photograph or image), all rays must pass through the lens centre, which is (ideally) stationary. Figure 6.11 shows an example where objects of varied heights are photographed in a stereo pair. Although, the top and base of a tower are at the same NE plan location, they appear at different positions on the photographs. This is due to relief displacement. It should be noted that: (a) the image displacement occurs in such a way that higher points on the ground are displaced radially away from the principal point; (b) the amount of image shift is related to the relief; and (c) the relief displacement decreases with increasing flying height, for which reason space photographs show less relief displacement than aerial photographs.

2. Relief displacement on line-scanner images: On line scanning images, relief displacements occur perpendicular to the nadir line, outward and away from the nadir point. Consider a tall object AB being imaged by a linear array scanning device (Fig. 6.12). The base B of the object is imaged at B' and the top A at A'. Although the points A and B occupy the same position in plan (map), they appear at different positions on the image, the top being displaced outwards and away from the nadir line in the scan direction. Figure 6.13 gives an example of IRS-1D PAN image from the Himalayas.

Fig. 6.13. Relief displacement in CCD line-scanning image. The scene (dated 28 November 1998) is an IRS-1D PAN image of a part of the Himalayas. The area has a high relief; the main river is the Bhagirathi (Ganges) river; in the NE corner is the Maneri dam and reservoir. Note the relief displacement of hill-tops and resulting partial shadowing of the Ganges valley at several places

Geometric Distortions 135

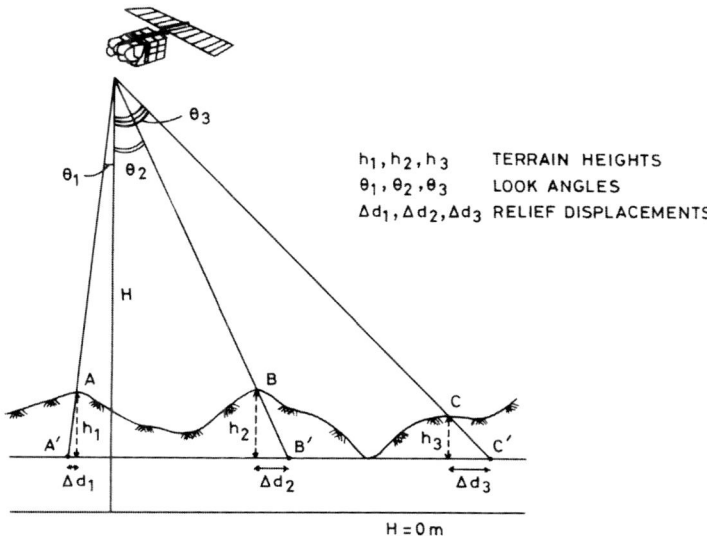

Fig. 6.14. Schematic terrain-induced displacement error for Landsat-TM-, SPOT-HRV-, and IRS-LISS/PAN-type image geometry

Figure 6.14 shows the typical geometrical configuration for Landsat-TM, SPOT-HRV and IRS-LISS/ PAN type of sensors. The terrain-induced relief displacement is broadly governed by two factors: off-nadir angle of look at the satellite and relative terrain height. Figure 6.15 gives a nomograph of relief displacement for Landsat TM. It is observed that in mountainous terrains such as the Himalayas and the Alps, where relative relief of about 2000 m could occur, Land-

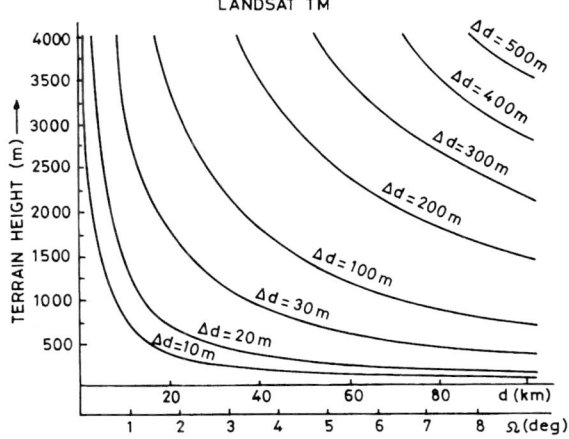

Fig. 6.15. Nomograph showing displacement in Landsat TM images induced by terrain height at various look angles. (After Almer et al. 1996)

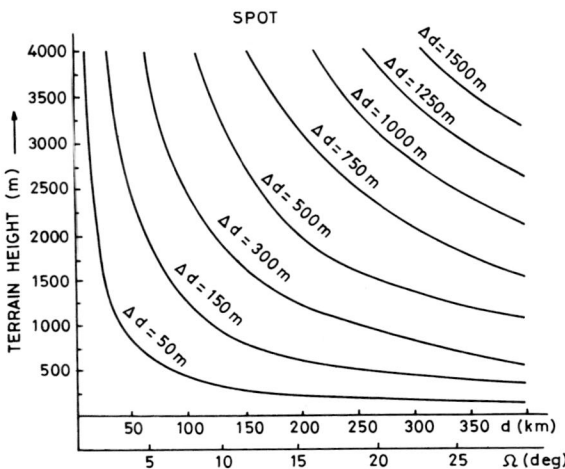

Fig. 6.16. Nomograph showing displacement in SPOT-HRV (off-nadir viewing) images induced by terrain height at various look angles. (After Almer et al. 1996)

sat TM data may possess relief displacements of the order of nearly 200 m (6–7 pixels!). Similarly, figure 6.16 shows a nomograph for SPOT-HRV (off-nadir viewing), where it is seen that the SPOT-HRV data in such areas (2000 m relative relief) may possess relief displacements of nearly 750 m (equivalent to distance of 75 pixels!). Therefore, for precision planimetric work, as also for geocoding and image registration, it is necessary to take relief displacements into account, particularly in mountainous terrain.

6.2 Stereoscopy

6.2.1 Principle

The aim of stereoscopic viewing is to provide three-dimensional perception (also now called 2.5 D, see Sect. 10.10). The principle of stereoscopy is well known (Fig. 6.17). If objects located at different distances are viewed from a set of two viewing centres, the viewing centres subtend different *perspective angles* (also called *parallactic angles*) at the objects (Fig. 6.17a). The angle depends on the distance or depth; the larger the distance, the smaller the angle. In other words, an idea of relative distances is conveyed by perspective angles. This implies that for relative depth perception, it is necessary that the same set of objects be viewed from two perspective centres.

Now, if an area is photographed from two stations, a set of two photographs can be used for stereoscopic viewing. The left photograph is viewed by the left

Stereoscopy

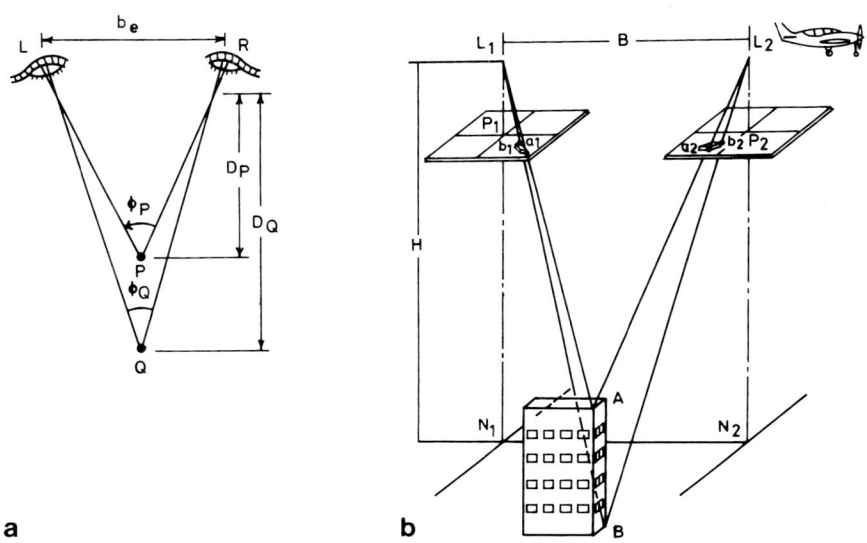

Fig. 6.17a,b. Principle of stereoscopy. **a** Relationship between object distance and parallactic angle. Two objects P and Q situated at distances D_P and D_Q are viewed by the two eyes L and R and subtend parallactic angles φ_P and φ_Q respectively. As $D_P < D_Q$, $\varphi_P > \varphi_Q$. **b** Photographs from two exposure stations sighting the same building. B: airbase; H: flying height; L_1 and L_2: lens centre positions; P_1 and P_2: principal points; N_1 and N_2: nadir points; a_1b_1 and a_2b_2: images of A and B on the two photographs respectively. (Wolf 1983)

eye and the right photograph by the right eye (Fig. 6.17b), commonly using stereoscopic instruments. When the two, left and right images, fuse or merge into each other, a three-dimensional mental model is perceived, called the stereoscopic model. This technique is called *stereoscopy*. The process virtually transposes the eyes in such a manner that the eye base (distance between the two eyes) is transposed to airbase (horizontal distance between two successive photographic stations). If per chance the relative positions of the photographs are interchanged (i.e. the left photo is placed on the right and the right photo on the left), the mental model shows inverted relief – the ridges appearing as valleys and vice versa – called *pseudoscopy*.

6.2.2 Vertical Exaggeration

In a stereoscopic mental model, almost invariably there is a geometric distortion, as the horizontal and vertical scales no longer match. The distortion is called *vertical exaggeration* (VE), as very often the vertical scale is greater than the horizontal scale. The distortion is primarily caused by the fact that the B/H ratio during photography does not match with the corresponding b_e/h ratio during stereoviewing. The vertical exaggeration can be written as

$$VE = \frac{B}{H} \times \frac{h}{b_e} \qquad (6.2)$$

where B = air base, H = flying height, b_e = eye base, and h = depth at which stereo model is perceived.

As a result of vertical exaggeration, the relief appears enhanced, and slopes and dips appear steeper in the mental model. Common values of VE range between 3–5. The VE is at times helpful in investigations, e.g. in a flat terrain, as minor differences in relief may get enhanced. However, it may be a problem in a highly rugged terrain. Rarely, a phenomenon called 'negative VE' occurs, in which the horizontal scale is larger than the vertical, and the relief becomes depressed (for more details, see e.g. Moffitt and Mikhail 1980; Wolf 1983).

6.2.3 Aerial and Spaceborne Configurations for Stereo Coverage

Remote sensing products from all types of platforms, i.e. aerial, space and terrestrial, can be used for stereoscopic viewing, provided the two photographs/ images form a geometrically mutually compatible pair. Some arrangements for stereo coverage are schematically shown in figure 6.18. The most common and typical arrangement is vertical photography (Fig. 6.18a) from an aerial or space platform with ca. 70% overlap for stereo viewing. The convergent type of photography, in which one camera is forward-inclined and another aft-inclined, to provide coverages of the same scene from two different perspective directions, is another configuration (Fig. 6.18b).

The coverages from Landsat sensors are such that successive paths have a small area of overlap, and these can also be used for stereo viewing (Fig. 6.18c). However, a specific problem occurring in space data for stereo studies is the low B/H ratio owing to very high altitudes. Typically, in aerial stereo photography the B/H ratio is around 0.4–0.6, which gives a VE of about 3–4. In contrast, in space sensors such as Landsat MSS and TM, the B/H ratio is around only 0.17–0.03, giving a VE of about only 1–0.2 (negative vertical exaggeration where relief gets depressed in the stereo vision!).

During the past decade, a lot of attention has been focused on improving the VE of stereo sensors from space. In order to improve the B/H factor in space image data, oblique-viewing spaceborne systems have been launched. For example, SPOT-HRV and IRS-PAN use tiltable sensors such that the line of sight can be tilted across-track to generate stereo coverage (Fig. 6.18d). Futhermore, configurations have been designed for dedicated spaceborne stereo programmes called 'in-flight' stereo capability sensors (e.g. JERS-OPS, stereo MOMS-02 and ASTER). Such systems include multiple CCD line arrays placed at the focal plane of the optical system, to acquire multiple-look (usually one vertical and two oblique) coverage, using either single-lens optics (Fig. 6.18e) or multiple-lens optics (Fig. 6.18f). These configurations generate along-track stereos.

Stereoscopy

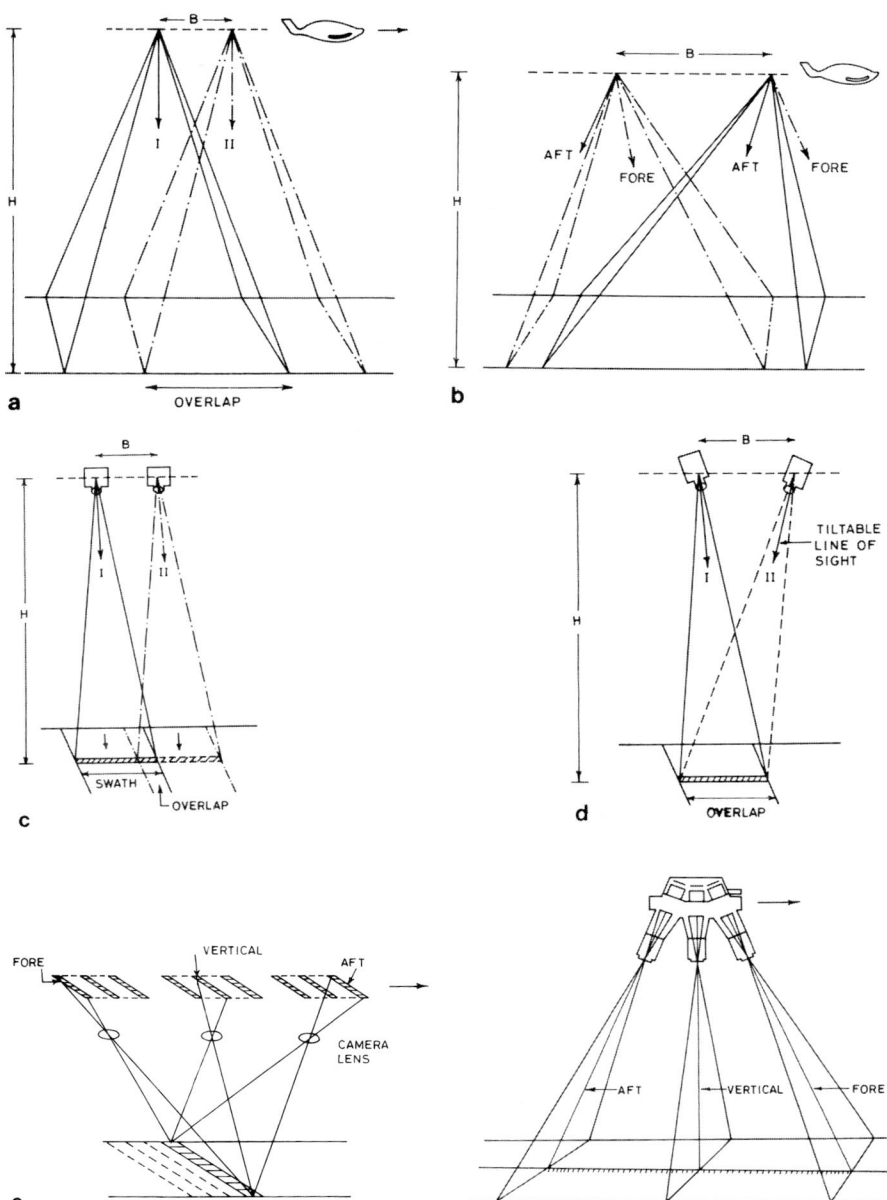

Fig. 6.18a–f. Configurations for stereoscopic coverage. **a** Vertical photography. **b** Convergent photography. **c** Stereoscopic coverage from Landsat MSS type images. **d** Stereoscopic coverage using a tiltable across track line of sight. **e** Stereoscopic coverage from multilook linear arrays using single optics and **f** using multiple optics

IKONOS and QuickBird-2 have the special capability to tilt line of sight in either along-track or across-track direction, to generate both along- and across-track stereos. Many of the earlier spaceborne systems such as SPOT-HRV, JERS-OPS and IRS-PAN had the limitation of low B/H ratio (approx. 0.3), due to smaller tilt angles of around 20°. IKONOS-1 has tilt angles of ± 45°, yielding stereos with a B/H ratio of 0.6 and VE of about 4.

In this work, the following examples of stereo pairs are included:
- aerial photographic stereo pair (Figs. 16.20, 16.44, 16.45, 16.51)
- aerial digital camera stereo pair (Fig. 16.16)
- Metric Camera photograph stereo pair (Fig. 8.5)
- IRS-PAN image stereo pair (Figs. 16.9, 16.31)
- ERS-SAR image stereo pair (Fig. 16.29)
- SRTM image stereo pair (Fig. 13.7)
- synthetic stereo pair (Figs. 10.40, 15.17, 15.18).

6.2.4 Photography vis-à-vis Line-Scanner Imagery for Stereoscopy

The mutual geometric compatibility of the photographs/images forming a stereo pair is important for stereoscopy. In general, distortions due to platform instability (roll, pitch and yaw), altitude–velocity variations, panoramic viewing, Earth curvature and oblateness affect the stereo compatibility.

There are some basic differences in the characteristics of photographs and line-scanner images relevant to stereo viewing.

1. The products from photographic systems and digital cameras possess central perspective geometry, as each frame is acquired with the lens centre at one position. The line-scanner images, on the other hand, are generated line by line, as the scanner keeps moving along the nadir line, and thus they lack the central perspective configuration. Therefore, a conventional stereogram is lacking in stereos generated from line-scanner images.

2. In photographs, as the entire scene is imaged simultaneously, there may be only interframe distortions; however, in scanner images intraframe distortions may creep in, as the images are generated line after line. This may complicate the geometry and hence the stereo compatibility, especially for photogrammetric applications.

6.2.5 Instrumentation for Stereo Viewing

Optical instruments used for stereo viewing differ in complexity and sophistication. The common viewing instruments are the lens stereoscope, mirror stereoscope, scanning stereoscope and interpretoscope.

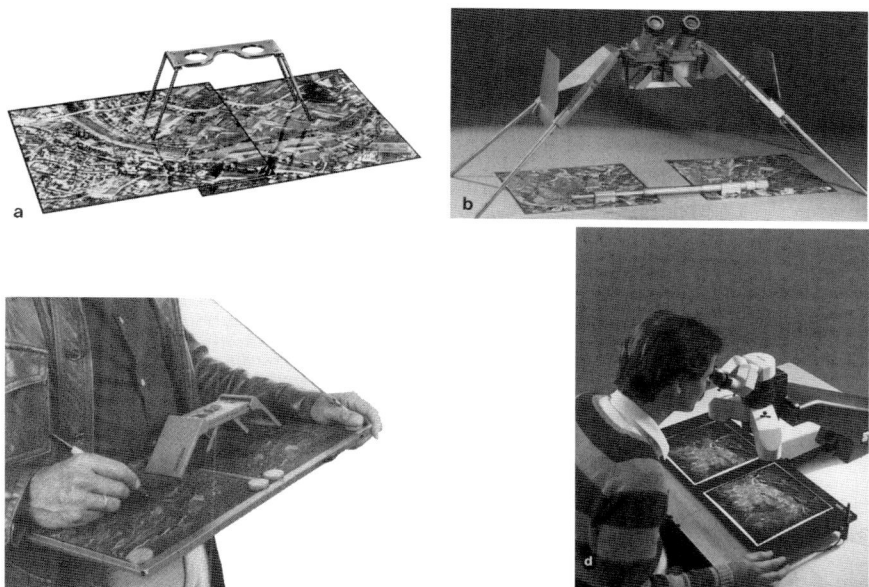

Fig. 6.19a–d. Stereoscopic instruments. **a** Lens stereoscope (courtesy of Carl Zeiss, Oberkochen). **b** Mirror stereoscope and a parallax bar (courtesy of Wild, Heerbrugg). **c** Portable mirror stereoscope (courtesy of Wild, Heerbrugg). **d** Interpretoscope (courtesy of Wild, Heerbrugg)

The *lens stereoscope* is a very simple instrument (Fig. 6.19a) consisting of a pair of lenses mounted on a common stand. The distance between the two lenses can be adjusted to suit the eye base of the viewer. Its main advantages are its light weight, low cost and portability. However, its limitations are the small field-of-view and a distorted view in peripheral areas. The *mirror stereoscope* (Fig. 6.19b) is technically more improved than the lens stereoscope and uses a set of prisms and mirrors. Its main advantage is its large field-of-view, enabling viewing of the entire stereo model at the same time. However, magnification is generally low, and binoculars are used to improve it, which in turn reduce the field-of-view. The *portable mirror stereoscope* (Fig. 6.19c) is a relatively new development, coupling the advantages of the light weight and portability of the lens stereoscope and the larger field-of-view of the mirror stereoscope. The *scanning mirror stereoscope* permits stereo viewing of the entire stereo model at variable magnifications. The *interpretoscope* (Fig. 6.19d) is a refined scanning stereoscope providing variable and high magnifications.

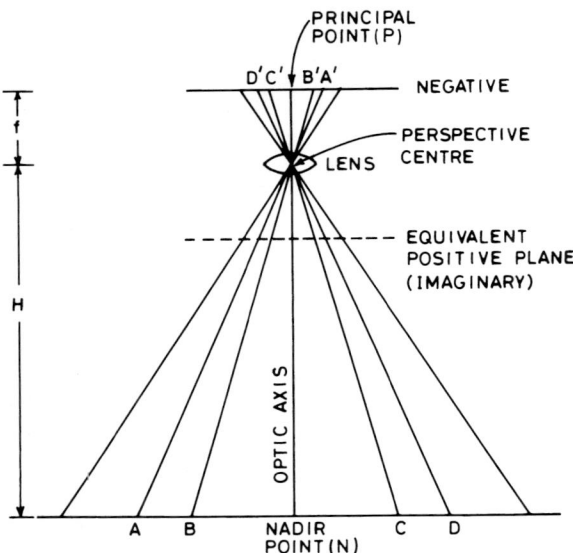

Fig. 6.20. Geometric elements of a vertical photograph and terminology

6.3 Photogrammetry

Photogrammetry is defined as the science and technique of making precise measurements on photographs. As such, the term photogrammetry is now extended to measurements on all remote sensing data products, whether photographs or scanner images. The basic purpose of photogrammetry is to determine distances, angles and heights from photo (or image) measurements, produce geometrically correct maps, and extract geometric information. Owing to its strategic and intelligence importance, photogrammetry has made rapid strides during the past few decades. In view of the fact that there are basic differences in their geometry, measurements on photographs and line-scanner images are discussed separately here.

6.3.1 Measurements on Photographs

6.3.1.1 Geometric Elements of Vertical Photographs and Terminology

Vertical photographs from aerial platforms have been extensively used for photogrammetric applications in the past. A vertical photograph is one in which the optical axis is vertical (the optical axis of a thin lens is defined as the line joining the centre of curvature of the spherical surfaces of the lens). The lens centre acts as the perspective centre through which all rays must pass. Thus the position of points on the image plane can be found by drawing rectilinear rays emanating from ground points and passing through the lens centre (e.g. ground points A and B are imaged at 'A' and 'B' respectively in Fig. 6.20).

The term *principal point* (P) denotes the geometric centre of the photograph. The principal point can be located with the help of marks appearing on edges or corners, called *fiducial marks*. If the optical axis is inclined, it is called an oblique photograph. In *high oblique photographs,* the horizon is visible, and in *low oblique photographs,* the horizon it not visible. The point on the ground vertically below the lens centre is called the *nadir point* (N) and the line joining successive nadir points is termed the *nadir line*; it gives the ground track of the flight path. In a remote sensing camera, the negative plate is placed at a distance f (focal length) above the lens centre. For all practical projection purposes, an equivalent positive plane can be imagined at a distance 'f', below the perspective centre (Fig. 6.20).

Scale: the scale of a map is defined as the ratio of the distance on the map to the corresponding distance on the ground. Consider points A and B lying on a flat ground covered in a vertical photograph and imaged at A'B' (Fig. 6.20). The scale of the photograph can be given as

$$\text{Scale} = \frac{A'P}{AN} = \frac{B'P}{BN} = \frac{f}{H} \tag{6.3}$$

The scale of a vertical photograph thus depends only on camera focal length and flying height above the ground. It does not depend on the angular location with respect to the principal point in a flat terrain. However, if the terrain has a variable elevation, the scale accordingly varies: the scale is larger for elevated areas and smaller for depressed areas. In areas of relief, it is often convenient to have an average scale, obtained by using the average height of the sensorcraft in Eq. (6.3).

The scale of a photograph can also be computed by comparing distances on the photograph to corresponding distances on a map of known scale. In this approach, some control points are first identified and the photo scale is computed as

$$\text{Photo scale} = \frac{\text{photo distance}}{\text{map distance}} \times \text{map scale} \tag{6.4}$$

In the above procedure, care should be taken to select control points having nearly same elevation, otherwise relief displacement would also affect the computed scale.

The scale of oblique photographs varies depending on the angle of inclination of optic axis. Such photographs are first rectified, and only then they can be used for photogrammetric applications (for further details, the reader may refer to standard works in photogrammertry, e.g. Moffitt and Mikhail 1980; Slama 1981; Wolf 1983).

6.3.1.2 Measuring Distances and Areas (Two-Dimensional Photogrammetry)

The simplest device for measuring distances on photographs is an interpreter's scale. It has both black and white markings for use on areas of light and dark photographic tones respectively. The distance is computed as

$$\text{Ground distance} = \text{photo distance} \times \text{scale factor} \tag{6.5}$$

Areas on photographs can be measured in several ways, e.g. by using an overlay grid, a planimeter or a digitizing table. In addition to these, there are other methods such as using optical analog systems, equidensity contour films etc., which depend on density slicing and finding the area occupied by a certain range of gray tone.

6.3.1.3 Relief Displacements and Three-Dimensional Photogrammetry

Objects of different heights photographed in a stereo pair are shown in figure 6.11. The relative shift in the image position of a point is called *parallax*. Its component in the X direction, i.e. parallel to the flight line, can be used to deduce elevation differences, and standard procedures for this are described in several works (e.g. Wolf 1983).

Thus, all the X, Y and Z coordinates of objects can be measured from a photostereogram. The data can be applied to compute dip, thickness of beds, displacement along faults, elevation differences, and to prepare maps of various types (see Miller and Miller 1961; Ray 1965; Ricci 1982; Pandey 1987). For example, from the Metric Camera stereo pair of an area in Iran, Bodechtel et al. (1985) measured dip data for structural analysis (see Fig. 16.25).

6.3.2 Measurements on Line-Scanner Images

6.3.2.1 Geometric Characteristics of Line-Scanner Images

Although line-scanner images have been available for quite some time, their photogrammetric applications have been somewhat constrained, owing to the following three main reasons.

1. Lack of central perspective geometry, which implies that a true conventional stereogram is lacking.

Photogrammetry 145

2. Presence of intraframe distortions of systematic and non-systematic types, which affect their geometric fidelity.
3. Photogrammetric methodologies developed during the past 70 years or so have focused on applications of camera photographic products. CCD line scanner stereo systems came onto the scene only in the mid 1980s. It took some time to establish the standard photogrammetric methodology and to develop suitable software for such remote sensing products, and this has only recently been made available commercially.

6.3.2.2 Measuring Distances and Areas on Line-Scanner Images (Two-Dimensional Photogrammetry)

The scale of an image is most easily computed by finding control points, and using the relationship as in Eq. (6.4). The images can be used to compute ground distances and areas in much the same way as photographs.

6.3.2.3 Three-Dimensional Photogrammetry

As mentioned above, the relief displacement on a line-scanning image occurs along the Y direction (i.e. perpendicular to the X direction or flight line). In order to obtain the X parallaxes required for the perception of terrain relief and height evaluation in the stereo mode, two approaches are employed as follows.
1. Images may be recorded from adjacent orbits by CCD line array cameras equipped with rotatable mirrors (e.g. SPOT-HRV and IRS-PAN; Fig. 6.18d). This type relies on perspective geometry to develop parallaxes.
2. Images may be recorded successively along the orbit path from fixed multilook fore-, vertical- and aft-pointing CCD line array cameras (e.g. JERS-OPS, Stereo MOMS-02; Fig. 6.18f). These types utilize parallel ray geometry to develop parallaxes. As the fore-, vertical- and aft-looking cameras pass over the terrain in succession, the top and base of, say, a pillar are imaged at different instances by the three cameras, the relative difference (i.e. parallax difference) being related to the relief. The parallax difference can be linked to the time interval and these data can be used to compute elevation differences (for details, see e.g. Goetz 1980; Welch 1980, 1983; Welch and Marko 1981; Hofmann et al. 1984).

In this way, using the CCD linear array image data, the X, Y and Z coordinates of various points in an area can be measured, and the data can be applied used for terrain modelling and mapping applications.

6.3.3 Aerial vis-à-vis Satellite Photogrammetry

Concepts and methods of photogrammetry developed around aerial vertical photography have been suitably adapted and extended to satellite image data. Satellite photogrammetry is of interest for the following main reasons: (1) satellite platforms are more stable in attitude; (2) as the satellite orbits around the Earth, its position can be predicted and accurately located; and (3) to cover a particular terrain, fewer control points are needed in satellite photogrammetry than in aerial photogrammetry. Satellite data – both photographic and scanner images – are fairly well suited to two-dimensional (planimetric) photogrammetric applications. The scale of a satellite image is in general more uniform in comparison to that of an aerial photograph, for the simple reason that in satellite images relief displacements are smaller, the platform is more stable, and angular distortions are only minimal. However, three-dimensional photogrammetric applications need stereo viewing with a reasonable VE. Aerial photographic remote sensing data have been routinely used for this purpose. As far as space-borne sensors are concerned, this requirement is likely to be fulfilled by the new-generation space-borne stereo sensors such as IKONOS. Thus, the research and application scene is poised for intense activity in this direction in the near future.

6.4 Transfer of Planimetric Details and Mapping

A variety of instruments are available for transferring planimetric details and mapping. They differ in sophistication and cost. A monoscopic zoom transferscope (for example an enlarger, a rectifier or mapograph, or a sketchmaster) is a simple type of instrument that permits variable magnification and allows the worker to view a photograph and a map simultaneously. Details from the photograph can be transferred onto the base map. Minor errors due to tilt can also be rectified by some instruments (e.g. sketchmaster and rectifier); however, relief displacements cannot be corrected. Nevertheless, this type of instrument has been found to be one of the most useful items of equipment for transferring planimetric details from Landsat-type images to base maps during visual interpretation.

On the higher end, the most sophisticated type of plotting equipment is the stereo plotter used for precise photogrammetric mapping. Such an instrument is used to produce topographical maps and digital terrain models from remote sensing data. An analytical plotter is a precision equipment in which coordinate measurements of various points in the stereo model are fed to a computer to generate a mathematical model of the terrain. Then, it is easy to carry out the various types of corrections and to derive the required information from the mathematical model.

Chapter 7: Image Quality and Principles of Interpretation

7.1 Image Quality

Image quality is a major factor governing the amount of information extractable from a remote sensing product. It is therefore necessary to first have an idea of the various factors affecting image quality, before proceeding to interpretation and applications. Basically, there are two aspects to image quality – radiometric and geometric. Both collectively govern the amount of extractable information. Whereas the geometric aspects were discussed in Chapter 6, we shall focus our attention here on the radiometric aspects of image quality.

During visual interpretation, objects on an image or photograph are discerned from one another primarily by relative differences in brightness (tone) or colour. It is a common experience that a bright object can be easily marked if located against a dark background. However, the same bright object may be difficult to locate against a bright background (Fig. 7.1). *Brightness contrast ratio*, i.e. the ratio of brightness of any two objects occurring side by side on an image or photograph, is an important factor in deciding to what degree any two features can be differentiated from each other by visual inspection (Fig. 7.2). Sometimes the term

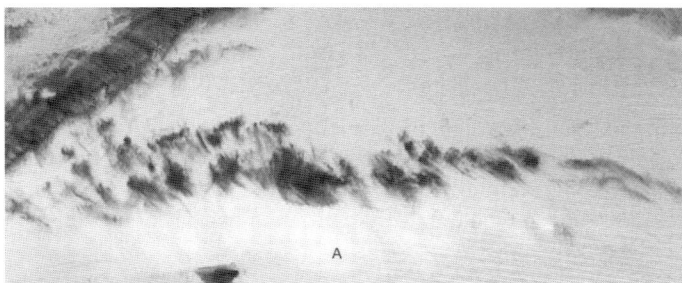

Fig. 7.1. Role of image contrast ratio in visual discrimination. This Landsat MSS image covers a part of the Sahara desert. In the background are very light-toned sand dunes and the presence of equally light-toned clouds is not readily detected on the image. The Presence of dark shadows with cirrus-type structure and the absence of a dune pattern in some places are the only indirect evidence of the existence of the clouds, such as at *A*. (Courtesy of R. Haydn)

148   Chapter 7: Image Quality and Principles of Interpretation

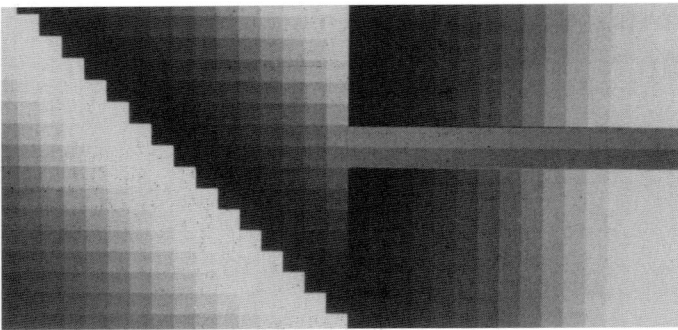

Fig. 7.2. Gray wedge illustrating the importance of image contrast ratio for visual discrimination. The boundary between the horizontal and vertical bars is sharp when the contrast ratio is high, and becomes less distinct as the contrast ratio decreases. Note further that the horizontal bar of the same gray tone appears to the eye to darker in the brighter background and brighter in the darker background

contrast ratio is used to denote the ratio between the maximum and minimum brightness in any one scene. For this, however, we prefer to use the term *dynamic range*. In general, a remote sensing product displaying good dynamic range is said to be of good image quality and an image with low dynamic range is termed flat or washed out. Digital processing techniques are available to enhance image contrast ratio and dynamic range in images (Sect. 10.5). Here, we shall discuss some of the more basic factors which govern the radiometric quality of images.

7.1.1 Factors Affecting Image Quality

The radiometric quality of photographs and images depends upon three main groups of factors: (1) ground properties, (2) environmental factors, and (3) sensor system factors (Table 7.1).

7.1.1.1 Ground Properties

Lateral variation in the relevant ground properties (namely, spectral reflectivity, thermal properties such as thermal inertia, emissivity etc., as the case may be) across the scene influences the radiometric image quality. Detecting such variations in ground across the scene is the crux of the problem and fashions the scope of remote sensing applications.

7.1.1.2 Environmental Factors

1. Solar illumination and time of survey. Remote sensing data should be acquired at a time when energy conditions are stable and optimum for detecting differences in ground properties. In the solar reflection region, energy conditions depend on the azimuth and angle of Sun's elevation and the time of survey. A noontime sur-

Image Quality 149

Table 7.1. Factors affecting radiometric quality of images and photographs (adapted after Moik 1980; Sabins 1987)

A. Ground properties
Lateral variations in relevant ground properties such as albedo, thermal properties, geothermal energy etc., including effects of topography and slope aspects

B. Environmental factors
1. Solar illumination and time of survey
2. Path radiance
3. Meteorological factors

C. Sensor system factors
1. Effects of optical imaging, image detection and recording systems
2. Shading and vignetting
3. Image motion
4. Striping

vey is generally preferred for obtaining images with uniform illumination and a minimum of shadow. In specific cases, however, a low-Sun-angle survey may be required, e.g. for enhancing structural-morphological features. Similarly, the thermal-IR survey is usually carried out in the pre-dawn hours when ground temperatures are quite stable.

2. Path radiance. Path radiance works as a background signal and tends to reduce image contrast ratio (Fig. 7.3). In the solar reflection region, scattering is the major source of path radiance and its effect can be minimized by cutting off shorter wavelengths during photography/imaging. In the thermal-IR region, the major

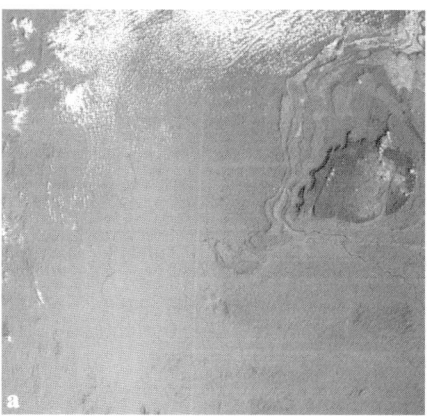
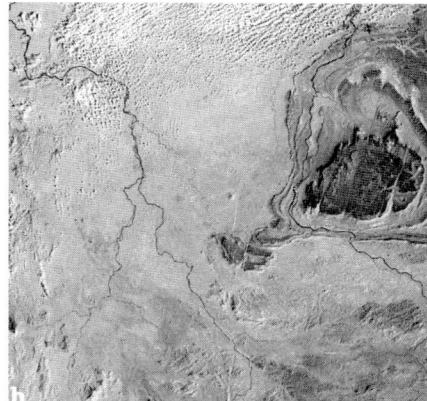

Fig. 7.3 a,b. Effect of path radiance on image quality. **a** Landsat MSS1 (green band) image and **b** Landsat MSS4 (near-IR band) of the same area in Venezuela. Note that the path radiance in the MSS1 band image has led to obscuring of many details and a poor image contrast ratio. (Courtesy of R. Haydn)

cause of path radiance is atmospheric emission; its effect can be minimized by confining sensing to atmospheric windows. Furthermore, atmospheric–meteorological models such as MODTRAN may be used to estimate the magnitude of atmospheric emissions. Digital techniques are also available for reducing the effects of path radiance and improving the image quality (Sect. 10.2.1).

3. Meteorological factors. Meteorological factors such as rain, wind, cloud cover etc. may significantly alter the ground properties and influence response in the optical region. For example, rain increases soil moisture, which alters ground albedo and thermal inertia. Clouds cast shadows and therefore alter the energy budget in the solar reflection and the thermal-IR region and also lead to restricted or poor coverage. Wind accelerates cooling – a factor which affects thermal response, and may also cause aerial platform instability. Therefore, meteorological factors may affect the image quality in three ways: (a) by changing the local ground properties, (b) by altering the energy budget and (c) by leading to poor ground coverage and platform instability.

7.1.1.3 Sensor System Factors

1. Effect of optical imaging systems. The optical imaging components, namely lenses, mirrors, prisms etc., are not absolutely perfect but real, and therefore minor diffraction, aberrations etc., are present. However, their effects are quite negligible.

2. Shading and vignetting. As view angle increases, the radiation intensity at the sensor surface decreases, theoretically as a function of $\cos^4$ or $\cos^3$ of the view angle; this variation is often termed 'shading' or '$\cos^4$ fall-off' (Slater 1980). Vignetting is absorption and blocking of more radiation by the lens wall, thus decreasing the radiation reaching the sensor. Normally, both shading and vignetting, in combination, produce significantly decreased image brightness towards the corners of images. To counter this effect, commonly an anti-vignetting filter is mounted on the optical lens system. Additionally, if the illumination fall-off characteristics in the lens system are known, digital techniques can be used to process and rectify the images. Yet another widely used technique is 'dodging', carried out during the printing stage (Fig. 7.4).

3. Image motion. The relative movement of the sensor platform with respect to the ground being imaged during the period of exposure or sensing leads to image motion. This results in the formation of streaks on the image. It is typically a problem in photographic systems where exposure duration are relatively long (about 1/30 to 1/100 s), and forward motion compensation (FMC) devices have to be used for better results. In digital cameras, although the exposure duration is relatively short (about 1/100 to 1/500 s), FMC devices may still be necessary in order to achieve high spatial resolution.

Handling of Photographs and Images

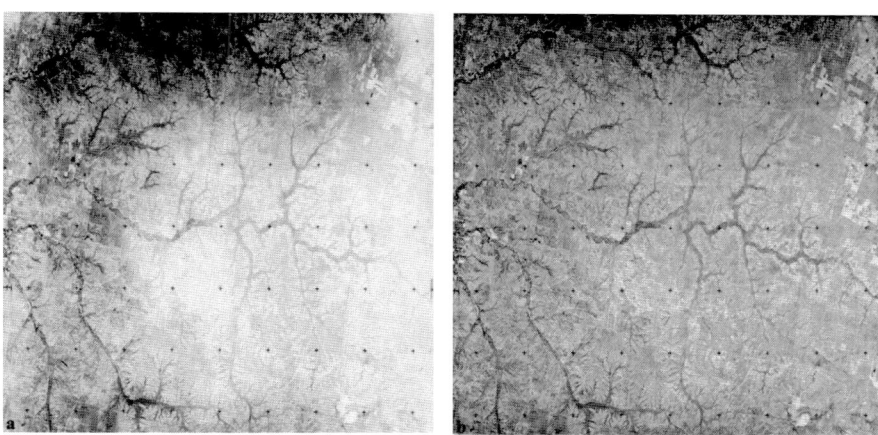

Fig. 7.4. a Vignetting effect in RBV image. **b** Vignetting effect having been rectified by 'dodging' at the printing stage (Courtesy of GAF mbH, Munich)

4. Striping. When a series of detector elements is used for imaging a scene (e.g. in the case of Landsat MSS and TM or in CCD linear or area arrays), the radiometric response of all the detector elements may not be identical. This non-identical response causes striping, and could lead to a serious degradation in image quality (de-striping is discussed in Sect. 10.2.5).

It should also be borne in mind that the radiometric quality of images and photographs is subject to the efficiency of the duplicating system, i.e. characteristics of the duplicating material and the photographic process (exposure, development and printing). Invariably there is a certain loss of information at each photographic regeneration stage and the amount of extractable information from the daughter products usually decreases, successively, at each next stage.

7.2 Handling of Photographs and Images

7.2.1 Indexing

Indexing is the first step carried out during the study of photographs and images. It aims at identifying the whereabouts of the area portrayed on the photographs and images. Generally, a small-scale topographical map (smaller than the photograph by a scale factor of 5–10) is used as a base map for indexing. Using control points, the area covered in each photograph or image is demarcated on the small-scale index map. Each image or photograph is assigned a certain index number, which can be used as an identification number.

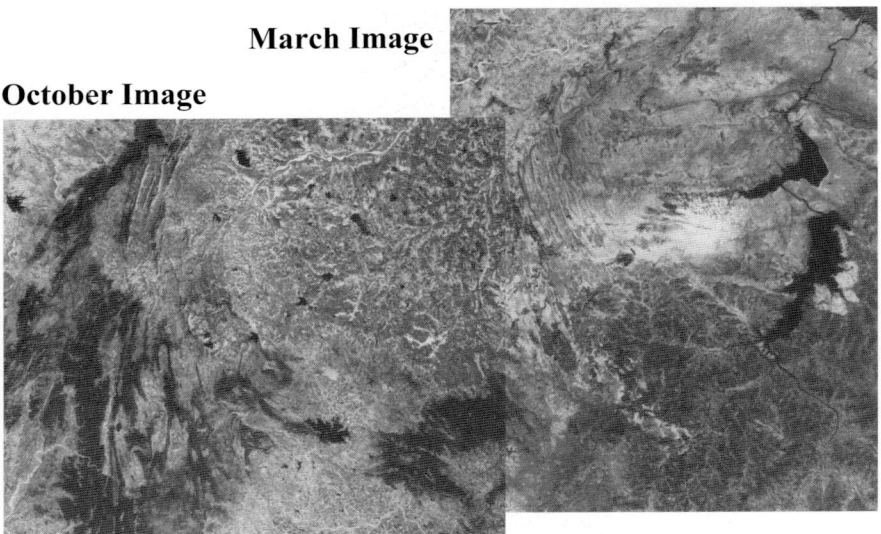

Fig. 7.5. Simple mosaic generated from Landsat MSS2 (red band) images. Note that the mosaic generated from images acquired in different seasons exhibits a break in photo tones

Indexing provides the following inputs, which are helpful during interpretation: (1) location of the area; (2) orientation, i.e. north; (3) scale, (4) geometric distortion in the images (approximate); and (5) regional setting. The index map is useful, for example, in planning reconnaissance study, stereoscopic or detailed study, and for collecting ancillary information.

7.2.2 Mosaic

A mosaic is a set of photographs (or images) arranged to facilitate a bird's-eye view of the entire area (Figs. 7.5 and 7.6). In a mosaic, adjacent photographs are arranged side by side so that there is no overlap, the adjacent boundaries match, and the features continue laterally. Often, photographs are cut and pasted together to generate a mosaic. Satellite images such as those from Landsat, IRS etc., in general, possess good geometric fidelity and are suitable for making reasonably scaled mosaics. These mosaics may exhibit sharp radiometric breaks at frame boundaries, particularly if the images pertain to different seasons or times (Fig. 7.5). The break in radiometric continuity can be removed by digital processing so that the radiometric levels of various objects in the two sets of image data match each other (Zobrist et al. 1983), and a larger homogeneous scene is generated (Fig. 7.6).

Handling of Photographs and Images 153

Fig. 7.6. A digitally processed mosaic generated from Landsat MSS images. The radiometric levels of various ground objects in the two sets of image data have been processed to match each other to create a homogeneous mosaic. (Courtesy of GAF mbH, Munich)

7.2.3 Scale Manipulation

The scale of photographs or images can be suitably altered by a variety of projection equipment, such as a photographic enlarger and rectifier. A photographic enlarger is the simplest device to enlarge transparent images or photographs. A rectifier is a more refined piece of equipment and is used for geometric rectification for angular (tilt and tip) distortions.

7.2.4 Stereo Viewing

The photographs and images are studied stereoscopically for 2.5-D (earlier called 3-D) perception and interpretation (see Sects. 6.2 and 10.10).

7.2.5 Combining Multispectral Products

The multispectral techniques of photography and scanning provide black-and-white products of the same scene in a number of spectral bands. For interpretation

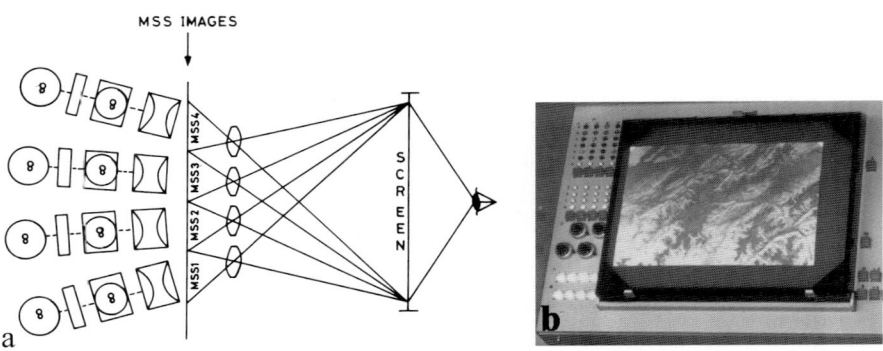

Fig. 7.7. a Working principle of additive colour viewer; **b** Additive colour viewer. (Courtesy of Carl Zeiss, Jena)

purposes, these can be studied one by one or collectively. For combining multispectral data products in visual/analog mode, the additive or subtractive theory of colours is used (Appendix 4.1).

Additive colour viewer (Fig. 7.7) is a simple instrument, which can combine three (or four) bands of multispectral photographs/images. In the 1970s, this was a very commonly used instrument for combining Landsat MSS images and generating false colour composites, although now, with the proliferation of PC-based processing devices, it has become obsolete. However, the working principle of displaying several images in colour mode used in this instrument is still valid, and the same principle is now applied in PC devices electronically for colour displays.

The basic technique is as follows: positive transparencies of different spectral bands are co-projected, and geometrically co-registered; each band is projected through one of the three primary additive colours (blue, green or red), using optical filters (Fig. 7.7a). The optical combination produces a colour composite which carries information from all of the three (or four) input images. The collective information is rendered in terms of colour – in varying hue, saturation and brightness across the scene.

Image handling and processing techniques have undergone a dramatic change during the past two decades. At the beginning of the Landsat era, most workers used only optical and photographic techniques to handle multiple images; digital processing was more or less restricted to larger research establishments. In the meantime, miniaturization, along with mass production of low-cost computer systems, has changed the scenario in favour of digital processing (see Chapter 10).

7.3 Fundamentals of Interpretation

Photo interpretation is the art and science of examining photographs to identify the objects portrayed on them and evaluate their significance. Principles of photo interpretation were initially developed for aerial photographic work. They are

Fundamentals of Interpretation 155

now extended to all remote sensing visual data products, including processed and unprocessed images. The concepts of photo interpretation have been well elucidated in various works on photo interpretation and photogeology (e.g. Colwell 1960; Miller and Miller 1961; Ray 1965; Allum 1966; Mekel 1970, 1978; Von Bandat 1983).

7.3.1 Elements of Photo Interpretation

A photograph or image is studied in terms of the following parameters: (1) tone or colour, (2) texture, (3) shape, (4) size, (5) shadow, (6) site or association and (7) pattern.

1. Tone or colour. Tone is a measure of the relative brightness of an object in shades of gray. The term is used for each distinguishable shade from black to white, such as dark gray, medium gray, light gray etc. The tone is an important parameter of photo interpretation and could be linked to various physical ground attributes, e.g. reflectivity in the solar reflection region, or radiant temperature in the TIR region. Colour products are obtained from colour photography or by colour coding of multispectral image data. The use of colour space dramatically increases the interpretability of data, as much subtler distinctions can be made (Appendix 4.1). Appropriate terms are used to describe the colour, such as deep red, pink, light blue etc.

2. Texture. Texture signifies the tonal arrangement and changes in a photographic image. It is defined as the 'composite appearance presented by an aggregate of unit features too small to be individually distinct'. It is a product of their individual colour, tone, size, spacing, arrangements and shadow effects (Smith 1943). Texture is dependent on scale; the same group of objects could have different textures on different scales. Texture is more important on larger-scale photographs. Various terms could be used to describe texture, e.g. fine, medium, coarse, speckled, granular, mottled, banded, linear, blocky, woolly, criss-cross, rippled, smooth, even etc. Some examples are given in figure 7.8.

3. Shape. Shape refers to the outline of an object. Many geomorphologic shapes are diagnostic, such as, alluvial fans, sand dunes, ox-bow lakes, volcanic cones etc. Further, structures such as dolerite dyke ridges and bedding strike ridges can also be identified by morphological shape.

4. Size. The size of a feature is also a significant parameter in photo interpretation. An idea of the size of an object can be obtained only after the scale of the image or photograph is known. Size, when considered in conjunction with shape and association, is a very useful parameter.

5. Shadow. Shadow cast by objects is at times quite informative, especially in the case of man-made objects. It gets more pronounced on low-Sun-angle images.

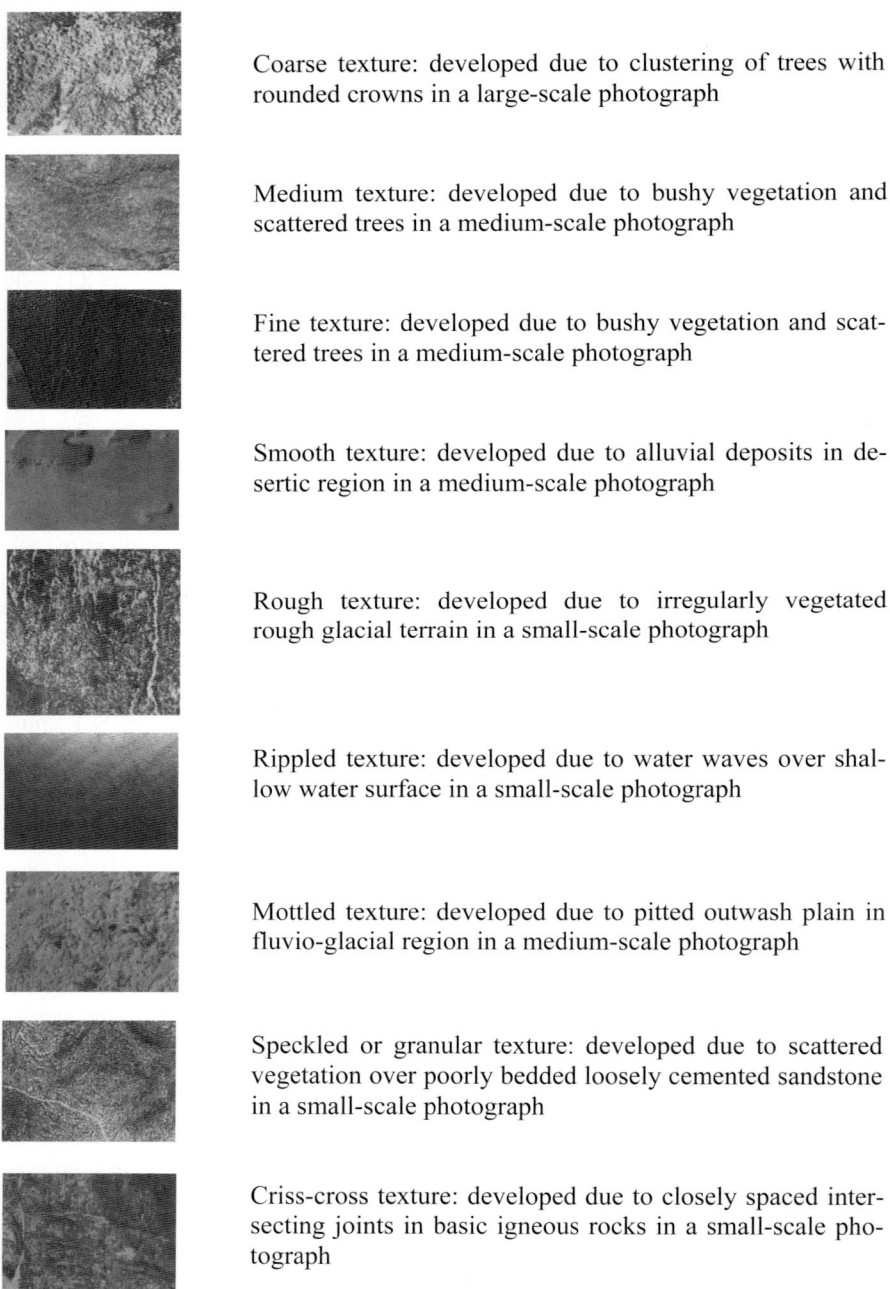

Fig. 7.8. Some typical photo textures. (Pandey 1987)

Fundamentals of Interpretation 157

This parameter should be considered together with shape and size, and direction of illumination.

6. Site or association. Certain features are preferentially associated with each other, and this mutual association of objects is one of the most important guides in photo interpretation; for example, extrusive rocks are associated with volcanic landforms such as cones, calderas, dykes, lava flows etc.; similarly, alluvial landforms include fans, meandering channels, ox-bow lakes, point bars etc.

7. Pattern. Pattern refers to the spatial arrangement of objects. It is an important parameter in photo interpretation, due to the fact that a particular pattern may have genetic significance and could be diagnostic. Patterns can be formed by different types of objects, such as rock outcrops, drainage, streets, fields, soil type etc. A specific term can be used to describe the spatial arrangement of each of these, e.g. linear, radial, rectangular, annular, concentric, parallel, en-echelon, checkerboard; other suitable terms may also be used as needed. It is important to note that pattern depends upon scale. The units which may be visible individually on a larger scale may coalesce or merge into each other on a smaller scale, and thus a group of objects forming a pattern on a larger scale may have to be described under the term texture on a smaller-scale photograph.

The above elements of photo interpretation are used to make observations on photographs and images. However, interpretations in terms of various physical attributes and phenomena have to be based on sound knowledge of the relevant scientific discipline. This becomes even more evident when we take into account the convergence of evidence approach. *Convergence of evidence* implies integrating all the evidence and interpretations gathered from different photo recognition elements, i.e. considering where all the evidence collectively leads to. The approach of convergence of evidence is very important for accurate geological interpretation, and for this a sound knowledge of geology is necessary.

7.3.2 Geotechnical Elements

The elements of photo interpretation are applied to study features on the Earth's surface such as landforms, drainage, vegetation, land use and soil. From the study and analysis of these surface features, which are referred to as geotechnical elements, significant information on lithology, structure, mineral occurrences and subsurface geology may be derived. In some photo interpretation studies, any one of these geotechnical elements could itself form the objective of study.

1. Landform. The shape, pattern and association of some landform features can be helpful in identifying geological features. For example, sand dunes have a peculiar typical pattern and shape, and are produced by wind action. Alluvial landforms such as ox-bow lakes, natural levees etc. are quite characteristic and typically produced by fluvial processes. Similarly, many marine landforms are distinctive in shape and pattern. Erosional landforms resulting in linear ridge-and-

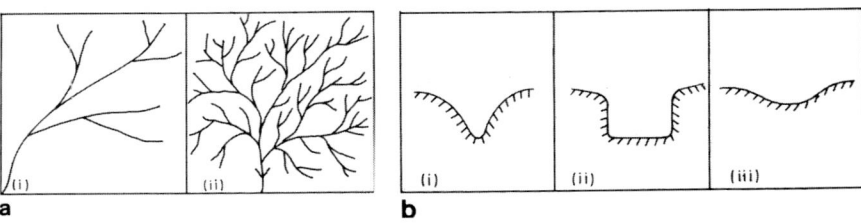

Fig. 7.9. a Drainage textures: (i) coarse and (ii) fine. **b** Typical valley cross-sections: (i) V-shaped, (ii) U-shaped and (iii) gently rounded

valley topography due to differential weathering are characteristic of alternating competent and incompetent horizons. Therefore, a systematic study of landforms is generally a pre-requisite in nearly all geological photo interpretation studies.

2. Drainage. Drainage is one of the most important geotechnical elements for geological photo interpretation. The study of drainage on photographs includes three aspects: (a) drainage texture, (b) valley shape and (c) drainage pattern.

The study of *drainage texture* comprises *drainage density* (ratio of the total stream length within a basin to the area of the basin) and *drainage frequency* (number of streams in a basin divided by the area of the basin). Drainage texture is primarily influenced by three factors: climate, relief and character of the bedrock or soil (i.e. porosity and permeability). Drainage density can be described as fine, medium or coarse. Drainage is said to be internal when few drainage lines are seen on the surface and drainage appears to be mostly sub-surface, e.g. commonly in limestones and gravels. External drainage refers to cases in which the drainage network is seen to be well developed on the surface. Low drainage density (coarse-textured drainage) implies porous and permeable rocks, such as gravels, sands and limestones. High drainage density (fine-textured drainage) implies impermeable lithology, such as clays, shales etc. Figure 7.9a illustrates coarse- and fine-textured drainage.

The shape of the valley may also vary and can give a good idea of the bedrock or soil. Figure 7.9b shows some typical types of valley cross-sections. Short gullies with V-shaped cross-sections often develop in sands and gravels, whereas U-shaped gullies develop in silty soils. Long gullies with gently rounded cross-section generally indicate clayey soils.

Drainage pattern is the spatial arrangement of streams. Drainage patterns are characteristic of soil, rock type or structure. Several authors have described drainage characters and classified them on the basis of genetic and geometric considerations (Zernitz 1932; Parvis 1950; Miller and Miller 1961; Howard 1967). Commonly, six drainage patterns have been considered as the basic types whose gross characteristics can be readily distinguished from one another. They are: dendritic, rectangular, parallel, trellis, radial and annular. A number of modified basic patterns have also been described (see Fig. 16.6). Their utility in geological interpretation is summarized in Table 16.1.

Fundamentals of Interpretation

3. Soil. The operation and interaction of natural agencies of weathering and erosion on the bedrock produce soil. The physical nature of soil therefore depends on the bedrock material and agencies of weathering.

Soils are classified as residual, transported or organic, depending upon their origin. On the basis of composition and physical characteristics, soils can be designated as clayey, loamy, silty, sandy, gravelly and combinations thereof. Broadly, they are called fine-textured, medium-textured or coarse-textured. Soils have characteristic hydrological properties, namely soil permeability and porosity, which govern the surface run-off vis-à-vis subsurface infiltration. Soils can be grouped as poorly drained, moderately drained, well drained and excessively drained. The coarse-textured soils, owing to their larger grain size, are invariably better drained than the fine-textured soils, in which infiltration of water is inhibited. These properties underlie the response of soils on photographs and images.

4. Vegetation. Vegetation in an area is controlled by climate, altitude, microclimate (local conditions), geological/soil factors and hydrological characteristics. The occurrence of plant association in different climate and altitude conditions is well known (see e.g. Von Bandat 1983). Commonly, alignment or banding of vegetation is observed on remote sensing photographs, which may be related to lithological differences or structural features. The height, foliage, density, crown, vigour and plant associations depend on the soil–hydrogeological conditions present. Therefore, tone of vegetation could be related to the bedrock, which may lead to broad vegetation bandings, parallel to the lithology (see e.g. Fig. 16.26). In some cases, structural tectonic features such as faults, fractures and shear zones produce water seepage zones, along which vegetation may become aligned. This alignment may be picked up on the remote sensing data, especially in semi-arid to arid areas with generally scant plant cover. Plants can thus reveal both structural and lithological features in a terrain. Recently, much work on geobotanical exploration using remote sensing data has been undertaken (Sect. 16.6.7).

Chapter 8: Interpretation of Data in the Solar Reflection Region

8.1 Introduction

As stated earlier (Chapter 2), the EM spectral region extending from 0.3 µm to approximately 3 µm is the solar reflection (SOR) region in terrestrial remote sensing. The Sun is the only source of energy in this spectral range, and the solar radiation scattered by the Earth's surface is studied for ground object discrimination and mapping.

This spectral region has been the most intensively studied region for remote sensing of the Earth, for the following reasons.

1. Remote sensing has evolved primarily from the method of aerial photo interpretation, the use of which is confined to the visible and near-infrared part of the solar reflection region. Even the earliest scanners and photo-detectors operated only in this part of the EM spectrum. Therefore, it is logical that the solar reflection range should have become the best investigated part of the EM spectrum.

2. The intensity of radiation available for sensing is highest in this region (Fig. 2.2); there is also a good atmospheric window (Fig. 2.4), permitting acquisition of good-quality aerial and space-borne remote sensing data.

3. The region includes the visible spectrum, in which the response of objects has been easy to interpret in terms of various directly observed objects and physical phenomena.

The interpretability and application potential of remote sensing data in the SOR region depends on image quality, which in turn is governed by a number of factors, grouped broadly into two sets: (1) sensor characteristics (discussed in Chapters 4 and 5) and (2) energy budget considerations (see Sect. 8.2).

The geometric quality of images and photographs were discussed in Chapter 6, and radiometric considerations in Chapter 7. In this Chapter, we first briefly review the parameters which govern the reflected radiance reaching the sensor, and then discuss the methodology for interpreting data in the solar reflection region. Numerous application examples are given in Chapter 16.

162 Chapter 8: Interpretation of Data in the Solar Reflection Region

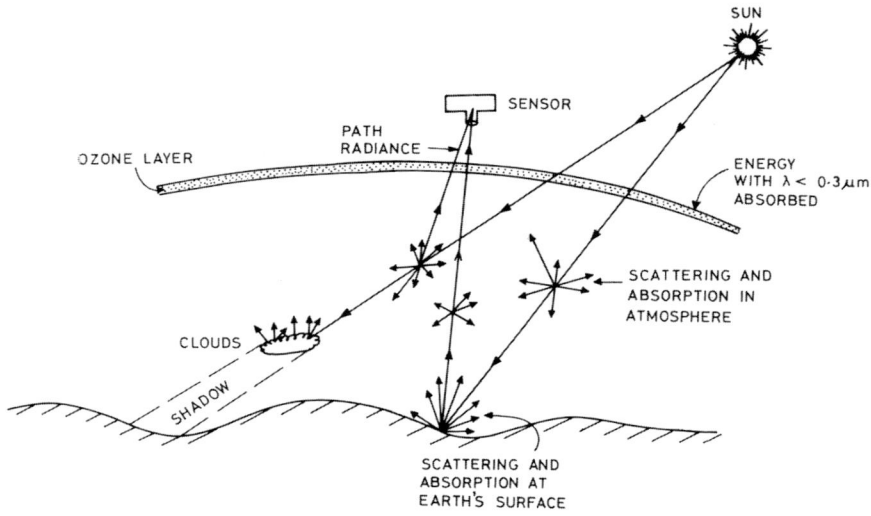

Fig. 8.1. Scheme of energy flow in the solar reflection region

8.2 Energy Budget Considerations for Sensing in the SOR Region

Figure 8.1 shows a schematic of energy flow in the solar reflection region. The Sun radiates EM energy, which illuminates the Earth's surface. As the radiation passes through the atmosphere, it gets modified due to atmospheric interactions (scattering and absorption). The radiation reflected from the Earth's surface again passes through the atmosphere, and again interacts with the atmosphere, before being collected by an aerial or space-borne remote sensor.

The total radiance received at the sensor is dependent upon five main groups of factors: (1) attitude of the Sun, (2) atmospheric–meteorological conditions, (3) topographical slope and aspect, (4) sensor look angle and (5) ground target characteristics (Table 8.1).

8.2.1 Effect of Attitude of the Sun

The Sun is the only source of energy for sensing in this spectral range. The magnitude of solar incident radiation reaching the Earth's surface is thus an extremely important factor and depends on the attitude of the Sun.

The Earth revolves around the Sun in a near-circular elliptical orbit. It is closest to the Sun in early January and farthest away in early July, although the variation

Table 8.1. Factors governing solar reflected energy reaching the sensor

	Primary variables	Secondary variables	Comments
1.	Sun attitude	– Time of the day – Yearly season, day of the year – Latitude – Earth–Sun distance	Vary with time and day, but constant within a scene
2.	Atmospheric meteorological factors	– Composition of the atmosphere – H_2O vapour, CO_2, O_3 concentrations etc. leading to absorption – Particulate and aerosol concentration leading to scattering and path radiance – Relative humidity – Cloud cover and rain	May vary within a scene, from place to place
3.	Topography and slope aspect	– Landscape slope direction – Landscape position – Goniometric aspects	Vary from place to place within a scene, depending on Sun–local topography relation
4.	Sensor look angle	– Sensor–target view angle	For space-borne systems, nearly constant within a scene; for aerial systems (OM scanners and panoramic cameras), varies within a scene
5.	Target reflectance	– Albedo of objects – Surface coating – Surface texture affecting – Lambertian vis-à-vis specular – Reflection pattern	Deciphering this attribute and relevant differences holds the clue in remote sensing

in the Earth–Sun distance is found to have little impact on the intensity of solar radiation reaching the Earth (Nelson 1985).

The angle of incoming solar radiation is one of the most important factors in reflected solar energy. The inclination of the Sun's rays is a function of latitude, yearly season or day of the year, and local time of day. The Sun's angle and direction can noticeably change the appearance of features on a scene (Fig. 8.2). Consequently, season (or day) of the year is important; summer and winter images may bring out different features.

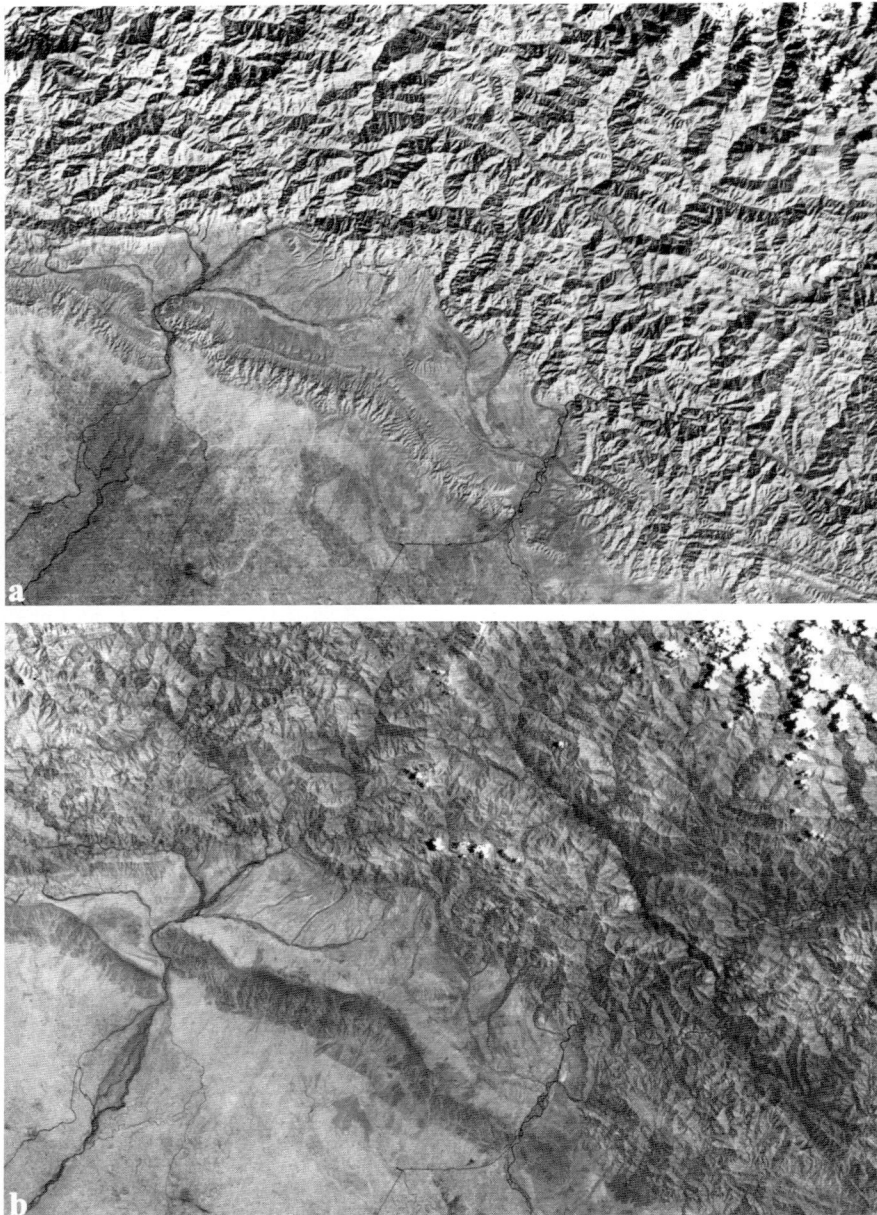

Fig. 8.2a,b. Landsat MSS4 (infrared) images of a part of the Himalayas. **a** Winter and **b** summer images show distinct differences in manifestation of landform, drainage and geological/structural features due to change in the solar illumination condition

A low Sun-angle setting enhances structural features in a direction perpendicular to the direction of sunrays, owing to shadows. On the other hand, a very low Sun elevation may lead to unduly reduced signal and extensive shadows. Therefore, image data sets with proper solar illumination condition must be carefully chosen for optimum results.

The effect of differences in the Sun's angle of inclination can be reduced by digitally transforming image data to a fixed Sun angle (see Sect. 10.2.2).

8.2.2 Effect of Atmospheric–Meteorological Conditions

Solar radiation, before being sensed by a remote sensor, has to travel twice though the atmosphere – first while incoming from the Sun, and then, after being back-scattered from the Earth's surface. In this process, it becomes modified through interaction with the atmosphere by scattering and absorption.

Atmospheric–meteorological conditions prevailing at the time of observations play an important role. Haze, aerosols and suspended particles in the atmosphere cause scattering and path radiance. Further, cloud cover blocks the solar radiation and casts shadows on the ground, limiting the effective ground area to be sensed and possibly increasing the atmospheric path radiance to some extent (Nelson 1985). These factors need be considered at the time of acquiring and interpreting image data. Digital processing methods for path radiance correction are discussed in Sect. 10.2.1.

8.2.3 Effect of Topographic Slope and Aspect

Uneven topography leads to a varying local angle of incidence. The ground illumination and intensity of back-scattering thus also varies accordingly (Fig. 8.3).

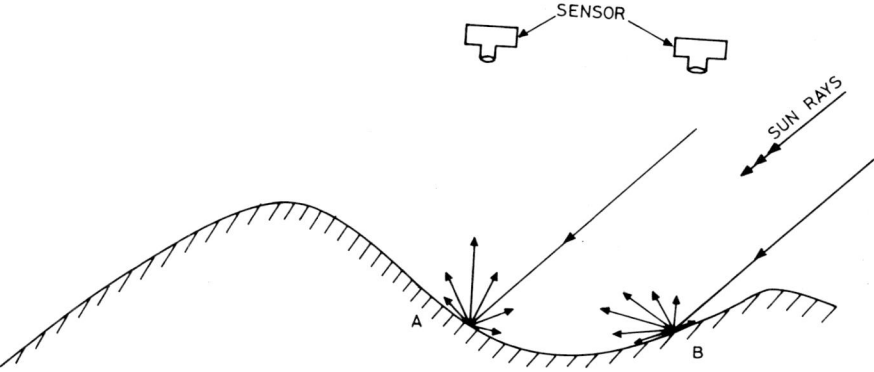

Fig. 8.3. Effect of topography on intensity of back-scattered radiation. Due to topographic orientation, the ground at *A* appears brighter than that at *B*, the images being acquired under a similar Sun angle and sensor configuration

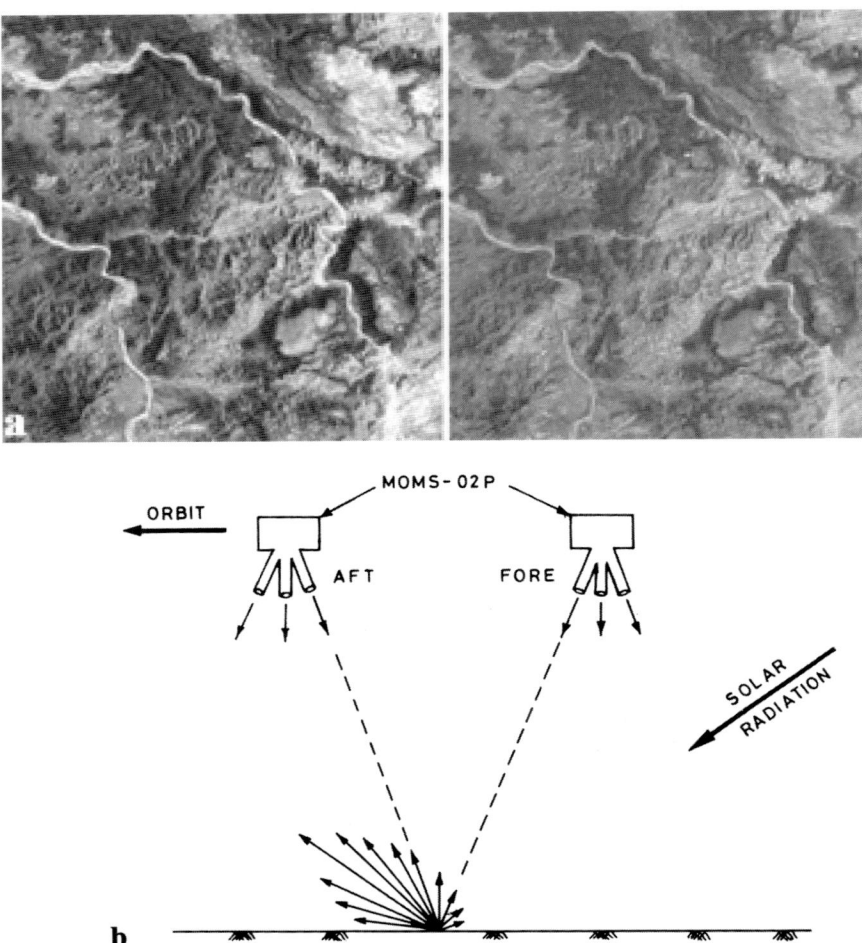

Fig. 8.4. a Fore and aft camera images from MOMS-02P acquired barely about 15 s apart. **b** Schematic explanation of the difference in radiometry of the two images in **a**. (**a** Courtesy of DLR, Oberpfaffenhofen)

Local landscape orientation (slope direction and magnitude) vis-à-vis Sun angle is extremely important in controlling illumination. Most natural terrestrial surfaces behave as a semi-diffuse reflector, being in between an ideal specular reflector and a Lambertian reflector. The radiation is scattered in various directions, the intensity of back-scattered radiation being maximum in a direction corresponding to the specular reflection. Thus, topographical slope (direction and magnitude) vis-à-vis Sun angle is one of the most important single factors affecting the intensity of incident illumination and back-scattered radiation (Stohr and West 1985).

The reflected radiance may vary at different times of the day. Topographical effects on reflected radiance data can be so severe, especially in areas of high relief, that correlation and extrapolation of photo units on the basis of simple tonal signatures may be quite impossible. Methods for topographic normalization are discussed in Sect. 10.2.3.

8.2.4 Effect of Sensor Look Angle

Most remote sensors operate with the optical axis nominally vertical (nadir-looking). However, in some cases, such as for stereoscopic applications, oblique-looking optical systems, are employed. As the look angle of the sensor (i.e. the angle made by the sensor's optical axis with the vertical) changes, the Sun–target–sensor goniometry also changes; this in turn changes the intensity of back-scattered radiation reaching the sensor (in terms of BRDF).

Figure 8.4a shows a set of MOMS-02P images, which were acquired from fore- and aft-looking cameras of the in-orbit stereo system The images exhibit different radiometry, although the two were acquired barely about 15 s apart. The differ-

Table 8.2. Albedo of various surfaces (integrated over the visible spectrum). (After Schanda 1986)

Surface	Percent of reflected light intensity
General albedo of the Earth	
Total spectrum	~ 35
Visible spectrum	~ 39
Clouds (stratus) < 200 m thick	5–65
200–1000 m thick	30–85
Snow, freshly fallen	75–90
Snow, old	45–70
Sand, 'white'	35–40 (increasing towards red)
Soil, light (deserts)	25–30 (increasing towards red)
Soil, dark (arable)	5–15 (increasing towards red)
Grass fields	5–30 (peaked at green)
Crops, green	5–15 (peaked at green)
Forest	5–10 (peaked at green)
Limestone	~ 36
Granite	~ 31
Volcano lava (Etna)	~ 16
Water: Sun's elevation (degrees)	
90	2
60	22
30	6
20	13.4
10	35.8
5	~ 60
< 3	> 90
Urban reflectance	~ 6–20

ence in radiometry is related to Sun–target–sensor goniometry for the two images (Fig. 8.4b). With the Sun illuminating from the right side, the aft (left) image gets more lighted (and exhibits greater contrast) than the fore (right) image.

8.2.5 Effect of Target Reflectance

Solar radiation incident on the ground surface may be absorbed or backscattered, as permitted by the spectral characters of the materials, discussed in detail in Chapter 3. The ratio of intensity of reflected to incident energy is called albedo. Albedo values for common natural objects are listed in Table 8.2.

The depth of penetration of the radiation in the SOR region is of the order of barely 50 μm. Therefore, surface properties, such as surficial coatings, moss, algae, clays, weathered products, oxidation and leaching, are highly important in the solar reflection region.

The degree of homogeneity and the density of the objects are other relevant variables. The surface texture influences the scattering characteristic of the target, i.e. the extent to which a surface will behave either as a specular surface or as a Lambertian surface.

With the knowledge that the above parameters influence the observed reflected radiance data, the aim of remote sensing investigations is to distinguish between various types of ground surfaces.

8.3 Acquisition and Processing of Solar Reflection Image Data

All types of sensors, including photography, opto-mechanical line scanners, CCD line scanners and digital cameras have been employed to acquire data in the solar reflection region. These were described in Chapters 4 and 5.

The sensors may be used from various platforms. From aerial platforms, conventional black-and-white photography has been the most common mode of acquiring data in the solar reflection region. From space platforms, multispectral scanners (e.g. Landsat MSS and TM, SPOT-HRV and IRS-LISS) have been extensively used. Lately, digital cameras have been finding greater applications from both aerial and space platforms (IKONOS and many new missions planned by various agencies).

Ground measurements aim to provide a reference base, necessary for reliable interpretations, and were discussed in Sect. 1.6.

Processing of the solar reflection image data is done for:
a) computing images of some physical attributes;
b) carrying out geometric and radiometric corrections;
c) image enhancement, transformation, and multiple image fusion etc.;
d) image classification.

Chapter 10 'Digital Image Processing' describes in detail the methods for geometric and radiometric corrections, image enhancement, transformation, fusion and classification. We mention here some concepts which can be used for computing images to bring out physical attributes in the solar reflection region.

1. *Radiance* is computed as (after USGS 1979)

$$\frac{DN}{DN_{max}}(L_{max} - L_{min}) + L_{min} \qquad (8.1)$$

where DN is the digital number; DN_{max} is the maximum digital number that can be recorded by the sensor (viz. 127, 255 etc.); L_{max} and L_{min} are saturation and threshold radiance levels respectively, for a given sensor band in milliwatts per square metre.

2. *Reflectance* is a measure of the percentage of light reflected from a given target. It can be computed from the radiance data by accounting for the strength of the incoming solar radiation and the general angle of incidence of radiation at the time of overpass (USGS 1979) (assuming that the target is Lambertian and the effect of the atmosphere is also Lambertian) using the following relation

$$\text{Reflectance} = \frac{\pi}{E[\sin(\alpha)]}(\text{Radiance}) \qquad (8.2)$$

where E is the solar constant for the spectral band under consideration at the top of the atmosphere (in mW cm^{-2}) and α is the solar elevation (90° minus solar–zenith angle). This provides an image with values understandable in physical terms (Robinove 1982) and removes the effects of differences on various images due to Sun attitude (time of day, season and day of the year etc.), and is found to be a reasonably efficient method for reducing scene variability (Nelson 1985).

8.4 Interpretation

8.4.1 Interpretation of Panchromatic Black-and-White Products

(Aerial panchromatic photographs, SPOT-PAN, IRS-PAN, ETM+ PAN, IKONOS-PAN, QuickBird-2 PAN etc.)

The application of the technique of photo interpretation panchromatic aerial photographs has been described in numerous standard publications (e.g. Leuder 1959; Miller and Miller 1961; Ray 1965; Allum 1966; Mekel 1970; Von Bandat 1983; Avery and Berlin 1985; Pandey 1987). Black-and-white panchromatic photographs and images of terrestrial features have also been also available from vari-

Table 8.3. Comparison of spatial resolution of selected space-borne remote sensing systems (panchromatic and multispectral bands in solar reflection region) (all data in m)

		Panchromatic	Blue	Green	Red	Near-IR	SWIR-I	SWIR-II
Landsat	MSS	–	–	79×79	79×79	79×79	–	–
	TM	–	30×30	30×30	30×30	30×30	30×30	30×30
	ETM+	15×15	30×30	30×30	30×30	30×30	30×30	30×30
IRS	LISS-I	–	72×72	72×72	72×72	72×72	–	–
	LISS-II	–	36×36	36×36	36×36	36×36	–	–
	LISS-III	–	–	23×23	23×23	23×23	70×70	–
	PAN	5.8×5.8	–	–	–	–	–	–
	WiFS	–	–	–	188×188	188×188	–	–
SPOT	HRV-PAN	10×10	–	–	–	–	–	–
	-Multi	–	–	20×20	20×20	20×20	–	–
	HRVIR-HR	–	–	–	10×10	–	–	–
	-Multi	–	–	20×20	20×20	20×20	20×20	–
JERS-OPS		–	–	18×24	18×24	18×24	18×24	18×24[a]
CBERS		20×20	20×20	20×20	20×20	20×20	-	-
ASTER		–	–	15×15	15×15	15×15	15×15	30×30[b]
IKONOS	PAN	1×1	–	–	–	–	–	–
	Multi	–	4×4	4×4	4×4	4×4	–	–
QuickBird-2	PAN	0.61×0.61	–	–	–	–	–	–
	Multi	–	2.5×2.5	2.5×2.5	2.5×2.5	2.5×2.5	–	–

[a] JERS-OPS has three bands in SWIR-II range.
[b] ASTER has five bands in SWIR-II range.

ous space-borne sensors, e.g. Metric Camera, Large-Format Camera, SPOT-PAN, IRS-PAN, ETM+ PAN, IKONOS-PAN and QuickBird-2 PAN. Products from these remote sensing missions differ in spatial resolution (Table 8.3). However, as far as radiometric aspects are concerned, data products from all the panchromatic band sensors are quite alike; therefore, interpretation of all the above products must follow the same stream of logic.

Pre-processed satellite sensor images/data and photographs form the bulk of products available for interpretation and applications. The interpretations are based on elements of photo recognition and geotechnical elements. Further, stereo viewing, which enables appraisal of relief, slope and 3-D morphology, is a special advantage generally available in panchromatic products in the SOR region (Fig. 8.5), in contrast to other spectral regions.

Fig. 8.5a,b. Metric camera stereo photo pair, French Alps; note the low Sun angle (Sun elevation 15°, azimuth 145°). (Processed by DLR, Oberpfaffenhofen)

On broad-band panchromatic black-and-white photographs and images, *snow* appears bright white due to high albedo. A *deep and clear water body* appears dark, as the solar radiation is either specularly reflected or penetrates the water body to a limited depth and is absorbed, with little or no volume scattering. On the other hand, a *turbid and shallow water body* appears in shades of light gray to gray, due to volume scattering and bottom reflection. Occasionally the sensor, Sun and water surface may be in such a geometric configuration that specular reflection is received at the sensor – this is called *glare*. *Vegetation* appears dark gray to light gray, the actual tone at the site being a function of the density and type of vegetation. Various photo parameters, such as shape and size of crown, density of foliage, stage of development, time of year, shadow etc., can be used to identify tree species, crop estimation and vegetation damage assessment. *Soils* appear in the various shades of gray, which may be related to the type of soil and its origin. On broad-band VNIR photographs and images, coarse-textured porous *sandy soils* of alluvial fans, natural levees and aeolian landforms are very light to almost white in tone; on the other hand, fine textured *clayey soils* of backswamps, flood plains and lakes are medium to dark gray. Local variation in tone may occur due to moisture content, organic matter, relief or grain size of the soil. In general, soils in arid climates are lighter-toned, because of the lack of vegetation and surface moisture, than in humid climates. *Calcareous soils* generally give a medium tone on VNIR data and show a pitted appearance or mottling due to variation in moisture content. *Alkaline soils* often show light tones, due to the presence of salts and scarce vegetation. *Rock surfaces* appear in shades of gray, the tone of the surface being dependent on the type of rock, its degree of weathering, soil cover, moisture and vegetation cover. Discrimination between rock groups could be based on converging evidence derived from a number of parameters including landform, soil, vegetation, drainage and structure. A number of *cultural features* like cities, townships and settlements, roads and railway tracks can often be recognized on the

Table 8.4. Salient response characteristics of the multispectral bands

Band name	Blue	Green	Red	Near–IR	SWIR–I	SWIR–II
Important spectral band characters	Very strong absorption by vegetation and Fe–O; good water penetration; high scattering by suspended atmospheric particles (Fig. 8.6a)	Some vegetation reflectance; good water penetration; scattering by atmospheric particles etc. (Fig. 8.8a)	Very strong absorption by vegetation; some water penetration; scattering by atmospheric particles (Fig. 8.8b)	High reflectance by vegetation and limonite; total absorption by water (Fig. 8.8c)	General higher reflectance; insensitive to moisture contained in vegetation or to hydroxyl-bearing minerals; absorption by water, snow (Fig. 8.7b)	High absorption by hydroxyl-bearing minerals, carbonates, hydrous minerals, vegetation leaves, water and snow (Fig. 8.9b)

Interpretation

Table 8.5. Response of common objects on multispectral bands

Band name / Object	Blue	Green	Red	Near - IR	SWIR-I	SWIR-II	Colour on CIR film and standard FCC
Forest							
- deciduous	Dark gray	Dark	Very dark	Light	Light	Dark gray	Deep red bright
- defoliated	Light gray	Light gray	Med. gray to light	Darker tone	Light	Light gray	Grayish to brownish red
Cropland	Gray	Gray	Med. gray	Light	Light gray	Light gray	Pinkish red
Water							
- clear and deep	Dark	Dark	Black	Black	Black	Black	Black
- silty and shallow	Light	Light	Gray	Black	Black	Black	Bluish
Soil							
- fallow fields	Light	Light gray	Light gray	Darker	Gray	Darker	Yellowish
- moist ground	Light gray	Light gray	Light gray	Very dark to black	Dark	Dark	Cyanish-light grayish
Snow							
- fresh	White	White	White	White	Very Dark	Very dark	White
- melting, dirty	White	White	white	Very dark	Very dark	Very dark	Cyanish white
Urban/industrial areas	Light	Light gray	Gray	Darker			Bluish gray mottled
Rocky terrain (bare)	Lighter	Lighter	Gray	Gray	Gray	Gray	Gray to dark colour

photographs and images, as these are marked by contrasting shapes, outlines and patterns.

8.4.2 Interpretation of Multispectral Products

(Landsat MSS, TM & ETM+, SPOT-Multi, IRS-LISS-I, -II & -III, JERS-OPS, ASTER, IKONOS, QuickBird-2)

Data from space sensors are available on a regular basis. The spectral distribution of the above sensors can be considered to be quite comparable (except for some bands in JERS-OPS and ASTER), although there are differences in spatial resolution specifications (Table 8.3). Interpretation of image data from these sensors would follow a common line of argument, and for this reason they have been grouped together. Multiband photographic products, although not available on a regular basis, can also be included under this discussion, since the spectral considerations are the same as for other sensors.

Spectral characteristics of the objects and the sensor wavelengths govern the response of objects in different channels of the sensor. The multispectral data can therefore help discriminate and identify different types of objects, depending upon their spectral attributes (Tables 8.4 and 8.5).

Fresh snow appears bright white in the VIS and NIR range and exhibits strong absorption in the SWIR range, due to the presence of water molecules (Fig. 8.7b).

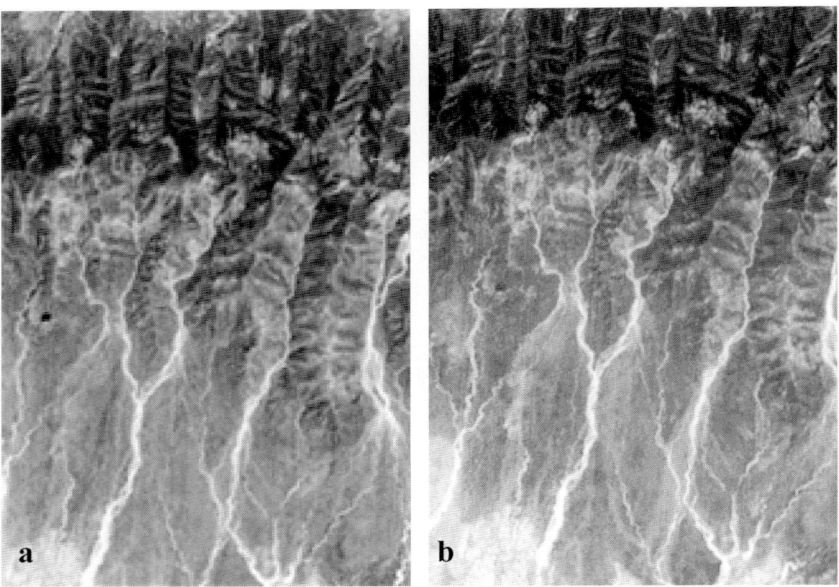

Fig. 8.6. a Blue-band and **b** green-band images from IRS LISS-II; the blue-band image exhibits higher scattering and path radiance than the green-band image

 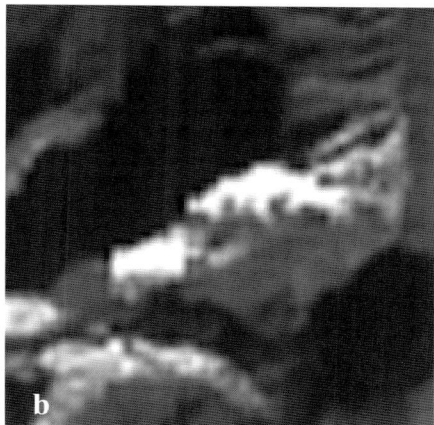

Fig. 8.7. a Near-IR band and **b** SWIR band images (IRS LISS-III). Note that cloud and snow are not distinguishable from each other on the near-IR image, but can be distinguished with the help of the SWIR band image

Melting snow appears bright white in the visible range and is dark in the NIR channels due to absorption of the radiation by the water film on the surface; in the SWIR, the melting snow is again dark. *Water* exhibits different types of responses, depending upon its silt content and the depth of the water body (Fig. 8.8). *Clear deep water* bodies are dark in the VIS, NIR and SWIR ranges. *Silted and shallow water* bodies strongly reflect the shorter wavelengths and therefore appear light-toned on blue, green and even red bands, the brightness gradually decreasing towards longer wavelengths (Fig. 8.8). In the NIR and SWIR, the water bodies, whether shallow or deep, silted or clear, appear black.

Forests in general appear medium dark in the visible and bright in the NIR. In the SWIR-I, they appear bright and in the SWIR-II again dark. *Coniferous forests* reflect less strongly in the NIR than deciduous forests. *Defoliated forests* appear brighter in the visible and darker in the NIR, owing to the absence of leaves. The response over defoliated forests also depends on the type of soil/bedrock. *Cropland* has generally medium density of leaves and vegetation and therefore appears medium gray in the visible channels (blue, green, red) and light in the NIR (Fig. 8.8) and SWIR. The response of cropland in the VIS, NIR and SWIR ranges is a mixture of the response of vegetation and soil. The cropland may also be marked by characteristic field pattern, which may be observed on suitable scales. *Soils and fallow fields* (dry) are light in the visible and medium gray in the NIR and SWIR bands. *Moist ground* is medium gray in the visible but very dark in the NIR and SWIR. *Rocky terrain* (bare) is usually brighter in the visible than in the NIR and SWIR, and is characterized by peculiar landform and structure. *Limonite* exhibits strong absorption towards UV–blue and is therefore very dark in the blue–green bands, and light-toned in the red, NIR and SWIR bands. *Clays*, on the other hand,

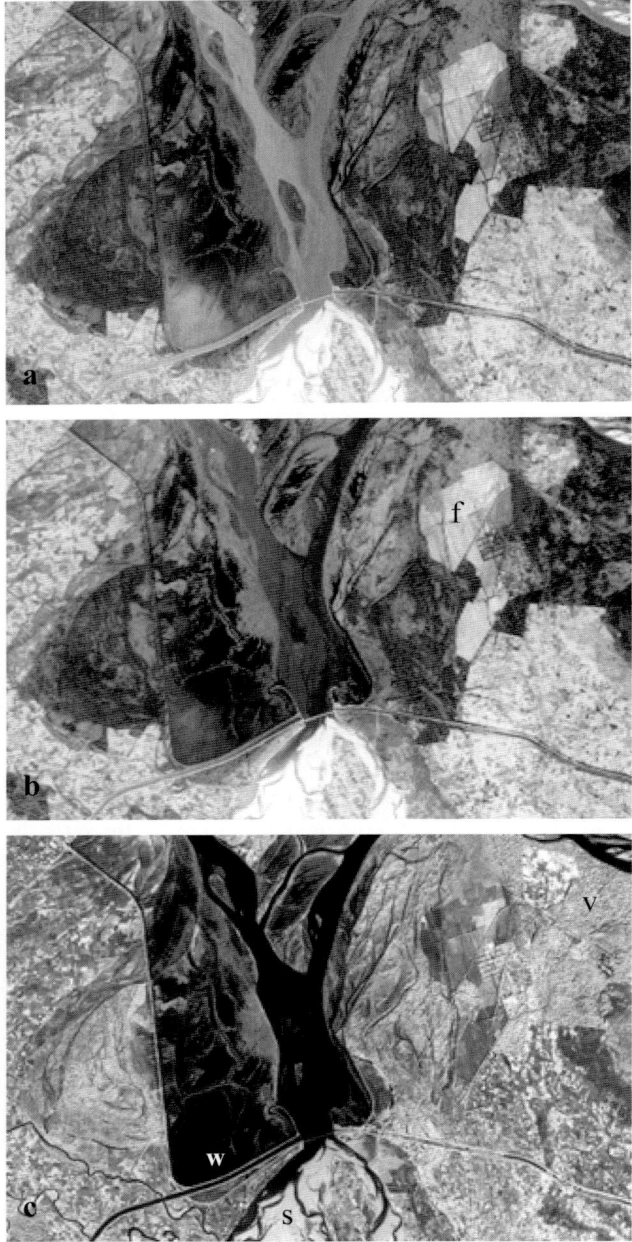

Fig. 8.8. Images in **a** green, **b** red and **c** near-IR bands of a part of the Gangetic plains, exhibiting differences in spectral response of various objects (IRS-LISS III sensor). w = water body; s = sand; v = vegetation (crop); f = fallow fields

Interpretation

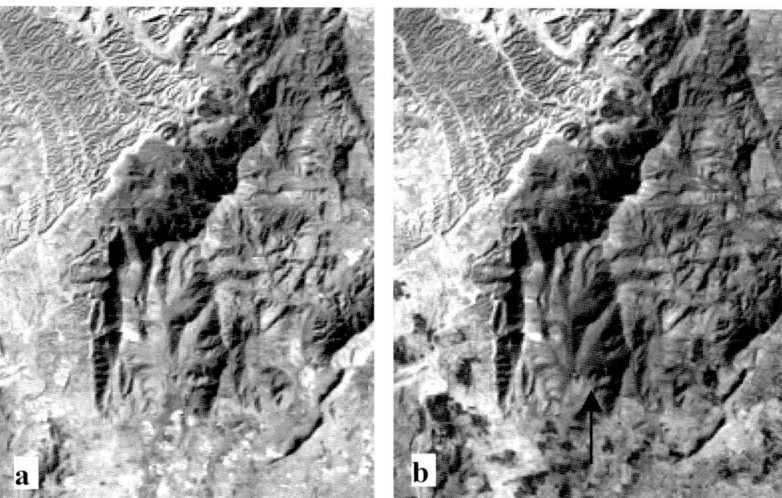

Fig. 8.9. a Near-IR and **b** SWIR-II band images (Landsat TM data of part of Khetri copper belt, India); *arrow* indicates the area of hydroxyl-bearing alteration zone in **b**

are light-toned in the visible, NIR and SWIR-I ranges, but are very dark in the SWIR-II range, due to strong absorption by the hydroxyl group of anions (Fig. 8.9).

The multispectral data from these sensors have opened up vast opportunities for mapping and monitoring of surface features and geological exploration, and a multitude of other applications.

8.4.3 Interpretation of Colour Products

8.4.3.1 Standard FCCs and CIR Film

The false colour composite (FCC) technique is extensively used for combining multi-spectral images through colour coding. In addition, colour infrared photographs have been routinely acquired from various aerial platforms and selected space missions. Several examples of such colour products are given in this treatment (e.g. Fig. 8.10). Interpretation of colour infrared film and standard FCCs follows a common line of argument, since the two are generated through a similar scheme of falsification of colours (i.e. coding spectral bands into colours)

response in blue wavelength	– cut off
response in green wavelength	– shown in blue colour
response in red wavelength	– shown in green colour
response in NIR wavelength	– shown in red colour.

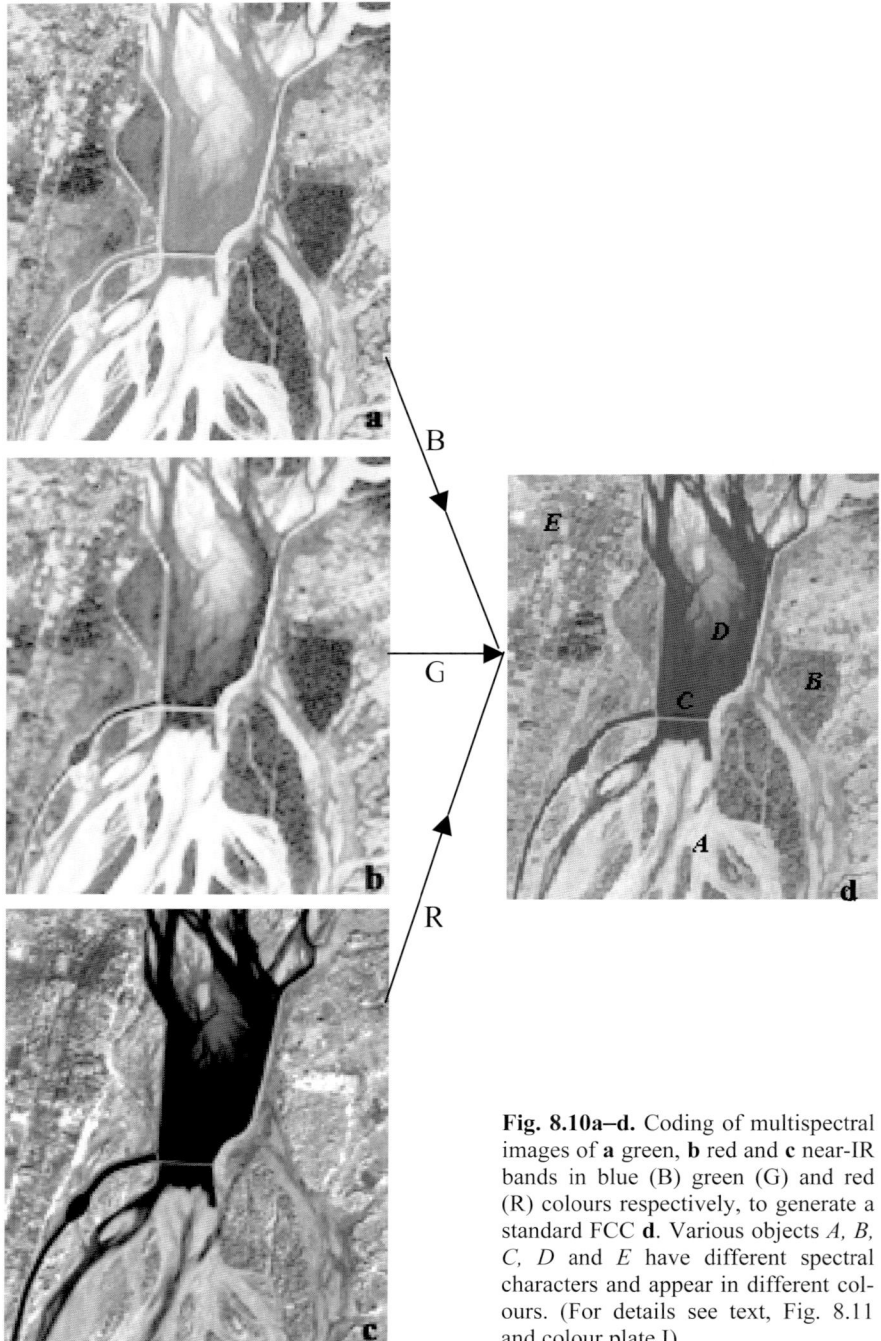

Fig. 8.10a–d. Coding of multispectral images of **a** green, **b** red and **c** near-IR bands in blue (B) green (G) and red (R) colours respectively, to generate a standard FCC **d**. Various objects *A, B, C, D* and *E* have different spectral characters and appear in different colours. (For details see text, Fig. 8.11 and colour plate I)

Interpretation

The responses of common objects such as forests, cropland, water bodies, snow, bare ground etc. on standard FCC/CIR film are also given in Table 8.5.

Before proceeding to interpretation, it is important to see how a standard FCC is generated. Figure 8.10 explains the methodology. A set of multispectral images of green, red and near-IR bands (Fig. 8.10a,b,c) are projected in blue, green and red colours, respectively. (This task is performed by selecting blue, green and red planes for displaying green band, red band and near-IR spectral images respectively, in a colour monitor. As the colour display planes are usually selected in the order red, green and blue, the term RGB is used.) Figure 8.10d shows the standard FCC so generated of an area. Note the following features.

A: Bare sandy soil/fallow fields appear very light to light gray, due to near-equal reflectance in all the bands.
B: Vegetation appears deep red, due to high reflectance in the near-IR band, which is coded in red colour.
C: Deeper water body is deep blue, due to some reflectance in the green band (coded in blue) and absorption in other bands.
D: Silted water body appears light cyanish due to reflectance in green and red bands (coded in blue and green respectively) and absorption in the NIR band (coded in red).
E: Township is light cyanish due to higher reflectance in green and red bands (coded in blue and green respectively) and low reflectance in the NIR band due to absence of vegetation (coded in red). It is also marked by the characteristic reticulate texture.

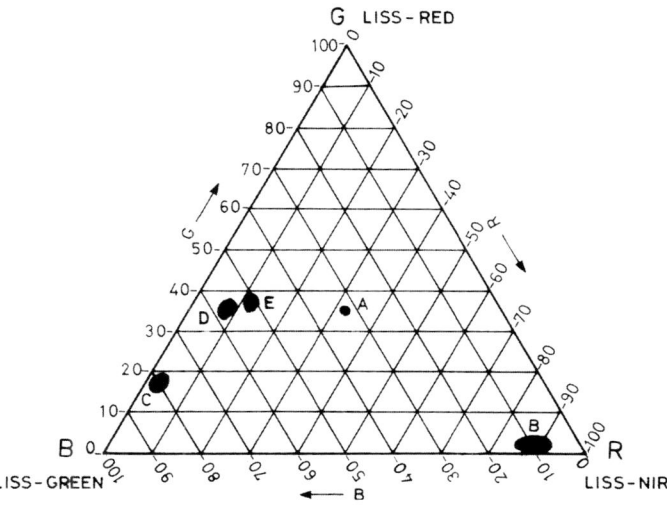

Fig. 8.11. RGB colour ternary diagram showing plots of colour features *A, B, C, D* and *E* in figure 8.10

The positions of different colours corresponding to *A, B, C, D* and *E* have been located in the RGB ternary diagram (Fig. 8.11), which yields the relative contributions of the three end members at the features. The information is indicative of the spectral character of the objects, from which the objects can be identified.

8.4.3.2 Other Colour Displays

The approach using an RGB ternary diagram is a practical means of interpreting all types of colour displays, including various types of (non-standard) FCCs, colour composites generated from thermal images, ratio images, and other spatial data sets, etc. Several examples are given (see e.g. figures 16.68, 16.69 and 16.72).

8.5 Luminex Method

The luminex method is a new development based on the detection of photo-luminiscence (Robbins and Seigel 1982). It is an active sensor like lidar and radar. The method is based on the phenomenon that certain minerals, when struck by UV radiation, exhibit luminescence. The sensor employs UV laser beams fired from airborne platforms, which cause active photo-luminiscence in certain key minerals such as scheelite, powellite, hydrozincite, autunite etc. These minerals are either themselves of interest or act as pathfinder minerals to certain deposits, e.g. tungsten and skarn deposits, molybdenum, tin, zinc, base metals, gold, uranium etc. The fired UV laser beam excites photo-luminiscence in minerals in its target area or foot-print. The light emanating from the target area is viewed by a telescope, spectrally separated and analysed to determine the presence of the mineral of interest.

8.6 Scope for Geological Applications

The scope and potential of solar reflection data in geological applications are so high that remote sensing has become almost an operational tool. Numerous examples are provided in Chapter 16.

In comparison to other data types, SOR data have some advantages: (1) the thermal-infrared data have relatively coarser spatial resolution and inferior spectral radiometric quality, and (2) radar data have relative disadvantages of angular looks and distortions. The chief limitation of the solar reflection data for geological application arises from the fact that the response in the SOR region is governed by barely the top 50 µm layer on the ground. In some cases, this may predominantly comprise lichen, moss, soil, vegetation, oxidation film and surficial coatings, which are invariably ignored by field geologists. However, systematic and detailed study of the geotechnical elements, together with the principle of convergence of evidence, can help unravel geological features in the region.

1. Landforms. Physiological features, drainage and relief are very well recorded in solar reflection data, due to high spatial resolution and contrasting albedos of the ground materials. Stereo pairs in particular serve as an excellent medium for landform studies. Further, landform features can at times be better delineated on multispectral images when subtle differences in moisture, vegetation etc. play a diagnostic role.

2. Structure. Structural features such as folds, faults, lineaments etc. can often be well detected on panchromatic black-and-white and multispectral data products in the SOR region, so much so that these have acquired the status of an essential technique in relatively less-explored areas. Vegetation alignments and variation in surface moisture mark many of the structural features, and these features are amenable to detection on multispectral images.

3. Lithology. Different rock types exhibit differences in landform, drainage, soil, vegetation etc. The cumulative effect of these may permit discrimination of different rock types. In addition, multispectral data and particularly high-spectral resolution remote sensing data have demonstrated the capability for mineralogic/lithologic identification (see Chapter 11).

4. Mineral exploration. Owing to its general utility in mineral exploration, remote sensing is considered as an efficient forerunner in all exploration programmes. In addition, remote sensing data are of proven utility for identification of limonite and hydroxyl minerals, which form significant guides to hydrothermal mineral deposits (see Sect. 16.6).

Additionally, data in the solar reflection region have been extensively applied for hydrocarbon exploration, groundwater investigations, engineering geology, geo-environmental surveys and a host of other applications (discussed in Chapter 16).

Chapter 9: Interpretation of Data in the Thermal-Infrared Region

9.1 Introduction

The EM wavelength region of 3–35 μm is popularly called the thermal-infrared region in terrestrial remote sensing. This is because of the fact that, in this wavelength region, radiation emitted by the Earth due to its thermal state is far more intense than solar reflected radiation (Fig. 9.1), and therefore any sensor operating in this region would primarily detect the thermal radiative properties of ground materials.

Out of the 3–35 μm wavelength region, the greatest interest has been in the 8–14 μm range, owing to the following three main reasons.

1. At ambient terrestrial temperatures, the peak of the Earth's blackbody radiation occurs at around 9.7 μm (Fig. 9.1), which indicates the highest energy available for sensing in this region.

2. An excellent atmospheric window lies between 8 and 14 μm, and poorer windows exist at 3–5 μm and 17–25 μm. Interpretation of the data in the 3–5 μm region is rather complicated, due to overlap with solar reflection radiation in day imagery, and the 17–25 μm region is still not well investigated. This leaves 8–14 μm as the preferred window for terrestrial remote sensing.

3. Prominent and diagnostic narrow spectral features (high-reflectivity or reststrahlen bands) occur, due to bending and stretching molecular vibrations in minerals in this region (see Sect. 3.4.3). These bands vary with composition and structure of minerals and can therefore be usefully applied to give information on mineral composition of rocks.

In view of the above, the 8–14 μm region has been of great interest for geological remote sensing, and the technique has made tremendous strides during the past nearly two decades.

Remote sensing in the TIR region has generally been of a passive-type, i.e. sensors collect data on the naturally emitted radiation. Active techniques, deploying

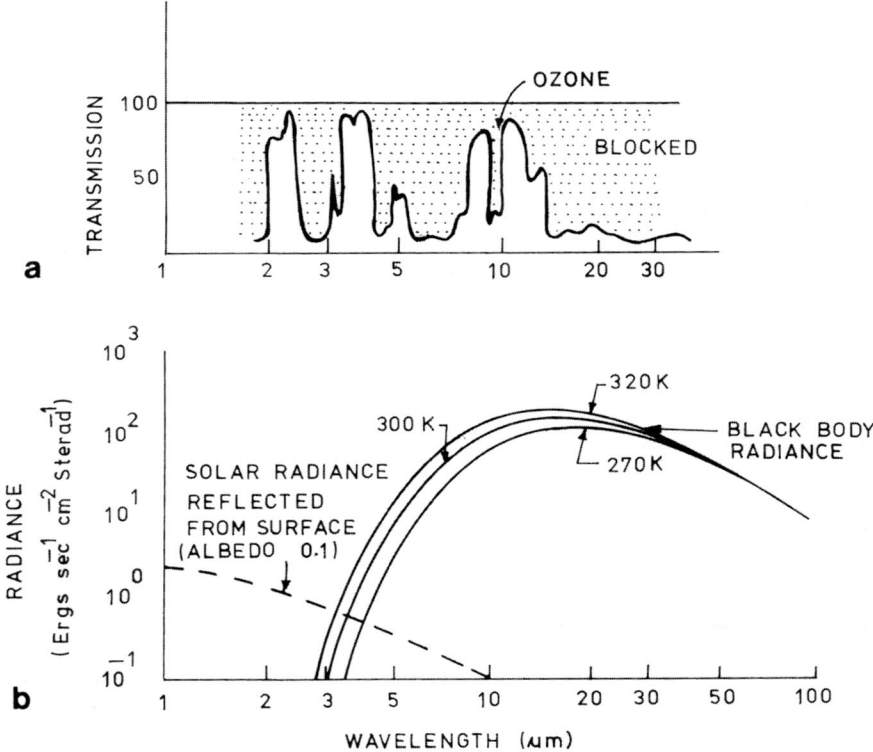

Fig. 9.1. a Atmospheric windows in the thermal-IR region; note the ozone absorption band at 9.6 μm. **b** Energy available for sensing; beyond 3–4 μm, blackbody radiation emitted by the Earth is the dominant radiation with a peak at around 9.7 μm

monochromatic wavelength laser beams (also called laser radar or LIDAR) have also been developed for some research investigations.

9.2 Earth's Radiant Energy – Basic Considerations

The atomic and molecular units within a body having a temperature above absolute zero (0 K or −273.1°C) are in agitated form, owing to which they interact, collide and radiate EM energy. How much energy an object on the ground radiates is a function of two parameters: surface temperature and emissivity. These parameters vary spatially and temporally.

Earth's Radiant Energy – Basic Considerations

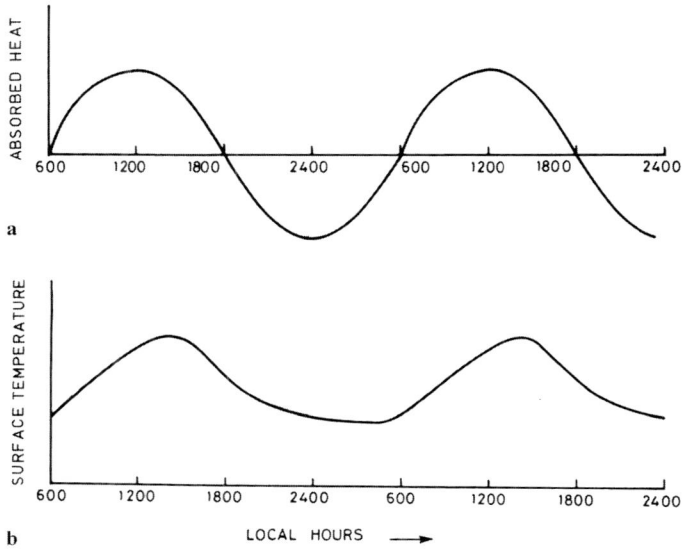

Fig. 9.2a,b. Bearing of solar heating cycle on the Earth's surface temperature. **a** Solar heating cycle. **b** Variation in surface temperature. (After Watson 1973)

9.2.1 Surface (Kinetic) Temperature

The surface temperature of the ground is called kinetic temperature. It is dependent on two main groups of factors: heat energy budget and thermal properties of materials. A detailed review of these parameters is given by Kahle (1980).

9.2.1.1 Heat Energy Budget

Heat energy transfer takes place from higher temperature to lower temperature, by radiation, convection or conduction. Changes in net thermal energy lead to variations in kinetic surface temperature. The following factors influence the heat energy budget.

1. Solar heating. The most important source of heat energy to the Earth's surface is the Sun. The solar radiation falling on the Earth's surface is partly absorbed and partly back-scattered. The absorbed radiation leads to a rise in the level of heat energy, and therefore surface temperature. It has been found that the thermal effect of the diurnal (day and night) cycle usually exists up to a depth of nearly 1-m, the most important being the top 10-cm zone. During the daytime, heat is transmitted from surface to depth, and at night the reverse happens.

In a generalized way, as the Sun rises in the morning, the Earth's surface temperature also starts rising (Fig. 9.2a,b). At noon, the Sun is at its zenith, after

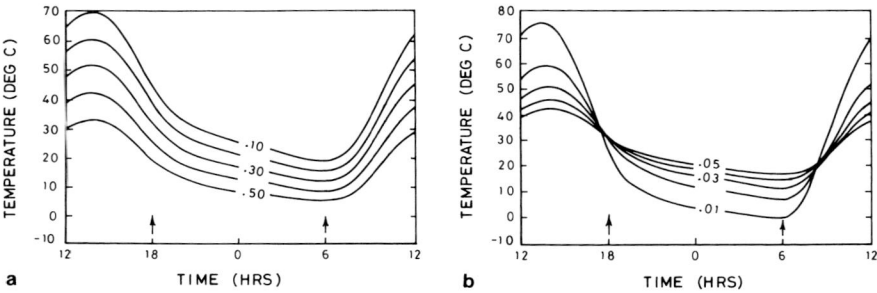

Fig. 9.3a,b. Diurnal temperature curves for varying values of **a** albedo and **b** thermal inertia. (Watson 1975)

which it starts descending; however, the surface temperature keeps rising and reaches a maximum in the early afternoon (around 14.00 h). After this, the surface temperature starts declining as the surface cools off, first at a rapid rate and later at a gentle rate. The surface temperature assumes an almost constant value from nearly midnight to just before sunrise.

The amount of solar energy incident on the Earth's surface depends on several parameters, such as solar elevation, cloud cover, atmospheric conditions, topographical attitude and slope aspect of the surface. The solar elevation (which depends on latitude and time of day and month) is a systematic variable and relatively easy to quantify (see e.g. Price 1977). Variables such as cloud cover and atmospheric conditions are accounted for by monitoring meteorological conditions. Topographical factors such as relief and slope aspect lead to unequal illumination, and terrain models may be used to account for such variations (Gillespie and Kahle 1977).

Since solar heating is the most important source of heat energy to the Earth's surface, a very important factor influencing heat budget is the solar albedo (A), i.e. the percent of solar radiation back scattered, the co-albedo (1−A) being absorbed and directly responsible for the rise in surface temperature. Materials having higher albedo generally exhibit lower temperatures, and those having lower albedo values exhibit higher temperatures (Fig. 9.3a) (Watson 1975). The solar albedo for different objects can be estimated from data in the VNIR region, which can help compute the fraction of solar energy absorbed by the ground surface.

2. Long wave upwelling and downwelling radiation. The longwave upwelling radiation corresponds to the radiation emitted by the Earth's surface. This component removes heat from the Earth's surface and depends on prevailing ground temperature and emissivity. The longwave downwelling radiation is the energy emitted by the atmosphere that reaches the ground and depends on the gases present in the atmosphere and prevailing atmospheric temperature. Empirical relations are used to estimate these components of heat energy (see e.g. Sellers 1965).

3. Heat transfer at the Earth–atmosphere interface. Heat transfer by conduction and convection takes place at the Earth–atmosphere interface. Further, the processes of evaporation and dew formation involve latent heat and affect net heat transfer on the ground. These heat transfers depend on the thermal state of the ground, the atmosphere and meteorological conditions. Empirical relations incorporating meteorological data can be used to estimate the amount of such heat transfers. Therefore, monitoring of meteorological conditions is important for modelling and interpreting thermal data.

4. Active thermal sources. Active geothermal sources such as volcanoes, fumaroles, geysers, etc. and man-made sources such as fire and thermal effluents etc., if any, introduce additional factors in the heat energy budget.

Heat balance. The various components of the heat energy fluxes give an estimate of the net heat flux conducted into the ground. The heat balance equation can be written as (after Kahle 1977)

$$E_s + E_r + E_m + E_i + E_a + E_g = 0 \qquad (9.1)$$

where
 E_s = net solar radiation flux absorbed by the ground
 E_r = net longwave radiation
 E_m = sensible heat flux between the atmosphere and the ground
 E_l = latent heat flux between the atmosphere and the ground
 E_a = heat flux due to the active source
 E_g = net heat flux conducted into the ground, which governs the rise in ground temperature.

9.2.1.2 Thermal Properties of Materials

Thermal properties of ground materials shape the pattern of distribution of the net heat energy conducted into the ground, and therefore govern ground temperatures. The material properties are influenced by mineral composition, grain size, porosity and water saturation. Important thermal properties are briefly described below (typical values listed in Table 9.1).

1. Thermal conductivity (K) is a measure of the rate (Q/t) at which heat (Q) is conducted by a medium through unit area of cross-section, under unit thermal gradient. It is expressed as cal cm^{-1} s^{-1} °C^{-1}. Thermal conductivity is dependent on porosity and the fluid filling the pores. In general, rock has a low value of K, water has a higher value of K, and that of steel is still higher (Table 9.1).

2. Specific heat (c) is a measure of the amount of heat required to raise the temperature of 1 g of substance through 1°C. Physically, a higher value of specific heat implies that more heat is required to raise the temperature of the material. Its units are cal g^{-1} °C^{-1}. Relatively speaking, water has much higher specific heat than rocks, and steel significantly lower (Table 9.1).

Table 9.1 Typical values of thermal properties of selected materials (Most data from Janza 1975)

Geological materials	K Thermal conductivity [cal cm^{-1} s^{-1} °C^{-1}]	ρ Density [g m^{-1}]	c Specific heat cal [g^{-1} °C^{-1}]	k Thermal diffusivity [cm^{-2} s^{-1}]	P Thermal inertia cal [cm^{-2} s$^{-1/2}$ °C^{-1}]
Igneous rocks					
Basalt	0.0050	2.8	0.20	0.009	0.053
Gabbro	0.0060	3.0	0.17	0.012	0.055
Peridotite	0.0110	3.2	0.20	0.017	0.084
Granite	0.0075	2.6	0.16	0.016	0.052
Rhyolite	0.0055	2.5	0.16	0.014	0.047
Syenite	0.007	2.2	0.23	0.009	0.047
Pumice, loose	0.0006	1.0	0.16	0.004	0.009
Sedimentary rocks					
Sandy soil	0.0014	1.8	0.24	0.003	0.024
Sandstone, quartz	0.0120	2.5	0.19	0.013	0.054
Clay soil	0.0030	1.7	0.35	0.005	0.042
Shale	0.0042	2.3	0.17	0.008	0.034
Dolomite	0.0120	2.6	0.18	0.026	0.075
Limestone	0.0048	2.5	0.17	0.010	0.045
Metamorphic rocks					
Marble	0.0055	2.7	0.21	0.010	0.056
Quartzite	0.0120	2.7	0.17	0.026	0.074
Slate	0.0050	2.8	0.17	0.011	0.049
Other materials					
Water	0.0013	1.0	1.01	0.001	0.037
Steel	0.030	7.8	0.20	–	0.168

3. *Density* (ρ). Mass per unit volume (gmcm^{-3}) is another physical property which comes into play in determining the distribution of temperature pattern. It is included in other parameters such as heat capacity, thermal inertia and thermal diffusivity.

4. *Heat capacity* (C = ρ.c) is the amount of heat required to raise the temperature of a unit volume of substance by 1°C. Its units are cal cm^{-3} °C^{-1}.

5. *Thermal diffusivity* [k = K/ (ρ.c)] is a measure of the rate at which heat is transferred within the substance. It governs the rate at which heat is conducted from surface to depth in the daytime and from depth to surface in the night-time. Its units are cm^{-2} s^{-1}. Water possesses high specific heat and therefore minor changes in moisture content have significant effects on thermal diffusivity of soils.

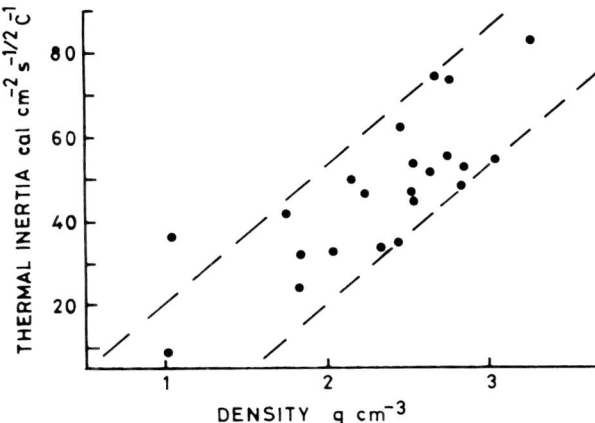

Fig. 9.4. Relationship between thermal inertia and density of rocks (the various values pertain to the data in Table 9.1)

6. *Thermal inertia* [$(P = (K\rho c)^{1/2})$] is a measure of the resistance offered by a substance in undergoing temperature changes. Its units are cal cm^{-2} s$^{-1/2}$ °C^{-1}. The thermal inertia increases if thermal conductivity (K), density (ρ) or specific heat (c) increase. This is quite logical, for, if K increases, then more heat is conducted to depth and the rise in surface temperature will be relatively less. Similarly, if ρ increases, then more material (in g cm^{-3}) is available to be heated, and accordingly more heat is required to raise the temperature (Fig. 9.4); if c is higher, then the material needs relatively more heat for the same rise in its temperature. Figure 9.3b shows surface temperature variations in a diurnal cycle for objects having different thermal inertia values, other factors remaining constant (Watson 1975). The temperature variations are found to be greater for objects having lower thermal inertia values and smaller for those having higher thermal inertia values.

9.2.1.3 Importance of Thermal Inertia in Remote Sensing

Thermal-IR sensing deals with the measurement of surface temperatures. It is shown that the variation in surface temperature of a periodically heated homogeneous half-space is dependent on a single thermal property called *thermal inertia* (Carlsaw and Jaegar 1959). Kahle (1980) made computations to show that if heat balance is constant, then for the same value of thermal inertia P, the variation in K, ρ and c can affect only the depth temperature profile, an not the surface temperature, which remains the same (Fig. 9.5). Therefore, thermal inertia becomes an intrinsic property of materials in the context of thermal remote sensing. Remote sensing methods for estimating thermal inertia are discussed in Sect. 9.3.5.

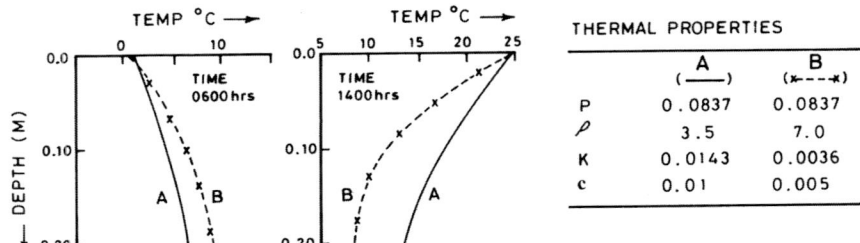

Fig. 9.5. Temperature profiles with depth at different time instances. Materials *A* and *B* have the same value of thermal inertia but differ in density and thermal conductivity. The surface temperature for *A* and *B* is the same, although the temperature profiles with depth are different in the two cases. (After Kahle 1980)

9.2.2 Emissivity

Emissivity is a property of materials which controls the radiant energy flux. Emissivity (ε) for a blackbody is unity and for most natural materials is less than 1, ranging generally between 0.7 and 0.95. If a natural body and a blackbody possess the same surface temperature, then the natural body will emit less radiation than the blackbody.

Emissivity (ε) depends on two main factors: composition and surface geometry. It is intimately related to reflectance or colour (spectral property). Dark materials absorb more and therefore emit more energy than light materials. Spectral absorptivity is equal to spectral emissivity (Kirchoff's Law). It has been shown that the percentage of silica, which is an important constituent in the Earth's crust, is inversely related to emissivity (in the 8–14 μm region) (Fig. 9.6). Therefore, the presence of silica, which has low emissivity, significantly affects the bulk emissivity of an assemblage. Moreover, smooth surfaces have lower emissivity than rough surfaces. In the case of broad-band thermal measurements, lateral emissivity variations are generally ignored. On the other hand, in multispectral thermal sensing attention is primarily focused on detecting lateral variations in spectral emissivity across an area, which in turn sheds light on rock composition.

9.3 Broad-Band Thermal-IR Sensing

In the case of sensors from aerial platforms, the entire 8–14 μm region is used for broad-band thermal-IR sensing, for the simple reason that it provides a high signal-to-noise ratio. On the other hand, as an ozone absorption band occurs at 9.6 μm, the bandwidth of broad-band thermal-IR sensors from space platforms is usually restricted to 10.4–12.6 μm.

Broad-Band Thermal-IR Sensing

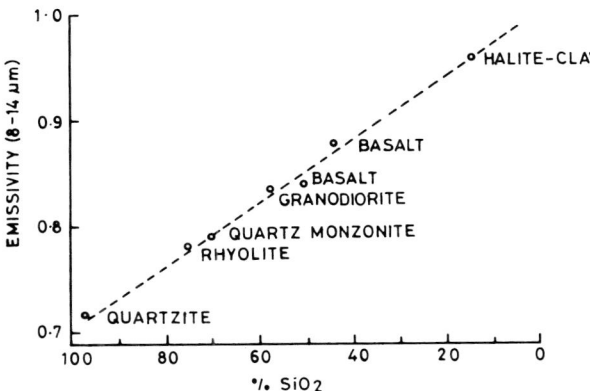

Fig. 9.6. Relationship between emissivity (8–14 μm region) and silica percentage in rocks. (After Reeves 1968)

Thermal-IR wavelengths lie beyond the photographic range and the thermal radiation is absorbed by the glass of conventional optical cameras. Thermal remote sensing data is collected by radiometers and scanners. The working principle of these instruments, including their operation, calibration and generation of imagery from scanner data were discussed in Chapter 5.

The thermal-IR image data can be displayed in real-time or recorded as per the requirements. they can also be combined with other spectral remote sensing data and/ or ancillary geo-data for interpretation.

9.3.1 Radiant Temperature and Kinetic Temperature

In thermal-IR sensing, radiation emitted by ground objects is measured. The *radiant temperature* (T_R) is defined as the equivalent temperature of a blackbody which would give the same amount of radiation, as obtained from a real body. The radiant temperature depends on actual surface or kinetic temperature (T_K) and emissivity (ε). It corresponds to the temperature actually obtained in a remote sensing measurement.

In the case of a non-blackbody, the total amount of radiation (W) emitted is given by the Stefan–Boltzmann Law as

$$W = \varepsilon \cdot \sigma \cdot T_K^4 \qquad (9.2)$$

$$= \sigma \cdot T_R^4 \qquad (9.3)$$

$$\text{where} \quad \varepsilon \cdot T_K^4 = T_R^4 \qquad (9.4)$$

This gives,

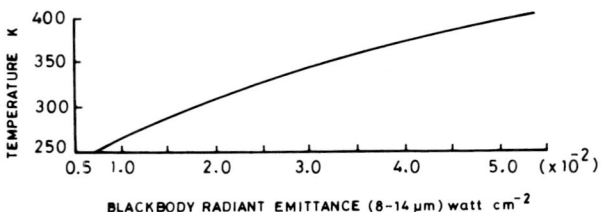

Fig. 9.7. Relationship between energy radiated by a blackbody in the 8–14 μm region (i.e. blackbody radiant emittance) and surface kinetic temperature

$$T_R = \varepsilon^{1/4} \cdot T_K \qquad (9.5)$$

Radiant temperature for a natural body will thus be less than that for a blackbody at the same temperature. This also implies that temperatures measured by remote sensing methods are less than the prevalent surface kinetic temperatures by a factor of $\varepsilon^{1/4}$.

The relation between temperature and radiant emittance for the temperature range 270–350 K in the spectral band 8–14 μm is illustrated in Fig. 9.7. It can be coarsely approximated by a linear function, although for precise measurements a better match is a curve with a T^4-relation (Scarpace et al. 1975; Dancak 1979).

9.3.2 Acquisition of Broad-Band Thermal-IR Data

9.3.2.1 Aerial TIR Data Acquisition

For aerial broad-band thermal sensing, the wavelength range 8–14μm is commonly used as a single band. A two-level calibration is provided on-board. The temperature data obtained in this way are still apparent data, the various possible errors being due to instrument functioning, calibration, atmospheric interference and unknown emissivity. Collectively these may lead to errors in the range of 1–2°C (Leckie 1982). However, absolute temperature data are only seldom needed, and relatively calibrated temperature data, as commonly obtained, are sufficient for most geological remote sensing applications.

Flight lines can be laid as single airstrips or in a mosaic pattern. Selecting the day of survey is always a critical decision in aerial thermal surveys. The predominant considerations are as follows.
(a) The meteorological conditions should be optimum; monitoring of atmospheric parameters may give a clue to the meteorological conditions.
(b) Ground conditions should permit maximum geological discrimination. The geological discrimination is largely affected by soil moisture; hence, soil moisture should preferably be different in different types of soils, and the soils be neither

Broad-Band Thermal-IR Sensing

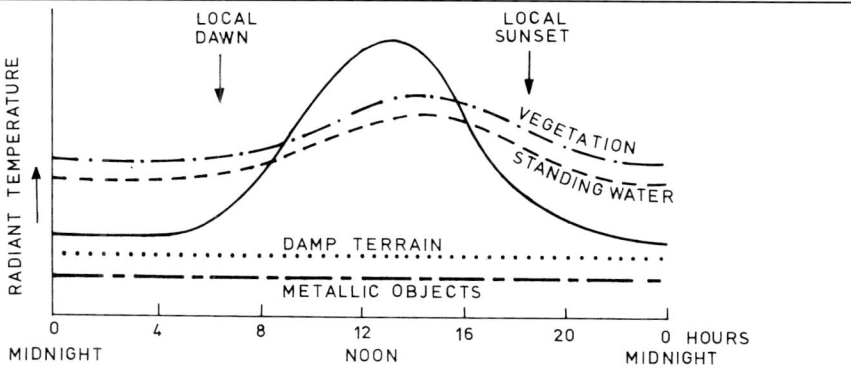

Fig. 9.8. Typical diurnal radiant temperature curves (idealized) for selected materials. (After Sabins 1987)

too wet nor too dry on the day of flight. An empirical method is to delay the flight after rains by up to a day, which should maximize soil moisture variation in the ground.

Ground objects exhibit a systematic variation in radiant temperature in the diurnal cycle (Fig. 9.8). Therefore, a set of day and night thermal data is usually used in thermal-IR investigations. It is seen that objects with differing thermal inertia values have similar temperatures in late afternoon and early morning (Figs. 9.3b and 9.8); This renders these hours unsuitable for thermal discrimination. The maximum contrast in surface temperatures is available at 13–14 hours (early afternoon). However, noon–early afternoon is usually windy and may increase the instability problems of the aerial platform. Further, rapid temperature changes occur in the noon hours with time. The pre-dawn time is generally considered as best for most thermal-IR surveys because (a) the effects of topography and differential solar heating can be avoided and (b) objects maintain steady temperatures from midnight to pre-dawn and therefore variations occurring due to logistic reasons can be minimized.

Aerial thermal-IR imagery may have many geometric distortions, e.g. panoramic distortion, distortions due to platform instability etc. These were discussed in Chapter 6 and methods to rectify them are given in Chapter 10.

9.3.2.2 Orbital TIR Data Acquisition

The use of broad-band thermal-IR channels from orbital platforms commenced with meteorological missions (e.g. TIROS, NOAA etc.), having a typical spatial resolution of 1–5 km. These data had only limited geological applications. A thermal-IR channel has been included in some missions for specific land resources applications, e.g. HCMM, Landsat-4, -5 TM and Landsat-7 ETM+ (Table 5.8).

For sensors on free-flying platforms, there are no navigational or planning considerations for acquiring data; these aspects are taken care of while designing the sensor (e.g. total field of view, swath width, etc.) and selecting orbital parameters.

The geometric characters of orbital imagery are more regular and uniform, as the platform is at a higher elevation and therefore more stable. The various types of geometric and radiometric distortions occurring in space-borne opto-mechanical scanner images are discussed elsewhere (see Chapters 6, 7 and 10).

9.3.2.3 Ground Measurements

The following ground parameters are usually monitored during a thermal-IR survey: (1) ground surface temperature at selected locations (temperature profile to a depth of about 1 m may also be useful), (2) wind speed, (3) air temperature, (4) sky temperature (radiative), (5) cloud cover, (6) humidity, (7) rainfall over the last few weeks or a month, (8) soil moisture at surface at selected locations, (9) groundwater level, (10) vegetation type and density, (11) type of soil or rock and (12) albedo. Determination of thermal properties such as emissivity and thermal inertia in the field (e.g. Marsh et al. 1982; Vlcek 1982) may help during data interpretation. It is generally useful to monitor the first six variables at least for a few days prior to the actual survey (Ellyett and Pratt 1975; Bonn 1977).

The purpose of ground investigations could be to provide a reference base for interpretation or to verify some anomalous signatures. Where the area to be flown is large, a significant time may elapse between start and finish points, and in such situations, monitoring of meteorological data could help in reducing the data to a common base for comparative interpretation. Further, field checking of some anomalies may be necessary to control the interpretation.

9.3.3 Processing of Broad-Band TIR Images

The most commonly used TIR image is a simple radiant temperature image, which may belong to the night-time or the daytime pass. For some investigations, temperature difference [$\Delta T = (T_D - T_N)$] image, and 'apparent thermal inertia' image (a computed parameter) may be better used. Additionally, thermal image data can also be used in any desired combination in multisensor integrated studies.

For image processing and data integration, it is necessary to register thermal images over other images. Some practical problems occur in registering thermal images, particularly because the temperature of the ground surface is highly variable and depends on time and meteorological atmospheric parameters (which are dynamic factors!), and also on such factors as albedo, thermal and emissive properties, slope, topography, moisture and vegetation. Therefore, locating stable ground control points (GCPs) is a difficult exercise, as even fixed features on the ground (e.g. topography) may be displayed differently on different day and night thermal images (Watson et al. 1982b).

Table 9.2. Physical factors affecting thermal (radiant temperature) data. (After Ellyett and Pratt 1975)

Variable	Physical properties	Ground and atmospheric factors	
1. Emissivity	a. Composition b. Surface geometry	Type of rock, soil, vegetation etc. Surface configuration of ground objects	
2. Kinetic temperature	a. Physical/thermal properties of materials	– Rock, soil (composition) – Grain size and porosity – Moisture content	
	b. Heat budget factors	– Solar heating	Season Latitude Cloud cover Solar elevation Time of day Topography and aspect Albedo (and co-albedo) Atmospheric absorption
		– Longwave radiation and heat transfer at the Earth–atmosphere interface	Ground temperature Emissivity Atmospheric temperature Wind speed Humidity Sky temperature Cloud cover Rain Topographical elevation
		– Active thermal sources	Fumaroles, geysers Fire, thermal effluents Volcanoes, etc.

Even then, handling daytime thermal-IR data is relatively easy as the accompanying VNIR images are usually available; registering night-time thermal-IR data over other images requires much more careful effort.

Another difficulty in registering thermal images is the resolution aspect, as thermal imagery generally has a coarser spatial resolution as compared to VNIR imagery, and lacks fine details. These difficulties may lead to mis-location. However, water bodies, if present, often serve as good control points, as they are quite distinct on the TIR images (and also on other data sets). Image processing of thermal-IR images in particular is addressed by Schott (1989).

9.3.4 Interpretation of Thermal-IR Imagery

As is obvious from the foregoing discussion, the radiant temperature (T_R) is dependent on two main factors: emissivity and kinetic temperature of the surface. The various physical–environmental factors to be considered during interpretation of radiant temperature image are summarized in Table 9.2.

A dedicated thermal remote sensing experiment typically uses a set of two passes – one pre-dawn (night) and one day (noon) pass. Qualitative image interpretation is based on the usual elements of photo interpretation (see Sect. 7.3.1). Here we discuss some typical features commonly seen on radiant temperature images.

1. Topography. Topographical features are enhanced on daytime thermal images due to differential heating and shadowing (Fig. 9.9b). The hill slopes facing the Sun receive more solar energy than those sloping away from it, and some of the hill slopes may lie in shadows, owing to rugged topography. These effects lead to local differences in thermal energy budgets and consequent differences in surface temperatures. However, on night-time images, the topography becomes subdued (Fig. 9.9c). Elevation is another variable influencing ground temperatures. The temperature elapses with elevation, the common lapse rate being 6.5°C per 1000 m. This effect may be more manifest on satellite TIR images covering large mountainous areas. In brief, therefore, it is always advisable to use topographical data conjunctively when interpreting thermal-IR data.

2. Wind and cloud cover. Wind trails can be seen on thermal-IR images of good spatial resolution. Wind causes dissipation of surface heat. Objects such as shrubs, boulders etc. act as barriers to wind and lead to formation of shadow trails with relatively higher surface temperature. Wind effects can thus be seen as alternating bright and dark parallel-curved lines. Cloud cover leads to differential heating and shadowing and hence a patchy bright (warm) and dark (cool) appearance on the image. Scattered non-uniform precipitation leads to unsystematic moisture levels in soils and also results in a mottled appearance on the image.

3. Land surface (rocks and soils). The land surface gets heated during the day and cools at night, thus showing temperature variation in a diurnal cycle (Fig. 9.8).

4. Standing water. Relative to land, standing water appears brighter (relatively warmer) on the night-time image and darker (cooler) on the daytime IR image (Figs. 9.8 and 9.10). Although thermal inertia values of rocks and water are nearly equal, the unique thermal pattern of water is related to convection, circulation and evaporative cooling over water bodies. In the daytime, as the temperature of the surface water starts rising, evaporation takes places and becomes stronger with increasing temperature. Due to evaporation, energy is transported from the water to the air and the water appears cooler. At night, as cooling of the surface water proceeds, convection brings warmer water from depth to the surface, decreasing the net drop in temperature of surface water, and the water appears warmer.

5. Damp terrain. Moisture content present in soils directly affects the thermal inertia values, and therefore influences the thermal pattern in a day-and-night cycle. A damp terrain has a very different thermal response to either standing water or land. The moisture present in materials leads to evaporative cooling and therefore the temperature of damp terrain is generally quite low and the thermal contrast in the day and night images of damp terrain is also less.

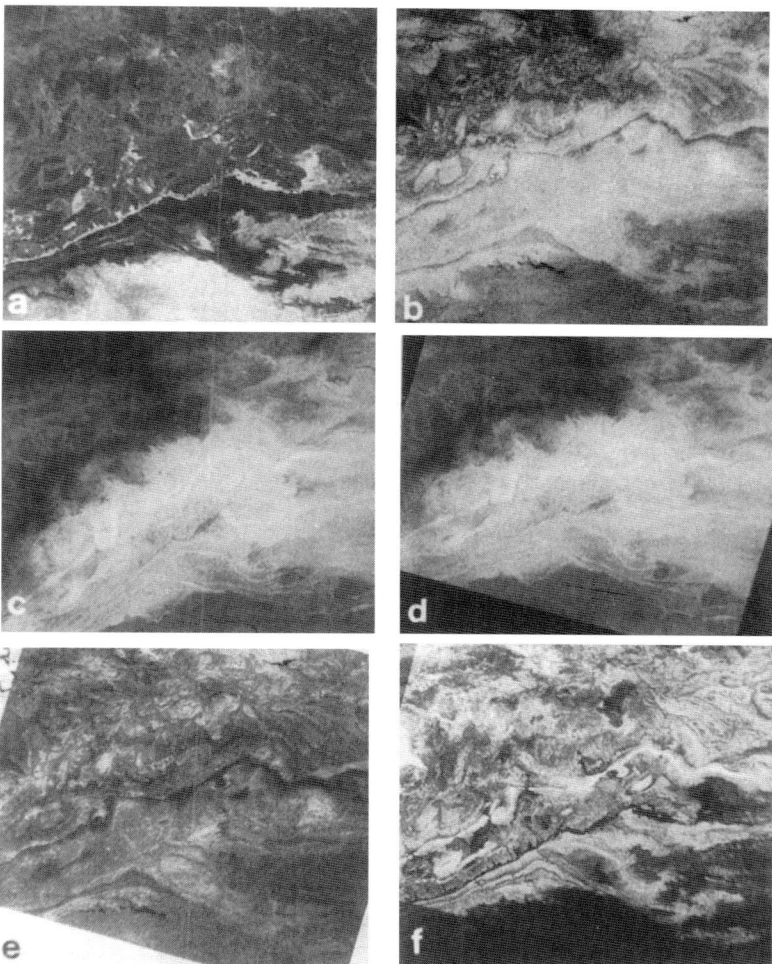

Fig. 9.9a–f. A set of HCMM images of the Atlas Mts. **a** Visible channel; note the higher spatial resolution and many geological features on the image; also seen are scanty clouds in the southern part. **b** Day thermal-IR image; topographical effects are enhanced by solar heating; water body in the NE, clouds in the S and hilly area (NW) appear cooler (darker). **c** Night thermal-IR image; note the difference in geometry as compared to **b**; topographical effects are subdued; the NW area being higher is, cooler; some geological details are seen on the night IR image but not on the day IR image. **d** Night TIR image after registration in geometric conformity to **b**. **e** Temperature-difference image (ΔT = day TIR − night TIR). **f** ATI image [$(1-A)/\Delta T$]; water has higher ATI and therefore appears bright; day clouds and snow are also bright; the wet sand in the SE part appears light grey; many geological details can be picked up. (**a–f** Courtesy of R. Haydn)

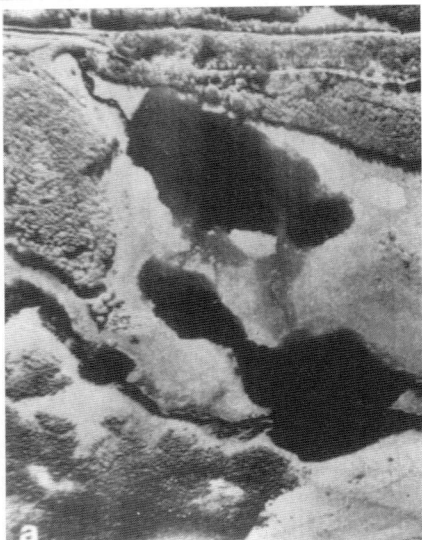 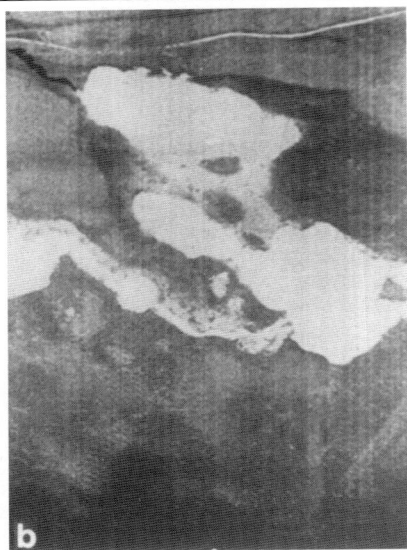

Fig. 9.10. a Daytime aerial photograph and **b** night-time TIR (8–14 µm region) aerial scanner imagery of the Oster Lakes, Bavaria, Germany. In the night, water body is warm and vegetation cooler. (Courtesy of DLR Oberpfaffenhofen)

6. Metallic objects. These have low emissivity (as they have high reflectivity = low absorptivity = low emissivity), and therefore exhibit low radiant temperatures and always appear dark (cool) on TIR images, day and night (Fig. 9.8).

7. Vegetation. In general, vegetation is warmer on the night-time and cooler on the daytime image, in comparison to the adjoining unvegetated land (Figs. 9.8 and 9.10). The lower daytime temperature is related to the transpiration process in the plants and the higher night-time temperature to the higher moisture content in leaves. Dry vegetation lying on the ground insulates the ground from the atmosphere and causes warmer night-time and cooler daytime responses. Extensive and thick vegetation cover acts as a barrier to geological mapping on TIR channels.

9.3.5 Thermal Inertia Mapping

Thermal inertia (P) is an intrinsic property of materials and is the single most important thermal property of materials, governing the diurnal variation in temperature of objects on the Earth's surface (Sect. 9.2.1.3). Materials with lower thermal inertia exhibit a greater range of temperature change in the diurnal cycle. Therefore, thermal inertia mapping can help discriminate various types of soils, rocks etc. on the Earth's surface. Thermal inertia values computed from remote sensing methods are only approximations of the actual thermal inertia (TI) values and are therefore called the *Apparent Thermal Inertia* (ATI) values.

9.3.5.1 Methods for Computing ATI

As defined earlier, thermal inertia is a measure of the resistance offered by an object to change in temperature. Conceptually, if the same amount of heat energy (Q) is given to the same quantity of materials with differing thermal inertia (TI), then the change in temperature (ΔT) will be different for different materials, such that:

$$\text{TI} \propto (1/\Delta T) \tag{9.6}$$

K. Watson (1971, 1973, 1975), a pioneer worker in the field of thermal-IR sensing, developed a simple thermal model assuming that the Sun causes periodic heating of the Earth's surface and that the ground losses of heat are only by radiative transfer. The solar energy incident on the Earth's surface is partly reflected and partly absorbed. If the solar flux S is incident on a surface of albedo A, then

$$\underset{\text{solar incident flux}}{S} = \underset{\text{reflected energy}}{A \times S} + \underset{\text{absorbed energy}}{(1-A) \times S} \tag{9.7}$$

where (1–A), called co-albedo, is responsible for the amount of heat absorbed [(1–A) × S], which causes a rise in surface temperature. However, the co-albedo varies spatially within a scene. In order to obtain an estimate of ATI, it is necessary to normalize for the variation in input heat energy [(1–A) × S]. Therefore

$$\begin{aligned}\text{ATI} &\propto \frac{1}{(\Delta T/1-A)\cdot S} \propto \frac{1}{S}\cdot\frac{(1-A)}{\Delta T} \\ &= N\cdot\frac{(1-A)}{\Delta T}\end{aligned} \tag{9.8}$$

where N is a scaling constant as solar flux can be considered to be uniform in a scene.

In order to compute an ATI image, three input images are required: broad-band panchromatic (VNIR), daytime TIR and night-time TIR images (Fig. 9.11). Pixel by pixel, albedo can be obtained from a VNIR image, and diurnal temperature variation (ΔT) can be computed from daytime (T_D) and night-time (T_N) coverages. In this way, apparent thermal inertia can be computed [Eq. (9.8)].

Pohn et al. (1974) used Watson's simple model and generated a family of curves giving a relationship between ΔT, ATI and A. Measuring ΔT and A from Nimbus data, they generated the first-ever ATI contour map for a part of Oman in the Arbian peninsula. Subsequent improvements upon the original Watson model were suggested to include the regional atmospheric parameters (Watson and Hummer-Miller 1981).

A generalized formulation to calculate ATI from global remote sensing data was given by Price (1977), which included various parameters such as diurnal temperature difference, albedo, solar constant, atmospheric transmittance at visi-

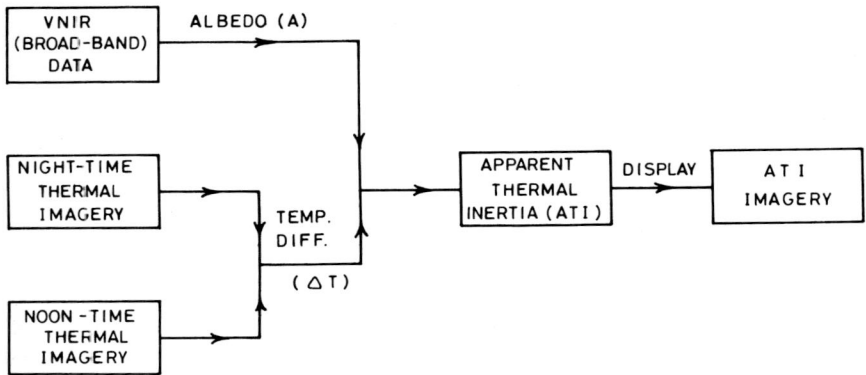

Fig. 9.11. Schematic for ATI calculation (based on Watson's simple periodic heating model)

ble wavelengths, the angular speed of the Earth's rotation, the ratio of the heat flux density transferred by the surface to the air to that transferred into the ground, and a parameter to account for solar declination and latitude. This was used in the HCMM (Heat Capacity Mapping Mission) programme in a simplified form as (Short and Stuart 1982)

$$ATI = N.C.(1-A)/\Delta T \qquad (9.9)$$

where
N = a scaling factor (= 1000)
C = a constant to normalize for solar flux variations with latitude and solar declination
A = apparent solar albedo
ΔT = diurnal temperature difference ($T_D - T_N$).

9.3.5.2 ATI Image Interpretation

Topography and solar illumination are very important factors in thermal-IR sensing. For example, poorly illuminated slopes may be computed to have erroneously high ATI! Therefore, topography/slope effects need proper correction and consideration during TIR data processing and interpretation.

Soil and alluvium commonly appear dark on the ATI image as they possess low ATI, due to lower density and thermal conductivity values. *Forests* are generally light-toned, as they have high ATI due to low albedo (A) and relatively small

ΔT, owing to moisture content and evapo-transpiration. *Agricultural areas* are depicted in varying shades of gray. *Water* and *snow* both possess very light tones on the ATI image (Fig. 9.19f), but for different reasons. Water has a light tone (high ATI) because the albedo is very low, and *snow* has it because although the albedo is high, the ΔT is very small. *Rock materials* are expressed in various shades of gray depending on their albedo and density. *Day clouds* are white (Fig. 9.9f) and *night clouds* appear black.

An ATI image and its corresponding ΔT image have, in general, an inverse mutual relation (Fig. 9.9e, f), owing to the fact that albedo values of many objects are similar, and ΔT appears in the denominator for calculating ATI. Therefore, higher values of ΔT would generally correspond to lower values of ATI and vice versa, although on detailed examination many exceptions may be found.

9.3.6 Scope for Geological Applications – Broad-Band Thermal Sensing

9.3.6.1 Geomorphology

Space-acquired night-time thermal images provide a good impression of regional landforms, owing to the fact that differences in night-time temperatures are related to elevation, soil moisture and vegetation. On the other hand, daytime thermal images give more information about topography and relief.

9.3.6.2 Structural Mapping

Thermal-IR images can be extremely useful in delineating structural features. Structural features such as folds, faults etc. may be manifested due to spatial differences in thermal characteristics of rocks. Bedding and foliation planes appear as sub-parallel linear features due to thermal contrasts of compositional layering. There are examples where structures have first been detected on the TIR image, not identified at all on the aerial photographs, and only subsequently located in the field (Sabins 1969). Faults and lineaments may be associated with springs or they may promote movement of groundwater to shallow depth; this would lead to evaporative cooling along a line or zone, producing a linear feature (Fig. 9.12). At times, daytime thermal images, on which topographical effects are enhanced, may also be useful in locating structural features.

9.3.6.3 Lithological Mapping

The spatial resolution of sensors is an important aspect in lithological mapping. The present-day space-borne TIR scanners provide rather coarse spatial resolution data. However, the viability of the technique for lithological discrimination has been shown by higher-spatial-resolution aerial scanner data. Both radiant tempera-

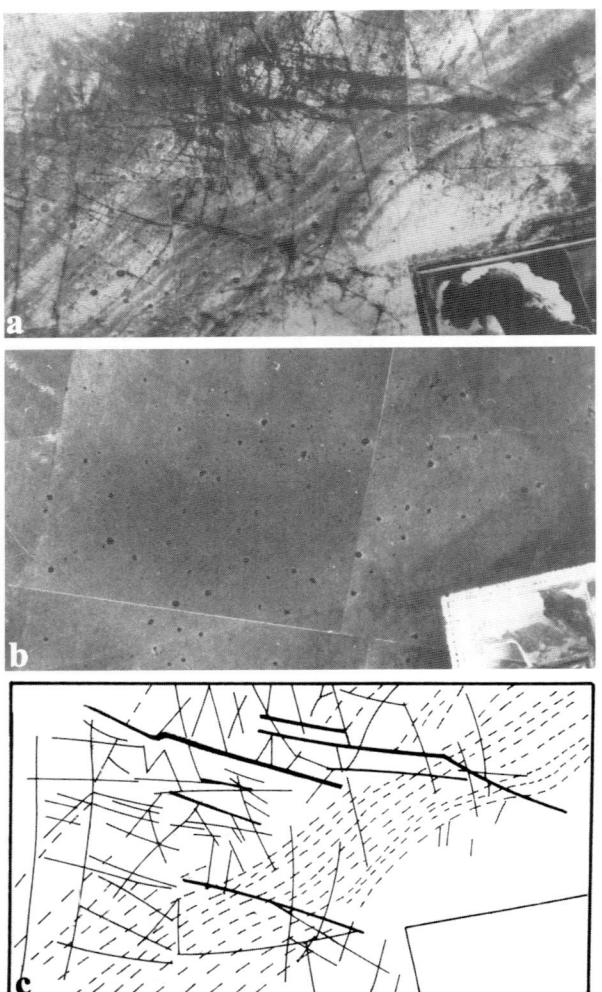

Fig. 9.12. a TIR image showing numerours structural features such as bedding and lineaments; **b** the corresponding aerial photograph, which is nearly featureless; **c** interpretation map based on the TIR image (Stilfonstein area, Transvaal, South Africa). (**a,b,** from Warwick et al. 1979)

ture images and ATI images (particularly the latter) can be used for lithological discrimination (see e.g. Kahle et al. 1976; Short and Stuart 1982: Watson 1982a; Abrams et al. 1984).

(a) Relevance of ATI in hard-rock areas. Discrimination on the basis of ATI appears to be easiest in the case of sedimentary rocks – orthoquartzites, dolomites, sandstones, limestones and shales have successively lower ATI (Fig. 9.13a).

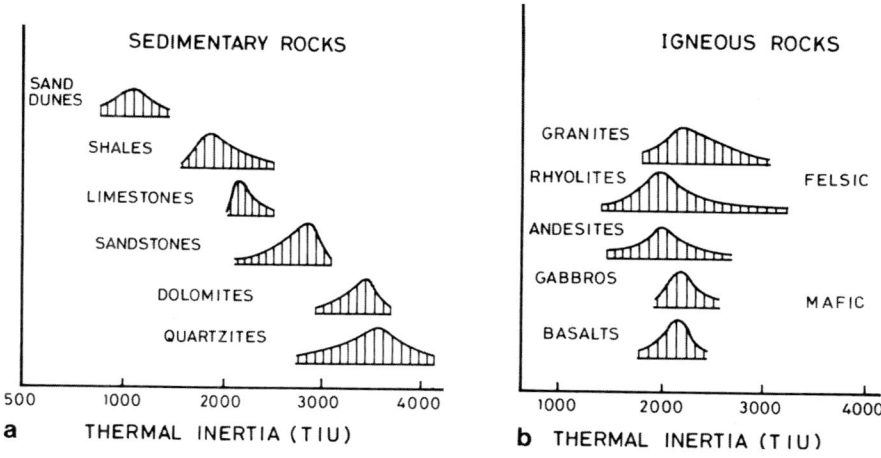

Fig. 9.13. ATI values for selected **a** sedimentary and **b** igneous rocks (plots vertically separated for clarity; compiled after Miller and Watson 1977; Short and Stuart 1982; Watson et al. 1982a). Note that the rocks with higher density and higher silica content possess higher ATI

This may primarily be due to the wide range of porosity and bulk density in the sedimentary rocks. For example, in an area of sedimentary sequence comprising shale, siltstone and sandstone, on the pre-dawn thermal image, sandstone appears relatively cool (high ATI), siltstone warmer, and shale still warmer (low ATI). Further, due to higher conductivity and specific heat, dolomites generally have higher ATI than limestones. Therefore, limestone and dolomite, which are hard to distinguish in the VNIR region, may be separated from each other by subtle temperature differences.

Distinction among igneous rocks on the basis of ATI values is more difficult, as compared to the sedimentary rocks, although smaller differences in ATI values of igneous rocks such as rhyolites, andesites, basalts and granites have been reported (Fig. 9.13b).

(b) Relevance of ATI in weathered areas. After weathering, different rocks would yield soil covers differing in bulk density, porosity and water saturation, which will lead to differences in ATI values. In this way, ATI images can help discriminate between weathered products derived from different parent rocks.

As many materials have overlapping ranges of ATI values, the ATI value may not be diagnostic for identification; however, it can be a useful data to complement the VNIR data for improved lithological discrimination.

It was mentioned earlier that the diurnal heat waves penetrate to a depth of nearly 1-m below the ground surface. The material in the top 10-cm zone is the most important region in thermal-IR remote sensing (Watson 1975; Byrne and Davis 1980). The top encrustation–oxidation soil–lichen layer, which is usually less than 1 mm thick, may have a negligible effect on the bulk thermal properties.

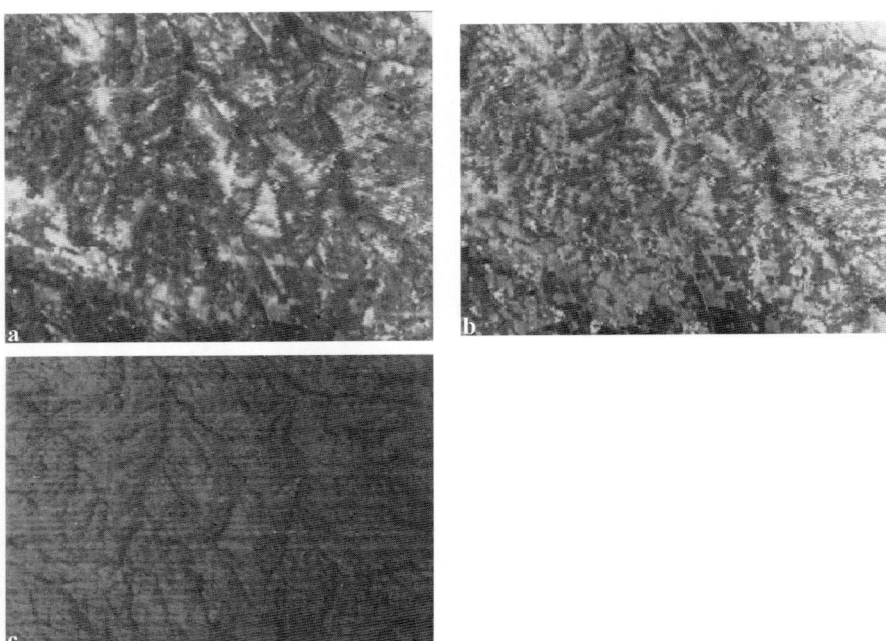

Fig. 9.14a–c. Aerial multispectral scanner images of a semi-arid region: **a** visible (red); **b** near-IR; and **c** daytime thermal-IR. In this area, the channel-ways are partly covered with vegetation and partly wet ground. On the TIR image, the drainage pattern is distinct, because wet ground and vegetation have near similar responses on the day TIR. On the other hand, as vegetation and wet ground have mutually contrasting responses on both VIS and near-IR channels, the drainage-ways fail to manifest clearly. (Courtesy of Space Applications Centre, Ahmedabad, ISRO)

This is in contrast to the case of solar reflection sensing, where the top few-microns-thick layer exerts a predominant influence over the spectral response. Therefore, thermal-IR image data are likely to yield better information on bedrock lithology than are the solar reflection data.

9.3.6.4 Hydrogeological Studies

The thermal-IR data are highly influenced by surface moisture, and this characteristic can be applied for hydrogeological investigations in a number of ways. For example, for (a) soil moisture estimation, (b) exploration of shallow aquifers or water-bearing fractures, (c) detection of seepage of water from irrigation canals, and (d) landslide studies, where mapping of moisture zones is important. Figure 9.14 gives an interesting example of the detection of drainage pattern on the TIR, whereas the same features are not observed clearly on the VIS and near-IR images.

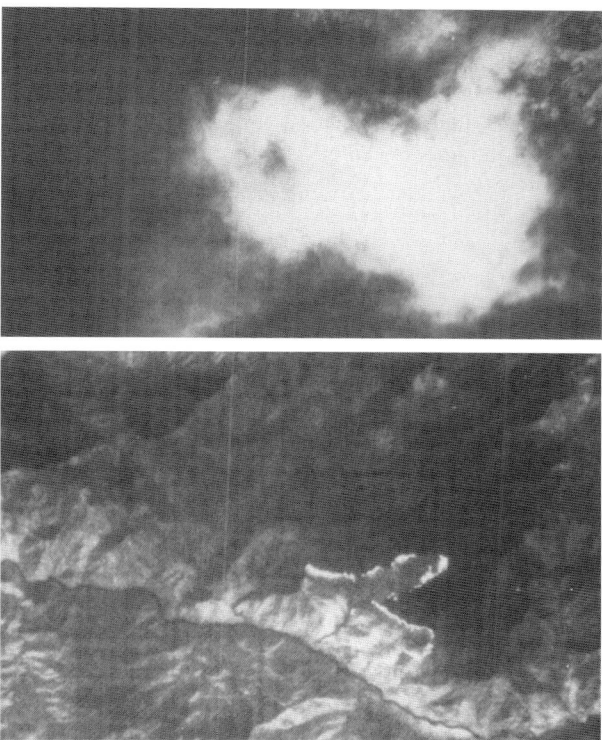

Fig. 9.15a,b. Penetration of smoke by thermal-IR radiation. **a** Visible band image; note that the fire front of the forest fire is engulfed by the smoke plume and is not observable. **b** Corresponding thermal-IR image showing the fire front. (Courtesy of NASA Ames Research Centre, in Sabins 1997)

9.3.6.5 Penetration of Smoke

For fighting forest fires, the essential information required is the exact location of the fire front. The smoke plume emanating from the fire may engulf the fire front completely. As visible/near-IR radiation is scattered by tiny suspended particles of smoke plume, VNIR images would not be able to penetrate the smoke plume to locate fire front. On the other hand, thermal-IR radiation, being longer-wavelength radiation, can penetrate the fire smoke and reveal the location of the fire front (Fig. 9.15).

9.3.6.6 Monitoring of Volcanoes, Geothermal Fields and Coal Fires

This group of applications deals with high-temperature features. Thermal-IR sensing has specific applications in the above problems of geohazards, which are discussed in Sect. 16.11 and 16.12 with examples.

9.4 Temperature Estimation

For estimation of temperature by remote sensing, the radiation intensity emitted from the target (heat source) is used. Planck's radiation equation can then be applied to convert measured spectral radiance to kinetic temperature. The Planck's radiation equation for a blackbody is (c.f. Eq 2.6):

$$L_\lambda = \frac{2\pi hc^2}{\lambda^5} \left(\frac{1}{\pi \left(e^{hc/\lambda kT} - 1 \right)} \right) \qquad (9.10)$$

where λ is wavelength (m), L_λ is spectral radiance, h is Planck's constant, k is Boltzmann's constant, T is temperature in kelvin, and c is speed of light. This can be rewritten as:

$$T = \frac{C_2}{\lambda \ln\left[\left(C_1 \lambda^{-5}/\pi L_\lambda\right)+1\right]} \qquad (9.11)$$

where, $C_1 = 2\pi hc^2 = 3.742 \times 10^{-16}$ W m^{-2}, and $C_2 = hc/k = 0.0144$ mK. This suggests that temperature (T) of a blackbody can be estimated from spectral radiance (L_λ) at a particular wavelength.

9.4.1 Use of Landsat TM Data for Temperature Estimation:

In a large number of investigations world wide, the Landsat TM data have been used for thermal studies. The spectral bands of Landsat TM together with their sensitivity limits are presented in figure 9.16. The thermal-IR band TM6 (10.4-12.5 µm) is sensitive from sub-zero to 68°C temperature range; therefore, it is useful for studying normal range (5° - 50°C) of thermal phenomena on the Earth's surface. The SWIR bands TM7 (2.08 – 2.35 µm) and TM5 (1.55 – 1.75 µm) together have the capability to measure temperatures integrated over the pixel in 160° – 420°C range. Hence, these bands are used for studying high temperature surface phenomena like volcanic vents and surface fires. Reviews and examples of estimation of surface temperature using Landsat TM data have been given by several workers (e.g. Rothery et al. 1988; Oppenheimer et al. 1993; Becker & Li 1995; Prata 1995; Prakash and Gupta 1999).

Temperature Estimation

Radiance from TM DN values can be computed from the general equation:

$$\text{Radiance} = \text{offset} + \text{gain} * \text{DN} \tag{9.12}$$

This can be rewritten as (Markham and Barker 1986)

$$L_\lambda = L_{min(\lambda)} + \frac{L_{max(\lambda)} - L_{min(\lambda)}}{Q_{calmax}} Q_{cal} \tag{9.13}$$

where L_λ is spectral radiance received by the sensor for the pixel in question, $L_{min(\lambda)}$ is minimum detected spectral radiance for the scene (i.e. L_λ at DN = 0), $L_{max(\lambda)}$ is maximum detected spectral radiance for the scene (i.e. L_λ at DN = 255), Q_{calmax} is maximum grey level (255), and Q_{cal} is the grey level for the analysed pixel.

The values of $L_{min(\lambda)}$ and $L_{max(\lambda)}$ are different for different bands of the satellite sensor and are subject to drift with time; these calibrations are provided from time to time by the agencies operating the satellite (NASA/USGS). A set of sample values of Landsat-4 TM bands 5, 6 and 7 are given in Table 9.3. Using Eq. (9.9) and applying the values of $L_{min(\lambda)}$ and $L_{max(\lambda)}$, any DN value can be converted into spectral radiance for the pixel.

Corrections to Spectral Radiance Values

The values of spectral radiance at the satellite sensor calculated as above may carry three main radiance components: (1) thermal emission from the ground element, (2) a component of atmospheric contribution, and (3) a component of solar reflection. Relative amounts of these components vary with wavelength, conditions and time of survey. It is important to carry out corrections to arrive at realistic estimates of the ground temperatures.

Estimation of the thermal emission from the ground is the aim of all corrections and the crux of the problem; correct estimates of this component would lead to

Table 9.3 Sample calibration values of spectral radiance for selected bands of Landsat-4 TM and Landsat-7 ETM+ used in temperature computation

	Landsat-4 TM[a]		Landsat-7 ETM+[b]			
			Low gain		High gain	
	L_{min}	L_{max}	L_{min}	L_{max}	L_{min}	L_{max}
Band 5	0.037	2.719	−1.0	47.57	−1.0	31.06
Band 6	0.1238	1.56	0.0	17.04	3.2	12.65
Band 7	0.015	1.438	−0.35	16.54	−0.35	10.80

[a] After Markham and Barker (1986)
[b] Source: htpp://ltpwww.gsfc.nasa.gov/IAS/handbook/handbook.htmls/chapter11/
Note: Applicability of this sample calibration data may be verified before use.

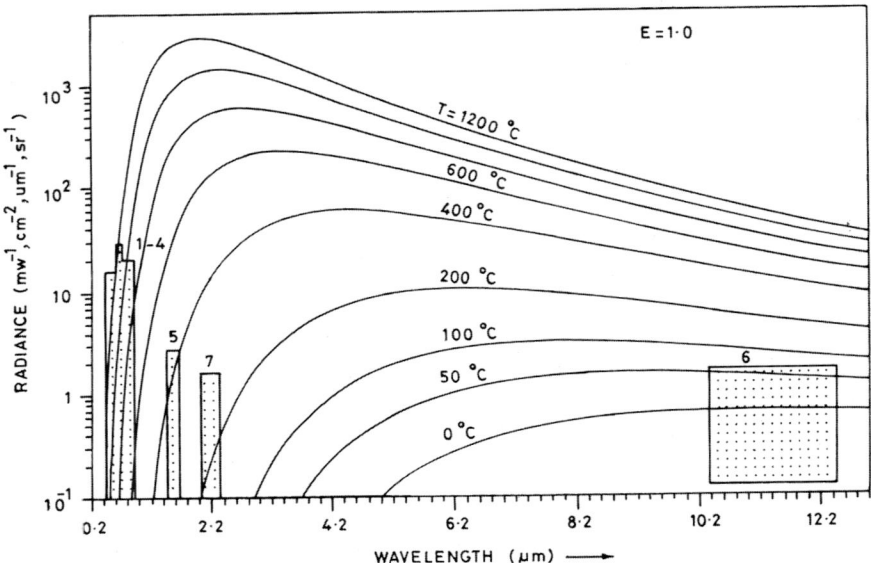

Fig. 9.16. Wavelength dependence of thermal radiance of blackbody (E = 1.0) plotted for a range of temperatures (0 to 1200°C); shaded boxes indicate operational range of radiance for Landsat TM bands 1 to 7. (Markham and Barker 1986)

better estimates of ground temperatures. Atmospheric correction of thermal data is discussed by a few workers (e.g. Price 1983; Li and Mcdonnell, 1988). Various models such as LOWTRAN7 and MODTRAN may be applied to provide estimates of the atmospheric contribution, which is subtracted from the at-sensor spectral radiance to obtain values of ground radiance. However, when working in atmospheric windows, the effect of the atmospheric contribution may be rather low. For example, Mansor et al. (1994) investigated the effect of atmospheric factors and found that there was no significant improvement in the strength of the TM thermal anomaly in the Jharia coal field after LOWTRAN7 processing. Thus, the influence of the atmospheric contribution in the TM data may be relatively minor, particularly for a comparative assessment of thermal anomalies and temperatures across an area (Prakash and Gupta 1999).

The solar reflection component is quite negligible in thermal-IR band TM6 (not to be confused with the effect of solar differential heating!). However, it is necessary to consider this factor for TM5 and TM7. Precise estimation of the solar reflected component is a difficult task. One empirical method is to compute the background solar reflected radiation by taking the average DN values of pixels adjoining the thermally radiant pixels; the effect of reflected solar radiation can be removed by subtracting the average background value from the anomalous TM7 and TM5 DN values (Prakash and Gupta 1999).

Estimating Radiant and Kinetic Temperatures

Once the corrected spectral radiance value (L_λ) for a pixel is known, it can be substituted in Eq. (9.7) to compute the temperature value. This temperature estimated from L_λ is the radiant temperature (T_R) as the effect of non-blackbodiness (spectral emissivity) is still to be taken into consideration. From the radiant temperature (T_R), the kinetic temperature (T_K) can be calculated using Eq. (9.5), as above. Commonly a spectral emissivity value of about 0.90–0.96 can be assumed for most terrain materials.

Utilizing the above concept and procedure, the temperature of various features and objects can be estimated from the Landsat TM data (for examples, see Sect. 16.11 & 16.12). The temperature so obtained represents the overall temperature of the pixel, and is called the *pixel-integrated temperature*. This can approximate the true surface temperature only in the case of surfaces where temperatures are uniform over large areas.

Sub-Pixel Temperature Estimation

In many cases, pixels consist of thermally mixed objects. For example, in the case of an active lava flow or surface fire, the source pixel is likely to be made up of two thermally distinct surface components (Fig. 9.17): (1) a hot component such as molten lava or fire with temperature T_h, occupying a portion 'p' of the pixel, and (2) a cool component (background area) with temperature T_c, which will occupy the remaining portion of the pixel (1–p). If the same thermally radiant pixel is concurrently sensed in two channels of the sensor, then we have two simultaneous equations

$$L_i = p\,L_i(T_h) + (1-p)\,L_i(T_c) \qquad (9.14)$$

$$L_j = p\,L_j(T_h) + (1-p)\,L_j(T_c) \qquad (9.15)$$

where L_i and L_j are the at-satellite spectral radiances in channels i and j, p is the portion of the pixel occupied by the hot source, $L_i(T_h)$ and $L_j(T_h)$ are the spectral radiances for the hot source in channels i and j respectively, and $L_i(T_c)$ and $L_j(T_c)$ are the spectral radiances for the cool source in channels i and j. Using the dual-band method developed initially for AVHRR data by Matson and Dozier (1981), the temperature and size of these two sub-pixel heat sources can be calculated if any one of the three variables T_h, T_c or p is known. Using data from SWIR bands TM7 and TM5, the method has been adapted to estimate sub-pixel temperature and size of hot areas (e.g. volcanoes and surface fires, see Sects. 16.11 and 16.12).

9.4.2 Use of Landsat-7 ETM+ Data for Temperature Estimation

The ETM+ sensor has been described in Section 5.5.1. For conversion of ETM+ DN values into radiance values, the same general method as applicable for

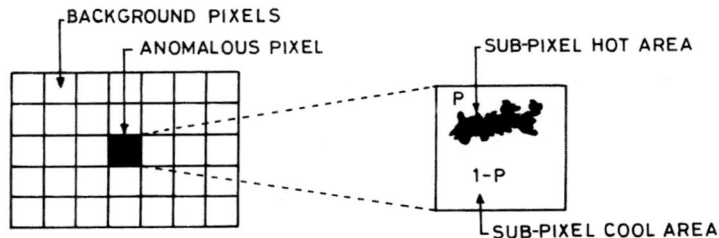

Fig. 9.17. Schematic representation of sub-pixel proportion of hot/cool areas

TM data can be used, the only difference being in the calibration values of the sensor. The ETM+ operates in two gain states: low gain and high gain. The $L_{min(\lambda)}$ and $L_{max(\lambda)}$ radiance values for ETM+ bands 5, 6 and 7 for the two gain states differ from each other. A set of calibration data (valid after 1 July 2000) is listed in Table 9.3. Using calibration data for the applicable gain state, DN values can be converted into radiance values and then into temperature values, as per the procedure described earlier.

For ETM+ thermal-IR band 6, radiance values can also be directly converted into temperature estimates by using the following simplified equation (web-site: htpp://ltpwww.gsfc.nasa.gov/IAS/handbook/handbook.htmls/chapter11/):

$$T = \frac{K_2}{\ln\left(\frac{K_1}{L_\lambda} + 1\right)} \quad (9.16)$$

where $K_1 = 666.09$ watts m^{-2} sr^{-1} µm^{-1} and $K_2 = 1282.71$ kelvin. This would give the radiant temperature, from which the kinetic temperature can be computed using Eq. (9.5) above.

9.5 Thermal-IR Multispectral Sensing

Thermal-IR multispectral sensing utilizes differences in spectral emissivity to discriminate between various minerals/mineral groups. The spectra of minerals and rocks in the thermal-IR region have been described earlier (Chapter 3). To briefly recapitulate:
1. The silicates, the most abundant group of minerals in crustal rocks, show prominent spectral bands in the TIR region. In fact, the Si-O stretching phenomenon in silicates dominates the 8–12 µm region. It is seen that the wavelength of the Si-O absorption band decreases from 11 µm to 9 µm in a uniform

succession for minerals with chain, sheet and framework structures (Fig. 3.10). This provides a distinct possibility for identifying minerals with these structures.
2. Hydroxides, carbonates, sulphates, phosphates and oxides are other important mineral groups frequently occurring in sedimentary and metamorphic rocks, and these also exhibit prominent spectral features in the TIR region (especially in the 8–14 µm region).
3. The mineral emission spectra are additive in nature, and therefore the spectra of rocks depend on the relative amounts of the various minerals present and the spectra of the individual minerals (Fig. 3.12).
4. The emission spectra are not significantly affected by surficial coatings (less than 1 mm thick) on mineral/rock surfaces.

Multispectral thermal-IR sensing exploits the presence of these subtle spectral emissivity differences in minerals. The technique has to use a multispectral sensor in order to pick up relative differences in emissivity at narrow spectral bands. Lyon and Patterson (1966) were the first to investigate the above idea of spectral emissivity differences in minerals and rocks for geological field investigations. Using field data from a spectrometer mounted on a mobile van, they found that the reststrahlen (low-emissivity) spectral band centre shifts to larger wavelengths and decreases in intensity with decreasing silica content (or corresponding increase in mafic minerals) in rocks (Fig. 9.18). Subsequent aerial surveys showed that the reststrahlen band of silicates was observable from aerial heights (Vincent and Thomson 1972; Vincent et al. 1972).

9.5.1 Multispectral Sensors in the TIR

Initially, Kahle and Rowan (1980) used six-channel data lying in the thermal-IR region from Bendix 24-channel aerial scanner data acquired over Utah. The study showed highly promising results for geologic discrimination. This led to the development of a dedicated aerial six-channel NASA-Daedulus Thermal-Infrared Multispectral Scanner (TIMS). The TIMS operates in six channels between 8–12 µm (Table 9.4) and has provided extremely interesting results (Kahle and Goetz 1983; Palluconi and Meeks 1985; Abbott 1990, 1991).

As far as a space-borne multispectral thermal-IR imaging system is concerned, ASTER is the only sensor to date. It operates in five spectral channels and has a spatial resolution of 90 m (Table 9.4) (see also Sect. 5.5.8).

The distribution of TIMS and ASTER channels in relation to the spectra of common rocks is shown in figure 9.18. It is obvious that the multispectral thermal-IR data from TIMS and ASTER have capabilities for mineralogical/litholigical discrimination and identification.

For ground truth, portable spectrometers are used to collect field data in the thermal-IR region (Hoover and Kahle 1987; Hook and Kahle 1996).

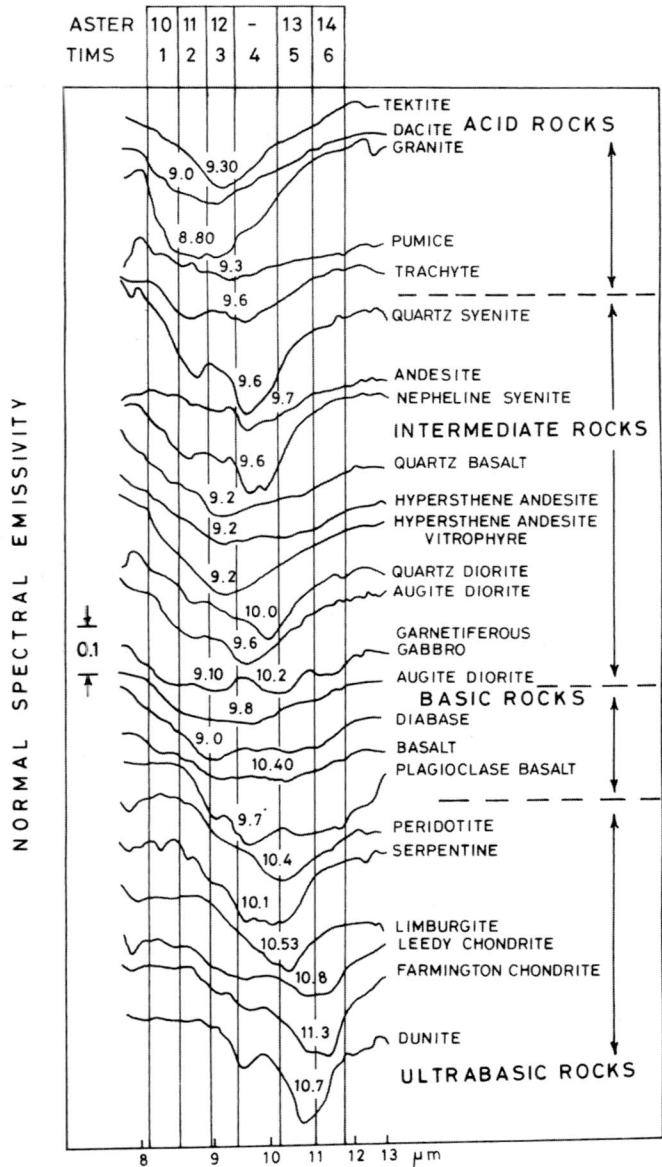

Fig. 9.18. Spectral normal emissivities of common terrestrial rocks. Spectra are arranged in decreasing SiO_2 content from acidic to basic types. *Numbers* refer to wavelengths of prominent minima. The positions of TIMS and ASTER channels (*vertical bars*) indicate that multispectral thermal sensing has tremendous potential in lithological mapping. (Spectral curves from Lyon and Patterson 1966)

9.5.2 Data Correction and Enhancement

The multispectral thermal-IR data require a special approach for data processing. The first step is calibration of the data. During scanning, the scanner sensors the blackbody, and these data are used for calibration. Then, the at-sensor radiance data needs to be corrected for atmospheric effects. Commonly this is accomplished by monitoring the atmospheric parameters and using a suitable model for atmospheric transmission and emission (e.g. MODTRAN model; Berk et al. 1989). This correction results in radiant temperature images.

After the above, the data are processed to derive spectral emissivity and surface kinetic temperature values, for which a number of approaches exist (Hook et al. 1992, 1999). A simple approach is the normalized emissivity approach (Gillespie 1986; Realmuto 1990). In this method, for a particular pixel, the highest of the set of temperatures in various channels is considered as the near-correct temperature of the pixel. The spectral emissivity in the chosen channel is assigned a value of 0.96, which represents a reasonable emissivity value for exposed rock surfaces.

This set of values (highest temperature and spectral emissivity value of 0.96) is used to compute the surface (kinetic) temperature of the pixel. From the kinetic temperature, spectral emissivity in each channel is computed using the corresponding radiant temperatures. In this way, spectral emissivity images are derived for various channels.

The various spectral emissivity images, and also the radiant temperature images in general, are very highly correlated. The contrast ratio in spectral emissivity is commonly less than 0.15, as compared to the contrast ratio of 0.5 or greater found in the VNIR data. Hence, the data need further processing for meaningful representation. A commonly used technique for enhancing correlated thermal-IR data is decorrelation stretch (Chapter 12). The decorrelated images can be colour-coded and displayed as a composite.

The spectral emissivity image data can also be subjected to principal component transformation (PCT). In such a case, the first PCA image (being a weighted average of the highly correlated input images) is almost entirely dependent on ground temperatures. In the noon hours, since ground temperatures are controlled

Table 9.4. Comparison of TIMS and ASTER (TIR channels)

TIMS		ASTER	
Channel	Range (µm)	Channel	Range (µm)
1	8.1– 8.5	10	8.1– 8.5
2	8.5– 8.9	11	8.5– 8.8
3	8.9– 9.3	12	8.9– 9.3
4	9.5–10.1	gap due to O_3 absorption band at 9.6	
5	10.2–10.9	13	10.2–10.9
6	11.2–11.6	14	10.9–11.6

Notes: 1. Spatial resolution of TIMS is 18 m and that of ASTER is 90 m.
        2. NEΔT of TIMS is 0.1 K and that of ASTER is 0.3 K.

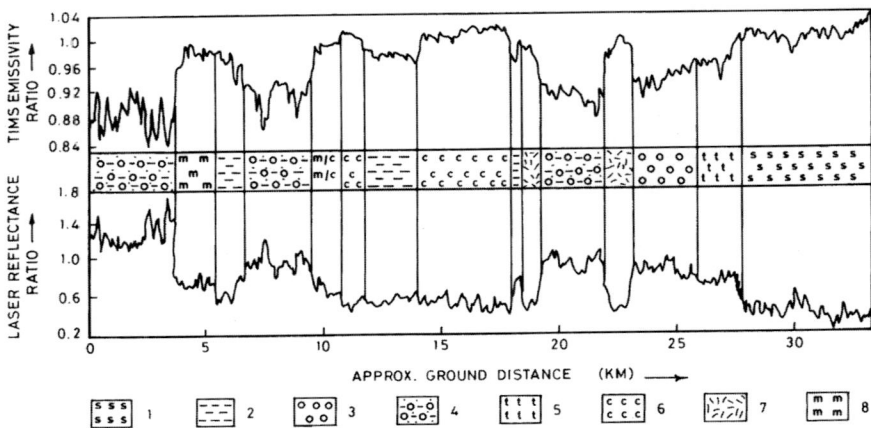

Fig. 9.19. Upper part of the figure shows TIMS emissivity ratio for bands 3/5 (8.9–9.3-μm waveband/10.2–10.9-μm waveband) for various surface materials in a profile in Death Valley, California. The lower part shows laser reflectivity ratio for wavelengths 9.23-μm/10.27-μm for the same profile for comparison. The two ratios have an inverse relation. *1* saline playa and lake sediments; *2* argillaceous sedimentary rocks; *3* fan gravels mixed (Blackwater, Tucki and Trail sources); *4* fan gravels of Blackwater and Train canyon; *5* fan gravels of Tucki wash; *6* Carbonate rocks and fans (predominantly dolomites); *7* basaltic lava; *8* breccia rocks and fans of mixed composition. (After Kahle et al. 1984b)

primarily by topography, the PC-1 image gives mainly topography. The lower-order principal components depend on spectral emissivity differences in different channels.

9.5.3 Applications

The multispectral thermal-IR imaging sensor TIMS has been flown over a number of test sites world-wide, yielding interesting geologic information. For example, application of TIMS data for lithological identification in Death Valley, California, is described by Gillespie et al. (1984) Using colour composites, various lithologic units could be differentiated and identified. Figure 9.19 shows the ratio of emissivity values in TIMS bands 3/5 in Death Valley, where a correlation with lithology is quite distinct.

TIMS data has been used for a number of other geologic studies. To mention just a few, for distinguishing between *pahoehoe* and *aa* lava flows in the Hawaii islands (Lockwood and Lipman 1987), for quantitative estimation of granitoid composition in Sierra Nevada (Sabine et al. 1994), for lithologic analysis of alkaline rock complex (Watson et al. 1996), and for mapping playa evaporite minerals and associated sediments (Crowley and Hook 1996). Some examples of TIMS image data as applied to lithologic mapping are presented in Sect. 16.4.5.

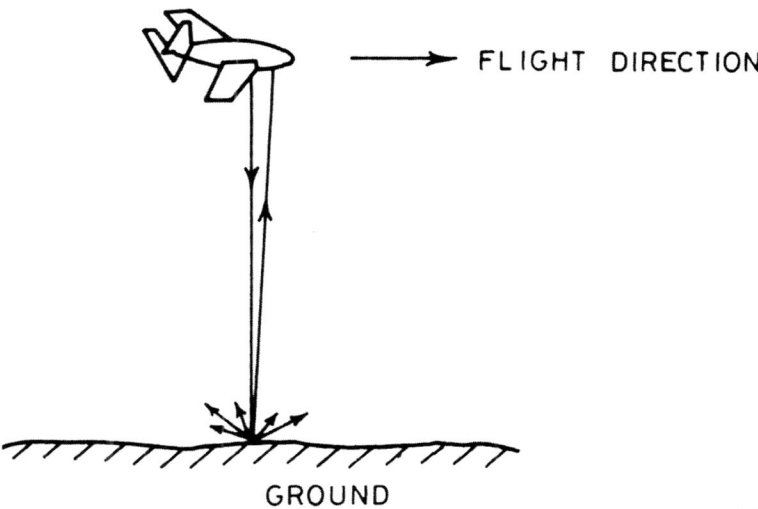

Fig. 9.20. Schematic arrangement of an active laser remote sensing system.

9.6 LIDAR Sensing

9.6.1. Working Principle

LIDAR (laser radar) sensing is an active sensor using an artificial laser beam to illuminate the ground. The laser rays, transmitted from the system on-board the remote sensing platform, impinge on the ground and are back-scattered (Fig. 9.20). The intensity of the back-scattered beam is sensed by the detector, amplified and recorded. The data can be used to discriminate ground objects. The technique uses differences in reflectivity in very fine, nearly monochromatic spectral bands. A number of laser wavelengths ranging from the visible to the thermal-infrared part of the EM spectrum are available.

Laser devices in the TIR region generally use a CO_2-laser transmitter and receiver system. Heterodyne detection techniques are employed to measure the laser back-scatter from the ground and to distinguish it from the background thermal radiation. The laser radiation is affected by atmospheric attenuation/absorption, for which appropriate corrections are needed. The LIDAR experiments have been flown in profiling mode from aerial platforms with a flying height of about 3 km, and an instrument foot size of about 3×15 m, the larger axis being oriented along the flight path.

9.6.2 Scope for Geological Applications

LIDAR sensing in the thermal-IR spectral region uses the presence of reststrahlen (high-reflectivity) bands in geologic material. With changes in the lattice structure and mineral composition, the high-reflectivity Si-O bands shift in position and intensity from 9 µm to 11 µm. Additionally, many other minerals, such as carbonates (at 11.3 µm), sulphate (at 10.2 µm and 9 µm), phosphates (at 9.25 µm) and aluminium-bearing clays (at 11 µm) also exhibit spectral features in the thermal-IR spectral region.

Kahle et al. (1984b) used a set of two laser wavelengths at 9.23-µm and 10.27-µm for a flight in Death Valley, California. Ground-track photography was used to locate the flight path and to compare it with ancillary data such as the geological map and other images. The lithological units covered included shale, limestone, quartzites and volcanics (rhyolitic tuffs and basalts).

The laser data were recorded on a strip chart and later digitized. A ratio plot of the intensity of the two lasers (9.23-µm/10.27-µm) was found to be very informative. It could discriminate between various rock types. To compare the reflectivity data with thermal emission data of the same area, Kahle et al. (1984b) registered the laser data over the TIMS image data, and found that the emissivity and reflectivity values are inversely related (Fig. 9.19). Cudahy et al. (1994) report similar work on using lasers for geological mapping in Australia.

9.7 Future

Future developments in thermal-IR sensing are expected in many directions, for example:
- understanding the emission spectra of natural surfaces and mixtures
- improved methods to incorporate environmental effects
- extending the use of the 3–5 µm region and the 17–25 µm region
- developing higher spatial resolution multispectral thermal-IR scanners from an orbital platform.

Chapter 10: Digital Image Processing of Multispectral Data

10.1 Introduction

10.1.1 What is Digital Imagery?

In the most general terms, a digital image is an array of numbers depicting spatial distribution of a certain field or parameter. It is a digital representation in the form of rows and columns, where each number in the array represents the relative value of the parameter at that point/over the unit area (Fig. 10.1). The parameter could be reflectivity of EM radiation, emissivity, temperature, or a parameter such as topographical elevation, geomagnetic field or even any other computed parameter. In this chapter, we deal with remote sensing multispectral images.

In a digital image, each point/unit area in the image is represented by an integer digital number (DN). The lowest intensity is assigned DN zero and the highest intensity the highest DN number, the various intermediate intensities receiving appropriate intermediate DNs. Thus, the intensities over a scene are converted into an array of numbers, where each number represents the relative value of the field over a unit area, which is called the picture element or pixel.

The range of DNs used in a digital image depends upon the number of bit data, the most common being the 8-bit type:

Bit number	Scale	DN-Range
7-bit or 2^7	128	0– 127
8-bit or 2^8	256	0– 255
9-bit or 2^9	512	0– 511
10-bit or 2^{10}	1024	0–1023
11-bit or 2^{11}	2048	0–2047
12-bit or 2^{12}	4096	0–4095

Advantages: There are several advantages of remote sensing data handling in digital mode (as compared to photographic mode), such as: (a) better quality of

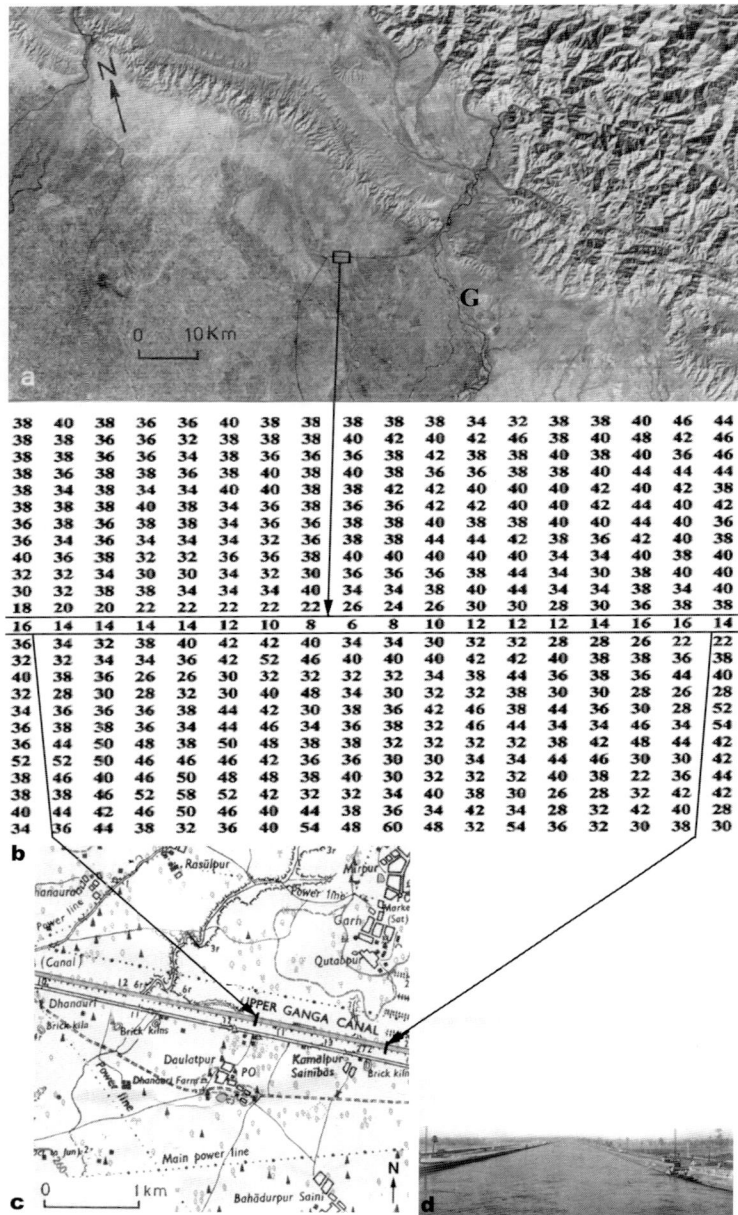

Fig. 10.1a–d. Structure of a digital image. **a** Landsat MSS4 image showing a part of the Himalayas and the Indo-Gangetic Plains (G = Ganga river). Note the prominent sharp canal oriented parallel to the scan lines in the boxed area. **b** DN output of the area marked in **a**. **c** Topographic map of the corresponding area. (Survey of India) **d** Field photograph

Introduction

data, as the entire DN-range (generally 256 levels) can be used, (b) no loss of information during reproduction and distribution, (c) greater ease of data storage, distribution and a higher shelf life, (d) amenability to digital image processing and (e) repeatability of results.

10.1.2 Sources of Multispectral Image Data

The various types of scanners (Chapter 5) used on aerial and space-borne platforms constitute the basic sources and provide the bulk of remote sensing multispectral digital imagery. In some cases, photographic products can also be scanned to generate digital image data. This could be particularly useful in cases where some of the input data for an investigation may be available only as photographic products, such as studies requiring use of old archival photographic records. In principle, this is simply the reverse of making analog images from scanner data (see Fig. 5.30). Common items of equipment for scanning photographic products include the microdensitometer and digitizing camera with a CCD array. However, as scanning of film involves an additional step at the data input stage, and every additional step means some loss of information, it is to be used only when necessary.

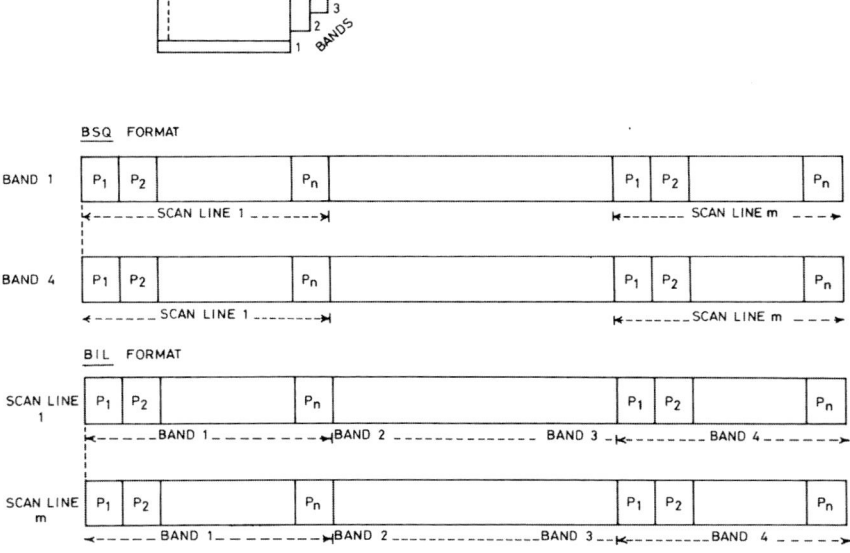

Fig. 10.2. Types of data formats: band sequential and band interleaved by line

10.1.3 Storage and Supply of Digital Image Data

A data disk is composed of tiny storage sites called bits. Each unit of information is called a byte. The alphanumeric characters are stored in various models such as ASCII or EBCDIC. The video data are commonly stored in binary mode, i.e. there are two alternatives for each bit – to remain unfilled or to get filled, also called no/yes, or 0/1.

Remote sensing data, as received from satellites by the data-receiving agency, is often first partially corrected, pre-processed and reformatted before it is supplied to users. A variety of media have been used. In 1970s–80s, computer-compatible tapes (CCTs) were largely used for this purpose. Flexible diskettes have been used for small sub-scenes. Now, with the rapid developments in electronics and requirement of large data sets, compact disks (CDs), optical disks, exabyte tapes and various types of cassettes with data storage capacity up to gigabyte values are used for distribution of remote sensing data.

A remote sensing data disk usually contains several files, such as header, ancillary, video data and trailer files. Each file may consist of several records. Header, ancillary and/or trailer files contain such information as location, date, type and condition of sensor, altitude, Sun angle and other data for calibration etc. The video data files contain image data, which may be arranged in a variety of ways, called formats. The two important formats used are band-sequential (BSQ) and band-interleaved by line (BIL) (Fig. 10.2). In the BSQ format, all the data for a single band covering the entire scene are written as one file, each line forming a record; in this way, there are as many video data files as there are spectral bands. In the BIL format, the video data for various bands is written line-wise (i.e. line 1 band 1, line 1 band 2, line 1 band 3, and so on). BIL is quite handy when multispectral data of a smaller sub-scene is to be picked out, and BSQ is particularly convenient when some of the spectral bands are to be skipped and only some are to be used.

On the basis of the degree of rectification carried out, remote sensing image data are broadly categorised as one of three types: (1) raw data, in which little correction has been carried out, (2) pre-processed data, in which some initial or basic corrections have been carried out, and (3) precision-processed data, in which rectification to a high degree has been accomplished and the data correspond to a certain standard map projection. Generally, raw data disks are not supplied to users, as the data contained therein is not ready for interpretation. The pre-processed type is the most commonly available product. Precision processing requires extra effort and has to be especially requested.

10.1.4 Image Processing Systems

The revolution in the field of computers during the last decade has made digital image processing widely accessible. The power of the workstations of yesterday is matched by the desktop PCs (Personal Computers) of today, and the change in

Introduction

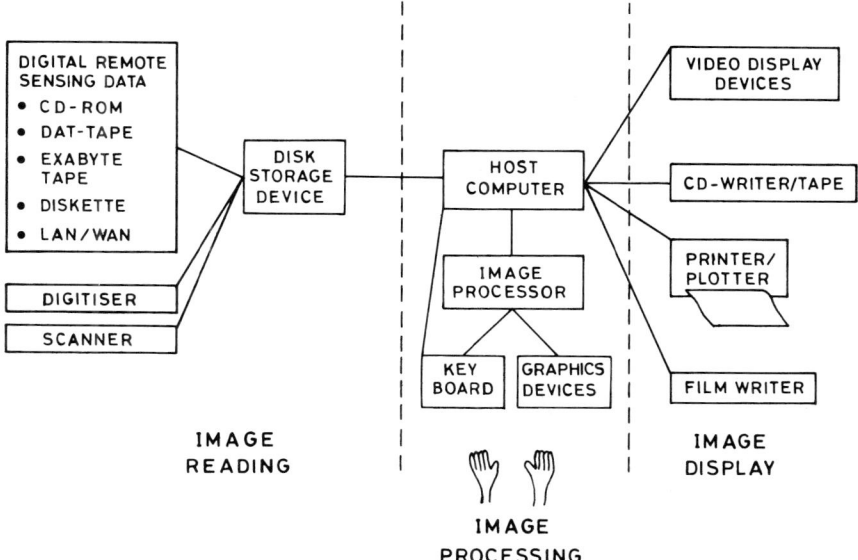

Fig. 10.3. Typical configuration of remote sensing image data processing system. (After Curran 1985)

scenario is a continuous on-going process. Therefore, it is best to talk in terms of only general system requirements, rather than capacity of computing systems.

An image processing system comprises hardware (physical computer components and peripherals) and software (computer programs). *Hardware* comprises the central processing unit (CPU), hard disk to store data, and random access memory (RAM) to store data and programs when the computer is working. Linked are a monitor, a keyboard and a mouse. Peripherals may include tape drives, audio and video systems, scanners, printers, photo-writers, CD-writers etc. (Fig. 10.3).

Software provides logical instructions that drive a computer. There are two basic kinds of software, called operating system software and applications software. *Operating system software* provides a basic platform for working and allows the running of applications software. *Applications software* performs applications of particular interest to the user. A list of selected image-processing software is given in Appendix 10.1.

A crucial component of the digital image processing system is the colour display device (monitor or cathode ray tube) which presents the results of image processing. The processor generates a signal driving the monitor. The display screen contains a phosphor layer. When high-energy electrons from the signal strike/excite the phosphor, visible light is emitted, the intensity of illuminated light being proportional to the current of the electron beam. Colour on the screen is displayed by using triads of phosphors: red, green and blue (the three primary additive colours), each one being independently excited by an electron beam. These are popularly called R, G, B electron guns.

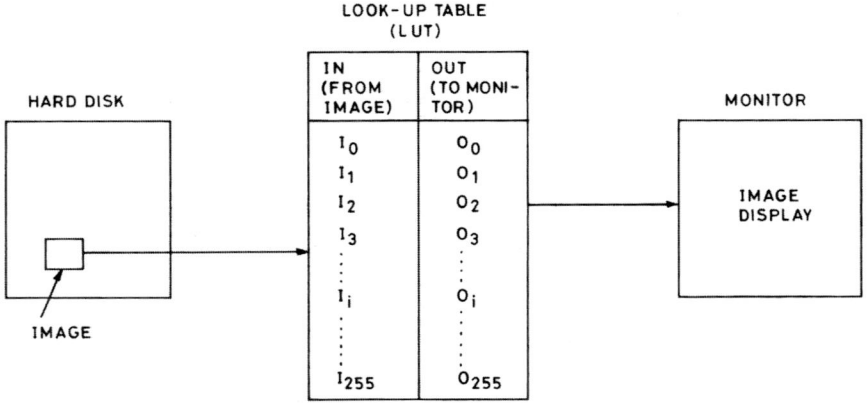

Fig. 10.4. Working concept for look-up table (LUT) of 8-bit data

Look-Up Tables (LUTs) provide data manipulation and display interactively, with a lot of flexibility. An image is displayed on the monitor through a set of three LUTs, one for each of the three R, G, B guns. A black-and-white image display would have identical LUTs for R, G, B guns. An LUT is used to maintain a specified selective relationship between the DN values in the input image and the gray levels displayed on the monitor (output) (Fig. 10.4). Digital image processing changes the LUTs, bringing about changes in the display accordingly, although the image data in the hard disk does not change, unless specifically saved. This allows flexibility in interactive working, by trial and error.

10.1.5 Techniques of Digital Image Processing

Digital image processing can be carried out for various purposes. Moik (1980), Hord (1982), Jensen (1986), Mather (1987, 1999), Gonzales and Woods (1992), Russ (1995), Schowengerdt (1997) and Richards and Jia (1999) have provided useful reviews on digital image processing of remote sensing data. Geological aspects of digital remote sensing analysis have been discussed by many, in particular Taranik (1978), Condit and Chavez (1979), Gillespie (1980), Drury (1993), Sabins (1997) and Harris et al. (1999).

Based on the objective and technique, digital image processing can be identified into various types. In this treatment, the discussion is divided under the following sub-headings.

1. Radiometric image correction – correction of the recorded digital image in respect of radiometric distortions.
2. Geometric image correction – deals with digital processing for systematic geometric distortions.

3. *Image registration* – superimposition of images taken by different sensors from different platforms and/or at different times, over one another, or onto a standard map projection.

4. *Image enhancement* – aims at contrast manipulation to enhance certain features of interest in the image.

5. *Image filtering* – extraction (and enhancement) of spatial-scale information from an image.

6. *Image transformation* – deals with processing of multiple band images to generate a computed (transformed) image.

7. *Colour enhancement* – use of colour space in single and multiple images for feature enhancement.

8. *Image fusion* – aims at combining two or more images by using a certain algorithm, to form a new image.

9. *2.5- dimensional visualization* – deals with the display of image-raster data as a surface from different perspectives.

10. *Image segmentation* – subdivides and describes the image by textural parameters.

11. *Classification* – classification of pixels of the scene into various thematic groups, based on spectral response characteristics.

Point vs. local operation. During digital processing of an image, old pixel values are modified and an image with new DN values is created, according to a certain scheme. In this respect, the various processing tasks can be classified into two types – point operations and local operations. In point operations, the DN value is

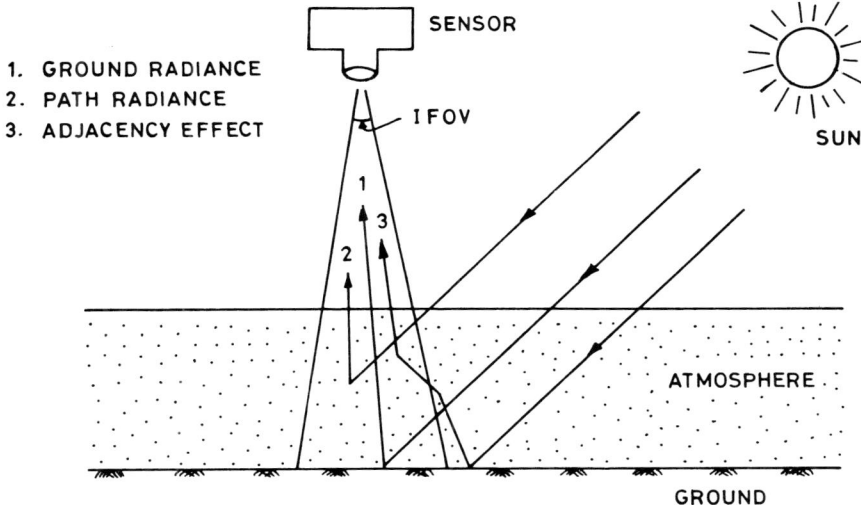

Fig. 10.5. Components of radiance reaching the sensor in the solar reflection region – ground radiance and path radiance; in addition, there could be some contribution due to the adjacency effect

modified at the particular pixel, irrespective of the surrounding DN values. In contrast to this, local operations modify the pixel values taking into consideration the adjoining DN values as well. Examples of these types of operations are presented at several places within this volume.

10.2 Radiometric Image Correction

The purpose of radiometric image correction, also sometimes called image restoration, is to rectify the recorded image data for various radiometric distortions.

10.2.1 Correction for Atmospheric Contribution

The effects of atmosphere on electromagnetic radiation were discussed in Sect. 2.3. The atmosphere leads to selective scattering, absorption and emission. The total radiance received at the sensor depends upon the ground radiance and the path radiance (Fig. 10.5) as

$$L_S = L_D \cdot A \cdot \tau_o + L_P \tag{10.1}$$

where L_S is the radiance received at the sensor
L_D is the total downwelling radiance
A is the albedo
τ_o is the atmospheric transmittance
L_P is the path radiance.

In the solar reflection region, scattering is the most dominant process leading to path radiance (see e.g. Turner et al. 1974; Dozier and Frew 1981). It is more pronounced in the shorter wavelength range (UV–blue), and may be quite negligible in the near-IR–SWIR region.

Path radiance affects remote sensing image data in two ways: (a) reduction in contrast, and (b) adjacency effect. Reduction in contrast takes place due to the masking effect, as a result of which dark objects appear less dark and bright objects appear less bright on the image. In addition, atmospheric scattering may direct some radiation from outside the sensor FOV towards the sensor aperture; this is called *adjacency effect* (Fig. 10.5). This leads to a decrease in spatial resolution of the sensor, and also reduced contrast.

Different methods and approaches are practised for atmospheric correction, which can be grouped under two categories: (1) dark-object subtraction method and (2) approaches utilizing atmospheric models.

1. Dark-object subtraction method. This is a simple empirical procedure, fast to implement. The method is based on the assumption that in every image there

Radiometric Image Correction

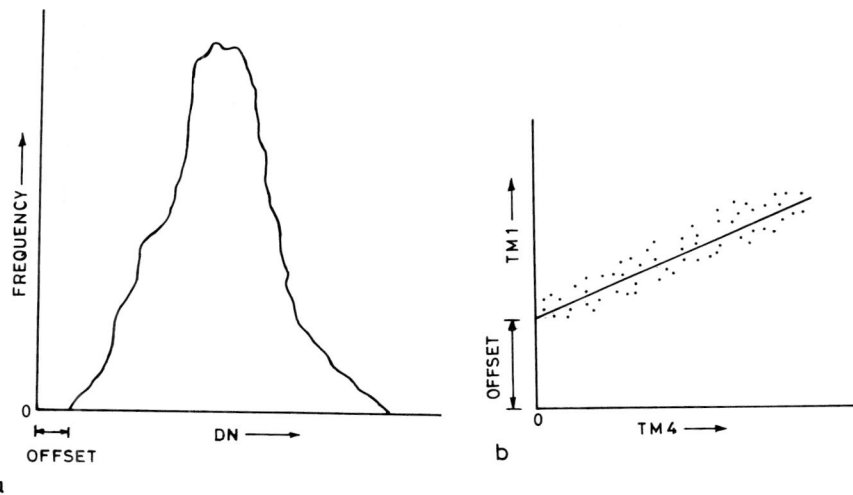

Fig. 10.6a,b. Correction for contribution due to atmospheric scattering by 'dark-object subtraction' method using **a** histogram and **b** scatterogram

should be at least a few completely dark pixels (0% reflectance); they may correspond to areas of deep, clear, open water bodies, deep shadows etc. Ideally, these pixels should have DN values of zero; however, because of the atmospheric haze, they record non-zero values. This characteristic is used in empirically estimating the haze component. In practice, the method involves first locating clear, open and deep water bodies or completely shadowed (dark) areas (where ground reflectances can be considered to be null and the DN values in infrared channels are also generally nearly zero). The minimum DN value in each channel over such pixels is taken as the path radiance and is subtracted from all other pixel values in the respective channel. Path radiance is very much reduced in the SWIR bands (Landsat TM5, TM7) and these bands are not normally corrected. Table 16.4 (Chapter 16) presents an example of 'dark-object subtraction' data.

Histogram method. The DN value to be subtracted from each band can also be estimated by plotting a histogram (Fig. 10.6a). The atmospheric contribution leads to an offset of the histogram for a given spectral band towards higher DN values by some amount. The minimum DN value can be assumed to be the haze component in the particular band.

Regression method. A scatterogram or regression method can also be used to estimate the haze component (Fig. 10.6b) (Crippen 1987). DN values of the visible band (blue/green/red) are plotted against the near-IR band. The line of best fit will intercept the short-wavelength axis at a DN approximating the haze component. This value can then be subtracted from all the pixels to remove the haze component. Chavez (1988, 1996) suggested improvements to the 'dark-object subtraction' method by incorporating a relative scattering model. It is based on the fact that Raleigh scattering is inversely proportional to the nth power of the wavelength, the value of n varying with the atmospheric turbidity condition.

2. Approaches utilizing atmospheric models. Remote sensing data can also be corrected for atmospheric contribution by using atmospheric models (codes), such as: 5S (Tanre et al. 1990), SMAC (Rahman and Dedieu 1994) and 6S (Vermote et al. 1997). In addition, Operational Atmospheric Correction (OAC) method has been developed for Landsat TM data (Ouaidrari and Vermote 1999).

All these models have a similar general approach. In order to be able to use the atmospheric model effectively, it is necessary to have measurements on atmospheric parameters at the time of imaging. The cost of such data collection (*atmospheric truth!*) is quite high, and the technical sophistication and infrastructure required is not always available. Hence, the use of atmospheric models for rectifying multispectral remote sensing data has generally been restricted to research purposes (although such corrections are imperative in hyperspectral data analysis; see Chapter 11).

10.2.2 Correction for Solar Illumination Variation

The solar illumination condition has a significant influence on remote sensing image data. It is important to normalize for the variation in the solar illumination condition so that spectral responses can be compared and interpretations derived. The variation in solar illumination occurring from scene to scene can be normalized by implementing a fixed Sun-angle transformation, as a first-order correction. For this, the azimuth direction is ignored; each DN value is multiplied by a factor derived from the Sun elevation angle, as follows:

$$DN_{new} = \left(\frac{\cos z}{\cos \theta}\right)(m \cdot DN + b) \qquad (10.2)$$

where DN_{new} is the new digital value, z is the fixed solar zenith angle, θ is the solar angle corresponding to the data to be rectified, DN is the digital number to be transformed, and m and b are linear constants. This computation provides DN values corrected for variation in Sun angle, and may also be carried out during the pre-processing stage.

10.2.3 Correction for Topographic Effects

Local topography influences the solar reflected radiance as recorded by a remote sensor in two ways.
(a) It causes differential solar illumination (i.e. some slopes receive more sunlight than others).
(b) Topographic slope and aspect govern the Sun–target–sensor goniometric geometry, which along with the BRDF (bidirectional reflectance distribution function) controls the magnitude of reflected radiance reaching the sensor.

Therefore, it is important to carry out corrections for topographic effects, so that spectral characters in the scene are brought out. Proy et al. (1989), Woodham (1989), Li et al. (1996) and Parlow (1996) provide reviews of the problem. Methods for topographic normalization include the following.

1. Minnaert correction. This provides a first-order correction for terrain illumination effects (Smith et al. 1980; Colby 1991). The method utilizes an empirical coefficient (Minnaert constant, k), estimated statistically for each image. The coefficient is related to the surface roughness describing the type of scattering. If k = 1, it implies a Lambertian surface; if k is between 0 and 1, it implies a semi-diffuse reflection. In a non-Lambertian reflection model, the following equation is used to normalize the brightness values in the image:

$$L_n = L \cdot \cos(E)^{k-1} \cdot \cos(I)^k \qquad (10.3)$$

where L is the measured radiance, L_n is the normalized radiance (i.e. equivalent radiance on a flat surface with incidence angle of zero), E is the slope (derived from DEM), and I is the incidence angle of solar radiation.

Although the method is useful in reducing topographic effects, a major problem is that it is scene dependent, and the Minnaert coefficient has to be determined uniquely for each set of Sun–sensor geometry for the scene.

2. Use of shaded relief model (SRM). A shaded relief model generated from a topographic map can be applied for topographic normalization of remote sensing data and enhancement of spectral characters. The method requires a digital elevation model (DEM) from which an SRM image is generated with a constant albedo. The SRM should correspond to the same illumination angle and direction as the image to be rectified. This SRM image gives brightness variation across the scene which is dependent solely upon topography, irrespective of the albedo variation. Ratioing of satellite sensor image and SRM image will yield an image with the spectral radiance of the ground cover.

However, the ratio image is quite noisy. Therefore, another approach would be to use a subtraction technique based on the following equation (Sundaram 1998):

$$DN_{new} = M\ (SAT_{DN} - SRM_{DN}) + A \qquad (10.4)$$

where M and A are the scaling constants (gain and bias).

Civco (1989) developed another method for topographic normalization of images. It uses an empirically derived calibration coefficient determined by comparing the spectral responses from large samples of an equal number of pixels falling on northern and southern slopes of the same category (e.g. deciduous forest). The difference in the spectral response (mean and variance) of same land-cover category pixels is calculated. The correction coefficient equilibrates the mean spectral response of the same category occurring on northern/southern slopes.

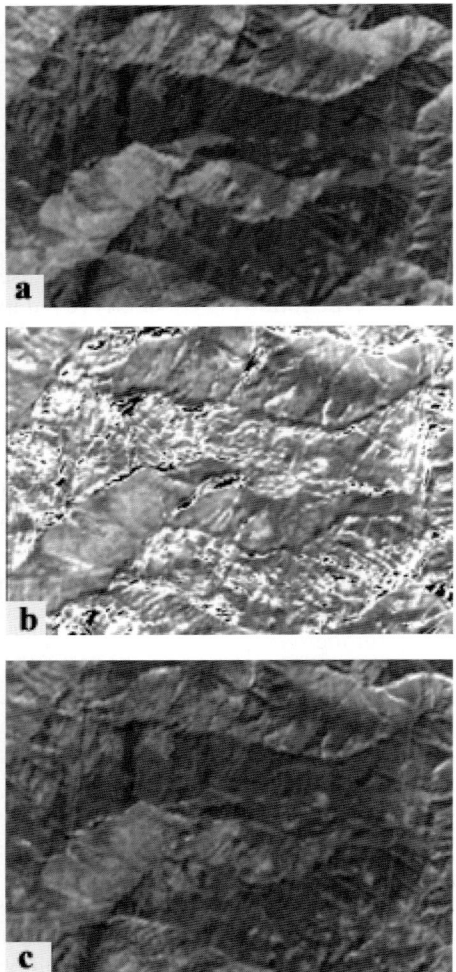

Fig. 10.7a–c. Topographic normalization. **a** Raw image (IRS-LISS-III SWIR band); **b** the image with Minnaert correction; **c** the image with Civco-method correction (for details, see text). (**a–c** Courtesy: A.K.Saha, S. Mitra and E. Csaplovics)

Figure 10.7a–c show the original, Minnaert-corrected and Civco-method-corrected images of a part of the Himalayas.

10.2.4 Sensor Calibration

Calibration of a remote sensor is necessary to generate absolute data on physical properties such as reflectance, temperature, emissivity etc. This type of data on physical properties is only seldom needed, such as while dealing with physical models where the physical data may be required to be input into the models. Aspects of sensor calibration and absolute data computation from satellite sensor data have been dealt with by a few workers (e.g. Price 1988; Hill 1991; Itten and Meyer 1993; Thome et al. 1993).

The radiometric response of a remote sensor is subject to drift with time, as the sensor may degrade/decay or its detector characteristics may change over a period of time. The task of sensor calibration is normally handled by the remote sensing agency operating the system, and calibration data are provided from time to time. For example, calibration values of Landsat-4 TM are given by Markham and Barker (1986), Landsat-5 TM by Thome et al. (1993), and the calibration values for Landsat-7 ETM+ are available at:
http://ltpwww.gsfc.nasa.gov/IAS/handbook/handbook. htmls/chapter11/.

In the field of geosciences, the Landsat TM/ETM+ data have a role in estimating temperature values of lava flows, volcanic vents, coal fires etc., for which details and examples are given in Sects. 16.11 and 16.12.

10.2.5 De-striping

Striping is a common feature seen on images, and arises due to non-identical detector response. When a series of detector elements is used for imaging a scene, the radiometric response of all the detector elements may not be identical. The dissimilarity in response may occur due to non-identical detector characteristics, disturbance with time or rise in temperature, or even failure of an element in an array. This leads to the appearance of stripes, called *striping*. Defective resampling algorithms may also lead to striping.

The phenomenon is quite commonly observed, e.g. in images from Landsat MSS, TM, IRS-LISS, SPOT-HRV etc. The various Earth resources satellites are placed in near-polar orbits. The OM line scanners scan the ground in an across-track direction; therefore striping in such products appears in an E–W direction or 'horizontal' on the image (Fig. 10.8a). On the other hand, the CCD-pushbroom scanners scan the ground in an along-track direction; therefore striping in their case appears oriented N–S or 'vertical' on the image (Fig 10.8b).

Removal of stripes, called de-striping, is sometimes carried out during pre-processing. However, even the pre-processed data, as available, are sometimes strongly striped and users have to rectify the data themselves. Further, the data must be de-striped before any image enhancement is attempted, otherwise striping is likely to get enhanced, leading to erroneous results. For this reason, de-striping is discussed in more detail here. Several procedures, often in combination, are used for de-striping.

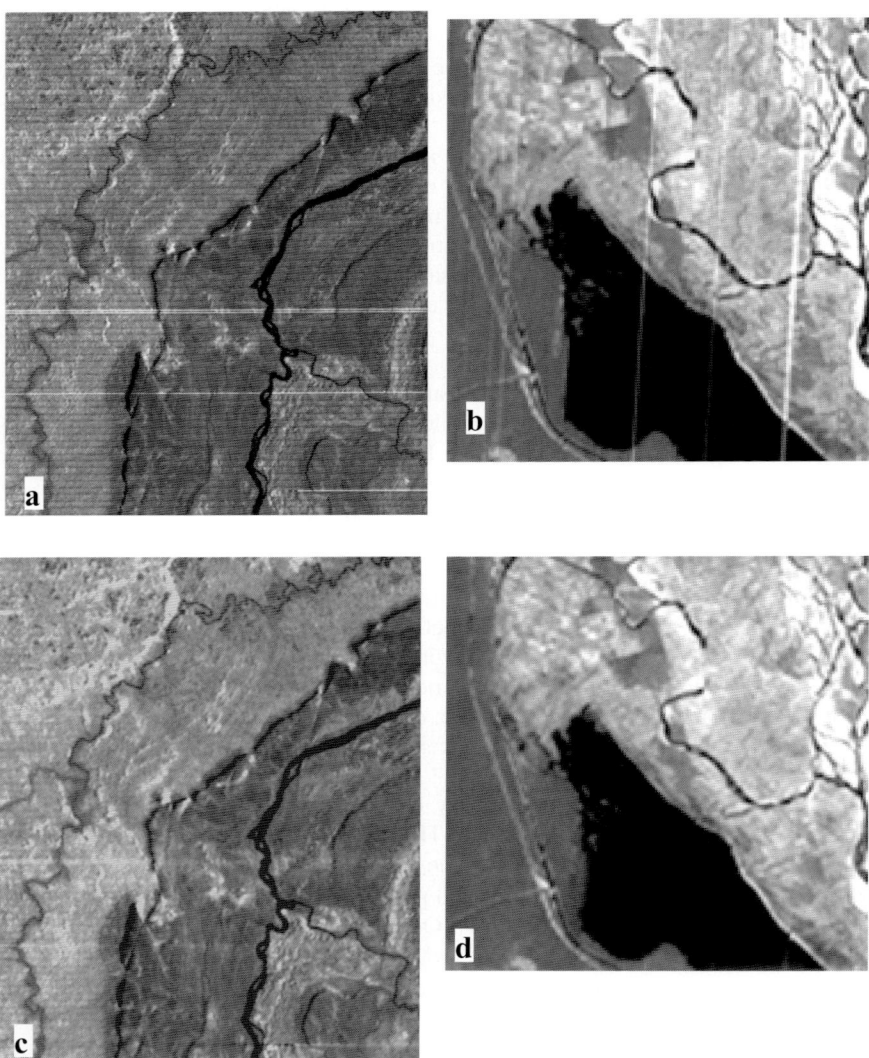

Fig. 10.8. a Horizontal scan lines and stripes observed in an opto-mechanical line scanner image (Landsat MSS); **b** vertical scan lines and stripes in a pushbroom scanner image (IRS-1C LISS-III); **c** and **d** are the respective images after de-striping by Fourier filtering

Radiometric Image Correction

1. Method of look-up tables (LUTs). Before the launch, and if possible also after the return of the sensor system, the radiometric responses of all the detector elements at different brightness levels are recorded. From this data set, relationships (LUTs) between DN value and intensity of radiation for each detector element are generated. This method was followed in several remote sensing experiments, e.g. in MOMS-01.

2. On-board calibration method. A major limitation of the LUT method is that relative changes occurring in the response characteristics of the detector elements after launch with time, or due to temperature fluctuations etc., are likely to remain unrectified. Therefore, in long-duration space missions, on-board calibration is provided. The detector units record flux from calibration lamps, once in each cycle of mirror rotation/oscillation/retrace. Ideally, this arrangement is best suited to the opto-mechanical type of sensors. The stability of the on-board calibration lamps is monitored from the ground. Data from different detector units can be brought to a common reference level (bias) and scale (gain). The method has been used in a number of sensors such as Landsat MSS and TM. The method is also used in CCD linear array scanners (e.g. IRS-LISS).

3. Histogram matching. Some striping is commonly found, even after implementing a look-up table or on-board calibration method. This may be removed by statistical histogram matching. Separate histograms, one corresponding to each detector unit, are constructed, and they are matched with each other. Taking one response as the standard, the gain (rate of increase of DN) and offset (relative shift of mean) for all the other detector units are suitably adjusted and new DN values computed and assigned. This yields a de-striped image where all DN values conform to a common reference level and scale.

4. Interpolation. Failure of a detector unit in an array of detectors may also lead to *missing lines* and striping. This can be rectified cosmetically, in two ways: (a) by interpolating using data of adjacent scan lines in the same spectral band, or (b) by interpolating data of the same scan line in adjacent spectral bands, using the concept of spectral slopes.

5. Fourier transform. In a CCD line scanner image there are several thousand columns, each corresponding to a detector element. De-striping and noise removal of such image data are better accomplished using Fourier transform techniques (Lei et al. 1996) (see Sect. 10.6.3). Figure 10.8c and d show examples of images after de-striping by Fourier filtering.

10.2.6 Correction for Periodic and Spike Noise

Periodic noise occurs at regular intervals and may arise due to interference from adjoining instruments. This is removed by various signal-filtering techniques. Spike noise may arise owing to bit errors during transmission of data or due to a temporary disturbance. It is detected by mutually comparing neighbouring pixel values. If adjoining pixel values differ by more than a certain threshold margin, this is designated as a spike noise and the pixel value is replaced by an interpolated DN value. Occasionally, a complete scan line may have to be rectified in this manner (see interpolation above).

Some of the above operations serve to render the image look cleaner and better and may also be considered as *cosmetic operations*.

10.3 Geometric Corrections

Geometric distortions are broadly divided into two groups: systematic and non-systematic (see Table 6.1). We discuss the correction for systematic distortions in this section. The non-systematic distortions are removed by the general technique of 'rubber-sheet stretching', as discussed under registration (Sect. 10.4).

10.3.1 Correction for Panoramic Distortion

Panoramic distortion arising due to non-verticality of the optical axis results in squeezing at image margins. A correction is necessary such that the horizontal distance is given by $X = H \tan \theta$, where H is the flying height/altitude, X is the horizontal distance and θ is the angle of rotation of the optical axis (see Fig. 6.2).

This correction is especially necessary in aerial scanner data as θ is typically about 50° (total angular field-of-view 100°). The correction is sometimes applied during pre-processing; the angle of rotation (θ) is related to time and therefore X can be related to a time-base to produce a geometrically rectified image. In spaceborne missions the angle θ is very small (e.g. in Landsat MSS, $\theta = 5.6°$) due to the very high altitudes involved, and the error can often be ignored.

10.3.2 Correction for Skewing Due to Earth's Rotation

This type of distortion is caused due to the rotation of the Earth underneath, relative to the line-imaging sensor as it scans the ground from above (see Fig. 6.10). In aerial scanner data, the effect of this type of skew is negligible. In satellite data, the skew is maximum in sensors orbiting in polar orbits. In a generalized way, the image skew is given by

$$\tan \varphi \times \sin \alpha = \left(\frac{2 \times R \cos \theta}{24 \times 60 \times 60} \right) \times \frac{1}{V} \tag{10.5}$$

where φ = image skew, θ = scene centre latitude, R = radius of the Earth (= 6367.5 km), V = velocity of the ground track of the satellite (in km s^{-1}) and α = inclination of the orbit plane from the equator. Once the amount of skew is computed, the

Registration 233

Fig. 10.9. a Landsat MSS image with aspect ratio and skew distortions. **b** The corresponding rectified image

scan lines are physically shifted at regular intervals, so that terrain features appear in proper geometric positions in relation to each other. Often, this type of rectification is also made during pre-processing.

10.3.3 Correction for Aspect Ratio Distortion

When the linear scales along the two rectangular arms of an image are not equal, aspect ratio distortion occurs. This can arise for many reasons, e.g. over-sampling/ under-sampling, or variations in the V/H ratio of the sensor-craft. The aspect ratio distortion due to design in sampling pattern is of a systematic type. For example, the Landsat MSS sensor, which provided global coverage in 1970s-80s, carried a systematic type of aspect ratio distortion. Here, due to over-sampling, the scale along x-axis was longer than that along y-axis by a factor of 1.38 (Fig 10.9). Aspect ratio distortion due to V/H variation in the spacecraft is of an unsystematic type and is rectified by the general technique of 'rubber-sheet stretching'.

10.4 Registration

10.4.1 Definition and Importance

Registration is the process of superimposing images, maps or data sets over one another with geometric precision or congruence, i.e. the data derived from the

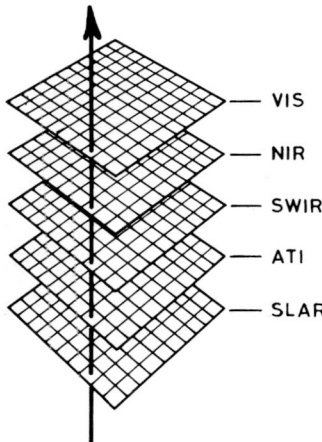

Fig. 10.10. Concept of image registration. The image data at each unit cell are exactly in superposition and geometric congruence

same ground element in different sensor coverage are exactly superimposed over each other (Fig. 10.10). The need for image registration is obvious. In many situations, we have multi-sensor, multi-temporal, multi-platform or even multi-disciplinary data. For integrated study, e.g. for general pattern recognition or change detection, the various image data sets must be registered over each other.

Registration of an image is carried out to a certain base. If the base is an image, it is called *relative registration*, and if the base is a certain standard cartographic projection, it is called *absolute registration*. The basic technique in both relative and absolute registration is the same. Absolute or map registration is widely used in GIS applications (Chapter 15). Here we discuss relative registration.

Multispectral images taken from the same sensor and platform position are easy to register. The problem arises when images to be registered are from different sensors, platforms, altitudes or look directions. In such cases, distortions, variations in scale, geometry, parallax, shadow, platform instability etc. lead to mismatch, if the various images are simply laid over one another. Therefore, there is a need for a method of digital image registration, whereby multi-images can be superimposed over each other with geometric precision.

10.4.2 Principle

Digital image registration basically uses the technique of co-ordinate transformation. A set of ground control points (GCPs) in the two images are identified and their co-ordinates define the transformation parameters. Typically, a set of polynomial or projection equations is used to link the two co-ordinate systems. An example of simple affine projection is as follows:

$$X' = a_0 + a_1 x + a_2 y + a_3 xy \qquad (10.6)$$
$$Y' = b_0 + b_1 x + b_2 y + b_3 xy \qquad (10.7)$$

where X' and Y' are the co-ordinates in the new system and x, y those in the old system. There occur eight unknown constants ($a_0, a_1, a_2, a_3, b_0, b_1, b_2, b_3$), which can be computed by using four control points (as each control point gives two equations, one for X' and one for Y'). However, only four control points may not be sufficient for a large image. Usually, a net of quadrilaterals is drawn, employing several control points over the entire scene, and a set of transformation equations for each quadrilateral is computed. These transformation equations are used for transposing pixels lying within the quadrilateral. This is also popularly called 'rubber-sheet stretching'.

The type of simple coordinate transformation outlined above may often meet the requirements for Landsat-type satellite data. Higher-order corrections, although giving better results, require more elaborate transformations and computer time.

10.4.3 Procedure

The basic concept in image registration is shown in figure 10.11. The various steps involved are as follows.

1. Selection of base. When only remote sensing images are to be mutually registered, it is often convenient to use any one of the images as base, and perform relative registration. Usually, the image with higher spatial resolution and better geometric fidelity is selected as the base. (On the other hand, when data from multi-disciplinary investigations, e.g. remote sensing, geophysical, geochemical, geological, topographic etc., are being collectively used, it is often better to use a standard cartographic projection as the base, see Sect. 15.4). For further discussion, let us call the base image as Image I (or master image) and the image to be registered as Image II (or slave image).

2. Selection of control points. A number of controls point or landmarks are identified which may be uniquely located on the two images. These points are also called *ground control points* (GCP's). Workers in digital images also refer them to as *templates*. The control points should be well distributed, stable, unique and prominent. Some of the features commonly used as control points are, for example, road-intersections, river bends, canal turns or other similarly prominent and stable features.

3. Matching of control points or templates. The next step is to match the selected control points on the two images. This is most commonly done visually, i.e. by computer-human interaction - the digital images are projected on the screen and the control points are visually matched. Further, the process of matching control points is basically iterative, and some other match points can always be used to check the earlier matching.

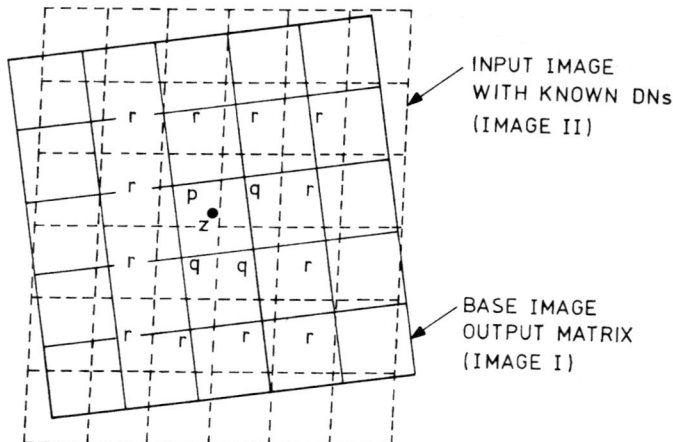

Fig. 10.11. Data resampling method for registration (after Bernstein 1976; Lillesand and Kiefer 1987). The value at z (the required pixel location) is given as that at p by the nearest-neighbour approach. In the bilinear interpolation approach, the value at z is computed by interpolating between p and q cells. In the cubic convolution approach, interpolation is made considering the values at p, q, and r

Another possibility is to match the templates or tie points digitally (only by computer); this is termed *template matching*. Template matching is a very critical operation, as it controls the accuracy of the entire registration operation. Commonly, once a template in one of the digital images has been defined, its approximate position in the other digital image can be estimated, which gives the search area. The most commonly used statistical measure of template matching is the correlation coefficient. The template is moved through the search area and at each position the correlation coefficient between the array of DNs in the two images is computed. The position giving the highest correlation coefficient is taken as the match position. When the template is being moved through the search area, only translational errors in the two digital images are taken care of. Differences in orientation due to rotation, skew, scale etc. hamper template matching. Other factors affecting template matching are real changes in scene, or changes in solar illumination and shadows etc. Therefore, if two digital images differ widely, the digital template matching may be a futile exercise.

4. *Computing the projection equation.* Once the matching of control points/ templates has been satisfactorily done, the next step is to calculate parameters for projections (polynomial or affine) from the coordinate data sets.

5. *Interpolation and filling of data.* Registration is accomplished by selecting a cell in the grid of the base image, finding its corresponding location in the slave image, with the help of the affine projections, and then computing the DN value at that point in the slave image by interpolation and resampling methods. This gives the DN value distribution of the slave image field, in the geometry of the master

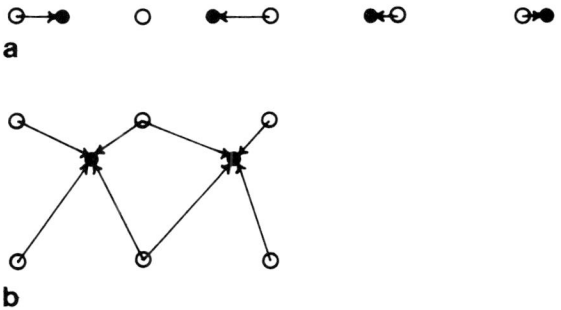

Fig. 10.12. Schematic representation of procedures of data resampling and interpolation. Open circles = input image matrix (locations of known DNs); filled circles = output image matrix (locations where DN values are desired). **a** Nearest-neighbour approach; **b** bilinear interpolation. (After Moik 1980)

image. The process therefore involves finding new DN values at grid points in the geometry of the base image (and not transferring old DN values to new locations!). If the entire image is registered, pixel by pixel, in the above manner, it takes a prohibitive amount of computer time. Therefore in practice, a few selected points are registered in the above manner and the space in between is filled by resampling/interpolation.

The commonly used methods of data interpolation and resampling are as follows.

Nearest-neighbour method. Here the point in the new grid simply acquires the DN value of that point in the older grid which lies closest to it (Fig. 10.12a). The DN value is not interpolated, but some of the old pixels are bodily shifted, leaving some pixels out. This technique is fast and quite commonly used. However, owing to sudden shifting of pixels, it may lead to the appearance of breaks or artificial lines on the images, as the local geometry may become displaced by half a pixel.

Bilinear interpolation. This is a two-dimensional extension of linear interpolation. It uses the surrounding four pixel values from the older (slave) grid. The procedure is first to interpolate along one direction, to obtain two intermediary DN values, and then to interpolate in the second direction to obtain the final DN value (Fig. 10.12b). It takes more computer time, but produces a smoother picture with better geometric accuracy.

Bicubic interpolation. This uses a polynomical surface to fit the DN dibstribution locally, and from this fitted surface DN values at the new grid points are computed. It gives a better visual appearance, but is quite expensive in terms of computer time. The image is also quite smoothened and local phenomena such as high spatial frequency variation etc. may also get subdued.

In addition, *krigging* is another interpolation method used for resampling during registration.

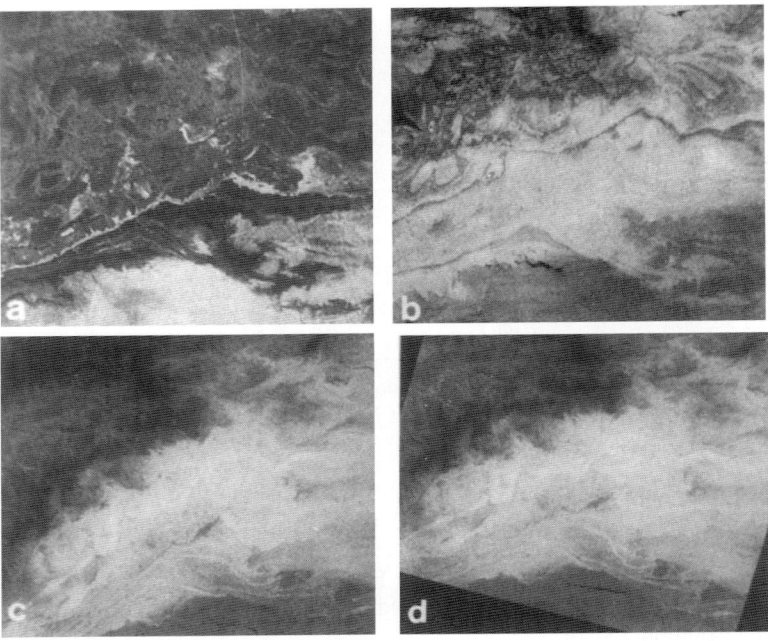

Fig. 10.13. a and **b** are visible- and thermal-IR-band daytime images of a part of the Atlas Mts.; **c** covers the same area in the thermal-IR night-time pass; note the difference in geometry between **b** and **c**. **b** (daytime TIR image) has been taken as the base image. **d** shows the night-time TIR image in registration with **b**. (HCMM data Processed at GAF mbH, Munich)

An example of image-to-image registration is given in figure 10.13. It is obvious that low-resolution and low-contrast images are relatively more difficult to register than sharp and high-contrast images. The accuracy of registration may thus depend upon the sharpness of features, overall image contrast and spatial resolution of component images.

10.5 Image Enhancement

The purpose of enhancement is to render the images more interpretable, i.e. features should become better discernible. This at times occurs at the expense of some other features which may become relatively subdued. It must be emphasized that digital images should be corrected for radiometric distortions prior to image enhancement, otherwise the distortions/noise may also get enhanced.

As a preparatory first step for image enhancement, the statistical data distribution is examined. A histogram describes data distribution in a single image and a scatterogram provides an idea of relative data distribution in two or more images.

Image Enhancement

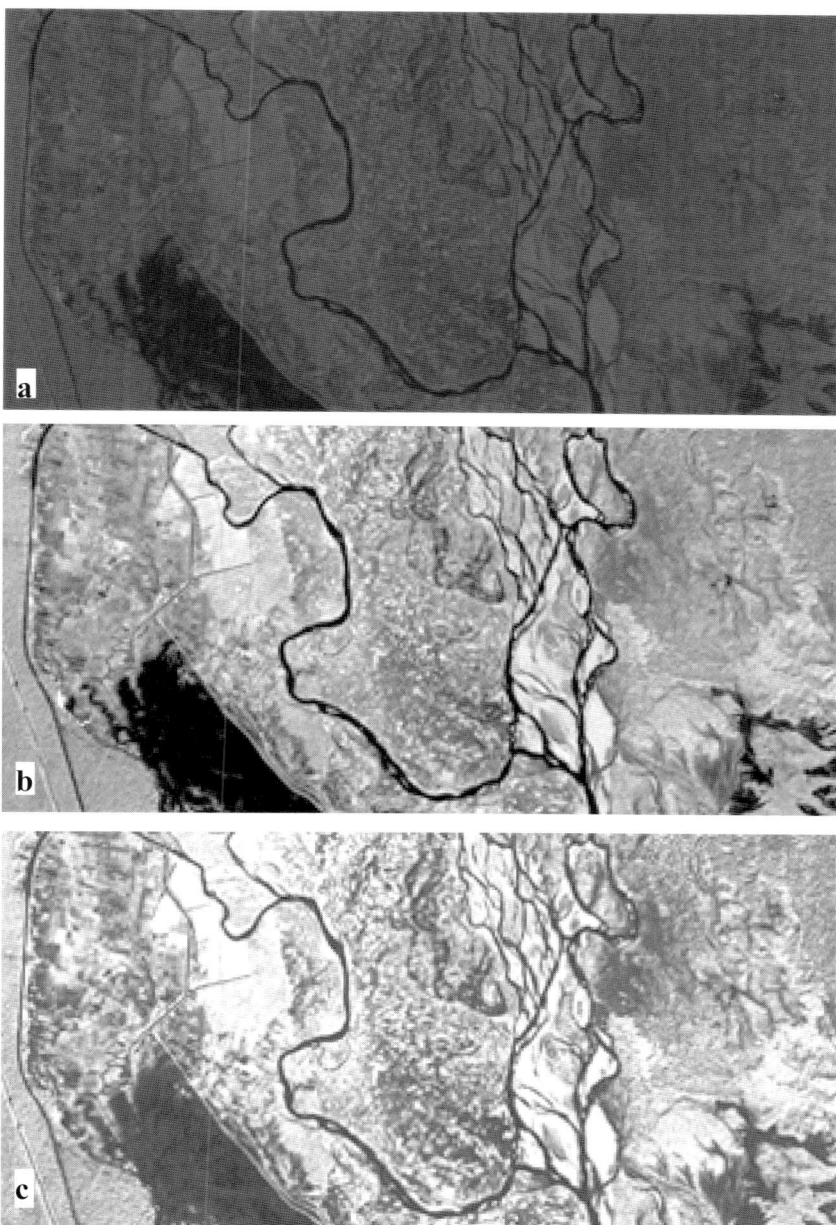

Fig. 10.14. Contrast stretching. **a** Raw image; **b** linear contrast-stretched image; **c** image with histogram equalization stretch; note the higher contrast in **c** than **b**. (IRS-1C-LISS-III near-IR band image covering a part of the Gangetic basin)

Techniques developed in the field of classical signal processing over many years are (e.g. Castleman 1977; Pratt 1978; Rosenfield and Kak 1982) advantageously adopted and applied to remote sensing images. In this section, we discuss the contrast-enhancement techniques.

Contrast enhancement deals with rescaling of gray levels so that features of interest are better shown on the image. In general, it so happens that the number of actually recorded intensity levels in a scene is rather low and the full dynamic range of the digital image (256 levels) is not utilized. As a result, typically, the image has a low contrast (Fig. 10.14a). This characteristic is indicated by the corresponding histogram in which the DN frequency is found to be segregated only in the lower DN range, with hardly any pixel occurrence in the higher DN-range (Fig 10.15). Details on such an image are scarcely visible. A rescaling of gray levels could improve the image contrast and allow better visibility of features of interest. Contrast enhancement is a typical point operation in which the adjacent pixel values have no role to play.

To perform the contrast-stretching operation, the first step is to plot a histogram of the DN values in the image. This itself may convey valuable information about the scene. A sharp peak would indicate no contrast in the scene, and a broad distribution would imply objects of high contrast. A multimodal distribution would clearly indicate objects of more than one type. This information could also be utilized for deciding on the type of contrast manipulation. As a common practice, gray levels with a frequency of less than 2% of the population, on either flank of the histogram, are truncated and only the intervening DN-range is stretched. This allows greater contrast in the entire image, although some of the gray levels on either extreme become squeezed or saturated. In any particular case, however, the actual cut-off limit should be based on the requirements.

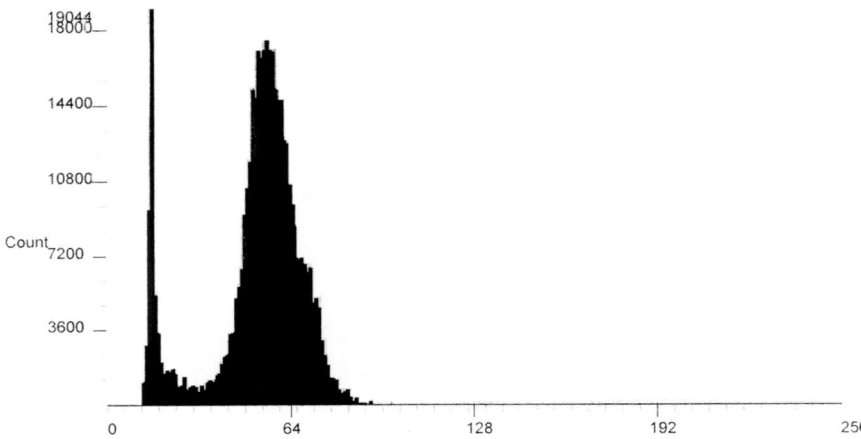

Fig. 10.15. Histogram of the raw image in figure 10.14a; note the segregation of DNs in the lower range, which is responsible for the low contrast in figure 10.14a

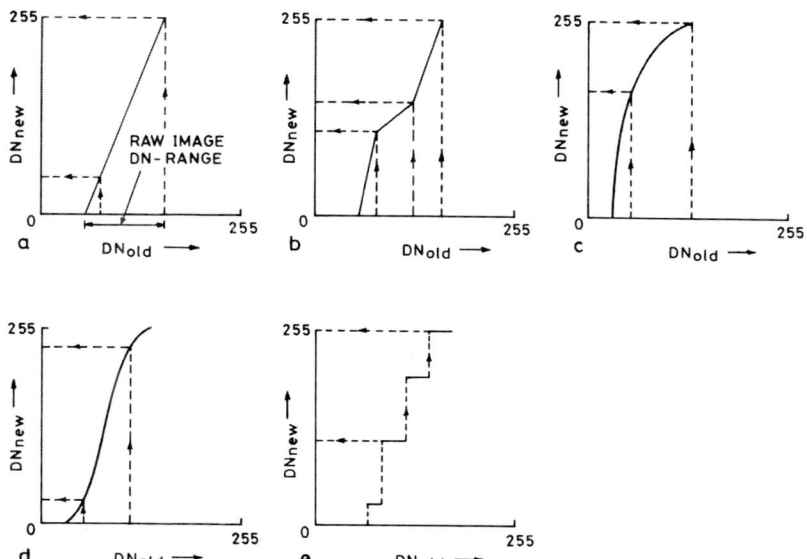

Fig. 10.16a–e. Transforms for contrast stretching. **a** Linear contrast stretch. **b** Multiple linear stretch. **c** Logarithmic stretch. **d** Histogram equalization stretch. **e** Density slicing

Some of the common methods for contrast manipulation are as follows.

1. Linear contrast stretching. This is the most frequently applied transform in which the old gray range is linearly expanded to occupy the full range of gray levels in the new scale (Fig. 10.16a). Figure 10.14b gives an example of a linearly stretched image, the raw image being shown in figure 10.14a.

2. Multiple linear stretch (piece-wise linear stretch). Different segments of the old gray range can also be stretched differently, each segment in itself being a linear stretch (Fig. 10.16b). This is a useful transform when some selected stretches of DN-ranges are to be enhanced and other intervening ranges are to be squeezed.

3. Logarithmic, power or functional stretch. The image data can be stretched using a logarithmic function, power function or any other function (Fig. 10.16c). The logarithmic stretch is useful for enhancing features lying in the darker parts of the original image, the result being an overall relatively brighter image. In contrast, the exponential function stretches preferentially features in the brighter parts of the original gray range, rendering the image generally darker.

4. Gaussian stretching. In this scheme of gray-scale manipulation, the new gray scale is computed by fitting the original histogram into a normal distribution curve. It renders greater contrast preferentially in the tails of the old histogram, which means that the cut-off limits in the old histogram become highly critical.

5. Histogram equalization stretching. This is also called ramp stretching, cumulative distribution function stretching or uniform distribution stretching. The new image has a uniform density of pixels along the DN-axis, i.e. each DN value becomes equally frequent (Fig. 10.16d). As pixels in the middle range happen to be most frequent, the middle range is substantially expanded and made to occupy a larger range of DN values in the new gray scale. The overall contrast in the image is thus significantly increased (Fig. 10.14c). This is a good transform when an individual image is to be displayed in black-and-white, but is not particularly suited for making colour composite displays.

6. Density slicing is also a type of gray scale modification. The old gray scale is subdivided into a number of ranges, and each range is assigned one particular level in the new gray scale (Fig. 10.16e). The steps can be of equal or unequal width. This is analogous to drawing a contour map for the parameter DN, and the contour interval could be equal or unequal. A density-sliced image is often colour coded for better feature discrimination (see pseudocolour enhancement, Sect. 10.8.2).

Thus, there are numerous possibilities to manipulate the image contrast to suit the needs of an investigation. Further, any type of simple or processed image can be subjected to contrast enhancement.

10.6 Image Filtering

Image filtering is carried out to extract spatial-scale information from an image. As it is a technique to enhance spatial information in the image, it can also be grouped under 'enhancement'. All filtering operations are typically local operations in which the DN values at the neighbouring pixels also play a role.

Image filtering operations are of two basic types: (1) spatial domain filtering, and (2) frequency domain (Fourier) filtering. Spatial domain filtering is carried out using windows, boxes or kernels (see below) and has dominated the remote sensing image processing scenario until now, for two main reasons: (a) it is simple and easy to implement and requires lower computational capabilities, and (b) it could meet the requirements of data processing for the Landsat MSS, TM type data. Spatial domain filtering will be discussed first and Fourier filtering later.

If we plot a profile of DN values from one end of an image to another, we find that the profile consists of a complex combination of sine waveforms. It can be broadly split into two: (1) high-frequency variations and (2) low-frequency variations (Fig. 10.17). Here, frequency connotes the rate of variation in the DNs. High-frequency variations correspond to local changes, i.e. from pixel to pixel, and low-frequency variations imply regional changes, from one part of the image to another. Correspondingly, there are two types of filtering techniques: high-pass filtering and low-pass filtering, to enhance one type of information over the other. Both of these types of filters (high-pass and low-pass) can be implemented through spatial-domain as well as frequency-domain filtering.

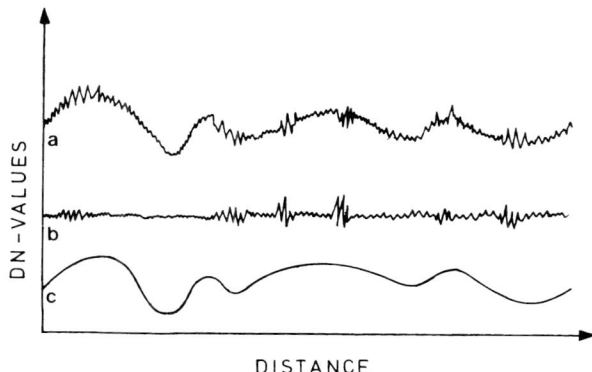

Fig. 10.17. High-frequency and low-frequency components in an image. **a** A typical profile of DN values consists of a complex combination of sine waveforms which can be split into **b** high-frequency and **c** low-frequency components

Spatial-domain filtering is carried out by using *kernels*, also called *boxes or filter weight matrices*. A kernel consists of an array of coefficients. Figure 10.18 shows some oft-used kernels. In order to compute the new DN, it is imagined that the kernel is superimposed over the old image-data array. The original DN values are weighted by the overlying coefficients in the kernel and the resulting total DN value is ascribed to the central pixel in the new image. The kernel is successively moved over all the pixels, in rows and columns, and the array of new DNs is computed.

The most common kernel size is 3×3, 5×5 or 7×7 being relatively less common. Kernel size can also be arbitrarily chosen, as per requirements in a certain case. There are no predefined kernels that would provide the best results in all cases. At times, the selection is based on trial and error. Odd-numbered kernel sizes are generally used so that the central pixel is evenly weighted on either side. Anisotropic kernels (e.g. 3×5) produce directional effects, and therefore can be used to enhance linear features in a certain preferred direction.

10.6.1 High-Pass Filtering (Edge Enhancement)

In a remote sensing image, the information vital for discriminating adjacent objects from one another is contained in the *edges*, which correspond to the high-frequency variations. The edges are influenced by terrain properties, vegetation, illumination condition etc. Thus, *edge enhancement* is basically a *sharpening* process whereby borders of objects are enhanced.

There are several ways of designing filter-weight matrices and enhancing edges in digital images (Davis 1975; Shaw 1979; Peli and Malah 1982; Jensen 1986; Russ 1995). Some of the more important methods are as follows.

$$\begin{vmatrix} 0 & 0 & 0 \\ 0 & 1 & 0 \\ 0 & -1 & 0 \end{vmatrix} \quad \begin{vmatrix} 0 & 0 & 0 \\ 0 & 1 & -1 \\ 0 & 0 & 0 \end{vmatrix} \quad \begin{vmatrix} 0 & -1 & 0 \\ -1 & 4 & -1 \\ 0 & -1 & 0 \end{vmatrix} \quad \begin{vmatrix} 0 & -1 & 0 \\ -1 & 5 & -1 \\ 0 & -1 & 0 \end{vmatrix}$$

a   b   c   d

$$\begin{vmatrix} 0 & 0 & 0 \\ 0 & 1 & 0 \\ 0 & 0 & -1 \end{vmatrix} \quad \begin{vmatrix} \frac{1}{16} & \frac{1}{8} & \frac{1}{16} \\ \frac{1}{8} & \frac{1}{4} & \frac{1}{8} \\ \frac{1}{16} & \frac{1}{8} & \frac{1}{16} \end{vmatrix} \quad \begin{vmatrix} \frac{1}{9} & \frac{1}{9} & \frac{1}{9} \\ \frac{1}{9} & \frac{1}{9} & \frac{1}{9} \\ \frac{1}{9} & \frac{1}{9} & \frac{1}{9} \end{vmatrix}$$

e   f   g

Fig. 10.18a-g. Some typical kernels (3 × 3 matrices) for filtering. **a** X-edge gradient. **b** Y-edge gradient. **c** Isotropic Laplacian. **d** 'Image with isotropic Laplacian' **e** Diagonal-edge gradient. **f** Image smoothing. **g** Image-smoothing 'mean image'

1. Gradient image. The gradient means the first derivative or the first difference. A gradient image is obtained by finding the change in the DN value at successive pixels, in a particular direction. The procedure is to subtract the DN value at one pixel from the DN value at the next pixel. Several variations exist. Enhancement parallel to X, called X-edge enhancement, can be obtained by proceeding along the Y-axis, i.e. taking differences of successive pixels along the Y-axis (DN_{new} = df/dy). Similarly, Y-edge enhancement is obtained by proceeding along the X-axis, (i.e. DN_{new} = df/dx). Figure 10.18a,b are the kernels used for enhancing edges by the gradient method in the X and Y directions respectively. In general, the first difference in the direction parallel to scanning has a smaller variation than that in the direction perpendicular to it, owing to striping.

The gradient image can be added back to the original image to provide a gradient-edge-enhanced image. The gradient method is often used in detecting boundaries, called line detection, as discussed later in image segmentation (Sect. 10.11), but not so frequently for edge enhancement, for which the Laplacian (given below) is generally considered to be better suited.

2. Laplacian image. The Laplacian enhancement is given by the second derivative, i.e. the rate of change. Basically, it would involve computing differences with respect to the two neighbouring pixels on either side, in a row or column. The isotropic Laplacian (kernel in figure 10.18c) can be written as

$$DN_L = d^2f/dx^2 + d^2f/dy^2 \tag{10.8}$$

DN_L has a high value whenever the rate of change of DN in the old image is high. Figure 10.19a and b give an example of the image and its Laplacian edge.

Image Filtering

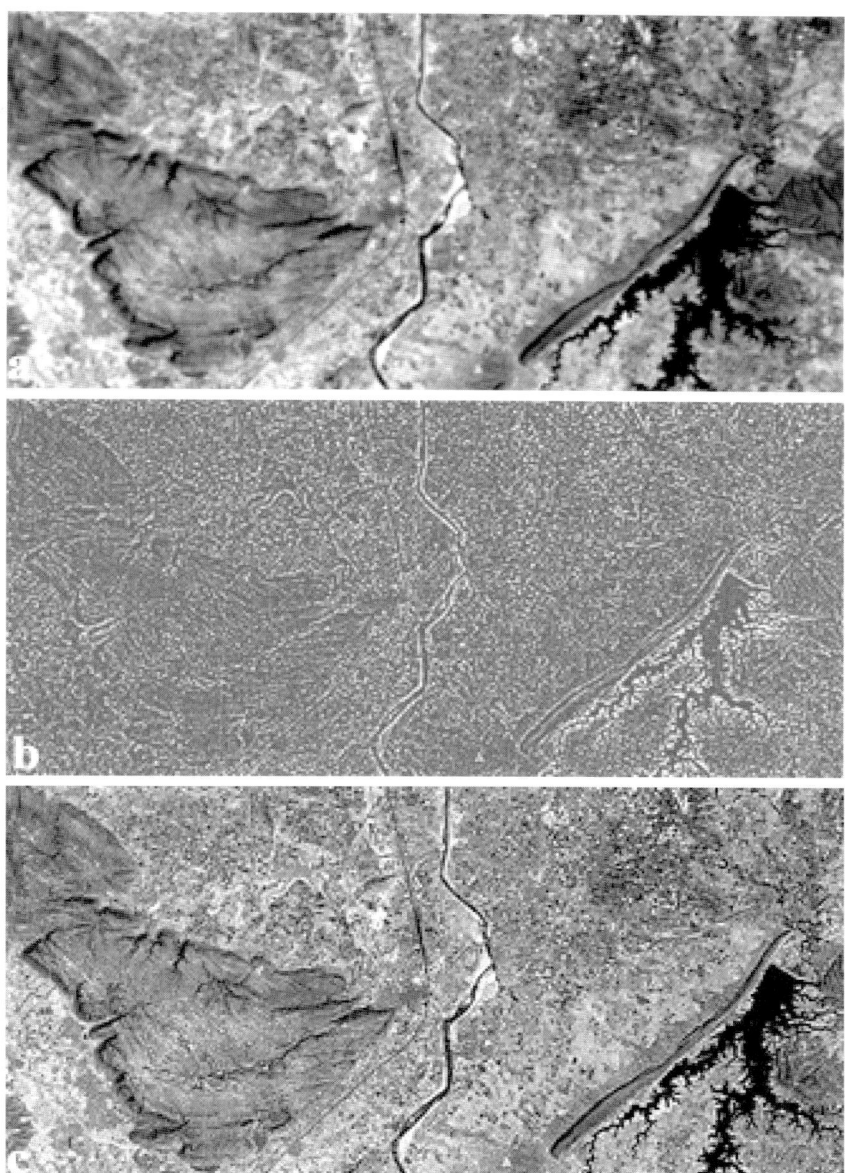

Fig. 10.19a–c. Edge enhancement. **a** Original image; **b** isotropic Laplacian image; **c** edge-enhanced image (produced by filter 'image with isotropic Laplacian' fig. 10.18d). The image is IRS-LISS-II NIR band covering a part of Rajasthan, India. The nearly NE-SW trending rocks occurring on the east are the Vindhyan Super Group, which are separated from the folded rocks of the Aravalli Super Group, occurring on the west, by the Great Boundary Fault of Rajasthan

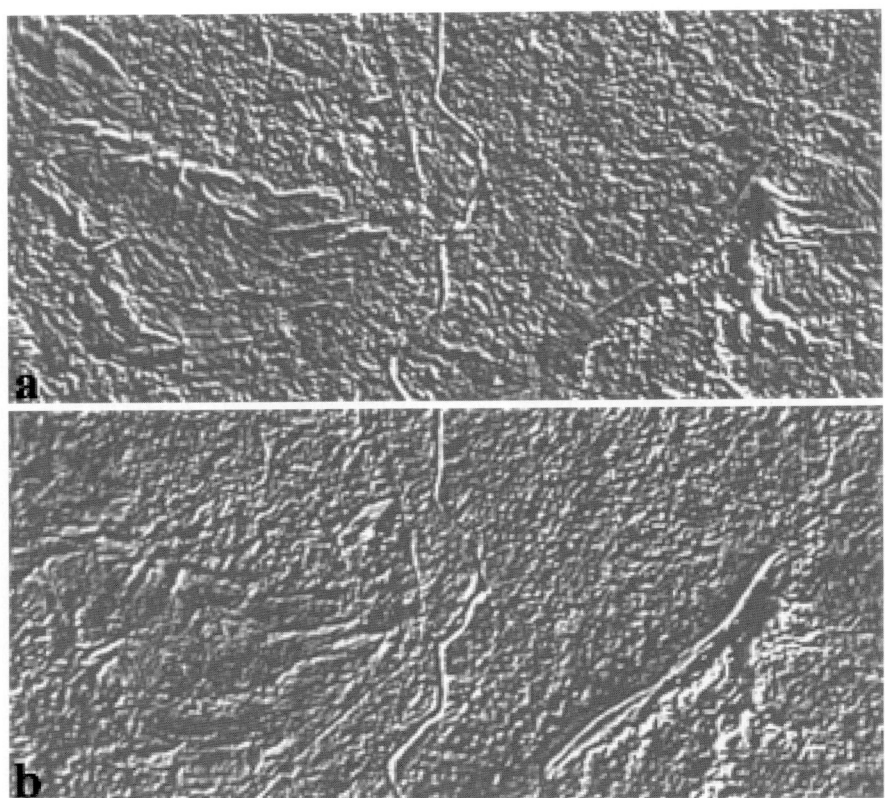

Fig. 10.20a,b. Examples of directional edge enhancement. **a** Diagonal edges with look direction towards SE. **b** The same with look direction towards SW

3. Low-frequency image subtraction. This is another particularly powerful method of edge enhancement. It is obvious from figure 10.17 that high-frequency information in an image can be obtained by subtracting a low-frequency image from the original image.

Invariably, an edge image is more susceptible to noise variations; therefore, it is added to the original image to give an edge-enhanced image (Fig. 10.19c). For this, a useful kernel is that shown in figure 10.18d.

4. Diagonal edge image. Diagonal edges can be enhanced by computing differences across diagonal pixels in an image (see e.g. kernel in figure 10.18e). The image can be generated for left-look (SE looking) or right-look (SW looking) directions (Fig. 10.20a,b). In this way, boundaries running NW–SE or NE–SW can be enhanced.

Some basic concepts in edge enhancement (i.e. image sharpening) have been outlined above. The subject matter encompasses a vast field where numerous other

Image Filtering

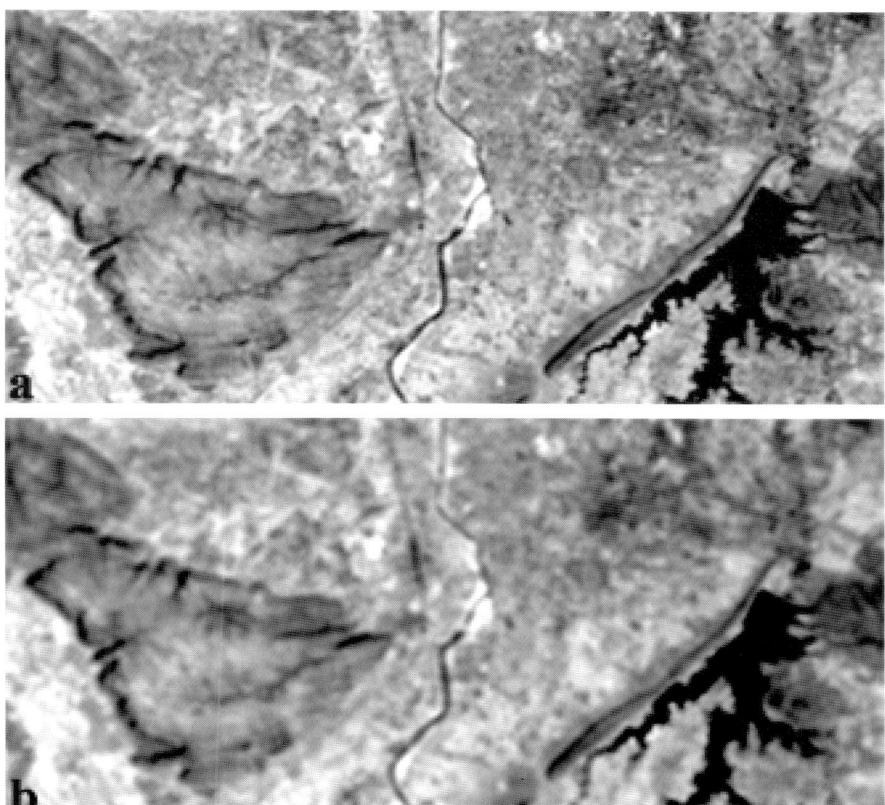

Fig. 10.21a,b. Image smoothing. **a,** resulting from a kernel 3 × 3, is less smooth than **b,** corresponding to a kernel 5 × 5

non-linear filters (such as Sobel, Robert's, Kirsch etc.) have been formulated and applied with varying success (for more details, refer to e.g. Russ 1992).

Special mention may also be made here of a subtractive box filter formulated by Thomas et al. (1981) for enhancing circular features on the Landsat MSS data.

Scope for Geological Applications

Edge enhancement brings out variations in DN values across neighbouring pixels, or high-frequency spatial changes; it is also called *high-pass filtering* or *textural enhancement*. The main applications of high-pass filtering in geological investigations are as follows.

(a) To obtain a sharper image showing more details for better interpretation in terms of local topography /landforms, lithology, vegetation, soil, moisture etc.

(b) To enhance linear systems (edges), to facilitate interpretation of fractures, joints etc; linear edge enhancement in specific directions can also be obtained by suitable anisotropic filters (although interpretation of such data warrants extra care).
(c) To reduce the effects of gross differences in illumination; on an edge-enhanced image, local variations become important so that details within a uniformly illuminated zone are better deciphered.

A limitation of the high-pass-filtered image is that only local variations with respect to the adjoining pixels become important and absolute DN values may have little significance, e.g. uniform large stretches of dark-toned and light-toned objects could show up in similar tones. The common practice is, therefore, to add/subtract the Laplacian from the original DN values, so that both original DN values and local variations are displayed together.

10.6.2 Image Smoothing

The main aim of image smoothing is to enhance low-frequency spatial information. In effect, it is just the reverse of edge enhancement. Typical kernels used for image smoothing are shown in figure 10.18f,g. Examples of image smoothing are given in figure 10.21. As image smoothing suppresses local variations, it is particularly useful when the aim is to study regional distribution over larger geological domains.

10.6.3 Fourier Filtering

The mathematical technique of *Fourier analysis*, which uses the frequency domain, is also applicable to remote sensing data. Fourier analysis operates on one image at a time. Figure 10.22 presents the various steps involved in Fourier filtering of an image.

Fourier analysis separates an image into its component spatial frequencies. Referring back to figure 10.17, the DN variation plotted along each row and column provides a complex curve with numerous 'peaks' and 'valleys'. The Fourier analysis splits this complex curve into a series of waveforms of various frequencies, amplitudes and phases. From the data on component frequencies, a *Fourier spectrum* can be generated, which is a two-dimensional scatter plot of the frequencies. Further, if the Fourier spectrum of an image is known, it is possible to regenerate the image through inverse Fourier transform.

Various frequencies corresponding to the original image can be identified in the Fourier spectrum. For example, low frequencies appear at the centre and successively higher frequencies are plotted outwards in the Fourier spectrum (Fig. 10.23).

The Fourier spectrum can be processed through filtering in the frequency domain. Practically, this involves designing a suitable filter (matrix) and transposing the earlier matrix by multiplying with the new (filter) matrix. By applying inverse Fourier transform, the filtered Fourier spectrum is used to generate the filtered im-

Image Filtering

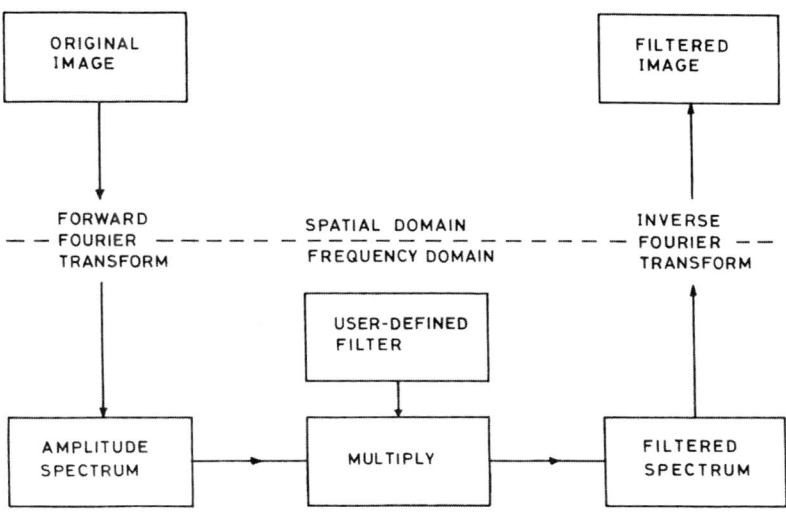

Fig. 10.22. Basic procedure in the frequency-domain filtering of a digital image

age. Figure 10.24 gives an example where a 'low-pass blocking filter' has been used to derive an edge-enhanced image.

As mentioned earlier, the use of Fourier analysis in remote sensing multispectral data processing by resources scientists has been rather limited, because of considerations of computer sophistication and computer time etc. (and therefore the technique of spatial-domain filtering has dominated the image processing scenario till now). Fourier transform provides higher flexibility in data processing; for example, an image can be filtered by blocking the low-frequency or high-frequency component. Further, modern remote sensors carry several thousand detector elements (such as in CCD line arrays and area arrays). For this type of data,

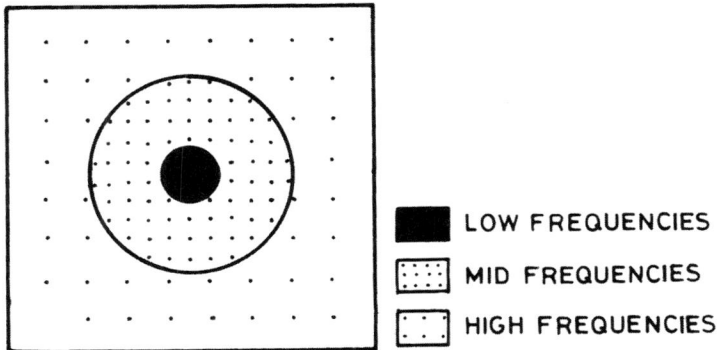

Fig. 10.23. Location of low-, mid- and high-frequency components in the amplitude spectrum

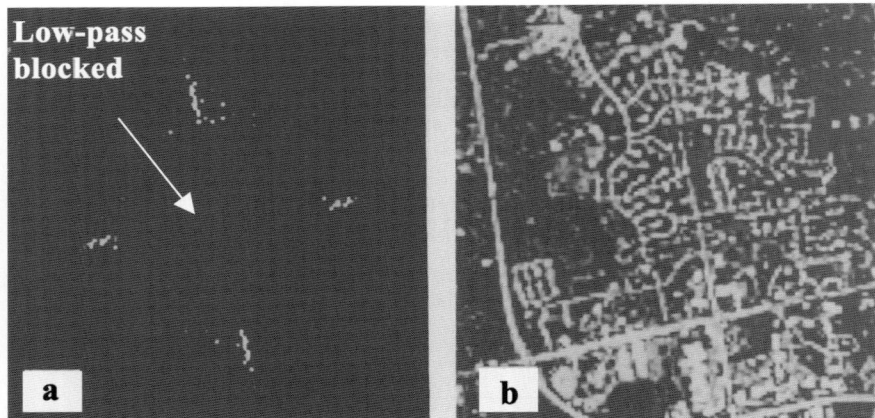

Fig. 10.24a,b. Fourier filtering. **a** Amplitude spectrum with low-frequency blocking filter; **b** edge-enhanced image from inverse transform of **a**. (**a,b** Lillesand and Kiefer 2000) (reproduced by permission, Copyright © 1999, John Wiley & Sons Inc)

it is extremely difficult to carry out noise removal (e.g. de-striping) with methods such as histogram matching etc., developed for OM line scanners. Fourier transform is a powerful means to rectify this type of image data (see Fig 10.8 and Fig. 11.19).

Therefore, with the availability of superior computing facilities (hardware/software), and also the growing need to handle data from CCD line and area arrays, it is expected that Fourier transform will find greater applications in remote sensing digital image processing.

10.7 Image Transformation

The enhancement techniques so far discussed are basically for single images. They are derived from the field of classical signal processing. A peculiar aspect of remote sensing is that it provides data in multispectral bands, which can be collectively processed to deduce information not readily seen on a single image. Moreover, images from different sensors, different platforms, or acquired at different times can also be superimposed over each other with geometric congruence. Image transformation deals with processing of multiple-band images to generate a new computed (transform) image.

In order to conceptually understand what is happening when the multispectral images are combined, it is necessary to become conversant with the concept of feature space and related terminology. Two sets of image data can be statistically represented through a two-dimensional plot called *feature space* – the *feature axes* here being the two spectral channels. The location of any pixel in this space is controlled by the DN values of the pixel in the two channels (features). This is

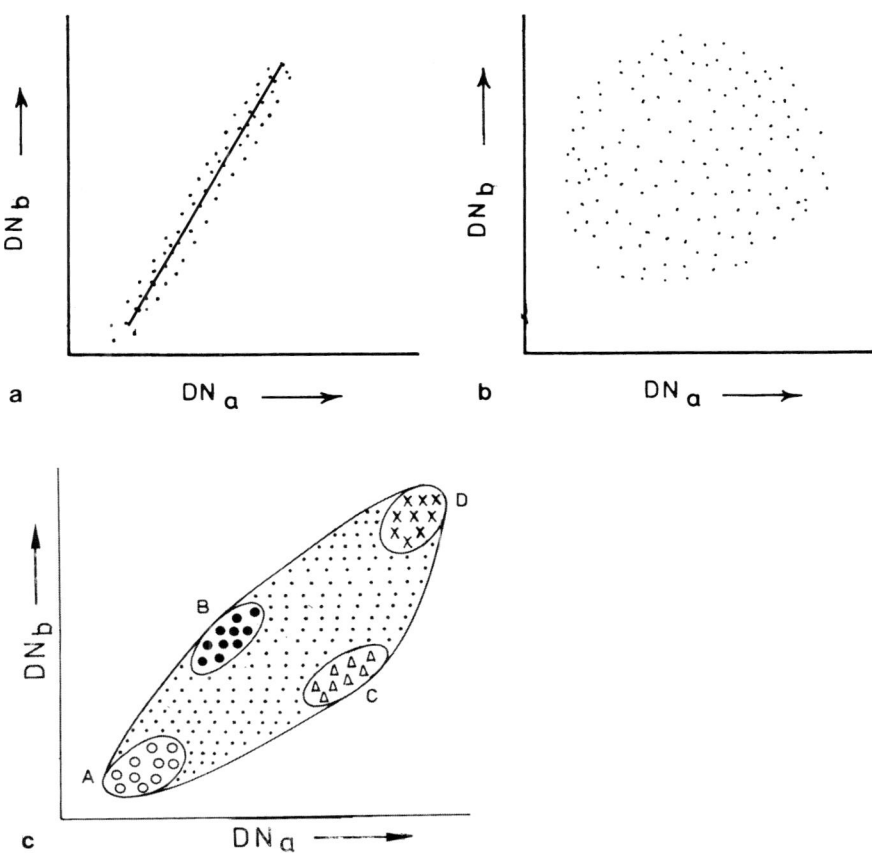

Fig.10.25a–c. Concept of scatterograms. **a** Extremely correlated feature axes. **b** Non-correlated feature axes. **c** A typical scatterogram using two Landsat MSS/TM channels; the pronounced elongated shape of the scatterogram indicates that the two channels are well correlated

also called a *scatterogram* or *scatter-diagram*. A purely visual inspection of such a diagram may itself be quite informative about the mutual relation of the two features. If the two features are highly correlated, a line approximates the scatterogram (Fig. 10.25a). Two non-correlated features produce a feature space plot in which points are scattered isotropically (Fig. 10.25b).

In general, the present Landsat TM, ETM+, SPOT-HRV and IRS-LISS spectral channels provide data which are well correlated. This fact is expressed by the pronounced elongated shape of the data points in a scatterogram (Fig. 10.25c). If some objects are spectrally unique, then they form clusters. The clusters can at times be separated from other data points. A spectral plot showing clusters of points is called a *cluster diagram*. The separability of clusters governs the dis

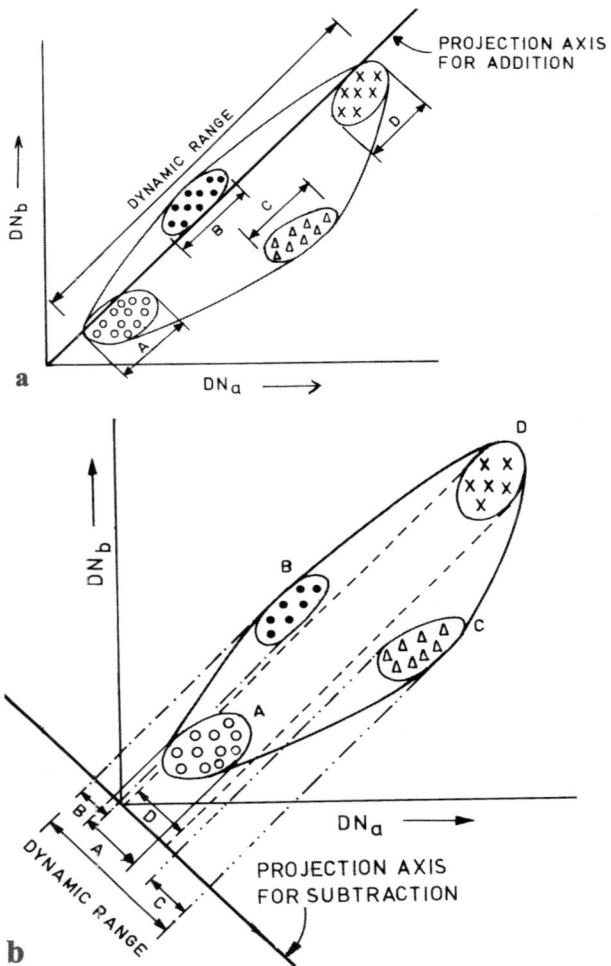

Fig. 10.26a,b. Transformation of data during addition and subtraction of images. **a** Simple addition of well-correlated multispectral images results in distribution of DN values as if the points were projected onto the line passing through the origin and having a slope of 45°. The new image would have a much larger dynamic range and therefore greater overall contrast; however, it does not necessarily mean better separability. For example, on the new image clusters A and D become more separated from each other, but clusters B and C would largely overlap. **b** Subtraction of well-correlated multispectral data would result in projection of points on the line perpendicular to the simple addition axis, passing through the origin and having a slope of 45°. The dynamic range of the new image is relatively small; the clusters A and D which were distinct on the addition image, appear overlapping and are now inseparable on the subtraction image; however, clusters B and C become separable from each other on the new image. (Note: scale along the projection axis is nonlinear and only the relative position of points/clusters is relevant here)

criminability of objects in the image. The meaning and utility of cluster diagrams is mentioned also in the section on classification (Sect. 10.12). The concepts of a feature or cluster diagram can now be easily adapted to understand the various multiple-image enhancement techniques, using simplified conceptual projection diagrams.

10.7.1 Addition and Subtraction

The simplest method to combine multispectral digital images is by addition and subtraction. In a most generalized way the linear combination can be written as:

$$DN_{new} = p \cdot DN_A \pm q \cdot DN_B + K \qquad (10.9)$$

where p and q are constants suitably selected to give weights to (A) and (B) input images, and K is a constant which helps in scaling the gray-scale.

If we form a simple addition of two spectral bands ($DN_{new} = DN_A + DN_B$), the DN value distribution in the new image would conceptually be given by projection of points on the line passing through the origin and having a slope of 45° (Fig. 10.26a). The resulting image is characterized by a much larger dynamic range and higher contrast than any of the originals, due to the high correlation of the two input images. However, it may not necessarily imply better separability of clusters; hand, there could be a situation in which the spectral differences become reduced between some categories on the addition image.

If we calculate the difference between the two bands ($DN_{new} = DN_A - DN_B$), the DN value distribution in the new image appears as if the data points were projected on the axis perpendicular to the simple addition axis (Fig. 10.26b). The image is characterized by a lower dynamic range (which can be taken care of subsequently by contrast manipulation). However, all those objects which do not possess a correlation pattern in conformity with the majority of data points show up strongly on such an image. Therefore, the technique is used for change detection in a multi-temporal image data set.

By introducing weighting factors on features (input images), a variety of linear combinations can be generated. The resulting new images may appear quite different in dynamic range and contrast distribution in differently weighted addition/subtraction cases – this may even confuse subsequent interpretations; therefore, for meaningful data handling and manipulation it is essential to formulate strategies beforehand. Careful stretching of spectral bands combined with properly weighted subtraction/addition (particularly the subtraction) may produce quite informative images. At times, the weighting factors may be derived from statistical approaches such as principal components (see Sect. 10.7.2).

Scope for Geological Applications

Figure 10.27 presents examples of simple addition and subtraction image processing. The following general points can be made.

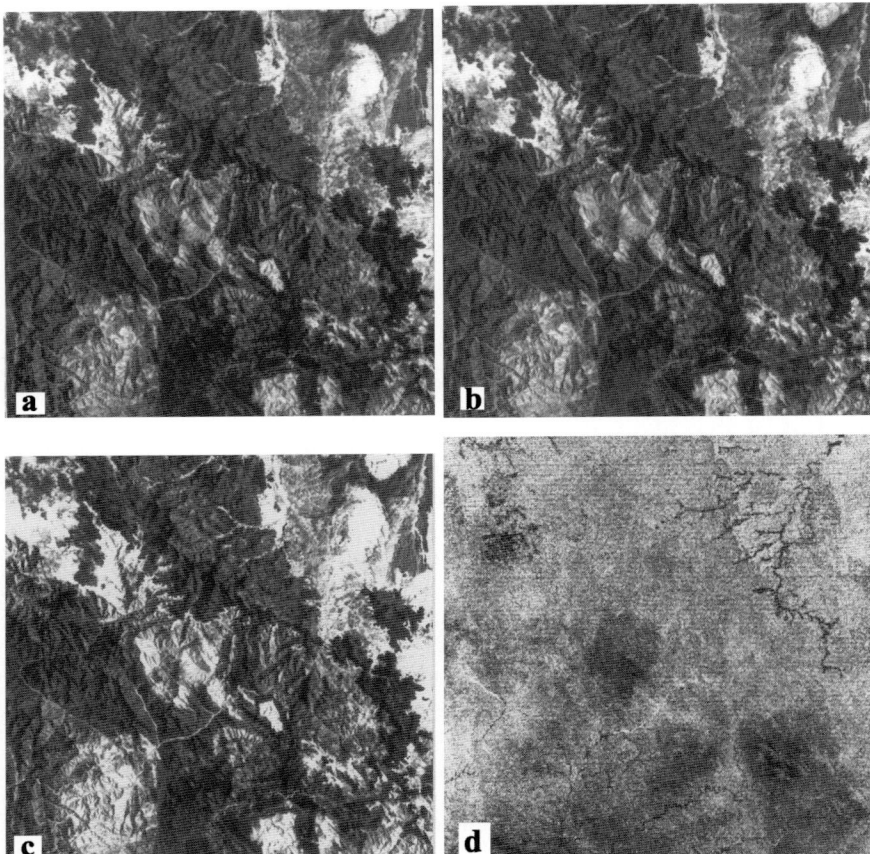

Fig. 10.27. a Landsat MSS5 and **b** MSS6 images of a part of Saudi Arabia used as input for making simple addition and subtraction images; **c** and **d** are the resulting addition and subtraction images respectively. Note the higher contrast in **c** in comparison to **d**. Also note the manifestation of the spectrally 'off-beat' (egg-shaped) feature in **d**. (**a–d** Courtesy of R. Haydn)

1. An addition image generated from a set of correlated images has a much larger dynamic range and higher contrast than any of the input images. Objects which have positive correlation in the two input bands become still better separable.

2. A subtraction image made from a similar data set has a smaller dynamic range and lower contrast. It is useful in picking up data points/objects which do not spectrally conform to the background, or are 'offbeat'. The technique is also used for change detection from repetitive images.

Image Transformation

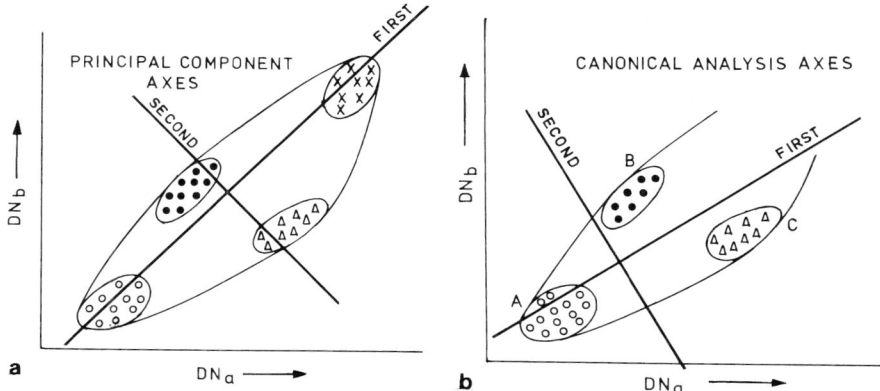

Fig. 10.28. a Concept of principal component axes. **b** Concept of canonical analysis axes

10.7.2 Principal Component Transformation

Principal component transformation (PCT), also called principal component analysis or Karhunen–Loeve analysis, is a very powerful technique for the analysis of correlated multidimensional data (for statistical principles, refer to any standard text, e.g. Davis 1986). Its application in digital remote sensing has been discussed by many workers (e.g. Kaneko 1978; Byrne et al. 1980; Haralick and Fu 1983). The n-channel multispectral data can be considered as n-dimensional data. As often happens, many of the channels are correlated, either negatively or positively. The PCT builds up a new set of axes orthogonal to each other, i.e. non-correlated. Most of the variance in the input image data can be represented in terms of new principal component (PC) axes, fewer in number.

To understand the PC transform conceptually, we again refer back to our concept of scatterograms. It was stated earlier that nearly unlimited possibilities exist to combine the two spectral channels in the form of weighted linear combinations. The PC transform provides statistically the optimum weighting factors for linear combinations, so that the new feature axes are orthogonal to each other (Fig. 10.28a). One digital image, called the PC image, can be made corresponding to each principal component axis (PCA), which can be contrast stretched or used as a component in other image combinations/enhancements etc.

The first PCA (PC1) has the largest spread of data, and when dealing with Landsat-type data it is generally found to be the weighted average filter of all the channels. As all the channels are represented in this image, it more or less becomes an intensity (or albedo in the solar reflection region) picture with little spectral information (e.g. Fig. 10.29a). The other PCAs are generally differences in various channels (e.g. the second PCA is the difference between the visible and infrared channels, and so on). These can be considered as spectral or colour information (Fig. 10.29b).

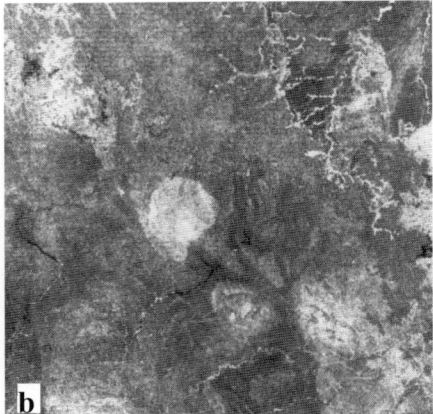

Fig. 10.29. a PC1 and **b** PC2 images from a Landsat MSS scene of Saudi Arabia. The PC1 image is a high contrast 'general-albedo' image; PC2 carries spectral information. Note the egg-shaped geological feature enhanced on the PC2 image (**a,b** Courtesy of R. Haydn)

A variation of PCT is canonical analysis. In this, new feature axes are computed in such a way that they enable maximum discrimination between some selected groups of objects, and not in the entire scene (Fig 10.28b), i.e. the statistics are computed not from the entire scene but only from some selected object types (training areas); the transformation to be carried out is based on these statistics. Thus, this provides maximum discrimination between the selected groups of objects.

Scope for Geological Applications

The PCT increases overall separability and reduces dimensionality, and is therefore highly useful in classification. A limitation of PCT is that the gray-tone statistics of a PC image are highly scene dependent and may not be extrapolated to other scenes. Further, geologic interpretation of PC images also requires great care as the surface information dominates the variation.

As indicated earlier, commonly the first principal component (PC1) image is a general albedo (intensity) image, whereas other PC images carry spectral information. The eigen-vector matrix gives the information on how each PCA is being contributed to by the various input spectral channels. The matrix-data form the basis to interpret the spectral information carried by each of the PC-axis images (see e.g. Crosta and Moore 1989). The method has been applied to map alteration zones (see e.g. Loughlin 1991). The following case study from Central Mexico provides an interesting example.

In a study to enhance the spectral response of hydrothermal alteration minerals in Central Mexico, Landsat TM data (solar reflection bands) were processed through the technique of PCT and interpreted by the 'Crosta technique' by Ruiz-Armenta and Prol-Ledesma (1998). Table 10.1 gives the eigen-vector matrix, i.e.

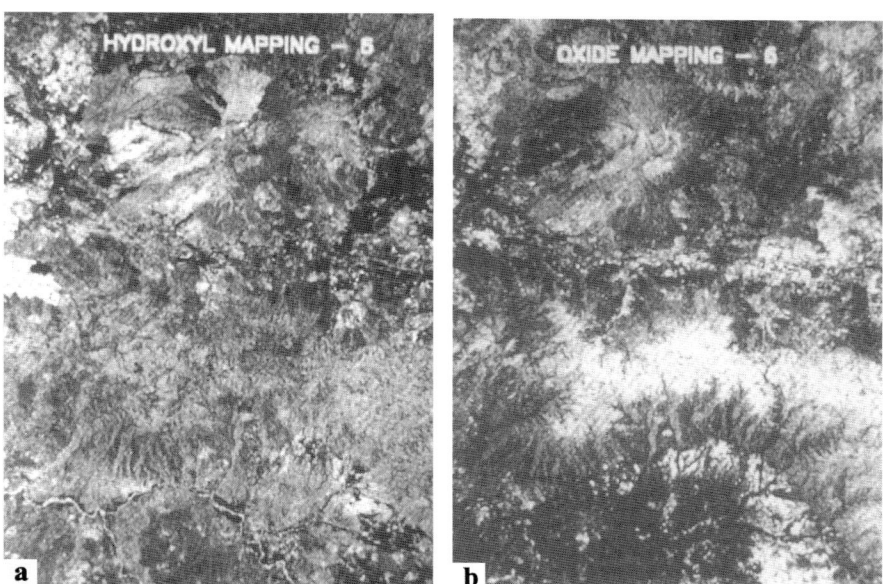

Fig. 10.30. a Inverse PC4 and **b** inverse PC5 images showing distribution of hydroxyls and iron oxides respectively (bright pixels in both cases), in an area in Central Mexico (for details, see text). (**a,b** Ruiz-Armenta and Prol-Ledesma 1998)

the loadings of each PCA in terms of the various TM channels. As is expected, PC1 is a general albedo (intensity) image carrying 87% of the scene variance. Other PCAs carry successively less scene variance, and carry negative/positive contributions from the six TM channels. Noteworthy is that PC4 has a high positive contribution (82%) from TM7 and a high negative contribution (−32%) from TM5. As hydroxyl minerals exhibit high absorption in TM7 and high reflectance in TM5, an inverse image of PC4 will represent hydroxyls as bright pixels (Fig. 10.30a). Similarly, in PC5 there is a high positive contribution (71%) from TM1

Table 10.1. Principal component analysis on six bands of the Ceboruco TM image (eigenvector loadings expressed in percentages) (Ruiz-Armenta and Prol-Ledesma 1998)

Inputs bands	Eigenvector matrix (%) of original bands						Eigen values (%)
	TM-1	TM-2	TM-3	TM-4	TM-5	TM-7	
PC1	23.38	16.50	33.07	24.19	75.65	42.15	87.29
PC2	−11.34	2.50	−1.86	92.64	−6.22	35.21	6.39
PC3	−54.18	−30.98	−51.79	−8.20	56.30	−13.52	5.32
PC4	−24.83	−11.91	−27.32	25.91	−32.01	82.45	0.57
PC5	71.20	1.75	−69.51	7.25	6.32	−1.15	0.38
PC6	−26.47	92.82	−25.31	−6.34	1.28	0.45	0.05

and a high negative contribution (−69%) from TM3. As iron oxide has strong absorption in TM1 and high reflectance in TM3, an inverse image of PC5 in this case will show iron oxide as bright pixels (Fig. 10.30b). In this way, careful interpretation of the PC eigen-vector matrix can allow deduction of thematic information from the PC images.

10.7.3 Decorrelation Stretching

Another modification of PCT is called decorrelation stretch, a highly useful technique for processing correlated multidimensional image data (Kahle et al. 1980; Campbell 1996). It involves the following steps: (1) a principal component transformation, followed by (2) a contrast equalization by Gaussian stretch, so that histograms of all principal components approximate a Gaussian distribution of a specified variance, resulting in a 3-D composite histogram of a spherically symmetric 3-D Gaussian type, and next (3) a co-ordinate transformation that is the inverse of the principal component rotation to project data in the original space. The inverting operation has the advantage of restoring basic spectral relationships. The decorrelation-stretched images can be used as components for making colour composites. Examples of decorrelation-stretched images are presented in figures 16.54 and 16.69.

10.7.4 Ratioing

Ratioing is an extremely useful procedure for enhancing features multispectral images. It is frequently used to reduce the variable effects of solar illumination and topography and enhance spectral information in the images (Crane 1971; Holben and Justice 1980; Justice et al. 1981). The new digital image is constructed by computing the ratio of DN values in two or more input images, pixel by pixel. The general concept can be formulated as

$$DN_{new} = m \left(\frac{DN_A \pm K_1}{DN_B \pm K_2} \right) + n \qquad (10.10)$$

where DN_A and DN_B are the DN values in A and B input images, K_1 and K_2 are the factors which take care of path radiance present in the two input images, and m and n are scaling factors for the gray-scale.

In a situation when DN_B is much greater than DN_A, use of the equation in the above form would squeeze the range too much, and in such a case a better alternative is to use the logarithmic of ratios (Moik 1980) or arctan functions ratios (Hord 1982). Images for complex ratio parameters including additions, subtractions, multiplications and double ratios etc. can be generated in a similar manner.

Image Transformation

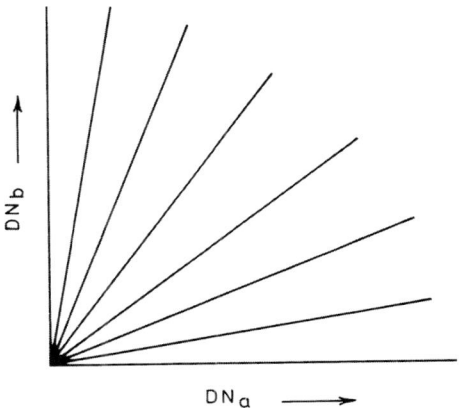

Fig. 10.31. The concept of ratioing. The figure shows a two-channel plot with lines of equal slopes; all points lying on a line have the same ratio and therefore acquire the same gray tone in the new ratio image, irrespective of the absolute albedo values

The resulting ratio image can again be contrast stretched or used as a component for colour displays etc.

To discuss what happens during ratioing and how it helps in feature enhancement, let us refer to the scatterogram concept. In figure 10.31, we have a two-

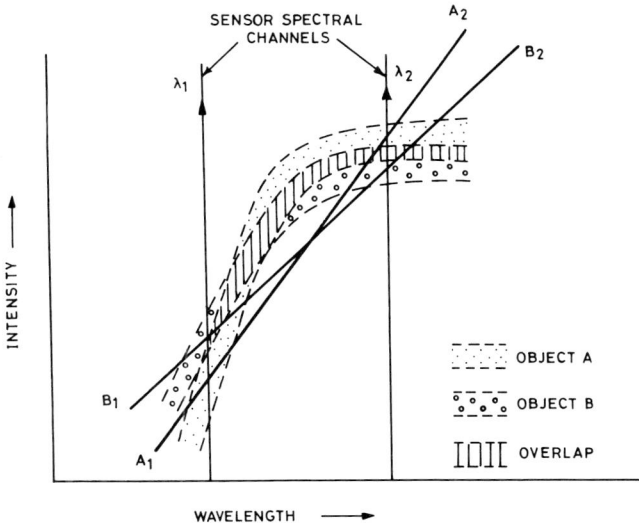

Fig. 10.32. The figure shows spectral curves of two objects A and B, λ_1 and λ_2 being the two sensor channels. The objects A and B have overlapping spectral responses in both λ_1 and λ_2 channels; therefore, no single channel is able to give unique results. However, if the ratio of the two channels is taken, the spectral slopes would be given by $A_1 - A_2$, and $B_1 - B_2$ lines, which make discrimination between the objects A and B possible

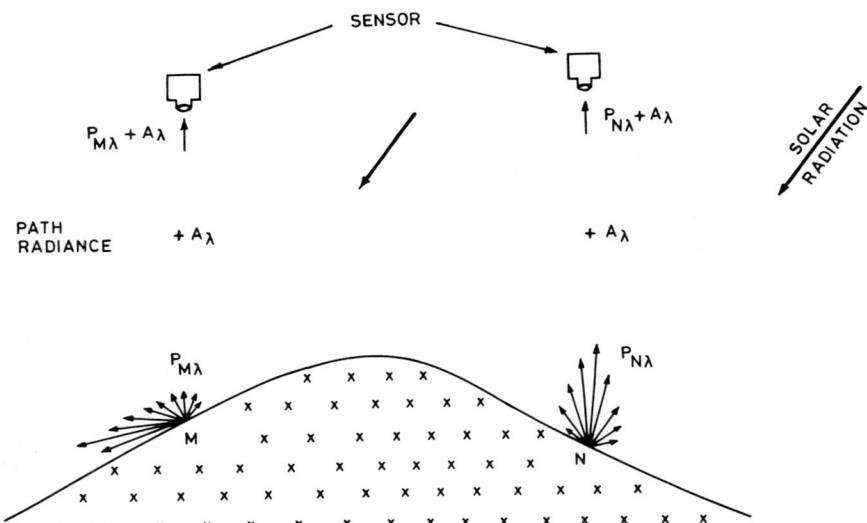

Fig. 10.33. Undulating terrain composed of a homogeneous rock (e.g. granite) illuminated from a direction and viewed from two sensor positions; ground segment N reflects more sunlight than the ground segment M towards the sensor. The atmospheric contribution (A_λ) is the same and is additive in both cases. The relative fraction of the signal due to path radiance is more at M than at N

channel plot with several straight lines of constant slopes. Each straight line corresponds to a particular ratio between channels A and B. Pixels with different DNs but the same ratio value in the two channel images would lie on the same line; therefore they would have the same gray tone on the ratio image. Thus, the various gray tones on the ratio image are controlled by slopes of the straight lines on which the ratio points lie (and not on the distance from the origin!). In this way, a ratio image gives no information on the absolute or original albedo values, but depends on the relative reflectance in the two channels. Therefore, basically, a ratio image enhances features when there are differences in spectral slopes (Fig. 10.32). In this respect, a ratio image can also be considered as a representation of spectral/ colour information.

In actual practice, the ratio values of images in the solar reflection region are influenced by path radiance to a high degree, including the skylight component. If we take the ratio values with shorter wavelength in the numerator and proceed from poorly illuminated to well-illuminated topographical slopes of the same surface material, the ratio values are found to decrease with distance, i.e. in poorly illuminated areas, the ratio values are higher in comparison to those in well-illuminated areas (c.f. Kowalik et al. 1983). This is because of two reasons:

(a) path radiance is higher towards shorter wavelengths;

Image Transformation

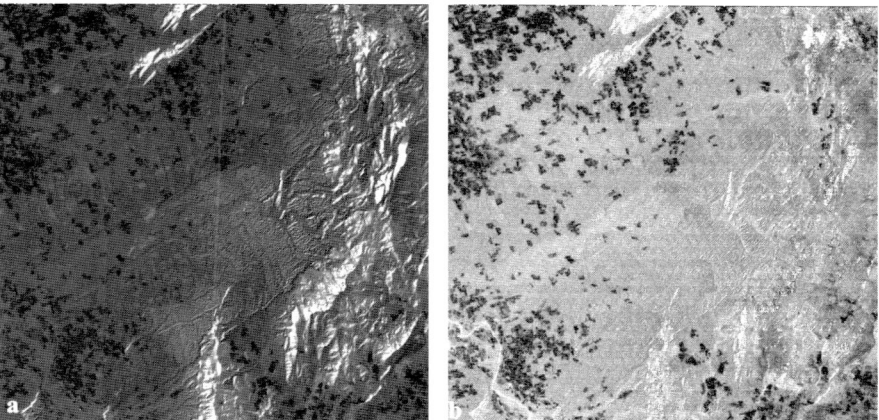

Fig. 10.34a,b. Examples of ratio images (TM1/3). **a** Ratio image without rectification for path radiance; the topographical effects are still seen on the ratio image. **b** Ratio image when path radiance is properly rectified; the resulting image is free from topographical effects. (Landsat TM sub-scene covering a part of the Khetri copper belt, India)

(b) the fraction of the total signal arising from the path radiance is greater in poorly illuminated areas than in well-illuminated areas (Fig. 10.33).

Ratio images derived from such data carry strong topographical effects; therefore, it is necessary to rectify for the path radiance component in the input images before constructing a ratio image. Figure 10.34 presents an example.

Scope for Geological Applications

The most important advantage of ratioing is that it provides an image which may be quite independent of illumination conditions. The pixels acquire the same DN value if the ground material has the same spectral ratio, quite irrespective of whether the ground happens to lie in a well- or poorly illuminated zone. Therefore, a properly made ratio image significantly reduces topographical effects (Fig. 10.34).

The Landsat TM data have been available on a global basis and have suitable spatial, spectral and radiometric resolutions. Some of the more useful ratio combinations and their applications are given in Table 10.2. Such processed images have found practical applications in mapping vegetation, limonite, clays and delineation of hydrothermal alteration zones. Examples are given in Chapter 16.

Table 10.2. Important TM/ETM+ band ratios

TM ratio	Application
7/5	Argillic versus non-argillic
3/4	Rocks versus vegetation
3/1	Fe-O versus non-Fe-O
(4−3)/(4+3)	NDVI (normalized difference vegetation index) for vegetation versus non-vegetation
(2−5)/(2+5)	NDSI (normalized difference snow index) for snow versus no snow

10.8 Colour Enhancement

10.8.1 Advantages

Colour viewing is a highly effective mode of presentation of multispectral images (Buchanan 1979; Haydn et al. 1982; Haydn 1985; Gillespie et al. 1986, 1987). It leads to feature enhancement owing to the following three main reasons.

1. Sensitivity of the human eye. Whereas the human eye can distinguish at the most only about 20–25 gray tones, its sensitivity to colour variation is very high and it can distinguish more than a million colours. Therefore, subtle changes in the image can be more readily distinguished by human interaction if colour is used.

2. Number of variables available. The black-and-white images carry information in terms of only one variable, i.e. tone or brightness (gray level). In comparison to this, a colour space consists of three variables – hue, saturation and brightness.

3. Possibility of collective multi-image display. In remote sensing, we often deal with multiple images, viz. images in different spectral bands, multi-sensor images, multi-temporal images, various computed images, etc. For a collective interpretation, the colour space offers a powerful medium. It stems from the colour theory (Wyszecki and Stiles 1967) that three input images can be viewed in colour space concurrently.

The display of wide-spectrum remote sensing data from gamma ray to microwave in the visible space is possible only through falsification of colours. Therefore, the colours seen on the false-colour image may not have any relation to the actual colours of the objects on the ground.

Colour Enhancement

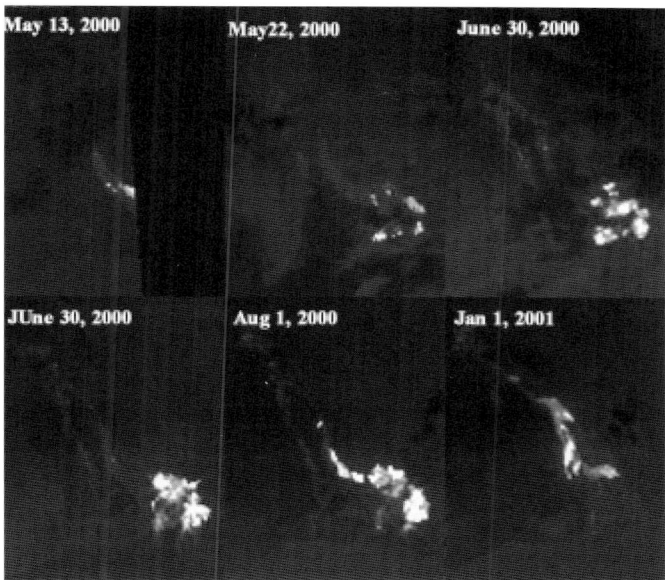

Fig. 10.35. Pseudocolour coding. A sequence of night-time thermal images (ASTER band 14) shows lava flows entering the sea, Hawaii Island. Colour coding from black (coldest) through blue, red, yellow, white (hottest). The first five images show a time sequence of a single eruptive phase; the last image shows flows from a later eruptive phase. (Courtesy of NASA/GSFC/MITI/ERSDAC/JAROS and US/Japan ASTER Science Team) (see colour Plate I)

10.8.2 Pseudocolour Display

This technique of colour enhancement is applied to a single image, i.e. one image at a time. In this, the gray tones are coded in colours, according to a suitable arbitrary scheme. Frequently, a density-sliced image (Sect. 10.5) is processed so that each gray tone slice is displayed in a particular colour, to permit better discrimination of features (also called *density-slicing colour coding*). Figure 10.35 presents an example where different temperatures over a lava flow surface are exhibited in different colours.

10.8.3 Colour Display of Multiple Images – Guidelines for Image Selection

As mentioned earlier, one of the chief advantages of colour display is that multiple (three) images can be displayed and interpreted concurrently. Often, there are many images to choose from, as any simple or processed image can be used as a component in a colour display. Screening and selection of bands may initially be done on the basis of dynamic range, atmospheric effects, knowledge of spectra of ground materials and objects of interest, followed by trial and error to some extent. However, as the number of available images increases the method of trial and error becomes more confusing and the selection more critical.

The purpose of selection is often to maximize information. In this respect, the choice of band combinations can be based on some statistical measures of the input images. For example, Crippen (1989) developed a statistical method to determine optimum bands for display so that the colour composite carries maximum information for discrimination. In this method, first a correlation matrix for the suite of images for the area is calculated; then a ranking index is computed from correlation coefficients for each pair-wise channel combination as:

$$\text{Ranking index} = (1 + 2abc - a^2 - b^2 - c^2)^{1/2} \qquad (10.11)$$

where a, b and c are the pair-wise correlation coefficients. A ranking index value of 1.0 indicates that the three bands in combination would exhibit maximum spectral variability, and an index value of 0.0 suggests the minimum variability. Another statistical measure could be the 'maximum volume of ellipsoid' for selecting three component bands for display in the colour space (Sheffield 1985).

The information content of the colour images depends largely on the quality of the input images. In general, meaningful and good-looking colour displays are obtained if: (a) each of the three component images is suitably contrast stretched and frequency histograms of all three input images are similar, (b) excessive contrast is avoided, as this may lead to saturation at one or both ends, and (c) the mean DN value lies at the centre of the gray range in each of the stretched images.

10.8.4 Colour Models

A number of colour models have been used for enhancing digital images (see Appendix 4.1). We discuss here two basic methods: (1) using the RGB model and (2) using the IHS model.

10.8.4.1 Using the RGB Model

The RGB model is based on the well-known principle that red, green and blue are the three primary additive colours. Display monitors utilize this principle for colour image display.

The RGB model is capable of defining a significant subset of the human perceptible colour range. Its field can be treated as a subset of the more basic and larger chromaticity diagram (Fig. 10.36). The position of R, G and B component-colours in the chromaticity diagram will define the gamut of colour that can be generated by using a particular triplet.

In RGB coding, a set of three input images is projected concurrently, one image in one primary colour, i.e. one image in red, the second in green and the third in blue. In this way, each image is coded in a (maybe false) colour. Variations in DN

Colour Enhancement

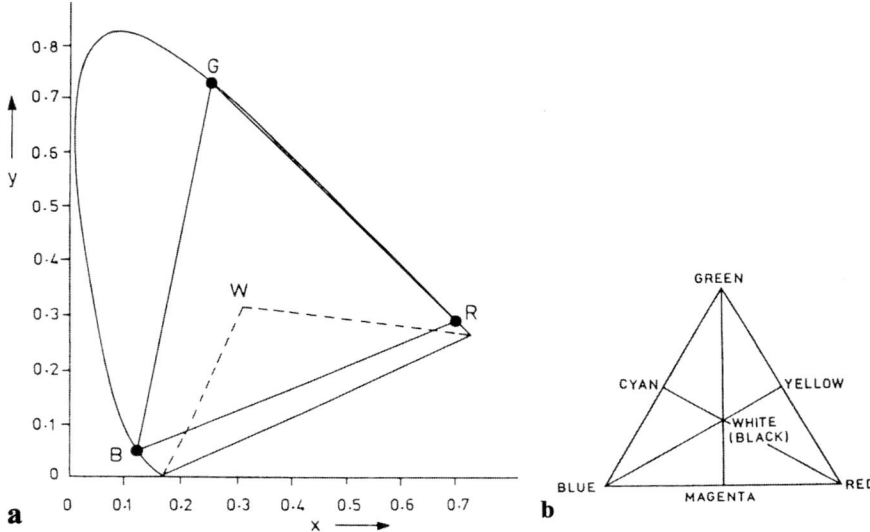

Fig. 10.36. a CIE chromaticity diagram with colour field generated by B, G and R primaries. **b** Schematic position of various colours in the BGR (or RGB) diagram

values at various pixels in the three input images collectively lead to variations in output colours on the colour display.

In principle, any image can be coded in any colour, although it is more conventional to code a relatively shorter wavelength image in blue, and a longer wavelength image in red. Various types of transformed and SAR images can also be used as input images.

Whatever the scheme, the R, G and B points in the chromaticity diagram define a triangle. Therefore, a simple way to understand relations between various colours is through the RGB colour ternary diagram. Several examples are given (see e.g. Figs. 8.10, 16.68, 16.69).

10.8.4.2 Using the IHS Model

In this model, colour is described in terms of three parameters: intensity (I), hue (H) and saturation (S). Hue is the dominant wavelength, saturation is the relative purity of hue, and intensity is the brightness (for details, refer to Appendix 4.1).

Conceptually, the IHS model can be represented with a cylinder, where hue (H) is represented by the polar angle, saturation (S) by the radius and intensity (I) by the vertical distance on the cylinder axis (Fig. 10.37). However, the number of perceptible colours decreases with decrease in intensity and the contributions of hue and saturation become insignificant at zero intensity. For these reasons, a cone seems to be a better representation. The mutual relationships between RGB - and IHS - transforms are discussed by many (e.g. Buchanan and Pendgrass 1980;

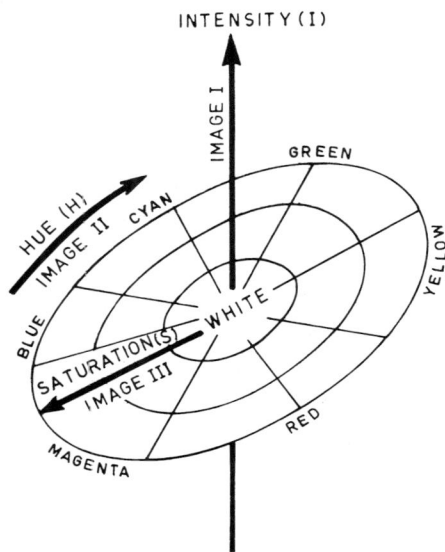

Fig. 10.37. Principle of colour coding in the IHS scheme. The first image is coded in intensity, the second in hue, and the third in saturation

Haydn et al. 1982; Gillespie et al. 1986; Edwards and Davis 1994; Harris et al. 1999).

The coding of remote sensing images using this system is called IHS transform. Each image in the triplet is coded in one of the three colour parameters:

1. first image: 0–255 DN values = intensity variation at pixels (i.e. dark to bright).
2. second image: 0–255 DN values = hue variation (blue–green–red–blue).
3. third image: 0–255 DN values = saturation variation at pixels (i.e. pure saturated colour to gray).

The scheme is inherently highly flexible. However, minor manipulations in hue axis lead to drastic changes in colour of the pixels (for example, suddenly from blue to green) on the display screen, and decoding of the colours has to be done very carefully.

When images of differing resolutions are being colour-displayed in IHS, it is useful to have the high-spatial-resolution image as the intensity image, as this becomes the base image and all other images can be easily registered to this. Further, hue is the polar angle of the conical/cylindrical colour space, and in principle, we can use the entire range blue–green–red–magenta–blue for coding. However, as this renders the same hue (blue) to the lowest and highest DNs, it is customary to restrict the hue range as blue–green–red for coding.

Examples of IHS coding are given in figures 16.113 and 16.121.

10.9 Image Fusion

10.9.1 Introduction

Image data can be acquired from a variety of aerial/space-borne sensors with differing spatial, temporal and spectral resolutions. Image fusion is the technique of integrating these images and other data to obtain more and better information about an object of study that would not be possible to derive from single-sensor data alone. It can be defined as 'a process of merging data from multiple sources to achieve refined information. Thus, it is a tool to combine multi-source imagery and data using advanced processing techniques (for a review on image fusion, refer to Pohl and van Genderen 1998). Fused data products provide increased confidence and reduced ambiguity in image interpretation and applications.

Image fusion techniques can be applied to digital images in order to:
- sharpen images
- improve geometric corrections
- enhance certain features not visible in either of the single data alone
- complement data sets for improved classification
- detect changes using multi-temporal data
- substitute missing information in one image with signals from another sensor (e.g. clouds, shadows etc.)
- replace defective data.

10.9.2 Techniques of Image Fusion

The techniques used for image fusion can be categorized into three types depending upon the stage at which the image data set are fused:

(a) pixel-based fusion
(b) feature-based fusion
(c) decision-based fusion

The first step in all fusion methods is pre-processing of the data to ensure that radiometric errors are minimized and the images are co-registered. The subsequent steps may differ according to the level of fusion. After image fusion, the resulting image may be further enhanced by image processing.

Fusing data at pixel level requires co-registered images at sub-pixel accuracy. The advantage of pixel fusion is that the input images contain the most original information; therefore, the measured physical parameters are retained and merged. In *feature-level fusion,* the data are already processed to extract features of interest to be fused. Corresponding features are identified by their characteristics, such as extent, shape and neighbourhood. In *decision-level fusion,* the features are classified and fused according to decision rules to resolve differences and provide a more reliable result to the user. For geological applications, most commonly the pixel-based fusion techniques are applied to remote sensing and ancillary image data, which will be discussed in the following paragraphs.

The pixel-based fusion techniques can be grouped into three categories:

(1) statistical and numerical methods
(2) colour transformations
(3) wavelet transform method

10.9.2.1 Statistical and Numerical Methods

A number of methods can be applied to statistically and numerically process the co-registered multiple-image data sets. For example, the following techniques generate image data with new information:
− principal component transformation
− decorrelation stretch
− addition/subtraction
− ratioing/multiplication.

These can also be treated under the general category of image fusion, as new computed images carry information from multiple input images.

Image fusion is frequently used to sharpen images. The problem is that often we have multiple images in different spectral bands, with different spatial resolutions, e.g. a panchromatic band image with high spatial resolution and several other coarser-spatial-resolution spectral bands. It would be good idea to generate a new image which exhibits the spectral character of a particular band but carries spatial information from the high-spatial-resolution image.

A simple technique to sharpen the image from a multi-resolution image data set using addition is as follows:

− first generate an 'edge' image from a high-resolution panchromatic image
− add this edge image to a lower-resolution image (Tauch and Kähler 1988).

The new image will exhibit the radiometric character of the lower-resolution image but the sharpened edges from the high-spatial-resolution image.

Bovery transform. This technique is of special interest in image sharpening and deserves specific mention. It maintains the radiometric integrity of the data while increasing the spatial resolution. The technique normalizes a band for intensity, and then the normalized band is multiplied by a high-resolution data set. Thus, the transformed band is defined as follows:

$$DN_{b1\,fused} = \frac{DN_{b1}}{(DN_{b1} + DN_{b2} ... DN_{bi} ... + DN_{bn})} \times DN_{Highres} \qquad (10.12)$$

where $DN_{b1fused}$ is the digital number of the resulting fused image in band1. The advantage of this technique is that it optically maintains the spectral information

Image Fusion

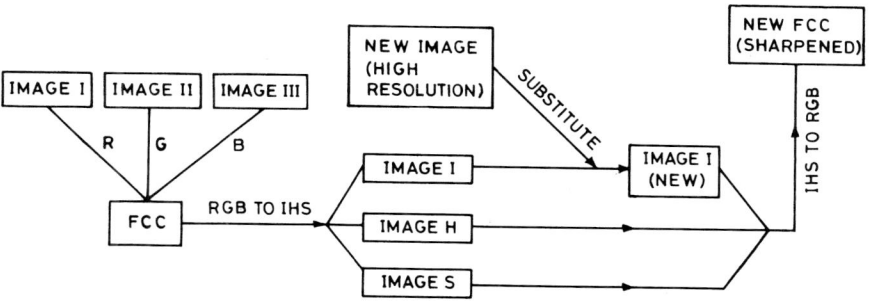

Fig. 10.38. Steps in image fusion and sharpening using RGB–IHS–RGB colour transformation

of the band whilst sharpening the scene. In this way, a set of three bands can be sharpened and then displayed in RGB.

10.9.2.2 Colour Transformations (RGB–IHS)

Image fusion by colour transformations takes advantage of the possibility of presenting multiple-image data in different colours. The RGB system allows the assigning of three different images to the three primary additive colours – red, green and blue – for collective viewing.

The most commonly used transform for multiband image data these days is the IHS transform that separates spatial (I) and spectral (H and S) from a standard RGB image. The IHS technique can be applied for image fusion either in a direct manner or in a substitutional manner. The first refers to the transformation of three image channels assigned to I, H and S directly. In the latter, an RGB composite is first resolved in to IHS colour space; then, by substitution, one of these three components is replaced by another co-registered image (Fig. 10.38). An inverse transformation returns the data to the RGB colour space, producing the fused data set. The IHS transformation has become a standard procedure in image analysis. The method provides colour enhancement of correlated data (Gillespie et al. 1986), feature enhancement (Daily 1983; Harris et al. 1990), improvement of spatial resolution (Welch and Ehlers 1987; Carper et al. 1990; Eosat 1994) and fusion of disparate data (Ehlers 1991). An example of a fused image using this method is given in figure 16.100.

10.9.2.3 Wavelet Transform Method

Given a set of images with different spatial resolutions and spectral content, this method allows improved spatial resolution of the coarser-resolution image, which can be increased up to the level of the highest-spatial-resolution image available.

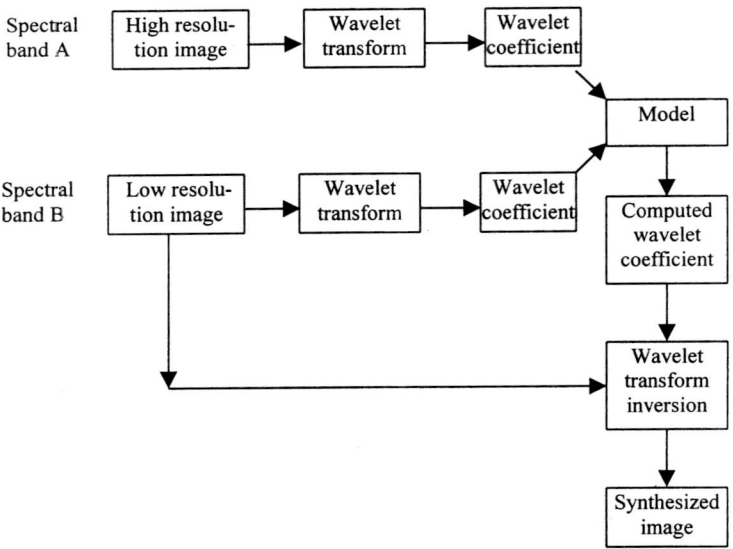

Fig. 10.39. Main steps in wavelet transform method for generating a synthesized image with spectral content of low-resolution image and spatial resolution of high-resolution image

The concept makes use of multi-resolution analysis (MRA) using wavelet transform (WT). WT is used to compute successive approximations of an image with coarser and coarser spatial resolutions, and to represent the difference of information existing between two successive approximations. Wavelet coefficients allow isolation of finer structures. In practice, wavelet coefficients from both low-resolution and high-resolution images are computed separately, and the data are fed into a model. From this, new wavelet coefficients are computed for generating a synthesized image, which corresponds to the spectral character of the low-resolution image but the spatial resolution of the high-resolution image (Fig. 10.39). For further details on this method see, for example, Alparone et al. (1998); Couloigner et al. (1998); Ranchin et al. (1998) and Zhou et al. (1998).

10.10 2.5-Dimensional Visualization

The term 2.5-dimensional visualization is used to imply that the display is more than 2-D but not fully 3-D. Raster images are typically two-dimensional displays in which the position of a pixel is given in terms of X, Y co-ordinates. In remote sensing image processing, the DN values can be considered as Z data, i.e. at each

X, Y location, a certain Z value can be associated. This means that the raster image data can be portrayed as a surface. However, in a true 3-D, it should be possible to provide more than one Z value at each X, Y location, to enable modelling of features in 3-D (e.g. an overturned fold or an orebody below the gossan). As remote sensing image processing generates a surface of varying elevation across the scene, the term 2.5-D is used since the display is more than 2-D but not fully 3-D.

There are different ways of visualizing 2.5-D raster data:
(1) shaded relief model
(2) synthetic stereo
(3) perspective view.

10.10.1 Shaded Relief Model (SRM)

A shaded relief model is an image of an area that is obtained by artificially illuminating the terrain from a certain direction. To generate a shaded relief model we need a digital terrain model (DTM), from which slope and aspect values at each pixel are calculated. Then, using an artificial light source (e.g. 'artificial sun'), and assuming a suitable scattering model, scattered radiance is calculated at each pixel, to yield an SRM. Changing the illumination direction can generate an infinite number of SRMs of an area (see also Sect. 15.6.3).

10.10.2 Synthetic Stereo

Another practical possibility for combining and enhancing multi-image data is by developing a synthetic stereo (Pichel et al. 1973; Batson et al. 1976; Sawchuka 1978; Zobrist et al. 1979; Haydn et al. 1982). It utilizes the simple geometrical relationship described by the standard parallax formula in aerial photogrammetry. The only difference is that now the height–parallax relationship is used to calculate parallax values, which are artificially introduced in the image, to correspond to the 'height' data.

Consider two images A and B to be merged by synthetic stereo; one of these images is to be used as the base image and the other as the 'height' image. Generally, the high-resolution image is taken as the base image on which data from the lower-resolution image is superimposed as height data. The two images are first registered over each other so that the 'height' values from the second image can be assigned to corresponding pixels in the base image. The geometric relationship for deriving the parallax shift is well known. Using DN values in the 'heighting' image, parallax values are calculated for each pixel which are to be assigned to the base-image pixels. The pixels in the base image are shifted by an amount corresponding to the parallax value. The processing is done pixel by pixel, line by line, to generate a set of new images. The resulting stereo pair can be analysed with an ordinary stereoscope. For example, figure 10.40 shows a synthetic stereo in which the base image is TM4 and the parallax is due to the image TM6 (thermal-IR

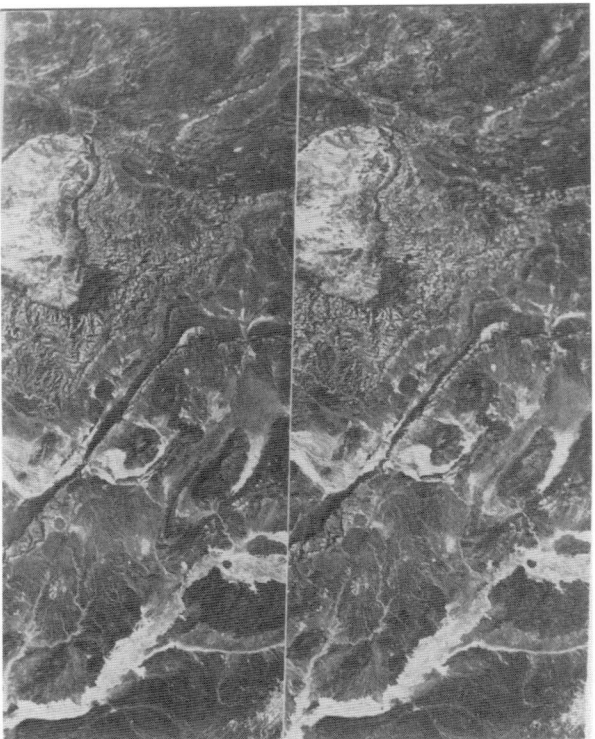

Fig. 10.40. Synthetic stereo formed by using TM4 as the base image and TM6 (thermal channel) as the height information. (Kaufmann 1985)

channel). Temperature variations appear like an artificial topography, where areas with higher temperatures appear elevated.

10.10.3 Perspective View

Perspective views are based on integration of remote sensing and DTM data. First, the remote sensing image is registered and draped over the topographic map. Then, the 3-D model can be viewed from any angle to provide a perspective view. Such models help in terrain conceptualization. Figure 10.41 shows a perspective view generated from MOMS-02P. The aft and fore cameras were used to generate the stereo model from which the DTM was derived. On this DTM, the high-resolution panchromatic channel was draped over.

2.5 Dimensional Visualization

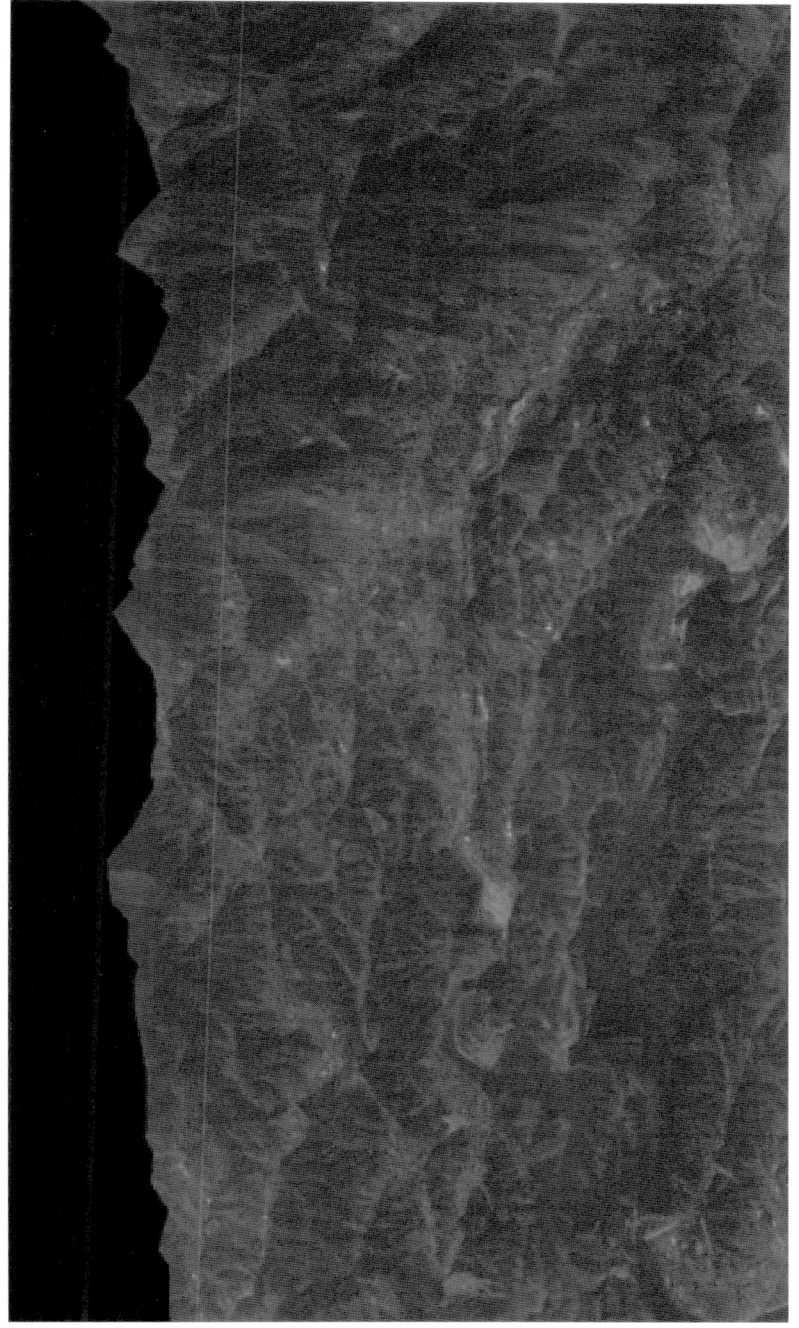

Fig. 10.41. Perspective view generated from MOMS-02P of an area in Korea. (Courtesy of DLR, Oberpfaffenhofen)

10.11 Image Segmentation

The purpose of image segmentation is to subdivide the image into zones which are homogeneous with respect to image properties. From a geological point of view, image segmentation could be of interest for automatic lithological mapping. However, several variables are involved in the end product, and the technique is still in the developmental stage. The reader may refer to other texts for details (e.g. Hsu 1978; Haralick 1979; Brady 1982; Rosenfeld and Kak 1982; Farag 1992; Pitas 1993).

The segmentation of an image into sub-areas involves two considerations: (1) the sub-area should be checked for homogeneity and (2) the presence of break or change characteristics between two sub-areas should be shown. Putting boundaries within any gray scale in order to subdivide the image is called thresholding and is a highly critical operation. Using only the gray level as the parameter for subdividing an image into uniform zones may be relatively easy. This is done by density slicing, which may be based on a certain statistical principle, such as multimodal distribution in a DN histogram. If the DN histogram has many modes, then each mode may be assumed to correspond to one category of objects, and troughs in the histogram can form boundaries for density slicing for image segmentation. The image can also be subdivided by gradient edges (Sect. 10.6.1), which are produced by sudden changes in DN values of successive pixels. This method may be used for finding sub-areas where the gray levels are more or less uniform, and for setting the boundaries where the gray levels suddenly change.

However, the present capability of these techniques is rather limited, as factors such as illumination, vegetation, moisture, topography etc. may predominantly affect the local gray-level values.

Texture is a relatively complicated parameter to quantify. It is related to spatial repetition of constituent elements and their orientation. Local image characters at various points, such as mean, variance and directional difference, can be used to compute a parameter which can possibly define texture. However, the technology in this direction is still under development.

10.12 Digital Image Classification

The purpose of digital image classification is to produce thematic maps where each pixel in the image is assigned on the basis of spectral response to a particular theme. The methods of image classification are largely based on the principles of pattern recognition. A pattern may be defined as a meaningful regularity in the data. Thus, the identification of pattern in the data is the job of the classification. In the context of remote sensing, the process of classification involves conversion of satellite- or aircraft-derived spectral data (DN values) into different classes or themes of interest (e.g. water, forest, soil, rock etc.). The output from a classification is a classified image, usually called a *thematic map* of the image.

Digital Image Classification

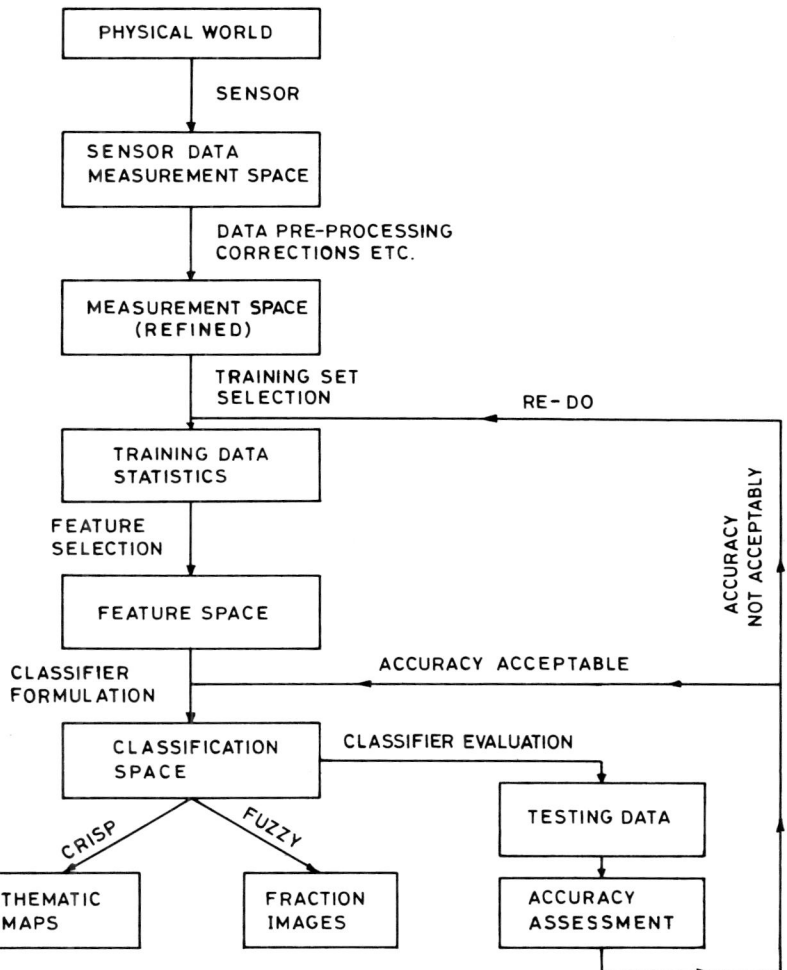

Fig. 10.42. Flow diagram of data in supervised classification approach

With the unprecedented increase in the availability of a large amount of multispectral digital data, the use of sophisticated computer-based classification techniques has become essential. Moreover, the digital thematic maps prepared from the classification can be used as direct input to a GIS.

There are two basic approaches to digital image classification: supervised and unsupervised. In a *supervised classification*, the analyst first locates representative samples of different ground-cover types (i.e. classes or themes) of interest in the image. These samples are known as *training areas*. These training areas are selected based on the analyst's familiarity with the area and knowledge of the actual ground-cover types. Thus, in a way, the analyst supervises the process of image classification, hence the name. The digital numbers of the pixels in all spectral bands representing these training areas are then used to generate some statistical parameters (also known as *training-data statistics* or signatures), depending upon the classification algorithm used. Finally, the decision rule developed on the basis of training-data statistics allocates each pixel of the image to one of the classes.

In *unsupervised classification* (also called *cluster analysis*), practically the reverse happens. The classes are first grouped spectrally on the basis of the digital numbers of the pixels. The analyst then confirms the identification of the groups to produce meaningful classes. A range of clustering algorithms may be used to produce these natural groups or clusters in the data. The analyst may, however, have to supply to these algorithms with some minimum information, such as the number of clusters, the statistical parameters to be used etc. Thus, unsupervised classification is also not completely free from the analyst's intervention. However, it does not depend upon a priori knowledge of the terrain.

A hybrid approach involving both supervised and unsupervised methods may generally be the best. For example, in the first stage, an unsupervised algorithm may be used to form the clusters, which can provide some information about the spectrally separable classes in the area. This could assist in generating training-data statistics. Then, in the second stage, using a suitable supervised classification algorithm, the pixels are allocated to different classes.

A large number of supervised and unsupervised statistical classification algorithms have been developed; a few of the commonly used algorithms will be described here.

10.12.1 Supervised Classification

A supervised classification involves three distinct stages: training, allocation and testing. A schematic representation of a supervised classification process is presented in Fig. 10.42.

10.12.1.1 Construction of Measurement Space

The physical world contains objects that are observed by a remote sensor. Data from measurements constitute the measurement space (i.e. each observed point represents a digital number in each of the n spectral channels in an n-dimensional measurement space). The sensor data may be corrected or processed (e.g. by enhancements, or multi-image operations etc.) before any classification is attempted. Although pre-processing may yield a refined measurement space in which it may be easier to recognize patterns, the spectral nature of the data gets changed; therefore, pre-processing operations ought to be performed very carefully, and only when really necessary.

10.12.1.2 Training

The training of a classifier is one of the most critical operations (Hixson et al. 1980). Training data are areas of known identity that are demarcated on the digital image interactively. The training data should be sufficiently representative of the classes in the area. As collection of training data is a costly affair, the size of the training data set ought to be kept small, but at the same time it should be large enough to accurately characterize the classes. The training data sets should therefore meet the following general requirements:
(a) training areas should be homogeneous (i.e. should contain pure pixels);
(b) a sufficiently large number of pixels should be available for the training data set;
(c) the training data set of each class should exhibit a normal distribution;
(d) training areas should be widely dispersed over the entire scene.

Typically, the sample size for the training data is related to the number of variables (viz. spectral channels/bands). Accordingly, the training sample size varies from a minimum of $10b$ per class to $100b$ per class, where b is the number of bands (Swain 1978). However, it is not a question of 'the bigger the better'.

Ideally, the training data should be collected at or near the time of the satellite or aircraft over-pass, in order to accurately characterize the classes on the image. This is, however, not always possible in practice, due to financial, time and resources constraints. Nonetheless, the time of training data collection is crucial to some studies, such as land-cover-change detection and crop classification.

In order to judge the quality of the training data, the following preliminary statistical study may be carried out.
(a) The histogram of the training data for each class may be examined. A normally distributed uni-modal curve represents good-quality training samples for that class.
(b) A matrix showing statistical separability (normalized distance between class means) may be computed for each spectral band to check whether or not any two classes are mutually distinguishable in any one or more spectral bands. The classes that have poor statistical separability in all the bands should be merged into each other.
(c) In order to obtain a better physical idea about the training data set, cross-plots or scatterograms between various spectral bands may also be plotted, to graphically depict the mutual relations of responses in different spectral bands for all of the classes.

After having decided about the quality, number and placement of training areas, the stage is now set for the next step, i.e. feature selection.

10.12.1.3 Feature Selection

Feature selection deals with the distillation process to decipher the most useful bands contained in the measurement space. It is performed to bin the redundant information, in order to increase the efficiency of classification.

As remote sensing data are collected in a large number of bands, some of the spectral bands exhibit high correlation. In such a situation, utilizing data from all of the spectral bands would not be advantageous for the classification process. The main purpose of feature selection is to reduce the dimensionality of the measurement space, minimizing redundancy of data, while at the same time retaining sufficient information in the data. The feature selection may be performed by utilizing various separability indices such as divergence, transformed divergence, Bhatacharya distance, J-M distance, or from principal component analysis etc. (Swain 1978).

Ratios and *indices* of spectral measurements can also be used as feature axes. For example, vegetation index (NDVI), snow index (NDSI), or various spectral ratios to map Fe-O, hydroxyl minerals etc. (Table 10.2) can also form feature axes for image classification. In addition, principal component analysis and canonical analysis, which have been already mentioned (Sect. 10.7.2), can also form feature axes in image classification.

10.12.1.4 Allocation

Once a suitable feature space has been created, the remote sensing image can be classified. The aim of the allocation stage is to allocate a class to each pixel of the image using a statistical classification decision rule. In other words, the objective is to divide the feature space into disjoint sub-spaces by putting decision boundaries so that any given pixel or pattern can be assigned to any one of the classes. The strategy of a statistical classifier may be based on:
(1) geometry
(2) probability
(3) a discriminant function.

1. Pattern classifiers based on geometry. This type of classifier uses a geometric distance as a measure for sub-dividing the feature space. Two such typical classifiers are the minimum distance to means (or centroid classifier) and the parallelepiped classifier.

The *minimum distance to means classifier* is the simplest one. The mean vector (as obtained from training data statistics) representing the mean of each class in the bands selected is used here to allocate the unknown pixel to a class. The spectral distance between the DN value of the unknown pixel and mean vectors of the various classes is successively computed. (There are a number of ways to compute the distance, e.g. Euclidean distance, Mahalobonis distance.) The pixel is assigned to a class to which it has the minimum distance. Figure 10.43 shows a two-dimensional representation of the concept of this classifier.

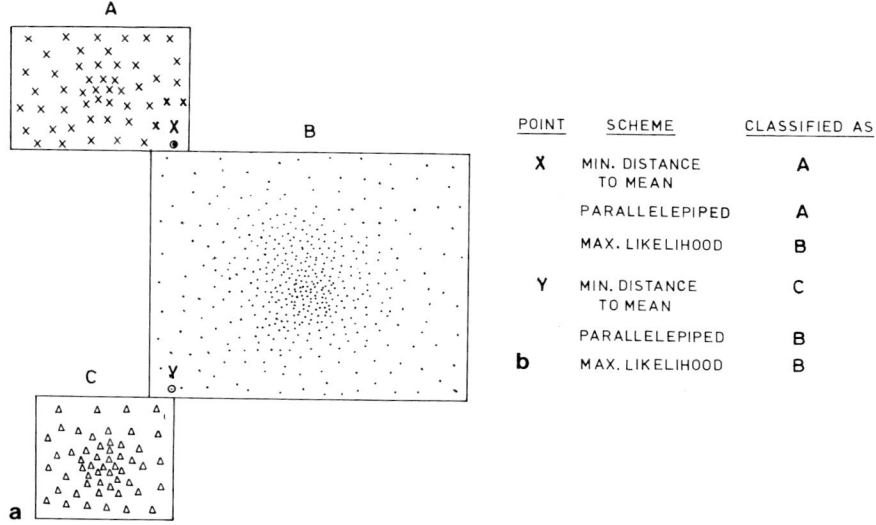

Fig. 10.43a,b. Comparison of some supervised classification schemes. **a** The three known class categories are *A, B* and *C;* unknown pixels *X* and *Y* are to be classified. **b** The categorization of the points *X* and *Y* as *A, B* or *C*, if different classification criteria are used

One major limitation of this classifer is that it does not take into account the variance of the different classes. Thus, using the above rule, an unknown pixel may be classified in a particular class, whereas it may go to another class if the scatter (or variance) of the class in a certain band is also considered.

The *parallelepiped classifier* is a type of classifier that may incorporate variance also. In this classifier, a range is usually determined from the difference between lower and upper limits of the training-data statistics of a particular class. These limits may be the minimum and maximum DN values of the class in a certain band, or may be computed as $\mu \pm \sigma$ (where μ is the mean of a class and σ is the standard deviation of the class in a band). Once the lower and upper limits have been defined for the classes under consideration, the classifier works on each DN value of the unknown pixel. A class secures the pixel if it falls within its range. For example, in a two-dimensional representation (Fig. 10.43), the classes possess rectangular fields and an unknown pixel lying within the field is classified accordingly. Parallelepiped classifiers are in general very fast and fairly effective. However, their reliability sharply declines if classes have large variances, which may lead to high correlation along the measurement axes. In addition, if a pixel does not fall within a particular range it may remain unclassified.

2. Probability-based classifier or maximum likelihood classifier (MLC). This type of classifier is based on the principle that a given pixel may be assigned to a class to which it has the maximum probability of belonging. A common strategy used is called the Bayes optimal or Bayesian, which minimizes the error of misclassifica-

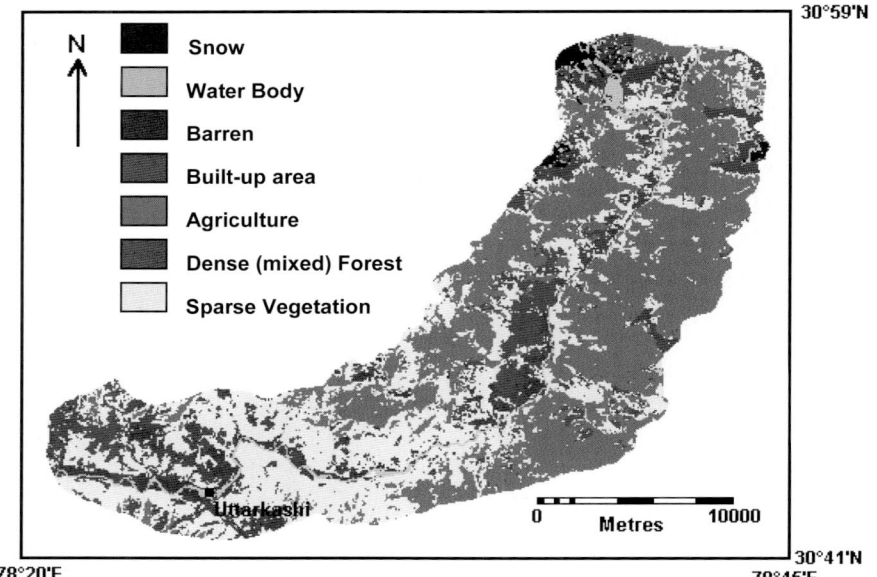

Fig. 10.44. Land-use/land-cover classification of remote sensing data (IRS-LISS-II) (using MLC) of a part of the Ganges valley, Himalayas. (Gupta et al. 1999; see colour Plate II)

tion over the entire set of data classified. It is another way of stating the maximum likelihood principle. The method is based on a PDF (probability density function) obtained from a Gaussian Probability distribution. The probabilities of a particular pixel belonging to each class are calculated, and the class with the highest probability secures the pixel (Fig. 10.43).

In practice there may be several classes, each having a different inherent probability of occurrence in the area, and this aspect may also be duly taken into account by considering the a priori probability of occurrence of the various classes. A priori probabilities may be generated from other data, such as by incorporating the effects of terrain characteristics (Strahler 1980) or by including data from a non-parametric process (Maselli *et al.* 1992).

Further, often a thresholding is also applied so that in the case that even the highest PDF is below a certain threshold limit (generally a value of two standard deviations), the pixel may be classified as unknown, lest the overall accuracy of the classification output may deteriorate. The MLC technique uses a fair amount of computer time, but has been most popularly used for classifying remote sensing data as it gives a minimum of errors.

Figure 10.44 shows an example of land-use/land-cover classification using IRS-LISS-II data in the Ganges valley, Himalayas.

3. Classifier based on a discriminant function. Such classifiers associate a certain function with each class and assign the unknown pixel to that class for which the value of the function is maximum. The determination of the optimum discriminating function or decision boundary is basically the training or learning in such a case. The technique is powerful if the classes can be linearly separated along any feature axis. The transformation involves the building up of a feature axis along which the groups are separated most and inflated least. In a comparative analysis between discriminant analysis and MLC (Tom and Miller 1984), it was found that the discriminant analysis algorithm offered distinct advantages in accuracy, computer time, cost and flexibility over the MLC in nearly all data combinations and the advantages increased manifoldly as the number of parameters was increased.

Overall, the maximum likelihood classifier (MLC) has been universally accepted as an effective method of classification. However, as with other statistical classifiers, problems exist with this classifier also. First is its normal distribution assumption: sometimes the spectral properties of the classes are far from the assumed distribution (e.g. in complex and heterogeneous environments). Second, this classifier is based on the principle of *one pixel one class allocation* (Wang 1990), thereby forcing each pixel to be allocated to only one class, although it may correspond to two or more classes (viz. mixed pixels). The maximum likelihood classifier shows marked limitations under these circumstances.

10.12.2 Unsupervised Classification

This classification method is applied in cases where a priori knowledge of the types of classes is not available and classes have to be determined empirically. It involves development of sets of unlabelled clusters, in which pixels are more similar to each other within a cluster, and rather different from the ones outside the cluster. Because of this characteristic of cluster building, it is also termed *cluster analysis*. In practice, the spectral distance between any two pixels is used to measure their similarity or dissimilarity. The simplest way to understand it is as follows: start with one pixel and assign to it the name cluster I; then take the next pixel, and compute its spectral distance to cluster I; if this distance is less than a certain threshold (say d), then group it in cluster I; alternatively put it in cluster II; then take the next pixel and classify it as a member of cluster I or II, or alternatively cluster III, according to its mathematical location. In this way, all the subsequent pixels may be classified.

Unsupervised classification is highly sensitive to the value of threshold distance d, which has to be very carefully chosen. The value of d controls the number of clusters that develop; it may happen that a particular d may yield clusters having no information value. The technique is very expensive by way of computer time, as each subsequent pixel has to be compared with all the earlier clusters, and therefore as d is reduced, the number of clusters increases and the computer time increases manifoldly.

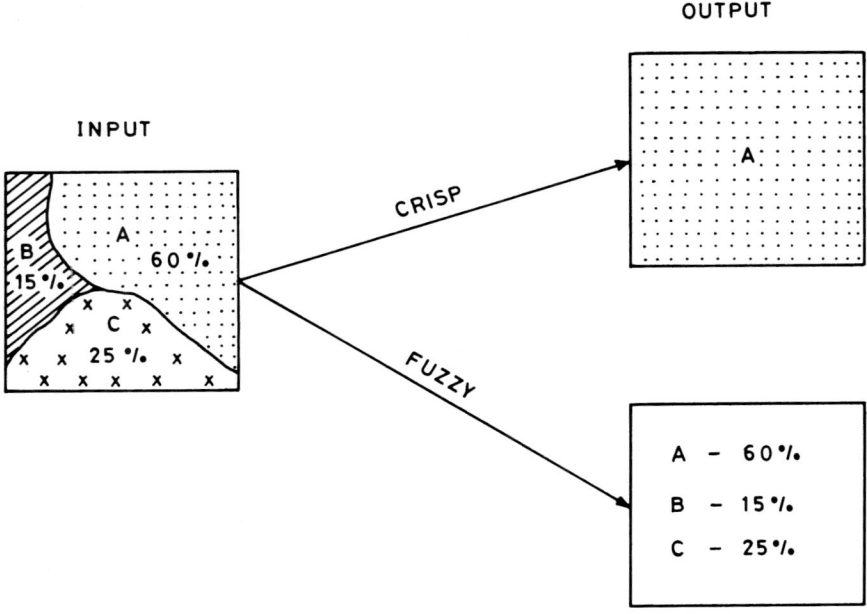

Fig. 10.45. The concept of crisp versus fuzzy classifiers. In a crisp classifier, a pixel is classified as belonging to the dominant category; in a fuzzy classifier, a pixel is assigned fractions of thematic categories

10.12.3 Fuzzy Classification

The classifiers mentioned so far seek to designate mutually exclusive classes with well-defined boundaries, and therefore are based on the 'one pixel, one class phenomenon' (Key et al. 1989; Wang 1990). These are termed *crisp classifications* (Fig. 10.45).

In practice, however, such an assumption may not be met by many pixels as they may actually record reflectance from a number of different classes within a pixel (mixed pixels!) (Chikara 1984; Fischer and Pathirana 1990; Foddy and Arora 1996). Therefore, other approaches are required to classify pixels, which may use the method of allocating multiple-class membership to each pixel. Such classifications are termed *fuzzy classifications*. The concept is based on fuzzy set theory, in which pixels do not belong to only one class but instead are given membership values for each class being constructed. These membership values range between 0 and 1, and sum to unity for each pixel. A pixel with a membership value of 1 signifies a high degree of similarity to that class, while a value near to 0 implies no similarity to that class.

Among the various fuzzy classifications, the fuzzy-c means (FCM) algorithm is more widely used (Bezdek et al. 1984). FCM is a clustering algorithm that randomly assigns pixels to classes and then, in an iterative process, moves pixels to other classes so as to minimize the generalized least-squares error. In addition, the well-known MLC can also be used as a fuzzy classifier, the reason being that the probabilities obtained from MLC have been found to be related to the proportion of classes within a pixel on the ground.

In a fuzzy classification, there will not be a single classified image but a number of fraction images or proportion images, equal to number of classes considered. For example, if there are five classes, there will be five fraction images.

10.12.4 Linear Mixture Modelling (LMM)

LMM is also a strategy to classify sub-pixel components. It is based on a simple linear equation, derived on the assumption that the DN value of a pixel is a linear sum of the DN values of component classes, weighted by their corresponding area (Settle and Drake 1993). Based on this, class proportions can be obtained, which sum to one. However, if the spectral responses of component classes do not exhibit a linear mixing, the results may become misleading. These aspects are discussed in greater detail in Chapter 11 (Hyperspectral Sensing).

10.12.5 Artificial Neural Network Classification

Another classifier is the artificial neural network (ANN) classifier, which is regarded as having enormous potential for classifying remotely sensed data. The major appeal in using neural networks as a classification technique is due to their probability-distribution-free assumption and their capability to handle data from different sources in addition to remote sensing data from different sensors.

A neural network comprises a relatively large number of simple processing units (nodes) that work in parallel to classify input data into output classes (Hepner et al. 1990; Foody 1992; Schalkoff 1992; Foody and Arora 1997). The processing units are generally organized into layers. Each unit in a layer is connected to every other unit in the next layer. This is known as a feed-forward multi-layer network.

The architecture of a typical multi-layer neural network is shown in figure 10.46. As can be seen, it consists of an input layer, a hidden layer and an output layer. The input layer merely receives the data. Unlike the input layer, both hidden and output layers actively process the data. The output layer, as its name suggests, produces the neural network's results. Introducing a hidden layer(s) between input and output layer increases the network's ability to model complex functions.

The units in the neural network are connected with each other. These connections carry weights. The determination of the appropriate weights of the connections is known as learning or training. Learning algorithms may be categorized as supervised and unsupervised, as in a conventional classification. Thus, the magnitudes of the weights are determined by an iterative training procedure in which the network repeatedly tries to learn the correct output for each training sample. The procedure involves modifying the weights between units until the network is able

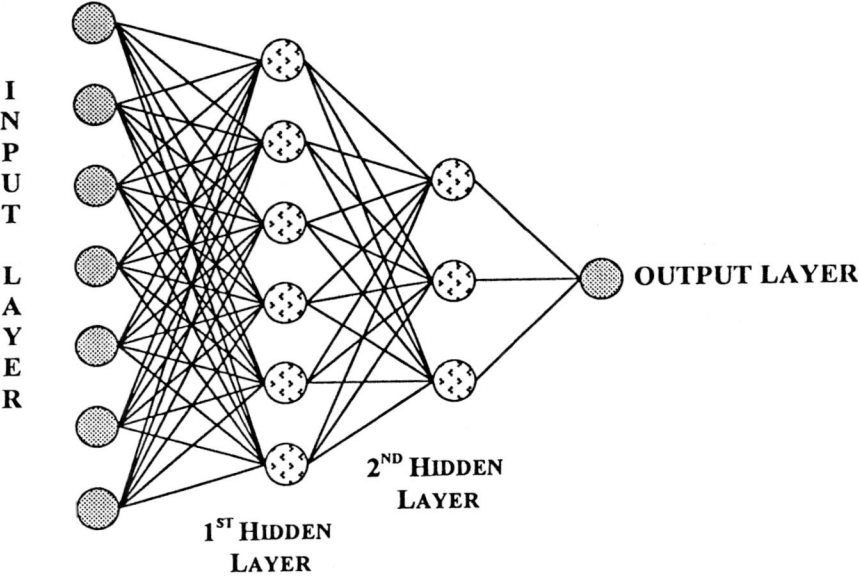

Fig. 10.46. Architecture of an artificial neural network (ANN) to characterize the training data. Once the network is trained, the adjusted weights are used to classify the unknown data set (i.e. the unknown pixels of the image).

In recent years, the neural network has been viewed as the most challenging classification technique after the MLC. An important feature of the neural network classification is that higher classification accuracies may be achieved with neural networks than with conventional classifiers, even when a small training data set is used. Furthermore, neural networks have been found to be consistently effective under situations where data from different sources other than remote sensing are considered (Foody 1995; Arora and Mathur 2001). The aspects of using ancillary data for classification are discussed further in Chapter 15.

10.12.6 Classification Accuracy Assessment

A classification is rarely perfect. Inaccuracies may arise due to imperfect training or a poor strategy of classification. Therefore, it is useful to estimate errors in order to have an idea of the confidence attached to a particular classification.

One of the most common ways of representing the classification accuracy is by defining it with the help of an error matrix (Congalton 1991). An *error matrix* is a cross-tabulation of the classes on the classified remotely sensed image and the ground data. It is represented by a $c \times c$ matrix (where c is the number of classes) (e.g. Table 10.3). The columns of the matrix generally define the ground data while the rows define the classified remotely sensed data (albeit the two are interchangeable). The error matrix has been referred to in the literature by different names, such as *confusion matrix, contingency table, misclassification matrix,* etc.

Digital Image Classification

Table 10.3. Error matrix for assessing classification accuracy

		Classified map								Total pixels	Accuracy
		Snow	Water body	Barren	Built-up area	Agri.	Sparse veg.	Mod. veg.	Dense forest		
Reference data	Snow	**104**	1	0	1	0	0	0	0	106	0.98
	Water body	1	**99**	1	3	0	0	0	0	104	0.95
	Barren	3	3	**75**	22	2	0	1	0	106	0.71
	Built-up area	0	7	11	**71**	5	6	0	1	101	0.70
	Agriculture	0	0	15	10	**69**	3	3	1	101	0.68
	Sparse vegetation	0	1	14	4	3	**63**	22	11	118	0.53
	Moderate vegetation	0	0	13	2	3	42	**32**	23	115	0.28
	Dense forest	0	1	1	0	0	20	11	**92**	125	0.74

Total testing pixels = 876   Average accuracy = 69.69%
Total diagonal elements = 605   Overall accuracy = 69.06%

To generate an error matrix, a set of testing-data are used which are usually collected from the ground data in the same way as the earlier used training-data. The testing area consists of pixels of known classes, which were not used in training the classifier, but were preserved and held back for the purpose of classifier evaluation at a subsequent stage. The resulting output from the classifier is compared with the ground truth by building up an error or confusion matrix. The feedback from the classifier evaluation could also be used to modify the strategy of classification or to improve feature selection.

Scope for Geological Applications

The digital classification approach to multispectral remote sensing data has found rather limited applications in geological–lithological mapping. Most of the world-wide available remote sensing data (e.g. Landsat MSS, TM, SPOT-HRV, IRS-LISS) pertain to principally the solar reflection region, which brings information from the top few-microns-thick surficial cover over the Earth's surface. The thermal-infrared data are related to the few-centimetres-thick top surface layer and, although better in this respect, are scarcely available with proper spatial resolution. In geological mapping, the geologist is interested in defining bedrock lithologies, irrespective of the type of surface cover, such as vegetation, scree, alluvium, cultivation etc., which may serve only as a guide at most. This could be a major reason why Landsat-data-derived classification maps are reported to have a low correspondence with field geological maps (c.f. Siegal and Abrams 1976).

It is in this context that enhancement has been a comparatively more rewarding approach for geological applications. In practice, the geologist can avail himself of an enhanced remote sensing product and interpret the image data suitably, considering the various surface features such as topography, drainage, land-cover, vegetation, alluvium, scree etc., for optimum lithological discrimination and mapping.

Chapter 11: Hyperspectral Sensing

11.1 Introduction

Hyperspectral sensing, also called imaging spectrometry, is a relatively new field that has rapidly grown during the last two decades. The term hyperspectral is used here to indicate a very large number of narrow spectral channels, e.g. 64 to about 200 channels at 10–20-nm interval, in comparison to the multispectral sensing in which we have typically 4–10 spectral channels at approximately 100–200-nm interval. The main aim in hyperspectral sensing is to image a scene in a large number of discrete contiguous narrow spectral bands. An almost continuous spectrum can be generated for each pixel, and the data are in image format, hence the name imaging spectrometry (Fig.11.1). A wealth of review information on imaging spectrometry technique and its geological applications can be found in Clark (1999), Kruse (1999), Mustard and Shine (1999) and van der Meer (1999).

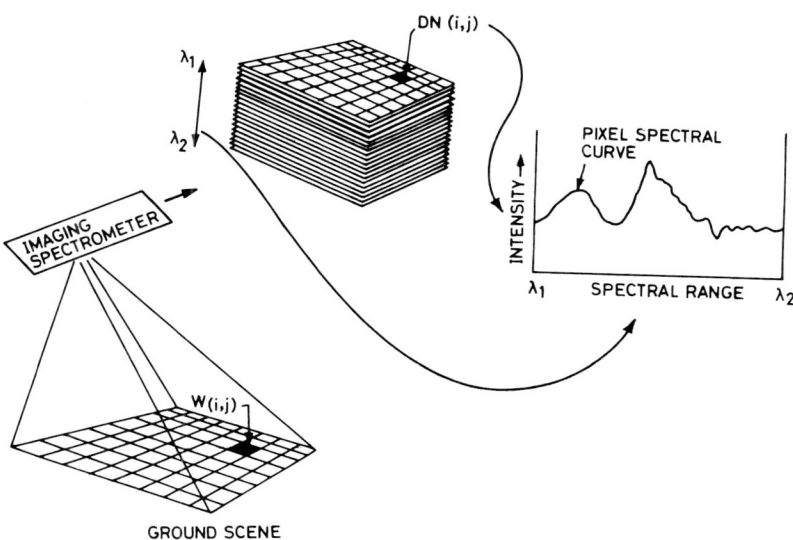

Fig. 11.1. Concept of imaging spectrometry

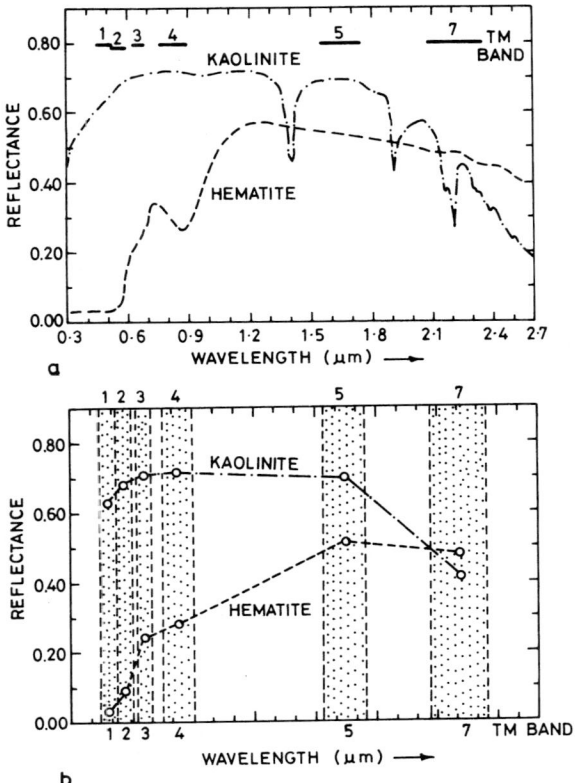

Fig. 11.2. **a** Laboratory spectra of kaolinite and hematite; **b** the same, as would be recorded by Landsat TM

Advantages and limitations. The most important advantage of the hyperspectral technique comes to the fore as we consider a relative limitation of the multispectral technique. Figure 11.2a shows spectra of two common minerals, kaolinite and hematite, along with band passes of the Landsat TM sensor, which has been the most widely used multispectral sensor. Figure 11.2b shows the spectral responses as would be recorded by the Landsat TM sensor. The coarse band passes of the Landsat TM allow recording of only broad variations in intensity values, and not the fine spectral features, which are lost. Hyperspectral remote sensing overcomes this limitation, as it allows generation of an almost continuous spectrum at each pixel by using hundreds of narrow contiguous spectral channels.

Imaging spectrometry possesses the capability to identify and map the distribution of specific minerals. The image spectral data, after adequate rectification and calibration, are compared to field/laboratory/library spectra of minerals in order to generate mineral maps. These maps may show details such as mineral combinations, their relative abundances and even mineral compositions in a solid-solution series.

Spectral Considerations 289

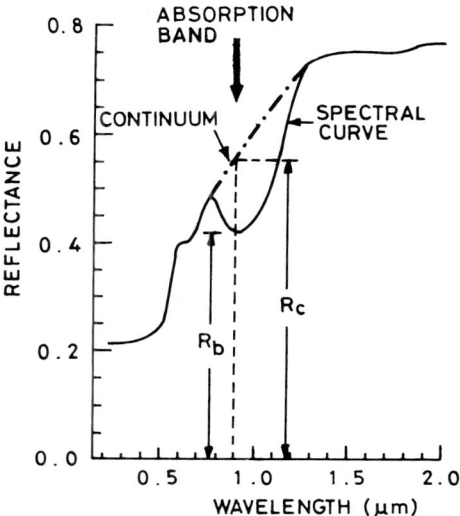

Fig. 11.3. Spectrum of goethite used to illustrate the terminology – continuum, absorption band and depth of absorption

Further, if a material is not commonly occurring but has diagnostic absorption features, its presence can be detected and distribution mapped. The main limitations of the technique are the high sophistication required in data acquisition and processing, and its high sensitivity to small changes in composition and structure of the surface material.

Most of the aerial and space-borne hyperspectral sensors operate in the solar reflection region (0.4–2.5μm), and their data processing and interpretation form the subject matter of this chapter. Some aerial hyperspectral sensors also carry 5–6 bands in the TIR range, similar to TIMS, which was discussed in Chapter 9.

11.2 Spectral Considerations

11.2.1 Processes Leading to Spectral Features

The basic information sought by imaging spectroscopic technique lies in the spectral features of objects. The various basic atomic–molecular processes [viz. electronic processes (crystal field effects, charge transfer effect, electronic transition, conduction bands) and vibrational processes (fundamentals, overtones and combinations of molecular motions)] governing spectra of minerals were discussed in Chapter 3. A large number of spectra showing broad spectral features were also presented there.

11.2.2 Continuum and Absorption Depth – Terminology

In a spectrum, an absorption feature can be considered to be composed of two components: the continuum and the individual feature. The background level is called continuum, on which a particular absorption feature is superimposed. The continuum is rather difficult to define. Qualitatively, it may considered as 'the upward limit of the general reflectance curve for a material' (Mustard and Sunshine 1999). Practically, for a particular absorption feature, continuum may be taken as the broad upward limit of the reflectance curve that would be obtained if the specific absorption feature was not present. Both absorption and scattering processes affect continuum. The continuum itself could be a wing of a larger absorption feature. For example, figure 11.3 shows the spectrum of goethite. The feature at 0.88 μm is the absorption band, for which the continuum is formed by a broader and stronger absorption feature in the UV.

Depth of an absorption band is defined relative to continuum

$$D = (R_c - R_b)/R_c = 1 - (R_b - R_c) \qquad (11.1)$$

where R_b is the reflectance at band bottom and R_c is the reflectance at the continuum, both being measured at the same wavelength (Clark and Roush 1984). The band depth of an absorption feature is related to the amount (degree of abundance) of the absorber, as well as to the grain size of the mineral.

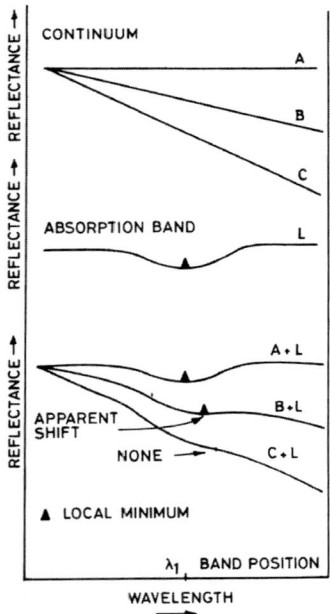

Fig. 11.4. Effect of sloping continuum on apparent position and slope of absorption feature

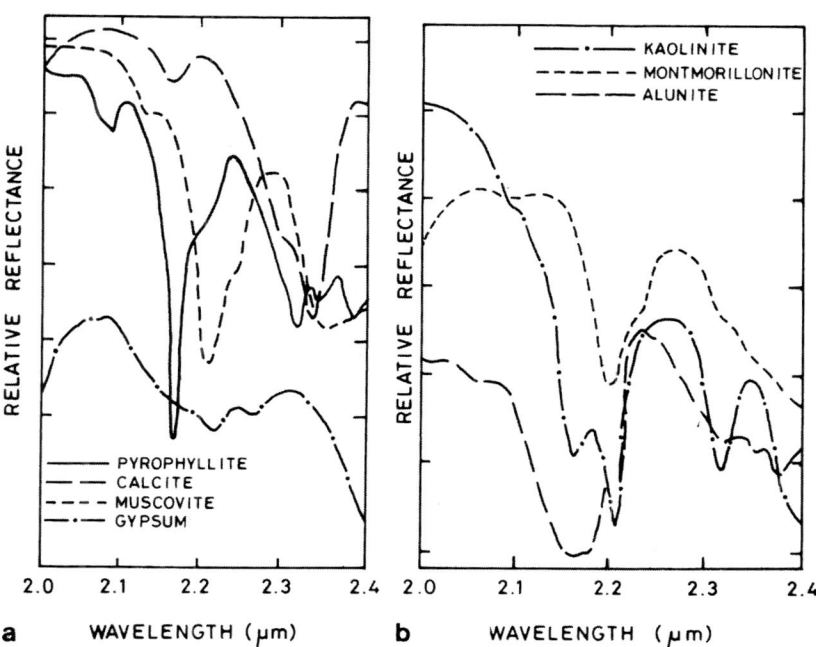

Fig. 11.5 High-resolution laboratory spectra of some common clay minerals in the SWIR region; each mineral exhibits diagnostic spectral features. (After Goetz and Rowan 1981)

The shape and slope of the continuum influence the appearance of the absorption feature, in terms of its position and depth (Fig. 11.4). If the continuum is sloping and an absorption feature is superimposed over it, then the local minimum may appear shifted, In case, the continuum is steeply sloping and a rather shallow and broad absorption feature is superimposed over it, then the resulting effect may appear as a change in slope with no local minimum. Therefore, it is necessary to remove the continuum, and the best way of doing this is by division. This normalizes the reflectance values and allows comparison of spectra for interpretation.

11.2.3 High-Resolution Spectral Features – Laboratory Data

It has been observed that changes in the chemical composition of minerals are characterized in terms of subtle changes in spectral absorption bands of minerals, i.e. subtle spectral features mark fine variations in the chemistry of minerals. Laboratory data show that these spectral features have in general a width of 10–40 nm. Hence, spectral sampling at a 10-nm interval is generally considered suitable for hyperspectral sensors which could allow obtaining information on chemistry of minerals. Some examples of hyperspectral laboratory data are given below.

Figure 11.5 shows high-resolution laboratory spectra of some common clay minerals. It is evident that various minerals like kaolinite, montmorillonite,

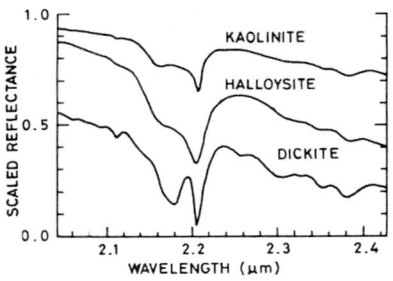

Fig. 11.6. Laboratory spectra of the kaolinite group of minerals (kaolinite, halloysite and dickite); all these have a common absorption band at 2.21 μm but each possesses a slightly different spectrum. (After Clark 1999) (Reproduced by permission, copyright © 1999, John Wiley & Sons Inc.)

Fig. 11.7. Shift in absorption band from calcite to dolomite (continuum-removed spectra). (After Clark 1999) (Reproduced by permission, copyright © 1999, John Wiley & Sons Inc.)

alunite, muscovite, pyrophyllite etc. have individually distinct and diagnostic spectral signatures. Even within a group of minerals, there could be subtle differences in spectra which discriminate one mineral from the other. For example, in the kaolinite group of minerals, kaolinite, halloysite and dickite, all of them have a

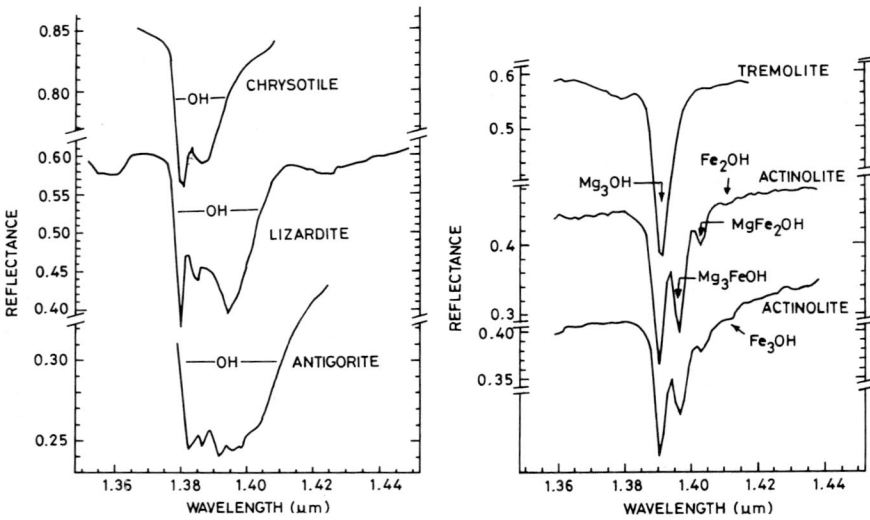

Fig. 11.8. High-resolution laboratory spectra of serpentine group of minerals (chrysotile, antigorite and lizardite). (After Clark et al. 1990)

Fig. 11.9. High-resolution laboratory spectra of tremolite and actinolite. (After Clark et al. 1990)

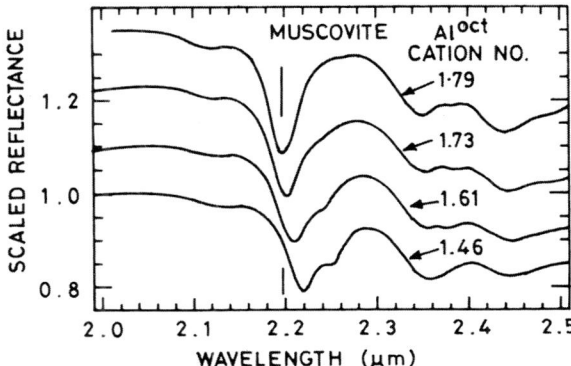

Fig. 11.10. Effect of substitution of aluminium in muscovite. (After Swayze 1997)

common absorption band at 2.21 μm, but have slightly differing spectra (Fig. 11.6).

Carbonates show diagnostic vibrational absorption bands in the SWIR at 2.30 to 2.35 μm (Fig. 3.8). The band position may vary with composition. Figure 11.7 shows the subtle shift in absorption band from calcite to dolomite. In the serpentine group, chrysotile, antigorite and lizardite are the three isochemical endmembers. Figure 11.8 shows that the OH-related absorption band can help distinguish the three minerals.

The high-resolution spectra are also sensitive to solid solution. Figure 11.9 shows spectra of tremolite and actinolite. The Mg_3OH absorption band in tremolite is sharply defined; in actinolite, due to the presence of Fe, the absorption band shifts and two additional smaller bands appear. The reflectance spectra can help estimate the ratio Fe/Fe+Mg (Clark et al. 1990b).

Elemental substitution is a common process in minerals. For example, it has been observed that substitution by aluminium in muscovite series leads to fine gradual, continuous changes in the spectral behaviour of muscovite (Fig. 11.10).

High- resolution spectral features of stressed vegetation. It is well known that in the spectra of plants, the red-edge region (around 0.75 μm) (Fig. 3.16) exhibits significant changes as the leaf transforms from the actively photosynthetic state to total senescence. However, even in actively photosynthetic plants, the slope and position of the red edge shows subtle variations, which are related to geochemical stresses. It has been reported that plants growing over soils containing toxic elements (metal sulphides) exhibit shifting of the red edge towards the blue end (shorter wavelengths) by about 7–10 nm (Fig. 11.11; Chang and Collins 1983). However, there is still some difference of opinion among scientists whether the shifting is always towards blue, or also towards near-IR in some cases. For example, Yang et al. (1999) report a shift in the red edge of wheat spectra growing over areas of hydrocarbon microseepages by about 7 nm towards the red edge (longer wavelengths). Hyperspectral sensors permit detection of such subtle spectral changes, and therefore could be of interest in mineral exploration.

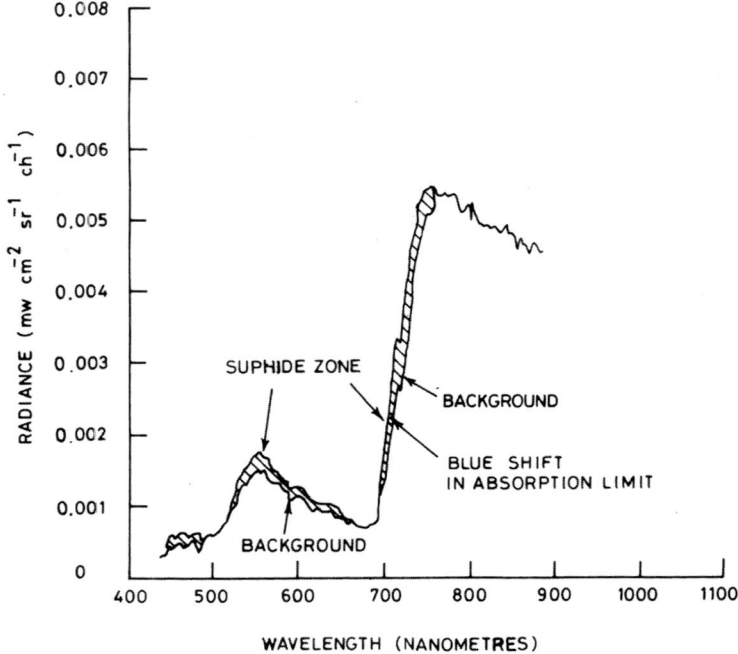

Fig.11.11. Blue shift of the red edge of the chlorophyll band in coniferous trees growing over copper sulphide rich soil zone. (Simplified after Chang and Collins 1983)

11.2.4 Mixtures

The field-of-view of a sensor may be filled with a variety of objects of varying surface extents. Spatial resolution of the sensor vis-à-vis surface dimension of the objects governs whether a pixel represents a pure pixel or a mixture of objects

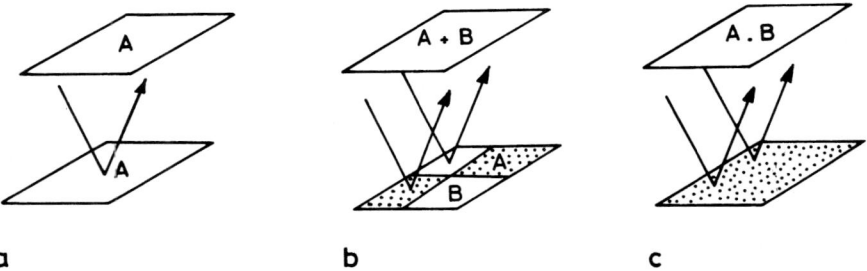

Fig. 11.12. Linear and non-linear spectral mixing. **a** A single homogeneous object forming a pure pixel. **b** Areal mixing (linear spectral combination). **c** Intimate mixing (non-linear spectral combination). (After Campbell 1996)

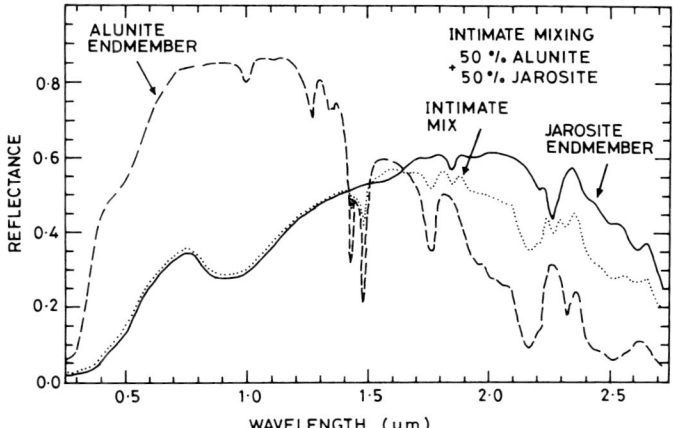

Fig. 11.13. Spectrum of an intimate mixture appearing as a highly non-linear combination of the two end-members. (After Clark 1999) (Reproduced by permission, copyright © 1999, John Wiley & Sons Inc.)

(Fig. 11.12). If the pixel covers a uniform ground area at the sensor resolution, it represents a pure pixel (Fig. 11.12a). In the case of mixtures, three types of physical mixtures are identified in the context of remote sensing spectroscopy.

1. Areal or linear mixture. This occurs when the pixel comprises two or more objects occurring in patches that are large relative to the resolution of the sensor (Fig. 11.12b). Thus, the constituent materials are optically separated from each other so that there is no multiple scattering between component materials. Spectral mixing occurs at the sensor and the signal from the mixture is a linear weighted average of the end constituents (Fig. 11.13).

2. Intimate or non-linear mixture. When different materials are in intimate contact with each other (Fig. 11.12c) (e.g. mineral grains in a rock), multiple scattering takes place on the contact surfaces. The resulting signal from the intimate mixture is a highly non-linear combination of the spectra of end-members (Fig. 11.13).

Intimate mixtures of light and dark objects (e.g. olivine and magnetite) deserve special mention. In such a mixture, photons reflected from the light objects have a high probability of striking and being captured by the dark objects. Thus, addition of even small amounts of an opaque or dark mineral to a reflecting mineral drastically reduces the albedo of the mixture (Fig. 11.14).

3. Coatings. Coating of one mineral over another is treated as a special type of physical mixture. In such cases, reflectance properties of the mineral forming the coating usually dominate or even mask those of the coated mineral.

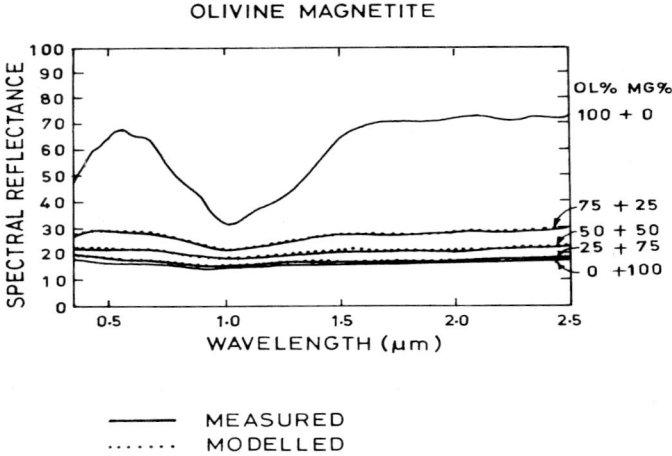

Fig. 11.14. Comparison of modelled (based on non-linear mixing model) and measured reflectance spectra of olivine–magnetite mixture series. (Johnson et al. 1983)

Effect of grain size. Grain size influences the amount of light scattered and transmitted by a grain. If the material is fine grained, a larger area is available for surface scattering. On the other hand, if the material is coarse grained, then a larger internal transmission path is provided to the radiation. Therefore, in the visible–near-IR region, in general, reflectance decreases as the grain size increases.

Effect of viewing geometry. The viewing geometry, including the angle of incidence and angle of view, influences the intensity of radiation received at the sensor; however, it does not have a spectral character, which means that its influence over the entire wavelength range is similar. Therefore, viewing geometry does not influence band position, shape and depth.

11.2.5 Spectral Libraries

Spectral libraries are the libraries holding spectral data/information. A large amount of data on spectra of various types of objects (minerals, rocks, plants, trees, organic substances etc.) has been generated in various laboratories worldwide and stored in libraries. Two such libraries available on the world-wide-web sites are those of the USGS (http://speclab.cr.usgs.gov) and NASA-JPL (http://asterweb.jpl.nasa.gov).

11.3 Hyperspectral Sensors

Imaging spectrometers are an advanced type of scanner, which generate digital images in a large number of contiguous spectral bands. Further, as radiation inten-

Hyperspectral Sensors

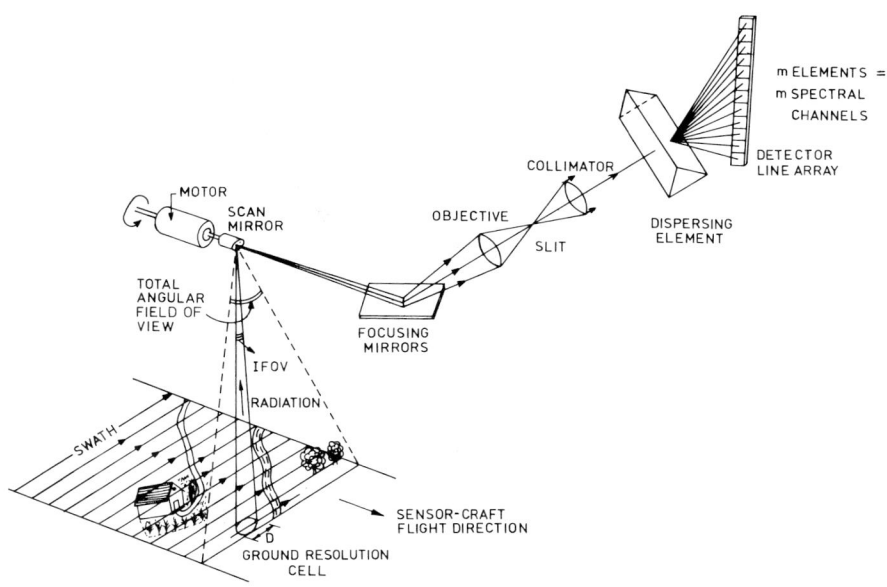

Fig. 11.15. Working principle of whiskbroom imaging spectrometer

sity is critical in interpretation, these sensors use 8-bit, 12-bit or 16-bit quantization levels.

11.3.1 Working Principle of Imaging Spectrometers

Depending upon the mechanism of working, two basic types of imaging spectrometers are distinguished: (1) the whiskbroom imaging spectrometer and (2) the pushbroom imaging spectrometer.

1. Whiskbroom imaging spectrometer. This is basically and opto-mechanical device, but produces image data in numerous (about 200) contiguous spectral channels. The working principle is illustrated in figure 11.15. A moving plane mirror directs radiation from different parts of the ground on to the radiation-sorting optics. The radiation-sorting optics comprises a dispersion device. The radiation dispersed and separated wavelength-wise is focused on the photo-detector. The heart of the device is the photo-detector, which consists of a CCD line array, aligned such that radiation of different wavelength ranges falls on different elements of the CCD line array. Thus, for each ground IFOV, radiation is dispersed and its intensity is recorded in as many spectral channels, as the number of detector elements in the line array. The rest of the principle is same as in a typical OM line scanner, i.e. imaging is carried out pixel by pixel, and line by line, as the sensor-craft keeps

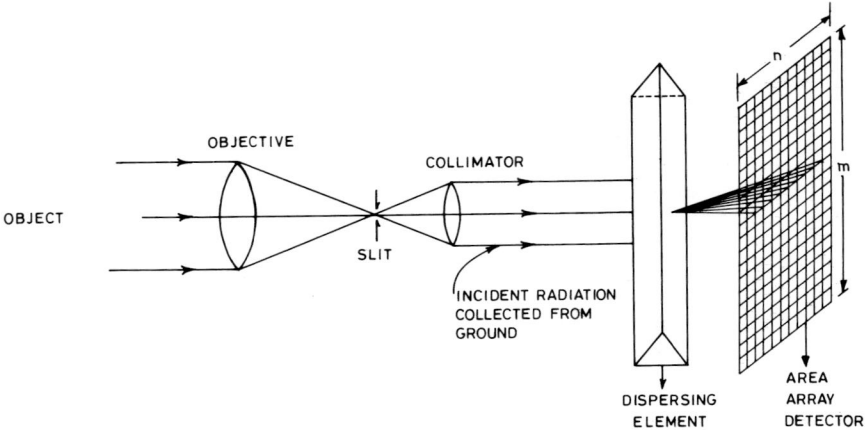

Fig. 11.16. Working principle of pushbroom imaging spectrometer

advancing. This leads to generation of images in numerous contiguous spectral channels.

For high spatial resolution, the whiskbroom mechanism in imaging spectrometers is suited only to an airborne sensor which flies slowly enough that the readout time of the detector array is only a fraction of the integration time. An example of this type of sensor is the Airborne Visible/Infra-Red Imaging Spectrometer (AVIRIS) (Vane et al. 1993) (Table 11.1), which has provided terrestrial hyperspectral sensing data over a number of sites.

2. Pushbroom imaging spectrometer. This type of sensor uses a two-dimensional CCD area array of detectors (instead of a CCD linear array!), located at the focal plane of the spectrometer. The working principle is shown in figure 11.16. The radiation is collected by the objective, passes through a slit, and is collimated onto a dispersing element. The radiation is separated according to wavelengths and focused onto an area array of detectors, for example $n \times m$ detector elements. In such a case, the imaging device measures radiation intensity in n number of spectral channels, over m across-track pixels. Thus, there is a dedicated column of n detector elements for each ground cell in the swath. The across-track width of the area array (m elements) is commensurate with the slit opening which determines the swath. The rest of the pushbroom scanning mechanism remains the same as in the CCD linear scanners, i.e. the radiation intensity is integrated for a certain period (dwell time) by a frontal photo-gate, after which the charge is read out. An example of a pushbroom imaging spectrometer is the Airborne Imaging Spectrometer (AIS) (Vane et al. 1983). Pushbroom imaging spectrometers obviate the requirement of OM-type scanning, and are ideally suited for use as space-borne imaging spectrometers.

Hyperspectral Sensors

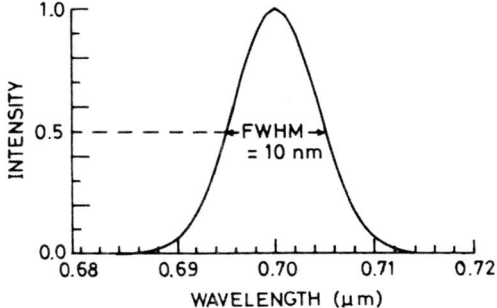

Fig. 11.17. Typical band-pass profile (gaussian) of spectrometer channel and the concept of 'full width at half maximum' (FWHM)

The above paragraphs have described the general working principle of imaging spectrometers. In detail, the design and instrumentation is highly sophisticated and complex, and undergoing continuous improvement.

11.3.2 Sensor Specification Characteristics

The characteristics of an imaging spectrometer are given in terms of three main parameters: (1) spectral range, (2) spectral bandwidth and (3) signal-to-noise ratio. *Spectral range* is the entire range of EM radiation over which the sensor operates, e.g. visible, NIR, SWIR etc. *Spectral bandwidth* means the width of the individual spectral channel in the sensor and is generally given in terms of nanometers. Hyperspectral sensors have narrower bandwidths than multispectral sensors, in order to be able to detect fine absorption features in the spectra of objects. It is observed that bandwidths greater than 20–40 nm are unable to resolve characteristic absorption features of minerals. Each spectral channel senses radiation of a certain wavelength band, the band-pass response most commonly being a gaussian one. The width of the band pass is usually defined as the width in wavelength at 50% response level of the function, and is called the 'full width at half maximum' (FWHM) (Fig. 11.17). In most spectrometers, spectral bandwidth and spectral sampling are the same, and this is also broadly called spectral resolution (although strictly speaking spectral resolution will be $2 \times$ FWHM).

Signal-to-noise (S/N) ratio for a spectrometer has the same connotation as for broad-band scanners [Chapter 5, Eq. (5.6)]. It depends upon detector sensitivity, spectral bandwidth, scene brightness, and transmittance through optics and atmosphere. Improvements in the design of imaging spectrometers have led to manifold increase in the signal-to-noise ratio during the last decade. Most manufacturers provide NER (noise equivalent radiance) specification.

11.3.3 Airborne Hyperspectral Sensors

The first imaging spectrometer was the NASA Scanning Imaging Spectroradiometer (SIS) built in the late 1970s, which provided image data in 32 contiguous spectral bands, each 15 nm wide, in the spectral range of 0.43–0.8 μm. Thereafter, an important development was the airborne spectroradiometer developed by the Geophysical Environment Research (GER) Company. It operated in one-dimensional profiling mode to acquire data in 576 narrow channels in the spectral range of 0.4–2.5 μm (Chiu and Collins 1978). This instrument, flown at an altitude of 650 m above the ground, could detect subtle changes, due to biogeochemical stresses, in the spectra of vegetation growing over mineral deposits (Collins et al. 1983), and thus generated a lot of interest.

A major advancement in instrumentation was the design of imaging systems with dispersing optics, which allowed acquisition of spectral data in a large number of discrete contiguous spectral channels. The Airborne Imaging Spectrometer (AIS) was designed and built at the NASA Jet Propulsion Laboratory (JPL) in early the 1980s. This instrument used an Hg-Cd-Te area array of 32 × 32 elements, and started operating in 1983 (Vane et al. 1983). It acquired image data in 128 spectral bands with a 9.3-nm spectral bandwidth, in the range 1.2–2.4 μm, with a small field-of-view (3.7°), providing a 32-pixel swath with pixel size of 11.5 m, from the designed flying height of 6 km. A subsequent version of this instrument, AIS-2, had a larger spectral range (0.8–2.4 μm region), spectral bandwidth of 10.6 nm and a broader swath (64 pixels).

The next important development was the Advanced Visible/Infrared Imaging Spectrometer (AVIRIS) which became operational in 1986/87 (Vane and Goetz 1988; Vane et al. 1993). This sensor has been flown over a number of test sites world-wide and is still operational. The AVIRIS images in 224 contiguous channels in the spectral range of 0.4–2.45 μm, with a spectral sampling of 9.4 nm. It is designed to operate at an altitude of 20 km to provide a spatial resolution of 20 m, over a swath of 11 km (550 pixels). It is a whiskbroom type of imaging spectrometer using CCD linear arrays. It uses silicon linear array detectors for the visible spectrum, and indium-antimonide line array detectors for the NIR and SWIR parts of the spectrum.

A number of private organizations have since recognized the immense potential of hyperspectral sensing. Among the more important commercial airborne imaging spectrometers are CASI, GERIS, HYDICE, MIVIS, HyMAP and SFSI. A list of selected aerial hyperspectral sensors, including their salient specifications, is given in Table 11.1.

11.3.4 Space-borne Hyperspectral Sensors

Among the hyperspectral sensors from space platforms, the Shuttle Multispectral Infrared Radiometer (SMIRR) was the first important experiment flown on the space shuttle STS-2 in 1981. It was a 10-channel profiling-mode spectroradiome-

Hyperspectral Sensors

Table 11.1. Selected airborne imaging spectrometer systems for geological remote sensing (after van der Meer 1999; and www.eol.ists.ca/documents/IS-Team-Canada 2000)

S. No.	Name	Full name	Country/ manufacturer	Availability	No. of channels	Spectral range (μm)	Bandwidth (nm)	IFOV mrad
1	AHS	Airborne Hyperspectral Scanner	Daedulus Enterprises		48	0.43–12.7	20–1500	2.5
2	AMSS	Airborne Multispectral Scanner	Geoscan Pty. Ltd.	1985	46	0.5–12.0	20–590	2.1–3.0
3	AVIRIS	Airborne Visible/Infrared Imaging Spectrometer	US	1987	224	0.4–2.45	9.4	1
4	CASI	Compact Airborne Spectrographic Imager	Canada	1990	288	0.43–0.87		
5	DAIS-7915	Digital Airborne Imaging Spectrometer	Germany	1994	79	0.5–12.3	15–2000	1.1–3.3
6	DAIS-21115	Digital Airborne Imaging Spectrometer	Germany		211	0.4–12.0		
7	GERIS	Geophysical and Environmental Research Spectrometer	Germany		63	0.4–2.5	16–120	2.5–.5
8	HYDICE	Hyperspectral Digital Imagery Collection Experiment	US	1994	206	0.4–2.5	7.6–14.9	0.5
9	ISM	Imaging Spectrometer Mapper	DESPA	1991	128	0.8–3.2	12.5–25	3.3–11.7
10	MAIS	Modular Airborne Imaging Spectrometer	China		71	0.44–11.8	20–800	3–4.5
11	MIVIS	Multispectral Infrared and Visible Imaging Spectrometer	Daedulus Enterprises/Italy	1993	102	0.43–12.7	8–500	2.0
12	SFSI	SWIR Full Spectrographic Imager	Canada		122	1.2–2.4	10	0.33
13	HyMAP (PROBE1)	Hyperspectral Mapper	Australia		100 200	0.45–2.5	10–20	2.5

ter, operating in the wavelength range 0.5–2.5 μm, out of which five channels were in the range 2.0–2.5 μm. This indicated for the first time that mineral identification from an orbit was possible (Goetz et al. 1982).

Following the success of AVIRIS, a number of space-borne imaging spectrometers have been planned and developed. The LEWIS Hyperspectral Imager (HSI) was the first such instrument, however it was lost a few hours after launch in July 1997. MODIS, the Moderate Resolution Imaging Spectrometer, is designed primarily for biological and physical processes on a regional scale. It has 36 bands in the VIS–NIR–SWIR–TIR range, with ground resolution ranging from 250 m to 1 km, and a swath width of approximately 2330 km.

The European Space Agency (ESA) has been developing two space-borne imaging spectrometers (Rast 1992): MERIS and PRISM (SPECTRA). MERIS, the Medium Resolution Imaging Spectrometer is proposed for launch on Envisat in 2002 and is designed mainly for oceanographic applications. SPECTRA, planned for Envisat-2, will cover the 0.4–2.4-µm wavelength range with a 10-nm contiguous sampling interval at a ground resolution of 32 m. The Australian Resources Information and Environmental Satellite (ARIES) is also proposed as a hyperspectral sensor. ARIES-1 will have about 32 contiguous bands in the 0.4–1.1-µm and about 32 contiguous bands in the 2.0–2.5-µm range; there will be other spectral bands to assist with atmospheric correction.

11.4 Processing of Hyperspectral Data

Processing of hyperspectral remote sensing data is quite different from that of multispectral data. There are hundreds of channels and the data may be of 12-bit or 16-bit type; therefore special processing strategies and high computational facilities are required.

A valuable overview on hyperspectral data processing is given by van der Meer (1999). The entire process of hyperspectral data processing can be grouped into three broad stages (Fig. 11.18).
1. *Pre-processing* – which aims at converting raw radiance data into spectrally and spatially rectified at-sensor radiance data.
2. *Radiance-to-reflectance transformation* – during which the influence of external factors (atmosphere, solar irradiance and topography) are removed from the data to convert at-sensor radiance into ground reflectance data.
3. *Data analysis for feature mapping* – involving spectral-curve matching with reference data, aimed at analysing the image data and generating mineral distribution maps.

11.4.1 Pre-processing

Pre-processing of hyperspectral image data is carried out to convert raw radiance into at-sensor radiance. This part is generally conducted by the agency operating the sensor, and the data released to users is commonly the at-sensor radiance data. However, at times some of the corrections have to be carried out by the users; therefore, it is pertinent to know the types and methods of the various corrections involved. The pre-processing includes spectral calibration, spatial rectification and noise adjustment.

Spectral calibration. The chain of spectral calibration comprises geometric spectral calibration, spectrometric calibration and radiometric calibration (Green 1992). Geometric spectral calibration is necessary as spectral response is not homogeneous when measured over the area covered by a pixel. The point-spread function describes the spatial variation of the measured signal. Similarly, a

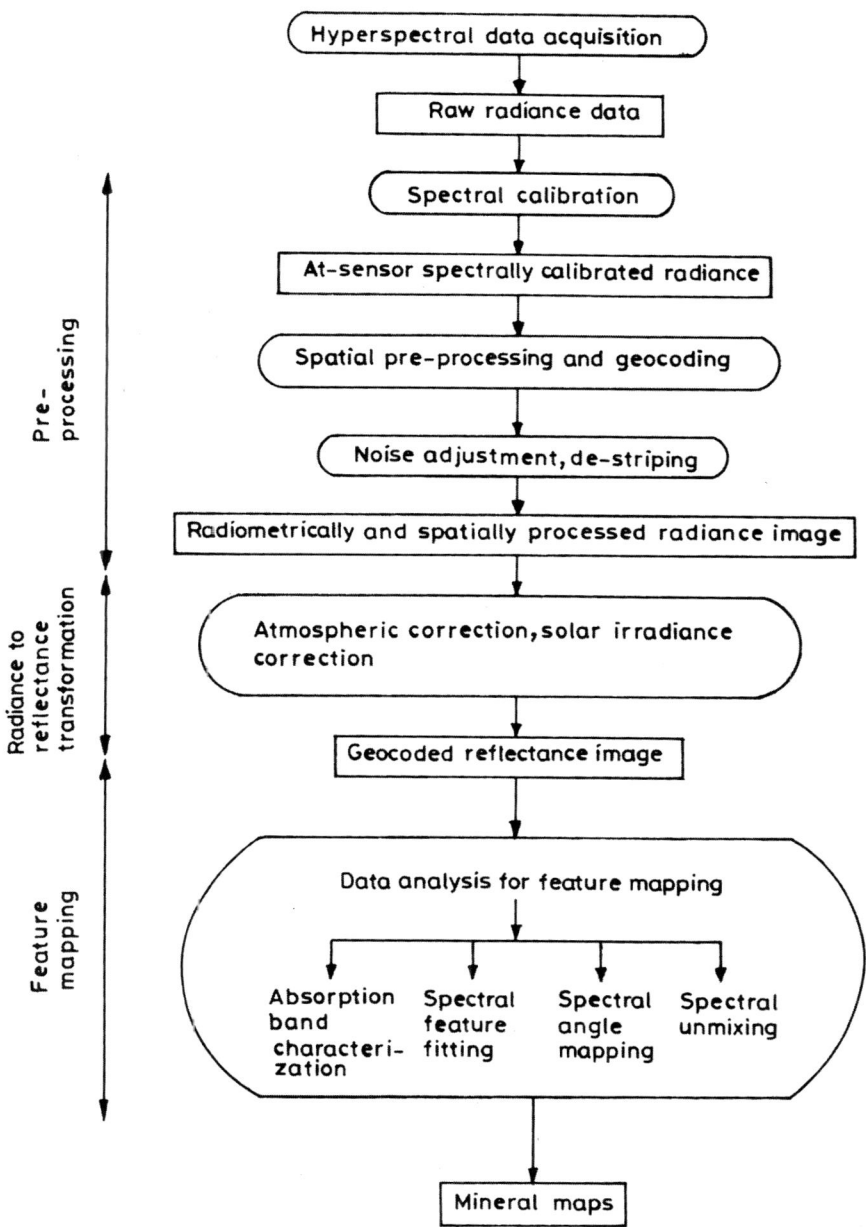

Fig. 11.18. Overview of scheme of hyperspectral data processing

spectral channel measures radiance in a wavelength band that stretches from a few nanometres lower to a few nanometres higher than the central wavelength of the channel (Fig. 11.17). The spectral response function is the curve describing the (gaussian) decline of radiance levels around the central wavelength for each channel, and is used for spectrometric calibration. Radiometric calibration is necessary to convert the measured digital numbers into meaningful spectral radiance values. For this a radiometric response function is used, which defines the relation between the signal (DN value) and incoming radiance, and is experimentally determined. With this, the raw radiance image data is converted into spectral radiance image data.

Geometric calibration and geocoding. Spatial distortions occur in the image data due to aircraft instability and ruggedness of the terrain. In modern times, the use of an on-board DGPS allows accurate location of the aircraft at image acquisition time. In addition, on-board gyros (on aerial platforms) are used to record angular distortions (roll, pitch and yaw). During spatial rectification of image data, the original observation geometry is reconstructed for each pixel, based on aircraft altitude, flight line, aircraft navigational information and surface topography (Meyer 1994). This generates a geocoded image data. Sometimes this geometric correction is not carried out on a routine basis but only on a specific request.

Signal-to-noise ratio and noise adjustment. Signal is the quantity measured by the sensor whereas noise is the random variability in the signal. Signal-to-noise (S/N) ratio is important in evaluating the performance of the sensor and may vary at different band passes for the same sensor. A higher S/N ratio leads to better mineral discrimination. The S/N ratio can be increased by improving design of the sensor and by reducing noise and retaining the signal during data processing (Clark & Swayze 1996). For noise reduction in the hyperspectral image data set, Green et al. (1988) developed a powerful method called minimum noise fraction (MNF) transform. It is a modification of principal component analysis (PCA). In this, first the noise covariance matrix is computed and then the reflectance data are rotated and scaled to make noise with unit variance in all the bands. The transformed data set is analysed with PCA. In this way, noise gets equally distributed among the bands, hence the method provides good discrimination of spectral features.

De-striping. The hyperspectral image data often contain striping, which has to be removed (Fig. 11.19a). Dickerhof et al. (1999) applied a combination of MNF and Fourier analysis (the sequential steps being: calculation of noise statistics, followed by MNF, forward FFT analysis, filtering, retransformation from frequency to spatial domain, low-pass filtering and finally reverse MNF transform). The resulting image data have increased SNR and are de-striped (Fig. 11.19c).

11.4.2 Radiance-to-Reflectance Transformation

The hyperspectral image data carry the influence of a number of external factors, which mask the fine spectral features of ground objects. These factors are: effects of the solar irradiance curve, atmosphere and topography. The *effect of the*

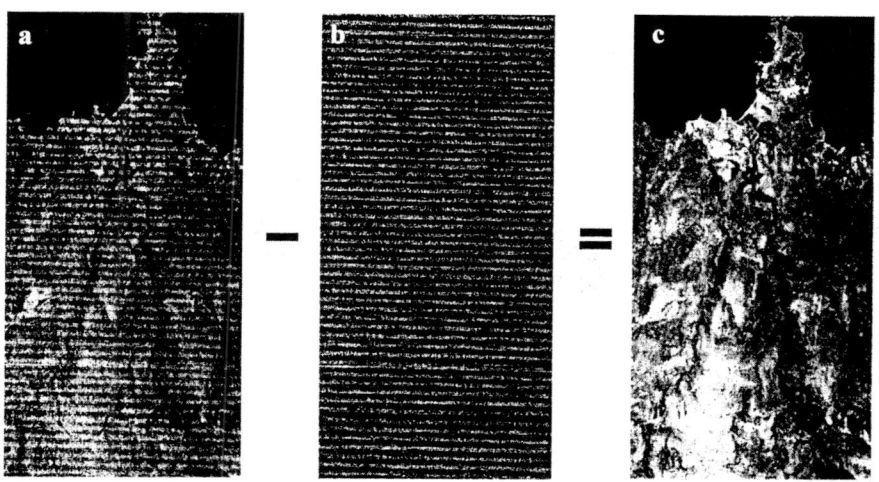

Fig. 11.19a–c. Destriping. **a** Original hyperspectral sensor image showing strong striping; **b** striping (noise) component; **c** de-striped image. (After Dickerhof et al. 1999)

solar irradiance curve arises from the fact that solar radiation intensity peaks at 0.48 μm and the radiation intensity drops off towards longer wavelengths (Fig. 2.2); therefore, the effect of solar irradiance is not uniform throughout. *Atmospheric effects* arise due to the fact that hyperspectral image data are collected over a wide wavelength range, which includes atmospheric windows as well as parts of the EM spectrum affected by atmospheric absorption and scattering. *Topographic effects,* in hyperspectral sensing, are similar in nature to those in multispectral sensing (Sect. 8.2.3). Spectral and spatial variations in at-sensor radiance data occur due to the above factors, which are external to the ground, and these must be adequately normalized in order to compute ground reflectance values in different channels.

A variety of techniques have been developed to remove the above artifacts. Rast et al. (1991) provide a comparative review. The various techniques can be categorized into two groups: (a) those using radiative transfer codes (atmospheric models) and (b) those using parameters derived from the scene (including flatfield correction, internal average relative reflectance correction and empirical line methods).

Correction method using atmospheric models. Figure 11.20 shows a typical atmospheric transmittance curve. Major absorption bands due to H_2O vapour are centred at 0.94, 1.14, 1.38 and 1.88 μm, O_2 bands at 0.76 μm and CO_2 bands near 2.01 and 2.08 μm. In addition, other gases such as ozone, carbon monoxide, nitrous oxide and methane produce absorption features in the 0.4–2.5-μm range. Atmospheric models include corrections for atmospheric effects, viz. absorption and scattering, as well as for illumination condition and topographic effects. First, the atmospheric water vapour is calculated from sensor data in the 0.94- and 1.14-μm bands. Then, the estimated water vapour content and solar and observational

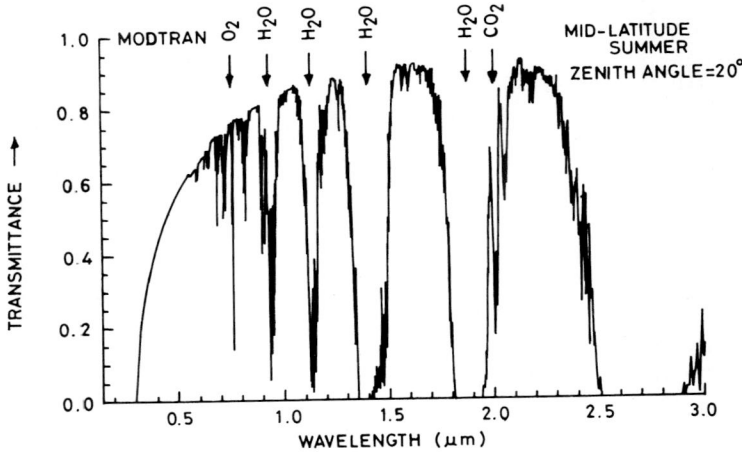

Fig. 11.20. Atmospheric transmittance curve (Berk et al. 1989)

geometry data are used to simulate transmission spectra of mixed gases. The atmospheric scattering effect is also modelled. The atmospheric models permit calculation of an atmospheric radiance spectrum, pixel-by-pixel, for a variety of atmosphere-types. Thus, through this correction, the at-sensor image data are transformed into ground reflectance data (see e.g. Richter 1996).

Flat-field (FF) correction method. In this type of correction, first a window with flat-field characters (morphologically and spectrally flat and homogeneous with high albedo) is identified in the image; for each band, the whole data set is divided by the mean value of the flat-field. This rectifies for the solar irradiance curve, major gaseous absorption features and system-induced effects, if any.

Internal average relative reflectance (IARR) correction method. When no flat-field calibration information is available, the IARR correction method can be used. In this method, an average pixel spectrum of the entire scene is used for normalizing the data sets, i.e. the reflectance at each pixel is divided by the average value of the entire scene. In cases where surface cover with strong absorption features is present, correction by IARR method could produce artifacts which may lead to erroneous interpretation.

Empirical line method. This method utilizes a-priori field knowledge of the scene. Two ground targets, differing widely in albedo values, are selected and their field spectra are collected. Windows of pixels associated with the two types of ground targets are located in the scene and extracted. Linear equations relating field and image spectra characterizing the same objects are computed for each band. The gain (slope) and offset (intercept) values are used to calibrate the image data and generate a relative reflectance image.

11.4.3 Data Analysis for Feature Mapping

After rectification as above, the hyperspectral sensor data take the form of reflectance image data in numerous contiguous bands. The large amount of image data has to be processed for a positive discrimination and meaningful interpretation. The general approach involves the following steps: characterization of the absorption features, comparison to ground truth (spectral libraries), and finally analysis. A number of techniques have evolved for handling hyperspectral data and feature extraction (e.g. Mustard and Sunshine 1999; van der Meer 1999). We consider here the basic concepts involved in some of the common techniques, viz.: (1) absorption band characterization, (2) matching complete shape of spectral feature, (3) spectral angle mapping, and (4) spectral unmixing.

1. Absorption band characterization (position, strength and shape). This method is based on quantifying absorption band characteristics. The spectral curve of a mineral species carries absorption bands with characteristic position, shape and strength. At wavelengths that do not exhibit absorption, reflection takes place. The continuous function is referred to as continuum or hull (Fig. 11.3) and gives in general terms the upward limit of the reflectance of a material. The factors influencing continuum are still not clearly understood. It is considered that continuum is influenced by non-selective absorption and scattering processes. As grain size influences diffuse scattering, continuum is also affected by grain size, in addition to mineral chemistry.

The method of absorption band characterization has been developed by Kruse et al. (1988), and integrated into analytical packages (Kruse et al. 1993). It involves the following main steps.

In order to isolate absorption features from the background signal, first continuum is defined; this can be done by defining high points in the spectrum (through slope and magnitude criteria) and drawing straight-line segments between the defined high points. This gives a model continuum.

To remove continuum from the reflectance data, the data set is divided by the model continuum at each channel. This normalizes the reflectance data and also

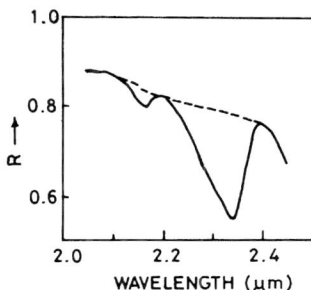
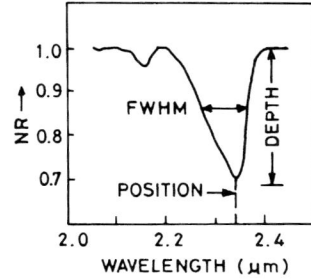

Fig. 11.21a,b. Absorption band characteristics – position and depth. **a** Original spectral curve; **b** the same after continuum removal. R = reflectance; NR = normalized reflectance

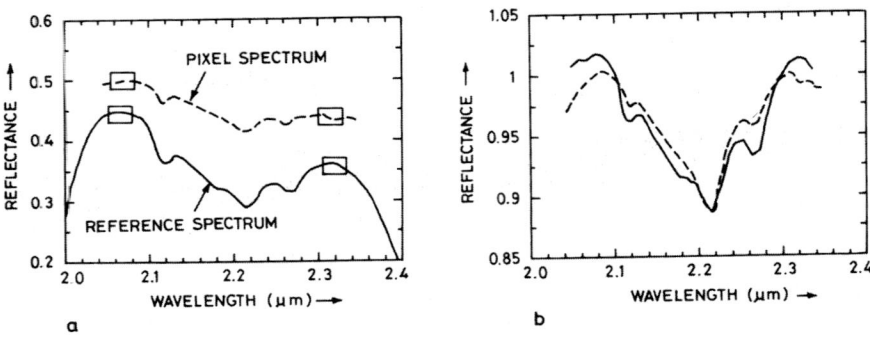

Fig. 11.22. Spectral feature fitting. **a** Reference and pixel spectral curves. **b** Continuum removed spectra are fitted to each other

provides a first-order correction for illumination effects, thus leaving only absorption features to remain.

The normalized reflectance curve is used to characterize the absorption feature in terms of various parameters – wavelength position of minimum, strength of absorption (depth) and shape (Fig. 11.21). The wavelength of band minimum is detected through the use of slope and magnitude (in a manner analogous to band maximum). The relative depth of band absorption 'D' is given in Eq. (11.1).

Band shape is given in terms of FWHM and asymmetry. FWHM (full width at half maximum) is the spectral width of the absorption band measured at a place where the depth of absorption is half of its peak (i.e. D/2). Asymmetry gives whether the absorption band is skewed to the left or to the right of the band minimum. Besides, slope of the continuum can also be taken as a parameter.

In this way, the information contained in the hyperspectral sensing data can by represented in terms of absorption band characteristics. This can be compared with similar absorption band characteristics of laboratory/field spectra. An expert system for analysing hyperspectral data which integrates band parameterization with spectral library data, has been developed by Kruse et al. (1993).

The method of absorption band characterization offers a rapid tool for processing the huge amount of data obtained in imaging spectrometry. The main advantage of the method is that the spectral information from a given band is condensed into four variables. However there are limitations, arising mainly from sensitivity to noise and the presence of sub-pixel mixtures, and also due to ambiguity cropping up in the case of materials with broad absorption features. The technique is better suited to sharp narrow absorption features.

2. Matching complete shape of spectral feature. The method of matching the complete shape of a spectral feature, also called *spectral feature fitting*, is more rigorous than the method of absorption band characterization. It is based on matching the complete shape of image spectra to library spectra, within a certain wavelength

Processing of Hyperspectral Data

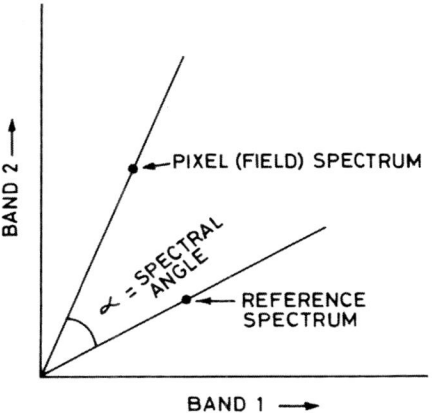

Fig. 11.23. The concept of spectral angle between reference spectrum and field (pixel) spectrum

range; the algorithm computes the degree of similarity between the two sets of spectra (Clark et al. 1990a; Clark and Swayze 1995).

To begin with, there must be a-priori knowledge of specific minerals or objects expected in the area. A certain spectral range is selected to define the spectral feature/absorption band. Continuum is removed separately from image data and library data (Fig. 11.22). The continuum-removed library spectrum is superimposed over the pixel spectrum, using simple linear gain and offset adjustment. A least-squares fit is calculated between the image (unknown pixel) and each of the reference members, separately. The root mean square (RMS) error of this fit yields an overall goodness-of-fit measure. At the same time, the band depth can be related to the abundance of the mineral. Therefore, statistical spectral similarity maps showing distribution and also relative abundance of minerals, based on spectral data, can be generated.

3. Spectral angle mapping. This is another method of generating a spectral similarity map between image data and library data (Kruse et al. 1993). First, the image and library data are processed as in the case of the earlier method of 'matching complete shape of spectral feature'. Then, instead of distance (RMS error), in this method the similarity is expressed in terms of average angle between the two spectra (Fig. 11.23). The spectra are treated as vectors in space, the dimensionality being equal to the number of bands used. Often the full spectral range is used. The angular difference between the pixel spectrum and the laboratory spectrum may range between zero to $\pi/2$. The spectra whose vectors are separated by small angles are considered mutually more similar, and in this way a spectral similarity map can be generated. A limitation of the method is that vector lengths are ignored, which amounts to de-emphasizing albedo differences.

4. Spectral unmixing. In most cases, a pixel is composed of mixed objects; in other words, there are many spectrally diverse objects present within a pixel. The

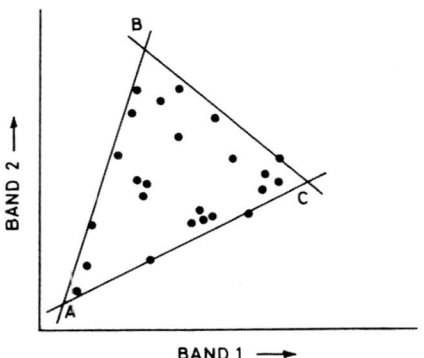

Fig. 11.24. Concept of spectral unmixing with scatteroplot of two bands; the end members A, B and C are defined by the scatteroplot

collective response of all the end members present in different proportions is recorded at the remote sensor. It is important to decipher these constituents and their relative proportions in a pixel. In the reflection domain, the spectrum of a mixture may be explained by linear or non-linear mixing of spectra of end members (Ichoku and Karnieli 1996). In simple cases, linear mixing models are used (e.g. Settle and Drake 1993). Mixture modelling or spectral unmixing uses these concepts and also a premise that in a particular area only a few minerals/materials are commonly occurring, such that their spectra are quite constant; varying proportions of the end members (common materials) are the cause of spectral variability in the scene.

In a simplified form, spectral unmixing considers that the mixed pixel reflection spectrum is a linear combination of the spectra of end members, weighted by their respective areal coverage in the particular pixel. The spectra of end members are known (either from a spectral library or through the application of statistical/procedures on the scene data) and the observed mixed pixel spectrum is known; the abundance/relative proportion values can be obtained by solving the equations.

Theoretically, the permissible numbers of end members are governed by the dimensionality of the data set, i.e. the number of spectral channels. The least-square approximation can be used to obtain the solution if the number of end members is equal to the number of spectral channels minus one (or less). However, as the hyperspectral data have a high channel-to-channel correlation, it is practically possible to use only a limited number (three to seven) of end members in the analysis.

The analysis starts by identifying the spectrally pure end members in the scene. Ideally, the data cloud should be bound by the end members. They may be selected on the basis of ground truth, spectral library, or from statistical procedures applied on the hyperspectral image data. For example, data processing can help to locate vertices of the intersections defining end members (Fig. 11.24).

The observed pixel spectrum is considered to be a mixture of the various end member spectra. Pixel-by-pixel analysis is carried out to calculate the amount (fraction) of each end member component in the observed pixel. This results in fraction images, which show the abundance and distribution of end member components in the scene. These images provide information on the composition of the surface.

In addition, techniques of matched filtering (Farrand and Harsanyi 1997) and cross-correlation spectral matching (van der Meer and Bakker 1997) have been developed for hyperspectral image data analysis.

11.5 Applications

In this section, we discuss a few applications of hyperspectral sensing data in the solar reflection region (VIS–NIR–SWIR). Some airborne hyperspectral sensors also carry spectral bands in the thermal-IR region which are useful for the identification of rock-forming silicate minerals. As these bands are similar to those of TIMS, the method of data processing and interpretation would be similar to that of TIMS/ASTER (discussed in Sect. 9.5).

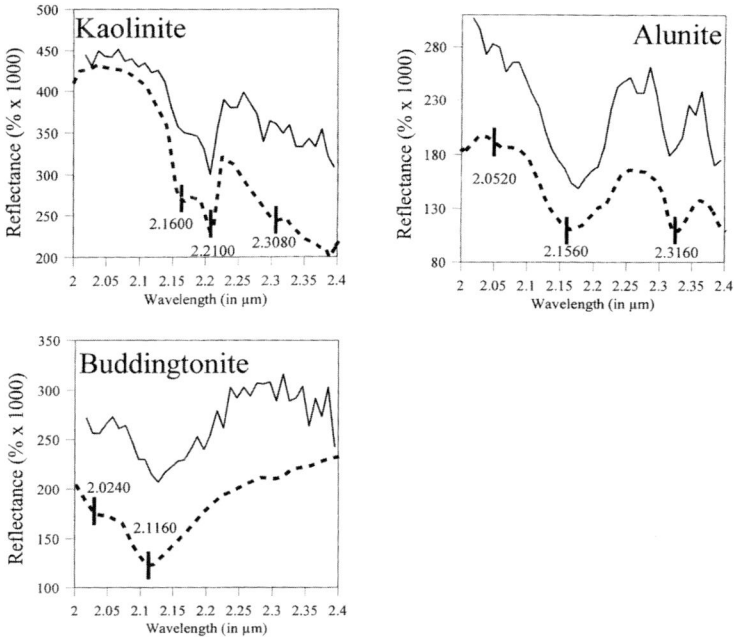

Fig. 11.25. Comparison of single-pixel AVIRIS spectra (*solid lines*) and corresponding laboratory reflectance spectra (*dashed lines*) of kaolinite, alunite and buddingtonite from the Cuprite mining district; vertical bars indicate centre positions of absorption features characterizing these minerals. (Van der Meer 1999)

A number of applications of hyperspectral sensing data in mineralogical studies have been reported. There are two main approaches: (a) mineralogical identification and (b) mineral species quantification.

1. Skarns. These are calc-silicate mineral assemblages that develop as a result of recrystallization and metasomatic processes around plutons of intermediate composition intruding into carbonate host rocks. Skarns are typically coarse-grained and zoned bodies. They constitute the principal source of tungsten, and also contain minor deposits of tin, cobalt, gold, arsenic, silver and many other metals. Besides, several industrial minerals are also associated with skarns, chiefly asbestos, graphite, wollastonite and magnesite.

As indicated earlier, limestone has an absorption band at 2.35 μm and dolomite at 2.32 μm (see Fig. 11.7). Using this spectral characteristics and Geoscan MK II imagery, Windeler and Lyon (1991) delineated a dolomitized limestone unit in the Ludwig skarn, Nevada.

2. Clay mineralogy. The Cuprite mining district, Nevada, is one of the most thoroughly studied geological test sites. Van der Meer and Bakker (1998) and van der Meer (1999) analysed AVIRIS data of the Cuprite area in order to identify distribution of various clay minerals: kaolinite, alunite and buddingtonite. Figure 11.25 shows a comparison of the spectra of the three clay minerals obtained from AVIRIS and laboratory samples. Based on the spectral data comparison (cross-correlogram approach), van der Meer and Bakker (1998) mapped the distribution of these minerals in the Cuprite mining district (Fig. 11.26).

3. Kaolinite content image. In central parts of Germany, there occur huge overburden dumps due to lignite open-cast mining, which was carried out on a large scale during the last century. Information on composition of dumps and its spatial variation is important in order to understand the hydro-geochemical effect of the dumps. Kaolinite forms an important alteration mineral in the overburden dumps.

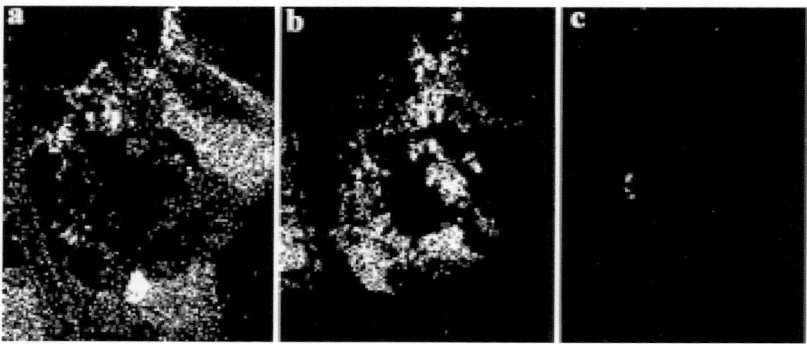

Fig. 11.26a–c. Alteration maps showing distribution of **a** kaolinite, **b** alunite and **c** buddingtonite in the Cuprite mining district, Nevada. (van der Meer et al. 2002) (Reproduced by permission, copyright © 2001, Kluwer Academic Publishers)

Applications 313

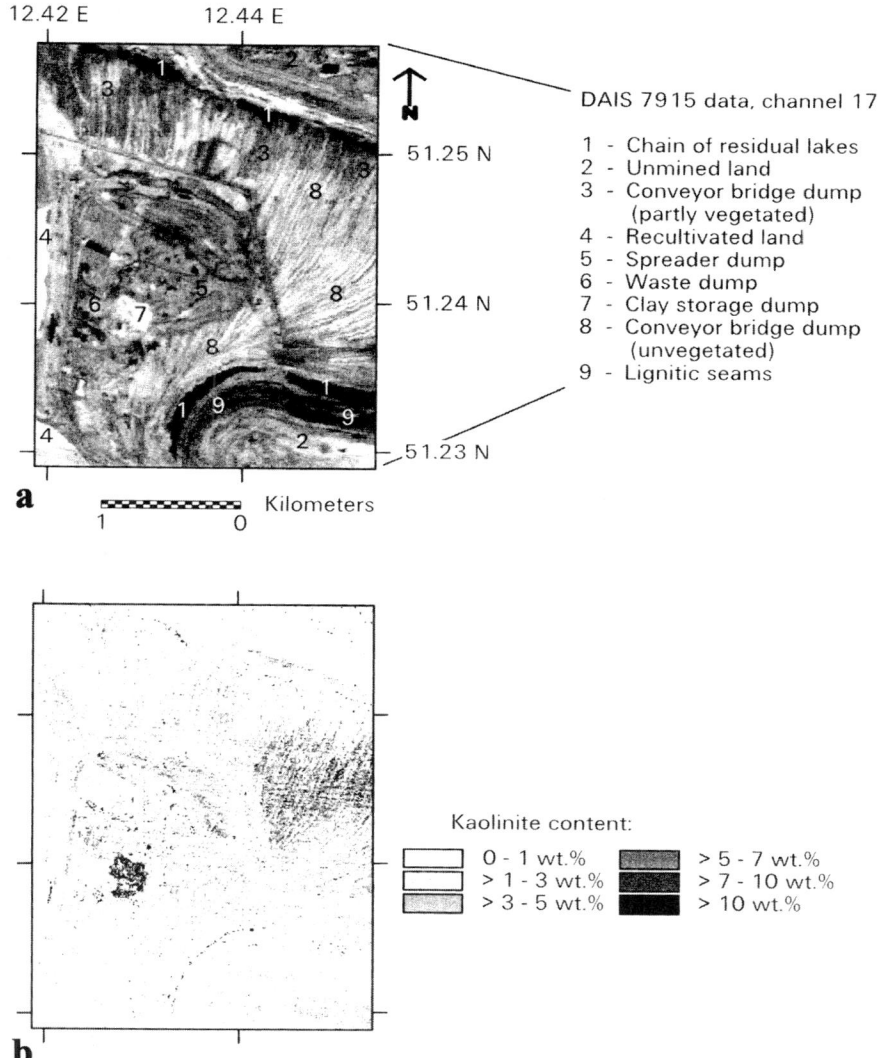

Fig. 11.27. a DAIS 7915 image (channel 17 = 0.754–0.788 µm) of the Espenhain open pit in the Central German Lignite Mining District; **b** kaolinite content image. (**a,b** Krüger et al. 1998)

Table 11.2. Examples of hyperspectral sensing studies

S. No.	Mineral/rock/theme	Data used	Authors
1	Identification of alteration minerals (alunite, kaolinite, buddingtonite, hematite and zeolites) and generation of mineral distribution maps	GERIS	Kruse et al. (1990)
2	Dolomitization of marble in a skarn deposit	Geoscan MKII	Windeler and Lyon (1991)
3	Identification of different iron oxides (hematite, goethite and jarosite) and their spatial distribution	GERIS	Taranik et al. (1991)
4	Detection of ammonium alteration (buddingtonite)	AIS	Goetz and Srivastava (1985)
5	Serpentinization in ultramafic rocks	GERIS	van der Meer (1994)
6	Detection of ammonium alteration (buddingtonite) and generation of quantitative ammonium concentration map	AVIRIS	Baugh and Kruse (1994)
7	Kaolinite, alunite, jarosite (mineral maps)	AVIRIS	Farrand and Seelos (1996)
8	Mineralogic mapping of mafic alkaline rocks, syenite, carbonatite	AVIRIS	Bowers and Rowan (1996)
9	Hypogene alteration vs. supergene alteration	GERIS	Agar and Villanuera (1997)
10	Hot spring mineral deposits	AVIRIS	Kruse (1997)
11	Kaolinite mineral map (in lignite overburden dumps)	DAIS-7915	Krűger et al. (1998)
12	Identification of kaolinite, alunite and buddingtonite and generation of mineral distribution maps	AVIRIS	van der Meer and Bakker (1998)

Krűger et al. (1998) analysed airborne imaging spectrometry (DAIS 7915) data. The depth of absorption at 2.2-μm-band was related to the kaolinite amount in the pixel. Using this approach, a kaolinite content image was generated. Figure 11.27a shows the DAIS image of the area, and figure 11.27b the corresponding kaolinite content image.

Fig. 11.28. Mineral alteration map based on Probe1 (HyMAP) hyperspectral data, Bluff test site, Utah; the red and brown colours represent the bleached zones related to hydrocarbon seepage. (van der Meer et al. 2001) (Reproduced by permission, copyright © 2001, Kluwer Academic Publishers) (see colour Plate II)

4. Hydrocarbon exploration. Hyperspectral sensing can form a very important tool for detecting onshore hydrocarbon micro-seepages (Yang et al. 2000). It is not only fast and low-cost, but can also detect even those traps which may be overlooked by seismic methods. Micro-seepages of hydrocarbons can result in several changes on the surface, such as:

- reduction of ferric oxide (bleaching)
- conversion of clays and feldspars into kaolinite
- increase of carbonate content
- anomalous blue shift of the red edge of vegetation spectra.

The above transformations on the surface are amenable to detection on hyperspectral remote sensing data. Van der Meer et al. (2001) give an example of the detection of bleached beds due to hydrocarbon seepage using Probe1/HyMAP data (Fig. 11.28).

As mentioned above, a number of applications of imaging spectrometry have been reported. A few selected examples are listed in Table 11.2.

Chapter 12: Microwave Sensors

12.1 Introduction

The EM spectrum range 1 mm to 1.0 m is designated as microwave. In the context of terrestrial remote sensing, this spectral region is marked by an excellent atmospheric window, i.e. the radiation traverses the atmosphere with minimal absorption and attenuation (Fig. 12.1). Therefore, this spectral range has aroused much interest for remote sensing applications.

The techniques and sensors for Earth observation in the microwave region can be divided into two broad types: passive, i.e. measuring the naturally available radiation, and active, i.e. illuminating the ground scene by an artificial source of energy and measuring the back-scattered radiation.

12.2 Passive Microwave Sensors and Radiometry

12.2.1 Principle

The passive microwave sensors, called radiometers, measure the intensity of naturally available radiation, the techniques having been adapted from the field of

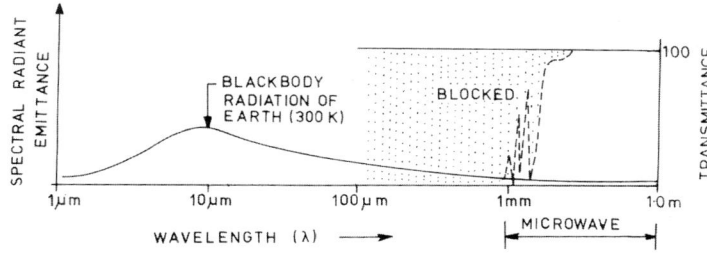

Fig. 12.1. Blackbody radiation emitted by the Earth and the atmospheric windows in the microwave region (1 mm–1.0 m)

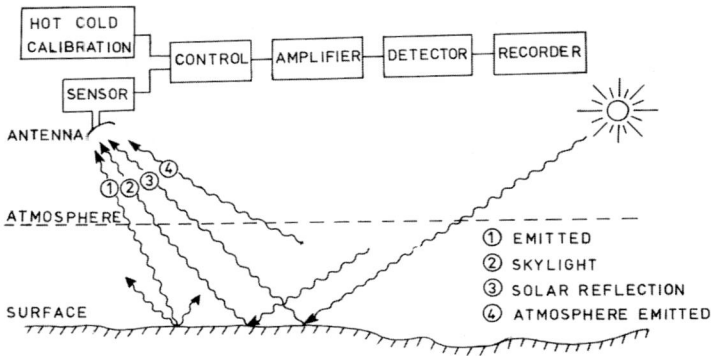

Fig. 12.2. Sources of passive microwave radiation and schematic for detection and recording

radio-astronomy. Commonly, passive microwave sensors operate in the wavelength range of 1mm–30 cm for Earth observations.

The blackbody radiation emitted by the Earth peaks at about 9.6 μm (Fig. 12.1) and falls off with increasing wavelength. At microwave wavelengths, although quite weak, it is still the most dominant source of naturally occurring radiation. The radiation emanating from some depth below the top ground surface is also transmitted through the overlying cover and also forms a part of the signal received by the microwave sensor (Fig. 12.2). Additionally, other radiation sources in this region include the reflected solar energy, the reflected skylight and atmospheric emission, all of which are very weak in intensity. Thus, as far as energy sources are concerned, the blackbody radiation emitted by the Earth forms the most important component.

It was discussed in Chapter 2 that at longer wavelengths such as in the microwave region, Planck's radiation equation is simplified to the Raleigh–Jean Law, which gives [(for abbreviations see Eq. 2.5)]:

$$w_\lambda \cong \frac{2\pi c k}{\lambda^4} \cdot T \text{ watts } m^{-2} \lambda^{-1} \qquad (12.1)$$

This means that the radiation intensity of a blackbody in the microwave region is directly proportional to the temperature of the radiating object. It also implies that temperature can be directly estimated from the radiation intensity. The temperature thus estimated in the real world is called *brightness temperature*.

12.2.2 Measurement and Interpretation

The study of microwave radiometer data is termed *radiometry*. In the microwave region, radiation is collected by the sensor antenna (instead of lenses and mirrors, as is done in the optical region). Owing to the weak intensity of the signal, the sensor employs a large antenna beamwidth, which leads to a lower spatial resolution. The signal received by the antenna from the ground is compared to that obtained from on-board calibration sources, alternately by a switching device (Fig. 12.2). The difference in ground radiation intensity is interpreted in terms of ground temperature difference. These sensors are most commonly operated in profiling mode and are popularly termed *microwave radiometers*. Sometimes they are operated in scanning mode to produce images, being called *scanning radiometers*. Multiband passive sensors in the atmospheric absorption bands (and not in the atmospheric windows!) have been the typical payloads of meteorological satellites for estimating water vapour, oxygen, atmospheric temperature etc.

The nature of passive microwave response is still not fully understood. It is related to a number of ground factors, such as ground temperature and emissivity, and electrical properties such as dielectric constant etc. The spectral emissivity depends on composition and surface roughness. Further, the dielectric constant is a measure of the manner in which EM radiation interacts with surface materials. If we consider that EM radiation is incident on an object, then a part of the radiation is reflected (R) and the remaining part is absorbed (A) (assuming R + A = 1 and transmission to be negligible). We know that the reflected radiation intensity (R) is related to the dielectric constant. Further, (1 − R) is the absorptivity, which in turn equals emissivity, and the spectral emissivity governs spectral radiance. Therefore, the spectral dielectric property becomes highly relevant in passive microwave radiometry.

Water and vegetation have dielectric constants which are very different from those of other ground objects. There is thus a distinct applicability for passive microwave radiometry in hydrological sciences and vegetation-related disciplines. In geology, the scope for passive microwave sensing lies in detecting seepage zones, springs, faults, soil moisture variations, vegetation banding etc. (see e.g. O'Leary et al. 1983).

There also appears to be scope for deciphering geological features present at depth, below the surface, since radiation from buried objects is transmitted through the overlying cover and recorded by the sensor. However, further work is needed to understand the response of ground features in order to accurately interpret passive microwave sensing data. Useful reviews on passive microwave remote sensing have been given by Schmugge (1980) and Ulaby et al. (1981, 1982, 1986).

12.3 Active Microwave Sensors – Imaging Radars

12.3.1 What is a Radar?

12.3.1.1 Definition and Development

Radar is an acronym for Radio Detection And Ranging. In the present context, however, the term radar is used to encompass all active microwave sensors applied for detecting the physical attributes of remotely located objects.

Basically, radar operates on the principle that artificially generated microwaves in a particular direction collide with objects and are scattered. The back-scattered energy is received, amplified and analysed to determine location, electrical properties and surface configuration of the objects.

Radar technology made tremendous strides during World Wars I and II, when this technique was used to detect ships and locate targets. In the early 1950s, the configuration of side-looking airborne radar (SLAR) came into existence, which allowed acquisition of continuous-strip image data, without actually flying over an area. In the 1960s and 1970s the SLAR technique saw rapid development and deployment for civilian purposes, e.g. the Radar Mapping in Panama (RAMP) Project and the Radam Project in Brazil (MacDonald 1969; de Azevedo 1971). These and other similar projects showed the tremendous potential in mapping large heavily vegetated and permanently cloud-covered virgin tracts on the Earth in a very short period by SLAR, due to the unique penetration capabilities of radar wavelengths. This paved the way for space-borne side-looking radar experiments (e.g. Seasat, SIR-A, B, C, ERS-1/2, Radarsat etc.). In the last two decades, there has been a considerable emphasis on the SLAR technique for planetary and Earth resources exploration programmes (Elachi 1980, 1983; MacDonald 1980; Elachi et al. 1982; Trevett 1986; Ford 1998; Henderson and Lewis 1998).

As explained above, SLAR throws artificially generated electromagnetic waves to illuminate the terrain and is therefore called an *active system*. This is in contrast to other, passive remote sensing methods that work on only naturally available energy, such as the sensors in the VNIR, SWIR and TIR region and passive microwave methods. The wavelength used is monochromatic and ranges from a few millimetres to about a metre. These wavelengths do not interfere with the atmosphere (Fig. 12.1). These factors make radar an all-time and all-weather capability, independent of solar illumination variables, which is responsible for the increased interest in this technology.

12.3.1.2 Basic Components of a Radar System

A typical radar system carries a pulse-generating device or transmitter, which is linked to an antenna (Fig. 12.3). The generated signal is radiated by the antenna in

Active Microwave Sensors – Imaging Radars 321

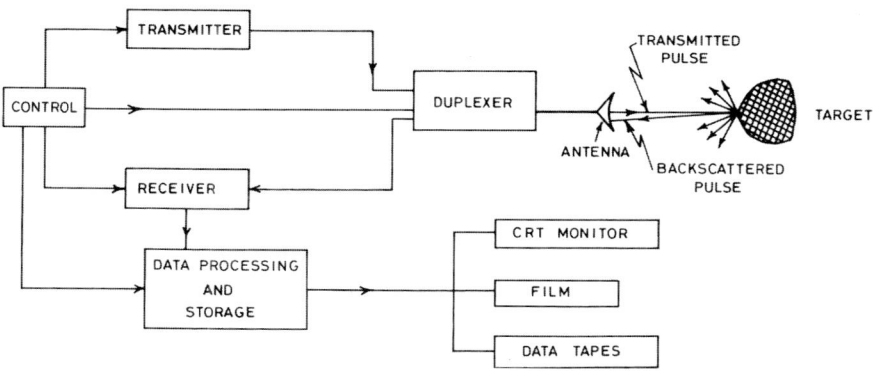

Fig. 12.3. Basic structure of a radar system

a given look direction, and scattered by the object. The back-scattered signal is sensed by an antenna. The systems which illuminate and observe objects from approximately the same location in space are called *monostatic* and those which illuminate from one location but observe from a substantially different location are called *bistatic*. In a bistatic system, two antennas are required. However, in usual radar configurations, the same antenna is switched to transmitting and receiving circuits alternately by a duplexer. A control unit serves as the radar brain and distributes timings signals and necessary commands to other components of the radar system. Finally, data processing and storage provide data output on a CRT terminal or in the form of film or data tapes, which could be used for processing and interpretation. The above are the basic functional components of a radar system (Fig. 12.3), the instrumentation being highly sophisticated in detail.

12.3.2 Side-Looking Airborne Radar (SLAR) Configuration

12.3.2.1 Working Principle

For imaging purposes, the radar is mounted on an aircraft such that the antenna (semi-cylindrical in shape, commonly about 5–10 m in length) is attached to the fuselage, aligned parallel to the aircraft axis. The antenna looks perpendicular to the flight line, and obliquely down upon the Earth on one side. The term SLAR is applied even when the system is space-borne. Sometimes two antennas, mounted one on each side of the aircraft, are used to scan the terrain on both sides of the ground track in a single flight.

In one scanning episode, the radar (SLAR) transmits a short pulse of monochromatic and coherent EM energy, illuminating narrow strips on the ground, perpendicular to the flight direction (Fig. 12.4a). The radar receiver records echoes (back-scatter) in order of arrival, which is related to ground distances (because of

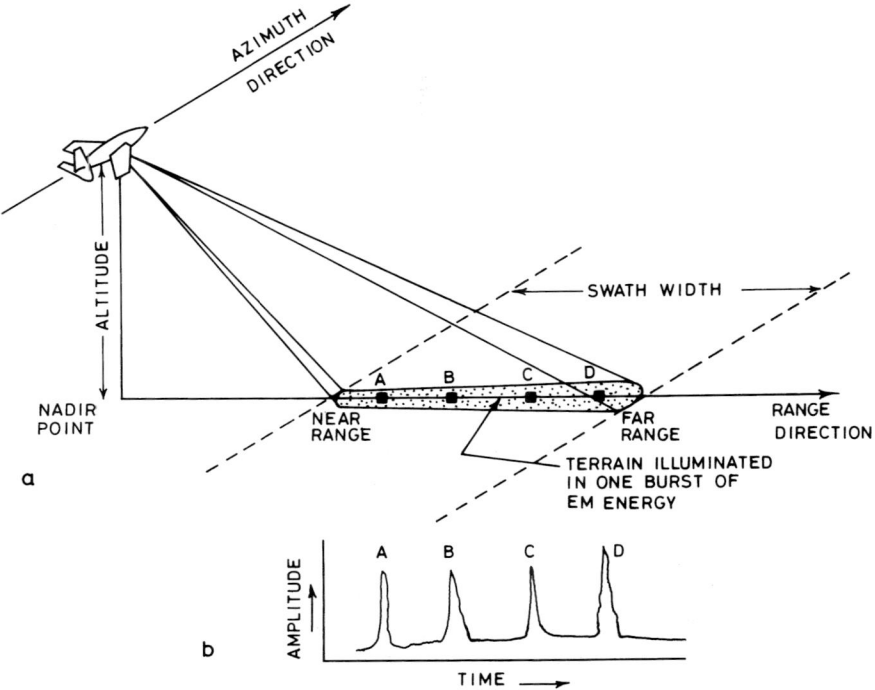

Fig. 12.4a,b. Working principle of side-looking airborne radar (SLAR). **a** Terrain illuminated in one burst of energy. **b** Conversion of target signals into time–amplitude signals

constancy of the speed of light). Thus, echoes from the objects situated closer to the ground track are received earlier, and those from farther away in the range direction, successively later (Fig. 12.4b). In this manner, target signals are converted into time–amplitude signals. After the last echo is received, determined by the swath width, a new pulse of energy is emitted by the SLAR. As the sensorcraft moves along its trajectory (azimuth direction), the radar beam likewise sweeps perpendicular to the flight path, and the entire terrain is scanned. The radar data result in an image (Fig. 12.5).

12.3.2.2 Data Recording and Output

The time–amplitude signal gathered at the radar receiver can be put on a line on film, image tube or data tape (Fig. 12.3). The earlier SLAR systems used films for recording SLAR data. However, digital recording is now widely used for its obvious advantages in handling and processing radar data.

Active Microwave Sensors - Imaging Radars

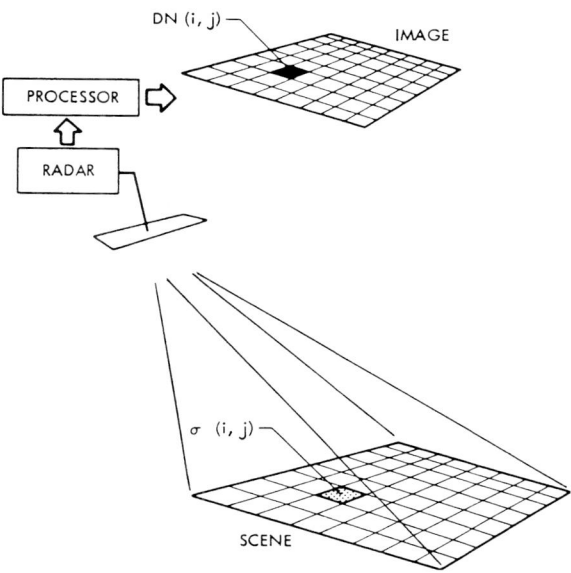

Fig. 12.5. Correspondence of radar image to ground scene. The radar data acquired over an area are processed to yield an image. The ground characteristic parameter (back-scattering coefficient σ) at the scene location (i, j) corresponds to the digital number (DN) at (i, j) in the image. (After JPL 1986)

12.3.2.3 SLAR Terminology

A jargon has evolved in the context of SLAR sensing (Fig. 12.6) and it is necessary to become conversant with it before proceeding further.

Range direction is the horizontal direction in which the aircraft-mounted antenna looks. It is generally perpendicular to the flight direction, except in case of *squint-mode* operation. Objects on the ground lying closest to the flight path (nadir line) are called *near-range* objects; they imply shortest travel time. *Far-range* implies far away from the aircraft in the range direction (largest travel time). *Range resolution* is the linear ground distance resolution in the range direction. *Slant range* denotes the direct linear distance between the antenna and the ground. *Ground range* pertains to the map distance in the range direction. This distance is also called *range distance* in most of the improved imaging radars. The total width of the ground imaged from one end to the other, i.e. from far range to near range, is called *swath width*. *Azimuth direction* is the horizontal direction of aircraft/spacecraft flight. The azimuth and range directions are generally mutually perpendicular. *Azimuth resolution* means the linear ground distance resolution in the azimuth direction. It varies according to the type of radar system (being coarser for real-aperture radars and finer for synthetic-aperture radar systems; see Sect. 12.3.3.2). *Pulse rectangle* corresponds to the unit area on the ground, sensed

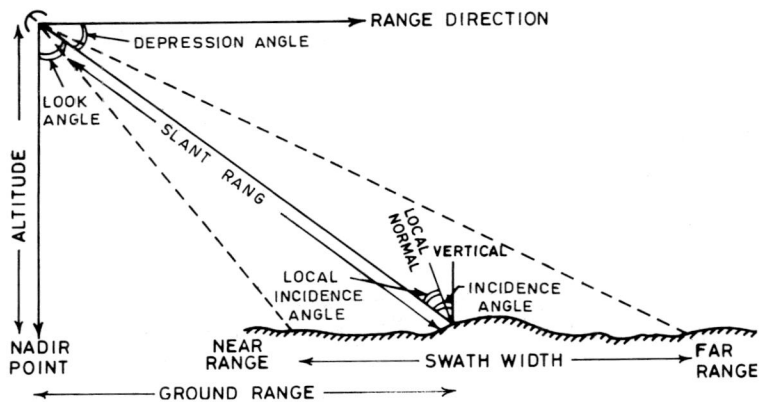

Fig. 12.6. Geometry of SLAR and terminology

by the SLAR (Fig. 12.7). It is also referred to as *spot size*. For real-aperture systems, it is the product of azimuth resolution (beam width) and range resolution (ground distance corresponding to pulse length).

Angular relations between antenna, incident ray and ground object (including surface topography) are vital factors in radar return. *Depression angle* is the acute angle that the transmitted ray (i.e. the line joining the antenna with the ground object) makes with the horizontal, as measured at the antenna in the vertical plane. It varies at different points in the swath and is smallest at far range and largest at near range. *Look angle* is the angle made by the transmitted ray with the vertical, as measured at the antenna in the vertical plane; it is complementary to the depres-

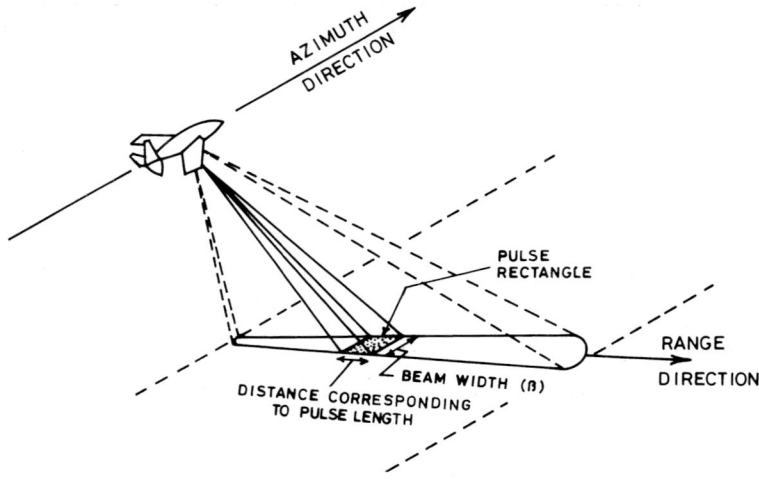

Fig. 12.7. Pulse length, beam width and pulse rectangle in a real-aperture SLAR

Active Microwave Sensors – Imaging Radars

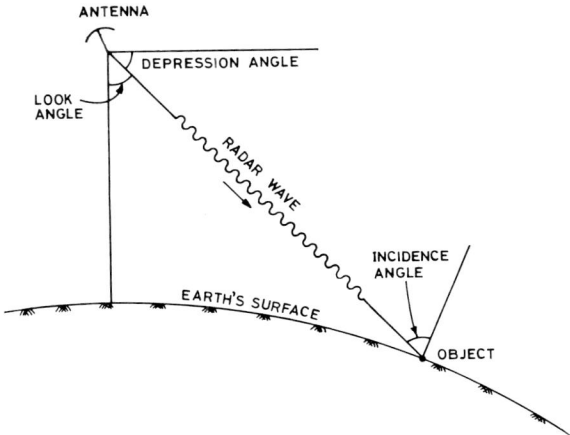

Fig. 12.8. In space-borne cases, the incidence angle (in a flat terrain) is always greater than the look angle at the antenna, owing to the curvature of the Earth's surface

sion angle. *Incidence angle* is that angle which the incident ray makes with the vertical on the ground, where the ground is assumed as essentially flat (or in other words, the vertical is drawn from the centre of the Earth). In the case of aerial SLAR sensing, the effect of the Earth's curvature can be ignored, and the incidence angle equals the look angle, and is complementary to the depression angle. However, in space-borne SLAR, the effect of the Earth's curvature is significant; the result is that the incidence angle is always greater than the look angle (by approximately 3–5 degrees in most cases of space-borne sensors) (Fig. 12.8). As radar return is governed by incidence angle, the incidence angle is generally used as a preferred specification to represent the SLAR viewing geometry (see Table 12.4, later). Another term, *local incidence angle,* is sometimes used for the angle made by the incident ray with the normal to the surface at a particular point (local topography).

In most cases, azimuth direction and range direction are mutually perpendicular. However, to produce images with different look directions (i.e. to produce squint-look images), the range/look direction may be altered by suitable manoeuvres at the sensor for imaging in *squint-look mode*. The angle between the azimuth direction and the range directional measured in the horizontal plane is called the *azimuth angle*. Complementary to this is the *squint angle* (angle between the across-track direction and squint mode look direction) However, as is obvious, in most cases the squint angle is zero.

12.3.3 Spatial Positioning and Ground Resolution from SLAR

From SLAR data, the position of objects is estimated from distances along two orthogonal directions – viz. the range direction and the azimuth direction.

12.3.3.1 Range Location and Resolution

A swath strip is illuminated by a single burst of EM energy and the time interval between the transmitted and received signal is used to give the position of the object in the range direction. If R_s is the slant range distance from the antenna to the object and t is the time interval between transmitted and received pulses, then

$$R_s = \frac{c \cdot t}{2} \tag{12.2}$$

where c is the speed of light. The ground range distance, R, can be computed from this as

$$R = \frac{R_s}{\cos \theta} = \frac{c \cdot t}{2 \cos \theta} \tag{12.3}$$

where θ is the depression angle (Fig. 12.6). The earlier SLAR images used slant range, which is actually the inclined distance, as the reference distance. This results in scale distortion, because of the cos θ factor. Most of the modern SLAR systems, however, use the computed horizontal ground range as the reference distance.

Range resolution is the ability of the radar to discriminate two targets situated behind each other in the range direction. A limitation on this is imposed by the pulse length or pulse duration. Two objects will be resolved if the received pulse from the first object ends before that from the second object starts. Thus pulses are made short-and now coded pulses are used to improve range resolution. If pulse duration is τ, then ground range resolution (R_r) is given as the range distance corresponding to the pulse length as

$$R_r = \frac{\tau \cdot c}{2 \cos \theta} \tag{12.4}$$

Therefore, range resolution is also dependent on the depression angle – for a smaller depression angle it is finer, and for larger depression angles it is coarser. This implies that objects situated the same distance apart may be resolved at far range, but sometimes may not be at near range. In general, systems at higher altitude have to use a longer pulse length than the corresponding systems at lower altitude, and this also influences range resolution.

12.3.3.2 Azimuth Location and Resolution

Broadly, azimuth location is determined from the position of the sensorcraft in the azimuth direction. Azimuth resolution is a very important factor in evaluating the performance of an SLAR system. Depending upon how azimuth resolution is obtained, two types of SLAR systems are distinguished: real-aperture radar (RAR) and synthetic-aperture radar (SAR).

1. Real-aperture radar. Real-aperture radars (RARs) were technically simpler and were used in the initial stages. They possess coarser azimuth resolution, which varies systematically from near range to far range. When an angular beam of energy is radiated from the antenna, it fans out in the range direction. Azimuth resolution (R_a) is simply the arc length or horizontal beam width (ß, see Fig. 12.7) at a particular place in the range direction, i.e.

$$R_a = B_\theta \cdot R_s \tag{12.5}$$

where B_θ is the angular beam width at the antenna and R_s is the slant range distance. The angular beam width B_θ can be approximated as

$$B_\theta = \frac{\lambda}{D} \tag{12.6}$$

where λ is the wavelength used and D is the antenna length. Therefore

$$R_a = \frac{\lambda \cdot R_s}{D} \tag{12.7}$$

This means that azimuth resolution becomes coarser with increasing slant range. It can be improved by using shorter wavelengths, smaller slant range and larger antenna length. Shorter wavelengths have limitations, as they are more attenuated in the atmosphere. Reduction in slant range improves azimuth resolution but decreases range resolution (see above). Further, as mentioned later, longer slant range, i.e. low depression angles, is useful for other purposes, e.g. for enhancing features in areas of low relief and reducing various image distortions etc. (Sect. 13.8). Antenna length can be increased within a certain physical limit. Thus, there are constraints to improving the azimuth resolution of RAR systems. It is generally of the order of 15–60 m, and varies with slant range. The unit cell (pulse length by beam width) is called the *pulse rectangle*.

2. Synthetic-aperture radar. Synthetic-aperture radar (SAR) systems are more sophisticated radars and use processing algorithms to yield finer azimuth resolution. Present-day aerial and space-borne systems use SAR technology. The basic limitation of the RAR is the insufficient or coarse azimuth resolution with increasing range distance. However, the fanning out of the radar beam width has another implication: that objects situated farther away are observed by radar for a longer duration, i.e. in a greater number of sweeps. In a synthetic-aperture SLAR, all the observations of objects are integrated, and successive antenna positions are treated as if they were individual elements of one long antenna array (Fig. 12.9). The Doppler principle is used to process the data and synthesize an antenna of a much larger length. The synthesized beam is narrow and has a constant width or resolution, irrespective of the range distance. The present-day aerial SAR systems have an azimuth resolution of about 1–10 m, depending upon the sophistication of processing algorithm.

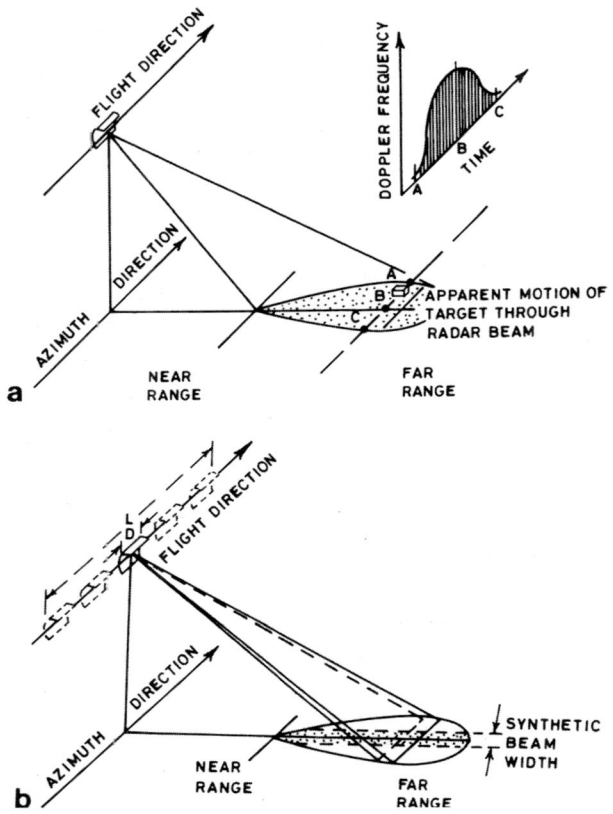

Fig. 12.9a.b. Working principle of a synthetic-aperture radar (Craib 1972). **a** Doppler frequency shift due to relative motion of target through radar beam. **b** Resolution of synthetic aperture radar. Note the synthetically lengthened antenna (L). The physical antenna length is D

12.3.4 SLAR System Specifications

A number of parameters are used to specify SLAR sensors, the important ones being the following.

1. Radar wavelength. Radar imagery is acquired at a fixed wavelength and this wavelength is one of the most important radar parameters. Ranges of wavelength used in radar remote sensing are listed in Table 12.1. They were coded alphabetically during the early classified stages of radar development, and the same nomenclature is followed to date, for harmony and convenience.

Table 12.1 Imaging radar bands

Radar band	Wavelength (cm)	Frequency (gigahertz, i.e. 10^9 cycles s^{-1})
Ka	0.8 – 1.1 (0.86)[a]	40.0 – 26.5
Ks	1.1 – 1.7	26.5 – 18.0
Ku	1.7 – 2.4	18.0 – 12.5
X	2.4 – 3.8 (3.1)	12.5 – 8.0
C	3.8 – 7.5 (5.7)	8.0 – 4.0
S	7.5 – 15.0 (15)	4.0 – 2.0
L	15.0 – 30.0 (23.5)	2.0 – 1.0
P	30.0 – 100.0 (50)	1.0 – 0.3

[a] Parentheses = commonly used radar wavelength.

2. Beam polarization. The plane of vibration of the electrical field vector defines the plane of polarization of the EM energy wave. In radar sensing, the transmitted wave train is always polarized in a particular direction (horizontal, H, or vertical, V). On interaction with the ground surface, it becomes partly depolarized. The radar antenna records back-scattered radiation only of a certain polarization, the like-polarized return (H–H or V–V) or the cross-polarized return (H–V or V–H). The H–H configuration yields strongest radar return and is therefore most widely used. An SLAR can also be operated with two receiving antennas – one for the like-polarized return and another for the cross-polarized return. In one radar image, however, the beam polarization is constant.

3. Look angle and swath width. Often several geometric parameters such as swath width, look angle and other sensor-linked characteristics are used as specifications. These parameters have been defined earlier.

4. Resolution. Resolution, in both azimuth and range direction, forms another important specification.

12.3.5 Aerial and Space-borne SLAR Sensors

A number of aerial and space-borne SLAR sensors have been available. Table 12.2 gives a comparison of the general characteristics of aerial and space-borne SLAR sensors.

Table 12.2. Comparison of general characters of aerial and space-borne SLAR sensing

Aerial SLAR sensing	Spaceborne SLAR sensing
1. The flying altitude is generally in the range of 3–12 km	1. The spacecraft altitude is in the range of 225– 800 km
2. Swath width is about 5–45 km	2. Swath width is about 25–100 km
3. Depression angle at the antenna is moderate to low (i.e. look angle is moderate to high)	3. Depression angle at the antenna is moderate to high (i.e. look angle is moderate to low). Low depression angles would imply very long slant range due to high altitudes, requiring high energy for the SLAR signal output
4. Look angle can be taken as equal to incidence angle (in a flat terrain)	4. Look angle is always smaller than incidence angle (in a flat terrain)
5. From the near range to far range, the change in look angle across the swath is substantial (about 20°). As applicability of SLAR image for geologic interpretation is influenced by look angle (see Sect. 13.8), it may differ from near to far range	5. The change in look angle across the swath is quite small (± 2–5°)
6. Platform instability may be caused due to atmospheric turbulence	6. The space-borne sensorcraft is comparatively more stable in attitude and predictable in position

12.3.5.1 Aerial SLAR Sensors

In the 1950s, the SLAR images and systems were made available for civilian applications. Since then, their use has greatly increased earlier on aerial and later on space-borne platforms. In the 1960s and early 1970s, rapid development in aerial radar sensing took place, and extensive aerial surveys for mapping purposes were conducted, particularly in the Latin American continent. Aerial radar technology gradually spread to other developing countries and other parts of the world.

The sensors used were initially K-band, RAR type. Later, as technology progressed, longer-wavelength, SAR-type, multifrequency, and multipolarization systems with variable depression angles have been used. A summary specification of selected aerial SAR sensors is presented in Table 12.3.

Generally, the airborne SLAR systems have a flying height in the range of 3–12 km, a swath width of about 5–45 km, and moderate to high look angles. For an aerial system, look angle equals incidence angle in a flat terrain. From near range to far range across the swath width, the change in look angle is significant (about 20°). This is important in view of the fact that the applicability of SLAR images for lithologic interpretation is influenced by look angle, being better for smaller look angles (see e.g. Ford 1998).

Table 12.3. Specifications of selected aerial SAR systems

System	Altitude (km)	Wavelength		Incident angle (deg)	Polarization	Swath width (km)	Resolution (m)	
		Band	λ (cm)				Slant range	Azimuth range
AirSAR (USA, JPL)	8	C L P	5.6 23.5 68.0	20–60	HH,VV,VH HV	7–13	7.5	2
C/X-SAR (Canada, CCRS)	6	X C	3.2 5.6	0–85	HH,VV,VH HV	18–63	6–20	1–10
E-SAR (Germany)	3.5	X C L	3.1 5.6 23.0	15–60	HH, VV	3	2	2
K-RAS (Denmark)	12.5	C	5.6	20–80	VV	9–48	2–8	2–8

12.3.5.2 Space-borne SLAR Sensors:

1. Seasat

Space-borne radar remote sensing began in 1978 with the launch of Seasat (Fig. 12.10a). The Seasat, basically an oceanographic survey sensor, carried a synthetic aperture radar (SAR) and a scatterometer, in addition to other sensors. The SAR operated in the L-band (23.5 cm) (Table 12.4). Owing to a large depression angle (about 70°), the Seasat images show high geometric distortion (see Sect. 13.2.2) and therefore could not be much used for geological applications.

2. Shuttle Imaging Radar (SIR) Series

NASA's Shuttle-Imaging-Radar (SIR) series of experiments (SIR-A, -B and -C) followed the Seasat experiment. The SIR-A was flown aboard Space Transportation System (STS)-2 in 1981 and used depression angles of 40° ± 3° (Table 12.4). Due to the relatively moderate depression angles, layover is largely absent on these images, shadows are also not so prominent, and therefore this has yielded quite interesting images. SIR-B, flown on STS-17 in 1984, had only limited success due to system malfunctioning.

SIR-C/X-SAR was an important programme, a joint venture between the USA (NASA/JPL), the German Aerospace Research Establishment (DLR) and the Italian Space Agency (ASI). The mission had capability to acquire SAR data concurrently at three wavelengths – L-band (23.5 cm), C-band (5.6cm) and X-band (3.1

Table 12.4. Salient Characteristics of selected space-borne SAR systems

Space-borne system / Parameter	Seasat (USA)	SIR-A (USA)	SIR-B (USA)	SIR-C/X-SAR (USA, Germany, Italy)	Almaz-1 (USSR/Russia)	JERS-1 (Japan)	ERS-1/2 (ESA)	Radarsat (Canada)	Envisat-1 (ESA)
Launch	1978	1981	1984	1994	1991	1992	1991/1995	1995	2002 (proposed)
Lifetime	100 days	2.5 days	8 days	Two flights, 11 days each	1.5 years	2 years	2–3 years each	5 years	5 years
Altitude (km)	795	259	225	225	300	568	777	~800	777
Orbit inclination (deg)	108	38	57	57	73	97.7	98.5	98.6	98.5
Wavelength (cm) Band	23.5 L-band	23.5 L-band	23.5 L-band	23.5, 5.7, 3.1 L, C, X	10 S-band	23.5 L-band	5.7 C-band	5.7 C-band	5.7 C-band
Incident angle (deg)	23	50	15–64	15–55	30–60	39	23	20–50	20–50 variable
Polarization	HH	HH	HH	HH, VV, VH, HV	HH	HH	VV	HH	HH, VV
Swath width (km)	100	50	10–60	15–60	20–45	75	100	10–500	100
Azimuth resolution (m)	25	40	17–58	30	15	18	28	9–100	25
Range resolution (m)	25	40	25	10–30	15–30	18	26	10–100	25
Processing	O, D	O	O, D	D	D	D	D	D	D
Recorder on-board	N	Y	Y	Y	Y	Y	N	Y	Y

O = optical; D = digital; N = no; Y = yes.

Active Microwave Sensors – Imaging Radars 333

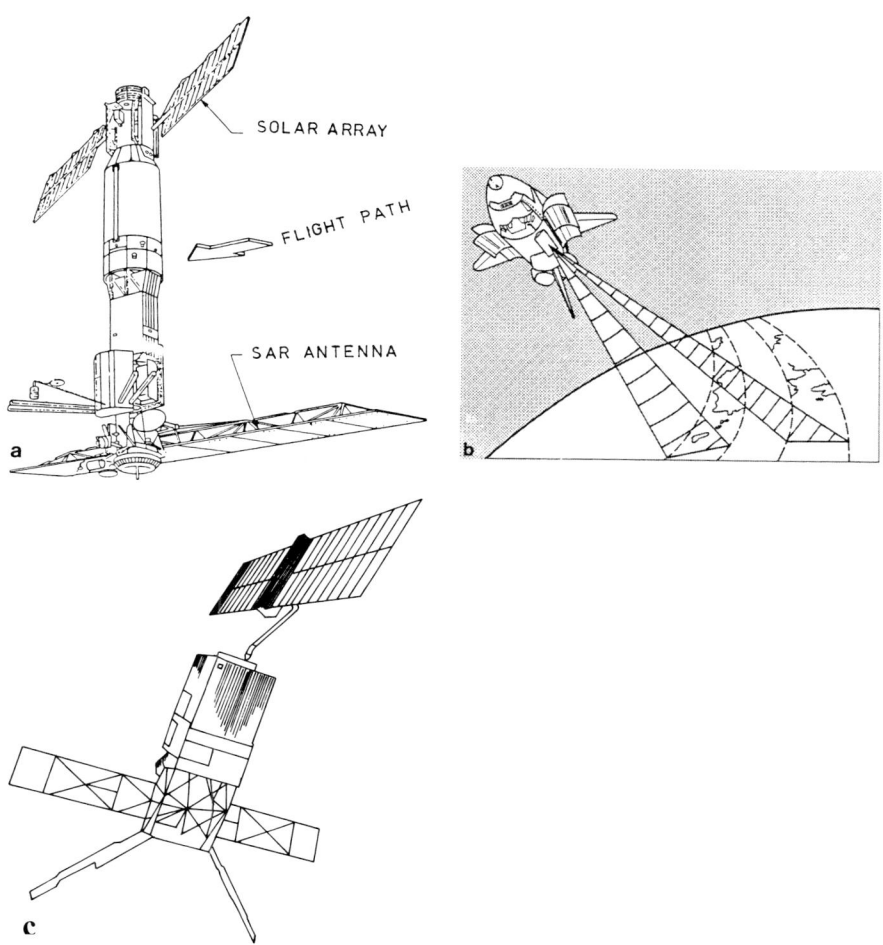

Fig. 12.10a–c. Selected SLAR-bearing satellites: **a** Seasat–the first SLAR-bearing satellite. **b** Shuttle Imaging Radar with a variable depression angle SLAR. **c** the ERS satellite, which has provided a wealth of SAR data, particularly for interferometry. (**a,b**) after NASA; **c** after ESA)

cm) – and in various polarization modes and at variable incidence angles (Fig. 12.10b) (Evans et al. 1997). The experiment was flown twice aboard the space shuttle 'Endeavour' in 1994. The first flight was made in April 1994 with an 11-day mission, and the second flight, also of the same duration, in October 1994. A wide choice of polarization, resolution, swath width and wavelength allowed several data collection modes from this experiment.

SIR-C web-site: http://www.jpl.nasa.gov/sircxsar

3. Almaz-1

Almaz-1 was a USSR/Russian satellite carrying an SAR sensor, which operated during 1991-92. Its S-band wavelength (10cm) and selectable incident angles make it different from other SAR missions. However, due to its low orbit (300 km), the on-board fuel depleted rather quickly, and this terminated the programme before the planned life.

4. JERS-1

The Japanese Earth Resources Satellite-1 (JERS-1) was launched in 1992 (Table 12.4). It carried an SAR sensor with L-band (23.5-cm wavelength) and HH polarization. The system was quite similar to the earlier Seasat except for the larger incidence angle of 39° in comparison to the 23° of Seasat. This has provided images with reduced geometric distortions arising due to topographic relief, and a better range resolution. With a tape recorder on board, the sensor has provided image data of several parts of the globe.

JERS web-site: http://hdsn.eoc.nasda.go.jp/

5. ERS-1 and -2

The European Remote sensing Satellites ERS-1 (1991–1996) and ERS-2 (1995–present) have been launched by the European Space Agency (ESA) (Table 12.4). These two satellites are identical to each other in payload and orbit. The on-board sensor is a C-band (5.7 cm wavelength) SAR sensor. A major application of ERS data has been in SAR interferometry, discussed in detail in Chapter 14.

ERS web-site: http://earth.esa.int/ersnewhome

6. Radarsat-1

Radarsat-1, the first Earth resources remote sensing satellite of Canada, was launched in 1995 with SAR (C-band, 5.6-cm wavelength) as the only imaging in-

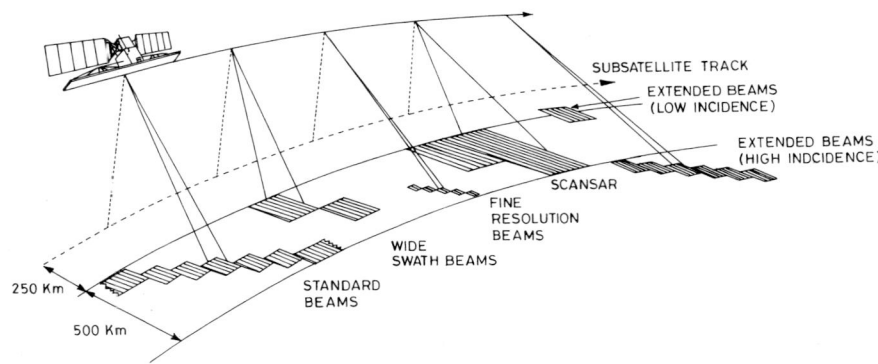

Fig. 12.11. Radarsat imaging modes. (After Radarsat Inc.)

strument on board (Table 12.4). Through ground command, it can provide data in a variety of beam selections – in terms of resolution, incidence angle and swath width. The various modes are called Standard, Wide Swath, Fine Resolution, Extended and Scan SAR (Fig. 12.11). For example, in Standard Beam Mode, it has a ground resolution of about 25 m × 28 m and a swath width of 100 km. In Fine-Resolution Beam Mode, it has a swath width of 45 km with a ground resolution of about 11(9) m × 9 m. In Extended Beam Mode, it can operate at high incidence angles (50–60°) or low incidence angles (10–23°). It carries two tape recorders on-board to store data from areas outside the visibility range of ground receiving stations. Further, it should be mentioned that Radarsat-1, with its orbit manoeuvring capability, is the only satellite to acquire data on Antarctica.

Radarsat web-sites: http://www.rsi.ca; www.radarsatinaction.com

7. SRTM

The Shuttle Radar Topographic Mission (SRTM) was a dedicated mission for satellite-borne single-pass SAR interferometry. It was an 11-day mission (February 2000) and carried a tape recorder. It used two sets of radar antennas for imaging in C-band and X-band. C-band covered any point on the equator twice (one ascending and one descending pass). It made global coverage (except areas more than 60° north and south) and has delivered the first continuous data set of this kind ever acquired. For more details, see Sect. 14.5.2.

8. Envisat-1

Envisat-1 is a mission proposed by the ESA as an advanced polar-orbiting Earth observation satellite. The payload is called Advanced SAR (ASAR). It features enhanced capability in terms of coverage, range of incidence angles, polarization, and modes of operation. The ASAR operates at C-band, 5.7-cm wavelength, similar to that of ERS-1and -2. Further, the orbit (35-day exact repeat coverage, 777-km altitude, 98.5° inclination from the equator) is also identical to that of ERS. In this way, it proposes to provide continuity of ERS-data. The spacecraft has a tape recorder for acquiring data from areas not covered by the Earth receiving system. The satellite is proposed for launch in 2002.

Envisat web-site: http://envisat.esa.int

Generally, the space-borne SLAR systems have an altitude in the range of 225–800 km, swath width of about 25–100 km, and moderate to high depression angles (or moderate to low look angles). Incidence angle (in a flat terrain) is always greater than look angle. From near range to far range across the swath width, the change in look angle is insignificant (about ± 2–5°). Further, the space-borne platforms are more stable in attitude and predictable in position in comparison to the aerial platforms.

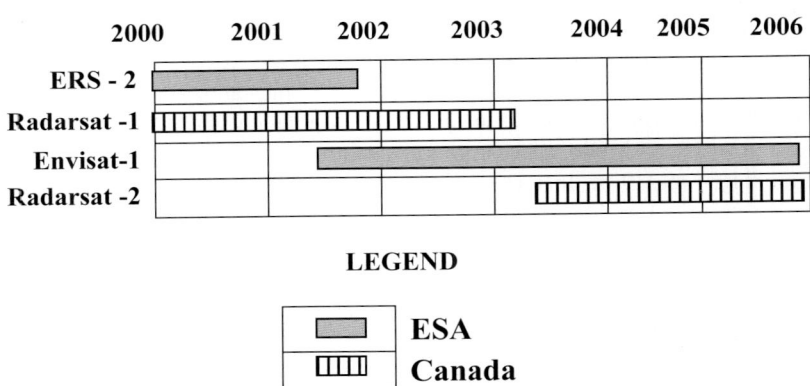

Fig. 12.12. Time distribution of important space-borne SAR sensors (2000–2006)

Figure 12.12 shows the time distribution of space-borne SLAR-bearing satellites which could provide data during the next five years.

Chapter 13: Interpretation of SLAR Imagery

13.1 Introduction

The technique of SLAR imaging and the various aerial and space-borne SLAR sensors were discussed in the preceding chapter. Briefly, the radar mounted on the base of the sensor platform looks sideways down on the Earth, transverse to the flight direction. It emits pulses of microwave energy, which illuminate long narrow stripes on the ground. The back-scattered signal, sensed by the antenna, is recorded in order of arrival time. This yields a radar image.

The interpretation and geological applications of SLAR data have been discussed by several authors (e.g. MacDonald 1969a,b, 1980; MacDonald and Waite 1973; Elachi 1980; Ford et al.1983, 1986; Sabins 1983; Trevett 1986; Buchroithner and Granica 1997; Ford 1998). The interaction of microwave EM energy with matter is governed by the wave nature of light. Both geometrical characteristics (shape, roughness, and surface orientation) and electrical properties (complex dielectric constant) are important. Broadly, the radar return is sensitive to decametre-scale changes in surface slope and centimetre-scale changes in surface roughness. In addition, the dielectric properties of the ground material (surface moisture and mineralogical composition) also influence the radar return.

The radar response opens up new avenues for discriminating and mapping Earth materials, as the radar signal provides a 'new look' at the ground. In this chapter we first discuss some SLAR image characteristics, in order to become conversant with the terminology. This is followed by a discussion on factors affecting radar return, and then interpretation and geological application aspects of SLAR sensing.

13.2 SLAR Image Characteristics

13.2.1 Radiometric Characteristics

The back-scatter at the radar, called radar return, is received by the antenna, amplified and recorded (Fig. 12.3). On an SLAR image, the intensity of radar re-

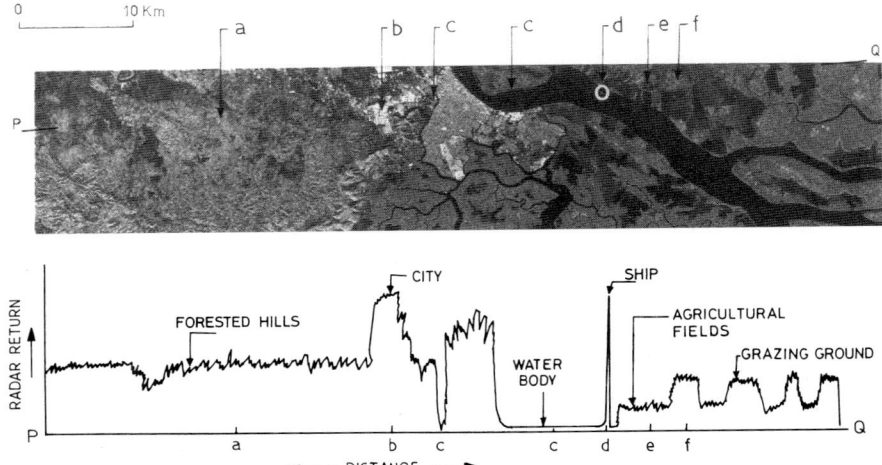

Fig. 13.1. a Radar image (L-band, SIR-B image of a region in Equador) showing some typical features. **b** Radar response profile drawn along P–Q on the image. The features seen are: *a* forested hills (diffused scattering); *b* city (Guayaquil, which forms the chief port and largest city in Equador) (corner reflection); *c* river (Guayas river; specular reflection); *d* ship (corner reflection); *e* agricultural fields (rice fields; near-specular reflection); *f* grazing ground (diffused scattering). (Image courtesy of C. Elachi and J.P. Ford, Jet Propulsion Laboratory)

turn is shown in shades of grey, such that areas of higher back-scatter appear correspondingly brighter. The most common types of responses on an SLAR image are the following (Figs. 13.1 and 13.2).

1. Diffused scattering. Most of the area on an SLAR image is dominated by diffused scattering caused by rough ground surfaces and vegetation (leaves, twigs and branches). These objects are also called diffuse scatterers and produce intermediate radar return.

2. Clutter is the term used for an intermediate, rapidly varying noise type of response, seen typically over sea surface and vegetation; this may be a hindrance to locating surface phenomena.

3. Hard targets. Some very strong responses are often seen on a radar image and are said to from hard targets, such as metallic objects and corner reflectors. Metallic objects produce high radar returns due to their high dielectric constant (discussed later). Bridges, automobiles, power lines, railway tracks and all metallic objects are generally easily identified on radar images (Figs. 13.1 and 13.2). Some hard targets may contain several scattering centres, e.g. in a ship or in an industrial complex, and the resulting image may exhibit a cluster of hard targets.

SLAR Image Characteristics

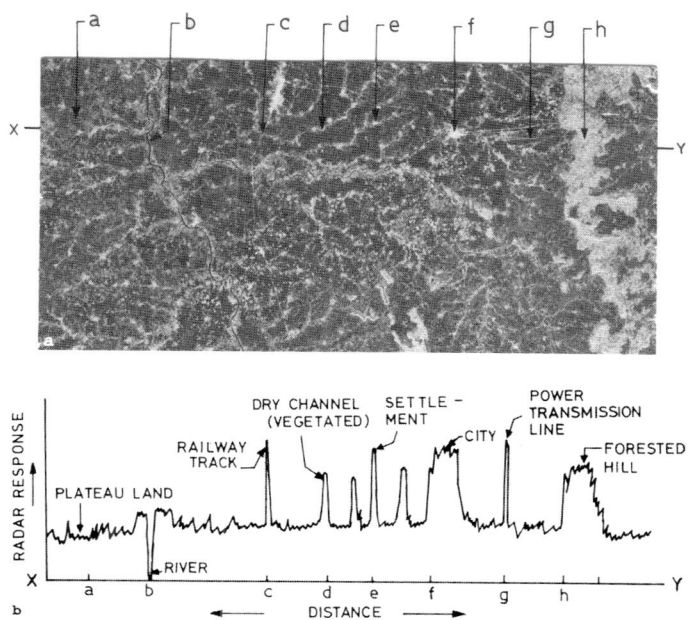

Fig. 13.2. a SIR-A image of a part of central India. **b** radar response profile along X–Y. Several typical features are seen such as; *a* plateau land with sparse vegetation (low-intensity diffused scattering); *b* river (Mahanadi river; specular reflection); *c* railway track (metallic object); *d* vegetated channel (diffused scattering); *e, f* settlement and city (corner reflection); *g* power transmission line (metallic object; note the corner reflection effect due to transmission line towers, which appear as small dots); *h* forested hills (diffused scattering and corner reflection). (Image courtesy of C. Elachi, Jet Propulsion Laboratory)

4. *Corner reflector* effect is produced when the object has a rectangular shape, such as a vertical wall joining with the ground or walls/roofs/ground at mutual right angles. This leads to echoes and high radar return (Fig. 13.3). Corner reflectors are formed typically by buildings, ships, sharply rising hills etc. Behind the corner reflector there often lies a shadow zone. As corner reflectors (CRs) appear prominently on SLAR images, they may serve as GCPs and therefore have practical use in locating tie points for SAR image registration. For this purpose, natural or artificial CRs (whose coordinates are very well defined, also see Sect. 14.5.3) may be used.

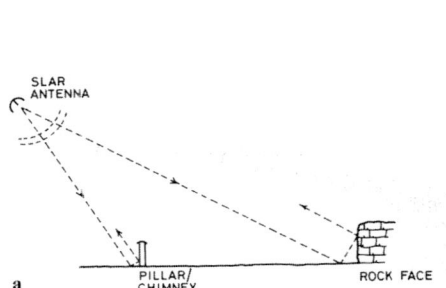

Fig. 13.3a,b. Corner reflection. **a** Mechanism of reflection of SAR waves straight back to the antenna (after Curran 1985); **b** aerial X-band SAR high-resolution imagery of Detroit, showing strong corner reflection effect; individual parking lot is clearly visible. (**b** Courtesy of MacDonald Dettwiler and Assoc. Ltd)

5. Specular reflection. This effect is produced by smooth surfaces, such as a quiet water body, playa lake, tidal flat etc. In the case of specular reflection, the SLAR beam is reflected in a small angular zone given by Snell's Law, and there occurs little or no return at the antenna, resulting in dark (black) tone (Fig. 13.l). A special case could be that in a certain area there may be a sub-pixel planar surface oriented nearly perpendicular to the incident beam (near-zero local incidence angle); due to quasi-specular reflection, this would generate very high radar return for the pixel. Other special cases of specular reflection are 'sea echo' and 'no show' in the surrounding sea clutter. *Sea echo* is a peculiar high signal recorded on the sea surface, brought about by reflection from sea waves. Further, oil films on the sea surface have a dampening effect on the waves and may produce 'no show' in the surrounding sea clutter.

6. Radar shadow. Imaging radar is a system which illuminates the ground from one side. In this configuration, some areas may not receive any radar pulse if they are located behind obstacles such as hills, buildings etc. This results in radar shadows. The extent of the shadow zone depends on the height of the sensor-craft, look angle, and elevation of the corner reflector. In radar sensing, the shadow zone is typically black, in contrast to the case of VNIR images, where some skylight may still faintly illuminate the shadow zone. Radar shadows are highly dependent on illumination geometry and ground relief. Higher look angles result in a greater amount of shadow (Fig. 13.4a). At times, radar shadows are helpful in detecting subtle topographical features. In a highly rugged terrain, on the other hand, radar shadows may be so extensive that they may render the image quite unsuitable for interpretation and application.

SLAR Image Characteristics

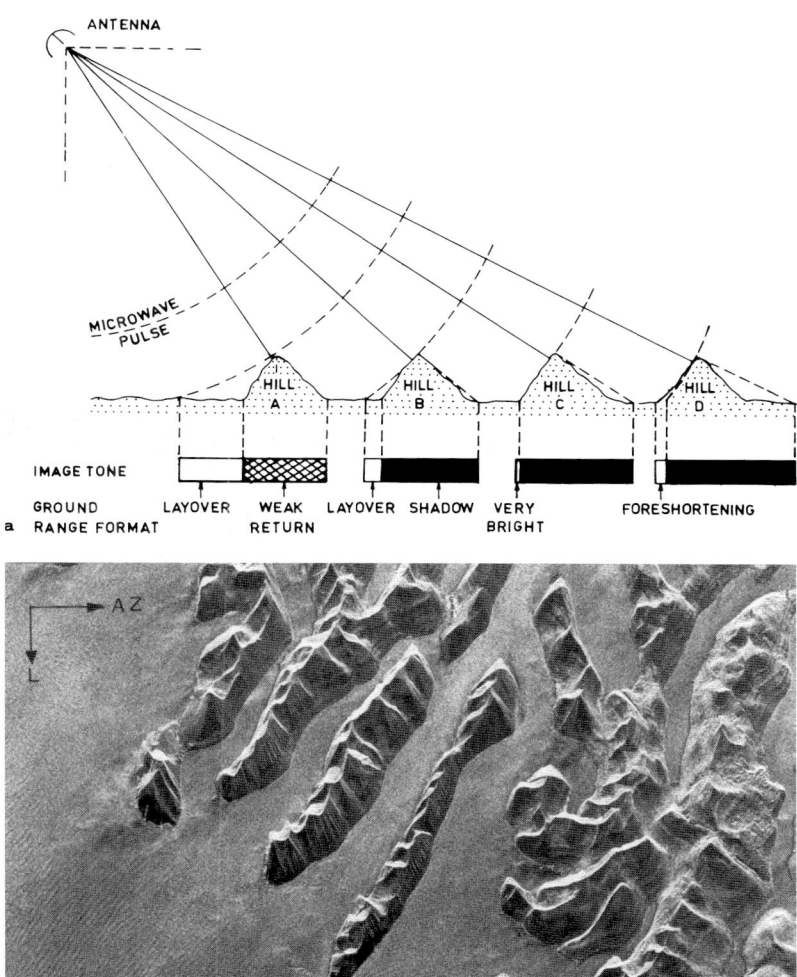

Fig. 13.4. a Layover, foreshortening and radar shadows (after Lewis 1976). **b** Seasat SAR image (18 August 1978) of parts of Iceland showing strong layover and foreshortening effects. (Processed by DLR; courtesy of K. Arnason) (AZ = azimuth direction; L = look direction)

7. Radiometric irregularities. At this juncture, we may also mention some unique radiometric irregularities seen on SLAR images. These types of irregularities are not seen on images of the optical EM spectrum. *Side-lobe banding* is the presence of brighter and darker bands, parallel to the azimuth direction, and these bands may sometimes be confused with terrain features (see Fig. 13.8b later). Subsidiary to the main pulse or lobe, some side lobes are emitted by the antenna, and are also reflected by the ground; reception of side lobes by the antenna leads to banding on the SAR images. The side-lobe banding is more prominent in the near range part of the image, whereas in the far range the banding is often absent, as the side lobe energy gets attenuated with distance. Further, this banding is often more prominent on cross-polarized images than on parallel-polarized ones, as the former require greater amplification of signal, resulting in simultaneous enhancement of the side lobe banding as well.

Another feature needing attention is the *speckle noise*, which accompanies data processing for higher azimuth resolution in SAR systems. The speckle noise occurs due to the coherency of the radar signal and the presence of statistically distributed reflecting targets inside one resolution cell. It produces a type of fine texture on the SAR images, similar to that observed when a scene is illuminated with a fine laser beam. Processing for higher azimuth resolution leads to greater image speckle noise. The speckle noise can be reduced by going through an averaging or smoothing filter. However, geologists in general appear to prefer a higher-resolution SAR image for interpretation (Ford 1982).

13.2.2 Geometric Characteristics

The geometry of radar imagery is fundamentally different from that of both types of optical sensor data (photographs and scanner imagery), owing to the basic difference that the location of objects on SLAR images is based on distances rather than on angles. We mention here a few important types of geometric distortions which affect interpretation and application of SLAR imagery (for details, see Leberl 1998).

1. Image displacement due to relief. The SLAR system basically works on slant-range distances. If the ground is flat, the image carries a regular non-linear geometric distortion, which is not difficult to rectify. However, if the ground has uneven topography, the images of objects get displaced, exhibiting foreshortening and layover (Fig. 13.4).

Foreshortening is a universal phenomenon on all SLAR images of undulating terrains. This distortion arises due to the fact that elevated points are displaced towards the sensorcraft (owing to decreased slant-range distance from the sensor craft), leading to relative shortening of all facets sloping towards the sensor-craft (Fig. 13.4a); this is called foreshortening. Complementarily, the hill facets sloping away are relatively lengthened.

Layover occurs in special situations when the slant-range distance to the top of a feature (e.g. a hilltop) is less than that to the base. In such a case, the top will be imaged earlier on the SAR image, and the base of the same topographic feature later. This is called layover (Fig. 13.4). This occurs when the terrain is high and rugged, topographical features rise sharply, and/or the look angle at the antenna is low. Layover renders physiographical conceptualization and interpretation difficult and geometric mapping altogether quite impossible.

SLAR Image Characteristics

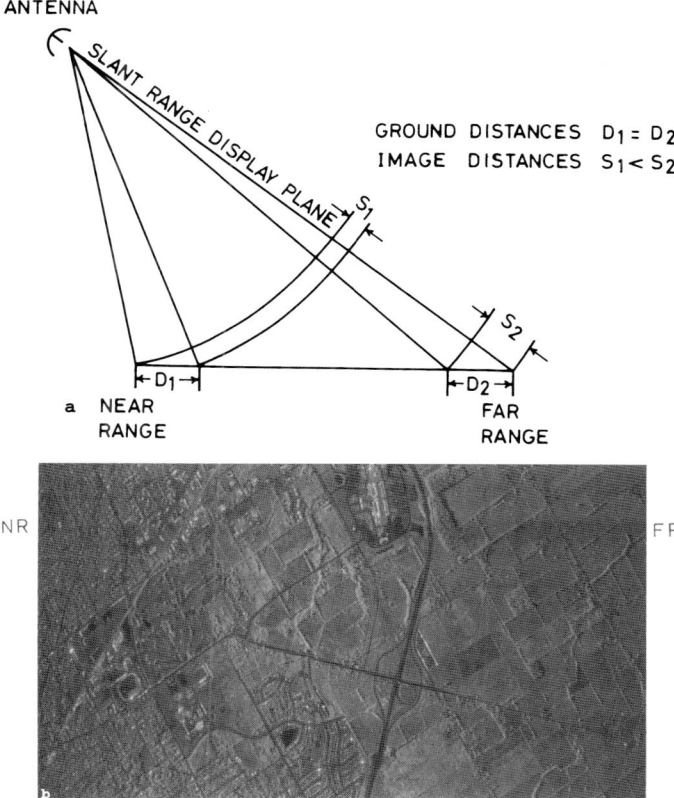

Fig. 13.5a,b. Scale distortion due to slant-range display. **a** Schematic showing that the ground distances D_1 and D_2 although equal, are represented on the image as S_1 and S_2 respectively, such that $S_1 < S_2$, which results in relative compression on the near-range side (after Sabins 1987). **b** Slant-range display image; note the compression of features on the near range (NR) side, in comparison to that on the far range (FR) side, resulting in curving of streets and rhombohedral fields. (Courtesy of Environmental Research Institute of Michigan)

2. Geometric distortion due to slant-range display. Some of the SLAR systems use slant range as the reference distance for image display. Although a set of objects may be equally spaced on the ground, the radar time intervals (related to slant range between the antenna and the objects) for these objects are unequal. The use of slant range as the base distance results in a scale distortion such that a relative compression of imagery at near range (or expansion of imagery at far range; Fig. 13.5) occurs. To counter this type of distortion, ground range (which can be computed from look angle) is used as the reference distance.

3. Look direction effects. It is clear that radar imagery is strongly direction dependent (e.g. Eppes and Rouse 1974). Owing to radar shadows, which get oriented parallel to the azimuth direction, features parallel to the azimuth direction are relatively enhanced and those parallel to the look direction are relatively suppressed. On an SLAR image, the position, orientation and extent of look-direction effects (shadows, layover and foreshortening) depend upon illumination geometry – i.e. relief in the area, altitude of the sensor-craft and look angle and look direction. The same area covered from differently oriented SLAR flights appears differently (see fig. 13.22 later). Due consideration must therefore be given to look direction effects while interpreting and mapping from a radar image.

4. Look angle effect. The geometry of an SLAR image is also dependent upon look angle (see Fig. 13.21 later). In the case of airborne SAR, the geometry changes rapidly across the swath due to look angle variation. On the other hand, in space-borne radar imaging, the effect of look angle variation across the swath on image geometry may be minimal.

5. Platform instability. Additionally, some distortions may arise in SLAR images due to sensor-platform instability. Space-craft are relatively more stable platforms and the images acquired from space possess better geometric fidelity than do those from aerial SAR systems.

Geometric Rectification

Raw SLAR images possess geometric distortions and characteristics so peculiar in nature that they render geographic location/recognition, and also superimposition and conjunctive interpretation with other data sets (such as photographs, scanner images and other standard maps), quite difficult. Therefore, it is necessary to carry out digital processing of SLAR image data for geometric rectification before any interpretation work can be taken up. This involves registration of SLAR image data to a base image by using tie points (for image registration, see Sect. 10.4). In a rugged terrain, image rectification using a digital elevation model (DEM) provides a better method (Naraghi et al. 1983). Ortho-rectified radar image products are provided by various agencies.

Radar Mosaic

Radar coverage is often used for large-scale regional surveys. For synoptic viewing, SLAR images can be put together strip by strip to form a mosaic. However, as mentioned earlier, radar images are highly direction dependent. Therefore, for mosaicking, it is advisable to have the entire area covered with the same look direction. If the area is imaged from opposite look directions, then shadows in adjacent strips become oppositely oriented, which makes conceptualization of the terrain difficult; hill ranges in one strip may look like valleys in another, owing to oppositely cast shadows. If on the other hand, a mosaic is generated from strips flown with the same look direction, the problem is that the far range of one strip falls adjacent to the near range of the next strip. This sudden change in

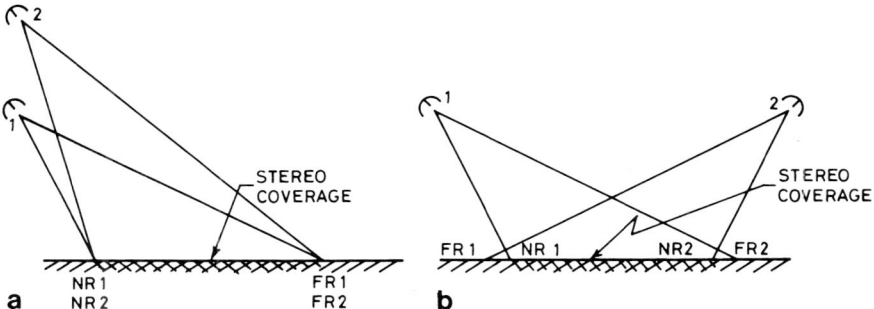

Fig. 13.6a,b. Flight arrangements for stereo coverage by SLAR with **a** parallel look direction and **b** opposite look directions

look angle creates an artifact that has no relation to the terrain features. In view of the above, radar mosaics have found rather limited application.

13.3 SLAR Stereoscopy and Radargrammetry

SLAR stereoscopy

Ideally, for stereo viewing, the two images in a stereo pair must be similar in thematic content and radiometry, and differ only in geometry caused by viewing perspective. However, any two SLAR images of the same area, acquired from two different stations, would possess differences in geometry as well as in radiometry, caused by the variation in look angle and direction. The perception of relief in the mental model is related to base/height (B/H) ratio, which governs vertical exaggeration (VE; see Sect. 6.2.2). In SLAR sensing, if the area is covered from two adjacent parallel flights with the same look direction, the B/H ratio is small (Fig. 13.6a), giving a small value of VE. If, on the other hand, SLAR images with opposite look directions are used (which certainly would increase B/H ratio and therefore VE) (Fig. 13.6b), images carry shadows in opposite directions; the resulting sharp differences in radiometry render image fusion difficult for stereoscopy. Therefore, stereo viewing of SLAR images has limitations.

Figure 13.7 presents an example of stereo radar images obtained from the SRTM mission (February 2000).

Radargrammetry

The technique of measurements from stereo radar images is termed radargrammetry or stereo radargrammetry. As raw/simple radar images have poor geometric accuracy, they were used in earlier times only for reconnaissance surveys. Now, with the availability of the global positioning system (GPS), differential GPS is used to control individual flight lines and locate the position of the im-

Fig. 13.7. Stereo image pair from SRTM. The region covered is NW of Bhuj, India. The elevation model is derived from the SRTM, over which Landsat-7 ETM+ image is draped to provide the land-cover information (printed in black-and-white from the colour image) (Courtesy of NASA/JPL/NIMA)

aging sensor at any instant of time. Using these data, stereo SLAR images are digitally combined to produce digital elevation models (for details, see Leberl 1998).

13.4 Radar Return

13.4.1. Radar Equation

The back-scattered signal received at the SLAR antenna is called radar return. The radar equation describes the dependence of radar return on various parameters, and is given as follows (see e.g. Lewis and Henderson 1998):

$$P_r = \frac{P_t \cdot G^2 \cdot \lambda^2}{(4\pi)^3 R_s^4} \cdot \sigma \tag{13.1}$$

where,
- P_r = power received (or radar return)
- P_t = power transmitted
- G = antenna gain
- λ = wavelength used
- R_s = slant range distance between antenna and target
- σ = effective back-scatter of the target, also called radar cross-section. It depends upon local incidence angle, surface roughness, complex dielectric constant, wavelength used and polarization. Local incidence angle, in turn, depends upon depression angle and terrain orientation. Surface roughness is also a function of wavelength and local relief. For all practical purposes, this is the single most important parameter influencing radar return.

Thus the radar return is governed by a complex interplay of factors, which can be grouped into two main categories: radar system factors and terrain factors (Table 13.1).

13.4.2 Radar System Factors

1. Power transmitted. The magnitude of power transmitted (P_t) directly affects radar return. Generally, radar flying at a higher altitude has to transmit a greater amount of power than one flying at a lower altitude; however, during a particular investigation, P_t can be taken as constant.

2. Antenna gain. Antenna gain (G) is a measure of current losses within the antenna material and is constant for a particular operation.

3. Radar wavelength. The radar return is directly related to the wavelength at which it operates. A particular SLAR imagery is acquired at a fixed wavelength and therefore this factor is constant. The various radar wavelengths used are listed in Table 12.1. The magnitude of the wavelength has to be taken into account while interpreting the imagery, as it primarily governs whether the surface would behave as a rough or a smooth surface (see Fig. 13.13 later). Further, smaller wavelengths are more prone to interaction with surficial features such as leaves, twigs and top soil, and carry greater information about surface roughness. On the other hand, larger wavelengths penetrate deeper, their signal carrying more information about the ground below the top cover of vegetation and soil.

Table 13.1 Factors affecting radar return

Group	Primary variable	Secondary variable
(A) Radar system factors	1. Power transmitted	
	2. Antenna gain	
	3. Radar wavelength	
	4. Beam polarization	
	5. Look angle	
	6. Aspect angle	Look direction and orientation of the general trend of the ground terrain
	7. Slant range	Flying altitude, look angle and ground relief
(B) Terrain factors	1. Local angle of incidence	Local topography (surface slope – amount and direction), look direction, look angle, flying altitude
	2. Surface roughness	Character of the ground material, radar wavelength, incidence angle
	3. Complex dielectric constant	Type of rock/soil, vegetation, moisture content
	4. Complex volume scattering coefficient	Object inhomogeneities, variation in dielectric property, radar wavelength, beam polarization

4. Beam polarization. The plane of vibration of the electrical field vector defines the plane of polarization of the EM energy wave. The back-scattered beam largely retains its polarization. However, a part of the beam may become depolarized after reflection, i.e. it may become polarized in some other plane. It is observed that depolarization depends on several factors, such as surface roughness, vegetation, leaf size, orientation etc. Based on earlier results (Dellwig and Moore 1966; Morain 1976; Ulaby et al. 1982; Evans et al. 1986; Zebker et al. 1987), it can be generalized that: (a) volume scattering promotes depolarization whereas direct surface scattering does not, (b) a higher degree of small-scale surface roughness, e.g. inhomogeneous soil or grass cover, leads to greater depolarization, (c) the orientation of features such as leaves etc. also appears to play a role in depolarization (also see Sect. 13.6).

Part of the depolarized wave-train can also be uniquely picked up by an antenna. In all, four polarization combinations can be used:
 1. Horizontal transmit – Horizontal receive (H–H)
 2. Horizontal transmit – Vertical receive (H–V)
 3. Vertical transmit – Vertical receive (V–V)
 4. Vertical transmit – Horizontal receive (V–H)
Figure 13.8 gives an example of like- and cross-polarized SLAR images.

Radar Return

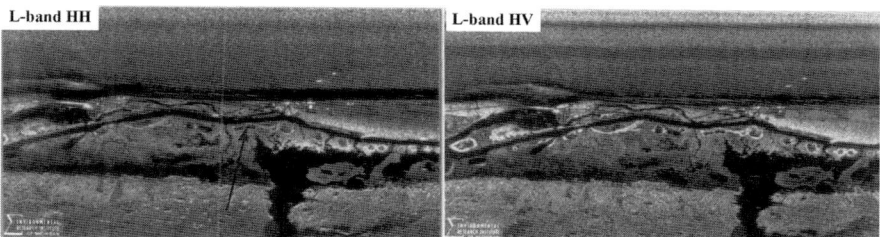

Fig. 13.8. L-band multi-polarization (HH and HV) SAR images; cross-polarized image has relatively weak signal, and also exhibits side-lobe banding. (Courtesy of Environmental Research Institute of Michigan)

5. *Look angle.* Look angle may vary from 0° to 90°. Commonly, in aerial surveys, the range of look angle is 60°–85°. In space-borne SLAR, as the altitude is high (about 300–800 km), lower look angles are generally used (see Table 12.2). The look angle influences radar return as it governs the angle of incidence. Lower look angle carries greater information about the features and their slope, but carries geometric distortion. As mentioned later, look angle is more important in lithologic discrimination.

6. *Aspect angle.* Look direction compared with the general terrain orientation is considered as the aspect angle, and influences radar return (MacDonald et al. 1969; Eppes and Rouse 1974). The same ground terrain may appear quite differently in terms of radar return when imaged from different directions (see Fig. 13.22 later). Aspect angle has to be duly considered while interpreting SLAR images and comparing a radar image with other images.

7. *Slant-range.* The slant-range distance depends on the altitude of the sensor-craft and the look angle. For aerial surveys, the typical slant-range distance is of the order of 5–20 km. For space-borne surveys, it may be of the order of 250–800 km. The radar return intensity decreases as the fourth power of slant range. Correction

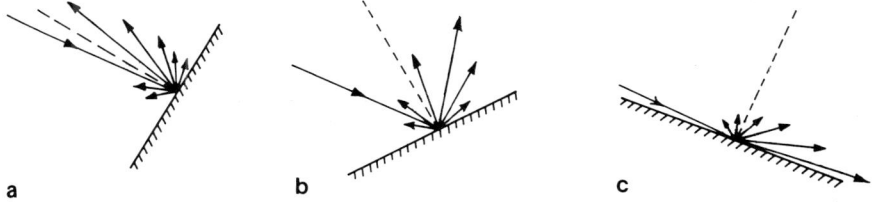

Fig. 13.9a–c. Variation in incidence angle caused by local topography, the orientation of the incoming radar beam being held constant. **a** Low incidence angle; **b** moderate incidence angle; **c** high incidence angle. This variation results in corresponding differences in radar return

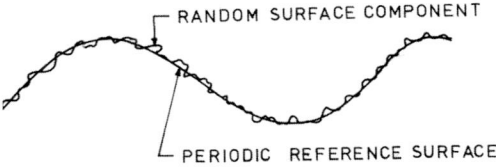
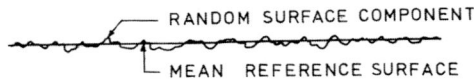

Fig. 13.10. Surface roughness, as conceived due to height variations. Random height variations superimposed on a periodic surface and that over a flat mean surface. (After JPL 1986)

for variation in slant range is generally applied as a radiometric correction during pre-processing of the data.

13.4.3 Terrain Factors

1. Local angle of incidence. Local angle of incidence (angle between the direction of the incident beam and the local normal to the surface) is one of the most important factors in radar return. A small incident angle leads to higher back-scatter (Fig. 13.9a). As the incident angle increases, the amount of back-scattered energy at the radar antenna decreases (Fig. 13.9b,c). When the incident angle is $\approx 90°$, the incident beam 'grazes' the plane and there is no radar return. Consequently, in general, if the slope is towards the sensor-craft, the radar return is high, and if it is sloping away the radar return is low.

On a very rough surface, e.g. a densely vegetated surface, the effect of surface roughness may be so intense that variation in topographical slope may not be of any consequence, and at incidence angles $> 30°$ the back-scattering coefficient may be almost constant.

2. Surface roughness. Roughness of the surface on which the radar beam is incident is very important. It may be considered as statistical variation of the random component in surface height relative to a certain reference surface. The reference surface could be a mean surface or could itself have a larger-wavelength periodic wave, such as the pattern of a row-tilled field (Fig. 13.10).

If the object surface is nearly smooth, the energy is totally reflected in a small angular region, and if the surface is rough, the energy is scattered in various directions (Fig. 13.11a,b). Therefore, for rough surfaces, the intensity of the radar return is nearly the same, quite irrespective of the look angle (Fig. 13.11d). On the other hand, for smooth surfaces, the radar return is high at low look angles and falls sharply with increasing look angle (Fig. 13.11c). All gradations occur between the two extremes of perfectly smooth and completely rough surfaces.

Radar Return

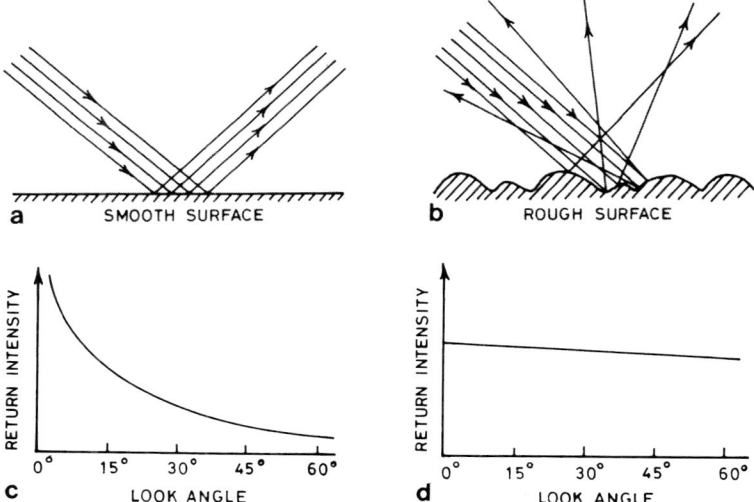

Fig. 13.11a–d. Radar reflection mechanism on **a** smooth and **b** rough surfaces; radar return for **c** smooth and **d** rough surfaces, as a function of look angle

Ground surfaces of the same roughness may behave as rough or smooth, depending upon the radar wavelength and look angle, and accordingly influence radar return. Therefore, the same object may appear dark or bright on different SLAR images. For example, a field could be rough (bright) for an X-band radar, but quite smooth (dark) for a P-band radar (Fig. 13.12).

A number of models have been proposed to explain the radar scattering in terms of ground surface roughness, and have been reviewed by Ulaby et al. (1982). For objective evaluation, criteria to quantify smoothness or roughness of a surface have been developed. One such criterion is the Raleigh criterion, which classifies a surface as rough if the root mean square of surface roughness (h_{rms}) has the following relation:

$$h_{rms} > \frac{\lambda}{8 \cos\theta} \tag{13.2}$$

where λ is the wavelength and θ is the incidence angle. Peak and Oliver (1971) modified the Raleigh criterion and proposed the following norms.

Fig. 13.12a,b. Dependence of radar return on ground surface roughness in relation to radar wavelength. The figure shows multi-wavelength (**a** X-band, 3-cm wavelength, HH; and **b** P-band, 72-cm wavelength, HH) aerial SAR images. Agricultural fields (crops) appear as diffuse reflector on the X-band and as specular reflector on the P-band. (Courtesy of Aerosensing Radarsysteme GmbH)

Radar Return

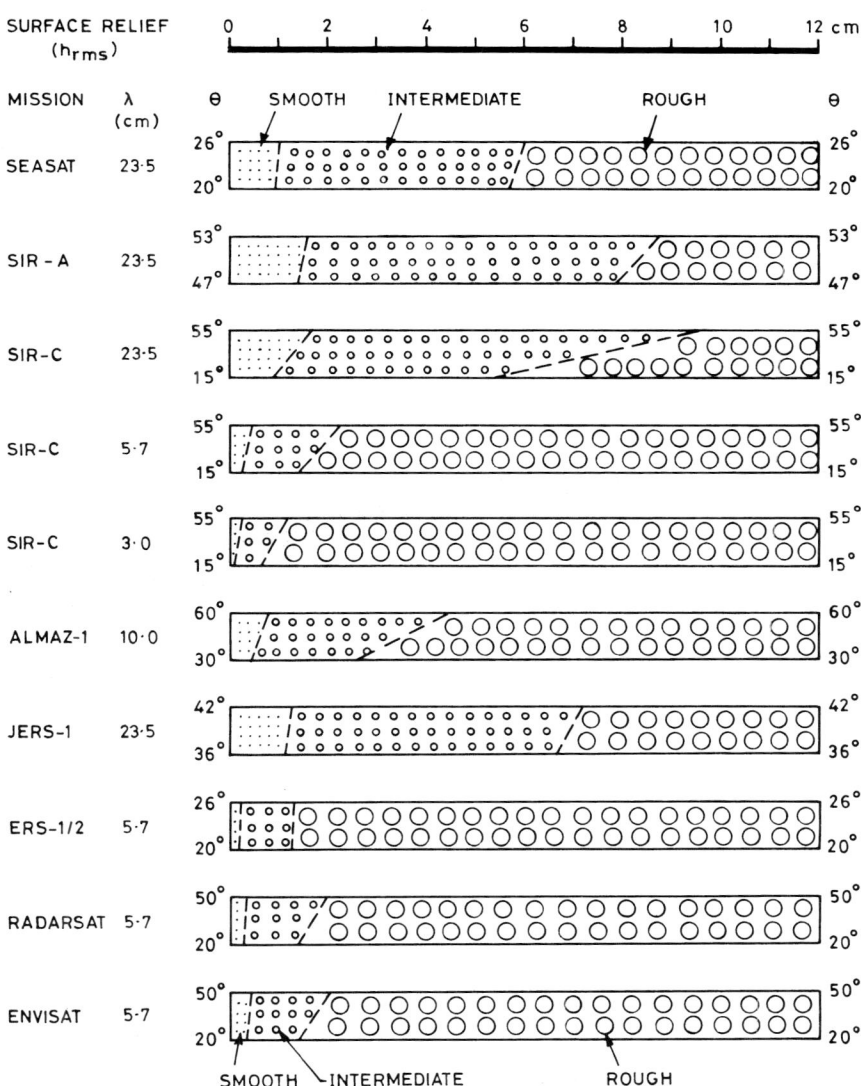

Fig. 13.13. Rough/intermediate and intermediate/smooth criteria limits for selected spaceborne SLAR configurations. The rough/intermediate limit corresponds to $h_{rms} = \lambda/(4.4 \cos \theta)$ and the intermediate/smooth to $h_{rms} = \lambda/(25 \cos \theta)$. The surface relief is also plotted in cm; λ = wavelength of the imaging radar; θ = incidence angle

The surface is

$$\text{smooth:} \quad \text{if } h_{rms} < \frac{\lambda}{25\cos\theta} \tag{13.3}$$

$$\text{intermediate:} \quad \text{if } \frac{\lambda}{25} < h_{rms} < \frac{\lambda}{4.4\cos\theta} \tag{13.4}$$

$$\text{rough:} \quad \text{if } h_{rms} > \frac{\lambda}{4.4\cos\theta} \tag{13.5}$$

These are presently the most widely used criteria. For various sensor configurations in space missions, based on the above criteria, the limits for rough/ intermediate and intermediate/smooth are schematically presented in figure 13.13. It is obvious that the same surface may behave as rough or intermediate or smooth, depending upon the radar wavelength and incidence angle. This opens up new avenues for mapping terrain features by using multiple-band SAR.

3. *Complex dielectric constant (δ).* As mentioned earlier, both geometrical characteristics and electrical properties govern the interaction of EM energy with matter. The most important electrical property of matter relevant in transmission and back-scattering of EM radiation is its spectral complex dielectric property (see

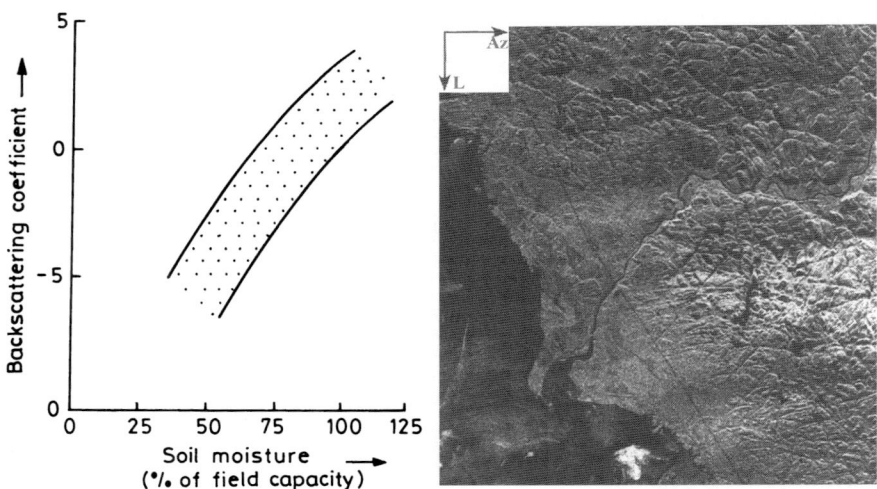

Fig. 13.14. a Increase in radar back-scattering coefficient with rise in soil moisture (redrawn after Bernard et al. 1984). **b** ERS-1 image showing high back-scatter (middle right of the image) due to a local storm. (**b** Copyright © ESA; courtesy of Canada Centre of Remote Sensing)

Table 2.2). Objects with lower dielectric constant reflect less energy than those with higher dielectric constant, other factors remaining the same. The reason for this is that a lower dielectric constant permits greater depth penetration, and as the energy travels through a larger volume of material, surface reflection becomes less. On the other hand, when objects possess higher dielectric constant, the energy gets confined to the top surface layers and back-scattering is higher.

At radar wavelengths, it is found that most dry rocks and soils have a complex dielectric constant of the order of barely 3–8, whereas water has the value of 80. Any increase in moisture content of soils/rocks results in a corresponding increase in dielectric constant of the mixture (e.g. Ulaby et al. 1983; Wang et al. 1983). For this reason, wet areas have a higher back-scattering acoefficient and appear brighter on radar images (Fig. 13.14a,b).

4. Complex volume-scattering coefficient. The concepts of surface and volume scattering were discussed earlier in Sect. 2.4. When a beam of EM energy is incident on a surface, a part of the energy becomes reflected/scattered on the surface, called surface scattering, and a part is transmitted into the medium. If the material is homogeneous, the wave is simply transmitted (Fig. 13.15a). If on the other hand, the material is inhomogeneous (layering, variation in textural and compositional characters, moisture etc.), the transmitted energy is further scattered (volume scattering) (Fig. 13.15c). A part of the volume-scattered energy may reach the sensor, carrying information about the subsurface/under-cover conditions. In nature, both surface and volume scattering occur concurrently, differing only in relative magnitude in different cases. The complex volume-scattering coefficient is a function of numerous variables such as wavelength, polarization of the incident beam, dielectric properties of the medium and its physical configuration.

Microwave energy interactions over water bodies are marked by predominantly surface scattering, owing to a very high complex dielectric constant (Fig. 13.15b), whereas vegetation cover is characterized by volume scattering (Fig. 13.15d). The multiple reflections from twigs, branches, leaves etc. resulting from volume scattering not only affect the intensity of the back-scattered signal but also cause depolarization. Volume scattering thus leads to a higher response on cross-polarized images. These peculiar radar responses may permit discrimination of vegetation types and densities.

13.5 Processing of SLAR Image Data

Simple SLAR images have commonly been used for visual image interpretation, and can be subjected to various image-enhancement procedures.

In areas of moderate to high relief, radiometric correction for topographic effects, to account for local change in incidence angle, is necessary before attempting thematic mapping from SAR data. Usually a transformation is carried involving a cosine function of the local incidence angle at each pixel. A prerequisite for this correction, therefore, is the availability of DEM, over which the SAR image

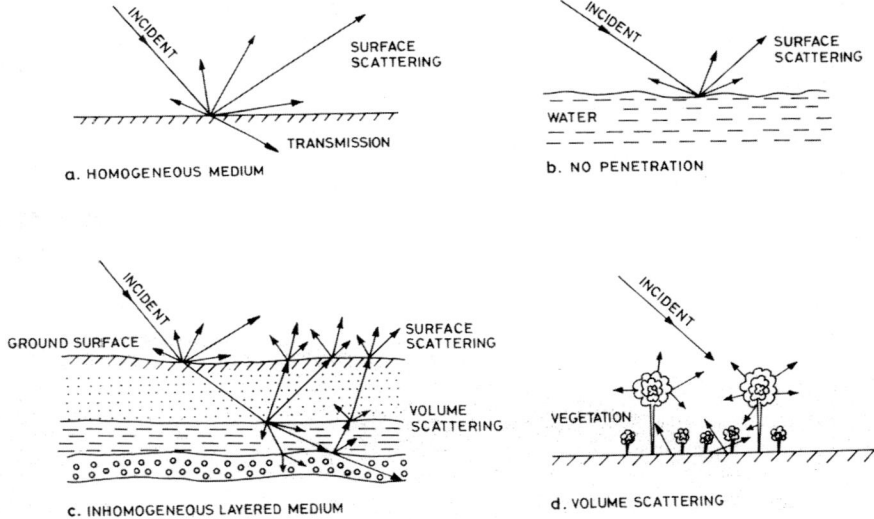

Fig. 13.15. a Surface scattering – an ideal case when the energy is incident on a homogeneous medium; the energy is partly scattered on the surface and partly transmitted; the transmitted ray simply travels further without any subsequent scattering, owing to the homogeneity of the material. **b** Surface scattering of microwave energy over water body; owing to high dielectric constant, nearly all the energy is scattered from the top surface. **c** Volume scattering, the incident energy is partly transmitted, and is further scattered owing to the inhomogeneity of the medium. **d** Volume scattering from vegetation; the energy suffers multiple reflections from crown, twigs, leaves, stem and the ground

can be registered. The effect of variable local incidence angle can also be reduced by ratioing different radar-band images acquired at the same time (Ranson and Sun 1994).

Speckle is a noise which affects the interpretability of SAR image data. Speckle reduction (through image filtering) may be carried out before attempting interpretation and digital classification of radar image data. More advanced and dedicated radar-image digital analysis aspects include texture extraction and image segmentation (see e.g. Ranson and Sun 1994).

The radar image data can be processed and combined with other data sets in several ways using digital image processing techniques.
1. Radar image data can be split into two components using a digital filter; the low-frequency component, possibly related to lithological contrast, and the high-frequency component, related to vegetation and sloping targets. These two components can be coded in different colours, or in an IHS scheme, which may lead to a better image interpretability (Daily 1983).
2. Radar data can be combined with other data, for example from the VNIR and TIR regions, or geodata such as geophysical data etc., in order to enhance geoscientific information (see e.g. Leckie 1990; Croft et al 1993).

Polarimetry 357

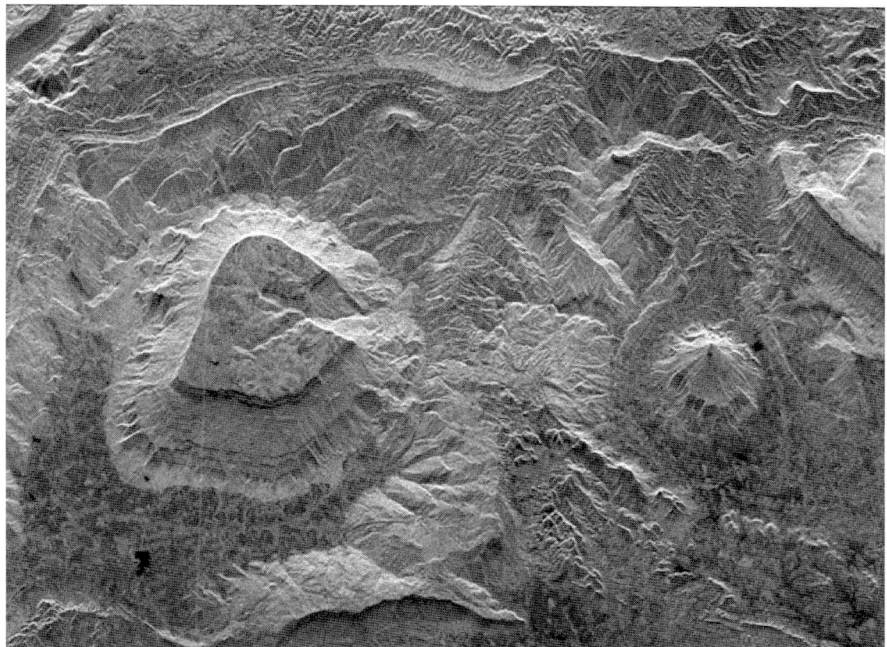

Fig. 13.16. Radar image showing landscape in the Hoei Range of north-central Thailand; the plateau terrain has been dissected by fluvial erosion; note the fine morphological details brought out by combining multiple SAR images. (SIR-C image; colour coding: R = L-band HH, G = L-band HV, B = C-band HV). (Courtesy of NASA/JPL) (see colour plate III)

3. Another interesting approach is to combine multi-channel and/or multi-polarization radar images in different colours. As different radar images with different bands and polarizations are sensitive to ground surface features of different dimensions, they collectively bring out greater geological–geomorphological detail. Figure 13.16 presents an example of the coding of multiple SAR images in RGB.

13.6 Polarimetry

Radar polarimetry is a relatively new technique that deals with the study of the polarization pattern of radar backscatter. It was mentioned earlier that scattering changes the electric field polarization of the EM wave. The change in polarization depends upon such factors as scattering mechanism, diversity of scatterers, orientation and shape of scattering elements, and multiple bouncing. Polarization diversity, therefore, carries information on the nature of the scatterer. For polarimetry, multi-polarization radar systems are used to yield data on polarization diversity,

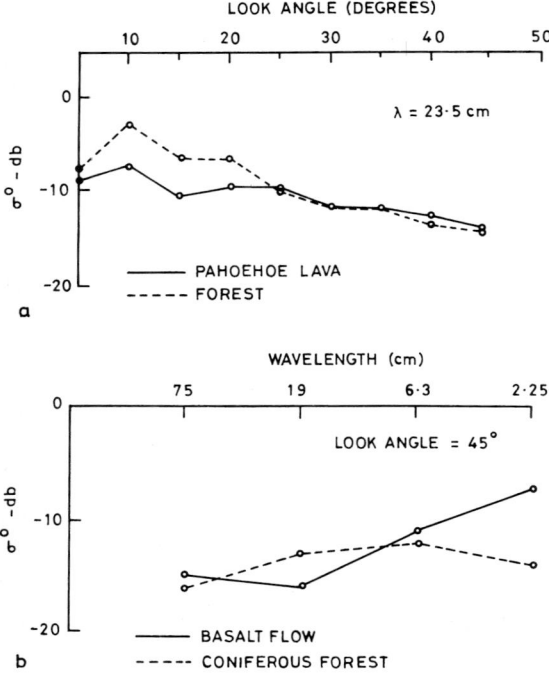

Fig. 13.17. a Field scatterometer data plot at a fixed wavelength (L-band 23.5 cm) and variable look angle; note that pahoehoe lava and forest are quite similar to each other except at look angles of around 10°. **b** Field scatterometer data plot at a fixed look angle (45°) but variable wavelength; note that the basalt flow and coniferous forest are quite similar except at a wavelength of 2.25 cm. (**a,b** simplified after Ford 1998)

which is compared to the transmitted wave. The change in polarization is analysed for each pixel. As volume scattering, commonly leading to depolarization, is more frequently associated with vegetation, these techniques have greater application potential in the field of forestry and agriculture (for details, see e.g. Elachi and Ulaby 1990; Boerner et al. 1998; Leckie 1998).

13.7 Field Data (Ground Truth)

Field data collected during an SAR image study are of two types:
1. Corner reflector
2. Scatterometer.

13.7.1 Corner Reflectors (CRs)

Corner reflectors are metallic tetrahedral bodies that are planted at well-known fixed positions in the field, during an aerial/space-borne SAR survey. As CRs are uniquely picked up on the SAR image data and their geographic position is well defined (to within centimetre accuracy), their data are used for geometric rectification of the SAR data (also see Sect. 14.5.3).

13.7.2 Scatterometers

Scatterometers are used to collect field data on back-scattering coefficients of various types of natural surfaces. They can be operated at various wavelengths and look angles. Usually the field scatterometer data are presented as curves/profiles. For example, figure 13.17a shows a scatterometer data plot at a fixed wavelength and varying look angle; figure 13.17b shows a data plot at a fixed look angle and variable wavelength. The scatterometer data help in understanding the behaviour of various natural surfaces at different wavelengths and look angles, which can be useful for SAR image interpretation and also the designing of better SAR sensors.

13.8 Interpretation and Scope for Geological Applications

For image interpretation, the well-known elements of photo interpretation are applied to SLAR images. These elements are tone, texture, pattern, shape, size, shadow, and association and convergence of evidence. Special advantages of SLAR sensing for geological applications stem from the fact that the SLAR return is influenced by the following:

(a) surface geometry – decametre-scale changes in surface slope
  – centimetre-scale changes in surface roughness
(b) complex dielectric properties
(c) illumination geometry.

Therefore, SLAR return provides information on:
- topography
- orientation of features in space
- surface roughness of the ground
- vegetation
- soil moisture
- drainage
- ground inhomogeneity
- metallic objects etc.

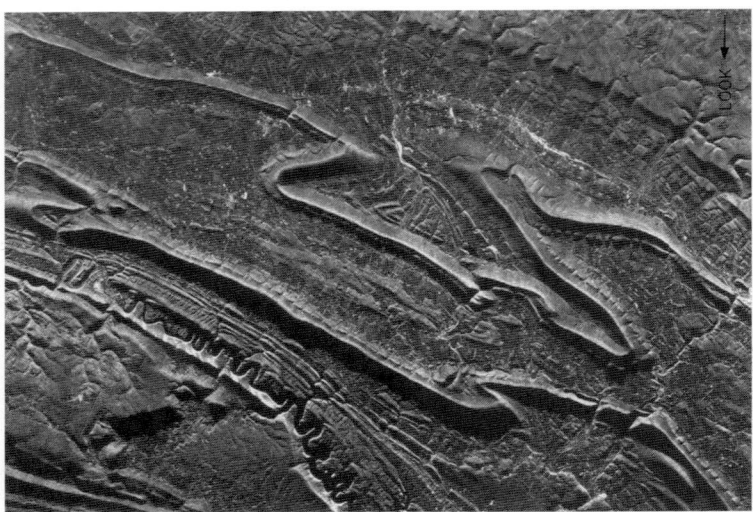

Fig. 13.18. Aerial SAR image of a structurally deformed terrain (Appalachian Mountains). X-band, swath width about 50 km (Courtesy of Intera Technologies Ltd.)

1. Geomorphology. SLAR images are of value in regional geomorphological studies, owing to the fact that minor details are suppressed on SLAR images. However, due to the fact that SLAR images provide 'oblique views', cartographic application is possible only from ortho-rectified radar images.

(a) Relief. General terrain relief/ruggedness is a very important type of information that can be obtained from SLAR images. The trend of physiographical features, i.e. hills and valleys, local relief, and look angle govern the manifestation of shadows and relief impression. In most cases, physiographical features run parallel to the structural trend or *grain*. If azimuth lines are aligned parallel to the physiographical trend, then alignment of hills (corner reflectors) and valleys (shadows) takes place perpendicular to the look direction, and this enhances manifestation of topographical features on radar images. Further, it should be noted that in areas of high relief, substantial areas are likely to remain unilluminated at high look angles, and strong geometric distortions would occur at low look angles. On the other hand, in regions of low relief, subtle relief impression on radar images may be revealed by high look angle radar imaging.

(b) Slope. For detailed interpretation of radar images, it is necessary to consider slopes and individual topographical facets. Based on the look angle, shadows and back-scatter at individual topographical facets, it may be possible to distinguish various slopes in terms of gradient (steep or gentle) and direction of slope (forward sloping or backward sloping).

Interpretation and Scope for Geological Applications 361

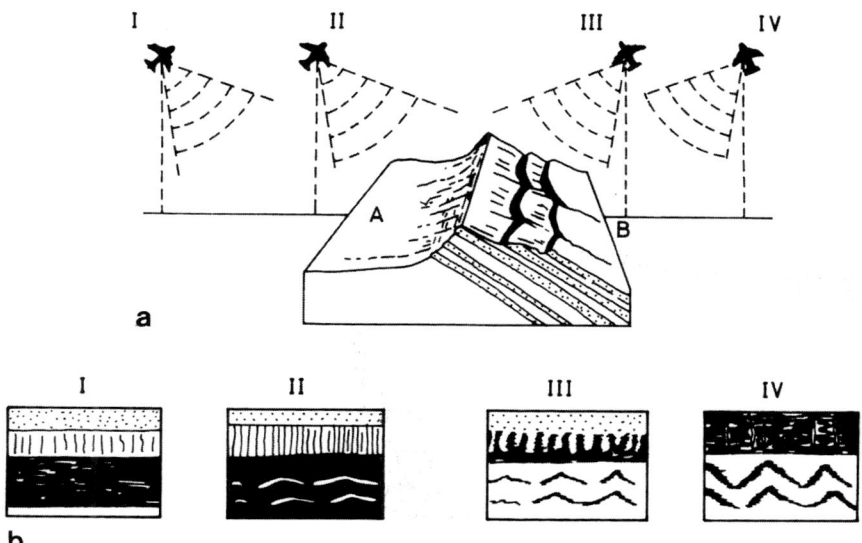

Fig. 13.19a,b. Influence of look angle and direction on representation of a dip slope on SLAR images. In **a** a dip slope and four different SLAR imaging positions I, II, III and IV are shown. **b** I–IV show the resulting images. (Modified after Koopmans 1983)

(c) Drainage. Study of the drainage network is often the first step in radar image interpretation. Types of drainage patterns and their relation to geology are well known (Table 16.1). Drainage channels are generally black to dark on radar images, as a water surface leads to specular reflection and there is no radar return at the antenna (Fig. 13.1). However, in semi-arid climates valley bases are often vegetated and appear bright on radar images (Fig. 13.2).

2. Structure. Radar images are found to be very useful for structural studies. Planar features such as bedding and foliation planes are often well manifested, in terms of variation in topography, relief, shadows, vegetation and soil. Extension of geological contacts can give the strike trend, and regional structures such as fold closures can be mapped (Fig. 13.18). Further, in a sedimentary terrain, morphological characteristics such as dip slope, slope-asymmetry, flatirons and V-shaped patterns of outcrop develop, and can be observed on radar images. The manifestation and pattern of these features on SLAR images is related to illumination geometry, as illustrated in figure 13.19.

In an area of low relief, higher look angle leads to shadows which may enhance outcrop pattern and structure. On the other hand, if relief is high, then low-look angle SAR data suffer from greater geometric distortions and high-look angle SAR data would have large shadows.

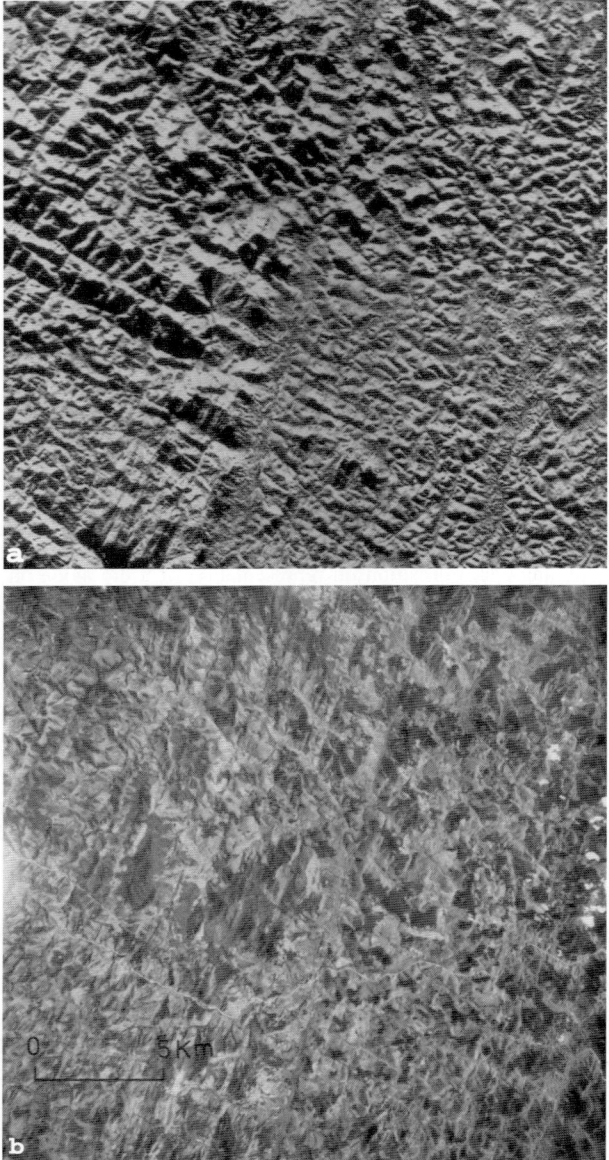

Fig. 13.20a,b. The structural details on radar imagery are better picked up due to microrelief than on the corresponding VNIR images and photographs. **a** X-band radar imagery of an area in Bahia, Brazil and **b** aerial photograph of the same area. (**a,b** Courtesy of A.J. Pedreira)

Interpretation and Scope for Geological Applications

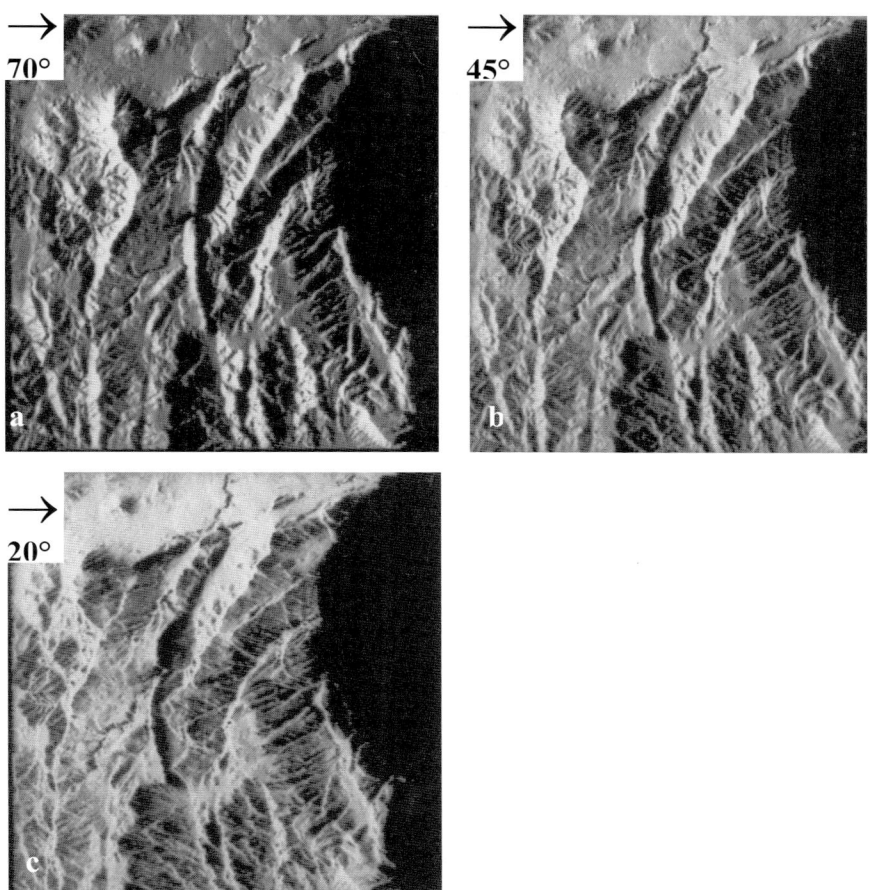

Fig. 13.21a–c. Effect of look angle on radar imagery. The three SAR images (**a–c**) of a test site in Italy have been acquired with a similar look direction but varying look angle: **a** 70°, **b** 45°, **c** 20°. Note the differences in manifestation of structural features on the three images. (**a–c** Courtesy of Institute of Digital Image Processing and Graphics, Graz)

Lineaments are extremely well manifested on SLAR images, generally much better than on the corresponding VNIR images and photographs (Fig. 13.20). In addition, several instances have been reported where structural features such as faults have been detected, extended or suspected, based on the SLAR image data (see e.g. Berlin et al. 1980; Sabins et al. 1980; Schultejann 1985).

It has been mentioned that look direction has a tremendous effect on the response and manifestation of features on SLAR images; figures 13.21 and 13.22 illustrate the role of look angle and direction respectively. Due consideration must be given to this fact while interpreting the relative dominance of SLAR linea-

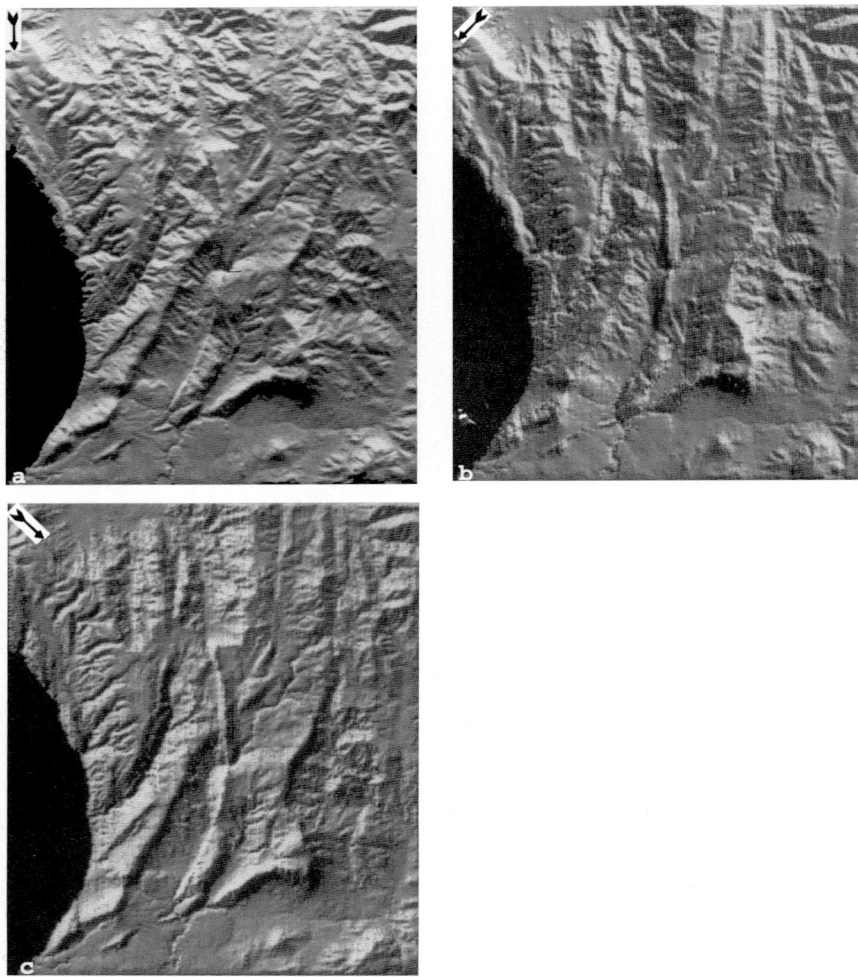

Fig. 13.22a–c. Effect of look direction on SAR imagery. The three radar images (**a, b, c**) of a test site in Italy have been acquired from different look directions, indicated by arrows on the images. In each image, note that the linear features aligned parallel to the look direction are relatively suppressed and those aligned perpendicular to the look direction are enhanced. (**a–c** Courtesy of Institute for Digital Image Processing Graphics, Graz)

ments. This phenomenon can also be used to locate, detect, or confirm suspected lineaments through specifically oriented flight lines. On the other hand, look direction is not relevant for studying features lacking directional character, such as natural forests and floods.

3. Lithology. On radar images, unique and direct lithologic identification is not possible. Interpretation for lithology must be based on indirect criteria such as sur-

Interpretation and Scope for Geological Applications 365

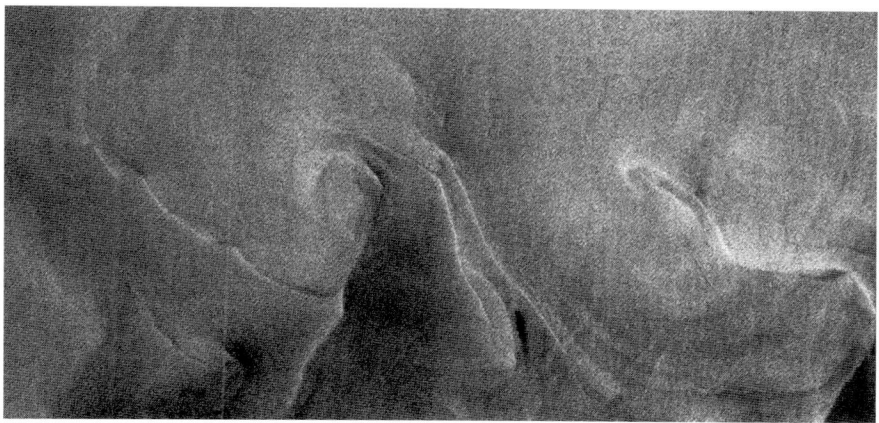

Fig. 13.23. Seasat image showing ocean features in the Gulf Stream, Western Atlantic Ocean). (Courtesy of MacDonald Dettwiler and Assoc. Ltd.)

face roughness, vegetation, soil moisture, drainage, relief, geomorphological features (sinkholes, flow structures etc.) and special features, contacts, etc. The radar image texture (Farr 1983) is also an important parameter in radar geology. Further, investigations have revealed good correlations between radar return and surface roughness of different types of lithologies (see e.g. Schaber et al. 1976). Based on scatterometer data, Blom et al. (1987) inferred that for separation of various lava flows, shorter wavelengths and smaller incidence angles are best; on the other hand, for sedimentary rocks, longer wavelengths and somewhat larger incidence angles are preferred. Figure 16.57 shows an example of a SEASAT image of Iceland, where lava flows of different ages can be discriminated owing to surface roughness differences; older lava surfaces are smother due to weathering and infilling with time, than are younger lava surfaces.

Further, it has generally been found that SAR images with low look angle possess higher sensitivity to slope changes and surface roughness, which is helpful in lithologic discrimination (see e.g. Ford 1998).

4. Soil moisture. In many investigations, soil moisture is a critical parameter. It has been mentioned that microwave response is governed by the complex dielectric constant (δ) of objects and that water has a very high δ, as compared to dry soil and rock. When moisture content in soil or rock is increased, a regular increase in δ of the mixture takes place (Fig. 13.14). This can be used to map soil moisture variation on the ground (Ulaby et al. 1983).

5. Depth penetration. Yet another important aspect in the context of geological application is depth penetration. Depth penetration increases with longer wavelengths as shorter wavelengths undergo a sort of 'skinning effect'. A number of conditions are necessary for depth penetration by SAR, and all of them must be fulfilled simultaneously (Blom et al. 1984; Ford 1998).

(a) The surface to be penetrated must be radar smooth.
(b) The cover to be penetrated must be extremely dry (moisture < 1 %), homogeneous, fine grained, and not too thick (up to two metres of penetration have been documented and up to six metres may be possible).
(c) The subsurface layer should be rough enough to produce backscatter to form an image.

Such conditions exist in aeolian sand sheets in desertic conditions. Further, from a single radar image, it is not possible decipher whether features observed on a particular SAR image are surface or subsurface. Therefore, complementary VNIR/SWIR/TIR image data are necessary for unambiguous interpretation.

6. Sea-bottom features. Lastly, some signatures of bottom features on radar images of the sea have also been reported in both shallow and deep water bodies (De Loor 1981; Kasischke et al. 1983; see Fig. 13.23). As water has a high dielectric constant at microwave frequencies, the microwaves cannot penetrate the water surface; it is considered that it must be the expression of the water surface 'morphology' (Bragg scattering of small gravity waves, currents, influencing wave pattern etc.) which 'reflects' the bottom topography on radar images.

Chapter 14: SAR Interferometry

14.1 Introduction

Synthetic Aperture Radar (SAR) interferometry, also called Interferometric SAR (InSAR, also IFSAR or ISAR), is a relatively new technique for producing digital elevation models (DEMs). It has made rapid strides during the last decade. As it provides a higher order of accuracy than the conventional stereo radargrammetry, the technique is on the verge of becoming operational as an alternative to the latter. The InSAR method combines complex images recorded by SAR antennas at different locations and/or at different times to generate interferograms. This permits determination of differences in the 3-D location of objects, i.e. generation of DEMs. The DEMs have wide application for geoscientific studies, e.g. for topographic mapping, geomorphological studies, detecting surface movements, earthquake and volcanic hazard studies, and several other applications.

Historical development. The use of SAR interferometry (InSAR) can be traced back to the 1960s, when the US military applied it for mapping Darien Province from aerial SLAR data. The first published report of the application of InSAR was by Rodgers and Ingalls (1969) for observing the surface of Venus and the Moon. Graham (1974) was the first to describe InSAR technique for terrestrial topographic mapping. Subsequently, Zebker and Goldstein (1986) applied the technique to aerial SAR data, and Gabriel and Goldstein (1988) to space-borne SIR-B data. Gabriel et al. (1989) introduced the technique of differential SAR interferometry (DInSAR) using data from three different Seasat passes. Until 1991, the research in to InSAR was constrained due to limited availability of suitable SAR data pairs. ERS-1 (launch 1991) started providing interferometric data sets. Subsequently, ERS-2 was launched (1995), and ERS-1 and ERS-2 were operated in tandem mode to provide SAR data sets of the same ground scene with a one-day interval, highly suitable for interferometry. The ERS-1/-2 tandem data sets have generated world-wide interest and provided a fillip to research in to this technique, bringing it to a near-operational stage.

14.2 Principle of SAR Interferometry

The radar signal is acquired as a complex signal comprising real (Re) and imaginary (Im) components (which is not so in optical data). These values contain

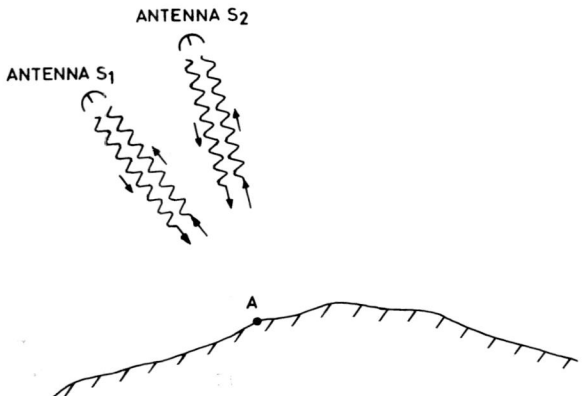

Fig. 14.1. The concept of phase and phase difference. Consider an object A being viewed from two SAR antenna positions S_1 and S_2. The emitted wave trains have the same phase at the antennas; however, the back-reflected wave trains differ in phase. Also, there is an ambiguity of many cycles

information about the amplitude (I) and phase (ϕ), which can be extracted from the real and imaginary components using the following relations:

$$\varphi = \arctan\left(\mathrm{Im}/\mathrm{Re}\right) \tag{14.1}$$

$$I = \sqrt{\mathrm{Im}^2 + \mathrm{Re}^2} \tag{14.2}$$

Thus, a unique feature of SAR data is that the phase of the complex signal is measurable directly, in contrast to the case of optical data (which has only intensity values). SLAR images in vogue (Chapter 13) utilize amplitude information, whereas phase information is used in SAR interferometry.

The amount of phase equals the two-way travel path divided by the wavelength, and therefore carries information on the two-way travel path to sub-wavelength accuracy (Fig.14.1). The phase of a single SAR image may not be of any particular utility. However, phases of two SAR images of the same ground scene, acquired from slightly differing angles, possess a phase difference. This forms the basic strength of the SAR interferometric technique – that the difference in slant ranges from the two antenna positions can be measured with fractional wavelength accuracy in terms of phase difference. The phase difference is related to slant-range difference and can be processed to derive height information, i.e. generate a DEM.

The theoretical aspects of the technique are well understood and reviewed by several workers (see e.g. Gens & van Genderen 1996; Madsen and Zebker 1998; Franceschetti & Lanari 1999). If we consider two SAR images taken from two slightly different viewing angles (Fig. 14.2), assuming no change in backscatter, the phase difference (ϕ_d) for the same surface element in the two SAR images is proportional to the travel-path difference ($\Delta r = r_1 - r_2$).

$$\varphi_d = \frac{4\pi}{\lambda}(r_1 - r_2) \qquad (14.3)$$

where λ is the radar wavelength. Figure 14.2 shows an idealized case, such that two SAR image are acquired from two antennas, S_1 and S_2, on parallel flight paths with flying height H. The baseline is B, and has a tilt angle α from the horizontal. Slant ranges from S_1 and S_2 to the same ground element are r_1 and r_2. With this geometry, the height (h) of the point above the datum plane can be represented as

$$h = H - r_1 \cos\theta \qquad (14.4)$$

where θ is the look angle. This can be rewritten as

$$h = H - r_1 \left(\cos\alpha \sqrt{1 - \sin^2(\theta - \alpha)} - \sin\alpha \cdot \sin(\theta - \alpha) \right) \qquad (14.5)$$

This means that h can be computed if we know H, α and ($\theta - \alpha$). H and α are directly known from the orbital flight data; the third parameter ($\theta - \alpha$) can be derived from range difference (r_d) and baseline length (B) [as ($\theta - \alpha$) = $\sin^{-1}$ ($\Delta r / B$); figure 14.2b]. Thus, this method offers a possibility to obtain elevation data of various points on the Earth's surface – by measuring travel-path differences.

The travel-path difference [2 ($r_1 - r_2$) = 2 Δr] is usually much greater than the wavelength λ; for example, in satellite sensing the travel-path difference is a few hundred kilometres and λ is in the range of centimetres. Therefore, the measured phase shows an ambiguity of many cycles (Fig. 14.1). However, for adjacent pixels, the relative difference in the two-way travel-path is smaller than λ and the phase difference is usually not ambiguous. Further, a simple relation between phase difference (ϕ_d) and relative terrain elevation 'h' can be derived [see Eq. (14.9)].

14.3 Configurations of Data Acquisition for InSAR

SAR interferometry can be considered to be of two main types: (a) single-pass type, in which two or more antennas simultaneously image the same ground scene, and (b) repeat-pass type, where separate passes over the same target area are used to form an interferogram. Single-pass interferometry can be in across-track mode or in along-track mode.

Across-track interferometry uses two SAR antennas mounted on the same platform for simultaneous data acquisition; the two antennas are separated from each

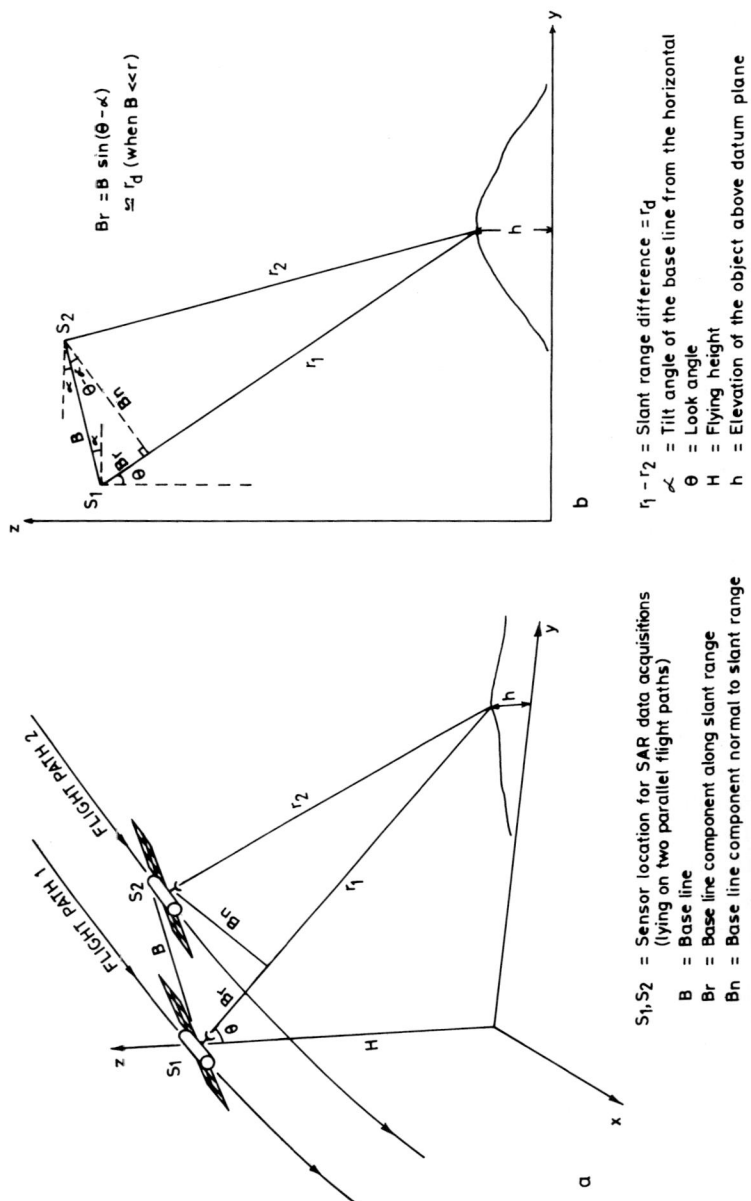

Fig. 14.2. a Geometry of SAR interferometry. **b** Simplified diagram showing angular relations (assuming orbital tracks mutually parallel and oriented perpendicular to the plane of the paper)

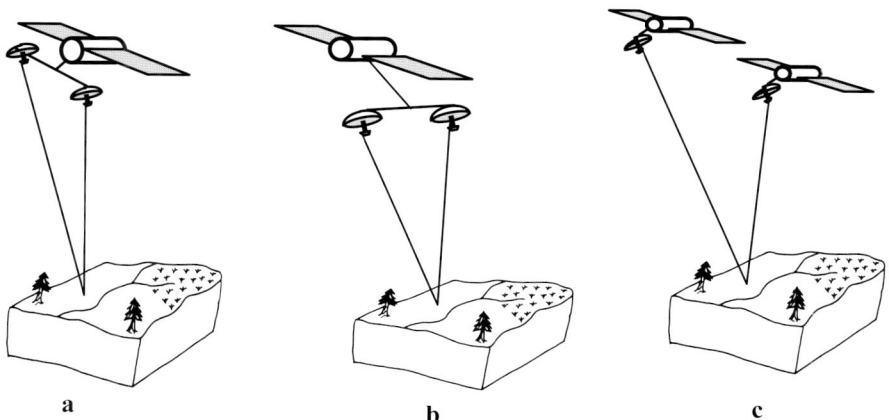

Fig. 14.3a–c. Configurations of data acquisition for InSAR. **a** Across-track interferometry; **b** along-track interferometry and **c** repeat-track interferometry

other by a fixed distance in a direction perpendicular to the flight path (Fig. 14.3a). The method has a certain spatial baseline and ideally zero temporal baseline (see Section 14.4). The across-track baseline allows us to measure target elevation, thus providing terrain topography or a DEM.

A number of aerial surveys have been carried out using this configuration. However, among the space-borne missions, only the Shuttle Radar Topographic Mapping (SRTM) mission (flown February 2000) falls in this category. A problem with airborne systems is that errors induced by aircraft roll cannot be distinguished from the effect of topographic slope. In view of the fact that satellite track is more stable than aerial flight path, this problem is less critical in satellite sensing.

Along-track interferometry employs two antennas on same platform, such that the two antennas are separated from each-other by a fixed distance in the flight direction (Fig. 14.3b). This approach has been employed from aerial SAR systems (Goldstein et al. 1989). This configuration is suited for mapping moving objects, e.g. water currents, vehicles or boats etc., and has importance also for surveillance. The phase difference between the two corresponding signals is caused by movement of an object in the scene. The velocity of the moving object can be computed from the relation

$$\varphi_d = \frac{4\pi}{\lambda} \cdot \frac{v}{V} \cdot B_x \qquad (14.6)$$

where
ϕ_d = phase difference
λ = wavelength of SAR
v = velocity of the moving object
V = velocity of the sensor platform
B_x = baseline component in the direction of motion of the sensorcraft (x direction).

In aerial sensors, with two antennas mounted on the same platform, the method allows measurement of the velocity component in the direction of line of sight, typically the velocity sensitivity being of the order of a centimetre per second.

Repeat-pass interferometry involves acquisition of SAR data with a single antenna covering the same area twice, each time with a slightly different viewing geometry (Fig. 14.3c). This method requires accurate information on the flight path (platform locations) and is therefore better suited for satellite systems than for airborne systems. Satellite systems repeat their orbit at regular interval of time; therefore, the method is also called *repeat-track interferometry* (RTI) or *multiple-pass interferometry*.

14.4 Baseline

Baseline is the most critical parameter in SAR interferometry. Broadly, baseline is a measure of the distance between the two SAR antenna locations. It can be given in terms of length and orientation angle (with respect to horizontal) of the line joining the two SAR antenna positions in space (Fig. 14.2b). Baseline can be described or resolved in terms of horizontal and vertical components. More commonly, it is given in terms of components resolved parallel and perpendicular to the slant range, which are called parallel-to-range baseline (B_r) and normal baseline (B_n) components respectively. The normal baseline component is the key element in InSAR for topographic modelling, controlling measurement of target elevation.

For interferometry, it is extremely important to have an SAR image pair with a suitable normal baseline (B_n) component. This factor alone determines whether a specific SAR image pair can be used for a particular application or not. A smaller baseline possesses higher sensitivity to height changes of ground objects. Solass (1994) has given potential applications of ERS–SAR interferometry as related to normal baseline lengths (Table 14.1).

For height estimation, it is important to estimate the baseline precisely (see e.g. Small et al. 1996). An error in baseline estimation would result in erroneous estimates of ground heights and the two cannot be distinguished. In areas of flat topography, the baseline can also be estimated from the local fringes in the interferogram. In general, satellite orbital data is used for estimating the baseline.

Larger baselines may result in decorrelation of the phase and reduced coherence between the two SAR images forming a pair. Critical baseline is the length at which the coherence between the SAR pair is completely lost.

As baseline is a key parameter in SAR interferometry, the configuration of SAR data acquisition has also been described using the baseline nomenclature, viz. spatial baseline, temporal baseline and mixed baseline.

A *spatial baseline* is one in which the target area is imaged from two different SAR antennas a certain distance apart, simultaneously (i.e. zero temporal baseline). This is the same as across-track configuration (Fig. 14.3a). *Temporal baseline* means that the SAR image pair is acquired from the same spatial position but at different times. A close approximation is the 'along-track' configuration (Fig. 14.3b).

Table 14.1. Potential applications for SAR interferometry for the ERS-1/-2 satellite as related to normal baseline component (Solaas 1994)

Application	Normal baseline (B_n)
Practical InSAR limit	$B_n < 600$ m
Digital terrain models	150 m $< B_n < 300$ m
Surface change detection	30 m $< B_n < 70$ m
Surface feature movement	$B_n < 5$ m

Satellite repeat-pass data sets (such as ERS-1/-2 etc.) often have *mixed baselines* – both spatial and temporal (Fig. 14.3c).

When the temporal baseline is large (i.e. data sets are acquired at substantial time intervals), the change and growth of vegetation etc. lead to differences in distribution of scatterers in the two images. This leads to loss of coherence, called temporal decorrelation, and renders the image pair unstable for interferometric processing. For example, in the case of ocean-surface monitoring, temporal baselines larger than a few seconds are undesirable and only lead to noisy interferograms.

14.5 Airborne and Space-borne InSAR Systems

14.5.1 Airborne Systems

CCRS-SAR. The Canada Centre for Remote Sensing started using the airborne SAR (CCRS-SAR) in 1986, and the system was employed for InSAR processing for topographic mapping in 1990, using the repeat-track method. Since 1992, it has been used with two across-track C-band (5.3 GHz frequency) antennas (baseline separation 2.8 m) for topographic mapping. Depending upon control points, the height/vertical accuracy is reported to be between 1 and 5 m. Later, in 1993, a set of double antennas for along-track interferometry was also introduced.

CCRS-SAR web-site: http://www.ccrs.nrcan.gc.ca/

AIRSAR. Airborne Imaging Radar Synthetic Aperture Radar (AIRSAR) is a system developed by the JPL/NASA for experimental applications of SAR (Lou et al. 1996). It operates in multifrequency and multipolarization modes (P, L, C bands in HH, VV, HV and VH) and records data in 12 channels (where a channel means single frequency, single polarization). The system has had several interferometry modes implemented. In 1991, the first airborne interferometric SAR for topographic mapping, called TOPSAR, was operated. Since then, TOPSAR mode has been flown over a number of test sites in the US. It uses both C- and L-band antennas, and yields output products with root-mean-square errors in height of the order of 1–2 m in flat terrain and 3 m in rugged terrain. AIRSAR also perform along-track interferometry, for which it uses separate sets of C- and L-band antennas. The system is equipped with devices to accurately determine position, attitude and motion compensation.

AIRSAR web-site: http://airsar.jpl.nasa.gov/

Further, in view of the immense potential of the technique for topographic mapping, numerous groups and private entrepreneurs have developed and acquired aerial InSAR systems aimed at commercial applications.

14.5.2 Space-borne Systems

SAR data from a number of satellite sensor systems have been used for interferometric processing, viz. Seasat, ERS-1/-2, JERS-1, Radarsat and SRTM. These satellite sensors were described in Sect. 12.3.5.2 (Table 12.4).

Seasat, launched by NASA in 1978, used 23.5-cm wavelength (L-band) radar. The first application of differential InSAR from space-borne SAR data was reported using Seasat (Gabriel et al. 1989). However, as InSAR processing was still not well developed at that time, only limited interferometric processing has been reported from Seasat data.

The *SIR-C/X-SAR* experiment was flown twice aboard the space shuttle 'Endeavour' in 1994 (Stofan et al. 1995). The second flight was made in October 1994, with a 24-hour near-repeat orbit, to allow acquisition of multiple-pass interferometry data. However, only limited interferometric results have been reported from SIR-C data.

JERS-1 (the Japanese Earth Resources Satellite-1) uses an SAR sensor with L-band (23.5-cm wavelength) and the spacecraft has a tape recorder to store data. The JERS-1 SAR has also been used with InSAR processing for earthquake-related ground deformation studies (see e.g. Murakami et al. 1995).

Radarsat-1, launched in 1995, uses a C-band SAR sensor (5.6-cm wavelength) and can gather data in a variety of beam selections. Further, it also has tape recorders on-board to store data of areas outside the visibility range of ground receiving stations. Radarsat-1 data has also been used for InSAR processing utilizing repeat track coverages.

ERS series, comprising ERS-1 (1991–1996) and ERS-2 (1995–present) have been launched by the European Space Agency (ESA) (Table 12.4). The ERS programme has yielded excellent data for InSAR processing and provided a big boost to the development and validation of the technique. There are two main reasons for this: (a) the two satellites have been operated in tandem mode to obtain SAR coverages with only one-day overpass interval; and (b) the ERS orbits have been well maintained, which allows precise computation of baseline. A number of agencies world-wide have used the ERS-1/-2 tandem data for interferometric investigations. Many of the published examples of InSAR applications are from ERS data.

ERS web-site: http://earth.esa.int/ersnewhome; http://earth.esa.int/INSI

Table 14.2 SRTM technical specifications

Shuttle

Launch date:	11 February 2000
Duration:	11 days
Altitude:	275-km orbit (approx.)

SAR sensors[a]

Parameter	C-band	X-band
Wavelength	5.6 cm	3.1 cm
Frequency	5.3 GHz	9.6 GHz
Look angle	30–60°	50–55°
Swath	225 km	50 km
Baseline separation	60 m	30 m

[a]Both SARs are left-looking.

SRTM. The Shuttle Radar Topographic Mission (SRTM) was the first experiment to provide satellite-borne single-pass SAR data for interferometry. SRTM was an 11-day programme, flown in February 2000. It was a co-operative effort between the US National Aeronautics and Space Administration (NASA), the US National Imagery and Mapping Agency (NIMA), the Italian Space Agency (ASI) and the German Aero-Space Agency (DLR).

SRTM was a unique mission exclusively dedicated to SAR interferometry. A two-frequency (Table 14.2) single-pass interferometric SAR was configured by simultaneously operating two sets of radar antennas separated by a baseline length of 60 m for C-band and 12 m for X-band. The configuration of the SRTM is shown in figure 14.4. The 60-m-long, deployable, stiff-boom structure perpendicular to the velocity direction of the space shuttle created the spatial baseline.

The look angles of the SRTM were fixed (C-band: 30–60 degrees, X-band: 50–55 degrees, off-nadir), both SARs being left-looking. The C-band swath width

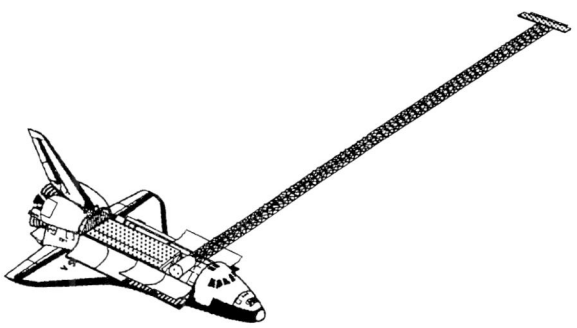

Fig. 14.4. The Shuttle Radar Topographic Mission (SRTM). (After NASA)

was 225 km and for X-band it was 50 km. The C-band covered any point on the equator twice (one ascending and one descending pass), whereas X-band coverage was small in swath. It has provided global coverage (except areas more than 60° north and south) and is the first continuous data set of this kind ever acquired.

The main goal of SRTM has been to generate very accurate and high-spatial-resolution DEMs, in general with about 30 m × 30 m spatial resolution (with C-band), and in special cases 10 m × 10 m spatial resolution (with X-band). The vertical height resolutions are expected to be better than 10 m for C-band and 6 m for X-band.

SRTM web-site: http://www.jpl.nasa.gov/srtm/

Envisat-1. This is proposed by the ESA for launch in early 2002 with a life period of 5 years. It uses SAR sensor wavelength and orbit identical to the ERS series (Sect. 12.3.5.2; Table 12.4), and this will provide continuity of ERS-type data which should be useful in interferometric studies.

14.5.3 Ground Truth and Corner Reflectors

For validation of any remote sensing results, it is necessary to have ground-truth data. In the case of SAR interferometry, the main task is to provide information on 3-D location of objects, which can be applied to various geoscientific themes. For this purpose, corner reflectors (CRs) are frequently installed at various sites in the field during an investigation, concurrently with aerial/satellite overpass. Usually, trihedral corner reflectors, made out of three triangular metallic plates, are used (Fig. 14.5). CRs serve as reference points in the SAR image and are also required for the calibration of the SAR system. Information on accurate location of the CRs is obtained through a differential GPS system and this helps in geometric calibration of the SAR data and validation of the DEM results.

Fig. 14.5. Corner reflector. (Courtesy of K. S. Rao and Y. S. Rao)

14.6 Methodology of Data Processing

The phase information required for interferometric processing is contained in single-look complex (SLC) and raw data products, and is not available in other types of SAR products such as multiple-look products etc. Therefore, for InSAR processing purposes, SLC (alternatively raw) SAR data products are required.

Important steps in SAR interferometry procedure are the following (Fig. 14.6):
(1) Selection of data sets
(2) Co-registration of the images
(3) Generation of interferogram
(4) Phase unwrapping
(5) Height calculation/DEM generation
(6) Geocoding.

1. Selection of data sets

The following guidelines are used for selecting a suitable pair of SAR images for interferometry.

Wavelength of the two SAR images should be the same, otherwise they will not form a pair for interferometric processing. For example, ERS-1/-2 can form a pair (for both $\lambda = 5.7$ cm) but ERS-1 ($\lambda = 5.7$ cm) and JERS-1 ($\lambda = 24$ cm) cannot form an interferometric pair.

Spatial baseline component (i.e. the normal baseline distance) should be within the limits for a particular application, as each type of application requires a certain normal baseline distance (see e.g. Table 14.1). If the baseline is longer, then it will result in spatial decorrelation between phases of the two scenes.

Temporal baseline, i.e. the time difference between the acquisition of two SAR images, should be suitable for interferometry. For DEM generation, zero temporal baselines is preferred. A longer time interval between the two scenes is usually accompanied by changes in backscatter character of the ground features (e.g. changes in vegetation etc.), which results in loss of coherence in the image pair (temporal decorrelation). For this reason, ERS-1/-2 were operated by the ESA in tandem mode, allowing acquisition of data with a one-day interval, and the pair covered the entire globe several times. Further, for applications of velocity mapping etc., suitable temporal baselines should be used to allow detection of features of interest.

2. Co-registration of SAR images

The two SAR images have to be co-registered very accurately and this operation forms the most critical and time-consuming step. The quality of the final interferogram is governed by the accuracy of co-registration. The concept and general procedure of co-registration are the same as discussed in Sect. 10.4, viz. calculating the field of the slave image in the geometry of the master image. Usually, first a coarse co-registration is carried out using data from satellite orbits, or using tie points selected in both images. After this, fine co-registration is implemented, for which various statistical methods specific to SAR images have been developed, such as:

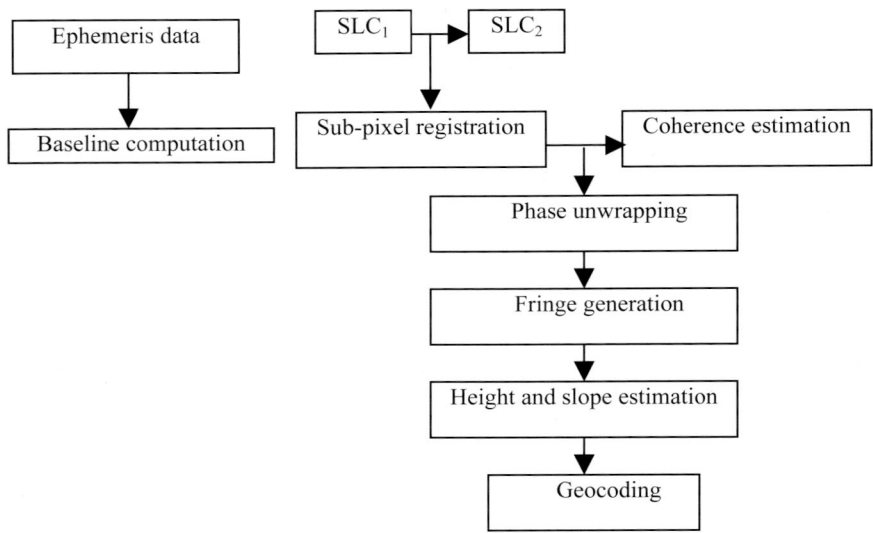

Fig. 14.6. Schematic of data flow in SAR interferometric processing

maximum value of coherence coefficient, cross-correlation of pixel amplitude, and minimization of average fluctuation of phase difference (see e.g. Franceschetti and Lanari 1999). Usually, the accuracies of co-registration obtained are in the range of 1/20 of a pixel. Any incorrect alignment, even on a sub-pixel level, causes reduction in coherence and may have a substantial adverse influence on the interferogram, as scatterers within the corresponding pixels are not same in the two images.

3. Generation of an interferogram

Interferogram is generated by multiplying the complex SAR values of the slave image with the complex conjugate of the corresponding master image. In essence, this means subtraction of one phase value from the other, at each pixel in the co-registered images, and averaging the amplitude values. The resulting difference image possesses phase values in the range of $-\pi$ to $+\pi$, which appear as fringes (Fig. 14.7).

At this stage, it is necessary to remove the effect of flat topography from the fringe image, this step being called 'flattening' (Fig. 14.8). The idea is that some regular fringes would have appeared even if the Earth's surface (screen) was essentially flat; therefore, it is necessary to eliminate the 'flat-topography fringes', in order to obtain fringes related to topographic variation only. Flattening is followed by a smoothing filter to remove the noise.

Methodology of Data Processing

Fig. 14.7. Interferometric fringes of Mt. Vesuvius (Italy), generated from ERS data (27 August and 5 September 1991). (Rocca et al. 1997)

4. Phase unwrapping

Phase unwrapping is a very important aspect of interferometry and leads to determination of the absolute phase from the measured phase. As the height of a terrain feature increases, the phase also correspondingly increases. However, as phase

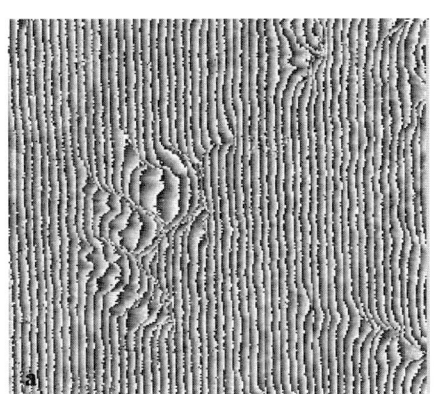 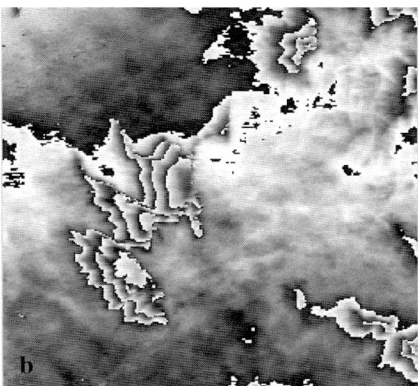

Fig. 14.8a,b. Fringe images generated from ERS-SAR data of the Panvel area near Mumbai. **a** Before flattening and **b** after flattening. (Courtesy of K. S. Rao and Y. S. Rao)

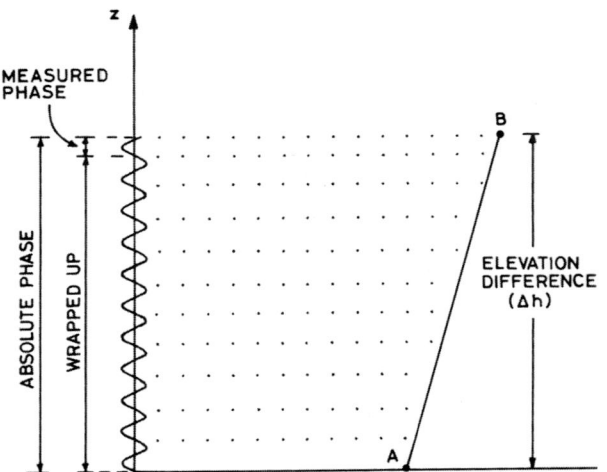

Fig. 14.9. The concept of phase wrapping. Consider elevation difference (Δh) between A and B. It can be represented in term of several full cycles of the SAR wavelength (which get wrapped-up) plus an incomplete cycle forming the measured phase component. Absolute phase consists of measured phase plus wrapped-up component

value is a periodic function of 2π, it gets wrapped up after reaching 2π. Therefore, the measured phase (ϕ_M, also called principal phase) values range between $-\pi$ to $+\pi$, irrespective of the elevation, and cannot be used directly to estimate terrain elevation. In order to obtain correct elevation data, the measured phase values (ϕ_M) must be unwrapped. The phase unwrapping may be considered as adding an integer number of cycles to each pixel to obtain the absolute phase. Figure 14.9 shows the concept of measured phase, wrapped phase and unwrapped phase, in relation to terrain height.

The wrapped and unwrapped phases are related as

$$\text{Unwrapped (absolute) phase} = \text{measured phase} \left(\varphi_M\right) + 2n\pi \qquad (14.7)$$

where n is an integer number. The correct value of n can be determined by a phase unwrapping process in the spatial domain. Assuming that the data are adequately sampled, the phase difference between adjacent pixels may be less than 0.5 cycles.

The various phase-unwrapping algorithms are classified into two major categories: (a) path-following algorithms and (b) least-square algorithms. The path-following algorithms use pixel-by-pixel operation to unwrap a phase and bring the phase differences in adjacent pixels to within the range $+\pi$ to $-\pi$ (within ± 0.5 cycles). The least-square algorithms minimize a global measure of the differences between the gradients of the wrapped input phase and the unwrapped solution (Ghiglia and Pritt 1998).

5. DEM generation

Height values at various points in the terrain (DEM) are to be derived from the phase values in the interferogram. The interferometric phase image is a representation of the relative terrain elevation with respect to slant-range direction. This coordinate system has to be transformed to the horizontal plane (ground-range and azimuth range axes) in order to generate a standard DEM. The correspondence between slant-range and ground-range is quite irregular, as the SAR image carries effects of foreshortening and layover. Further, very accurate estimation of the baseline is essential for accurate DEM generation.

6. Geocoding of the DEM

The InSAR DEM generated as above is in a co-ordinate system related to the SAR geometric configuration. Conversion is required to present the data/DEM in the universal cartographic grid, called geocoding. It requires computation of the absolute position of a pixel in the specific/standard Cartesian reference system. The transformation is usually carried out using a reference DEM. In the case that a reference DEM is not available, geocoding transformation can also be achieved by accurate estimation of the baseline and imaging geometry.

14.7 Differential SAR Interferometry (DInSAR)

Differential SAR interferometry is a new and powerful technique used to detect relative changes, of the order of a few centimetres or even less, occurring in the vertical direction on the Earth's surface (Gabriel et al. 1989). The technique utilizes three or more SAR images. The basic concept is illustrated in figure 14.10. The three satellite passes are processed to yield two interferograms, for example one from passes 1 and 2, and another from passes 1 and 3. Then, the two interferograms are differenced to generate a differential interferogram. This depicts surface changes and movements that have taken place in the intervening period, e.g. those resulting from earthquakes, volcanic activity, landslides etc. If no surface change has occurred in the intervening period, the differential interferogram shows near-zero values throughout. The relation between differential phase change ($\Delta\phi_d$) and surface elevation change Δh is given as

$$\Delta\varphi_d = \varphi_2 - \left(\frac{B_2}{B_1}\right)\varphi_1 = \frac{4\pi}{\lambda} \cdot \Delta h \qquad (14.8)$$

where ϕ_1 and ϕ_2 are the phase values in the two interferogram images, and B_1 and B_2 are the respective parallel baseline components.

Experimental proof of the sensitivity of DInSAR was obtained through field experiments by the ESA (Coulson 1993). A large number of corner reflectors were installed in the field and two of them were secretly moved in the middle of the ex-

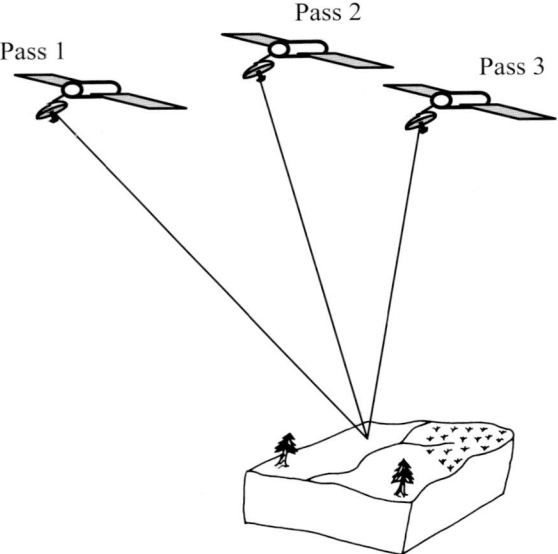

Fig. 14.10. Principle of differential SAR interferometry

periment by 1cm. Repetitive SAR coverages processed by the DInSAR technique could detect the movement of those particular CRs.

In general, three SAR passes with high temporal coherence and suitable baselines have been difficult to obtain for DInSAR processing. If two SAR passes are available, then sometimes an existing good-quality DEM can be used as one of the data sets, for some of the applications of DInSAR. However, the quality of DEM has a significant impact on the results. For local small-scale changes such as landslides, the DEM cannot be used as the third data layer. For changes occurring over large areas, such as those as a consequence of a major earthquake, high-quality DEM could possibly be used as the third data layer for DInSAR. For example, Massonnet et al. (1993) used two SAR passes and an existing DEM to generate a differential interferogram for the Landers earthquake of 1992.

It should be noted that, basically, the DInSAR technique can detect the component of motion in the line of sight. The full amount and direction of motion can be estimated by combining data from ascending and descending satellite passes, or may be derived from models that describe the motion for the region of interest.

14.8 Factors Affecting SAR Interferometry

A number of factors affect interferometric SAR data processing. An in-depth study of quality assessment of SAR interferometric data has been carried out by Gens (1998). Some of the more important factors are the following.

1. Baseline determination. In generating DEM it is very important to know the baseline and viewing geometry correctly. The exact orbit/position of the satellite is influenced by numerous factors, including the Earth's gravitational field, the Moon, the Sun, other planets etc. Usually, data on orbit-state vectors are taken to provide information on position and velocity of the satellite, and to compute the baseline.

2. Atmosphere. The atmospheric interaction with the SAR wavelength leads to refraction (causing mis-registration) and artifacts in the phase difference. This effect is more pronounced if the atmosphere is highly heterogeneous, i.e. spatially varying. The interferograms produced from two-pass interferometry exhibit a greater influence of atmospheric heterogeneity than do those from three-pass interferometry.

3. Temporal decorrelation. Physical changes in the character of the terrain surface (i.e. vegetation growth, change etc.) occurring between two SAR data acquisitions lead to decrease in phase coherence. This causes temporal decorrelation of the data sets and is undesirable for InSAR processing.

4. Baseline decorrelation. The length of the spatial baseline controls viewing geometry. Longer baselines imply greater relative change in viewing geometry, which results in decreased coherence and greater decorrelation. This factor is particularly relevant in considering the applicability of repeat-pass satellite data for various InSAR applications.

14.9 Applications

SAR interferometry has applications to a number of geoscientific themes, e.g. generating a DEM (which forms a basic item of information for many tasks), study of earthquake effects, and monitoring of volcanoes, glacier ice sheets, landslides, land subsidence etc. (see e.g. Dixon 1995; Massonnet and Feigl 1998; Rosen et al. 1999).

The resolution requirements for different applications are considered to be different (Fig. 14.11). As most of these requirements can be met from SAR interferometry, the technique has aroused much interest world-wide.

1.Digital elevation model (DEM)

High-resolution topographic data are a useful input for a number of geoscientific terrestrial ecosystem applications for the very simple reason that topography is a very important factor influencing nearly all surface processes – solar radiation, rainfall, water runoff, microclimate, vegetation distribution, soil development, to name just a few. The foremost application of SAR interferometry lies in the generation of DEM. The various salient intermediate steps, including generating a phase-difference image, phase unwrapping, converting it into relative height difference and geocoding, have been discussed earlier.

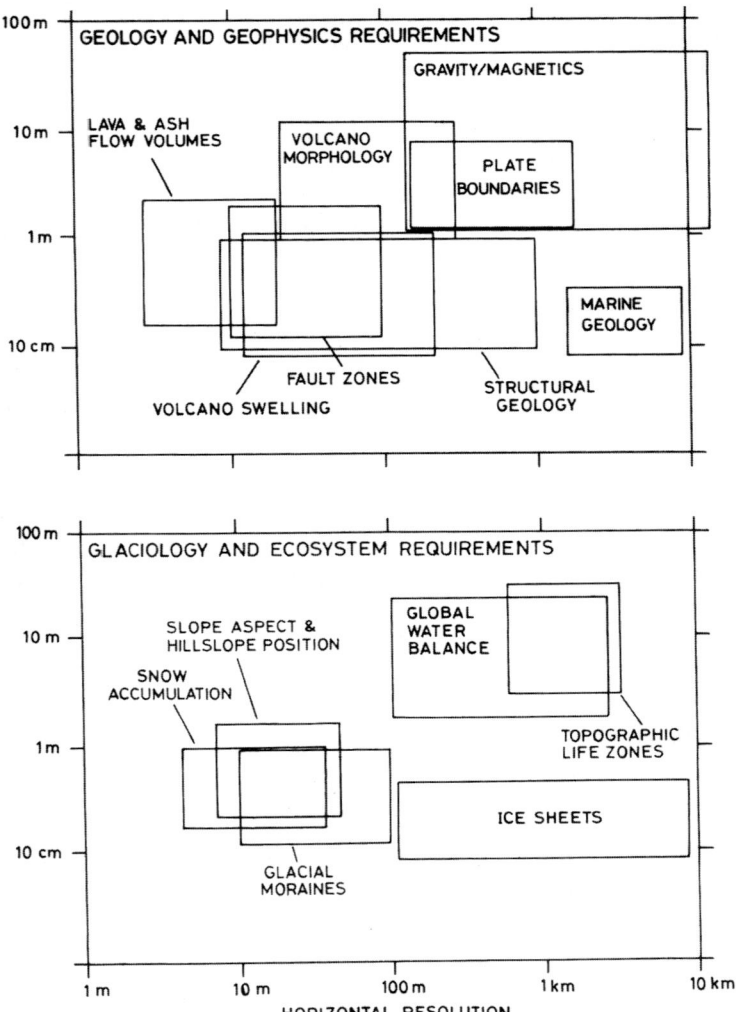

Fig. 14.11. Vertical and horizontal resolution requirements for various geoscientific applications (modified after Zebker et al. 1994)

The relation between relative change of height of the terrain (h) and phase difference (ϕ_d) can be written as

$$h = \frac{\lambda\, r_1 \sin\theta}{4\pi B_n} \cdot \varphi_d \tag{14.9}$$

where
- r_1 = the range distance
- θ = look angle
- B_n = component of normal baseline
- ϕ_d = phase difference.

2. Earthquakes

Massonet et al. (1993) demonstrated for the first time the applicability of differential InSAR for detecting minute co-seismic surface changes, with an accuracy of up to a centimetre or better. Thereafter, several other applications of DInSAR materlized, such as studies of land subsidence, volcanoes, landslides and glaciers.

The region of Southern California was struck by an earthquake, called Landers earthquake (Mw = 7.3), on 28 June 1992. It led to rupturing over a complex fault system spread over an 85-km length. There were also many pre- and post-earthquake shocks, mainly commencing from 23 April 1992 and continuing through 1993. The main shock (28 June 1992) caused right-lateral slips reaching up to 4 m and 6 m along some of the fault branches, as inferred from geodetic and seismological data. As three data sets are required for DInSAR, Massonnet et al. (1993) used ERS-1 SAR passes of 24 April 1992 and 7 August 1992 (as pre- and

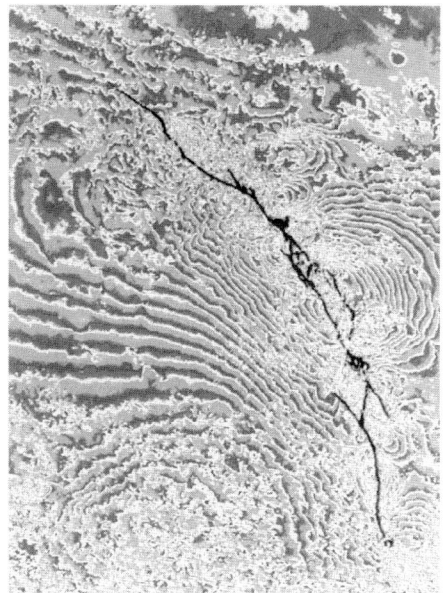

Fig. 14.12. Co-seismic interferogram of the Landers earthquake (28 June 1992) generated from ERS SAR images. The *dotted black lines* denote known faults; *solid white lines* show the surface ruptures mapped in the field. One cycle of colour represents one fringe equivalent to 2.8-cm vertical change. (Massonnet et al. 1993) (see colour plate III)

post-earthquake data sets) and an existing good-quality DEM. By applying the DInSAR technique, they obtained a spectacular image (Fig. 14.12), showing co-seismic ground deformation with an accuracy of 9 mm. Further, in a follow-up study using several repetitive ERS-SAR coverages of 1993, Massonnet et al. (1994) delineated sequential surface changes due to the series of aftershocks. The DInSAR image provided greater insight into the earthquake phenomenon – e.g. distribution of strains, rupture process and fault segmentation, as occurring in different phases of pre- and post-major shock. These results gave a big boost and increased confidence in the technique.

Following this, several studies of this type have been carried out world-wide, which have validated the potential of SAR differential interferometry for determining co-seismic displacements. Most of the reported results have agreed well with the conventional geodetic and seismological observations. For example, a part of Turkey was struck by a devastating earthquake (Mw = 7.4) on 17 August 1999. The epicentre was located on the eastern coast of the Marmara Sea near Izmit. This seismic activity (hypocentre 17 km deep) took place on the North Anatolian fault system, with movements of a pure right-lateral strike-slip type. Due to a rugged topography, the area exhibits large layover effects. Fig. 14.13a shows the fringe pattern due to the earthquake (interferogram) and 14.13b shows the fringe traces superimposed over the shaded relief map of the area for a better appreciation. Several slip planes can easily be identified.

3. Land subsidence.

Land subsidence monitoring is another important area of application of DInSAR, in which the mining industry and environmentalists are particularly interested. DInSAR can detect vertical movements of about 1 cm in a horizontal resolution of about 20 m. Monitoring of land subsidence due to natural gas extraction in the Groningen area has been carried out using DInSAR (Halsema et al. 1995). In the Selby coalfield, UK, the method gave estimates of 8 cm land subsidence during an interval of 35 days (Stow and Wright 1997). Similarly, Fielding et al. (1997) reported land subsidence values of about 25 cm accruing over a period of eight months over an oilfield in California.

4. Landslides.

Differential SAR interferometry has been used to study some of the major landslides. For example, the Saint-Etiene-de-Tinee landslide in southern France was studied by Rocca et al. (1997) using ERS-1 interferometric data (Fig. 14.14). The study indicated a movement of about 1-cm per day and this result was in good agreement with the field data. However, it seems that generally the technique is suited to the study of only larger landslides.

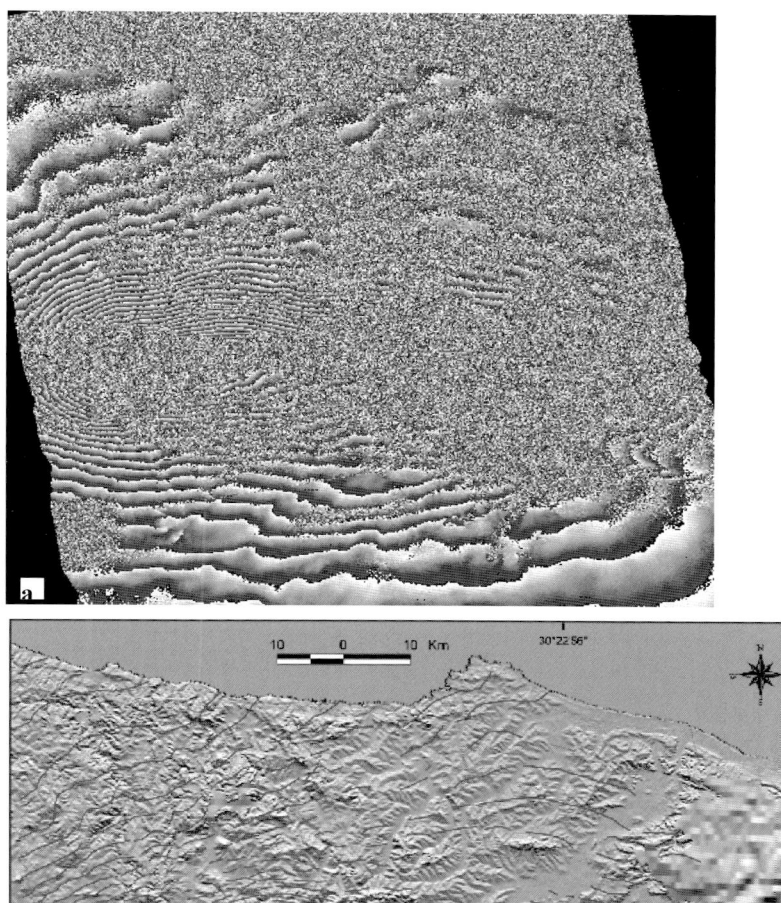

Fig. 14.13. a. Interferogram for the Izmit earthquake, Turkey, 17 August 1999 (generated from 13 August 1999 and 17 September 1999 ERS-2 image pair, with normal baseline of 65 m). The event occurred along the North Anatolian Fault with right-lateral displacement; **b** shows the fringe traces as superimposed over a shaded relief map for a better appreciation of the situation. (**a, b** courtesy of S. Stramondo) (see colour plate IV)

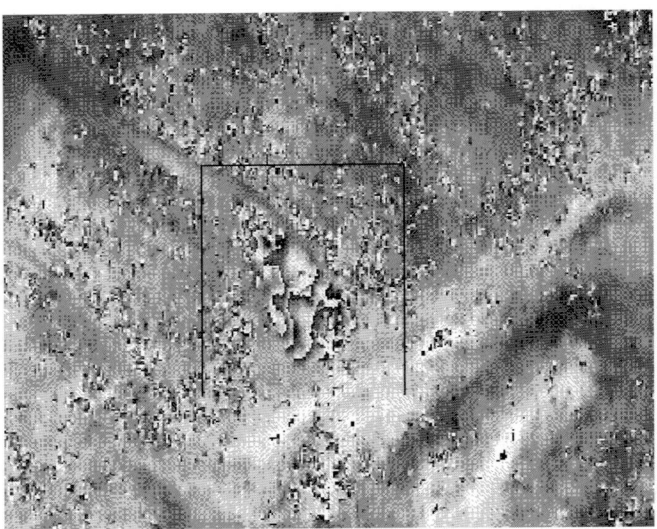

Fig. 14.14. ERS-1 SAR interferogram for landslide in France generated with 8-day interval. Landslide velocity of 1 cm per day has been estimated (Rocca et al.1997)

5. Ice and glacier studies.

Glacier and ice cover a significant part of the Earth and influence global climate and water resources. The InSAR technique can help estimate glacier/ice flow velocity and changes in topography. Figure 14.15 shows an interferogram example from Bagley Ice Field where the phase fringes exhibit influences of both topography and motion.

As glacier topography is spatially variable and the object is in motion, it poses a difficult problem to estimate both variables through InSAR. The strategy requires having InSAR data baselines with widely different sensitivities to displacements and topography, so that when one is being estimated the data set are insensitive to the other variable.

Goldstein et al. (1993) first used ERS-1 SAR repeat-pass data for interferometry to monitor ice stream velocity in Antarctica. As satellite orbit is almost repeated near to the poles, Goldstein et al. (1993) obtained SAR coverages with a 4-m spatial baseline. As such small baselines become quite insensitive to height (see Table 14.1), they could use the SAR pair for estimating ice flow velocity. Their estimate (390 m per year) corresponded closely with the field data.

Joughin et al. (1998) estimated both topography and ice sheet motion for Ryder glacier, Greenland. If the ice flow velocity is constant over a certain time interval when the SAR over-passes are being made, double-differencing pairs of interferograms can be used to cancel the effect of object motion. Joughin et al. (1998) used this technique to generate ERS-1 SAR DEMs of ice sheet topography with an error of 2 m.

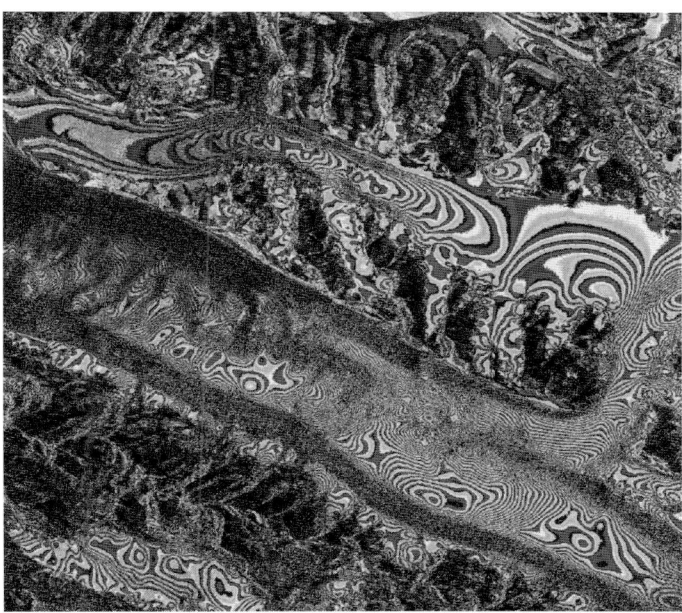

Fig. 14.15. Interferogram of Bagley Ice Field, Alaska; phase fringes show both topography and motion influences. (D.R. Fatland, http://www.asf.alaska.edu/step/insar/absracts/ fatland.html) (see colour plate V)

6. Surface manifestation of subglacial geothermal activity.

An interesting example of surface manifestation in terms of topographic change of subglacial geothermal activity is given by Jonsson et al. (1998). Iceland is marked by ice-glacier covered areas and also geothermal activity in some places. In such regions, ice cauldrons are created by melting in subglacial geothermal areas. The melt-water accumulates in a reservoir for a couple of years, until it drains in a jokulhlaup (a sudden release of water). In such events, the ice surface over the depression drops down by several tens of metres. However, as ice flows from the adjacent areas into the depression, the surface topography rises again, until the next jokulhlaup occurs. In such areas, the monitoring of changes in surface topography is very important for disaster mitigation and forewarning. Jonsson et al. (1998) used ERS-1/-2 SAR data over an area in the Vatnajokull ice cap, Iceland. From DInSAR processing of several data sets, they inferred uplift rates of 2–18 cm per day. An example is shown in figure 14.16.

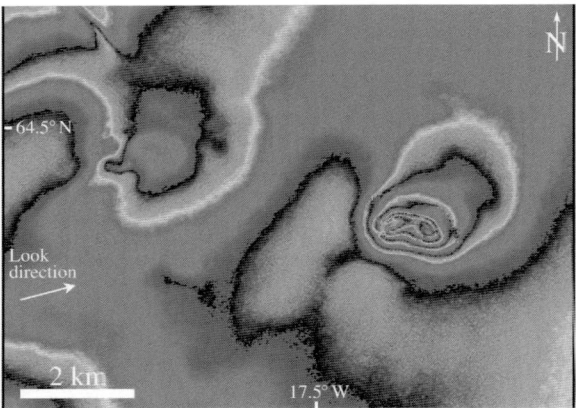

Fig. 14.16. Differential interferogram generated from ERS-1/-2 data (March 27–28, 1996, ascending orbit) over cauldrons in the Vatnajokull ice cap. Fringes indicate uplift in ice surface during an interval of one day. (Data copyright ESA, processing by DLR, Jonsson et al. 1998) (see colour plate V)

7. Volcano monitoring.

The monitoring of volcanic hazards is another important area of application of SAR interferometry, in which it is likely to emerge as a forewarning tool. Also relevant in this context is the all-time, all-weather capability of SAR, which is essential for a warning system.

Volcanoes have been one of the most frequently used test sites for InSAR validation studies (see e.g. Evans et al. 1992; Massonnet et al. 1995; Briole et al. 1997). State of the art and perspectives for volcano monitoring using InSAR have been reviewed by Puglisi and Coltelli (1998).

Around an active volcano, two broad categories of surface changes can be identified: (a) topographic, which are rather coarse changes due to accretion of pyroclastic cones, collapse of craters, lava flow and emplacement of domes etc., and (b) ground deformation and shape changes, which are finer changes associated with magma movement within the volcano, fault movements, lava flow cooling etc. It is necessary to distinguish between the two.

The topographic application of SAR is hindered by the steep topography in volcanic terrains (layover and foreshortening effects). Further, topographic application, including monitoring of lava flows and domes for the purpose of quantifying volume of emitted lava, while the eruption is actively flowing is entirely ruled out through the InSAR–DEM route, owing to the simple fact that continuous changes on the volcanic surface lead to loss of coherence of the interferometric pair. On the other hand, the loss-of-coherence parameter could be used to classify areas where lava flows or domes are moving! Therefore, coherence maps have value for change detection in an otherwise stable background. Further, assuming a

Applications 391

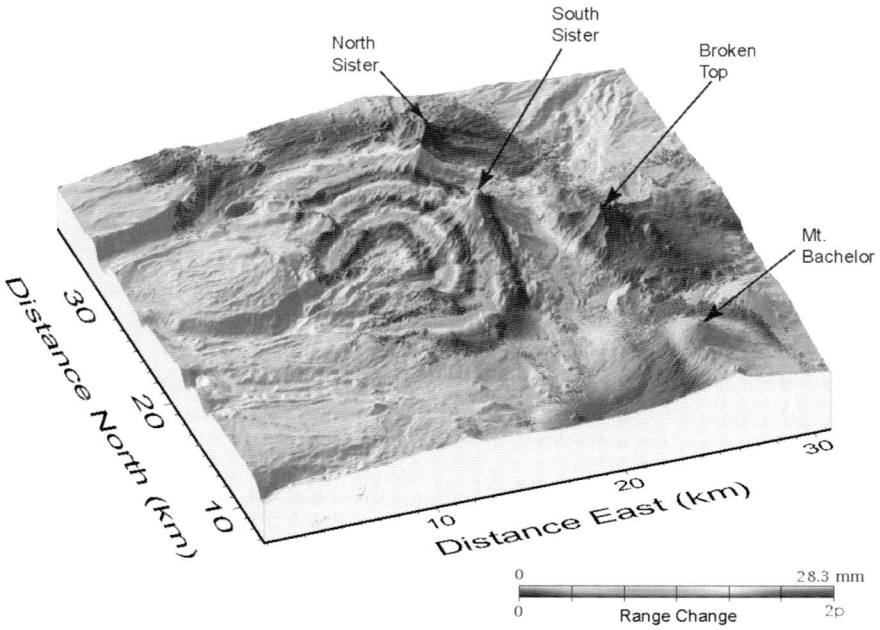

Fig. 14.17. Differential InSAR from satellite SAR data (passes in 1996, 2000) of the Three Sisters volcano region, Oregon. A broad uplift of the ground surface over an area of about 15–20-km diameter, with maximum uplift of about 10 cm at its centre, is detected. (Courtesy of USGS; interferogram by C. Wicks; http://vulcan.wr.usgs.gov/volcanoes/sisters/) (see colour plate VI)

certain thickness of the lava flow, volumetric estimates of the lava outpour could be obtained from coherence maps in a volcanic terrain.

The possibility of monitoring ground deformation through InSAR looks more promising and has the potential of becoming a tool for disaster forewarning. Concurrently with the emplacement of dykes and a magma reservoir at shallow depth, which may be precursors to volcanic eruption, some ground deformation around the volcano takes place. These fine surface deformations and movements can be detected through InSAR. For example, ERS-InSAR technique has been used for monitoring small surface changes due to volcanic inflation (Massonnet et al. 1995) and flank deformation due to lava emplacement (Briole et al. 1997).

The Three Sisters region is a volcanic area in Oregon, USA. Using satellite SAR data of 1996 and 2000, the USGS detected uplift of the ground surface over an area of 15–20 km diameter. The uplift is maximum (10 cm) at the centre (Fig. 14.17). It is believed to be a result of the intrusion of a small volume of magma at depth.

Thus, differential SAR interferometry has the potential for monitoring the dynamics of a volcano, particularly in the mid–long term. This can provide information about the precursors of the eruption and assist in disaster mitigation. A high accuracy would be required in order to make InSAR operational for this application, which essentially means eliminating atmospheric effects and also linking/inverting mono-dimensional along-line-of-sight DInSAR observations to 3-D displacements (Puglisi and Coltelli 1998).

14.10 Future

The future of SAR interferometry looks extremely bright. The technique has already passed experimental tests and is set to enter the operational stage. It has many important applications – generation of high-resolution DEM, investigations of geohazards such as co-seismic effects of earthquakes, monitoring of volcanoes, landslides, land subsidence and glaciers. It is likely to become operational in the near future for volcano monitoring and forewarning,.

Many countries have plans for future radar satellites. As mentioned earlier, Envisat-1 from the ESA is scheduled for launch in 2002 and carries an advanced SAR system (Table 12.3). The Japanese ALOS (Advanced Land Observation Satellite) is projected for 2003. The Canadian Radarsat-2 will carry an advanced SAR system in comparison to Radarsat-1 and is projected for 2003. The French 'Interferometric Cartwheel' is a recent new proposal dedicated to SAR interferometry using a simultaneous baseline. In addition, there are proposals to provide InSAR processing on-board satellites, and to down-communicate InSAR fringe patterns in real-time. All this speaks volumes for the future of SAR interferometry.

Chapter 15: Integrating Remote Sensing Data with Other Geodata (GIS Approach)

15.1 Integrated Multidisciplinary Geo-investigations

15.1.1 Introduction

The purpose of integrated multidisciplinary investigations is to study a system or phenomenon using several approaches and as many attributes as possible/required, in order to obtain a more comprehensive and clearer picture. The main advantage of such an approach is that ambiguities, which may arise from the use of only one type of data, can often be resolved by combining several data sets.

Undoubtedly, the growth in computing and data-processing capabilities, particularly the geographic information system (GIS), has played an important role in developing an integrated geo-exploration approach. As such, the integrated approach need not necessarily include remote sensing data; for example, Campbell et al. (1982) demonstrated the successful identification of porphyry–molybdenum deposits using pre-drilling exploration data and an artificial intelligence program. However, remote sensing data are almost invariably used as basic data in geo-investigations, forming a very important data source in GIS. In the context of remote sensing data processing, multidisciplinary geodata are often referred to as *collateral* or *ancillary data*.

15.1.1.1 Advantages

The chief advantages of the strategy of combining several data sets are twofold.
1. Using multidisciplinary data, the number of attributes or channels of information are increased, and this should correspondingly enhance the capability of discrimination and/or identification.
2. Interpretation of all the data sets collectively should result in a coherent analysis.

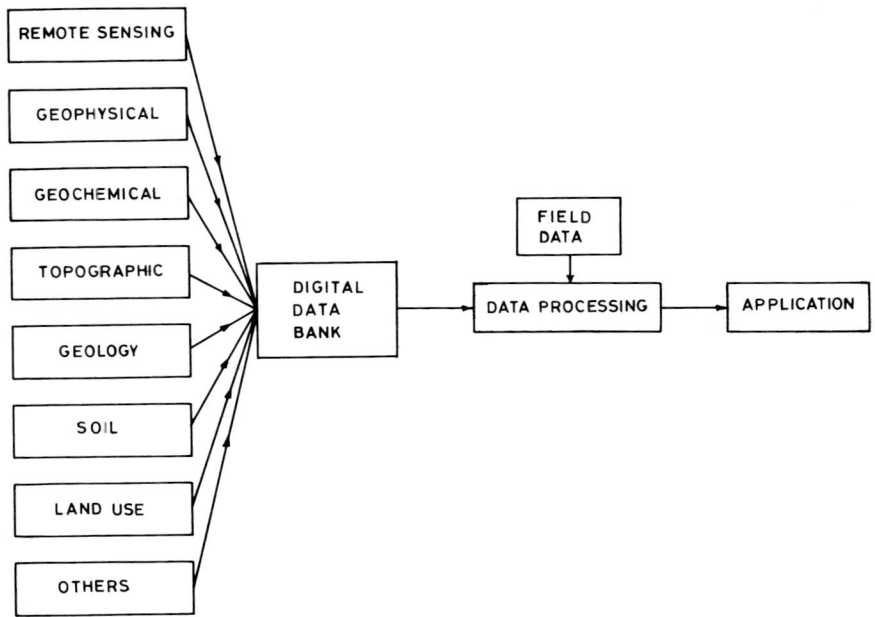

Fig. 15.1. Concept of an integrated multidisciplinary geo-investigation

15.1.1.2 Limitations

We encounter several limitations/difficulties while trying to integrate multidata sets.

1. Combining data sets of different types involving categorical and continuous attributes (see Sect. 15.2.3) is work of a special nature, requiring unique statistical treatment. Often the analytical classification process has to be split into two or more hierarchical stages.
2. Many of the collateral data are derived from maps, which have to be digitized, requiring specific instrumentation facilities.
3. A different base map may have been used for each of the multidata sets; this means that the data must be resampled and geometrically projected to a common base, using control points etc.
4. The spatial resolution and geometry of each of the multidata sets may be different. Further, there may be subjectivity errors, where on one map the boundary between any two units may appear at one place, but in another data set the same level boundary may be at another place. The problem is acute in situations describing categorical attributes occurring with gradational contacts, e.g. in cases of gradational rock types, soil types, vegetation types etc. In such cases, the place-

ment of boundaries could be arbitrary and would affect the mutual geometric compatibility of different data sets.
5. The reliability of some data sets may be questionable; this may adversely influence the reliability of the entire processing, and therefore due care has to be exercised in selecting the data sets, keeping the possible noise factor in mind.
6. Excessive information/data may also be a problem in handling and processing, and therefore optimization has to be duly considered.

Nevertheless, the approach of combining multidata sets is gaining wide application, simply because it improves image/data interpretation. Figure 15.1 shows the general working concept in a multidisciplinary data analysis.

15.1.2 Scope of the Present Discussion

The techniques developed around multiple-image processing and data handling are immediately applicable to assembling and working with multidata sets. In a multidisciplinary geo-investigation, the sources of data could be satellite or aerial photographs, field surveys, geochemical laboratory analyses, geophysical data etc., and may be available in the form of maps, profiles, point data, tables and lists etc. (Fig. 15.2). If GIS methodology is not used, then integrating such a variety of data sets would involve elaborate manual exercises, in order to extract relevant information. On the other hand, utilizing GIS techniques, the requirements of multiple-data integration can easily be fulfilled by the available hardware and software tools. In this chapter, we deal with the handling of multidata sets, as developed around mainly image (raster data) processing techniques.

15.2 Geographic Information System (GIS) – Basics

15.2.1 What is GIS?

The geographical information system, also called 'geobased information system' (GIS), is a relatively new technology. It is a very powerful tool for processing, analysing and integrating spatial data sets (see e.g. Aronoff 1989; Star & Estes 1990; Maguire et al. 1991; Bonham-Carter 1994; Heywood et al. 1998; Longley et al. 1999). It can be considered as a higher-order computer-based system which permits storage, manipulation, display and output of spatial information. This technology has developed so rapidly during the last two decades that it is now being regarded as an essential tool for numerous applications, a nearly indispensable tool for handling spatial information for Earth resources exploration, development and management.

GIS technology is aptly suited to integrate data in a multidisciplinary geo-investigations for the following main reasons.

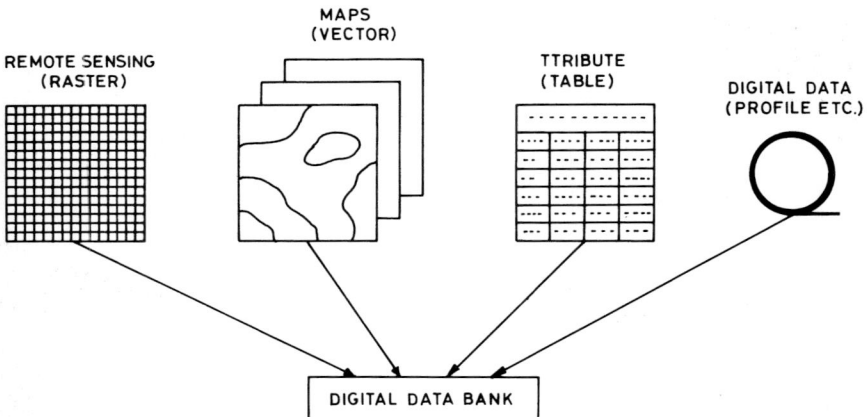

Fig. 15.2. Source data of various types and formats required to be input into a digital data bank

1. Concurrent handling of locational and attribute data. Invariably, we are required to deal with geodata comprising both locational (where it is) and attribute (what it is characters). Such a capability is available only in GIS packages, not in other types of packages.

2. Variety of data. Investigations often comprise diverse forms and types of data. such as: (a) topographic contour maps, (b) landform maps, (c) lithological maps, (d) structural maps, (e) geophysical and geochemical profile data/maps, (f) tables of various observations and data sets, and (g) point data, for example GPS locations etc. (Fig. 15.2). GIS packages offer methods for integrating the above variety of data sets.

3. Flexibility of operations and concurrent display. Modern GIS are endowed with numerous functions for computing, searching for, processing and classifying data, which allow analysis of spatial information in a highly flexible manner with concurrent display.

In addition, as a GIS is computer based, there are the advantages of speedy and efficient processing of large volumes of data, with repeatability of results.

Components of a GIS

A GIS is made up of hardware and software. The hardware comprises a basic computer system (viz. CPU, storage devices, keyboard and monitor), a digitizer and/or scanner for inputting spatial data, a colour monitor for displaying spatial data in image mode, a plotter for production of maps, and a printer for printing tables, data, raster maps etc.

Geographic Information System (GIS) – Basics

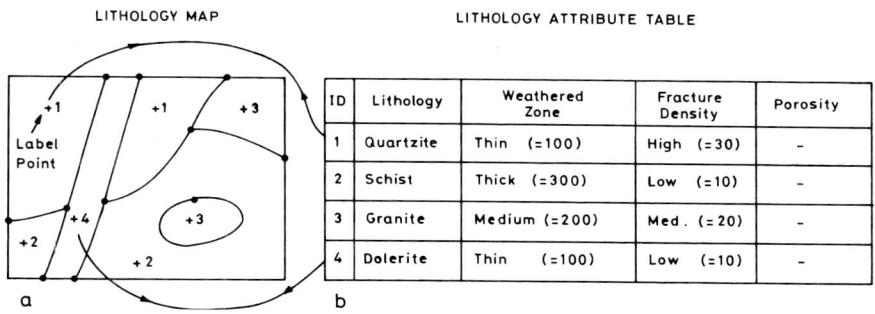

Fig. 15.3a,b. The two fundamental components in a GIS: **a** map and **b** attribute table; GIS maintains a link between the map feature and the corresponding tabular information

The software component of a GIS enables data input, storage, transformation, processing and output. There are a number of GIS software packages available on the market. Some of the more widely used GIS packages are (in alphabetical order): Arc-Info, Idrisi, Ilwis, Geomedia, Grass, Mapinfo and Spans.

15.2.2 GIS Data Base

A data base is a collection of information about things and their relationship to each other. In GIS, the data base is created to collate and maintain information. The geographical or spatial information has two fundamental components (Fig. 15.3) as follows:

(1) location (position) of the feature (where it is), e.g. location of a mine, city or power plant;
(2) attribute character of the feature (what it is), e.g. lithology type, topographic elevation, or landform type etc.

Data pertaining to the above two aspects are explicitly or specifically recorded in a GIS.

1. Location data. The location (or spatial position) is given in terms of a set of latitude/longitude, or relative coordinates. From a geometrical point of view, all features on a map can be resolved into points, lines (segments) or arcs and polygons. The location of a smelter plant, power plant or dam is a typical example of point data. Lineaments, including surface traces of faults, joints, shear zones, bedding planes (on plans) and roads, canals etc., are typical examples of linear data. Maps showing topographical contours, geophysical or geochemical anomaly contours or lithological distribution are examples of polygon data. All features, whether points, lines or polygons can be described in terms of a pair of co-ordinates: points as a pair of x–y coordinates; lines as a set of interconnected points in a certain direction; and polygons as an area enclosed by a set of lines.

Table 15.1. Types of attributes and measurement scales in GIS (after Davis 1986)

Type of attribute/property	Type of scale	Remark	Example
Categorical	Nominal	Mutually exclusive categories of equal status	A, B, C, D or quartzite, schist, marble etc.
	Ordinal	Hierarchy of states in which all intervening lengths are not equal	Drainage density – low, medium, high
Continuous	Interval	Possess lengths of equal increment but no absolute zero	A linear contrast-stretched image
	Ratio	Possess lengths of equal increment and also a true zero	Temperature scale in °C

2. *Attribute data.* Attribute data are the information pertaining to what the feature is, i.e. whether the point indicated is city or a power plant or a mine, or that the information at the specified location pertains to lithology etc.

In GIS, the thematic information is stored in data layers, frequently called *coverages* or *maps*. A coverage consists of a set of logically related geographic features and their attributes. For each map layer (coverage), there is one attribute table providing a description of various items on the map. The attribute data are stored as tables in a data file. The various data are managed and maintained in DBMS (data base management system).

15.2.3 Continuous vs. Categorical Data

In a multidisciplinary investigation, the various types of attribute data could be of two basic types, categorical and continuous (Table 15.1). Different scales are used to measure the above two different types of attributes. The categorical attributes are measured at nominal and ordinal scales. A *nominal scale* classifies observations into mutually exclusive categories of equal status, e.g. quartzite, schist, marble etc. An *ordinal scale* uses the hierarchy of states, e.g. drainage density – low, medium and high. On such a scale, although the categories can be encoded in numerals (e.g. 1 = low, 2 = medium, 3 = high) the intervening lengths, i.e. increments, are usually non-equal. The continuous attributes are measured at interval and ratio scales. An *interval scale* possesses lengths of equal increments but arbitrary zero, e.g. a linear contrast-stretched image where zero has been arbitrarily set, but intervals between successive gray levels are equal. A *ratio scale* possesses lengths of equal increments and also a true zero, e.g. a temperature scale in degrees Celsius or the size of an area in m^2.

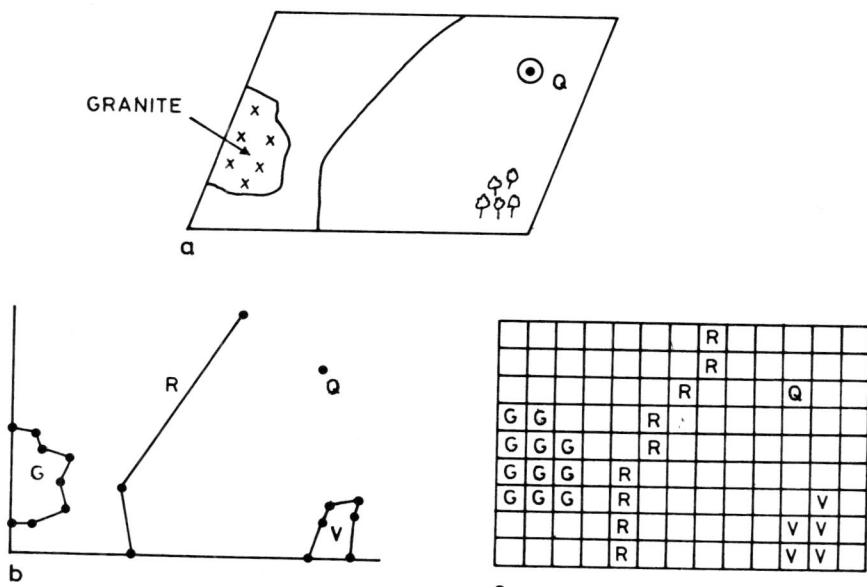

Fig. 15.4a–c. Basic data structures in GIS. **a** A map, and the same in **b** vector format and **c** raster format (G = granite; R = road; Q = quarry; V = vegetation)

15.2.4 Basic Data Structures in GIS

Two basic types of data structures exist in GIS: (a) vector and (b) raster (Fig. 15.4). In vector format, the features are defined by their positions with respect to a coordinate system (Fig. 15.4b). Every feature on a map has a unique position, whether it is a point, a line or a polygon. A point (e.g. the location of a mine) is represented by a single position, whereas a line (e.g. a road) is a set of points. A polygon (e.g. an area underlain by granites) is bounded by a closed loop, joined by a set of line segments. In vector format, *topology* (i.e. mutual relations between various spatial elements) is specifically defined. The data in vector format is geometrically more precise and compact. The method of cartographic manual digitization, which is very widely used, employs vector mode. However, the vector data are not amenable to digital image processing, and they are also relatively tedious for performing certain GIS operations such as overlay, neighbourhood etc.

A raster has a cellular organization (Fig. 15.4c). Remote sensing data are the most typical example. The location of a feature is represented in terms of row and column positions of the cells occupied by the data. Limitations in raster structure arise from degradation in information due to cell size. On the other hand, the raster structure is simple, easy to handle and suitable for performing image processing as well as GIS operations.

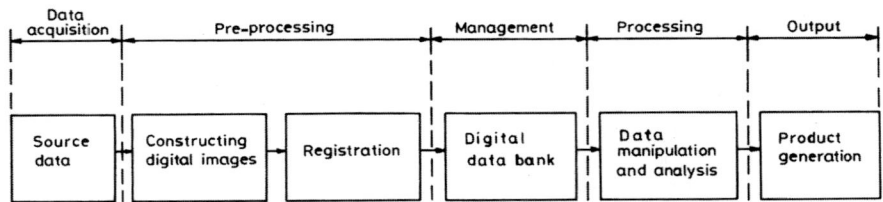

Fig. 15.5. Main segments of a GIS

In this presentation, the examples of geo-data processing utilize mainly a raster data structure.

15.2.5 Main Segments of GIS

Broadly, a GIS comprises five main segments or stages (Fig. 15.5): (1) data acquisition; (2) data pre-processing; (3) data management; (4) data manipulation and analysis, and (5) data output. These are discussed in the following sections.

15.3 Data Acquisition (Sources of Geo-data in a GIS)

15.3.1 Remote Sensing Data

Remote sensing has become one of the most important input data sources in GIS. Remote sensing data acquisition in various spectral ranges (UV, VIS, NIR, SWIR, TIR and microwaves) has been discussed in detail in earlier chapters. Platforms for remote sensing include space-borne, aerial and ground-based types (Table 15.2).

15.3.2 Geophysical Data

Geophysical methods rely on measurements of physical properties of geological materials to discriminate between different types of objects. Details of geophysical methods can be found in standard texts (e.g. Parasnis 1980). Wherever necessary, the geophysical data can be inducted into the GIS.

The *magnetic methods* measure anomalies in local geomagnetic fields in order to infer intensity of magnetization in rocks. The intensity of magnetization depends on magnetic susceptibility, which is used as a physical attribute in geo-exploration. Various types of magnetometers are used to measure the magnetic field, and the magnetic anomalies are expressed in gamma or nano-Tesla. The survey takes the form of profiles in a base grid. The data at different stations

Table 15.2. Acquisition of various types of geodata from different platforms

		Ground-based platform	Aerial platform	Space-borne platform
1	*Remote sensing* spectral data (UV, VIS, NIR, SWIR, TIR and microwaves)	✓	✓	✓
2	*Geophysical*			
	a. Magnetic data	✓	✓	✓
	b. Gravity data	✓	P	P
	c. Electromagnetic	✓	✓	–
	d. VLF induction data	✓	✓ L	–
	e. Electrical data	✓	–	–
	f. Seismic/ SONAR	✓	–	–
3	Gamma ray	✓	✓	–
4	Geochemical data	✓	✓	–
5	*Geological*			
	a. Structural data	✓	✓	✓
	b. Lithological data	✓	✓	✓
6	Topographical	✓	✓	✓
7	*Other thematic data*			
	a. Vegetation/forestry	✓	✓	✓
	b. Land use	✓	✓	✓
	c. Soil	✓	✓	✓
	d. Hydrological	✓	✓	✓
	e. Meteorological	✓	✓	✓

✓ = Yes; P = Possible (under research and development); L = Limited height

along the profile are displayed as point data or as profiles. If the profiles are quite closely spaced, the data can be interpolated to generate contour maps. The magnetometers were initially deployed as ground-based instruments. After World War II, improvements led to the development of *airborne magnetic methods*, which have become a widely used tool for regional exploration, especially for structure, geological mapping, basement topography, mineral exploration etc.

The launching of MAGSAT (Magnetic Field Satellite) initiated the use of magnetic methods for the study of the Earth from space (Fischetti 1981; Fig. 15.6a). The MAGSAT orbited the Earth for about 8 months during 1979–80. It carried a scalar and a three-axis vector magnetometer, which possessed an accuracy of ± 3

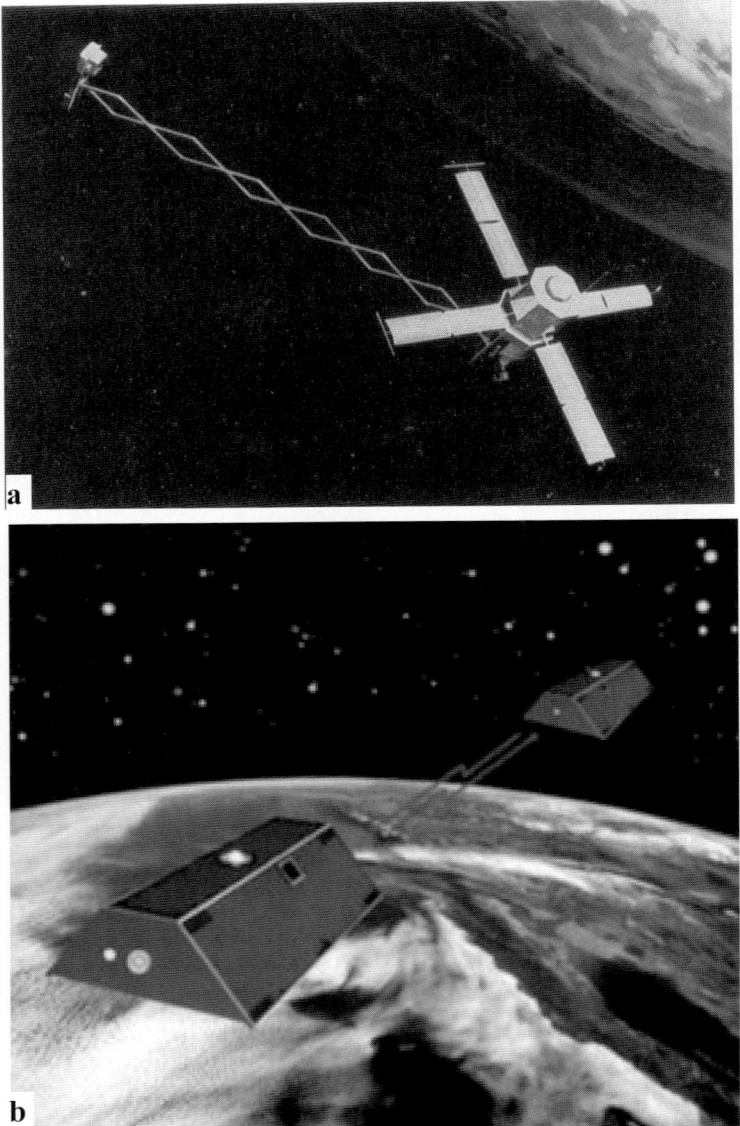

Fig. 15.6. **a** MAGSAT (1979–80); **b** the twin GRACE satellites for gravity measurements (proposed for launch 2002) (**a,b** Courtesy of NASA)

gamma in total field, and ± 6 gamma in each component in vector measurements. The MAGSAT data have been used to develop models of the main field for the 1980 Epoch and maps of crustal anomalies.

Gravity methods utilize measurements of the gradient of the Earth's gravitational potential, i.e. the force of gravity (g). Gravity anomalies yield information about the variation in density of the material and, hence, the material at depth. The anomalies are measured by instruments called gravimeters and are expressed in milliGal. Gravity surveys, like magnetic surveys, are carried out along profiles, which are referred to a base grid. Gravity surveys have been essentially ground based. Airborne gravity-gradiometer is under research and development.

The NASA's GRACE (Gravity Recovery And Climate Experiment), proposed for launch in early 2002, will aim to map tiny variations in the Earth's gravitational field from space. GRACE will use a set of two low-altitude satellites in the same orbit, one satellite 220 km ahead of the other (Fig. 15.6b). As the pair of satellite pass over zones of stronger gravity, the lead satellite will be affected first, affecting the distance between the two. Satellite-to-satellite tracking will be done with extraordinary precision (1-µm accuracy). It will yield variations in velocity, i.e. acceleration (web-site: www.csr.utexas.edu/grace/).

As in the case of magnetic data, gravity data are usually presented as point data, profiles or contoured maps.

Electromagnetic induction methods provide data on differences in electrical conductivities for differentiating between ground materials. The method was initially developed as a ground-based technique, but its most spectacular success has been from airborne platforms. *VLF* (very low frequency) method is a type of EM induction method that utilizes EM waves from radio stations. The VLF-EM signal carries subsurface information, and the method has been used from ground-based and airborne platforms.

There are also other geophysical exploration methods, for example electrical methods, which measure electrical properties such as resistivity, self-potential etc., and seismic methods, which measure elastic wave propagation properties. These are essentially ground-based techniques, and their data are shown as depth-profile data. A modification of the seismic technique involving audible frequencies is SONAR, used in ocean bathymetric surveys.

The types of platforms that are used for geophysical data acquisition are summarized in Table 15.2.

15.3.3 Gamma Radiation Data

The natural radioactivity of surface materials also constitutes an important attribute. Of the three natural radioactivity emissions (alpha, beta and gamma rays), only γ-rays can be used for remote sensing. Although the γ-radiation is basically EM radiation possessing very high energy, it is discussed here separately from remote sensing for the following two reasons:

(1) the technique is restricted to ground-based and very-low-altitude aerial survey (about 100–150 m above the terrain), owing to the rapid attenuation of the γ-ray intensity with altitude;

(2) γ-ray surveys are carried out in non-imaging profiling mode, similar to many geophysical methods, such as magnetic, EM induction etc., in contrast to the imaging mode normally used in remote sensing.

The main sources of γ-radiation in the crustal rocks are potassium (^{40}K), uranium (^{238}U) and thorium (^{232}Th). The intensity of γ-radiation at different ground locations is measured using instruments such as the Geiger-Mueller counter and the scintillation counter, which give total counts. Further, the γ-radiation associated with different sources, i.e. ^{40}K, ^{238}U and ^{232}Th, differs in energy level. Multichannel γ-ray spectrometers can be used to provide relative concentrations of these constituents in an area. The data can be expressed as single or composite (ratio) parameters and presented as point data, profiles or maps.

The γ-ray technique has potential for application in snow-pack studies, soil moisture measurements, environmental surveillance, geological mapping, and mineral exploration, especially of radioactive minerals and hydrothermal deposits (Bristow 1979; Duval 1983).

15.3.4 Geochemical Data

Geochemical data are often conjunctively used with geological and geophysical data. The data may consist of the distribution of major elements, minor elements, ionic complexes or the relative distribution of some constituents. Geochemical methods have been, by and large, ground based. The technique involves laying of base lines and grid, and then sampling of rocks/soil/water, followed by their chemical analysis. Further, it has been found that the air contains aerosols, particulate matter and vapour, which may be representative of the underlying terrain. Air sampling at heights of 60–150 m has been reported for exploration of certain deposits, e.g. mercury-bearing deposits (Barringer 1976). Further, a higher concentration of iodine vapours in the atmosphere has been linked to oil fields and petroleum source rocks. However, the air-sampling technique for geochemical exploration has found only limited use, as its application depends upon the pattern of dispersion and meteorological processes.

15.3.5 Geological Data

For integrated geo-investigations, geological data forms an important input. The data may comprise lithological or structural description at points or in the form of maps.

15.3.6 Topographical Data

For most investigations, topographical data would constitute a primary information. Often, topographical maps form the basis on which all other geodata are co-registered, in order to develop a digital data bank for the GIS. Data on elevation, slope and aspect (i.e. slope direction) also form vital collateral information for interpretation of other types of data.

15.3.7 Other Thematic Data

Attribute data from any discipline can, if necessary, also be used as input in GIS. For example, thematic maps giving the following information may be used:

1. Vegetation
2. Forest
3. Land use
4. Soil
5. Hydrology
6. Metereology etc.

As such, there is no limit to the data sets which could be incorporated in GIS and used in integrated geo-investigations. Only the researcher can define the most useful data sets for a particular investigation.

15.4 Pre-processing

Sources and acquisition of collateral geodata have been discussed in Sect. 15.3. Most of the collateral data to be combined with remote sensing data are usually available in the form of maps showing distribution of either the continuous or categorical type of attribute. Pre-processing is almost invariably required to convert the collateral data sets into a form suitable for storage in GIS data bank (GIS data base), so that the data are amenable to integrated analysis. It can be a simple to a fairly complex exercise. The main pre-processing operations in the raster-based GIS are: (a) data input, (b) interpolation, (c) black-and-white image display, and (d) registration (Fig. 15.7).

1. Data input. This means the encoding of data into a computer-readable form. The maps, tables etc. obtained from various sources must be digitized for inclusion in GIS. It involves two types of methods, usually in combination: (a) digitization of maps for entering locational data and (b) keyboard entry for computerizing the attribute data.

Keyboard entry involves manually entering the data at a computer terminal. It is mainly used for entering the attribute data, called tables in GIS. Feature labels, to identify points, lines and polygons are also entered through the keyboard.

For map digitization, several types of systems are used. A semi-automatic type, called a co-ordinate digitizer (or cartographic manual digitizer), is one of the most widely used systems. It uses an electronic digitizing tablet on which the map to be

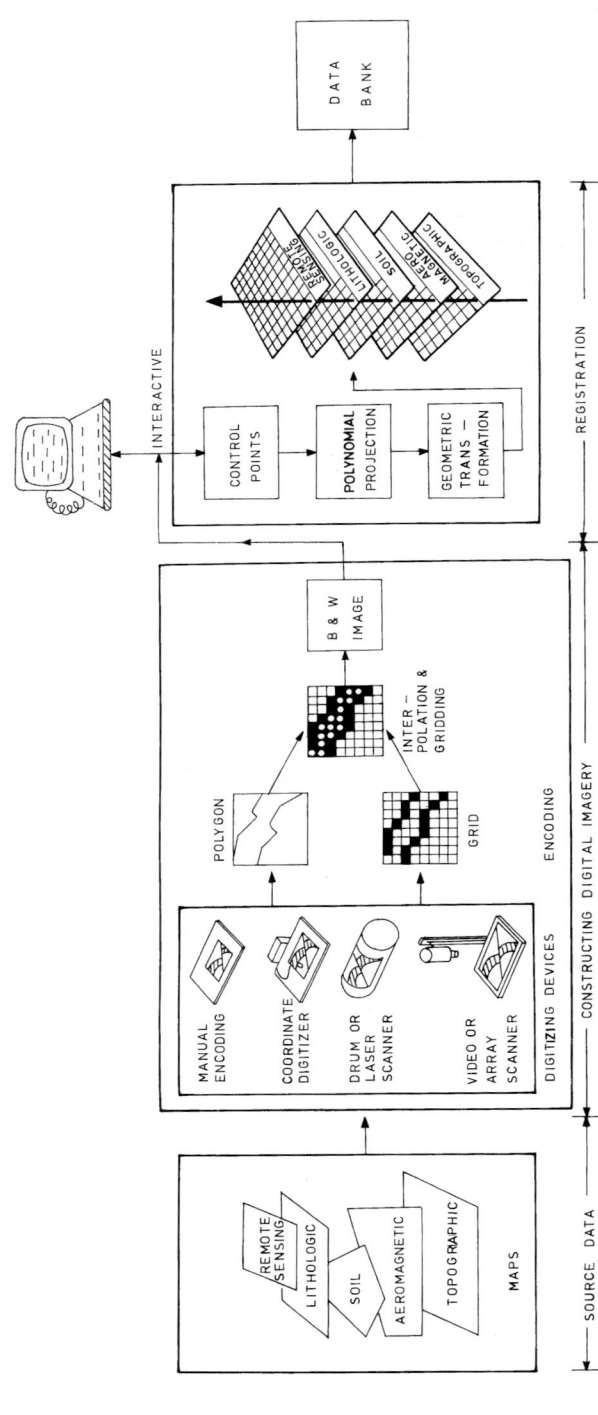

Fig. 15.7. Main steps in generating raster-based GIS

Pre-processing 407

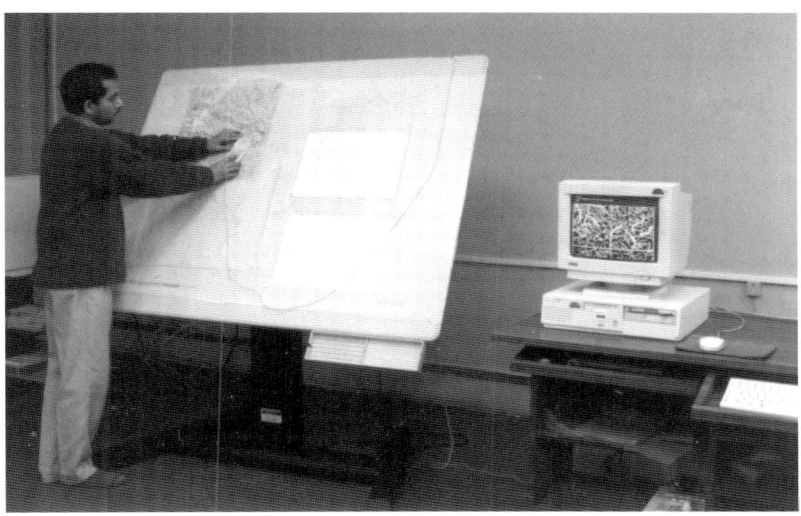

Fig. 15.8. A0-size digitizer

digitized is spread out and the tablet is connected to a microcomputer (Fig. 15.8). A hand-held pen or cursor is used to locate points on the map or to trace contours. The co-ordinates of various points on the map are successively read and stored in the microcomputer (vector mode digitization). By manual interaction, the relative attributes, viz. relative contour values, identification of different categorical units etc., are assigned to various map positions. An interactive display-and-edit device is used for carrying out concurrent monitoring and corrections. A vector-to-raster conversion yields data in raster format.

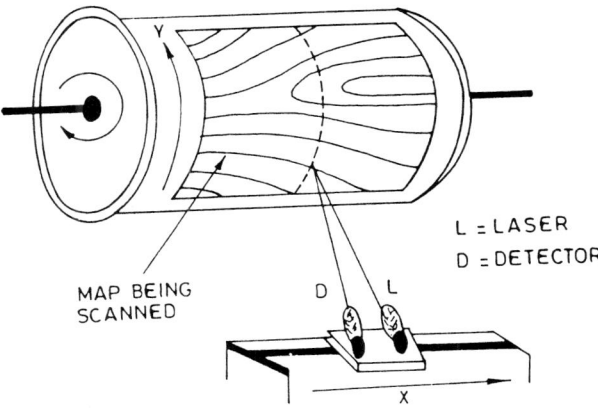

Fig. 15.9. Drum scanner – working principle. (After Burrough 1986)

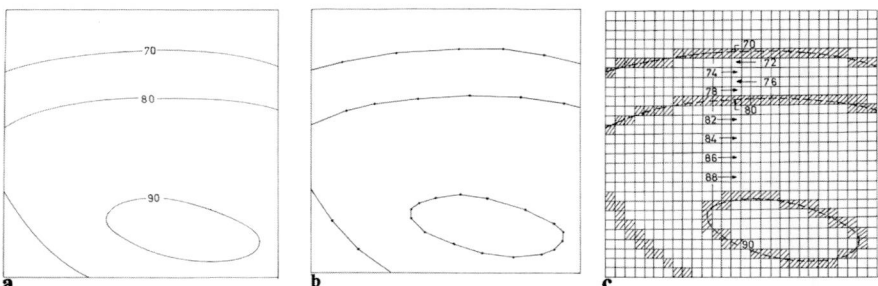

Fig. 15.10a–c. Schematic of constructing digital imagery for continuous type of data. **a** Input map of continuous type data; **b** polygon format encoding; **c** grid-cell format encoding and interpolation

There are also automatic scanning systems which are faster than the semi-automatic devices (Leberl and Olson 1982; Boyle 1984). These are of two types: (a) the rotating drum type, on which the map to be digitized is mounted and the map is scanned as the drum rotates (Fig. 15.9), and (b) the CCD array type, which uses a large number of photo cells to image/scan the map spread out on the table. The relative attributes or values are again assigned by human interaction. These devices yield data directly in raster format.

2. Interpolation. At this stage, some selected cells in the output grid, defining lines/boundaries, have been filled up, the rest of the grid cells being vacant. These vacant grid cells need to be filled up with values in order to generate the full image. The method of filling up vacant cells depends upon the type of data – whether continuous or categorical.

In the case of the continuous type of attributes, the raster cells through which the contour lines pass are assigned appropriate relative values (e.g. 400 ppm, 500 ppm, etc. or 50 milliGal, 60 milliGal etc.). Filling up of the vacant raster cells is

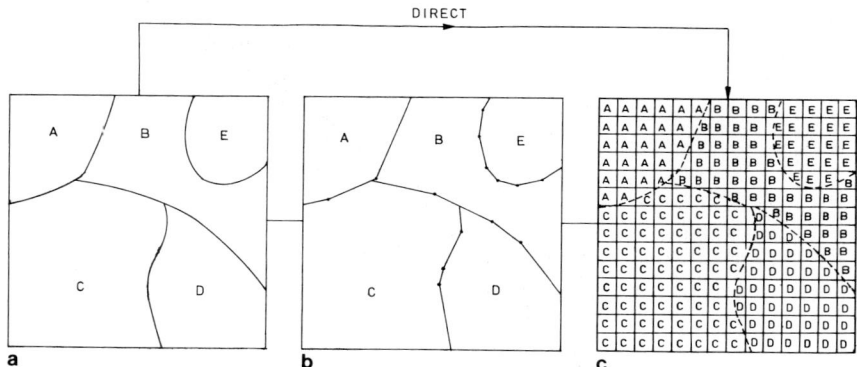

Fig. 15.11a–c. Schematic of constructing digital imagery for categorical type of data. **a** Input map; **b** polygon format encoding; **c** grid-cell format encoding

done by the process of interpolation, which basically deals with predicting unknown values at a point from known values in the vicinity. It is a type of neighbourhood operation. Interpolation programmes may employ a wide range of statistical methods for computation. In this way, each grid cell acquires a digital value which represents the intensity/magnitude of the field/parameter at that point (Fig. 15.10).

In the case of categorical attributes, the intervening grid cells are filled up by the 'principle of extension'. The lines serve as boundaries and the zones between

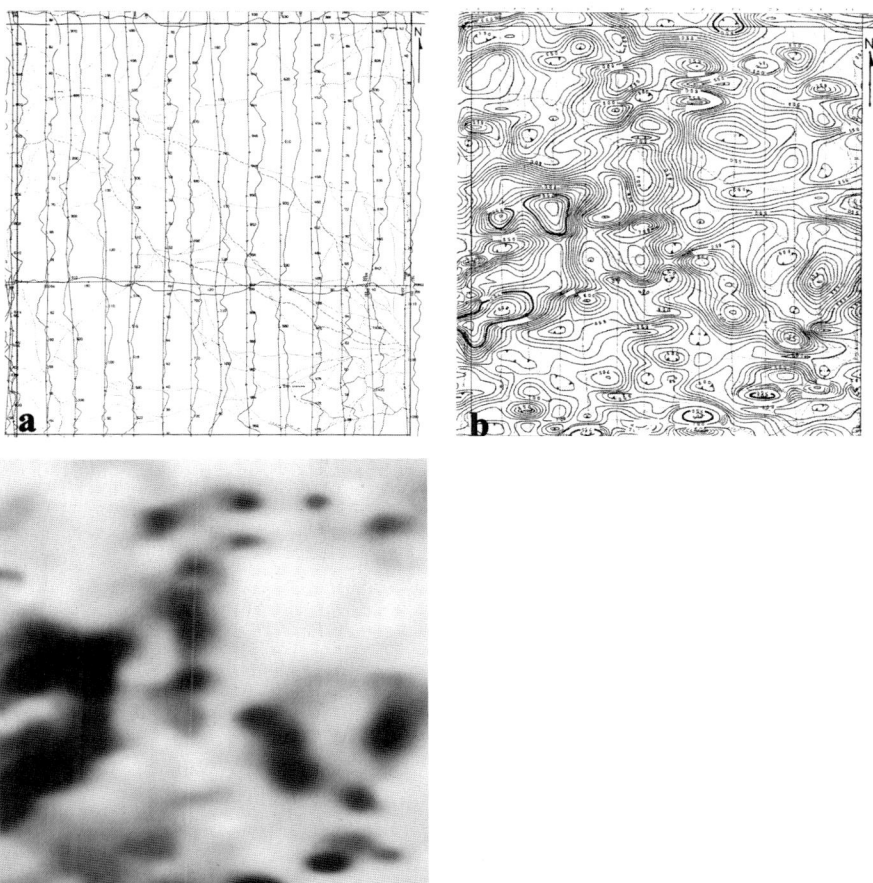

Fig. 15.12. a Aeromagnetic (total field) profile data (Lethlakeng area, SE Botswana; courtesy of Geological Survey Department, Lobatze). **b** Contour map prepared from the interpolated profile data (contour interval: 10 nTesla). **c** Image generated from **b**. Note that the bright areas on the image correspond to higher anomaly values and the darker parts to the lower anomalies. (**a–c** Courtesy of GAF mbH, Munich)

the boundaries are given identification marks or values. In each zone, a unit cell is assigned a particular value and all the adjoining cells, up to the boundary on either side, acquire the same value (Fig. 15.11).

3. Image display. The spatial data matrix can be presented in the form of black-and-white images by selecting a suitable gray scale. For a continuous type of data, normally the lowest value is given the darkest tone (0 DN) and the highest value the brightest tone (255 DN), the intermediate values getting appropriate tones. This is similar to the method of generating remote sensing images. Figure 15.12 a–c show the stages in generating an image from typical profile data of a continuous type of attribute.

The categorical type of data can also be displayed as a black-and-white image by choosing an appropriate gray scale, i.e. each categorical unit is displayed in one tone (a particular DN value) (Fig. 15.13). In such geo-images, the different DN values or gray tones may not necessarily have any relative significance (Fig. 15.14). In some cases, however, the DN values may be linked to a parameter of interest, such as age of the rocks, sand/shale ratio etc.

In general, such geo-images of categorical attributes may have rather specific or limited utility, for instance in examining the entire area in binary mode, e.g. areas of sandstone/no sandstone, or areas of marble/no marble etc. (Fabbri 1984). This approach could be useful in stratification of data sets aimed at understanding and integrating other parameters.

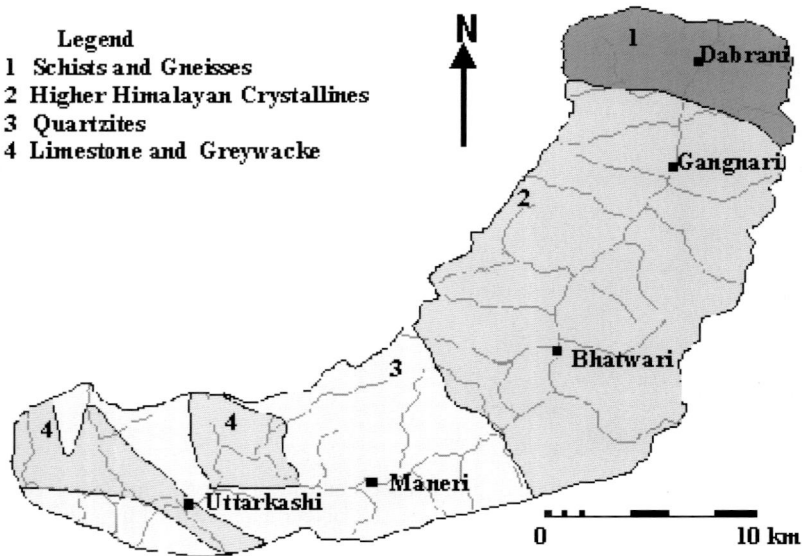

Fig. 15.13. Representing a geological map in image mode; various units are represented in shades of gray

Pre-processing

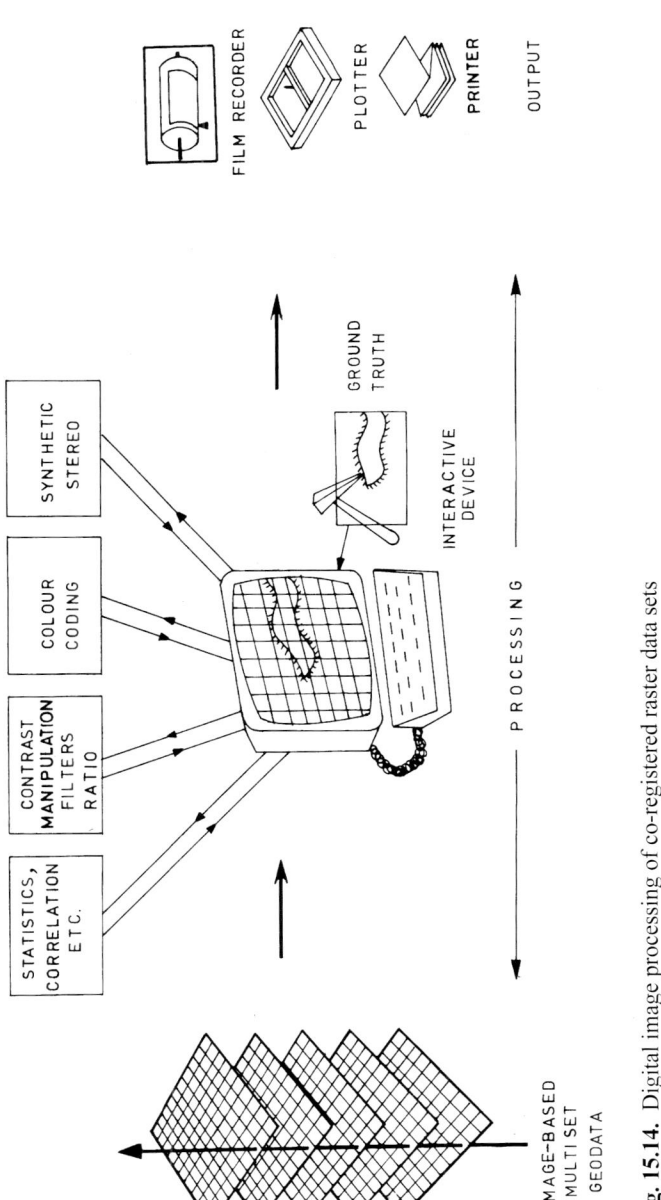

Fig. 15.14. Digital image processing of co-registered raster data sets

4. Registration. The next step, before integrated interpretation of image data sets can be attempted, is to geometrically superimpose the multidata sets, i.e. registration. The registration process can be broadly classified into two types: image-to-image (discussed in Sect. 10.4) and image-to-map registration. Cartographic maps are prepared using a specific standard projection such as the Lambert Conformal Conical projection, Transverse Mercator projection etc. Image-to-map registration implies that all the image data are transferred to the standard map projection. When registering remote sensing image data onto cartographic maps, a number of factors must be appreciated (Catlow et al. 1984).

(a) The map data are invariably subjected to generalization by the cartographer, involving simplification of shape and combining of features.
(b) The thematic boundary drawn on maps, in attempting to represent the spatial distribution of continuous variables, may be artificial, often involving considerable interpolation.
(c) Temporal variations may have occurred during the time interval of two data-set acquisitions.

Basically, the registration procedure involves the following: selection of a base projection; selection of control points; and performing geometric transformation.

Selection of a base projection. The choice of the base projection is generally guided by two considerations: relative spatial resolution levels of various multi-data sets, and the number of data sets available on a common projection. It is often worthwhile to use the image with higher spatial resolution as the base, so that minimal loss in overall resolution takes place. In addition, if several data sets possess a certain common projection, e.g. a standard cartographic projection, it could be worthwhile to use that particular projection as the base, so that the job of registration is simplified and the accompanying loss in information due to resampling is minimized.

Selection of control points. The black-and-white images, original maps etc. are studied to locate prominent and easily defined points, such as sharp bends in the river, prominent topographical features, or railroad intersections etc. These serve as control points for registration.

Performing geometric transformation. Using the control points identified above, geometric transformation is carried out; the image data are resampled and interpolated to yield co-registered images. The procedure is same as described in Sect. 10.4.3.

Once the multidata sets acquired from different sources have been co-registered to a common geographic base, this forms a digital data bank of the GIS (see Fig. 15.7).

15.5 Data Management

The GIS software packages are built around Data Base Management Systems (DBMS) which comprise a set of programs to manipulate and maintain data in a data base. The DBMS in GIS use relational data models, and hence are also called Relational DBMS. They provide advantages in operations such as controlled ordering and organization, sharing of data, data integrity, flexible user-preferred views, search functions etc. The DBMS acts as a controller to provide interaction between the data base and the application program.

15.6 Data Manipulation and Analysis

Taking in to consideration the pattern of most commercially available software packages, the discussion here is divided into three parts: (a) image processing operations, (b) classification, and (c) GIS analysis. However, it must be stated that any boundary between these raster-based operations is purely artificial.

15.6.1 Image Processing Operations

Once the multiple geo-data are co-registered in a raster GIS, wide possibilities exist for data processing, enhancement and analysis. The whole range of digital image processing modules (Chapter 10) is available for data processing (Fig. 15.14).

1. Black-and-white images. A black-and-white image is typically a single-parameter image, i.e. one image displays variation in a single parameter across the scene, in shades of gray. The various digital processing techniques for single and multiple images (viz. contrast enhancement, filtering, transformation etc.) can be applied to enhance information and detect local and regional features. For example, figure 15.15a is a groundwater-table image; figure 15.15b is produced by gradient filtering of 15.15a; it shows groundwater-table gradients across the area. Figure 15.15c is a difference image generated using topographic elevation image minus water-table-level image; it depicts depth to water table from the surface. A number of examples of groundwater data image processing are given by Singhal and Gupta (1999).

2. Colour coding. The generation of colour composites of co-registered multidata sets is a commonly used technique for collective interpretation. Either of the colour coding schemes (RGB or IHS; see Sect. 10.8) can be applied. The resulting colour composites can be suitably interpreted for thematic mapping. Many examples of this type of multidata compositing have been reported. For example, Con-

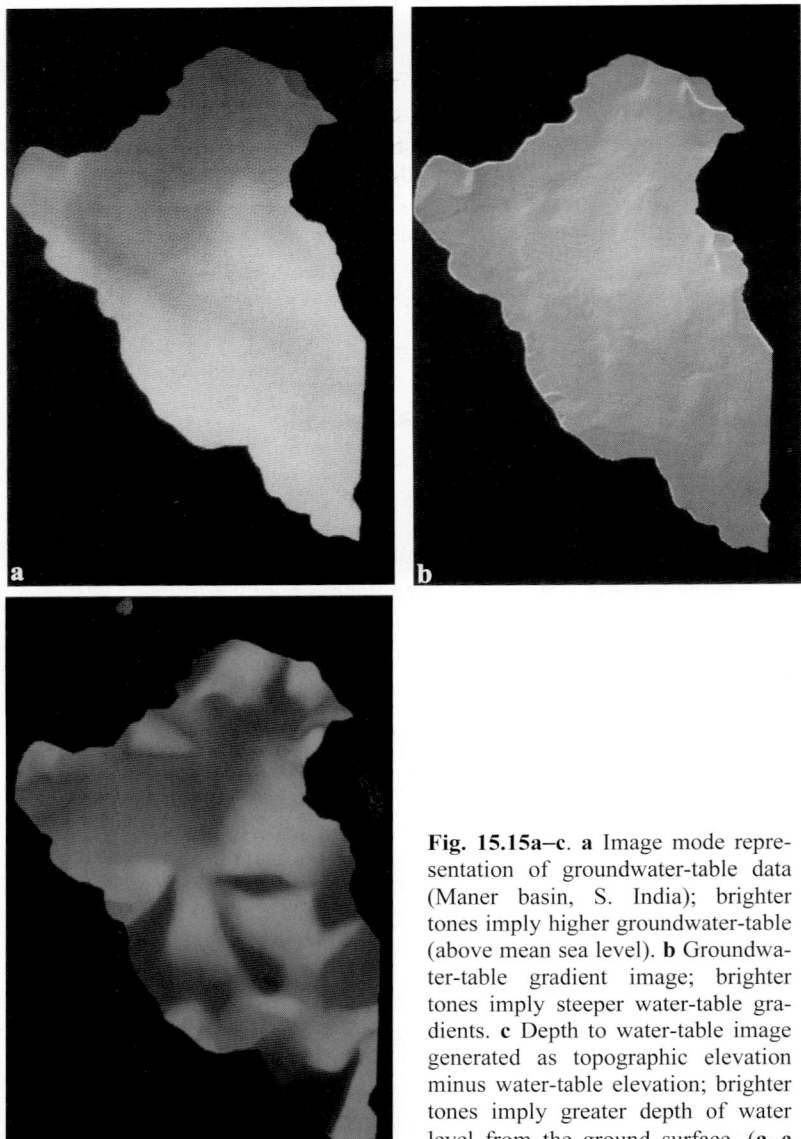

Fig. 15.15a–c. a Image mode representation of groundwater-table data (Maner basin, S. India); brighter tones imply higher groundwater-table (above mean sea level). **b** Groundwater-table gradient image; brighter tones imply steeper water-table gradients. **c** Depth to water-table image generated as topographic elevation minus water-table elevation; brighter tones imply greater depth of water level from the ground surface. (**a–c** Courtesy of N.K. Srivastava)

radsen and Nillson (1984) utilized Landsat MSS, geochemical and geophysical data (aeromagnetic) over an area in South Greenland, and generated a com-

Data Manipulation and Analysis 415

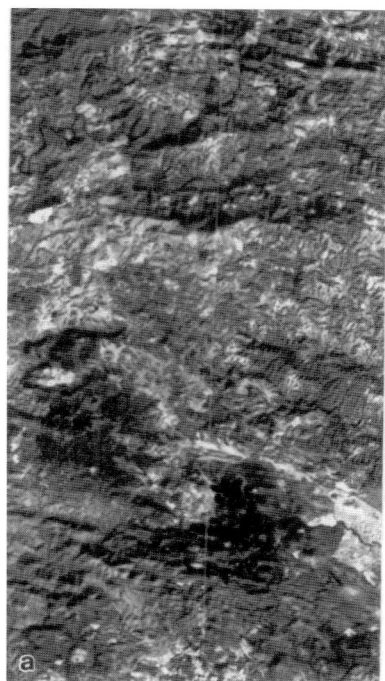 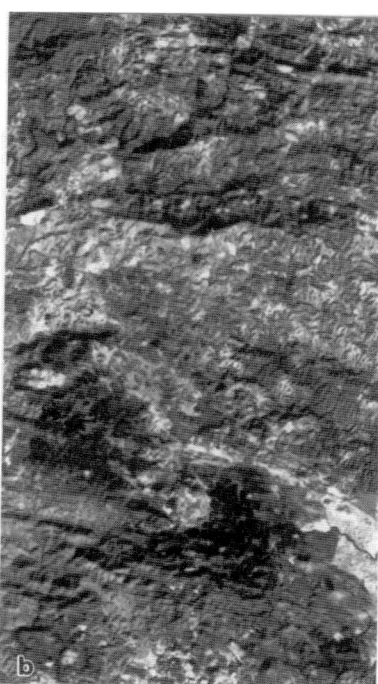

Fig. 15.16. Synthetic stereo pair (Tharsis Mining Complex, SW Spain). The TM4 has been used as the base image and aeromagnetic (total field) values correspond to the parallax. The maximum parallax range = 60 nT. (Volk 1986)

posite colour image by coding Landsat MSS7, aeromagnetic value and Fe value as I, H and S respectively, for visual thematic mapping.

3. *Synthetic stereo.* The technique of 2.5D visualization involving synthetic stereo, described earlier (see Sect. 10.10), can also be applied for feature enhancement and visual display of co-registered multidata sets. Frequently, the remote sensing image is used as the base, over which the collateral data image is incorporated as parallax (Fig. 15.16).

4. *Colour-coded synthetic stereo.* A still more interesting and informative approach for feature enhancement can be through integration of the colour coding and synthetic stereo techniques. In this way, four-dimensional data can be pictorially presented by using three variables in the colour space and one as parallax (Volk et al. 1986; Harding and Forrest 1989). Figure 15.17 shows an interesting example, and demonstrates the potential of pictorially integrating multidisciplinary data for visual interpretation.

The image shows part of the Mayasa Concession, Spain, known for its rich and extensive polymetallic-pyrite belt (Strauss et al. 1981; Ortega 1986). Geologically, at Almaden, Hg mineralization is found to occur in the vicinity of the Almaden syncline and mostly at the contact of basic rocks with a certain (Criadero) quartzite. The area shown in Fig. 15.17a,b constitutes the strike extension of the known Hg deposits at Almaden. The remote sensing, geochemical and geophysical data have been combined to form a synthetic colour stereo. The colour coding has been carried out in the IHS scheme (I = TM4, H = Hg values, and S = constant), gravimetric data are represented as parallax, and Pb values are depicted as white contours. It is interpreted that some areas (see Fig. 15.17c) are marked by the association of quartzite band, Palaeozoic rocks, high gravity anomaly corresponding to basic rock, significant Pb values, and high Hg values, and therefore can constitute potential target areas. Thus, in this manner, multidisciplinary geo-exploration data can be combined and coherently interpreted for in order to define targets for further exploration.

15.6.2 Classification

As far as classification is concerned, the basic aim of incorporating collateral data with remote sensing data is to improve discrimination/classification accuracy. A number of methods have been developed to incorporate additional spatial data, such as data on soil type (Foody 1995), data from digital elevation models (Franklin 1994), data from aerial photographs, and geophysical data (Gong 1996). Conjunctive use can be made at any one of the following three stages (Hutchinson 1982): (1) pre-classification stage for stratification, (2) classification stage for classifier operations, and (3) post-classification stage for further sorting.

1. Pre-classification stratification. At the pre-classification stage, the collateral data can be used for identification of homogeneous areas or strata, i.e. stratification. In a natural environment, the physical attributes of objects may vary laterally and the purpose of stratification is to subdivide the study area into smaller homogeneous sub-areas. The main advantages of stratification are two-fold; upon subdivision, the units become easier to handle, and homogeneous units with low variance are identified, which increases the accuracy of classification. For example, in a quantitative study utilizing co-registered Landsat, Seasat and SIR-A (digitized) data sets in a part of Northern Algeria, Rebillard and Evans (1983) used geological maps as a first step to stratify and locate areas of homogeneous characteristics. They identified different lithological units (e.g. clay, white sand, Pliocene outcrop etc.) and subsequently used a linear discriminant analysis program in a supervised approach for classification. Whilst this technique provides the convenience of working with a smaller data set, it does not seem appropriate in complex environments (Harris and Ventura 1996).

2. Integration at the classification stage. Remote sensing data is of the continuous type, whereas collateral data can be of the continuous or the categorical type. Combining continuous and categorical attributes is a particular problem and calls for special statistical approaches. Thus, if:

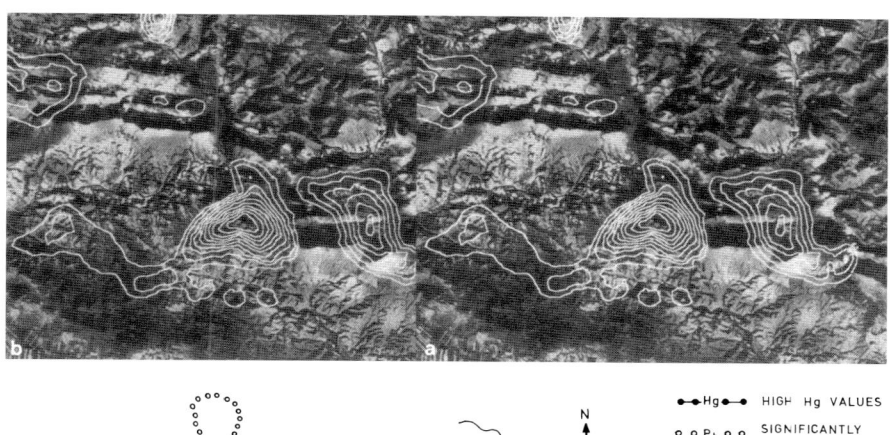

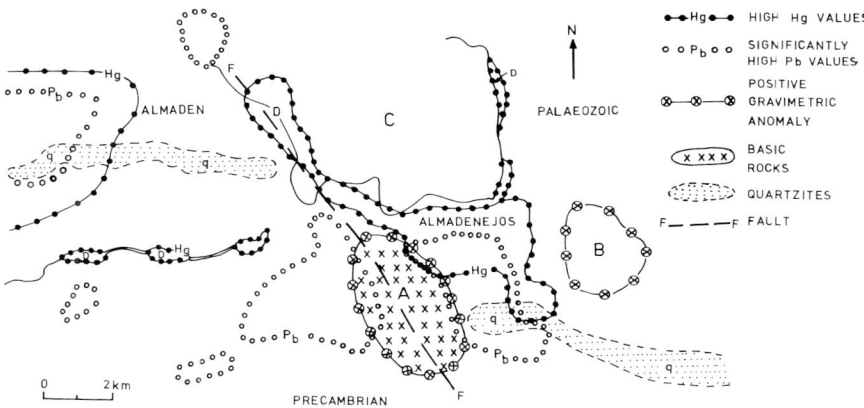

Fig. 15.17a–c. Colour synthetic stereo pair demonstrating integration of remote sensing, geochemical and geophysical exploration data in the Mayasa Concession, Spain. The colour coding scheme uses IHS, where intensity (I) = TM4, hue (H) = Hg values in ppm (so that low values correspond to the blue end and high values correspond to the red end of the colour wheel), and saturation (S) = constant. The gravimetric data is represented as parallax, so that higher positive gravity anomaly is shown as higher positive relief. The Pb values in ppm are indicated by the white contours. **c** Interpretation map of above. From the image stereo pair, a strong association of the quartzite ridge with high Hg and Pb values is evident near Almaden. The quartzite ridge appears as a dark-coloured band on the image. It is displaced by a transverse-oblique fault F–F, along which a basic intrusive rock (high gravity anomaly) has been emplaced. The region *A* is marked by the quartzite band, the Palaeozoic rocks in the north, high gravity anomaly due to basic intrusives, significant Pb values and high Hg values, so that this constitutes a suitable area for further exploration. At *B* (not in stereo) and *C,* there occur an unknown gravity high and a low respectively. At *D,* the hues (green, yellow and red) indicate the presence of high Hg values, which can be related to Hg dispersion along the streams. (Courtesy of Minas de Almaden y Arrayanes Sa, Madrid) (see colour plate VII)

(a) collateral data are of continuous type, they can be easily incorporated as additional channels or attributes in a multidimensional classification (see e.g. Peddle 1993). The statistical procedures are the same as used in other classifications, except that pre-standardization of data is a requisite to enable comparison and mutual compatibility of multidisciplinary data (see e.g. Batchelor 1974).

(b) collateral data are of categorical type, some possibilities of combining them with remote sensing data have been reviewed by Strahler et al. (1980). One use of collateral data can be in computing a priori probabilities which are used in a maximum-likelihood classifier. Strahler et al. (1978, 1980) applied this technique in a forest cover classification study. They linked topographical data to the occurrence of forest cover types and from this estimated a priori probabilities at different elevations and aspects (i.e. geographic orientation with reference to north) and found that applying these a priori probabilities to remote sensing data could improve classification accuracy by as much as 27%.

3. Post-classification sorting. At the post-classification stage, the collateral data can be used to sort out classes which are spectrally similar, but amenable to discrimination on the basis of collateral data. Hutchinson (1982) gave an example of this approach using Landsat MSS and ancillary data from a study in a desert area in California. As a few object classes were spectrally similar on the Landsat data, he started by initially grouping such classes (e.g. basalt, desert varnish surface, al-

Table 15.3. Example of the post-classification sorting approach (Hutchinson 1982)

Initial class assignment based on Landsat MSS	Discrimination rule based on topographical data	Final class assignment
Active sand dunes	Slope < 1%	Dry lake
	Otherwise	Active sand dunes
Stabilized sand	Slope < 1%	Dry lake
	1% < slope < 15%	Stablized stand
	Otherwise	Active stand dunes
Dissected cobbly alluvial fans	Slope < 3%	Low gradient undissected basalt alluvial fans
	3% < slope < 15%	Dissected cobbly alluvial fans
	Otherwise	Mountain scrub
Shadow and basalt	Slope aspect N or NW	Shadow
	Slope < 3%	Highly dissected alluvial fans
	3% < slope < 8%	Slightly dissected basalt alluvial fans
	8% < slope < 15%	Dissected basalt alluvial fans
	Otherwise	Basalt mountains

luvial fans and shadows) together, and later used topographical data to sort out and further separate the problem classes from each other (Table 15.3), finally observing that this approach is quick, easily implemented and efficient. Moreover, several types of ancillary data can be incorporated in framing decision rules for sorting (see e.g. Joria and Jorgenson 1996).

The drawback in using ancillary information in classification is the difficulty in dealing with data of different formats, measurement units and scales. All the data from different sources may first have to be brought onto the same platform before these methods are applied. Nevertheless, once the ancillary information has been incorporated effectively, the accuracy of classification may be increased.

In addition, the conjunctive use of ancillary data may be more useful in knowledge-based classifiers (Peddle *et al.* 1994). However, these techniques are based on more complex heuristic rules which sometimes are difficult to compose and understand, and also are very subjective and data dependent.

Table 15.4. Important GIS functions (modified after Aronoff 1989)

1	Maintenance and analysis of the spatial data	– Format transformations – Geometric transformations – Editing functions etc.	
2	Maintenance and analysis of the attribute data	– Attribute editing functions –Attribute query functions	
3	Integrated analysis of spatial and attribute data	– Retrieval/classification/ measurement	– Retrieval – Classification – Measurement
		– Overlay operations – Arithmetic/Boolean	
		– Neighbouring operations	– Search – Topographic functions – Thiessen polygons – Interpolation – Contour generation
		– Connectivity functions	– Contiguity – Proximity – Network – Spread – Seek – Perspective view
4	Output formatting	– Map annotation, text, labels, graphic symbols, patterns etc.	

15.6.3 GIS Analysis

GIS analysis functions are unique as they can concurrently handle spatial as well as non-spatial (attribute) data. In this treatment, five broad types of GIS analysis functions are discussed: retrieval, measurement, overlay, neighbourhood and connectivity (Table 15.4).

1. Retrieval functions. As mentioned earlier, in GIS, a coverage consists of a set of logically related geographic features and their attributes. The information stored can simply be retrieved by selective search on spatial and attribute data. The output display will show selectively retrieved data in their proper geographic locations. The criteria for selective retrieval could be based on attributes, or Boolean logical conditions (see later) or classification. A common example could be: select pixels lying in a specific rock type. As the selective search can be operated on both spatial and attribute data, this becomes a powerful function in handling and processing data in the GIS environment.

2. Measurement functions. Some measurement functions are commonly included in GIS software, such as those to measure distances between points, lengths of lines, perimeters and areas of polygons or number of points/cells falling in a polygon etc. Sample applications could be: find the number of cells of a particular class, or the area of an exploration block, or measure the distance between a mine and the smelter plant.

3. Overlay functions. This forms possibly the most important function in a GIS. Often, we have to deal with several data sets of the same area; overlay functions perform integration in a desired manner. There are two fundamental types of overlay operations: arithmetic and logical. Arithmetic operations include addition, subtraction, division and multiplication of each value in one data layer by a value at the corresponding location in the second data layer. The logical operations involve

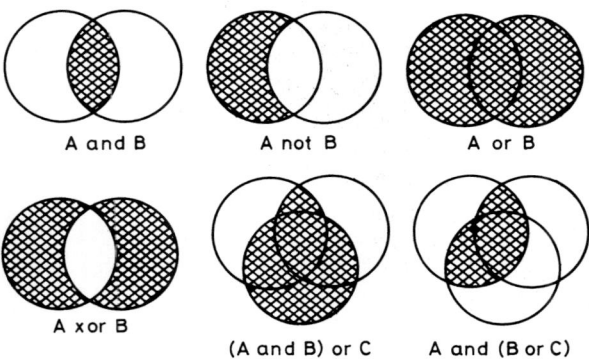

Fig. 15.18. Concept of Boolean conditions

Data Manipulation and Analysis 421

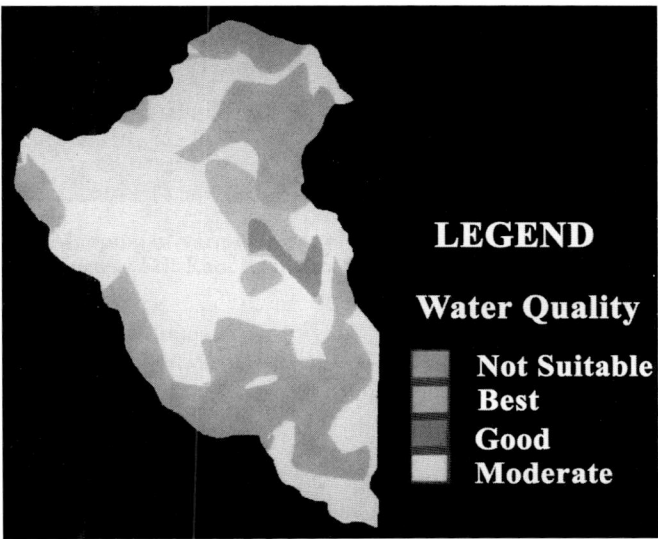

Fig. 15.19. Classification of groundwater. The groundwaters have been classified with respect to suitability for drinking purpose per WHO norms, using data on chemical quality of groundwater (TDS, Cl, Na + K, and Ca + Mg), and applying Boolean logic. (Courtesy of N.K. Srivastava) (see colour plate VI)

determining those areas of interest where a particular condition is fulfilled (or not fulfilled). These are generally performed using Boolean operators: AND, OR, XOR, NOT (Fig. 15.18).

Overlaying can be done in both raster and vector structures. However, in vector format, overlaying is rather tedious as it leads to creation of new polygons of various shapes, sizes and attributes which become difficult to handle. In contrast, the overlaying operation in raster structure proceeds cell by cell, and therefore is relatively simple. Figure 15.19 gives an example of an overlay operation.

4. Neighbourhood functions. These functions are required to examine the characteristics of an area surrounding a specific location. They are useful in determining local variability and adjoining information. Neighbourhood functions commonly include search, topography and interpolation.

a. Search function is a frequently used neighbourhood function. A suitable search area (e.g. a rectangle, square or circle) can be defined. Sample application could be: search pixels of limestone within 10-km distance of a cement plant.

b. Topographic functions. A raster data set can be represented in terms of a digital elevation model (DEM). The term topography here refers to the characteristics of such a DEM surface. The topographic functions include slope and aspect, which are typical neighbourhood functions. Slope is defined as the rate of change of elevation, and aspect is the direction that a slope faces.

c. An *interpolation* module is commonly provided in GIS packages. This is a typical neighbourhood operation and involves predicting unknown values at given locations using the known values in the neighbourhood. Interpolation is required to be carried out during registration and generation of a co-registered image data bank, and also in DEM generation.

Besides, the various 'local operations' in the field of digital image processing can also be considered as neighbourhood operations, e.g., high-pass filtering, image smoothing, etc.

5. *Connectivity functions.* These functions operate on the inter-connections of three basic elements (i.e. points, lines and polygons). They operate by creating a new data layer and accumulating data values in the new layer derived from data values over the area being traversed. Connectivity functions are grouped into contiguity, proximity, network, spread, stream and perspective-view functions.

a. Contiguity. Areas possessing unbroken adjacency are classed as contiguous. What constitutes broken/unbroken adjacency in a particular case can be prescribed depending upon the problem under investigation. Sample application could be: check for contiguity of a polluting tank from the adjoining water body.

b. Proximity function. Proximity is a measure of the distance between two features. The notion of distance could be a simple length, or a computed parameter such as travel time, noise level etc. Using the proximity function, a buffer zone is created around a feature. A buffer zone is defined as an area of a specified width, drawn around the map location. Buffering can be done around points, lines or polygons (Fig. 15.20).

c. Network function. A set of interconnected linear features forming a pattern is called a network. This function is commonly used in analysis where resources are to be transported from one location to another. It can be applied in environmental studies and pollutant dispersion investigations.

d. A *spread function* helps evaluate characteristics of an area around a particular entity. It is endowed with characteristics of both network and proximity functions. In this, a running total of the computed parameter is kept, as the area is traversed by a moving window. This is a very powerful function, particularly for environmental impact assessment and pollution studies.

e. Stream function. The job of the stream function (also known as the *seek* function) is to perform a directed search outward in an incremental manner, starting from a specified point and using a decision rule. The outcome of a stream function is the delineation of paths from the start point until the function halts.

Data Manipulation and Analysis 423

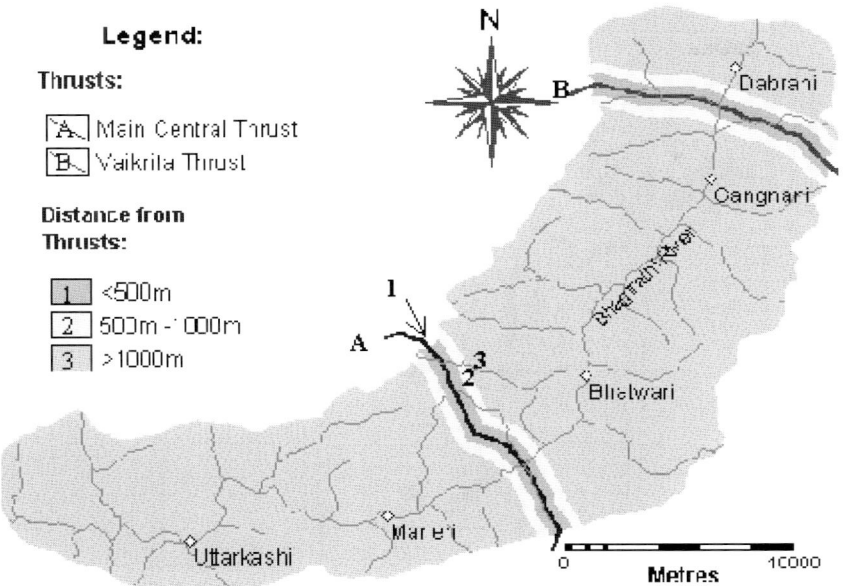

Fig. 15.20. Example of buffer zones along thrusts; *arrow* indicates the closest (< 500 m) zone

Applications could be: to trace the path of water flow or the path of a rolling boulder (rock fall) (Fig. 15.21).

f. Perspective view. Raster image data can be displayed as a 2.5-D surface, where the height corresponds to the value at that pixel (see Sect. 10.10.3). Various enhancements such as shaded relief modelling and perspective view can be applied on this type of data. This generates views valuable for understanding a pattern, as the human mind can easily perceive shapes and forms. Further, an additional raster data set can be superimposed over the perspective view model by draping.

g. Classification is a procedure of subdividing a population into classes and assigning each class a name. In vector data, each polygon may be assigned a name as an attribute. In raster structure, each cell is assigned a new numerical value for class identification.

Sources of Error in GIS

Errors of several types may creep in at different stages in GIS and affect the data quality. The two basic categories of errors are: (a) inherent and (b) operational. Inherent error is that which comes from the source data. Operational error is introduced during the working of the GIS.

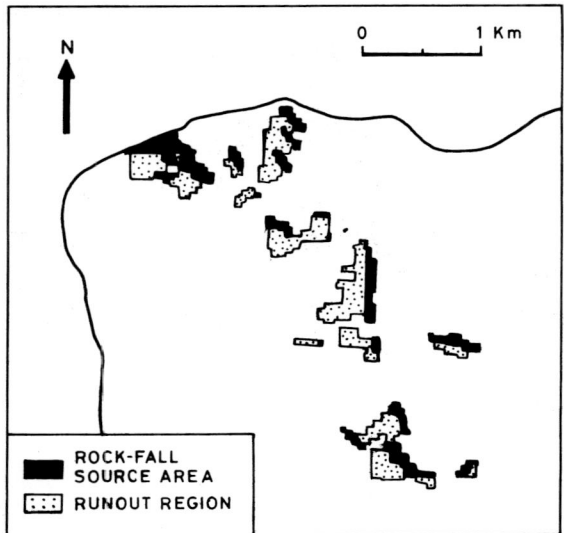

Fig. 15.21. Application of seek function to delineate rock-fall runout distance. (Courtesy of J. Mathew and A. Ranjan)

It is not possible to avoid errors completely; however, they can be managed to be kept within permissible limits. Therefore, an understanding of the types and sources of errors is necessary for better job management.

15.7 Applications

GIS methodology has found applications in almost all branches of natural resources investigations – mineral exploration, hydrocarbon exploration, groundwater, forestry, hydrology, soil erosion, environmental studies, urban planning, various natural hazards etc.

We discuss an example on GIS-based landslide hazard zonation in order to illustrate the general scheme of working.

Landslide hazard zonation using GIS

Landslides cause widespread damage in the Himalayas every year. Mitigation of disasters caused by landslides can be investigated only when knowledge about the expected frequency of mass movements in the area is available. Landslide hazard zonation is a process of ranking different parts of an area according to the degrees of actual or potential hazard from landslides (see e.g. Varnes 1984).

Applications

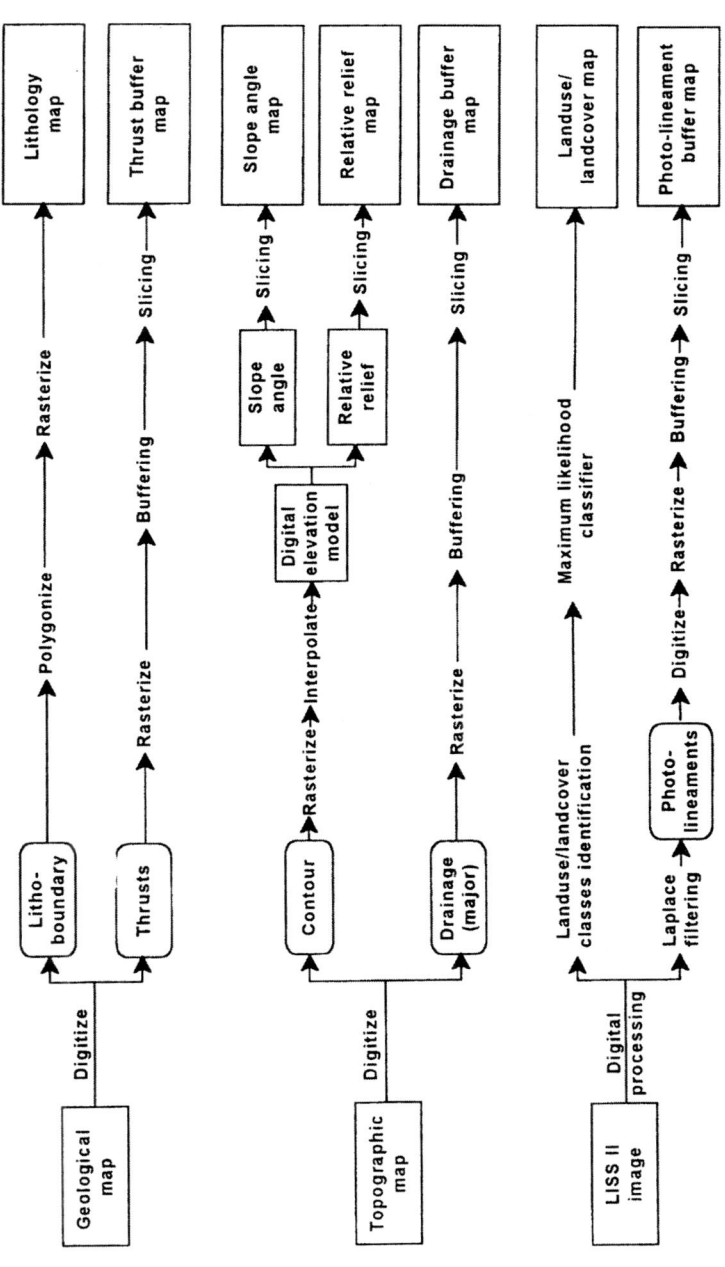

Fig. 15.22. Overview of methodology of data pre-processing for landslide hazard zonation

The evaluation of landslide hazard is a complex task as the occurrence of a landslide is dependent on many factors. In the last few decades, several field based hazard zonation studies with manual integration of data have been carried out in the Himalayas (see e.g. Anabalagan 1992, Pachauri and Pant 1992). However, these approaches have several limitations: for example, the extent of the area covered is generally small, and manual overlay of a thematic map is tedious and has poor integration capability.

With the advent of remote sensing and GIS technology, it has become possible to efficiently collect, manipulate and integrate a variety of spatial data, such as geological, structural, surface cover and slope characteristics of an area, which can be used for landslide hazard zonation (Gupta and Joshi 1990; van Westen 1994; Nagarajan et al. 1998).

The following case study on landslide hazard zonation is from a part of the Bhagirathi Valley, Himalayas (Gupta et al. 1999; Saha et al. 2002). The study utilized different types of data, including Survey of India topographic maps, IRS-1B and -1D multispectral and PAN satellite sensor data, and field observations. The processing of multi-geodata sets was carried out in a raster GIS environment. The following data layers were prepared:

- buffer map of thrust faults
- buffer map of photo lineament
- lithology map
- land-use/land-cover map
- buffer map of drainage
- slope angle map
- relative relief map
- landslide map.

An overview of the methodology of data layer preparation is given in figure 15.22.

Landslides are caused by mutual interaction of various factors, whose relative importance can be estimated from field knowledge/data. Using the ordinal scale and relative weighting–rating system, the various data layers were integrated to yield a landslide hazard index (LHI) map. Figure 15.23 shows the methodology of GIS data integration.

Applications 427

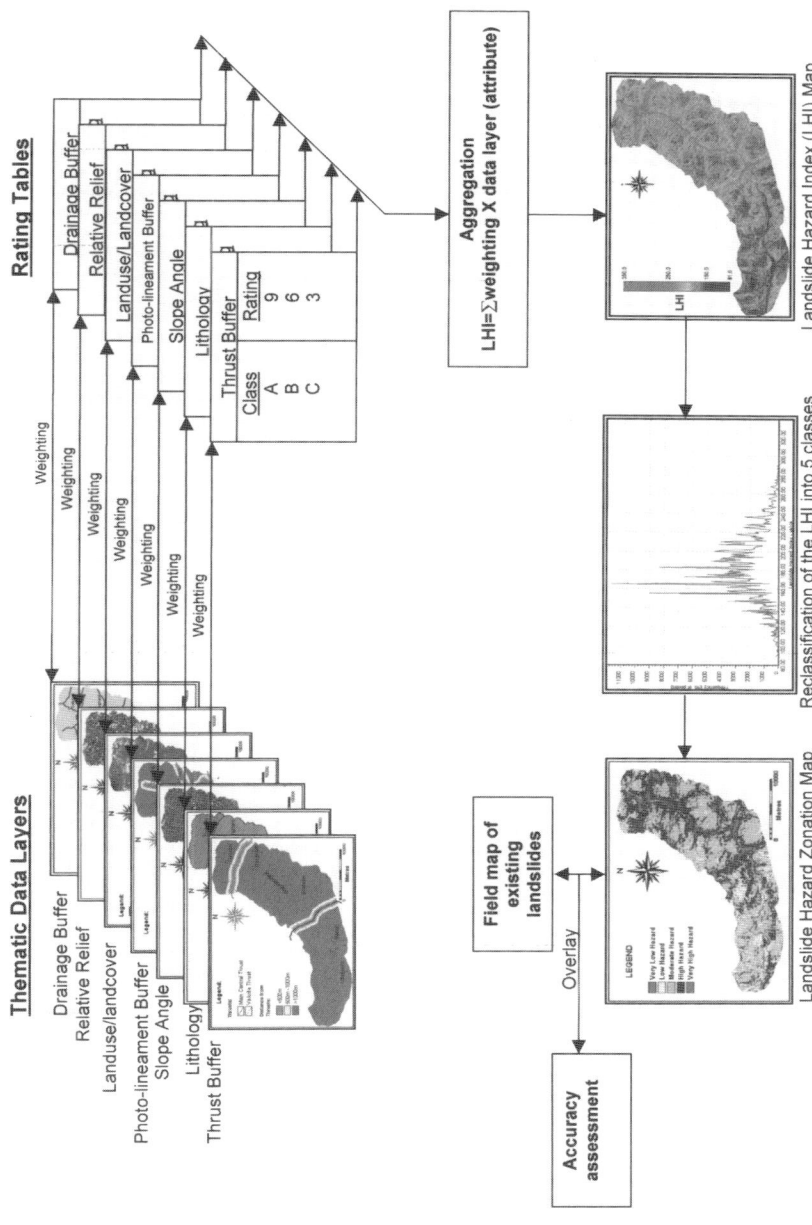

Fig. 15.23. Scheme of data integration in GIS for landslide hazard zonation

The LHI frequency diagram was used to delineate various landslide hazard zones, namely, very low, low, moderate, high and very high. Field data on landslides were employed to evaluate and validate the landslide hazard zonation map.

There have been numerous applications of GIS. For example, Bonhan-Carter et al. (1988) integrated geological data sets in GIS for gold exploration. Goosens (1991) used Landsat TM, aeromagnetic and airborne radiometric data to map granitic intrusions and associated skarns. Miranda et al. (1994) integrated SIR-B and aeromagnetic data for reconnaissance mapping. Rowan and Bowers (1995) integrated Landsat TM data with SAR data and a data base of known mines and prospects in a GIS study for mineral exploration. Jiang et al. (1994) report integration of Landsat TM-derived lineaments with aeromagnetic and geochemical data for exploration of porphyry copper deposits. Brainard et al. (1996) used GIS for assessing risk in transporting hazardous waste. Bruzewicz (1994) discusses GIS for emergency management. There is virtually no end to the range of applications GIS has found.

Chapter 16: Geological Applications

16.1 Introduction

Multispectral remote sensing data have shown tremendous potential for applications in various branches of geology – in geomorphology, structure, lithological mapping, mineral and oil exploration, stratigraphic delineation, geotechnical, ground water and geo-environmental studies etc. The purpose of this chapter is to review briefly the parameters involved in various thematic applications and present a few illustrative examples using mainly satellite data.

Rock attributes (i.e. structure, lithology, rock defects etc.) and physical processes (i.e. climatic setting, weathering and erosion agencies) operating in a region over a period of time govern the nature and appearance of landscape, i.e. topography, drainage, soil and vegetation. (Fig. 16.1). These in turn, influence photocharacters. The main task in remote sensing image interpretation is to decipher geological parameters from observations on elements of photo interpretation and geotechnical elements.

It should be appreciated that even when soil and vegetation cover is heavy, remote sensing data have their value in providing information on subsurface geol-

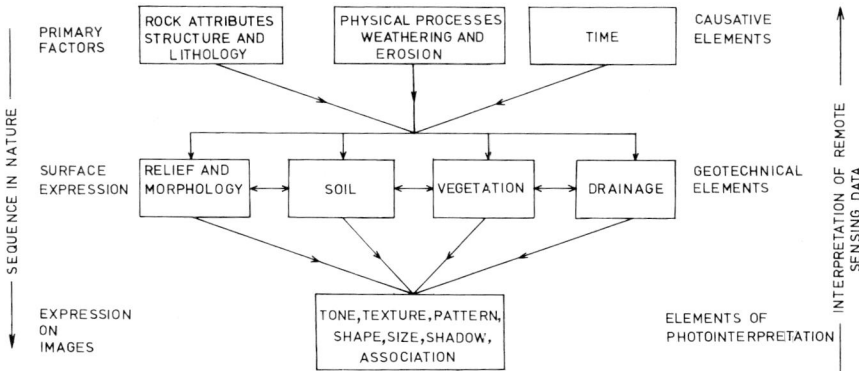

Fig. 16.1. Conceptual diagram showing the bearing of rock attributes and surface processes on geotechnical elements and elements of photo interpretation

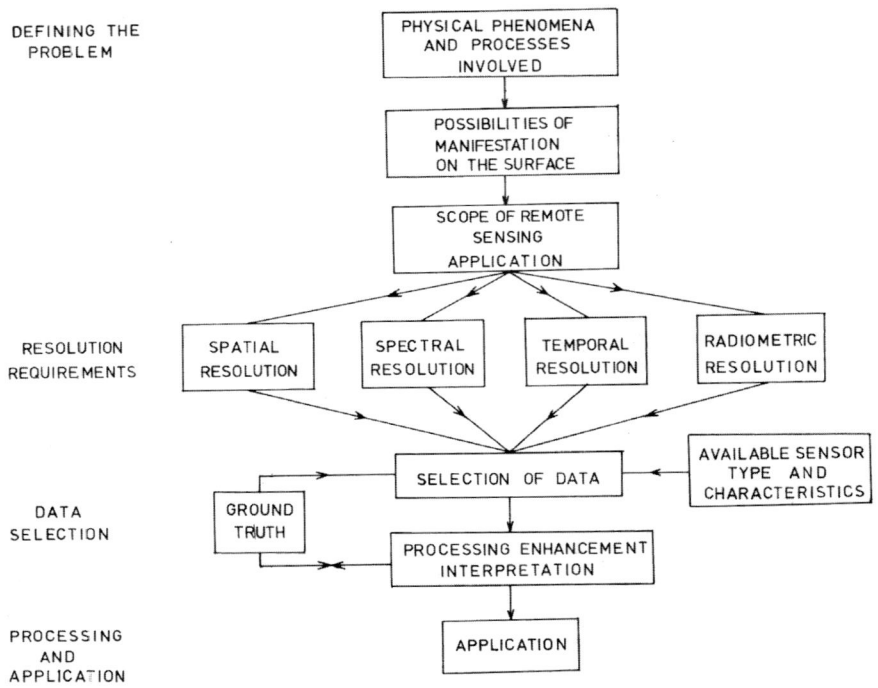

Fig. 16.2. Different stages in a typical remote sensing programme comprise: defining the problem, understanding the resolution requirements, selecting data sets, and finally processing and application

ogy, at least to some extent. The type of bedrock and structure control the type of soil, soil moisture and vegetation, which can give indirect information on the geology of the area.

Remote sensing investigations should not be considered as an alternative to field investigations. On the other hand, remote sensing data interpretation must be supported by field data – including field observations, sampling, analysis and even subsurface exploration – for reliable inferences. Remote sensing investigations may be said to have a two-fold purpose: (a) to allow viewing of the ground features in a different perspective, on a different scale, or in a different spectral region, and (b) to reduce the amount of field work involved in covering the entire study area.

A remote sensing application programme typically passes through several stages, which include defining the problem, assessing resolution requirements, selecting data sets, and finally data processing, interpretation and application (Fig. 16.2).

1. Defining the problem. The first and the most important task in any application assignment is to define the problem, i.e. to identify the various physical features,

Geomorphology

processes, and phenomena involved, so as to understand the possibilities of manifestation of the phenomena on remote sensing images. In a nutshell, for a particular geological application, we should be able to outline precisely what we should look for on the remote sensing data. For example, in a study on geological structure, vital clues are given by tectonic landforms and various types of trends and alignments. Similarly, in a problem on soil erosion, the important physical parameters will be: areas of steep topography, landslides, poor vegetation, higher surface runoff, drainage characteristics, and types of soils and rocks. Therefore, it is of utmost importance to physically conceive and define the problem.

2. Resolution requirements. Once the problem has been defined, the next step is to estimate the resolution requirements, i.e. what spatial, spectral, radiometric and temporal resolution of the remote sensing data will be sufficient to detect and/or identify the above physical parameters of interest. Different types of tasks have different resolution requirements. For example, in an investigation to map landslides in a hilly terrain, broad-band panchromatic VNIR remote sensing data with high spatial resolution will be necessary; on the other hand, for delineating mineral or rock types, data with essentially high spectral resolution will be required. Similarly, temporal resolution (repetivity) requirements depend upon the dynamics of the features of interest.

3. Selection of data. Depending upon the resolution requirements, and sensor types and characteristics available, the remote sensing data sets are to be selected for a particular application task. Often, care is required to ensure that the atmospheric-meteorological conditions existing at the time of remote sensing coverage are optimum, with regard to cloud cover, dust, haze, rain and solar illumination. Together with remote sensing data, related ground truth and ancillary information, e.g. geological structural data, topographical, soil, vegetation maps etc. and other ground information, are gathered as required.

4. Data processing, interpretation and application. After selection, the data are processed, transformed, rectified, enhanced, superimposed over other data sets and interpreted for features of interest. The interpretations are controlled by ground and ancillary information. Finally, the results of data interpretation are transferred to application groups.

In the following pages, we discuss the application of remote sensing data in various sub-disciplines of geology thematically, giving a few examples.

16.2 Geomorphology

Geomorphology deals with the study of landforms, including their description and genesis. Landform is the end product resulting from interactions of the natural surface agencies and the rock attributes. It depends upon three main factors: (a) climatic setting, including its variation in the past, (b) underlying bedrock (rock type and structure) and (c) the time span involved (Fig. 16.1). One of the widest applications of remote sensing data has been in the field of geomorphology, due to three reasons.

1. Remote sensing data products (aerial photographs and satellite images) give direct information on the landscape – the surface features of the Earth, and therefore geomorphological investigations are most easy to carry out based on such data.
2. Landform features can be better studied on a regional scale using synoptic coverage provided by remote sensing data, rather than in the field.
3. Stereoscopic ability permits evaluation of slopes, relief and forms; vertical exaggeration in stereo viewing brings out morphological details.

1. Spatial resolution. Geomorphology involves the study of a number of parameters, namely: extent and gradient of the slopes; their variations, shape, size, pattern; whether the slopes are barren or covered with soil or vegetation; type of surface material; whether the slope is stable or unstable; and mutual relations of the slopes. Whereas local landforms are best studied on large-scale to medium-scale stereo photographs, the regional setting of landforms extending over several kilometres and their mutual relationships can be better evaluated on coarser-resolution space images. The study of landforms on satellite images is also sometimes referred to as *mega-geomorphology*.

2. Spectral resolution. Data in the VNIR broad-band range have been used extensively for geomorphological investigations, as they provide higher spatial resolution and are able to bring out differences in topography, vegetation, soil, moisture and drainage. The application potential of thermal-IR data appears to be limited due to coarser spatial resolution. Radar may be advantageously used for gathering data on micro-relief, including surface roughness, vegetation, soil moisture and drainage.

3. Temporal resolution. Although land surface features are stable, their manifestation and detection on remote sensing images are predominantly influenced by temporal surface parameters, namely, soil moisture, vegetation, land cover and drainage (dry/wet channels). Therefore, it is important that the remote sensing data are acquired at a time which provides adequate discrimination between features of interest.

In many cases, landforms are characteristic of a particular type of terrain, e.g. (1) sink holes and collapse structures mark limestone terrain, (2) landslides, soil creep, rockfalls etc. indicate mass wasting and unstable slopes, (3) terraces, natural levees, fans, point bars and oxbow lakes indicate a fluvial environment, (4) moraines, drumlins and broad U-shaped valleys point towards a glacial environment, (5) sand dunes and loess mark aeolian terrain, (6) deltas, spit bars, lagoons and beaches etc. indicate marine processes, (7) volcanic cones, calderas and volcanic flows characterize an igneous environment, and so on. The various landforms are described in detail in standard works on geomorphology (e.g. Bloom

Geomorphology

1978; Thornbury 1978). As landforms are directly observed on remote sensing data products, it is important that the photo interpreter must have a sound knowledge of geomorphological principles and processes.

Valuable review contributions on geomorphological applications of remote sensing, particularly aerial photography, have been made by Tator (1960), Miller and Miller (1961), Ray (1965), Verstappen (1983) and Von Bandat (1983). An outstanding presentation on mega-geomorphology has been made by Short and Blair (1986).

The following describes the salient features of various landforms on remote sensing images, the description having been organized using genetic classification. Many, or rather most, landforms in nature are a result of multiple processes, and therefore the categorization made here may appear arbitrary in places.

16.2.1 Tectonic Landforms

Tectonic landforms may be defined as structural landforms of regional extent. W.M. Davis in 1899 considered that structure, processes and time constitute the three most significant factors shaping the morphology of a land. Of the three, structure, i.e. the deformation pattern, has the most profound control and this idea

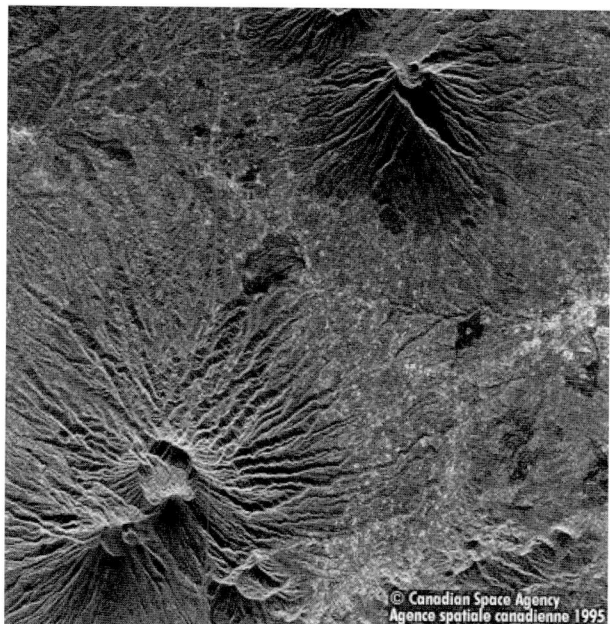

Fig. 16.3. Strato-volcanoes south of Jakarta, Indonesia. The look direction is from the top. The large circular caldera (bottom right) is more than 1 km wide. Radial drainage is well developed. (Courtesy of Radarsat Inc.)

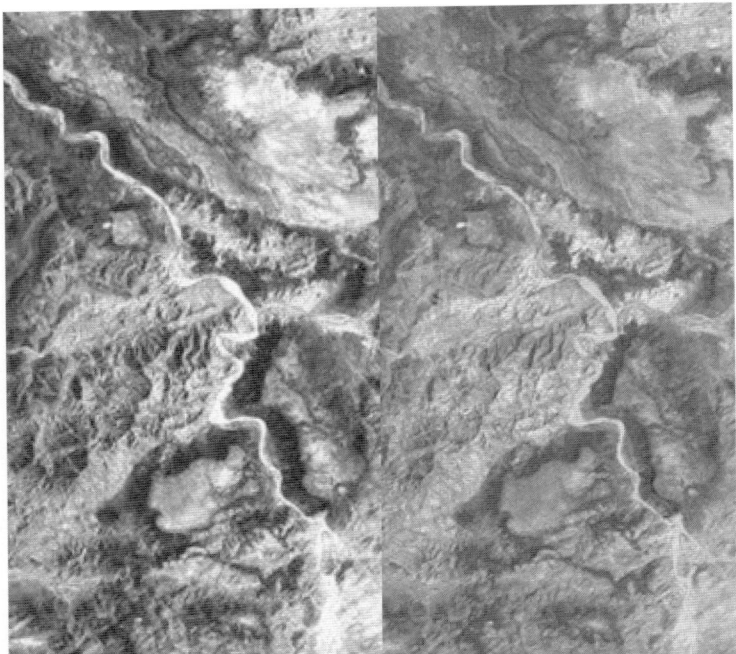

Fig. 16.4. MOMS-02P panchromatic stereo pair showing volcanic plateau landform in Somalia. (Courtesy of DLR, Germany)

led to the concept of morphotectonics. In almost all cases, the structure of the rock has an intrinsic influence on landforms due to selective differential erosion and denudation along structurally weaker zones. Everett et al. (1986) provide numerous examples. Many examples given here (see e.g. Figs.16.26, 16.28, 16.29) could also be considered as landforms of tectonic origin.

16.2.2 Volcanic Landforms

Volcanic landforms are primarily constructional, and result from extrusion of magma along either vent centres or fractures on the Earth's surface. Central-type neo-volcanic eruptions are confined to plate boundaries, most being concentrated on the convergent margins around the Pacific Ocean. They result in landforms such as conical mountains (Fig. 16.3). Fissure-type eruptions create sheets of flows forming plateaus (Fig. 16.4). Basaltic weathered surfaces are frequently marked by black-cotton soil and high-density dendritic, trellis and rectangular drainage patterns (Fig. 16.5). Short (1986) has reviewed and classified the various volcanic landform types. Interestingly, orbital repetitive remote sensor data can occasionally capture the process of active volcanism, and thereby add to our understanding of the endogenic phenomenon (see Fig. 16.104).

Geomorphology

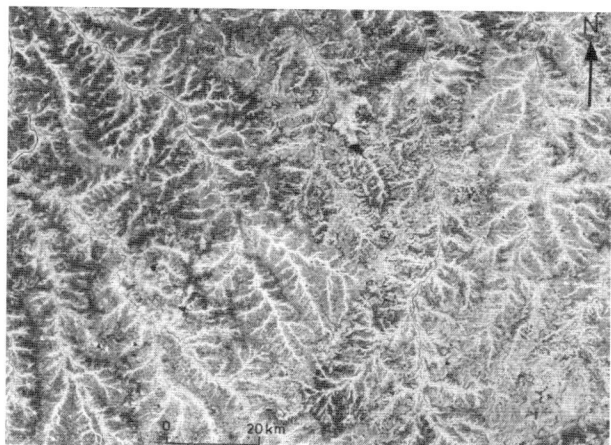

Fig. 16.5. Landform formed by flood basalt – Deccan Plateau, India. The topography is flat-topped plateau, generally dark-toned in the VNIR range, due to black-cotton soil. Vegetation is sparse and the terrain is marked by high-density dendritic drainage, monotonously extending over large tracts. (Landsat MSS4 infrared image)

16.2.3 Fluvial Landforms

Running water is one of the most prominent agents of landform sculpturing, whose effects are almost everywhere to be seen. Huge quantities of sediments or rock material are removed, transported from one place to another and dumped by rivers, thus modifying the land surface configuration (Baker 1986). The fluvial landscape comprises valleys, channel ways and drainage networks.

The *drainage pattern* is the spatial arrangement of streams and is, in general, characteristic of the terrain. Different drainage networks possess geometric regularity of different types, which reveal the character of the geological terrain, and also help in understanding the fluvial system. Howard (1967) has summarized the geological significance of various drainage patterns (Table 16.1; Fig.16.6). Six types of drainage patterns have been considered as basic, i.e. with gross characteristics readily distinguishable from other basic patterns, namely *dendritic, rectangular, parallel, trellis, radial* and *annular*. A number of modified basic patterns have also been described.

In addition to the above, several special drainage/channel patterns have also been identified. Some examples are given in figure 16.6. Each of these drainage patterns indicates a specific geological characteristic of the terrain. *Meandering* is the most common channel pattern. Such rivers take on a sinuous shape, developing alternating bends with irregular spacing along the length. These rivers have relatively narrow deep channels and stable banks and this pattern is most widely developed in flood plains. A *distributory pattern* consists of several branching channels, originating from the same source. It indicates the spreading of water and sediments across the depositional basin and develops over alluvial fans and deltas.

436 Chapter 16: Geological Applications

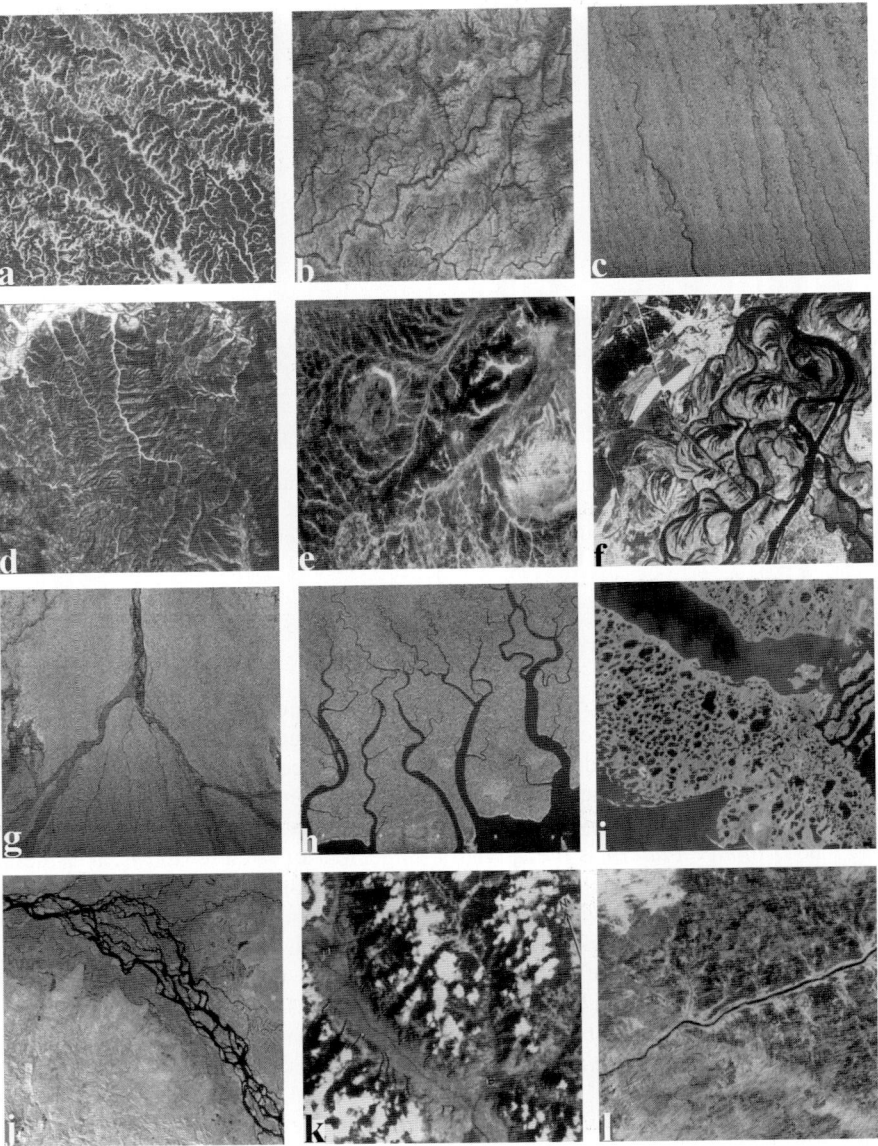

Fig. 16.6a–l. Important types of drainage patterns: **a** dendritic; **b** rectangular and angulate; **c** parallel; **d** trellis; **e** annular and sub-radial; **f** meandering; **g** distributary; **h** anastomotic; **i** deranged; **j** braided; **k** barbed; **l** rectilinear

Table 16.1. Common drainage patterns and their geological significance (see Fig. 16.6)

Type	Description	Geological significance
Dendritic	Irregular branching of streams, haphazardly, resembling a tree	Homogeneous materials and crystalline rocks; horizontal beds; gentle regional slope
Sub-dendritic	Slightly elongated pattern	Minor structural control
Pinnate	High-drainage-density pattern; feather-like	Fine-grained materials such as loess
Rectangular	Streams having right-angled bands	Jointed/faulted rocks, e.g. sandstones, quartzites etc.
Angulate	Streams joining at acute angles	Joints/fractures at acute angles to each other
Parallel	Channels running nearly parallel to each other	Steep slopes; also in areas of parallel elongate landforms
Trellis	Main streams running parallel and minor tributaries joining the main streams nearly at right angles	Dipping or folded sedimentary or low-grade metasedimentary rocks; areas of parallel fractures
Radial	Streams originating from a central point of region	Volcanoes, domes, igneous intrusions; residual erosion features
Centripetal	Streams converging to a central point	Depression, crater or basin, sink holes
Annular	Ring-like pattern	Structural domes

An *anastomotic pattern* comprises multiple interconnecting channels, separated by relatively stable areas of flood plains. A *deranged pattern* is a disorderly pattern of haphazard and erratic short streams and ponds found in swampy areas and glacial moraines and outwash plains. A *braided pattern* is controlled by the load carried by the stream and is marked by a shallow channel separated by islands and channel bars. A *barbed pattern* is one in which the tributaries join the main stream in bends pointing up stream. It indicates strong structural and tectonic control and uplift. Several other types of drainage patterns related to the geometry have also been described in the literature. A *palimpsest pattern* is one which includes traces of an older pattern, which forms the background for the development of the present pattern. A *drainage anomaly* is a local deviation from the regional drainage pattern, e.g. rectilinearity, local appearance/disappearance of meanders, anomalous ponds, marshes, fills, turns, gradients, piracy, rapids etc., and implies certain local geological phenomena.

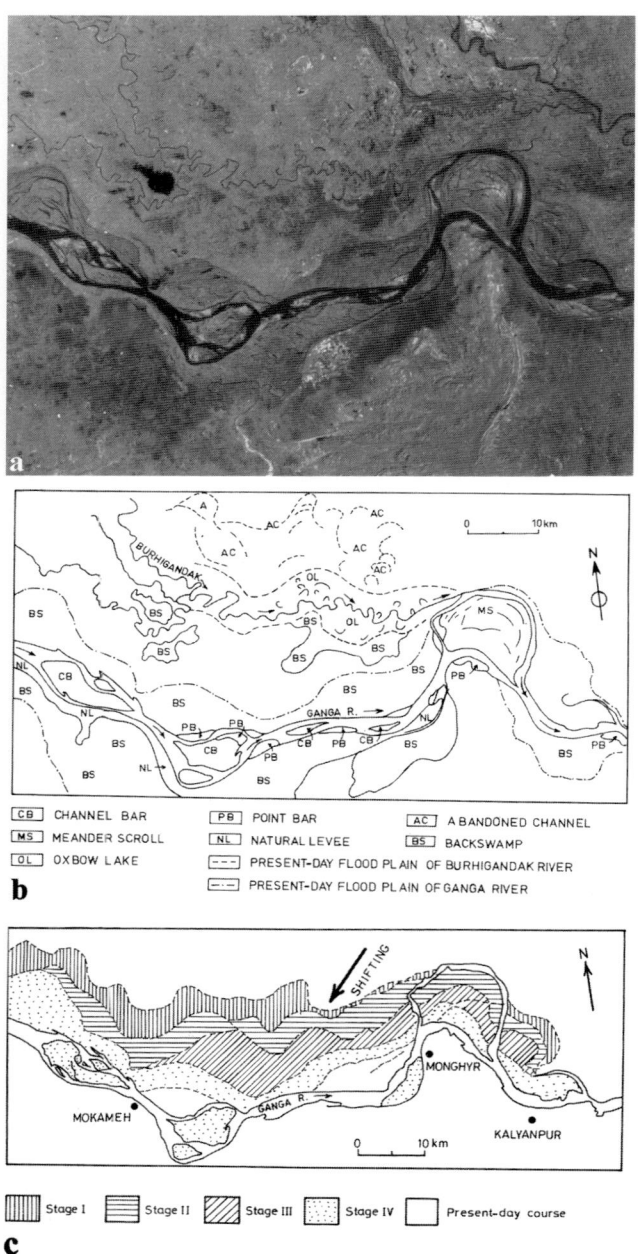

Fig. 16.7. a Landsat MSS4 image of a part of the Middle Ganges basin. **b** Interpretation map showing various fluvial landforms. **c** Successive southward migration of the Ganges River in several stages as interpreted from Landsat images. (Philip et al. 1989)

Geomorphology 439

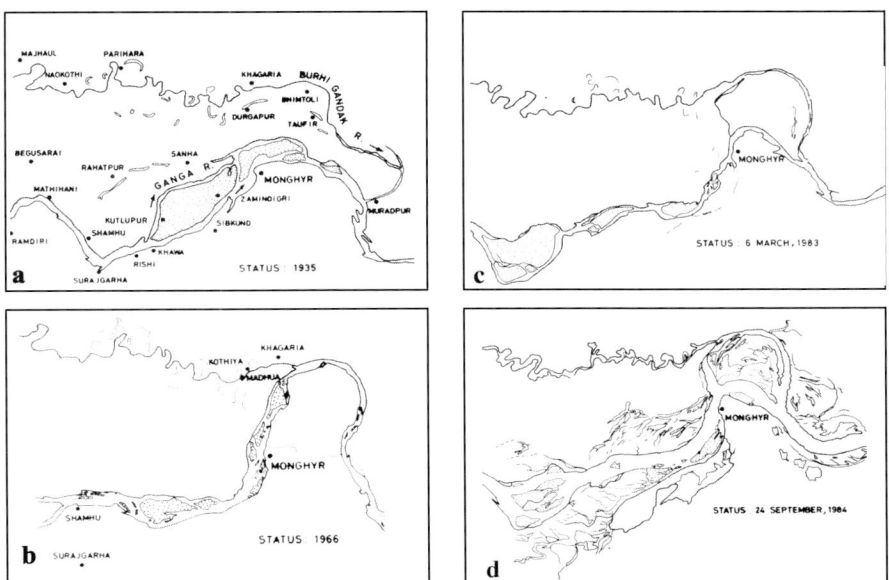

Fig. 16.8a–d. Channel planform pattern of Ganges and Burhigandak rivers, and changes through the years 1935–84. Patterns based on **a** topographic maps (1935), **b** aerial photographs (1966), and **c, d** Landsat images (1983) and (1984). (After Philip et al. 1989)

The landforms associated with fluvial erosion are gorges, canyons, V-shaped valleys, steep hill slopes, waterfalls, pediments etc. Typical depositional landforms include fans, cones, alluvial plains, flood plains, natural levees, river terraces, meander scars, channel fills, point bars, back swamps and deltas. Depending upon the dimensions involved, the landforms can be identified on aerial and satellite remote sensing data. An important application of repetitive remote sensing data is the study of dynamic features, such as changes in planform and migration of rivers (Figs. 16.7 and 16.8), and delineation of palaeochannels and palaeohydrology (see Sect. 16.8.3).

Stereoscopic analysis of remote sensing data can be of great help in geomorphic studies, as illustrated in figure 16.9. The stereo pair shows a part of Central India where the area is covered with basalt flows, and the drainage is controlled by mainly fractures/lineaments. This area shows an example of river piracy (or capture) (Rao and Subramanian 1997). River piracy develops when a river flowing at a lower level encroaches, due to headward erosion, upon a stream flowing at a higher level, diverting part of the water of the higher stream to the lower stream. This type of information can be better derived from 3-D (now called 2.5-D) stereoscopic views.

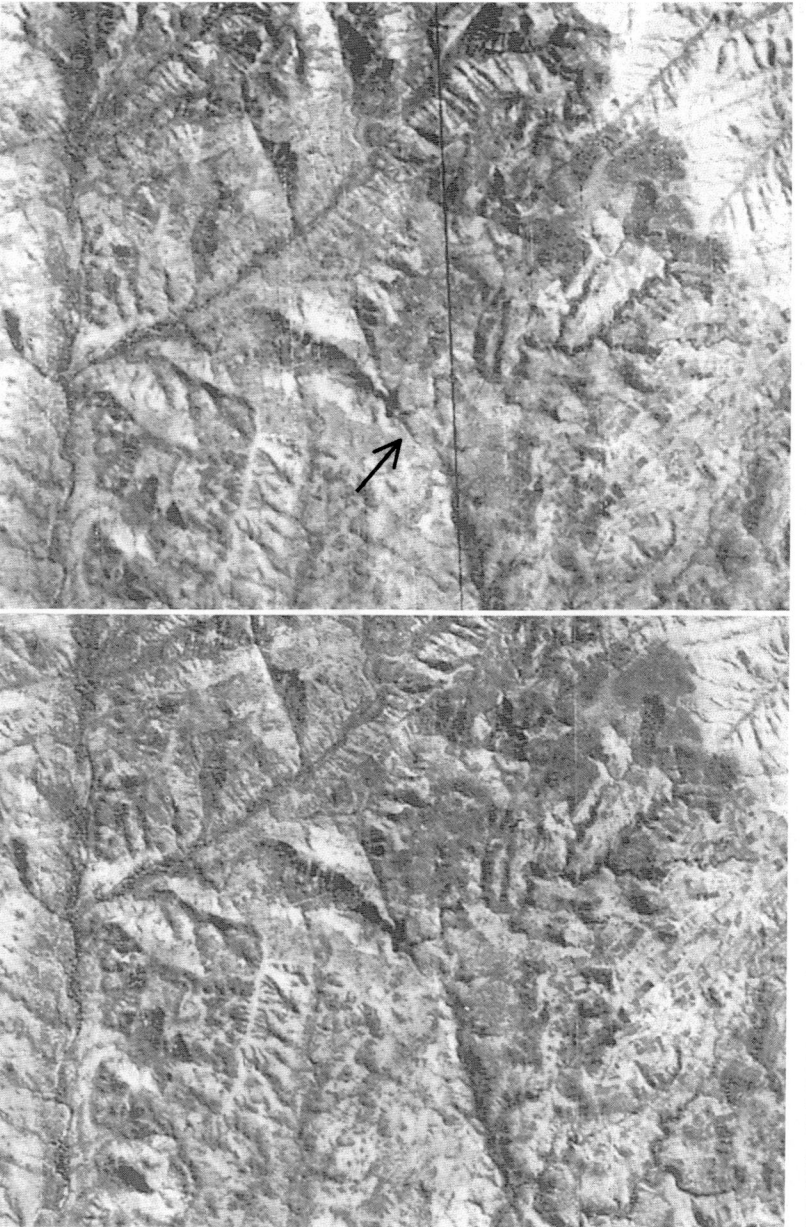

Fig. 16.9. IRS-PAN stereo images showing river capture (*arrow*) in Central India. (Rao and Subramanian 1997)

Geomorphology 441

Fig. 16.10. The Bissan river delta in Guinea. The river carries enormous amounts of sediments (black-and-white image printed from Landsat MSS standard FCC). (Courtesy of P. Volk)

16.2.4 Coastal and Deltaic Landforms

The oceans cover a major part of the Earth and surround the continents. A coastline is the boundary between land and ocean. In a general sense, the coast refers to a zone of indefinite width on both sides of the coastline. Coastal landforms are those which are influenced and controlled by proximity to the sea. Several types of coastal erosional landforms, such as cliffs, terraces, benches, shelves, caves, islands etc., and depositional landforms, such as beaches, spits, bars, tidal flats and deltas, can be identified on aerial photographs and satellite images, depending upon the dimensions involved and the scale provided by the sensor. Selected examples are given by Bloom (1986) and Coleman et al. (1986).

Rivers transport huge quantities of sediments from the land to the seashore, and a variety of landforms may develop, commonly grouped under the term delta (see Figs. 16.10, 16.11 and 16.12). Sometimes gigantic deltas are formed by great rivers. Deltas include distributary channels, estuaries, bars, tidal flats, swamps, marshes, etc. and are often marked by an anastomotic drainage pattern .

Many of the coastal landforms are composite products, having evolved as a result of multiple processes, due to eustatic-level changes. The shoreline of emergence is an exposed portion of the sea floor, subjected to sub-aerial agencies. Similarly, the shoreline of submergence is simply a drowned portion of a sub- aerial landscape, now subjected to submarine agencies. For example, figure 16.28 could also be considered as a coastal landform of composite type.

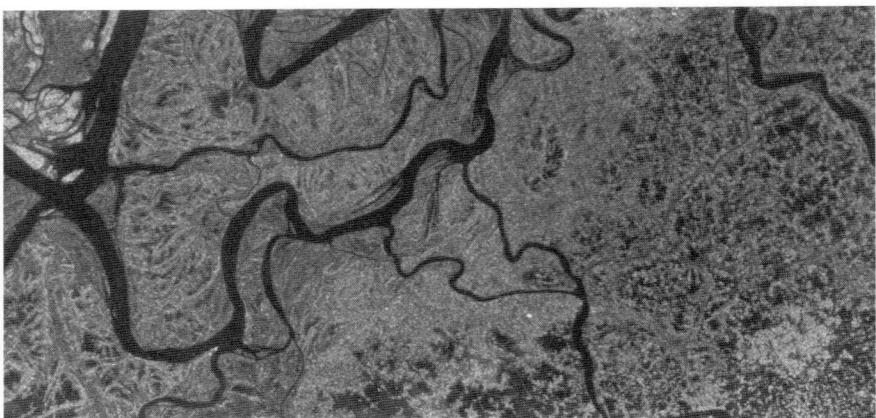

Fig.16.11. SIR-B image covering part of the Ganges flood plains, Bangladesh. The mottled grey and black areas on the right are cultivated fields connected by extensive irrigation and drainage channels. The uniform grey areas on the left correspond to the flood plain, susceptible to recurring floods and major reworking. (SIR-B image courtesy of JPL)

16.2.5 Aeolian Landforms

Deserts, where aeolian activity predominates, cover a significant part of the land surface. These are generally remote, inhospitable areas. The distribution of deserts in the world is not restricted by elevation, latitude or longitude. Erosion, transportation and deposition create the landforms in deserts, chiefly by wind action.

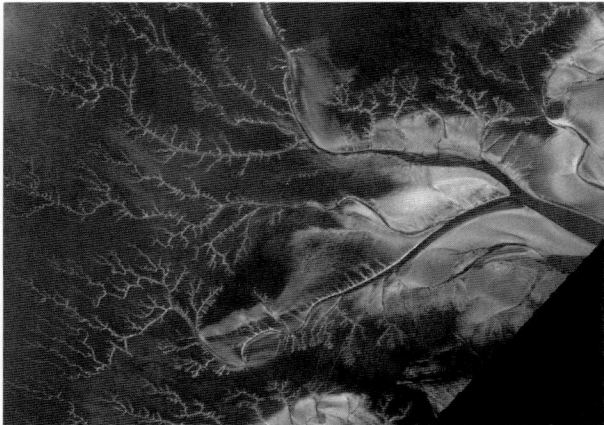

Fig. 16.12. Tidal flats in northern Germany (X-band aerial SAR image). Note the high-density dendritic drainage and variation in surface moisture over the tidal flats. (Courtesy of Aerosensing Radarsysteme GmbH)

Geomorphology 443

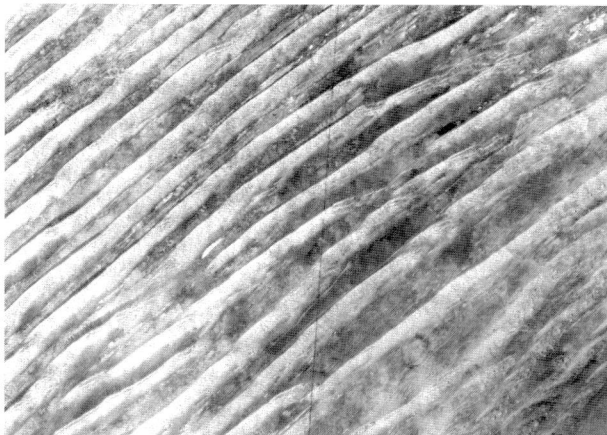

Fig. 16.13. Longitudinal sand dunes of the famous Rub' al Khali, which forms the world's largest continuous sand desert, Saudi Arabia. (ASTER image, printed black-and-white from colour composite) (Courtesy of NASA/GSFC/MITI/ERSDAC/JAROS, and US/Japan ASTER Science Team)

The aeolian terrain is marked by scanty or no vegetation and little surface moisture. Therefore, on VNIR photographs and images, the area has very light photo tones. Active dunes have no vegetation and stabilized dunes may have scanty grass cover. The various landforms can be distinguished on the basis of shape, topography and pattern (see e.g. Walker 1986).

Aeolian erosional processes lead to the formation of a variety of landforms such as yardangs (sculptured landform streamlined by wind), blow-outs (deflation basins), desert pavement (stony desert) and desert varnish (dark shiny surficial

Fig. 16.14. Star sand dunes in part of Saudi Arabia. In the south-east corner are circular irrigation fields due to sprinkler irrigation. (MOMS-02P image, printed black-and-white from colour composite) (Courtesy of DLR, Obepfaffenhofen)

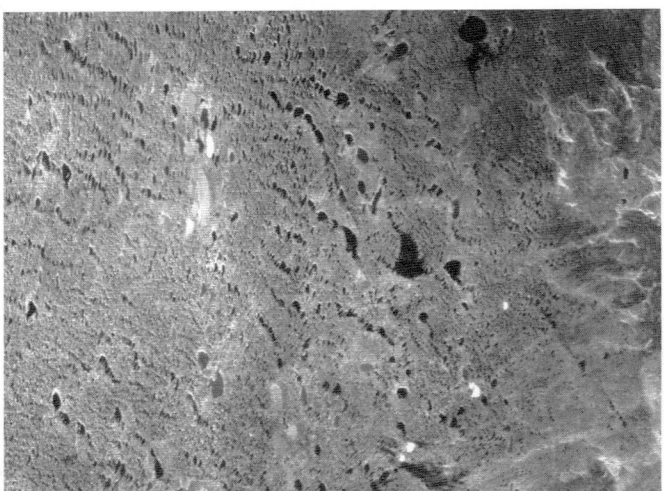

Fig. 16.15. Part of the Stuart desert, Australia, where a sudden storm has led to widespread flash floods and the formation of numerous lakes. (Landsat MSS4 image) (Courtesy of R. Haydn)

stains). The transportation action removes loose sand and silt particles to distant places. Dust storms are aeolian turbidity currents. Loess deposits are homogeneous non-stratified and unconsolidated wind-blown silt. They are susceptible to gullying and may develop pinnate and dendritic drainage patterns. Dry loess slopes are able to stand erect and form steep topography. Aeolian deposition leads to sand sheets, various types of dunes such as crescent dunes, linear dunes, star dunes, parabolic dunes, and complex dunes and ripples (Figs. 16.13 and 16.14). Other landforms in deserts could be due to fluvial activity, such as fans, dry river channels and lakes. Desert lakes (playas) are generally salty, shallow and temporary, and constitute sources of mineral wealth such as salts formed by evaporation.

There is little vegetation in desertic terrains, common types being xerophytes (drought or salt-resisting plants), succulents (which store water in their system) and phreatophytes (with long roots that reach the water table). The plants that are present hold soil, inhibit deflation, check surface velocity and provide sites for deposition. Rainfall may seldom occur in deserts, but when it does, it may be violent and lead to widespread fluvial lacustrine processes (Fig. 16.15).

Remote sensing data can help to monitor changes in deserts, their landform, movement etc., and to locate oases and buried channels (see Sect. 16.8.3).

16.2.6 Glacial Landforms

Glaciers are stream-like features of ice and snow, which move down slopes under the action of gravity. Glaciers occur at high altitudes and latitudes, and

Fig. 16.16. Stereo image pair of the near-IR spectral band showing aligned lakes due to glacial scouring (and possibly bedrock structure) in the Canadian shield of eastern Ontario, Canada. (Aerial digital camera images acquired using a Kodak DCS 460 CIR camera, with 3060 × 2036 × 12-bit format; the pixel spacing is 60 cm) (Courtesy of Doug King)

about 10% of the Earth's land surface is covered with glacial ice. The areal extent of glaciers is difficult to measure by field methods, and remote sensing data images provide information of much practical utility in this regard. Further, multispectral data can help delineate different zones in a glacier (see e.g. Hall and Martinec 1985; Williams 1986).

Typical erosional landforms of glacial origin are broad U-shaped valleys, hanging valleys, fords, cirques and glacial troughs. Figure 16.16 shows aligned lakes due to glacial scouring (and possibly bedrock structure). The huge moving masses of ice and snow erode and pick up vast quantities of fragmental material and transport these varying distances before deposition. The glacial deposit is typically heterogeneous, consisting of huge blocks to fine silt or rock flour, and is called till matrix. The depositional landforms include moraines, drumlins, till, glacial drift etc. Below the snow line (line of perpetual snow), the ice melts and gives rise to streams. In this region, up to a certain distance downstream the landforms have characteristics with both fluvial and glacial properties, and they are called fluvio-glacial. Typical fluvio-glacial landforms include outwash plains, eskers, fans and deltas and glacial lacustrine features (Fig. 16.17).

Broadly, glacial landforms produce gently rolling or hummocky topography with a deranged or kettle-hole drainage pattern. Images exhibit a mottled pattern due to varying soil moisture and the presence of a large number of ponds and lakes.

Special landforms associated with specific geological rock types, such as karsts in limestones, intrusives in igneous rocks, etc. are described with their respective lithologic types.

16.3 Structure

1. Scope. Remote sensing techniques have found extensive application in structural studies to supplement and integrate structural field data, the aerial and space-acquired data providing a completely new dimension in terms of synoptic view. In-

Fig. 16.17. Seasat-SAR image of part of Iceland showing prominent glacial features such as broad valleys and flowing ice; many of the fluvio-glacial features, such as outwash plains, streams with suspended material and dammed lakes, are also seen. (Processed by DLR) (Courtesy of K. Arnason)

formation on features can be most usefully gathered on a scale appropriate to their size. For example:

Dimensions of feature	*Optimum tool/scale of observation*
μm/mm	Microscopy
mm/cm/dm/km	Field data
dm/km/tens of km	Aerial platform data
tens of km/hundreds of km	Satellite platform data

It was in this perspective that aerial photography galvanized regional structural analysis in the late 1940s–early 1950s, and the same phenomenon was repeated by satellite data in early 1970s.

2. Basis and purpose. The basis for deriving structural information from remote sensing data emanates from the concept of morphotectonics – that rocks acted upon by erosional processes result in landforms which are related to both internal characteristics (rock attributes, i.e. lithology, and structure) and external parameters (types and intensity of erosional processes) (Fig. 16.1). This implies that: (a) structures have significant influence or landform development, and (b) erosional landforms carry imprints of structural features of rocks.

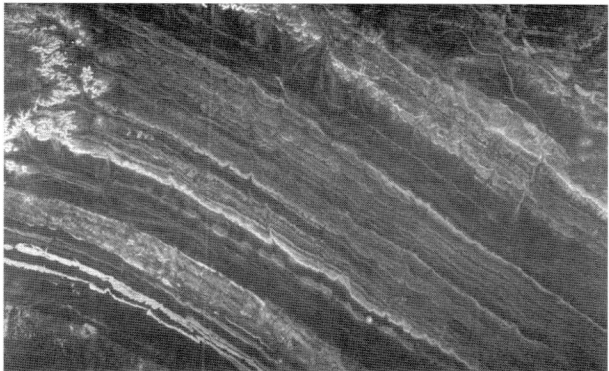

Fig. 16.18. Manifestation of bedding as prominent and regular, linear features marked by contrasting topography, tone, texture and vegetation. (SIR-A image) (Courtesy of JPL)

Most commonly, structural-geological studies commence by deciphering planar discontinuities in the rocks, with a view to understanding their characteristics, disposition and spatial relations. A *planar discontinuity* is marked by contrasting physicochemical conditions in rocks, in terms of mineral composition, chemical weathering property, mechanical strength and erodibility. The chances of identification of discontinuities are related to the resolution of the sensor (i.e. scale of observation) and the dimension of the discontinuity. Discontinuities can be of various types, e.g. bedding, foliation, faults, shear zones, joints etc. Bedding is the primary discontinuity and other structures constitute secondary discontinuities.

3. Manifestation of discontinuities. Discontinuities are expressed as differences in topography, slope, relief, tone and colour of the ground, soil and vegetation, and combinations of these. The discontinuities may be observed on simple photographs and images, or remote sensing data may be processed specifically to enhance certain discontinuities (see Sect. 10.6.1). Further, manifestation of discontinuities on remote sensing images may be direction dependent, i.e. a remote sensor may be configured so as to enhance certain discontinuities in comparison to others.

4. Vertical vs. horizontal discontinuities. In general, the remote sensing techniques, such as aerial photography, shuttle-based photography, and the most widely used Landsat MSS, TM, SPOT data, provide plan-like information, the data being collected while the sensor views the Earth vertically from above. On these images and photographs, vertical and steeply dipping planar discontinuities are very prominently displayed. On the other hand, gently dipping or subhorizontal structural discontinuities are comparatively suppressed and are difficult to delineate, especially in rugged mountainous terrains. Such discontinuities have a strongly curving or irregular pattern, and can be better identified on the basis of accompanying field/ ancillary information.

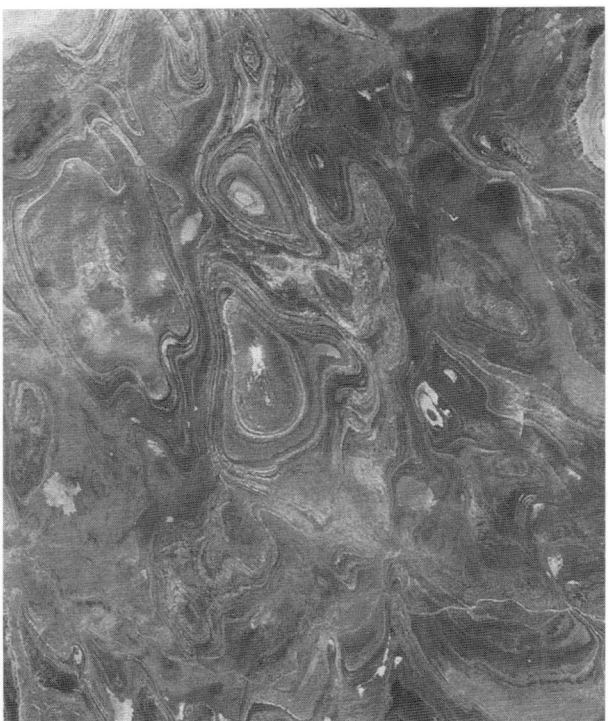

Fig. 16.19. Sub-horizontal beds resulting in an outcrop pattern in the shape of concentric loops and ellipses. The area forming a part of Tanezrouft, Algeria, is very dry, completely barren of vegetation, quite flat and is mainly a gravel desert. The Palaeozoic sedimentary rocks, which form the bedrock, are mildly deformed. Wind erosion and deflation have carved out the peculiar outcrop pattern. (Landsat MSS4 image) (Courtesy of R. Haydn)

16.3.1 Bedding and Simple-Dipping Strata

Bedding is the primary discontinuity in sedimentary rocks, and is due to compositional layering. The alternating sedimentary layers may differ in physicochemical properties, and this leads to the appearance of regular and often prominent linear features, marked by contrasting topography, tone, texture, vegetation etc., on images and photographs (Fig. 16.18). Linear features due to bedding are long, even-spaced and regular, in contrast to those produced by foliation or joints.

Orientation of bedding is important, for the main aim in structural interpretation is to delineate the attitude of beds (i.e. strike and dip) and deduce their structural relations. The beds may be sub-horizontal, inclined or vertical and may form segments of larger fold structures.

Structure 449

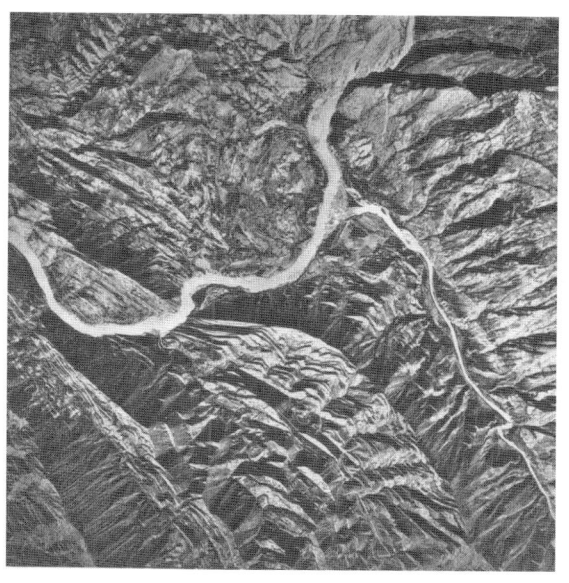

Fig. 16.20. Inclined/steeply dipping beds. The aerial photographic stereo pair shows competent (sandstone) and incompetent (shale) beds, which are intercalated. The sandstones form strike hogback ridges with dip slopes on one side and steep talus slopes on the other; the shales form strike valleys. The sandstone outcrop, as it meets the river, makes a V; the notch points towards the down-dip direction. (Stereo pair courtesy of Aerofilms, London)

1. Flat-lying beds are recognized by a number of features, such as banding extending along topographical contours or a closed-loop pattern, dendritic drainage on horizontal beds, and mesa landforms. As the beds are sub-horizontal, erosional repetition becomes a common feature (Fig. 16.19).

2. Inclined beds commonly form elongated ridges and valleys due to differential weathering (Fig. 16.20). The outcrop pattern is often determined by the structure and relief in the area. Common landforms are: hogbacks, cuestas, dip slopes, strike valleys and sometimes trellis drainage. Strike direction of beds is given by the trend of ridges, vegetation bands, tonal bands and linear features corresponding to lithological layering. The rule of 'V's can often be successfully applied for the determination of dip direction (Fig.16.20).

3. Vertical beds are identified by straight contacts, which run parallel to the strike of the beds, irrespective of surface topography; trellis drainage is also quite common.

16.3.2 Folds

A fold can be delineated by tracing the bedding/marker horizon along the swinging strike, and the recognition of dips of beds. Broad, open, longitudinal folds are easy to locate on satellite images. On the other hand, tight, overturned, isoclinal folds are difficult to identify on satellite images, owing to the small areal extent of hinge areas (which provide the only clues of their presence); therefore, such folds need to be studied on appropriately larger scales, such as aerial photographs. Some interesting examples of fold structures are given below.

1. Richat structure, Mauritania. The Richat structure is a classic example of the potential of remote sensing data in structural mapping (Fig. 16.21). This structure came into the limelight through the Gemini-4 photographs, although it was found subsequently that some French investigators had known about it even earlier. Within the Great Sahara desert, the Richat structure lies in a remote part of Mauritania and is located on a plateau about 200 m above the adjacent desert sands. The adjoining terrain consists of sub-horizontal sedimentary rocks of Ordovician age, resembling an extensive mesa. The Richat structure consists of series of concentric quartzite ridges, separated by concentric valleys underlain by shales. The rounded outcrop pattern has a gigantic 'bull's eye' shape of about 40 × 30 km. Annular and radial drainage patterns are present, although poorly developed owing to the arid conditions in the area. The origin of the structure has been a standing geological enigma, various views being: (a) an impact origin (now largely discarded), (b) intrusion of magma at depth, possibly as a plug, and (c) diapiric intrusion of shales (Short et al. 1976; Everett et al. 1986).

Fig. 16.21. The Richat structure located in Mauritania. It consists of concentric ridges (quartzite) and valleys (shales), and is about 40 × 30 km in dimension. Its origin has been a long-standing geological riddle. (Landsat MSS4 image) (Courtesy of R. Haydn)

2. Structures in parts of the Anti-Atlas mountains, Morocco. The Anti-Atlas mountains (Morocco) exhibit broadly folded doubly plunging complex structural features. The rocks comprise various types of sedimentary strata, such as sandstones, siltstones, clay-shales, limestones, dolomites and conglomerates, and some bedded volcanics. The Landsat TM ratio 4/7 image and its structural interpretation are shown in figure 16.22. The TM ratio 4/7 (near-IR/SWIR-2.2-µm) enhances various lithological units marked by differences in spectral character, due to variation in composition, surface moisture, soil and vegetation. A number of structures, such as plunging anticlines, synclines and doubly plunging, broad open folds, have been deciphered which were not so well shown on any published geological map of the area (Kaufmann 1985).

3. Fold structures in parts of the Aravalli hills, India. Some typical large-wavelength open folds are developed in the area around Alwar, Aravalli hills, Rajasthan, India (Fig. 16.23). The area lies in the north-western part of the Precambrian Indian shield and the rocks comprise predominantly quartzites, phyllites and schists, belonging to the Delhi Super Group. The quartzites form strike ridges and the associated phyllites and schists form slopes and valleys.

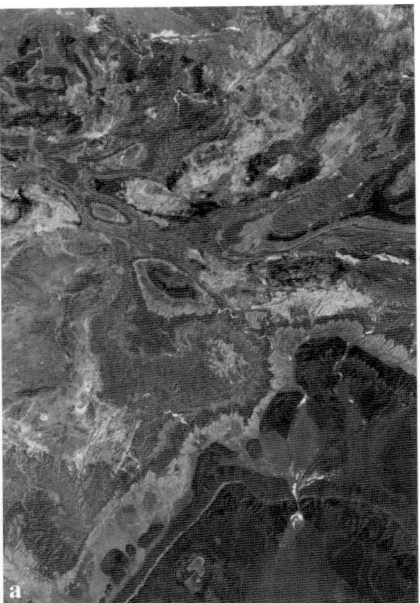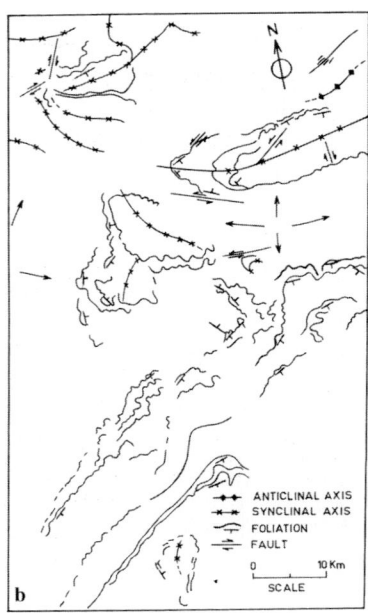

Fig. 16.22. a TM 4/7 ratio image of a part of the Anti-Atlas mountains, Morocco. **b** Interpretation map showing various structural features, viz. anticlines, synclines, doubly plunging folds and faults. (**a,b** Kaufmann 1985)

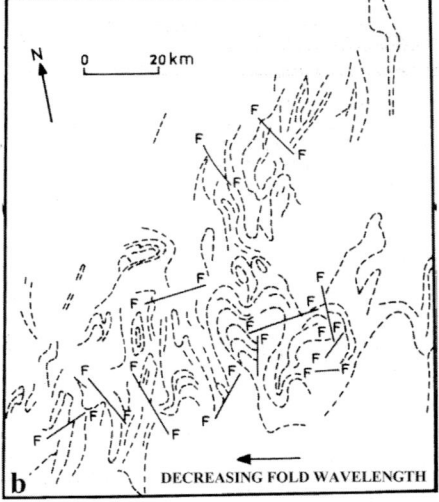

Fig. 16.23. a Landsat MSS4 (infrared) image showing the fold pattern in the rocks of Delhi Super Group near Alwar, Rajasthan; a number of faults are also observed. **b** Interpretation map of **a**; note the regional variation in wavelength of folds across the area

Structure 453

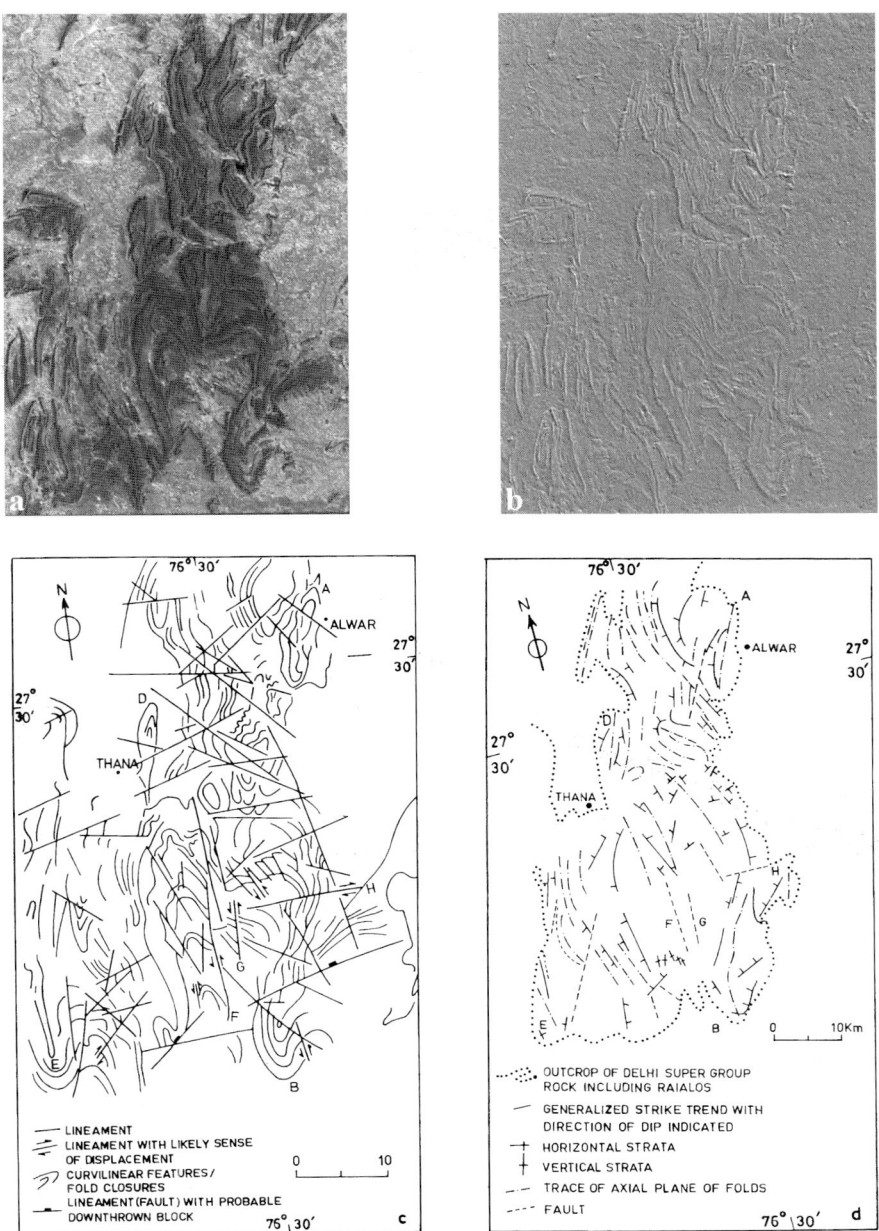

Fig. 16.24a–d. a Sub-scene of Fig. 16.23a (window marked). b Shows 'edges' of the sub-scene in a. c Structural interpretation from the images shown in a and b. d Structural map of the corresponding area based on field investigations (after Gangopadhayay 1967). For further discussion, see text

Numerous structural features – folds and faults – are seen in Fig.16.23a. The general strike of the rocks is NNE-SSW and the folds exhibit closures towards north and south. A regular variation in wavelength of the folds is conspicuous, the folds being broad and open on the east, and gradually becoming tighter towards the west (Fig.16.23b) – a fact reported by Heron (1922) after extensive field investigations and readily shown on the synoptic views provided by the satellite data.

An enlarged sub-scene of the above (Fig. 16.24a) and the 'edges' on this sub-scene (16.24b) show the fascinating fold pattern and many transverse structures. The interpretation map (Fig. 16.24c) from the Landsat images and the field structural map of the area (Fig. 16.24d, redrawn after Gangopadhyay 1967) are also compared. The example demonstrates the utility of satellite imagery in delineating structural features of such dimensions.

4. Photogrammetric measurements for structural analysis – Zagros mountains, Iran. The principles of deriving structural orientation data from stereo photography have been discussed elsewhere (Chapter 6). Bodechtel et al. (1985) made quantitative measurements on small-scale (1:820,000) Metric Camera stereo photographs for structural evaluation in parts of the Zagros Mountains, Iran. Geologically, this terrain consists of Mesozoic–Tertiary sedimentary sequences, deformed into long-wavelength doubly plunging folds (Fig. 16.25). The characteristic sedimentary layering is very well depicted, with dip slopes and fold closures. The main structure in the example area is a doubly plunging anticline, identified as Kuh-e-Gashu (Fig. 16.25a,b). The opposite dips on both the flanks are indicated by V-shaped structures and flat irons, formed by the outer resistant lithological horizon. The fold axial trace runs almost E–W and the axis plunges eastwards on the east and westwards on the west. An oval-shaped salt dome has intruded the anticlinal core and locally disturbed the fold axis. Further to the north lies a doubly plunging syncline, which is succeeded by another doubly plunging anticlinal structure, identified as Kuh-e-Guniz. Its long eastward closure with prominent outcrop pattern is distinct on the photograph. The axial zones of folds are marked by minor faults and fractures (lineaments).

Structural data for statistical evaluation was derived by Bodechtel et al. (1985) by applying photogrammetric methods on the Metric Camera stereo photographs, using a precision instrument. Orientation of the bedding plane was measured using the three-point method (i.e. the X, Y, Z co-ordinates of three points located on the same bedding plane), and these measurements provided strike and dip data of beds (Fig. 16.25b). On projection, this gave the statistical orientation of the bedding planes, their intersections being β or fold axis. The computed data were found to be in close correspondence with field data.

Application of similar methodology on SPOT-HRV-PAN data for structural geological measurements was reported by Berger et al. (1992).

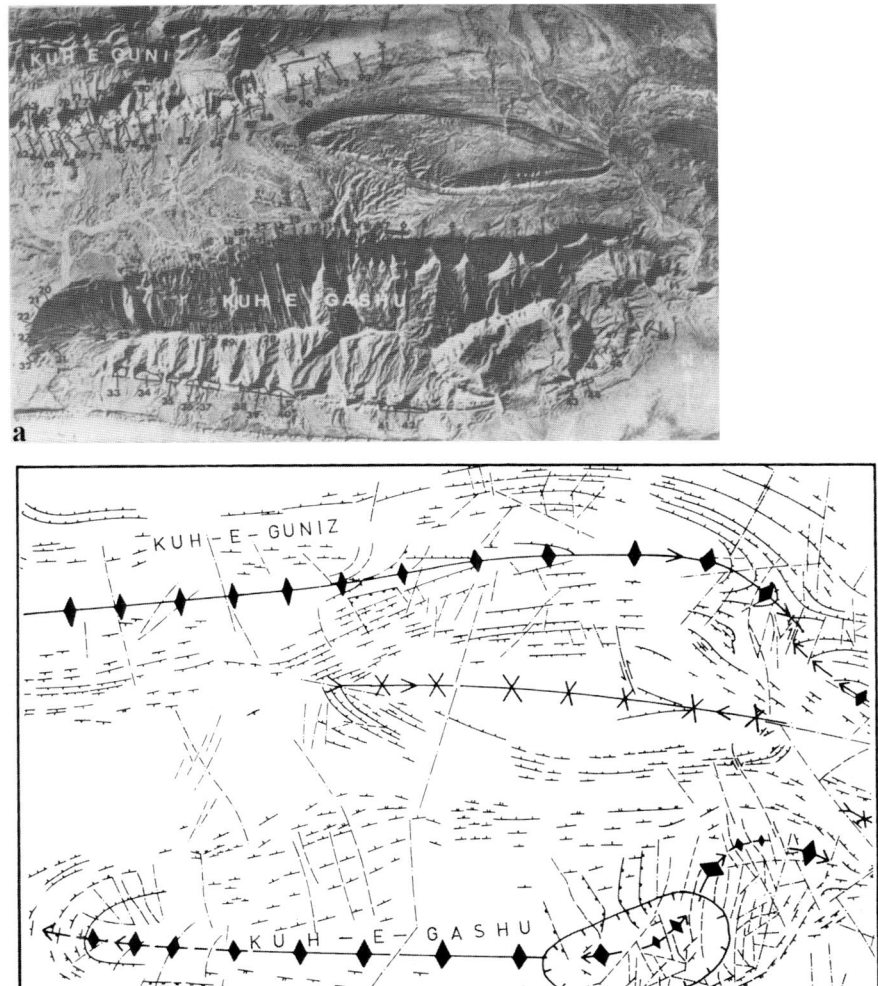

Fig. 16.25. a Metric Camera photograph of part of the Zagros mountains, (Iran) covering the Kuh-e-Gashu and Kuh-e-Guniz anticlines. **b** Structural map based on photogrammetric investigations of the Metric Camera stereo photographs. (**a,b** Bodechtel et al. 1985)

Fig. 16.26. Landsat MSS (red band) image showing the Yamuna fault. The Siwalik group of rocks border the Himalayas on the south, and are well bedded, comprising mainly sandstones and shales; the beds are severed and displaced by a fault (Yamuna fault). Large-scale drag effects, with a left-lateral sense of displacement, are distinct

16.3.3 Faults

One of the greatest advantages of remote sensing data from aerial and space platforms lies in delineating vertical to high-angle faults or suspected faults. These are indicated on the images and photographs by one or more of the following criteria: (1) displacement of beds or key horizons, (2) truncation of beds, (3) drag effects, (4) presence of scarps, (5) triangular facets, (6) alignment of topography including saddles, knobs etc., (7) off-setting of streams, (8) alignment of ponds or closed depressions, (9) spring alignment, (10) alignment of vegetation, (11) straight segment of streams, (12) waterfalls across stream courses, (13) knick points or local steepening of stream gradient, and (14) disruption of valley channels.

Low-angle faults are difficult to interpret, since the images provide planar views from above. Such faults have strongly curving or irregular outcrop and can be inferred on the basis of discordance between rock groups, e.g. with respect to attitude of beds, degree of deformation, degree of metamorphism etc.

Some examples of faults observed on multispectral remote sensing images are described below.

1. Faulting marked by dislocation and drag effects – the Yamuna fault, Sub-Himalayas, India. The river Yamuna emerges from the Himalayas along a fault, called the Yamuna fault or the Paonta fault (Fig. 16.26). Geologically, the terrain comprises the Siwalik Group of rocks, containing well-bedded clastics (sandstones, silstones, shales and conglomerates) of Miocence–Pleistocene age. These deposits represent the molasse sequence of the Himalayas and border the Himala-

Structure 457

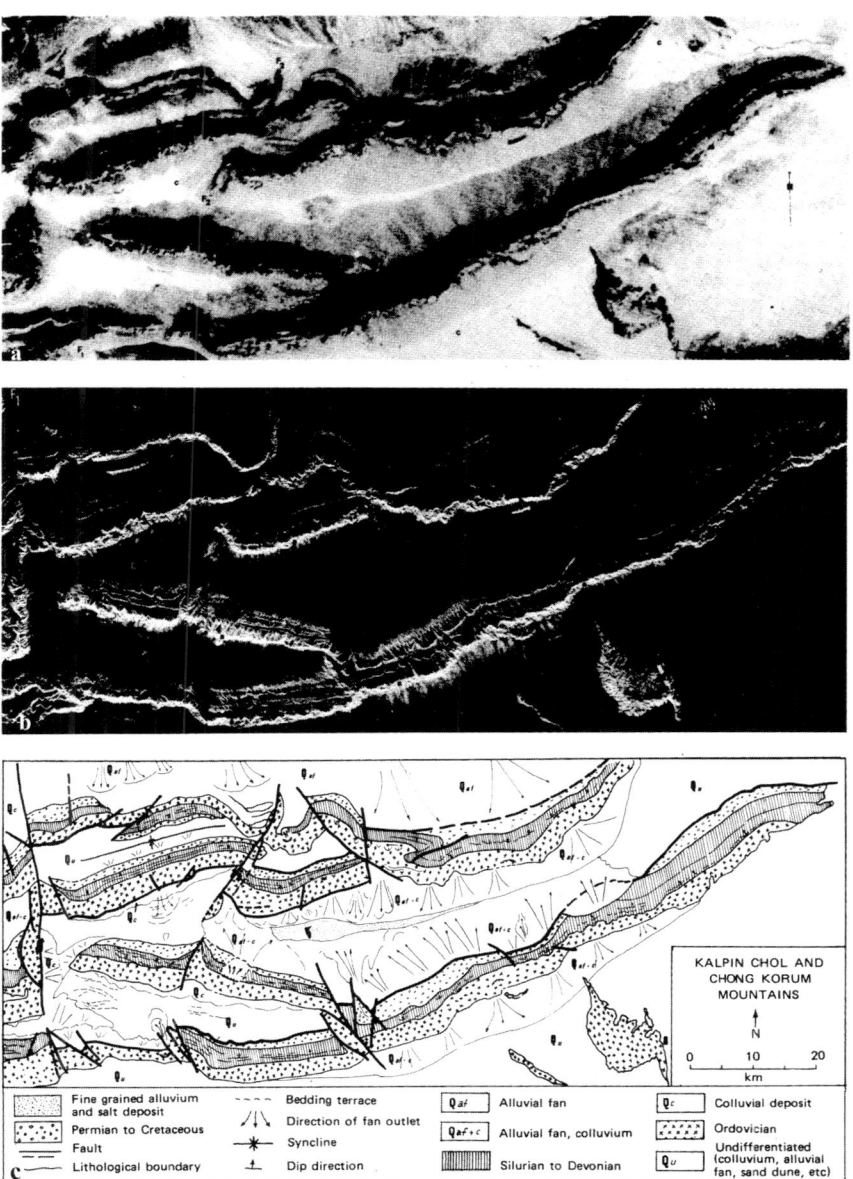

Fig. 16.27. a Landsat MSS (infrared) image of the Kalpin Chol and Chong Korum mountains. **b** The corresponding SIR-A imagery. Note that the various rock units exhibit greater contrasting characters on the SIR-A imagery and therefore, structural features such as folds and faults are better delineated on the SIR-A image than on the corresponding Landsat MSS image. **c** Interpretation map showing lithological units, folds and faults in this area. (**a–c** Woldai 1983)

yas to the south all along their E–W strike length. The Siwalik Group is severed by the Yamuna fault, being truncated and displaced (strike slip > 10 km). Large-scale drag effects of the order of a few kilometres are seen. Prior to the Landsat studies, the direction of displacement along this fault was considered to be right-lateral on the basis of field and aerial photographic interpretation (Rao et al. 1974). However, the Landsat image, with its synoptic view, clearly indicates the left-lateral direction of displacement along this fault (Gupta 1977b; Sharma 1977).

2. Structures enhanced by textural differences – the Kalpin Chol and Chong Korum Mountains, China. The Kalpin Chol and Chong Korum Mountains form the sourthern extension of the Chinese side of the Tien Shan ranges. The Landsat MSS and SIR-A coverages provide interesting information on the structure of this area (Woldai 1983, Fig. 16.27). The area includes a wide complex of rocks ranging in age from Precambrian to Cenozoic. The well-bedded sedimentary sequences, ranging in age from Ordovician to Permian, form the dominant rocks. These rocks generally trend E–W, and dip gently towards the north. The Landsat image shows predominantly shadows, owing to the fact that the beds are gently dipping northwards, and the solar illumination is from the SE. On the other hand, the SIR-A imagery exhibits a high radar response, since the beds are oriented such that they face the radar look direction (which is southwards). The fine gullies and surface roughness differences are picked up prominently on the SIR-A image. In general, the various important rock groups can be better delineated on the SIR-A image than on the corresponding Landsat image. Based on continuity of units, a number of folds and major/minor faults (lineaments) were identified which were not previously accounted for in any geological or tectonic map of this area (Woldai 1983).

3. Fault with vertical displacement – the Nagar-Pakar fault, Indo-Pak border. A prominent E–W trending fault occurs at the northern boundary of the Rann of Kutch, and is called the Nagar-Pakar fault (Fig.16.28). This fault runs for a few hundred kilometres' strike length, and limits the Kutch rift to the north (Biswas 1974). On the Landsat image, the fault forms an exceedingly conspicuous straight boundary where any evidence of lateral movement is lacking. The southern block is full of playas and salinas, and indicates frequent invasion by coastal waters, i.e. it has subsided. The northern block is covered with extensive desert dunes. A river (the Luni), entering this region from the east, takes a turn and is confined to the southern block, although in its natural course it appears that it ought to have been flowing on the northern block, again indicating that the northern block has relatively gone-up.

Several other examples of faults are given elsewhere (see e.g. Figs. 16.30, 16.34–37 and 16.93–99).

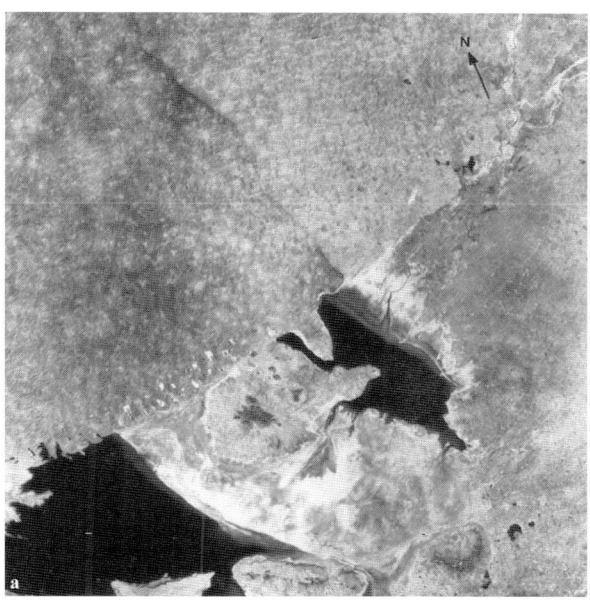
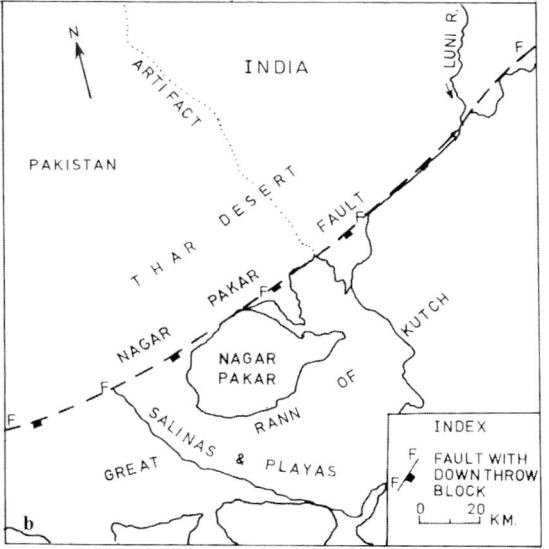

Fig. 16.28. a Landsat MSS7 image (infrared) showing the nearly E–W-trending prominent Nagar-Pakar fault. **b** Interpretation map. The block on the south of the fault is frequently invaded by coastal waters, playas and salinas and forms the downthrow side. The northern block is covered with dunes. The terrain is dry, bare of vegetation and quite uninhabited. The Precambrian inlier at Nagar Pakar (Pakistan) forms highlands. Also note the artificial Indo-Pak boundary running through the dunes

16.3.4 Neovolcanic Rift Zone

Divergent plate boundaries are frequently located under oceans as mid-oceanic ridges. Much interest is attached to these boundaries in the search for a proper understanding and validation of the existing concepts on plate tectonics. Deep-sea drilling programmes on some of the mid-oceanic ridges are a testimony to the scientific importance of these natural features.

The Atlantic mid-oceanic ridge is exposed in Iceland. A part of this neotectonic–neovolcanic rift zone has been covered by Seasat-SAR from two different but parallel orbits, located only 20 km apart. Compared to the sensor's altitude of about 800 km, the set of two views provides a moderate stereo effect (Fig. 16.29a). The images show that the region is marked by some distinct structural and morphological features. Figure 16.29b is a simplified geological map of the corresponding area.

Due to the rugged topography of the terrain and the low look angle of the SAR, the effects of layover and foreshortening are distinct, particularly in the stereo view. The hill slopes facing the radar beam appear bright and shortened, and those sloping away are dark and elongated. The scene is viewed at an angle of about 20° to the strike of the neovolcanic zone, marked by the general direction of faults and palagonite ridges. The palagonites seem to have piled up during subglacial fissure eruptions in the Upper Pleistocene. The typical rugged broken relief of the palagonite renders it easily mappable on the radar imagery.

Many of the significant faults can also be traced, especially those with a considerable downthrow on the NW (i.e. towards the illuminating radar pulse), which are distinct on the radar images. On the other hand, faults with the downthrown block away from the sensor are subdued.

The Holocence (not glacially eroded) lava flows, which have come mainly from the prominent shield volcano in the upper middle part of the scene, show up as a medium gray toned unit of generally uniform fine-grained texture. The effect of differing surface roughness on the radar response is clearly demonstrated in Lake Thingvallavatn. On the right-hand image, the stronger wind has ruffled the water surface, which results in stronger backscatter, making it difficult to distinguish the lake from the adjacent lava flows.

16.3.5 Lineaments

1. Definition and terminology. The term lineament has been extensively used recently, and often with differing shades of meaning. The photo-linears, i.e. linear alignments of features on photographs and images, are one of the most obvious features on high-altitude aerial and space images, and therefore the use of the term lineament has proliferated in remote sensing geology literature in recent years.

This term has also been applied to imply alignment of different geological features, such as (1) shear zones/faults; (2) rift valleys; (3) truncation of outcrops; (4) fold axial traces; (5) joints and fracture traces; (6) alignment of fissures, pipes, dykes and plutons; (7) linear trends due to lithological layering; (8) lines of significant sedimentary facies change; (9) alignment of streams and valleys; (10)

Structure

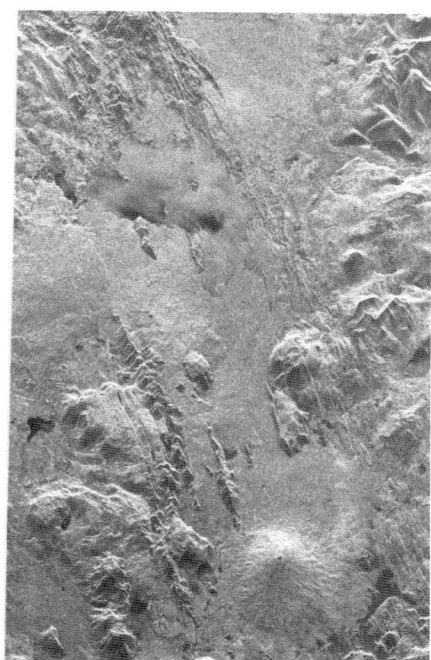

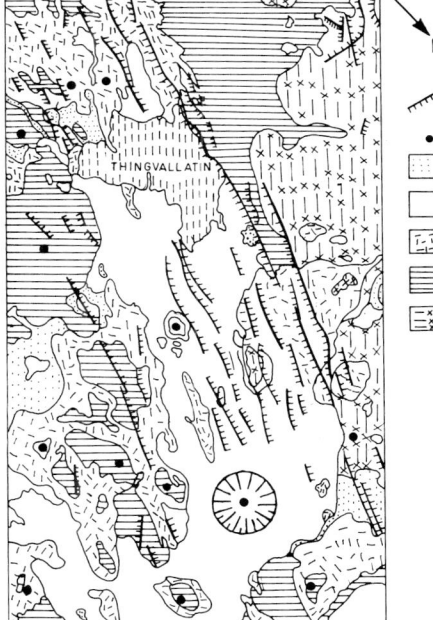

Fig. 16.29. Stereo pair acquired by Seasat-SAR from two adjacent and parallel orbits. It shows the neovolcanic rift zone in SW Iceland; the rift zone is characterized by numerous parallel faults, palagonite ridges, and Holocene lava flows. On the right image, note the disturbed water surface of Lake Thingvallavatn, which has led to a strong backscatter. A corresponding geological interpretation map of the area is also shown. (SAR images processed by DLR; interpretation map courtesy of K. Arnason)

Fig. 16.30. Mapping of various types of lineaments on IRS-1C LISS-III image of Cuddapah region, India. (Courtesy of D.P. Rao, A. Bhattacharya and P.R. Reddy)

topographic alignments – subsidences and ridges; (11) alignment of oil and gas fields; (12) occurrence of geysers, fumaroles and springs along a line; (13) linear features seen on gravity, magnetic and other geophysical data; (14) vegetation alignments; (15) soil tonal changes etc.; and (16) natural limits of distribution of certain features of the Earth's surface.

Hobbs in 1904 first used the term lineament to define a "significant line of landscape which reveals the hidden architecture of rock basement". O' Leary et al. (1976) reviewed the usage of this term and defined lineament essentially in a geomorphological sense as "a mappable simple or composite linear feature of a surface whose parts are aligned in a rectilinear or slightly curvilinear relationship and which differs distinctly from the pattern of the adjacent features and presumably reflects a sub-surface phenomenon". Hence, this category includes all structural alignments, topographical alignments, natural vegetation linears and lithological boundaries etc., which are very likely to be the surface expression of buried structures. This definition seems to be the most practical in the context of remote sensing image interpretation.

2. Scale and manifestation. The manifestation of a lineament is dependent on the scale of observation and dimensions involved. Lineaments of a certain dimension and character may be more clear on a particular scale, for which reason tectonic features of the size of hundreds of kilometres need to be studied on smaller-scale images.

Lineaments occur as straight, curvilinear, parallel or en-echelon features (Fig. 16.30). Generally, lineaments are related to fracture systems, discontinuity planes, fault planes and shear zones in rocks. The term also includes fracture traces described from aerial photographic interpretation. Dykes and veins also appear as lineaments. The pattern of a lineament is important on the image; lineaments with straighter alignments indicate steeply dipping surfaces; by implication, they are likely to extend deeper below the ground surface.

At times, the relative sense of movement along the fault/lineament may be apparent on the image; for example, displacement across some of the lineament features is clearly seen in figure 16.30. Stereo view, if available, is generally useful in giving a better understanding of the lineament's features (Fig. 16.31).

On a certain photo or image, both major and minor lineaments are invariably observed (see Fig. 16.33a). The major lineaments may correspond to important shear zones, faults, rift zones and major tectonic structures or boundaries. On the other hand, minor lineaments or micro-lineaments may correspond to relatively minor faults, or joints, fractures, bedding traces etc. These are expressed as soil–tonal changes, vegetation alignments, springs, gaps in ridges, aligned surface sags and depressions, and impart the textural character in a larger scene.

The ground element corresponding to a lineament would depend on the scale of the remote sensing data. On regional scales (say 1:250,000), lineament features may be more than ca. 5 km in length, representing long valleys and complex fractured zones. On larger scales (e.g. aerial photographs or IRS-1C/D PAN images, say 1:20,000 scale), shorter, local drainage features, individual fracture traces etc. may appear as lineament features.

3. Mapping of lineaments. Mapping of lineaments can be done on all types of remote sensing images: stereo panchromatic photographs, multispectral and thermal-IR images and SLAR images. The panchromatic, NIR SWIR, and thermal-IR images contain near-surface information, whereas SLAR images may provide limited depth penetration (of the order of a few metres at best) in arid conditions. Manifestation of lineaments is related to ground conditions and sensor spectral band.

Lineaments can be mapped on simple data products, as well as on processed/enhanced images. Further, different types of digital techniques, aimed at enhancing linear features, can also be applied in the form of isotropic and anisotropic filters. Anisotropic filters enhance linear features in certain preferred directions;

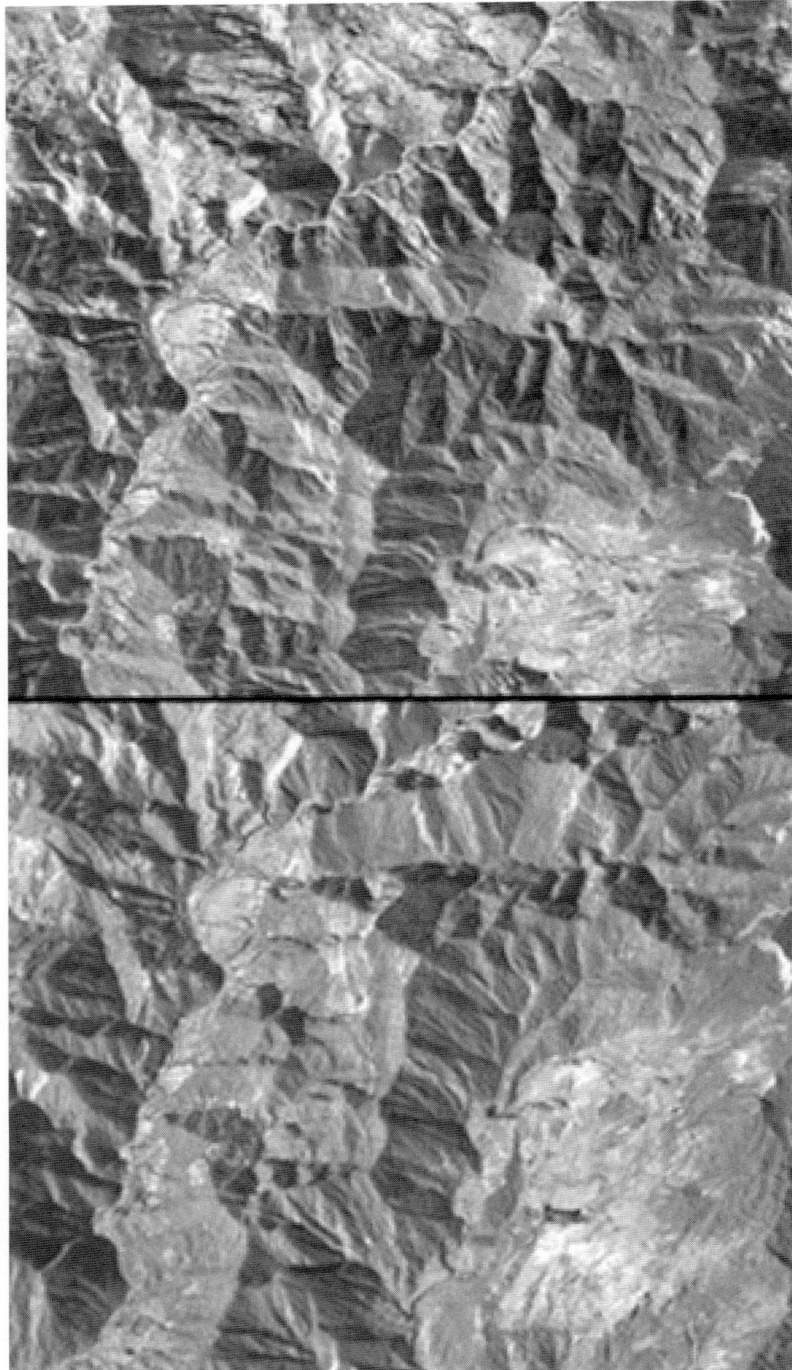

Fig. 16.31. Stereo pair (IRS-1D PAN) of a region in the Himalayas. The metasedimentary rocks possess E–W strike. Note the lineament followed by a river, running nearly N–S transverse to the general strike of rocks, with a likely right-lateral sense of displacement

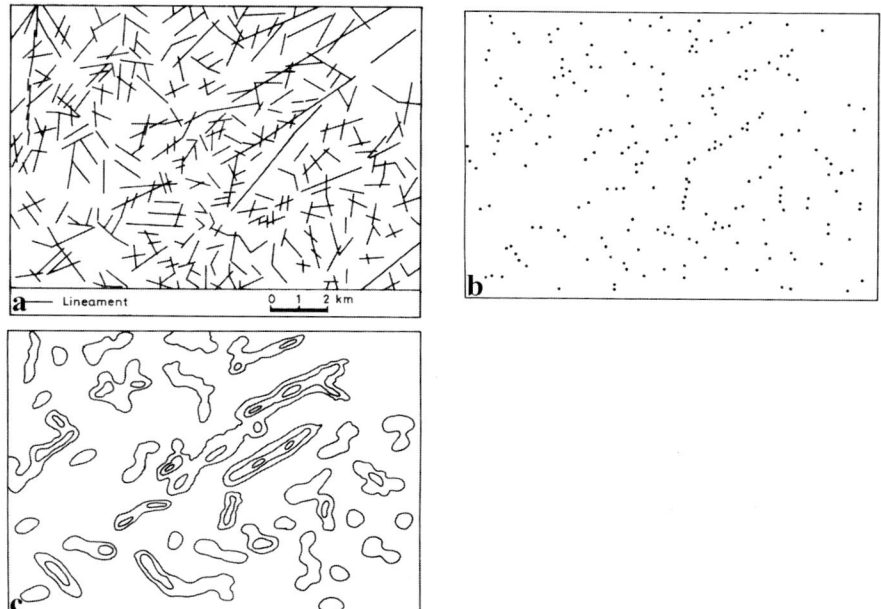

Fig. 16.32. a Lineament interpretation map of the area shown in figure 16.78; **b** lineament intersection point diagram; **c** contour diagram from **b**

however, the related artifacts are a common problem and render the interpretation of such enhanced products quite difficult.

4. Visual vs. digital interpretation. On an image, the lineaments can be easily identified by visual interpretation using tone, colour, texture, pattern, association etc., i.e. the elements of photo interpretation. Alternatively, automatic digital techniques of edge detection can also be applied for lineament detection. Edge detection techniques, with numerous possible variations, lead to many artifacts or non-meaningful linears, which may crop up due to illumination, topography, shadows etc. Therefore, visual interpretation technique is generally preferred and extensively applied.

Visual interpretation of lineaments involves some degree of subjectivity, i.e. the results may differ from person to person (Siegal 1977; Burns and Brown 1978; Wise 1982; Moore and Waltz 1983). Putting information in vector form, a method considering the location, direction and length characteristics of lineaments was developed by Huntington and Raiche (1978) in order to compare lineament interpretations of different workers. The questions of reproducibility of geological lineaments and their measurement by co-efficient of association (i.e. a measure of the number of times a given pixel is classified as being on a lineament) have been discussed by Burns et al. (1976).

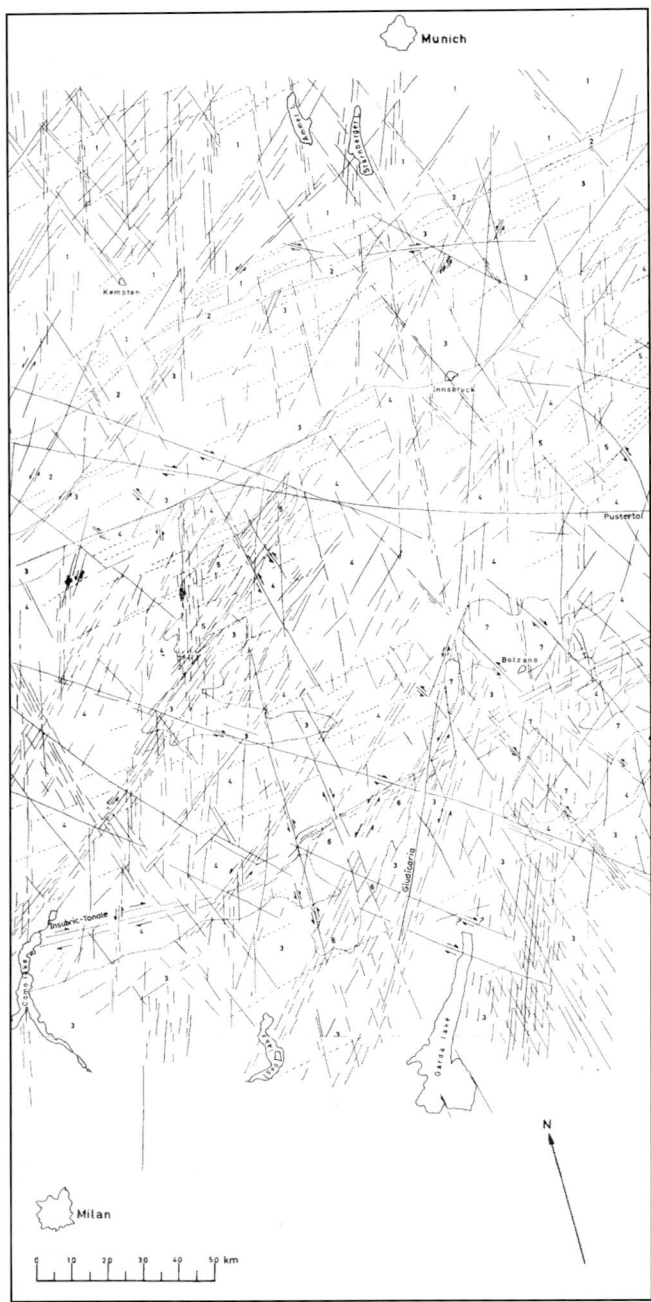

Fig. 16.33a. Lineament map of the Eastern Alps as interpreted from Landsat MSS images. (Gupta 1977a)

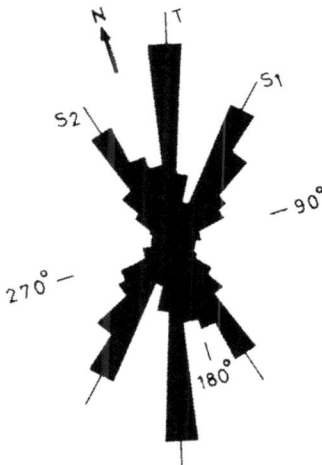

Fig. 16.33b. Rose diagram of lineaments in figure 16.33a. Clearly three major directions stand out: (1) N45–225° (S_1); (2) N15–195° (T); and (3) N345–165° (S_2). (Gupta 1977a)

5. Statistical analysis. Lineaments mapped on a remote sensing image possess spatial variation in trend, frequency and length. For statistical analysis, lineaments are often grouped in ranges of angles (commonly 10° interval) in plan. The lineament data can be processed manually, which is quite tedious, or alternatively the lineament map could be digitized. The lineament map could be mounted on a digitizing tablet, linked to a computer, and the end points of each lineament could be successively read using a hand-held movable pen or cursor, to digitize each lineament. Once the data of all the lineaments was collected, it could be reformatted and processed to provide the statistical information.

There are a number of ways to assess the statistical distribution of lineaments in an area. One is by considering the number of lineaments per unit area; the second, by measuring the total length of lineaments per unit area; and the third, by counting the number of lineament intersections per unit area. The method of lineament intersection per unit area (density) is generally faster and more convenient (Fig. 16.32). The intersections of two (or more) lineaments are plotted as points. The number of points falling within a specified grid area is counted. The data are contoured to give the lineament intersection density contour map. This gives zones of different degrees of fracturing in the area. Further, the lineament-intersection density maps can be rasterized and converted into lineament-intersection density images (see e.g. Fig. 16.40 later). In the same manner, it is possible to create lineament-length images and lineament-number images. Such processed data products provide statistical information on the distribution of fractured zones.

6. Discriminating between different genetic types of lineaments. In rocks, fractures or joints originate in different ways. There are two main types of fractures distinguished: (a) shear fractures, which originate due to shear failure in rocks, and (b) dilational fractures, which are of tensile (extensional) origin. A third 'hybrid' type exhibit features of both shear and dilational origin (Price and Cosgrove 1990). For some applications, it is important to distinguish between the genetic types of fractures/ lineaments; for example, the dilational type is more open and more productive for groundwater. The genetic distinction between the various lineaments can sometimes be made on remote sensing data, on the basis of two considerations: (a) relative movement along an individual discontinuity, and (b) orientation and statistical distribution of lineaments. It is always desirable that both these types of evidence are mutually supportive. Sometimes, relative movements along lineaments can be seen to indicate faults and shear fractures (see Fig. 16.30); in contrast, lineaments related to dilational fractures do not show any relative displacement. Further, in some areas statistical analysis of lineament trends could distinguish between fractures of shear origin and those of dilational origin, in a generalized way. For example, figure 16.33a presents a lineament interpretation map of a Munich–Milan section in the eastern Alps. A large number of lineaments are seen; their statistical trends are given by the rose diagram in figure 16.33b. Based on mutual angular relationships, it can be inferred that S_1 and S_2 sets are shear fractures and the T set is of a dilational type.

Applications

Lineament studies have found applications in numerous fields of Earth sciences – global tectonic studies, delineation of major structural units and contacts, analysis of structural deformation patterns and identification of geological boundaries/ domains/provinces etc. Further, lineaments are zones of deformation and fracturing, which implies that they are zones of higher secondary porosity. As these zones become significant channel-ways for migration of fluids, lineaments constitute important guides for exploration of mineral deposits, petroleum prospects, groundwater etc. A few examples to illustrate the types and scope of lineament studies are given below.

1. Lineament corresponding to the subduction zone – the Indus suture line, Himalayas. The Himalayas constitute one of the youngest and most active mountain systems on the Earth, formed as a result of the collision of two crustal plates – the Eurasian plate on the north and the Indian plate on the south. The collision zone is named the Indus suture zone, after the Indus river, which follows the inter-plate boundary for quite a distance. The Landsat image (Fig. 16.34) provides a regional perspective of the tectonic feature, where it is seen to occur as a major NW–SE-trending lineament zone that can be traced as a straight line for more than 100 km. The dark tones in the image along the Indus suture zone are due to an ultramafic ophiolite suite of rocks emplaced along the subduction zone.

2. Lineaments associated with tectonomagmatic activity – Um-Ngot lineament, Shillong Plateau, India. Deep-seated tectonic and magmatic activities often leave their footprints on the terrain. Due to their large dimensions (a few tens to hundreds of kilometres), such features are best identified on satellite remote sensing

Structure 469

Fig. 16.34. The NW–SE-trending prominent lineament is the Indus suture zone (ISZ), which forms the subduction zone in the India–Eurasia plate collision and constitutes a major tectonic boundary in the Himalayas. It extends for more than a hundred kilometres as a straight line in the rugged terrain of the Himalayas. The Indus river follows this zone for a considerable length. The dark-toned bodies are ophiolitic intrusives along the suture line. (Landsat MSS infrared-band image)

images, especially with multispectral data, which can pick up subtle differences in spectral characters of surface features.

An interesting example of this type occurs in the Um-Ngot lineament zone, Shillong Plateau, which was first identified on Landsat images (Gupta and Sen 1988). The Shillong Plateau consists mostly of an Archean gneissic complex and Proterozoic meta-sedimentaries and is marked by granitic intrusives and a deformation pattern characteristic of schistose–gneissic rocks. A unique and striking feature is the occurrence of a major lineament zone, called the Um-Ngot lineament (Fig. 16.35). It is about 50 km long and 5–10 km wide, cuts across the general strike of the rocks, and is evidently post-Precambrian. The Precambrian trends are bent or truncated against this tectonic feature. The lineament zone contains the Sung Valley alkaline-ultramafic-carbonatite complex (Fig. 16.35b,c), described from the Shillong plateau and considered to be of Cretaceous age (Chattopadhyay and Hashimi 1984; Krishnamurthy 1985). The alkaline ultramafic complex is marked by relatively poor vegetation (Fig. 16.35b) in a terrain of generally fairly good vegetation (pine trees). The vegetation anomaly is evidently due to the pres-

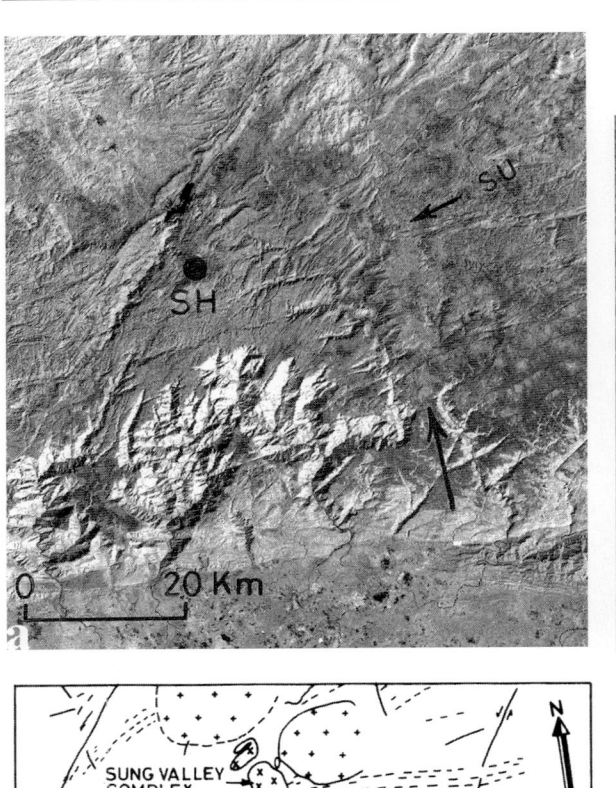
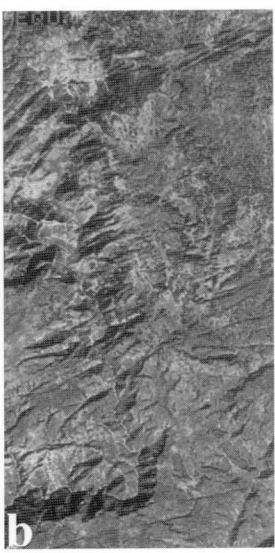
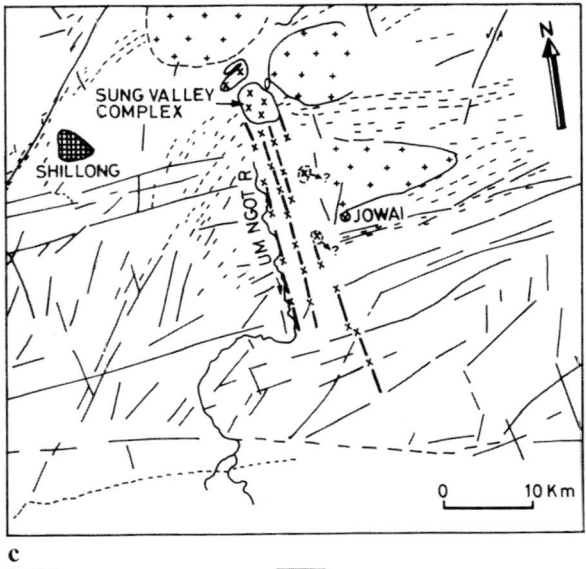

Fig. 16.35. a Landsat MSS infrared image of the Shillong plateau showing the presence of a major N–S-trending lineament zone (indicated by *arrow*). (*SH* Shillong). b Part of the same area covered by IRS-LISS-II sensor. The circular feature in the north is the Sung valley carbonatite intrusive, bare of vegetation. c Lineament tectonic interpretation map. (a,c Gupta and Sen 1988)

ence of toxic elements in the ultramafic suite of rocks. A few other circular to semi-circular features are also seen in the lineament zone. It is inferred that the N–S-trending lineament zone and the associated carbonatite suite of rocks formed as a result of up-arching of the mantle plume, and are genetically related to the Ninetyeast Ridge in the Indian Ocean (Gupta and Sen 1988).

Several examples of such lineaments have been described in the literature. The neovolcanic rift zone in Iceland (see Fig. 16.29) could also be considered as an example of this type.

3. Lineament zone with lateral displacement – Kun Lun–Tien Shan Fault, China. At a particular scale, some of the tectonic features may appear sharp, well-defined, although very extensive (e.g. the Indus suture line, Fig.16.34). On the other hand, other tectonic features may appear as a wide zone, characterized by numerous parallel en-echelon to overlapping lineaments, with a similar sense of displacement, spread over a wide zone, e.g. the Kun Lun–Tien Shan fault (Fig. 16.36).

The Pamirs form a central knot of the great Asian mountain system, from which various ranges such as the Himalayas, Karakorum, Kun Lun, Tien Shan and Hindukush splay out in various directions. The Tarim basin is enclosed between the Tien Shan and Kun Lun ranges. The Kun Lun ranges are separated from the Tarim depression by the Kun Lun–Tien Shan fault (Desio 1974).

In the area shown in Fig. 16.36a, the Kun Lun range extends in a NW–SE to NNW–SSE direction. The dislocation zone separating the Kun Lun range from the Tarim basin is very conspicuous as a zone of numerous overlapping, parallel lineaments with a strong right-lateral sense of displacement (Fig. 16.36b). The lineament zone is about 20–30 km wide and is nearly vertical, indicating its deep-seated origin.

There are a number of ways to assess the statistical distribution of lineaments in an area. One is by considering the number of lineaments per unit area; the second, by measuring the total length of lineaments per unit area; and the third, by counting the number of lineament intersections per unit area. The method of lineament-intersection density is generally faster and more convenient. Intersections of two (or more) lineaments are plotted as points, and the number of points falling within a specified grid area is counted and contoured, to give the lineament-intersection density contour map (e.g. Fig. 16.36c). This shows the location of maximum tectonic disturbance in the lineament zone.

4. Delineation of buried lineament structures – Hoggar mountains, Algeria. The southern part of Algeria is dominated by arid–hyperarid climatic conditions. Figure 16.37a,b show a Metric Camera photograph (panchromatic) and an SIR-A image, respectively, of a part in the southern Hoggar Mountains, Algerian desert. The terrain is typically arid with dry channels, bare slopes and sands. Morphologically, the area depicted in the photograph is characterized by hills trending NE–SW and a flat plain covered with sands, located between the two dominant ridges. The panchromatic photograph shows the structural trend of the formations (NE–SW) and the presence of several lineaments in the hilly terrain; however, lit-

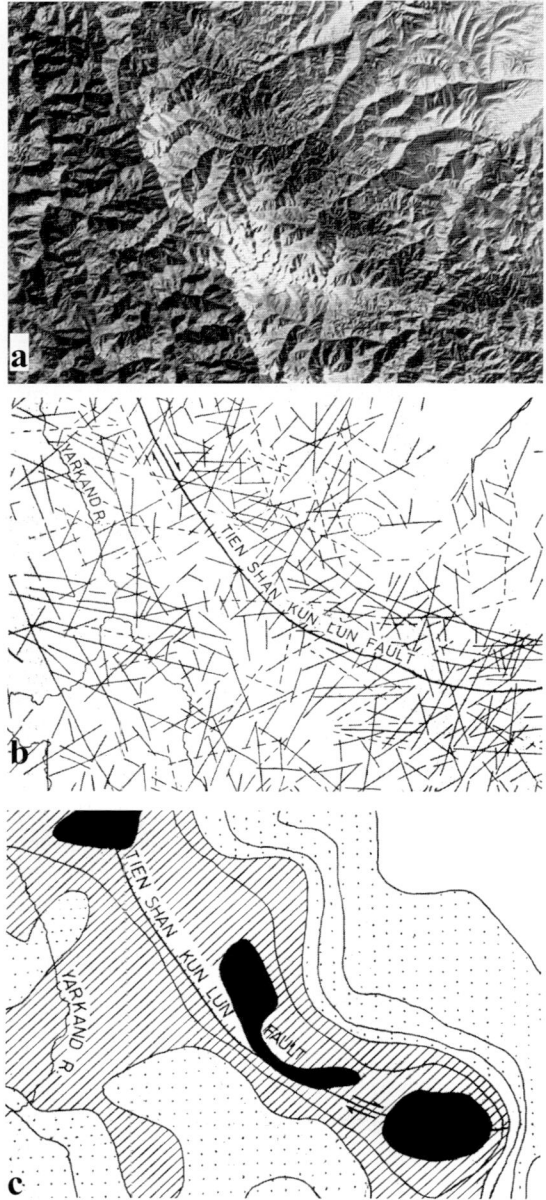

Fig. 16.36. a Landsat MSS (infrared) image of the border area between the Tarim basin and Kun Lun ranges. Development of numerous lineaments with right-lateral sense of movement is conspicuous, and the movement is distributed over a large zone. **b** Lineament–tectonic interpretation map of the above area. **c** Contour map showing the lineament intersection density and zones of differing tectonic disturbance

Structure 473

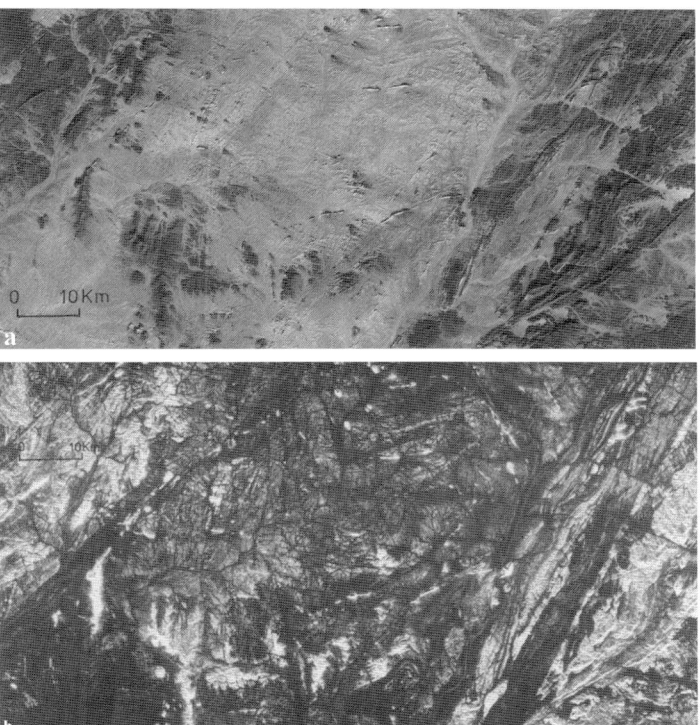

Fig. 16.37a,b. Comparison of a Metric Camera photograph (panchromatic) and SIR-A image of the southern Hogger mountains in Algeria. **a** The Metric Camera photograph shows structural details in the mountainous areas; however, no information is provided on the flat plain covered with dry sand. **b** The corresponding SIR-A image; note the presence of numerous linear structures buried under the sand cover, brought out on the radar image. (European Space Agency)

tle information is obtained in the sand-covered flat area. The corresponding SIR-A image shows that the radar return is influenced, collectively, by topographical relief, overlying sand cover, sub-surface topography of the bedrock and possibly variation in soil moisture. Due to the depth-penetration ability of the active microwave sensors in arid–hyperarid conditions, numerous buried structures can be delineated on the radar imagery.

5. Lineament corresponding to inferred fault separating zones of differing metamorphism – the Banas lineament, Aravalli ranges, India. The significance of some lineaments may not be clear at a first glance; our lack of understanding is responsible in part for the scepticism associated with the lineament type of work. Although lineament interpretations are somewhat subjective, not all lineaments may be spurious, and only detailed ground data of, at times, different types, may reveal the true significance of lineaments. Further, it may be recalled that it is relatively easier to identify steeply dipping faults marked by strike-slip displacement than many other discontinuities which may be low-dipping and/or may represent facies or lithologic differences.

The Aravalli mountain ranges constitute the north-western part of the Indian penisular shield, and extend for about 600 km with a general strike of NNE–SSW. The rocks have undergone polyphase metamorphism and deformation (Heron 1953), which has led to a complex tectonic deformation pattern.

Figure 16.38a shows a Landsat MSS4 (infrared) image of part of the Aravalli hills. A peculiar and conspicuous feature is the nearly E–W-trending lineament running transverse-oblique to the general strike of the rocks and extending for a strike length of about 100 km (Bharktya and Gupta 1983). The Banas River follows this lineament for some distance in the north-east. The lineament is marked by strong tonal and textural alignments, but there is little, if any, evidence of strike-slip movement. Figure 16.38b shows the geological map of the area based on a study of metamorphism (Sharma 1988). On comparison, it is evident that the lineament identified on the Landsat image corresponds to the inferred fault separating regions of sharply differing metamorphism. On the two sides of the inferred fault, the broad metamorphic characters of the rocks are as follows (Sharma 1988):

Greenschist facies
1. Greenschist facies followed by thermal metamorphism in places
2. Temperature = 450° ± 50°C
3. Pressure = 3–4 kbar
4. Deformation D_1, D_2, D_3

Amphibolite facies
1. Upper amphibolite facies; local partial melting
2. Temperature = 700–900°C
3. Pressure = 7–8 kbar
4. Deformation D_1, D_2, D_3

It is obvious that the inferred fault separates regions of vastly different levels and types of metamorphism. Assuming a high geothermal gradient of about 50–60°C km^{-1}, the difference in depth comes to about 5–6 km. It appears that the lineament is the manifestation of a major structural tectonic discontinuity, with dominantly dip-slip movement, developed at depth during the dying phase of tectonic deformation and metamorphism.

6. *Micro-lineament pattern related to lithology – Eastern Alps.* The term 'microlineament' is used here for the very small linear features frequently observed on an image. As mentioned earlier, micro-lineaments on Landsat-type images correspond to relatively minor faults, fractures, joints and bedding traces in the rock. These are expressed as alignments of local depressions, ponds, gaps in ridges, springs, soil and tonal changes, vegetation etc. and are responsible for the textural characteristics in a larger scene.

Figure 16.39a is a Landsat MSS4 (infrared) image of a part of the Eastern Alps, and Fig. 16.39b shows a lineament interpretation map of part of the above image. The general strike of the formations is ENE–WSW. The distribution of length (dimension) and number (frequency), i.e. texture, of the micro-lineaments exhibits spatial variations, so that four broad zones can be distinguished in this area. On

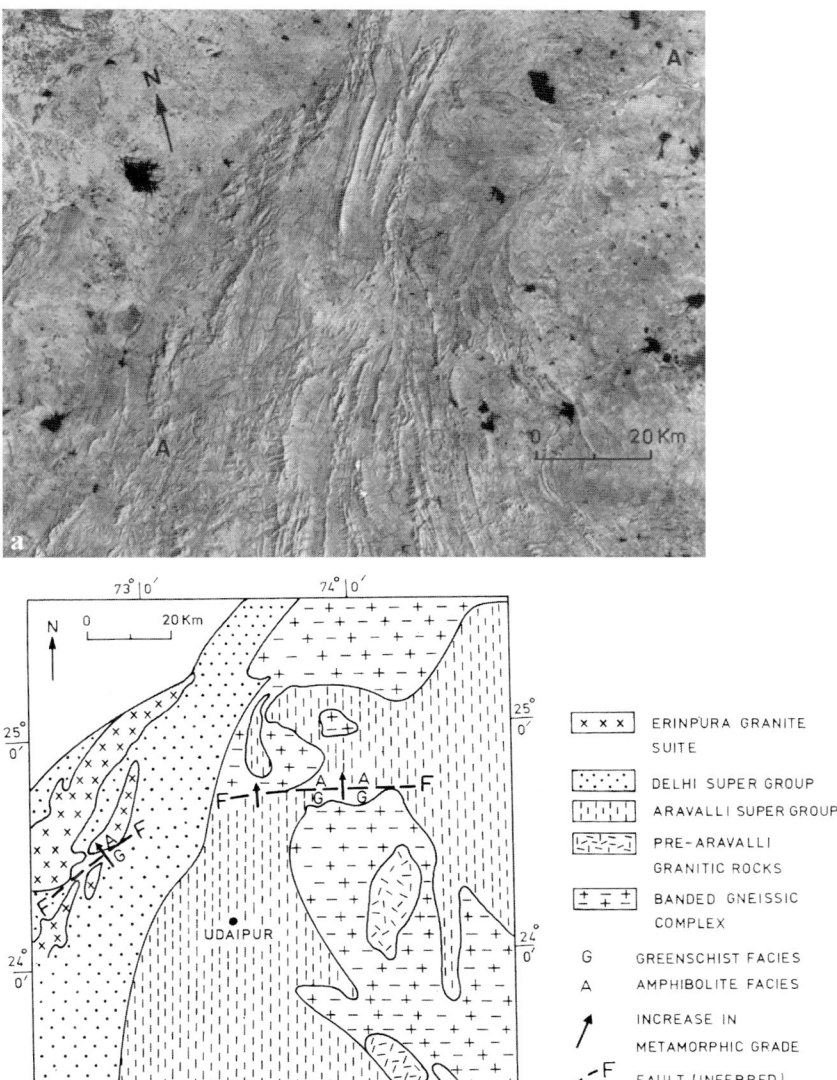

Fig. 16.38. a Landsat MSS (infrared) image of part of the Aravalli hills, India. The area comprises metamorphic rocks generally trending NNE–SSE. Note the presence of a major transverse-to-oblique, nearly ENE–WSW-trending lineament (marked A-A). **b** Geological map of the area prepared independently by Sharma (1988); a fault is inferred on the basis of sharply differing metamorphic grades. A comparison indicates that the lineament corresponds to the inferred fault

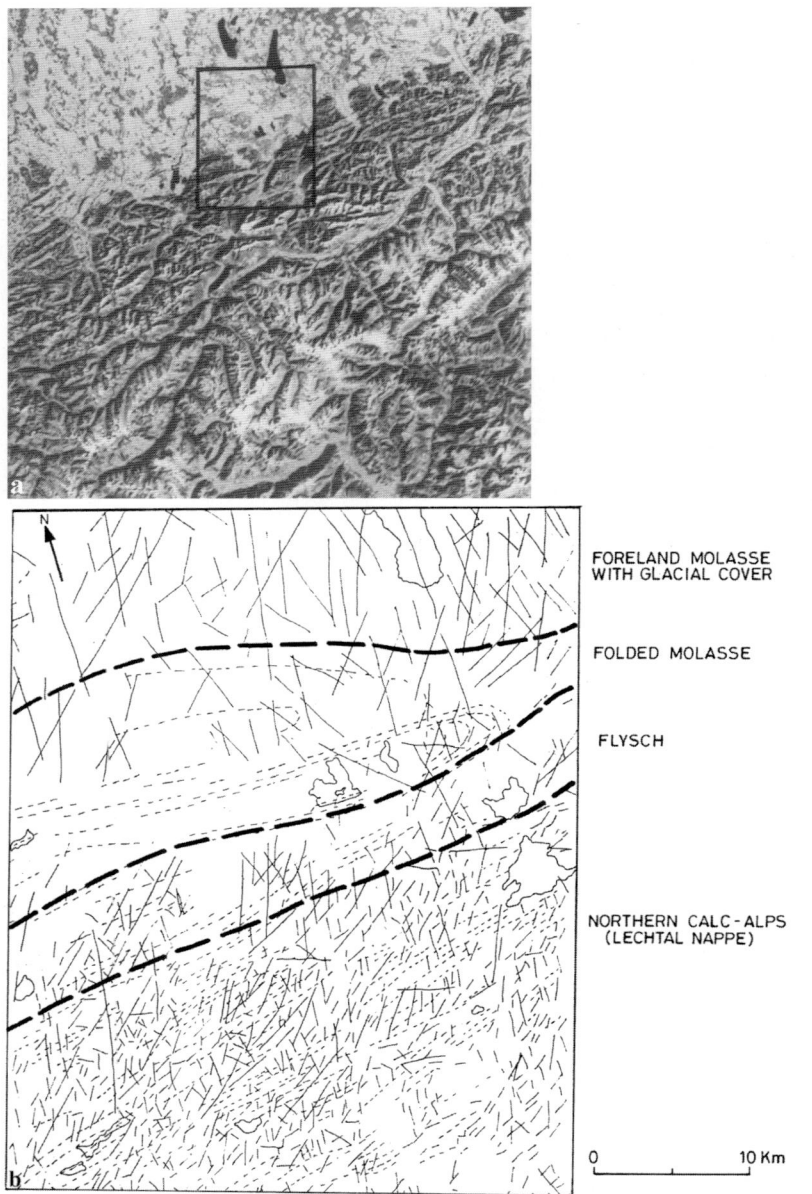

Fig. 16.39. a Landsat MSS (infrared) image of part of Eastern Alps. **b** Micro-lineament map corresponding to boxed area in **a**. The micro-lineaments are related to minor faults, fractures and joints etc. Based on their distribution (number, length and density), four broad zones can be distinguished. This indicates that the texture imparted by micro-lineaments can be used for distinguishing broad lithological groups in a certain terrain

comparison with a published geological map, these zones are found to correspond to: (a) foreland molasse with glacial cover, (b) folded molasse, (c) flysch and (d) northern calc-Alps (Lechtal nappe). It is well known that the development of joints and fractures, including their frequency and extent, is linked to the competency of the rock. Different rocks exhibit a different degree of jointing under the same stress conditions. Therefore, the micro-lineament pattern can be linked to the lithological characteristics – at least for distinguishing different lithological groups – in a particular terrain.

7. Micro-lineament density image – Iberian pyrite belt, Spain. The geological significance and manifestation of micro-lineaments on aerial and space images have been discussed above. During an exploration study for polymetallic mineralization in the area north-west of Tharsis, Iberian belt, Spain, Volk (1985) carried out an interesting image manipulation of the micro-lineaments mapped on aerial panchromatic photographs.

The mineralization in the Iberian pyrite belt is considered to be synsedimentary, dominantly strata-bound, in a volcanic sedimentary complex sequence of Lower Carboniferous age. The hanging wall consists of schists and quartzites of Upper Devoninan age, whereas the footwall consists of schists–graywacke of Lower to Upper Carboniferous age. The formations are generally well bedded and foliated. A large number of micro-lineaments are developed parallel to the strike, due to contrasting lithologies; in comparison to this, transverse lineaments are few. Since the mineralization is strata-bound, structures parallel to the strike possess greater significance for exploration. The micro-lineaments were mapped on aerial photographs and those oriented within 15° with respect to the general strike trend were separated from the rest (Fig.16.40a). The above map was rasterized to give a contour map showing lineament density (Fig. 16.40b). This was converted into a digital image (Fig. 16.40c). The lineament-density image was registered and combined with the TM5 gradient image (pseudoplast) (Fig. 16.40d). The resulting image shows the various important lithological contacts and minute structural details. (The techniques of constructing images from contour maps and registration of multidata sets are discussed in Sect. 15.4.)

16.3.6 Circular Features

Circular features often preferentially catch the geologist's eye on remote sensing images. This is due to the special importance attached to circular features in geology. A special subtractive box filtering, succeeded by a histogram equalization stretch can be used to enhance circular features on remote sensing data (Thomas et al. 1981).

Circular–quasicircular features may be associated with: (1) intrusives; (2) structural domes and basins; (3) volcanoes; (4) tensional ring fracturing; and (5) meteorites. Their salient characteristics are as follows:

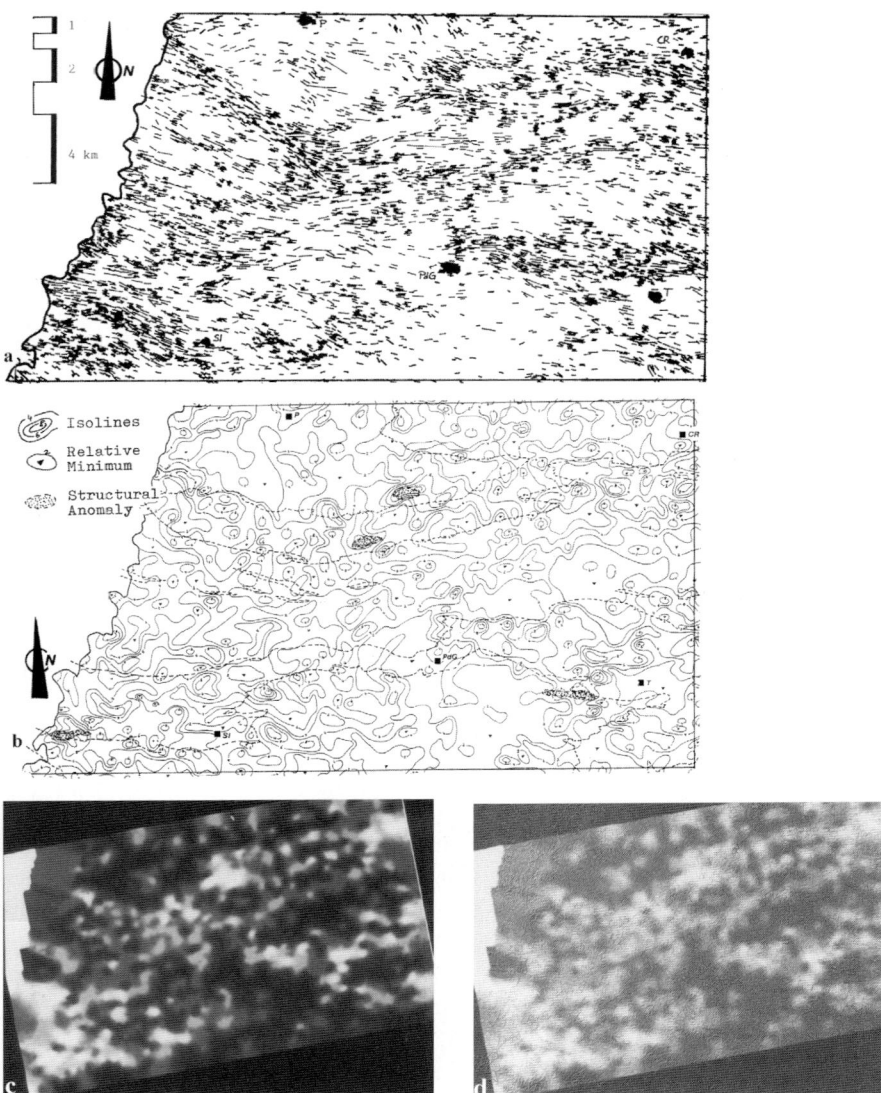

Fig. 16.40. a Micro-lineaments, nearly parallel to the strike trend of the rocks, based on aerial photographic interpretation (central SW Iberian pyrite belt, Spain). **b** Lineament density contour map from **a** (contours indicate the lineaments per raster cell). **c** Lineament density image after digitization and interpolation of **b** (high density = white; low density = black). **d** Additive combination of **c** and the TM5 gradient image (pseudoplast). The various important lithological contacts and minute structural details are well depicted on this image. (**a–d** Courtesy of P. Volk)

Structure

Fig. 16.41. Impact craters in Sahara desert, northern Chad. The concentric ring structure left of centre is the Aorounga impact crater, about 17 km in diameter. A possible second crater, similar in size to the main structure, appears as a circular trough in the centre of the image. The dark streaks are deposits of wind-blown sand. (Printed black-and-white from colour composite, SIR-C/X-SAR image) (Courtesy of NASA/JPL)

1. *Intrusive* circular to near-circular bodies are commonplace in metamorphic belts. These may exhibit a 'shoving-aside' pattern of the host rock and truncate the latter (see Figs. 16.25, 16.35, 16.38, 16.47).
2. *Structural domes and basins* formed by cross-folding can be recognized from structural patterns (see e.g. Figs. 16.21–16.25).
3. *Volcanoes* have prominent morphological expressions and frequently occur in clusters (see e.g. Figs. 16.3, 16.29, 16.49, 16.50).
4. *Ring-like fractures* due to tensional forces may also lead to circular features on images; their genetic type must be identified from the tectonic setting of the area.
5. *Meteoritic impact craters* are marked by random distribution, variability in size and form, but regularity in morphology (Gold 1980), and lack any type of tectonic association (Fig. 16.41). Field evidence of ultra-high pressure or shock metamorphism, shatter cones and impact melt may be associated. Remote sensing data have shown that there are many more circular structures possibly caused by meteoritic impacts on the Earth than was earlier believed.

480   Chapter 16: Geological Applications

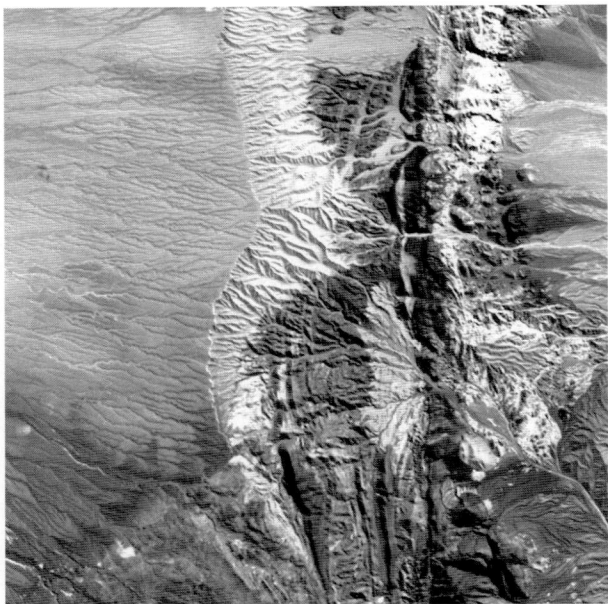

Fig. 16.42. Angular unconformity, Chile Altiplano. On the right side of the image are Cretaceous sediments dipping at an angle of about 50° eastward. On the left side occur flat-lying volcanic pyroclastic deposits. (ASTER image, printed black-and-white from colour composite) (Courtesy of NASA/GSFC/MITI/ERSDAC/JAROS, and US/Japan ASTER Science Team)

16.3.7 Intrusives

Igneous intrusions may occur in a variety of forms and dimensions, such as batholiths, laccoliths, lopoliths, sills, dykes, plugs etc. These are identified on the basis of their shape, form, relations with host rocks and dimensions. Examples are given later in figures 16.47 and 16.48. Salt domes are generally massive, lack bedding and truncate existing structural features, and are quite prominent on remote sensing data products (see Fig. 16.25).

16.3.8 Unconformity

Identification of unconformity on remote sensing data is based mostly on indirect evidence, such as truncation of lithological units and structural features or differences in degree of deformation. The expression of an angular unconformity (Fig. 16.42) may be quite similar to that of a low-angle fault.

16.4 Lithology

There are two main approaches to lithological/mineralogical mapping from remote sensing data: (a) mapping of broad-scale lithologic units, and (b) identification of mineral assemblages (including quantification of specific minerals).

1. *Mapping of broad-scale lithologic units* is based primarily on the principles and techniques of photo interpretation. This method employs mainly panchromatic aerial stereo photographs. Various multispectral satellite data and radar images can also be used in the same way at appropriate scales.

2. *Identification of mineral assemblages* utilizes the existence of characteristic absorption bands to help recognize specific mineral assemblages, e.g. clays, carbonates, Fe-O minerals etc. Multispectral and hyperspectral sensor data are required for this type of application. In some cases, detailed analysis of absorption characteristics (absorption band depth etc.) of data derived from hyperspectral sensors can help estimate the quantity of specific minerals.

As such, there is no sharp boundary between data inputs for the above two approaches. For example, aerial photographic and Landsat MSS/TM data have long been used to map broad lithologic units. Further, TM-type data is also used to identify specific minerals such as Fe-O, clays etc. TIMS multispectral data and hyperspectral data have been used to recognize the existence of specific mineral groups. Further, the analysis of TIMS thermal-IR data and hyperspectral data has enabled quantification of mineral contents in certain cases.

As hyperspectral sensor data are of a specific nature and have rather limited availability, their interpretation and applications are included in Chapter 11. In this section, we deal with panchromatic and multispectral data applications for lithologic–mineralogic interpretations (although this segmentation is just for convenience).

16.4.1 Mapping of Broad-Scale Lithologic Units – General

Mapping of broad-scale lithologic units can be carried out on panchromatic and multispectral satellite sensor data and radar data. Broad lithological information is deduced from a number of parameters observed on remote sensing images, viz. (1) general geologic setting, (2) weathering and landform, (3) drainage, (4) structural features, (5) soil, (6) vegetation and (7) spectral characteristics. The above parameters are also interdependent, and interpretation is generally based on *multiple converging evidence*; however, even a single parameter could be diagnostic in a certain case.

Over the same lithology, the spectral response may be quite variable, being a function of several factors: state of weathering, moisture content, soil and vegetation. Therefore, spectral enhancement followed by visual interpretation is generally preferred for geological–lithological discrimination purposes, for this type of approach.

Weathering pattern and products may carry a lot of information about the bedrock and need to be carefully studied. For example, Macias (1995) studied weathered products on TM data to distinguish between mafic and ultramafic rocks in an area in Australia. Rencz et al. (1996) could identify kimberlite plugs in Canada due to their negative relief (depressions) leading to accumulation of water which resulted in 'cooler' signatures on the thermal-IR images. Other examples of this type of interpretation and application can also be found in the literature (see e.g. Glikson and Creasey 1995).

Differences in lithological units may be obvious on simple black-and-white remote sensing photographs and images and/or on false-colour composites. Alternatively, image manipulation, such as ratioing, hybrid compositing, principal component transformation etc. may also be used to facilitate distinction between different lithologies.

On the following pages we discuss some important image characteristics of the main types of lithological units with examples. For practical applications, sensor characteristics and resolution parameters vis-à-vis ground dimensions also have to be duly considered.

16.4.2 Sedimentary Rocks

Sedimentary rocks are characterized by compositional layering (Fig. 16.43). The layers of different mineral assemblages possess differing physical attributes, and this results in the appearance of regular and often prominent linear features on images. Banding on remote sensing images may arise due to the following: (1) different compositional bands possessing different spectral characteristics; (2) differences in susceptibility to erosion and weathering for different bands, resulting in differential erosion between hard and soft rocks, i.e. competent layers stand out in relief over the incompetent layers; (3) differing moisture content, depending upon mineral composition; and (4) lithological layering associated with vegetation banding.

Frequently, all the above features collectively lead to banding on photographs and images. The resulting linear features are long, even-spaced, and few in number (in comparison to those produced by foliation in metamorphic rocks), and constitute rather continuous ridges and valleys. This type of banding is the most diagnostic feature of sedimentary terrains (Fig. 16.43; also see Figs. 16.18, 16.20 and 16.26).

1. **Sandstone.** *Weathering* – generally resistant in both humid and dry regions, excepting when poorly cemented or contains soluble cement. *Landform* – tends to form hills, ridges, scarps and topographically prominent features; inclined strata

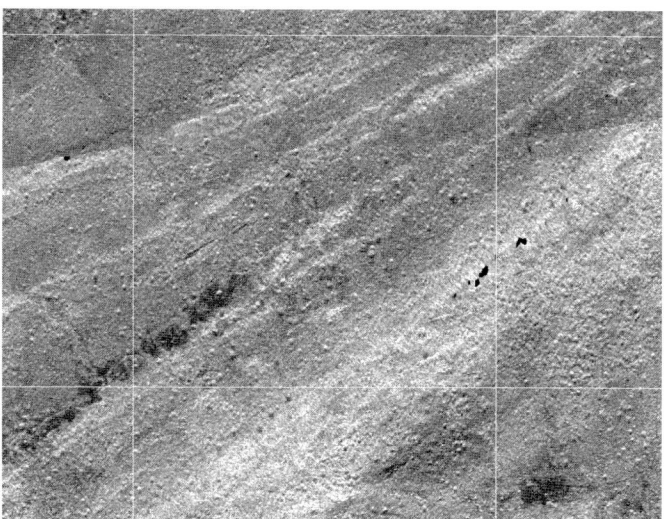

Fig. 16.43. Typical regular sedimentary layering. (X-band SAR image of an area in Brazil. (Courtesy of Aerosensing Radarsysteme GmbH)

form cuestas, hogbacks and cliffs; sub-horizontal beds form mesas (Fig. 16.44). *Drainage* – low to medium density due to good porosity and permeability and steeper slopes; partly internal drainage; frequently rectangular and angulate drainage patterns due to jointing; sub-parallel pattern in inclined strata; sub-dendritic in homogeneous sub-horizontal beds; trellis pattern in intercalated sequence. *Bedding* – often shown by compositional layering as fine lineations on the photographs/images; massive pure sandstones may lack manifestation of bedding. *Jointing* – invariably very prominent; several sets may be developed; coarse-grained sandstone and conglomerate show more widely spaced and less regular jointing. *Soil cover* – variable; scant soil cover over pure sandstone; in the case of impure sandstone, thicker soil cover may develop. *Vegetation* – sandstone generally supports good vegetation due to good porosity and soil cover; in arid–semiarid regions, it may support bushes and tree growth, whereas the adjoining clay shales may contain only grass; pure sandstone is barren of vegetation. *Spectral characteristics* – in VIS-NIR-SWIR images, the bare slopes of pure sandstone are generally light-toned; in the TIR range, it often appears cooler (dark-toned) due to low emisssivity and higher topographical location; low spectral emissivity at ≈ 9 µm; the overall spectral response over the rock may be highly variable, depending upon the presence of other minerals (e.g. limonite, carbonate, clays), weathering, soil cover, vegetation and orientation. *Similarities* – sandstone may appear similar to limestone in arid terrains, especially when massive, on VNIR data; however, limestones have a broad absorption band in the SWIR (2.35 µm); quartzites are similar photo units, but are located in metamorphic settings.

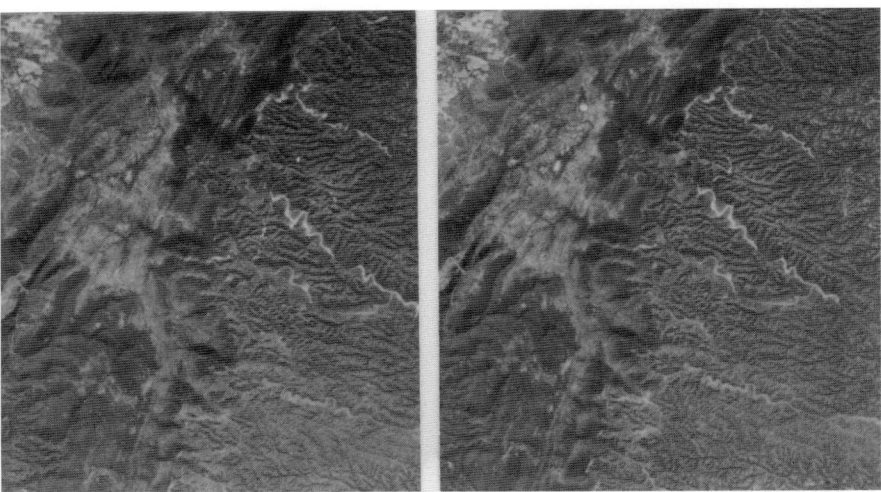

Fig. 16.44. Stereo aerial photos showing sandstone and shale marked by differences in landform and drainage. (Courtesy of A. White)

2. **Shale.** *Weathering* – generally incompetent; easily eroded. *Landform* – tends to form low grounds and valleys; in humid climates, it may form gently rounded hills; erosion is more intense in arid and semi-arid regions (Fig. 16.44). *Drainage* – chiefly external drainage due to its impervious nature; high drainage density, generally well developed, fine-textured, uniform drainage; most commonly dendritic to sub-dendritic drainage owing to homogeneity; gullies tend to be long and gently sloping with gentle V-shaped cross-sections; in loose unindurated clayey sediments, the gullies have steeper cross-sections and form badland topography. *Bedding* – rarely seen. *Jointing* – rarely exhibits prominent jointing. *Soil cover* – often thick cover; moisture-rich in humid areas due to its impervious nature and associated lower elevations; dry in arid–semiarid regions. *Vegetation* – in semi-arid areas, clay shale may have poor vegetation due to its impervious nature and low water content; in arid areas, shale may be nearly barren of vegetation; in humid areas, vegetation bandings may mark the lithology; shale and clay grounds are often used for agricultural purposes. *Spectral characteristics* – in VIS-NIR-SWIR images, dry bare shales appear light-toned, except in the 2.1-2.4 µm region, where they are dark due to the absorption band; wet shales are dark-toned in solar reflection images; in TIR images, shale generally appears light-toned (warmer), although the tone may be highly variable due to other variables and surficial cover. *Similarities* – shale appears similar to other soft rocks such as schists/phyllites but differs in regional setting.

3. **Limestone and dolomite.** *Weathering* – highly susceptible to dissolution by water; dolomite is harder than limestone and also less susceptible to solution activity; the carbonates are resistant rocks in arid regions. *Landforms* – in humid areas,

Lithology 485

Fig. 16.45. Stereo aerial photographic pair showing the limestone terrain characterized by depressions, dry valleys and mottled surface due to variation in surface moisture. (Courtesy: Aerofilms Ltd.)

karst topography, subsidence and collapse structures, sink holes, trenches, caverns, subsurface channels, etc. due to action by surface and subsurface water; these features are often elongated in the direction of prominent joint, bedding or shear planes; sudden or gradual disappearance of surface streams may be observed (sinking creeks!); in arid regions, the carbonates form ridges and hills (Fig. 16.45). *Drainage* – marked by low drainage density in both arid and humid terrains; internal drainage high in humid areas; in arid regions, low drainage density due to poor availability of water; valleys tend to be U-shaped. *Bedding* – often only weakly shown owing to the chemical origin of the carbonate rocks; intercalations of shales may enhance manifestation of bedding. *Jointing* – generally well developed; the joints provide sites for water action and control the shape and outline of solution structures. *Soil cover* – light-coloured calcareous soil often develops on carbonates; as a result of the removal of carbonates in solution, the insoluble constituents such as limonite and clays are left as a red-coloured residue, called terra-rosa. *Vegetation* – variable, depending upon weathering and climate; in humid climates, vegetation may be quite dense; vegetation may be sparse in karst landforms; carbonate terrain is also suitable for cultivation. *Spectral characteristics* – in the VIS-NIR-SWIR images, bare slopes of limestones generally appear light toned (except at longer SWIR wavelengths, where they may be dark due to the carbonate absorption band at 2.35 μm); very frequently, mottling due to variation in moisture content; terra-rosa appears as dark bands along joints or as patches, and may exhibit Fe-O absorption bands in the blue–UV region; SLAR images may exhibit surface roughness, topographical variations (collapse and subsidence

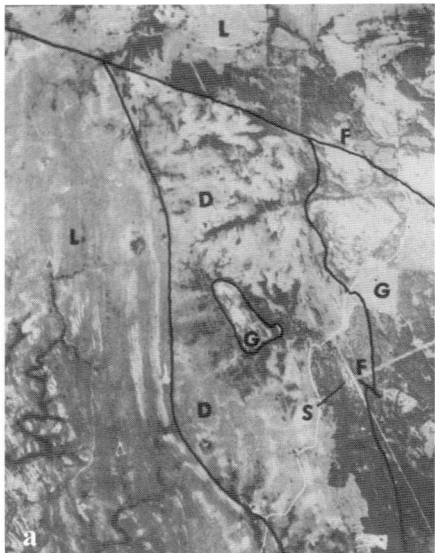

Fig. 16.46a,b. Discrimination between limestone and dolomite. **a** Aerial photograph of an area in Oklahoma, USA; rock types consist of limestone (L), dolomite (D), and granite (G); these rocks appear similar in tone in VNIR and exhibit limited textural differences. **b** Thermal-IR image (pre-dawn); discrimination between these rocks is possible on the TIR image; dolomite has higher thermal inertia than limestone and granite, and therefore is brighter (warmer) than the other two rocks; granites appear dark (cool) (K. Watson in Drury 1987)

etc.); limestones may be distinguished from dolomite on the TIR images due to differences in thermal inertia (Fig. 16.46). *Similarities* – carbonates may appear similar to sandstones in arid regions but the latter exhibit compositional banding; further, carbonates possess an absorption band in the SWIR (at 2.35 µm) and lack one at ≈ 9 µm.

16.4.3 Igneous Rocks

Igneous rocks are characterized by the absence of bedding or foliation. *Intrusive igneous rocks* are generally massive, isotropic and homogeneous, characteristics which can be easily observed on the remote sensing images. The intrusive igneous rocks may occur in different shapes and dimensions, e.g. batholiths, laccoliths, dykes, sills etc., and this may also help in identification (Figs. 16.47 and 16.48). *Extrusive igneous rocks* can be delineated with the help of the associated volcanic landforms, lavaflows, cones, craters, volcanic necks, dykes etc. The flows have rough surface topography and discordant contacts with the bedrock. Lava flows may be interbedded with non-volcanic sediments, and a number of flows collectively may impart the impression of rough sub-horizontal bedding. Older extrusive rocks have thicker soil and vegetation cover due to prolonged weathering, which renders their identification more difficult; younger

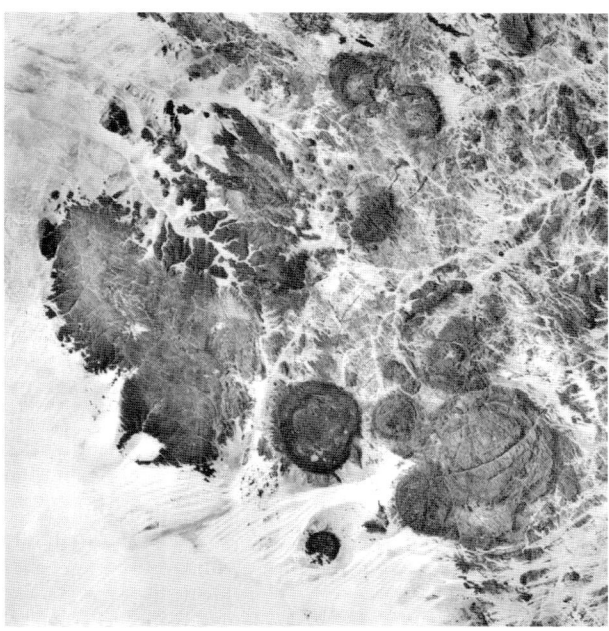

Fig. 16.47. The Air Plateau forming the southern extension of the Hoggar Mountains, North Africa, is a Precambrian massif lying on the border of the Sahara desert sands. Numerous large alkaline granitic intrusions (Late Palaeozoic or Mesozoic age) are seen. Several extensive faults with lateral movements are also conspicuous. The area contains deposits of uranium, wolfram, copper etc. (Courtesy of R. Haydn)

weathering, which renders their identification more difficult; younger extrusive rocks are relatively easy to demarcate.

1. Granites. *Weathering* – weathering characteristics of granites differ greatly; in humid warm climates, granites are more prone to weathering than in cold dry climates. *Landform* – granites occur as bodies of gigantic dimensions formed either as intrusive bodies or as products of migmatization; in warm humid climates, they typically exhibit smooth rounded shapes due to spheroidal weathering; topography is generally low lying; sometimes woolsack weathering is shown, in which isolated outcrops protrude through thick weathered mantle; huge boulders may be observed in valleys; in arid and semiarid regions, steep, sharp, jagged forms develop; sharp and steep forms may also develop due to rapid erosion in the tropics. *Drainage* due to rapid erosion in the tropics, generally low to medium density dendritic drainage; sometimes sickle-shaped drainage develops; rectangular and angulate drainage patterns may develop when well jointed. *Bedding* – absence of compositional layering. *Jointing* – three–four sets of joints often seen; the granites may exhibit sheeting (sub-horizontal extensive joints). *Soil cover* – may be quite thick in warm humid climates. *Vegetation* – variable from poorly to thickly vegetated; the weathered slopes are suitable for cultivation. *Spectral characteristics* – on VIS-NIR-SWIR images, granite is light- to medium-toned; on TIR images granites appear cooler (darker); low-emissivity bands due to quartz and feldspar occur at $\cong 9$ μm; surface moisture, soil and vegetation may significantly influence spectral response over granites. *Similarities* – all large-sized intrusive bodies may appear quite similar; basic and ultrabasic rocks are darker in the VIS region, smaller in size, have relatively poor vegetation cover, and display low emissivity bands at relatively longer wavelengths (10–11.5 μm) in the TIR region.

2. Basic and ultrabasic intrusives. *Weathering* – basic and ultrabasic rocks (gabbro and peridotite) are highly susceptible to weathering and alteration, especially under humid conditions; they commonly yield lateritic residual deposits and montmorillonitic soil. *Landform* – these bodies have smaller dimensions than the acidic intrusives and occur as laccoliths, lopoliths, plugs etc.; they commonly form undulating rolling topography in humid warm climates; in arid–semiarid regions, sharply rising rough and jagged forms are common; domal structures or circular depressions may indicate intrusion. *Drainage* – commonly coarse dendritic drainage patterns; rectangular and angulate patterns in jointed rocks; locally, radial and concentric patterns associated with domal upwarps; fine dendritic drainage over weathered soil (montmorillonitic) cover. *Bedding* – absent; layers formed by primary differentiation may resemble bedding. *Jointing* – often well developed; the joints vary in orientation with the form of the igneous body; altered masses may not exhibit a joint pattern. *Soil cover* – in warm humid climates, soil is well developed; the soil cover is thicker over ultrabasic rocks. *Vegetation* – scanty over basic rocks; ultrabasic rocks may even be barren of vegetation owing to the toxic effects

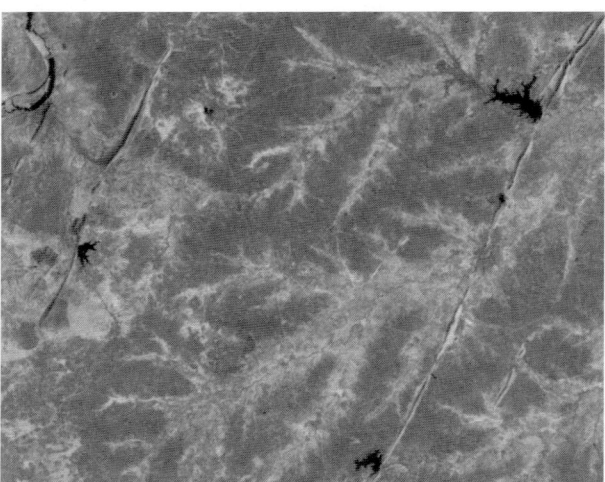

Fig. 16.48. Dykes forming linear features. (Landsat TM FCC of a part of Central India, printed black-and-white)

of certain metals; plant selectivity may occur, i.e. preferential growth of certain plants over a specific lithological unit. *Spectral characteristics* – in the VNIR range, basic–ultrabasic rocks are rather dark-toned; alteration products may comprise clays which possess absorption bands in the SWIR region (2.1–2.4 µm); silicate absorption bands occur in the TIR region (10–11.5 µm). *Similarities* – as discussed in the case of granites.

3. Dolerite. Dolerites are intrusive basic rocks of hypabyssal type, occurring as dykes and sills. Therefore, except for the mode of occurrence, most of the characters of dolerite are similar to those of gabbro. Dolerite sills are difficult to identify on remote sensing data, as these bodies can be mistaken for layering in the host rocks. On the other hand, dykes can be easily demarcated by their discordant relationship with respect to the host rocks, and a relatively thin sheet-like shape (Fig. 16.48); in relief, dykes commonly stand out above the country rocks, appearing as walls, but may also occasionally form trenches.

4. Acidic extrusive rocks *(rhyolite, pumice and obsidian)*. *Weathering* – highly susceptible to weathering, which means obliteration of characteristic landforms and features in older flows. *Landforms* – acidic lava is viscous and therefore restricted in extent, oblate in outline, has a rough topography and an irregular hummocky surface; volcanic landforms are associated. *Drainage* – absent or very coarse due to high porosity in newer flows; older lavas display higher drainage density and dendritic drainage; radial drainage is associated with volcanic structures. *Bedding* – absent. *Jointing* – may be faintly developed in some lava flows. *Soil cover* – may be thick in older flows. *Vegetation* – sparse in the case of younger flows; older flows frequently support good vegetation. *Spectral characteristics* – generally light-toned on VIS-NIR-SWIR images; weathering, soil cover and vegetation may have significant influence; 'clay bands' in the SWIR may be present in the case of altered flows.

5. Basalt. *Weathering* – highly susceptible to weathering; original landforms are difficult to identify after weathering. *Landforms* – basaltic lava has low viscosity and is commonly of a ropy type; presence of flow structures; oblate outline with flat and rough topography; basalts of central-eruption type exhibit gently sloping cones; those of fissure-eruption type are more flat near the centre, become serrated along the periphery, and are associated with dykes. *Drainage* – may be almost absent in the initial stages due to high permeability; fine dendritic drainage develops over weathered plateau basalts due to low permeability; radial and annular drainage may be seen associated with cones. *Bedding* – absent; successive flows interbedded with non-volcanic sediments may resemble coarse bedding shown on scarp faces. *Jointing* – typical columnar jointing often present. *Soil cover* – dark-coloured montmorillonitic soils develop as a result of weathering. *Vegetation* – sparse on younger flows; weathered areas may support vegetation and cultivation. *Spectral characteristics* – dark in the VIS range, especially at the blue end; medium tones in the NIR-SWIR ranges; mottled appearance due to variation in mois-

Fig. 16.49. AIRSAR image showing Kilauea volcano, Hawaii. The main caldera is seen on the right side of the image. Lava flows of different ages exhibit differences in backscatter due to variation in surface roughness. (Multiband image with colour coding: R = P-band HV; G = L-band HV, B = C-band HV) (Courtesy of NASA/JPL) (see colour plate VIII)

ture; absorption bands of mafic minerals in the TIR range; soils are very dark in the VIS-NIR range and exhibit clay bands in the SWIR range. *Similarities* – the basalts appear somewhat similar to acidic extrusives and may be differentiated on the basis of landform, weathering and spectral properties.

Figure 16.49 shows an AIRSAR image of Kiauea volcano, Hawaii. Basaltic lava flows of varying ages which possess differences in surface characteristics can be distinguished. Figure 16.50 is another image showing various volcanic features and lava flows.

16.4.4 Metamorphic Rocks

Metamorphic rocks are marked by foliation and some stratification. The foliation is manifested as photo-lineations, which are short and numerous, parallel to one another. Rocks of regional metamorphism are generally deformed (Fig. 16.51). The metamorphic derivatives of intrusive rocks, i.e. meta-intrusives, may appear quite similar to the intrusive rocks on remote sensing images.

1. Quartzite. *Weathering* – highly resistant both in humid and dry climates. *Landform* – forms hills, ridges, scarps and topographically prominent features. *Drainage* – low to medium density because of the steep slopes, even though permeability is very low; rectangular and angulate drainage patterns frequent; trellis pattern in intercalated sequences. *Foliation* – quartzites are most commonly mas-

Lithology 491

Fig. 16.50. ASTER FCC image of part of the Andes. The scene is dominated by the Pampa Luxsar lava complex (mostly upper right of the scene). Lava flows are distributed around remnants of large dissected cones. On the middle-left edge are the Olca and Paruma stratovolcanoes appearing in blue due to lack of vegetation. (Courtesy of NASA/GSFC/MITI/ERSDAC/JAROS, and US/Japan ASTER Science Team) (see colour plate VIII)

sive and lack foliation. *Jointing* – very prominently developed, often three to four sets. *Soil cover* – massive quartzites have scant soil cover; impure quartzites weather to yield good soil cover. *Vegetation* – massive pure quartzites are barren of vegetation, whereas weathered impure quartzites support good vegetation. *Spectral characteristics* – generally light-toned on VIS-NIR-SWIR images; low emissivity bands due to quartz/feldspar are prominent at ~ 9 μm in the TIR region; limonite, clays, and vegetation may significantly influence spectral response. *Similarities* – may sometimes resemble marble in arid regions; however, marble is a highly deformed rock and has absorption bands in the SWIR region; may also resemble sandstone, but sandstone is located in sedimentary settings.

2. Marble. *Weathering* – in arid and semi-arid regions, marble is quite resistant to weathering; in warm humid climates, it may be susceptible to solution activity. *Landform* – forms ridges and hills in arid climates; may display karst features in humid climates, often shows smooth rounded surfaces. *Drainage* – generally coarse density. *Foliation* – the intercalated bands of amphibolites and schistose rocks may indicate the trend of foliation; marbles may show a highly deformed,

Fig. 16.51. Aerial photo stereo pair of a typical metamorphic terrain; note the development of folds, faults and fractures. (Courtesy of A. White)

folded pattern due to plastic deformation. *Jointing* – often well jointed. *Soil cover* – thin soil cover may be present. *Vegetation* – variable depending up on composition, weathering and soil cover. *Spectral characteristics* – in general, marbles are light-toned on VIS-NIR-SWIR images; variation in moisture content leads to mottling; exhibit absorption band in the SWIR region (2.35 µm). *Similarities* – as discussed in quartzites and limestones.

3. Schist and phyllite. *Weathering* – generally incompetent rock. *Landform* – constitutes valleys and lower hill slopes; develops rounded forms in humid climates and relatively steeper slopes in arid climates (Fig. 16.51). *Drainage* – dendritic drainage often well developed; high drainage density; occasionally drainage may be controlled by foliation. *Foliation* – strongly developed but may be masked under soil cover. *Jointing* – generally well jointed and fractured. *Soil cover* – often thick. *Vegetation* – schist and phyllite support fairly good vegetation in humid climates, whereas in arid climates the vegetation may be sparse; suitable for cultivation. *Spectral characteristics* – generally moderate to light tones; iron-rich minerals such as biotite, chlorite and amphiboles may produce medium tones; prominent absorption bands occur in the SWIR region (2.1–2.4 µm). *Similarities* – schists and phyllites are similar to gneisses, which are some what less foliated and coarsely layered; weathering, landform and spectral characteristics of schists may resemble those of shales, which occur in sedimentary settings.

4. Slate. Slates are similar to schists and phyllites in photo characteristics, except that they are very dark in the visible region and strongly foliated and cleaved.

5. Gneiss. *Weathering* – possess greater resistance to weathering than schists and phyllites, and less resistance than quartzites. *Landform* – generally low-lying undulating terrains; rounded smooth surfaces. *Drainage* – high density; sub-parallel; sub-dendritic, rectangular drainage patterns. *Foliation* – gneisses typically show foliation; they are commonly interbedded with other lithological layers such as mica schists, amphibolites and quartzites; this leads to the development of a banded pattern on photographs and images. *Jointing* – well developed. *Soil cover* – highly variable. *Vegetation* – generally good vegetation cover. *Spectral characteristics* – highly variable, depending upon mineral composition, as the rock comprises layers of mainly quartz + feldspar and mica + amphiboles + chlorite; spectral banding (dark and light bands), is generally prominent. *Similarities* – at times gneisses may appear similar to acidic intrusive rocks, and can be distinguished from the latter on the basis of foliation, banded pattern and metamorphic setting; they may also appear similar to schists and phyllites, from which they can be distinguished on the basis of coarser foliation and faint tonal/lithological layering.

16.4.5 Identification of Mineral Assemblages

Identification of mineral assemblages is based on the broad spectral characteristics of minerals as described in Chapter 3. Figure 16.52 gives an overview of the important mineral-absorption bands. The discussion can be divided into two parts:

(1) solar reflection region (VIS-NIR-SWIR)
(2) thermal emission region (TIR)

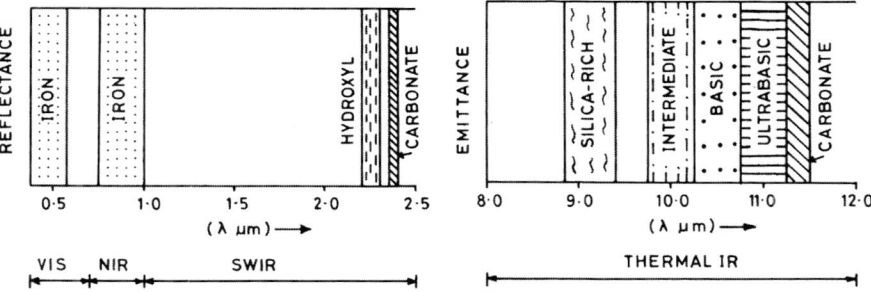

Fig. 16.52. Absorption bands in the optical region, which enable remote sensing mapping of mineral assemblages and rocks. Bands in the VIS-NIR-SWIR correspond to low reflectance, and those in the TIR to low emittance

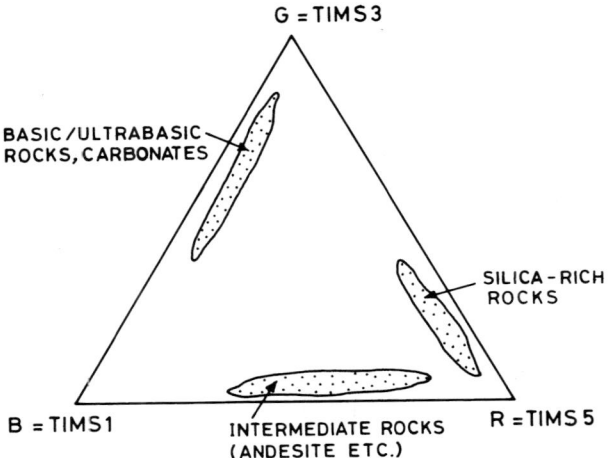

Fig. 16.53. Colour ternary diagram for TIMS data FCC (after decorrelation stretch); colour coding: R = TIMS5, G = TIMS3, B = TIMS1. On such an FCC, silica-rich rocks appear red to red–orange, intermediate rocks appear in shades of purple, and carbonates and basic/ ultrabasic rocks appear blue–green (also see Fig. 16.54)

16.4.5.1 Solar Reflection Region

In the solar reflection region, there are three main mineralogical assemblages that can be identified from multispectral remote sensing data: iron-bearing, hydroxyl-bearing and carbonate-bearing (Fig. 16.52). As minerals of these groups occur in hydrothermal alteration zones and are of vital significance in exploration, their identification is discussed under mineral exploration (see Sect. 16.6).

16.4.5.2 Thermal-IR Region

In the thermal-IR region, the various important rock-forming minerals exhibit the characteristic Reststrahlen (= low-emissivity) bands (Sect. 3.6.2). The position of low-emissivity bands systematically shifts from ≈ 9 µm in granites to ≈ 11 µm in peridotites (Fig. 16.52). This makes it possible to identify mineralogic assemblages using multispectral TIR data. A number of applications using TIMS data (for TIMS channels, see Table 9.4) for litholgical studies have been reported. A few examples are listed in Table 16.2.

In Death Valley, California, Gillespie et al. (1984) mapped alluvial fans using thermal-IR multispectral (TIMS) data. In this area, the various rocks exposed are quartzites, carbonates, shales, basalts, andesites and rhyolites. These rocks also form alluvial fans. It was found that the freshly produced debris (fans) and the original rocks in the provenance have a similar spectral response (colour on the

Lithology

Table 16.2. Examples of application of TIMS data for lithological studies

	Geology/theme	Data used	Authors
1.	Alluvial fans – their mapping and compositional study	TIMS	Gillespie et al. (1984)
2.	Basalt flows (aa, pahoehoe flows of different types and ages)	TIMS	Lockwood and Lipman (1987)
3.	Limestone vs. dolomite Silica-rich vs. silica-poor rocks	TIMS	Watson et al. (1990)
4.	Carbonatite–alkalic igneous rock complex – various rock units	TIMS + AVIRIS	Rowan et al. (1993)
5.	Leucogranite to anorthosite – their discrimination and creation of images of silica content	TIMS	Sabine et al. (1994)
6.	Playa and evaporites – different types and mineralogy	TIMS	Crowley and Hook (1996)

TIMS FCC) owing to similarity in mineral composition. Differentiating between the fan and in situ rock is possible based on the texture and shape of the image unit. Further, in cases where the original composition is modified by subsequent differential erosion and weathering (e.g. by removal of minerals in solution or suspension from the original fans), the spectral response of the material changes; this characteristic permits relative age mapping of various geological units.

Frequently, the TIMS data are presented as FCCs after decorrelation stretch. On such a colour composite, temperature variations lead to intensity differences;

Fig. 16.54. FCC generated from TIMS images after decorrelation stretch (R = TIMS5, G = TIMS3, B = TIMS1). Note the prominent compositional layering; red to red–orange are silica-rich, and blue–green are areas rich in carbonates; also see Fig. 16.53. (Lang et al. 1987) (see colour plate IX)

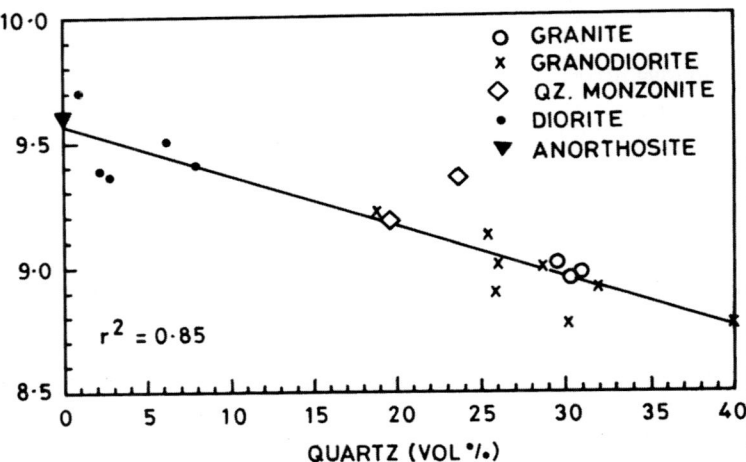

Fig. 16.55. Linear correlation of quartz content of plutonic rocks with wavelengths of emissivity minimum derived from TIMS spectra. (Simplified after Sabine et al. 1994)

compositional differences imply change in spectral emissivity of the material, and this is manifested in terms of colour (hue and saturation) variation (Gillespie 1992). Further, on a TIMS FCC (channels 5, 3, 1 coded RGB), silica-rich rocks (e.g. granite, sandstone etc.) appear red; intermediate rocks (e.g. andesite) appear purple and basic/ultrabasic rocks and/or carbonates blue–green (Fig. 16.53). Figure 16.54 is an example of a TIMS FCC showing compositional variation in rocks represented in terms of colour variation.

Quantitative Estimation

The thermal-IR multispectral data can also be used for quantitative estimation of specific minerals. In a study, Sabine et al. (1994) compared TIMS emissivity spectra (for method of computing emissivity, refer to Sect. 9.5.2) of igneous rocks ranging from leucogranite to anorthosite. They found that the emissivity minimum varied linearly with SiO_2 content of the rocks (Fig. 16.55). This relationship could be used to create images of the SiO_2 component. In another study, Reinhackel and Kruger (1998) used 6-band TIR data of DAIS-7915 to estimate quartz content in an area of lignite overburden dumps and found the accuracy of quantitative estimation of quartz (4.2 wt%) to be of the same order as obtained from laboratory analysis of field samples.

16.5 Stratigraphy

Stratigraphy deals with the deduction of the age relations and geological history of rock units in an area, i.e. broadly, the history of formation of rocks. Studies on stratigraphy attempt to derive information on the following (Lang 1999).

1. *Sequence:* i.e. characterizing vertical changes in lithology, documented in stratigraphic columns.

2. *Correlation:* determination of correspondence of strata from different locations and/or ages.

3. *Facies:* identifying rocks possessing similar depositional/environmental/age characters.

4. *Geometry:* i.e. 3-D form of strata, as determined through cross-sections, fence diagrams etc.

Mapping is the basic source of stratigraphic information; therefore this type of data is best available in sections in the field and on large-scale maps. However, in some cases, remote sensing data, owing to its special advantages of synoptic view and multispectral approach, can provide new and valuable input by revealing the nature of contacts, trends and outcrop patterns on a different scale.

Further, stereo photo analysis and DEM-based methods allow perception of elevation and relief; therefore, these methods can also be usefully exploited in deriving stratigraphic data.

The ability to define stratigraphic details (e.g. Member/Formation/Group/ Super-Group) is based on the scale of observation and mapping. Therefore, prime factors of concern are the resolution and scale of remote sensing images and photographs. The image-pixel size directly governs the details that can be mapped on the image, implying the map scale that can be obtained. Images with coarser pixels can be utilized only for smaller-scale regional maps, whereas image data with smaller pixel size can allow mapping on large scales (Fig. 16.56).

In some cases, relative ages of rocks may be indicated by an attribute more readily observed on multispectral data, such as radar surface roughness. Figure 16.57 is a Seasat-SAR image showing a group of Holocene basaltic lava flows in the neovolcanic zone of south-central Iceland. The entire region is very dry, and almost unvegetated, as all precipitation infiltrates directly into the porous ground; only sparse vegetation grows over one of the lava flows. On the VNIR imagery (Landsat MSS, TM or aerial photographs), these lava flows are impossible to discern from each other, and from the adjoining alluvium (ash fields), due to their quite similar VNIR spectral reflectances. On the other hand, because of the different surface roughnesses, the various lava flows and ash fields can easily be separated from each other on the radar image (Fig. 16.57). The original surface roughness of the lava flows falls into the category 'very rough', with regard to the

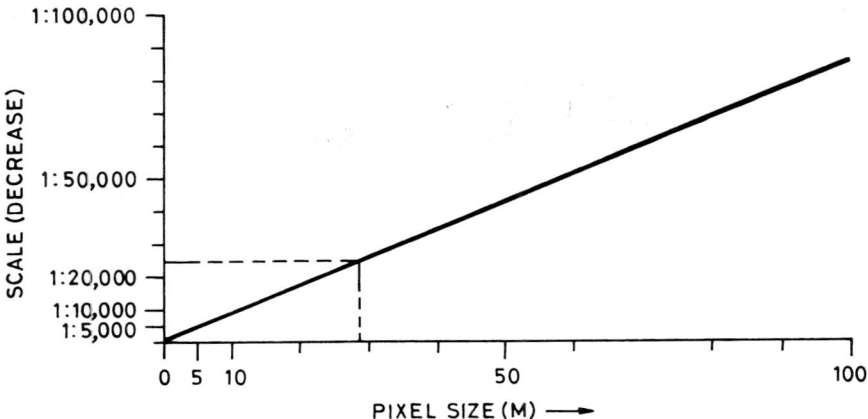

Fig. 16.56. Nomograph relating spatial resolution (pixel size) to the maximum image scale that is potentially useful for photo-stratigraphic interpretation. (Lang 1999)

wavelength of Seasat-SAR (height of surface irregularities » 10 cm); hence recent the new lava flows have a high radar backscatter. However, with time, due to infilling of aeolian sand, the lava surface gradually becomes smoother, resulting in a gradually darker tone for these lava flows with age on the radar image. Owing to the constancy of other factors governing the backscatter (humidity, vegetation etc.) in this area, it is possible to determine the relative age of the lava flows from the radar signal. Lava flow Nos. 486, 366, 363 and 068 (between 100 and 4000 years old) and 056, 058 and 062 are increasingly older, and these areas also become increasingly dark-toned in this succession on the radar image. The regularity in the relation age-vs.-backscatter is somewhat disturbed where a glacier river crosses the lava No. 068 (Fig. 16.57).

16.6 Mineral Exploration

16.6.1 Remote Sensing in Mineral Exploration

Remote sensing techniques play a very significant role in locating mineral deposits and effectively reducing the costs of prospecting and exploration. The various useful elements and minerals occur in a vast variety of genetic associations; however, the commercial deposits of minerals are limited in genetic type and mode of occurrence. This forms the basis of concept-based prospecting. Remote sensing data, by virtue of its synoptic overview, multispectral and multi-temporal coverage, can help to rapidly delineate metallogenic provinces/belts/sites

Mineral Exploration

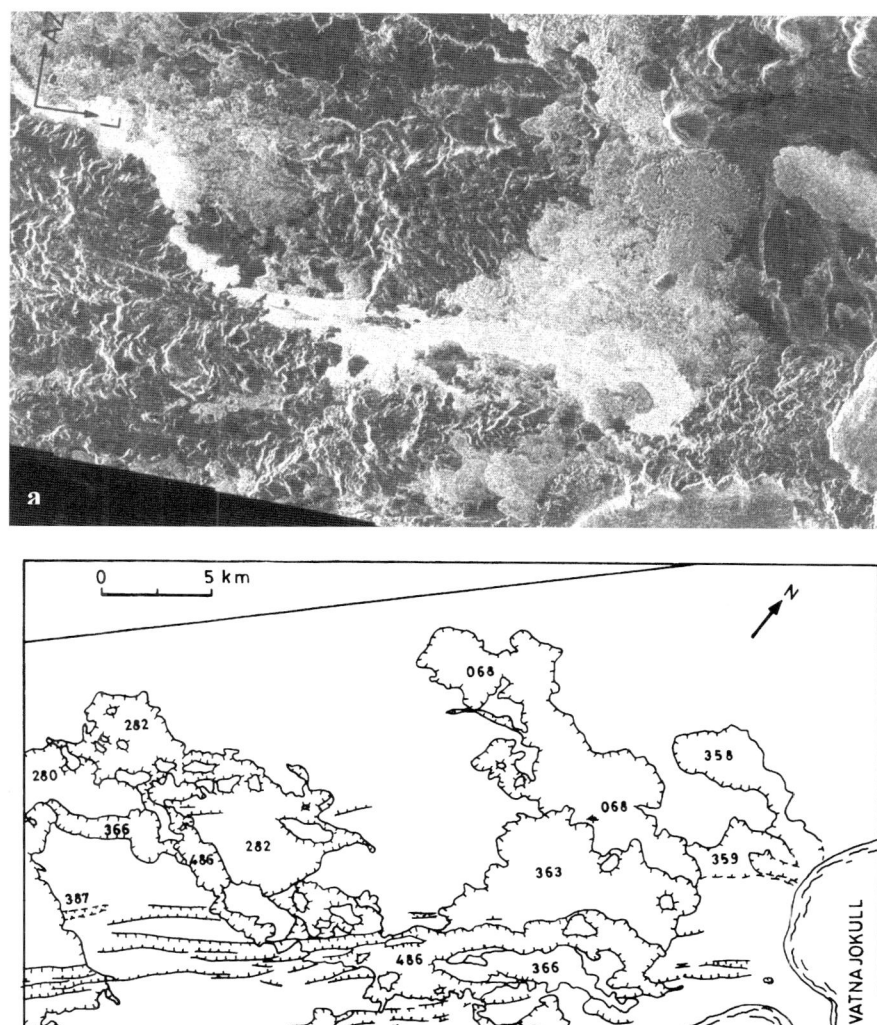

Fig. 16.57a,b. Example showing stratigraphic discrimination using SAR data. **a** Seasat-SAR image of a part of Iceland. **b** Geological map of the corresponding area (for details, see text). (**a** Arnason 1988; **b** Jakobson 1979)

and mineral guides over a larger terrain, based on known models of commercial ore occurrences. This can help to isolate potential areas from non-interesting areas for further exploration.

Most commonly, an exploration programme is marked by four stages: (1) prospecting stage, (2) regional exploration stage, (3) detailed exploration stage and (4) mine exploration stage.

The *prospecting stage* includes *reconnaissance* and *preliminary investigation stages*. The aim is to define 'targets'. At this stage, small-scale (1:500,000 to 1:100,000) satellite sensor data, supplemented with limited larger-scale (1:50,000) aerial photographs or satellite images and airborne geophysical surveys, form the most powerful data sources for defining targets.

During the *regional exploration stage,* extensive surface geological mapping (scale of 1:50,000 to 1:25,000) is carried out, and a few selected sites are surveyed using geophysical and geochemical techniques. During this stage, detailed analysis of remote sensing data provides an extremely useful input for completing surface geological mapping. Further, establishing a data base in a GIS approach, integrated data processing and enhancement of indicators could greatly help in the exploration programme. The outcome is delineation of significant anomalies or refined targets for detailed exploration.

During the next stage, i.e. *detailed exploration,* investigations of target areas are carried out at much larger scales (1:10,000 to 1:5000). The high-spatial-resolution remote sensing data and the GIS-based image-processing techniques can be valuable at this stage.

After this, the *mine exploration stage* starts to define the mineral deposit at depth, which may finally pave the way to development, mining and exploitation programmes. The high-spatial-resolution remote sensing data can be also used for monitoring open-cast mining and the effects of subsurface mining activities during this stage (see Sect. 16.13).

It should be appreciated that there is one important limitation of remote sensing data in mineral exploration – the depth aspect. Most of the mineral deposits occur at a certain depth and are not localized on the Earth's surface. Remote sensing data have a depth penetration of approximately a few micrometres in the VNIR region, to a few centimetres in the TIR and some metres (in hyper-arid regions) in the microwave region. Therefore, in most cases, a remote sensing data interpreter has to rely on indirect clues, such as general geological setting, alteration zones, associated rocks, structure, lineaments, oxidation products, morphology, drainage, and vegetation anomaly, since only rarely is it possible to directly pinpoint the occurrence and mineralogy of a deposit based solely on remote sensing data.

In this perspective, the hyperspectral sensors, which aim at defining mineralogy, are expected to play a greater role in mineral exploration in the future, by helping to delineate ore minerals or their path-finders (see Chapter 11).

Mineral Exploration

Table 16.3. Main types of mineral deposits and their surface indications observable on remote sensing data

Genetic type and form/mode		Minerals (examples)	Salient surface indications
Magmatic	– Segregation and differentiation	Chromite, magnetite, ilmenite, platinum group	Intrusive bodies, concordant/discordant relations, drainage and landform features
	– Pipes	Diamond	
	– Pegmatites	Gemstones, rare earths	
Sedimentary/volcano-sedimentary/metamorphic	– Bedded/layered	Banded iron formations, phosphorites, manganese, volcanogenic massive sulphide deposits, coal	Bedded/layered terrain, stratigraphic (age) aspects, structural controls etc.
Hydrothermal and related	– Greisen	Tin deposits	Alteration zone, mineralogy
	– Skarn	Minerals of wolfram, molybdenum, lead–zinc, tin, copper	Host rock, lithology, intrusive contact, calc-silicate minerals, alteration
	– Porphyry	Copper–molybdenum deposits	Batholithic intrusives, alteration, structural and lithological controls, gossan
	– Veins and lenses	Base metals, gold	Structural controls, alteration zone, gossan
Placer	– Mechanically concentrated by fluvial, aeolian and marine action	Diamond, monazite, gold, platinum etc.	Suitable landforms of deposition
Supergene enrichment	– Chemical leaching and deposition	Base metal sulphides	Gossan, alteration zone, oxidation and leaching
Residual enrichment	– Laterization	Bauxite, laterite, manganese minerals	Landform and drainage

16.6.2 Main Types of Mineral Deposits and their Surface Indications

The method and applicability of remote sensing techniques in prospecting workable deposits of valuable minerals are intimately linked to the mode of formation and occurrence of a mineral deposit. Therefore, a brief review of the major types of mineral deposits is relevant (Stanton 1972; Smirnov 1976; Guilbert and Park 1986) (see Table 16.3). These deposit types are frequently marked by certain surface indicators, some of which could be observed on remote sensing data products, and the same are also mentioned in Table 16.3.

A number of principal geological criteria or guides have been distinguished for mineral prospecting (Mckinstry 1948; Kreiter 1968: Peters 1978). Of these, those that can be observed on remote sensing data are: (1) stratigraphical–lithological, (2) geomorphological, (3) structural, (4) rock alteration and (5) geobotanical. In addition to these, geochemical and geophysical anomalies and other ancillary data can also provide valuable inputs during prospecting.

16.6.3 Stratigraphical–Lithological Guides

Stratigraphical (age) criteria refer to the geological setting and the stratigraphical position of the geological unit (e.g. beds or intrusives). As some types of mineral deposits are confined to certain age groups/lithologic horizons (e.g. deposits of coal, iron, manganese, phosphorites etc.), these criteria serve as useful guides during exploration. An idea of the type of terrain, namely igneous, sedimentary or metamorphic, and the stratigraphical position of major geological units can be obtained from remote sensing data on a suitable scale. Thus, attention can be focused on sub-areas of greater prospect.

Some mineral deposits are preferentially confined to a particular type of rock, and the rock may form a useful lithological guide. The deposits may be syngenetic (forming an original part of the rock mass) or epigenetic (introduced into the rock). Syngenetic sedimentary deposits are typically regular and extensive, e.g. banded iron formations, bauxites, coal and phosphorites. Syngenetic igneous deposits are relatively less regular and occur in differentiated intrusives, e.g. chromite, mangnetite, etc. The lithologically bound epigenetic deposits are formed due to strong preference for particular host rocks, e.g. carbonates, volcanic flows, metapelites etc., by the migrating mineralizing fluids. Remote sensing data of adequate spatial and spectral resolution can help locate the occurrence of lithological guides under suitable conditions, by virtue of synoptic overviews and the multispectral approach. Several examples of this type of application are known in the literature. For instance, using Landsat MSS data and supervised classification, Halbouty (1976) located the likely extension of known strata-bound copper deposits in the Tertiary Totra sandstones of Bolivia, into the adjoining territory of Peru.

16.6.4 Geomorphological Guides

Geomorphological guides are particularly important in prospecting for mineral deposits which are products of sustained weathering and erosion. Deposits formed by residual and supergene enrichment can be indicated by geomorphological criteria, e.g. hills, ridges, plateaus and valleys, in areas of sustained weathering and leaching. All these deposits are sought in the Quarternary cover. Patterns of relief, drainage and slopes are vital; information on all these features can be obtained from remote sensing data. Similarly, placer deposits (e.g. diamonds, gold, monazite etc.) are formed as a result of the mechanical concentration of fluvial, aeolian, eluvial and marine processes. Suitable sites for their deposition and occurrence

Mineral Exploration

503

Fig. 16.58. SIR-C radar image of Namibia diamond field, S. Africa. The diamond deposits, derived from the Kimberley pipes, occur as placers along the palaeochannels of the Orange river. The area is covered with thick layers of sand and gravel under which the paleochannels lie buried. Some of the ox-bow lakes can be seen on the radar image (printed black-and-white from false-colour composite). (Courtesy of NASA/JPL)

can be better located on the remote sensing data, e.g. in the case of fluvial placers, by delineating buried channels, abandoned meander scars and scrolls.

Figure 16.58 presents an SIR-C image of Namibian diamond deposits in South Africa, which form an area of active mining for diamonds. The diamonds occur as placer deposits along the paleochannels of the Orange river, having been derived from the famous Kimberley pipes. The exploration and mining are focused on palaeochannels, which lie buried under the thick layers of sand and gavel. As radar has the capability to penetrate sand sheets in desertic areas, this type of remote sensing data is of interest for exploration in such cases.

16.6.5 Structural Guides

The structural controls of ore formation, which eventually become structural guides during exploration, can be of different dimensions and scales. The structure can govern: (a) the distribution of metallogenic provinces within orogenic belts or platforms, (b) the distribution of ore-bearing regions and fields within the metallogenic provinces and (c) the localization of deposits in a particular ore field (Kreiter 1968). Remote sensing data on suitable scales can provide useful inputs in investigations on the relationship of global, mega, and minor structural features with ore deposits.

Deduction of information regarding localization of mineral deposits by certain types of geological structural belts, shear zones, faults, fractures, contacts, folds,

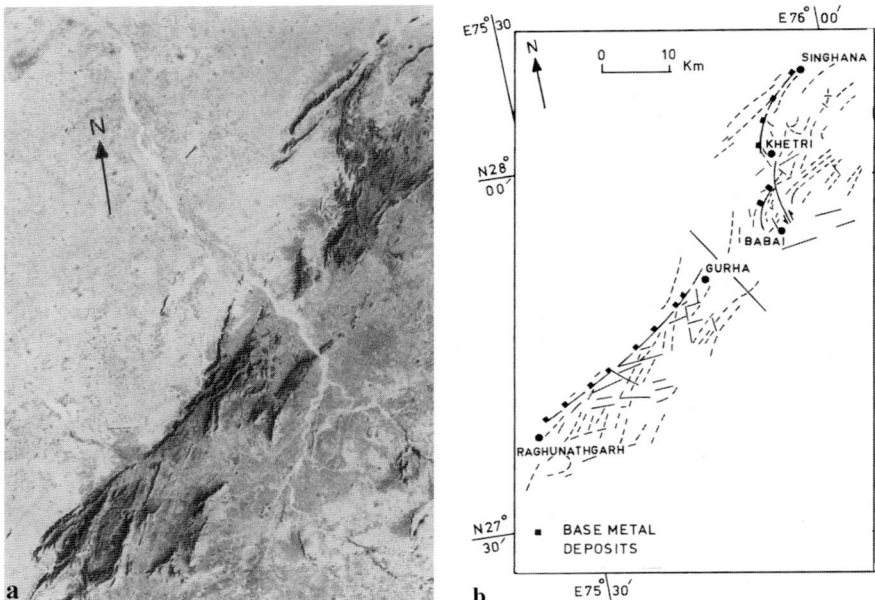

Fig. 16.59. a Landsat MSS4 (near-IR) image of the Khetri Copper belt, India. b Structural interpretation map of the above image. Distribution of major sulphide occurrences is also shown. A regional control on ore localization is obvious

joints or intersections of specific structural features is vital in planning exploration strategy (Cox and Singer 1986). Guild (1972) reviewed the relationship of global tectonics and metallogeny. Offield et al. (1977) using satellite data discovered a significant E–W-trending ore-controlling linear feature in South America.

Epigenetic mineral deposits are formed by deposition of mineral-bearing solutions in voids/fracture spaces and replacement of the host rocks, and therefore commonly exhibit a strong structural control. The Khetri copper belt (Rajasthan, India) furnishes an interesting example. Here the rocks belong to the Precambrian Alwar and Ajabgarh Groups and possess a regional strike of NNE–SSW. The polymetallic sulphide mineralization in the area is found to be localized close to the contact between the Alwar and Ajabgarh Groups (Roy Chowdhary and Das Gupta 1965; Bharktya and Gupta 1981). This provides structural–stratigraphical control of sulphide mineralization in the area. On the Landsat image, the contact is manifested as a regional lineament striking nearly NNE–SSE to NE–SW (Fig. 16.59).

Numerous studies of this type using all types of remote sensing data have been carried out with varying success. There are difficulties in integrating lineament maps with mineral exploration models, which could be due to the following reasons (Rowan and Bowers 1995): (a) some of the features mapped as lineaments may not be of structural–geologic nature, and (b) it may not be possible to distinguish between post-mineralization and pre-mineralization structures. Rowan and Bowers (1995) integrated lineament structures derived from Landsat TM and SAR

Mineral Exploration

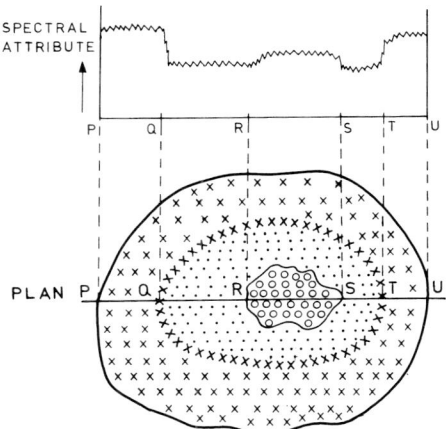

Fig. 16.60. Ringed and zoned targets formed by alteration haloes – an idealized schematic illustration showing spectral detection of zoned targets

data with a data base of known occurrences in GIS for a more fruitful interpretation of lineaments for mineral exploration. Hutsinpiller (1988) correlated lineament intersection density to alteration and observed that lineament intersection density was nearly twice as dense in altered zones as compared to unaltered zones.

16.6.6 Guides Formed by Rock Alteration

Alteration zones often accompany mineral deposits and constitute one of the most important guides for mineral exploration. These are of particular significance in the case of hydrothermal deposits, which include metals such as copper, lead, zinc, cobalt, molybdenum etc. The alteration zones usually have an abundance of minerals such as kaolinite, montmorillonite, sericite, muscovite, biotite, chlorite, epidote, pyrophyllite, zeolites, quartz, albite, goethite, hematite, jarosite, metal hydroxides, calcite and other carbonates, actinolite–tremolite, serpentine and talc. The alteration may affect the host rock as well the deposit. Often, the alteration zones constitute ringed or zoned targets, i.e. the ore-body may be surrounded by a halo of altered rock so that there is a variation in the spatial distribution of minerals. The degree of alteration may differ from core to margin. The appearance of relative amounts of different mineral groups, such as the propylitic, phyllic, opalitic, quartz and limonite groups may mark different zones or rings of alteration. Remote sensing data with adequate resolution (spectral, spatial and radiometric) holds promise for identifying such alteration zones or rings for exploration (Fig. 16.60).

Table 16.4. Typical DN-values of Landsat TM bands over altered and unaltered rocks and vegetation (Khetri copper belt, India)

	TM1	TM2	TM3	TM4	TM5	TM7
1 Raw data						
– Unaltered rocks	62	32	44	52	72	51
– Altered rocks	56	25	36	46	56	22
– Vegetation	53	22	21	72	43	18
2 Path radiance	41	13	10	09	–	–
3 Corrected data						
– Unaltered rocks	21	19	34	43	72	51
– Altered rocks	15	12	26	37	60	22
– Vegetation	12	9	11	63	43	18

16.6.6.1 Spectral Characteristics

The alteration minerals can be broadly categorized into four groups: (1) hydroxyl-bearing minerals, including clays and sheet silicates, (2) iron oxides, (3) carbonates, and (4) quartz-feldspars (framework silicates).

1. Hydroxyl-bearing minerals form the most widespread product of alteration. An abundance of clays and sheet silicates, which contain Al-OH- and Mg-OH-bearing minerals and hydroxides in the alteration zones, implies that absorption bands in the 2.1–2.4-μm range (= TM7) become very prominent (see Fig. 3.6). Most minerals and Earth materials reach their peak reflectance at about 1.6 μm (= TM5). Thus, a ratio of 1.6 μm/ (2.1–2.4) μm (= TM5/TM7) would yield very high val-

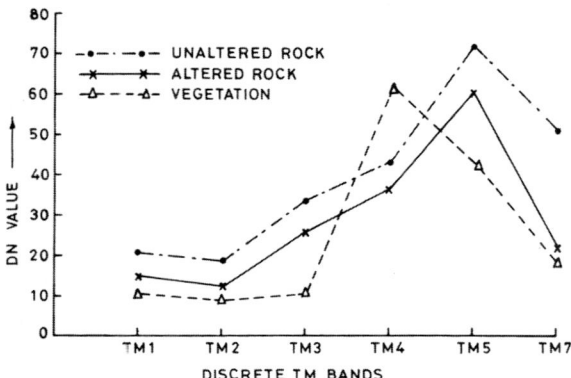

Fig. 16.61. Data plots of Landsat TM DN values for altered rocks, unaltered rocks and vegetation (Khetri copper belt, India)

Mineral Exploration 507

Fig. 16.62. TM7/TM5 ratio image of a part of the Khetri copper belt, India; dark areas (*arrows*) are hydroxyl-mineral-bearing alteration zones

ues for altered zones comprising dominantly hydroxyl-bearing minerals. This characteristics of phyllosilicates has been used in numerous mineral exploration investigations (Abrams et al. 1977, 1983, 1985; Podwysocki et al. 1983; Abrams & Brown 1985; Kaufmann 1988).

As an example, Table 16.4 gives typical values of altered rocks, unaltered rocks and vegetation in TM bands in the Khetri copper belt, India. Figure 16.61 presents the corresponding TM spectral curves. Note the steep spectral slope between bands TM5 and TM7 for altered rocks. In this sub-scene in the Khetri copper belt, the ratio of TM7/TM5 for unaltered rocks is found to be 0.7 and that for altered rocks 0.37. Figure 16.62 gives an example of a TM7/TM5 image depicting the alteration zone.

It should be mentioned here that principal component analysis (discussed in Sect. 10.7.2; see Fig. 10.30) can also be employed for alteration mapping (as also Fe-O mapping).

2. Iron oxide. Iron oxide is quite a common constituent of alteration zones associated with hydrothermal sulphide deposits. The presence of iron oxide (limonite) leads to strong absorption in the UV–blue region, affecting the slope of the reflectance curve in the UV–VIS region (see Fig. 3.3). Therefore, the ratio of blue/green (TM1/TM2 type) or blue/red (TM1/TM3 type) yields very low values for limonite-bearing zones. This spectral characteristics has been extensively applied in one form or another for limonite mapping and exploration of hydrothermal sul-

Fig. 16.63. Limonite zones (bright) on Landsat MSS ratio image of an area in Greenland. (Inverted ratio image of MSS green band/red band, which makes limonite appear in light tones; printed black-and-white from colour composite; adapted after Conradsen and Harpoth 1984)

phide deposits (see e.g. Rowan et al. 1974, 1977; Conradsen and Harpoth 1984). Figure 16.63 presents an example.

Limonite comprises three minerals: jarosite (which is pale yellow), goethite (yellow–orange) and hematite (red–orange). A closer look at the spectral curves of jarosite, goethite and hematite shows that hematite has a substantially lower reflectance in the red band (TM3) than jarosite or goethite (Fig. 16.64), the reflectance of hematite in the red band being quite close to that of vegetation. Therefore, on a TM3/TM1 type ratio, although jarosite and goethite would appear as clear bright pixels, hematite is likely to get mixed up with vegetation. This suggests that in certain areas where hematite is the dominant Fe-O and there are mixed objects (vegetation + rock), TM3/TM1 would not yield unambiguous results.

In such a situation, MSS-type data with two near-IR bands (MSS3 and MSS4) has utility. For limonitic areas, the reflectance in MSS4, which operates in the

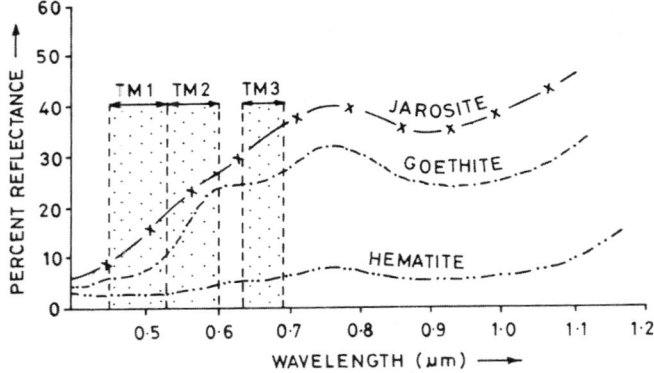

Fig. 16.64. Spectral reflectance curves of jarosite, goethite and hematite. Note the much lower reflectance of hematite than of jarosite and goethite in TM3

Mineral Exploration

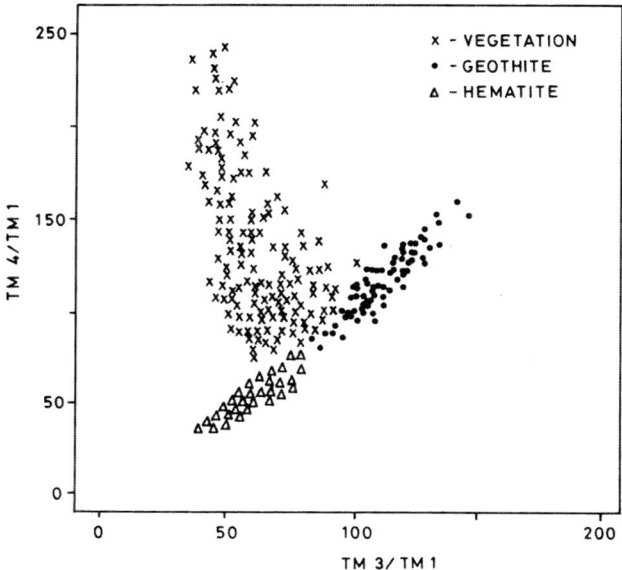

Fig. 16.65. Scatter plot of TM3/TM1 vs. TM4/TM1 with fields of goethite, hamatite and vegetation. (Data from the Newman area, Western Australia; redrawn after Fraser 1991)

0.8–1.1-μm band, is reduced due to the ferric-ion absorption band at 0.88 μm (see Fig. 3.3). This produces relatively high values for limonite-bearing zones in the ratio of two NIR bands, viz. MSS3/MSS4. Segal (1982, 1983) indicated that a compound or double ratio (MSS1/MSS2)/(MSS3/MSS4) is diagnostic in mapping limonite-bearing zones, where limonitic areas would have very low values in such a double-ratio image.

The present-day multispectral imaging sensors (e.g. TM, ETM+, HRV-multi, LISS-II/III) carry only one near-IR band. For this type of satellite remote sensing data, the method utilizing ratio of the two NIR bands cannot be implemented.

For TM data, Fraser (1991) adopted the Directed Principal Component Technique (DPCT) for discriminating ferric oxide (hematite and goethite) from vegetation. The DPCT is based on applying PCA (principal component analysis) to two input band images, which are selectively chosen for a certain purpose (Fraser and Green 1987; Chavez and Kwarteng 1989). In this way, both the correlated and uncorrelated information is enhanced on the resulting DPCT-I and DPCT-II images respectively. Ratio images are better used as input bands for DPCT, as the effects of illumination geometry and per-pixel brightness are already removed in ratio images (Fraser and Green 1987). Fraser (1991) used TM3/TM1 and TM4/TM1 as the two input bands for DPCT to discriminate between ferric oxide and vegetation. Figure 16.65 shows the scatter plot of such a two-axis ratio data from the Newman area, Western Australia. Hematite–goethite form one axis and vegetation the other. Therefore, such DPCT images provide discrimination between ferric-oxide pixels and vegetation pixels. If the scene variance is dominated by vegetation, then DPCT-I will contain vegetation and DPCT-II ferric oxide; on the other hand, if the variance is dominated by ferric oxides, then DPCT-I will contain ferric oxides and DPCT-II vegetation.

Ruiz-Armenta and Prol-Ledesma (1998) report a similar strategy for mapping goethite, hematite and vegetation, but using TM3/TM1 and TM4/TM3 as the two input ratio bands for DPCT in an area in Central Mexico.

It may be mentioned here that Fe^{3+} absorption features are present both in true gossan (limonite formed as a result of decomposition of sulphides) and pseudo-gossan (laterite formed as a result of weathering of ferromagnesian minerals). Distinction between the two, although critical for exploration, is not possible on the basis of the spectral character of Fe^{3+} alone. Thus, a limitation of the TM-ratio method is that it may produce false alarms, and therefore the method must be controlled through field data. Further, some of the deposits where Fe-O minerals are weakly developed, e.g. in a reducing environment, can be overlooked if we depend too heavily on this criterion alone.

3. *Carbonates.* The presence of carbonates leads to a general increased reflectance in the VNIR region. However, carbonates have absorption features in the SWIR (2.35 µm) and TIR (11.5 µm) regions, which can be used for identifying this mineral group (see Fig. 3.8).

4. *Tectosilicates* (quartz + feldspars) do not have absorption features in the solar reflection region, and therefore their presence also leads to a general increased reflectance in the VNIR–SWIR region. These minerals have absorption features in the thermal-infrared, and multispectral TIR data can be applied for differentiating between different types of silicates (see Sect. 3.4.3).

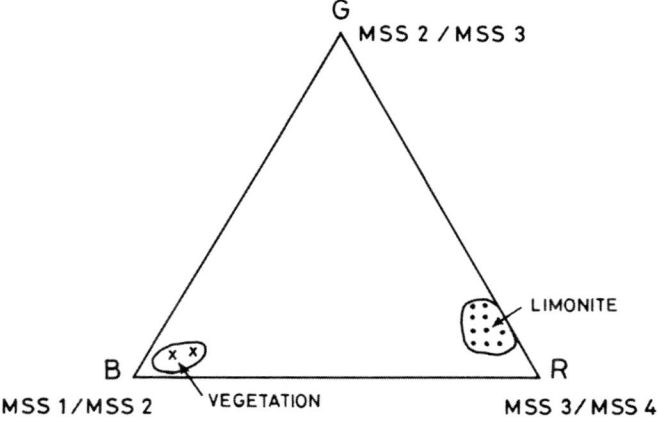

Fig. 16.66. Colour ternary diagram using MSS ratio parameters: MSS1/MSS2 = blue, MSS2/MSS3 = green, MSS3/MSS4 = red. On this CRC, limonite appears red–orange; vegetation is predominantly blue

16.6.6.2 Colour Ratio Composites

By far the most important and widely applied image enhancement technique for mineral exploration using multispectral satellite remote sensing, has been the colour ratio composite (CRC). This involves generation of three ratio images, which are colour coded in RGB for visual interpretation.

Rowan et al. (1974) were the first to show that a composite of ratios MSS1/MSS2, MSS2/MSS3 and MSS3/MSS4 (new band terminology used here) provides a powerful means for discriminating hydrothermally altered areas from regional rock and soil units. In their scheme, band ratios MSS1/MSS2, MSS2/MSS3 and MSS3/MSS4 were assigned the colours blue, green and red respectively (Fig. 16.66). On this CRC, vegetation appears blue (high MSS1/MSS2 value = B, low MSS2/MSS3 value = G, low MSS3/MSS4 value = R); iron oxide appears orange-red (high MSS3/MSS4 value = R, moderate MSS2/MSS3 value = G, low MSS1/MSS2 value = B). Other rock/soil materials have varying colours on the CRC, corresponding to their spectral characteristics.

The above approach of CRC generation has been successfully adapted for Landsat TM data and used in a number of regions. An example of the TM CRC colour ternary diagram is presented in figure 16.67. It uses the following colour scheme: TM5/TM7 = R, TM3/TM1 (or TM3/TM2) = G, and TM4/TM3 = B. On this CRC, clays appear red (high 5/7 ratio), limonite appears green (high 3/1 or 3/2 ratio), and vegetation appears in blue–purple (high 4/3 ratio).

16.6.6.3 Application Examples

1. Porphyry copper deposits, Arizona, USA. Porphyry copper deposits are huge disseminated deposits formed by hydrothermal activity. The residual copper-bearing hydrothermal solutions derived from the cooling intrusive magma lead to

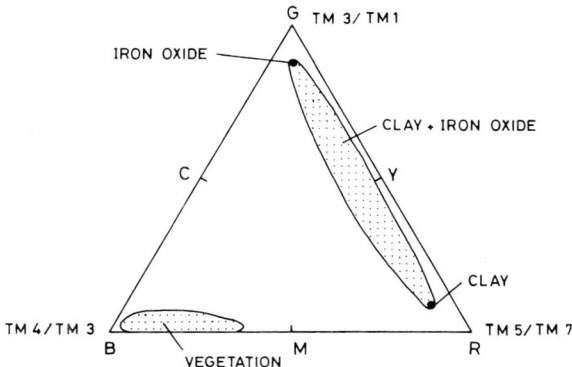

Fig. 16.67. Colour ternary diagram for Landsat TM CRC, using TM4/TM3 = blue, TM3/TM1 (or 3/2) = green, TM5/TM7 = red. On this CRC, limonite appears green, clays appear red and vegetation in shades of blue

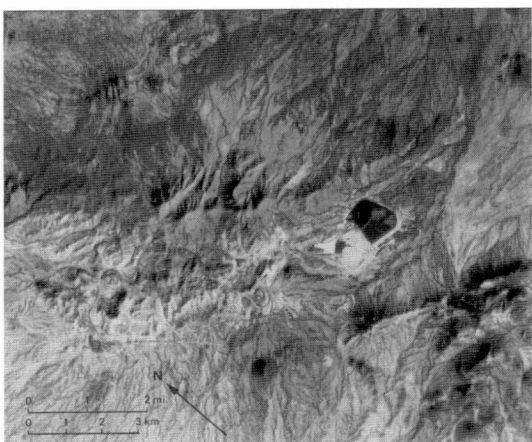

Fig. 16.68. The CRC of a part of the Silver Bell porphyry copper deposit (see text). The arcuate yellowish-orange band that trends from left to right through the centre of the area is the alteration zone comprising dominantly iron oxide and clays; field information is in agreement with these data. (Abrams and Brown 1985) (see colour plate IX)

extensive changes in mineralogy and chemistry of the host rock and the intrusion, and bring about the mineralization. In the south-west United States, several porphyry copper deposits occur, some of which were selected for the NASA–Geosat test case project (Settle 1984; Abrams et al. 1985).

At the Silver Bell test site, the host rocks are predominantly volcanic in nature, and the surface exposure of the alteration zone is extensive, vegetation being rather sparse. It has been established that specific rock alteration associations and sequences occur in the prophyry-type deposits in this terrain; for example, propylitic–argillic–phyllic–potassic alteration zones occur successively from outer to inner zones. Multiplatform and multisensor remote sensing data (Landsat MSS, aerial TM simulator and Bendix Modular Multispectal Scanner) were analysed by Abrams and Brown (1985) for this area. Figure 16.68 shows a colour ratio composite generated from the aerial TM simulator, such that 0.65/0.55 μm = green, 0.82/1.15 μm = blue and 1.65/2.2 μm = red. In this CRC scheme, iron-oxide minerals appear in green (high 0.65/0.55 band ratio). Similarly, a local concentration of clay minerals appears red (high 1.65/2.2 band ratio). The co-occurrence of iron-oxide and clay minerals is expressed in various shades of yellow. Field data on the limits of alteration compare very well with the alteration zone marked on the CRC (Abrams and Brown 1985).

2. Escondida porphyry deposit, Chile. The Escondida (or La Escondida) deposit lies in the Atacama desert, Andes mountains, Chile. It is a porphyry-type deposit and produces copper–gold–silver. The mineralization is related to a period of faulting, folding and igneous activity, which accompanied intrusion of the Andean batholith during Late Mesozoic–Early Tertiary times. Widespread alteration of

Mineral Exploration 513

Fig. 16.69. ASTER image showing the open-pit Escondida mine, Chile. ASTER-SWIR bands 4, 6 and 8 are coded in RGB respectively. The Al-OH-bearing minerals appear in shades of blue–purple, Mg-OH-bearing minerals appear in shades of green–yellow and Al-Mg-OH-bearing minerals appear in shades of red (for details on colour interpretation, refer to figures 16.70 and 16.71). (Courtesy: NASA/GSFC/MITI/ERSDAC/JAROS, and US/Japan ASTER Science Team) (see colour plate X)

both hypogene and supergene type has taken place. The hydrothermal alteration mineral zones include propylitic, phyllic and potassic zones. A high-grade supergene cap overlies the primary sulphide ore.

The Escondida deposit is being mined by open-pit method and commenced production in 1990. Figure 16.69 is ASTER image from SWIR radiometer bands 4, 6 and 8 coded in R, G and B respectively. It depicts the mine and shows lithologic–mineralogic variation at the surface. Interpretation of colours is facilitated by the colour ternary diagram (Fig. 16.70). Aster4 operates in the spectral range 1.6–1.7 µm, which is a general high-reflectance band, and is coded in red. Aster6 (2.225–2.245 µm) is absorbed by the Al-OH minerals, whereas Aster8

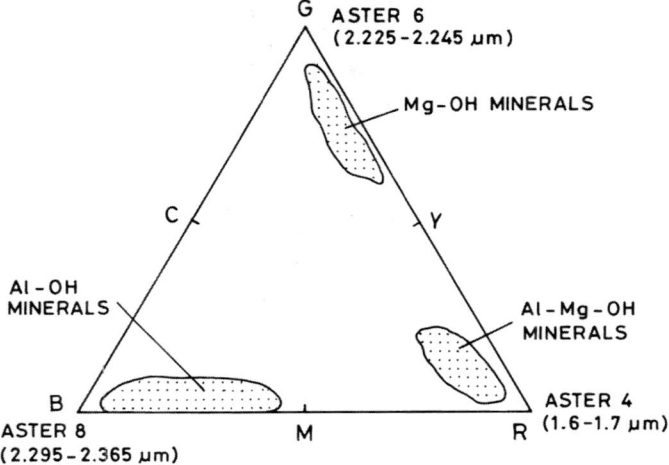

Fig. 16.70. Colour ternary diagram for ASTER FCC in figure 16.69

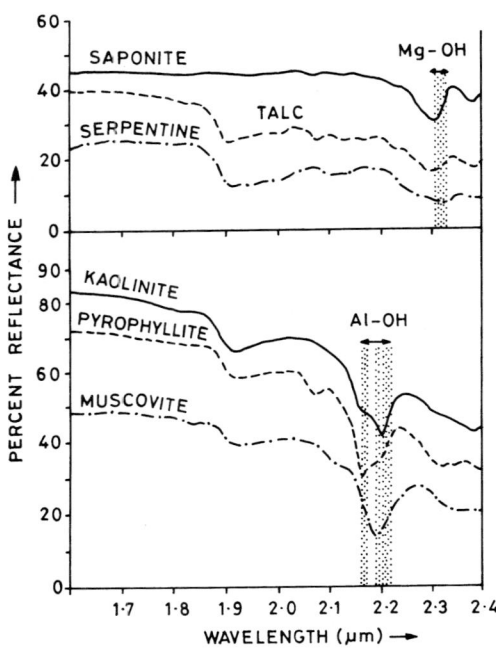

Fig. 16.71. Spectral curves of Al-OH- and Mg-OH-bearing minerals

(2.295–2.365 µm) corresponds to the absorption by Mg-OH minerals (and carbonates, if present) (Fig. 16.71). Therefore, in this colour coding scheme, Al-OH-bearing minerals appear in shades of blue–purple (high B, low G and intermediate R); Mg-OH-bearing minerals appear in shades of green–yellow (low B, high G, intermediate R); and Al-Mg-OH-bearing minerals appear in shades of red (high R, low B, low G).

3. Uranium exploration. Uranium is found to occur in a variety of geological settings. It is highly mobile and soluble in an oxidation environment and is precipitated as soon as the uranium-rich solutions enter a reducing environment. However, the reverse happens with iron, which is precipitated in an oxidation environment as ferric (Fe^{3+}) iron and becomes mobile in a reducing environment as ferrous (Fe^{2+}) iron. Therefore, the uranium mineralization may be marked by bleaching (decoloration), due to the removal of iron from the adjacent rock formations – as happens in the Lisbon Valley, Utah, which has been investigated as the NASA–Geosat joint test study by Conel and Alley (1985).

In the Lisbon Valley, uranium mineralization occurs as strata-bound pod-like deposits within a sedimentary sequence of sandstones, mudstones and conglomerates. The mineralization is restricted to a lithological member called the Moss Back Member, which forms a part of the Triassic Chinle Formation. The Chinle Formation is scantily exposed in plan and is generally covered by the overlying red Wingate sandstone. A very close spatial association between the uranium mineralization in the Moss Back Member and the bleached sections of the overlying Wingate Formation has been known in the area. This is attributed to the removal of iron from the overlying Wingate Formation under the reducing condition, which must have accompanied the deposition of uranium. From a remote sensing exploration perspective it is thus easier to detect the bleached Wingate sandstones, whose exposures are more extensive, than to directly map the uranium-bearing formations, whose exposure in plan is very restricted (Fig. 16.72).

It is further implied that local presence of reducing conditions, which may be associated with other mineral deposits, including hydrocarbon seepage etc., can also be detected on remote sensing data.

16.6.7 Geobotanical Guides

The study of vegetation differences in geological investigations is called geobotany. In geobotanical remote sensing, we discuss the spectral behaviour of vegetation in order to decipher geological information. Important reviews and contributions in the field of geobotanical remote sensing have been made by many, including Brooks (1972), Siegal and Goetz (1977), Lyon et al. (1982), Mouat (1982), Collins et al. (1983), Horler et al. (1983), Milton et al. (1983), Von Bandat (1983), Sellers (1985) and Rock et al. (1988).

In geological studies, vegetation has generally been considered to be a noise or a hindrance, masking the geological information. Most of the world's mineral production is obtained from low to moderately vegetated land surface. However, a

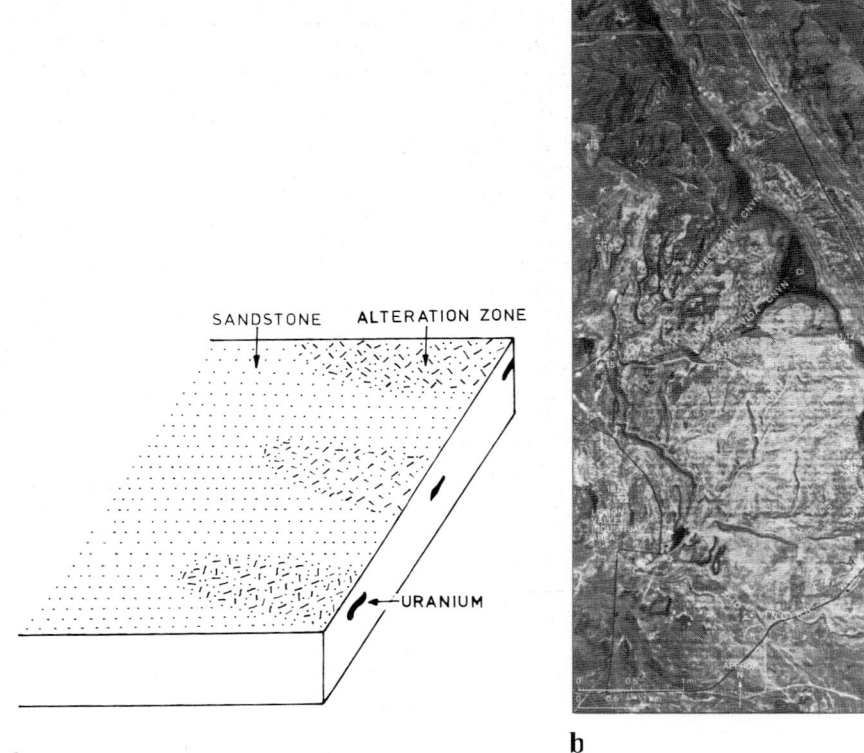

Fig. 16.72. a Schematic showing alteration–bleaching zone in the overlying Wingate sandstone associated with the uranium mineralization at depth. **b** False-colour image of the Lisbon valley, generated from canonically transformed aerial scanner data. Red–olive green colour = bleached parts of overlying Wingate Formation; shades of grey and beige = unbleached Wingate Formation. (**b** Conel and Alley 1985) (see colour plate X)

major part of the land surface on the Earth is moderately to heavily vegetated, and the largest mineral deposits yet undiscovered, probably now remain in areas of vegetation cover. Therefore, botanical guides for mineral exploration, which can easily be observed from remote sensing platforms, are gaining importance.

The spectral behaviour of vegetation is responsive to differences in soil–lithology characteristics (Fig. 16.73). For example, Bell et al. (1989) found that in a region in north-west Ontario there is a higher proportion of coniferous vegetation over metavolcanic rocks and more deciduous trees over metasediments.

Metals present in soil may lead to vegetation stress. For the purposes of mineral exploration, the geobotanical changes in vegetation are of three types: structural, taxonomic and spectral (Mouat 1982).

Mineral Exploration 517

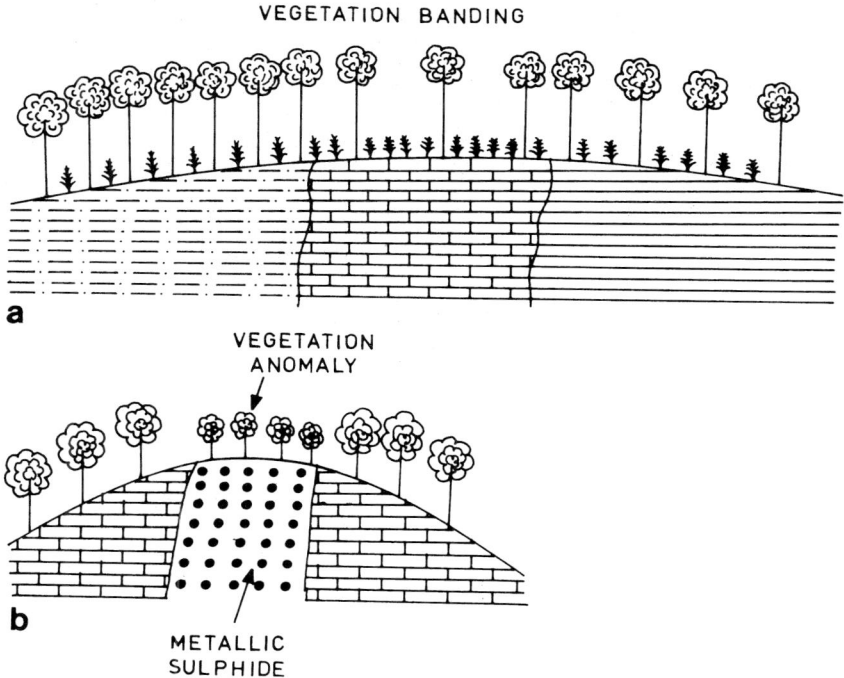

Fig. 16.73a,b. The concept of geobotanical guides. **a** Vegetation banding – the density of vegetation is related to lithology. **b** Vegetation anomaly as related to the presence of toxic metals; the growth of trees is stunted by toxic metals in soil derived from the bedrock

1. Structural changes in vegetation mean changes in morphology, i.e. vegetation density, mutation of leaves, flowers, fruits and phenological changes (changes within timing or seasonality of physiological events). For example:

(a) Chlorosis or the loss of chlorophyll pigments in green vegetation and the consequent yellowing of leaves is a common response due to the toxic effects of metal in the soil.
(b) Abnormality of form may result from radioactive minerals or boron in the soil, and these may even lead to changes in the colour of flowers.
(c) Vegetation may exhibit dwarfism or stunted growth, due to changes in geochemical conditions and, rarely, even gigantism when bitumen or boron are present.
(d) The density of total plant biomass may change, depending upon the soil characteristics; for example, the vegetation growing on serpentine-bearing soil has been described as sparse, dwarfed and xerophytic, whereas in adjoining areas the same species may be mesophytic; similarly, vegetation density differences have

been noted over copper porphyry by Elvidge (1982) and over hydrothermally altered andesite by Billings (1950).

(e) The timing of physiological events such as flowering, fruiting and senescence may shift and occur at a premature or delayed time due to vegetation stress. On plants growing over metallic sulphide deposits, it was found that the buds opened later and leaves were smaller than in those growing on soils with background levels. Further, senescence set in relatively earlier in such stressed plants than in other normal plants. This phenomenon provides 'autumn and spring windows' for detection of soil conditions and bedrock chemistry. It implies that the growing season for vegetation on metal-rich soil is shorter than that for vegetation growing on soils with background concentrations of metals. This phenomenon can be utilized to map mineral deposits with remotely sensed imagery (Masuoka et al. 1982; Labovitz et al. 1985).

2. Taxonomic differences in vegetation mean the presence or absence of some plant species (indicator plants) or changes in community structure, i.e. relative abundance. Some of the plants act as 'universal indicators' and are found only on mineralized soils. For example, the calamine violet (*Viola calaminaria*) grows only on soils with anomalous zinc content; the copper flower (*Becium homblei*) grows over copper deposits, and *Astragalus pattersoni* and *A. preussi* are used to indicate selenium mineralization, usually associated with uranium. There could additionally be 'local indicators', i.e. plants which grow preferentially over mineralized grounds in one region, but may grow over non-mineralized grounds in other regions. For example, the Mexican poppy (*Eschscholtzia mexicana*) indicates copper mineralization in Arizona but not necessarily in other areas.

3. Spectral differences refer to changes in spectral characteristics. It has been reported that vegetation (conifers) growing over mineralized areas exhibit a subtle change in spectral characteristics due to vegetation stress, such that the position of the red edge shifts by about 7–10 nm, although there is still a difference of opinion as to whether the shift is always towards the blue or the red end (see Sect. 11.2.3). It is also reported that the red-band absorption (≈ 0.67 μm) is more characteristic of concentrated heavy metals in the soil, rather than the red edge, the absorption band being regularly shallower and narrower over mineralized areas (Singhroy and Kruse 1991).

The morphological and taxonomic differences in plant associations can be picked up by good-spatial-resolution remote sensing data, e.g. broad-band panchromatic photography and images from low-altitude sensors. In order to detect phenological differences, remote sensing data should be collected at frequent temporal intervals (high temporal resolution) so that the phenomenon may be detected when it is active. Repetitive low-altitude aerial photography has been recommended for this purpose (Lyon et al. 1982). Hyperspectral remote sensing techniques can be used to detect spectral differences induced due to vegetation stress.

16.7 Hydrocarbon Exploration

Oil and gas pools occur at great depth, of the order of a few km from the surface, and are localized in geological features called traps. The strategy for hydrocarbon exploration relies heavily on the delineation of suitable traps, in a general oil-bearing terrain. The traps could be of structural or of stratigraphic type. Structural traps consist typically of structural features such as folds, faults and unconformities. Stratigraphic traps are commonly generated by facies variation during the sedimentation process.

Remote sensing techniques have been usefully applied for hydrocarbon exploration for quite some time (e.g. Berger 1994). These techniques aim at identifying suitable targets for further exploration by geophysical and drilling methods. From a remote sensing point of view, *surface geomorphic anomalies* have a special significance in hydrocarbon exploration. As oil and gas occur at great depth, it is through such features that targets for exploration can be identified in the first instance. Halbouty (1980) related the significance of surface geomorphic anomalies for 15 giant oil and gas fields.

Surface geomorphic anomalies may exhibit two characteristics: (a) morphostructural, and (b) tonal, sometimes both concurrently, in which case they possess a high correlation with hydrocarbon reservoirs. The anomalies appear as generally circular to oblong features, which are discernible on synoptic satellite sensor images. They may differ from the adjoining (background) area in terms of topography, tone, vegetation (natural) etc., with boundaries often hazy/blurred. Some of the morphostructural anomalies may be marked by drainage patterns, reflecting adjustments to subsurface shallow-buried structures. A few examples are given below.

1. Drainage anomaly, Banskandi, Assam. Numerous oil and gas pools occur in the Tertiary sequence of Assam and adjoining areas, north-east India. Therefore, this region has been subject to extensive surveys dedicated to hydrocarbon exploration. On remote sensing images of the area, a number of circular features and anomalies could be delineated, many of which were subsequently proved by geophysical surveys and drilling (Agarwal and Misra 1994). One such feature is the Banskandi anomaly (Fig. 16.74), which was first observed on the Landsat MSS image.

In the Banskandi area there are no geologic exposures on the surface, the area being covered with soil and vegetation. The drainage (Barak River) exhibits a striking circular anomaly (Fig. 16.74a), which is evidently controlled by subsurface structure. Field data show the presence of nearly N–S running longitudinal synclines on either side of the anomaly, and an anticline is exposed to the south of the anomaly area (Fig. 16.74b). Based on the above indications, it could be inferred that the Banskandi drainage anomaly represents a shallow-buried anticlinal–domal structure, which could be a potential site in this region of hydrocarbon-bearing structures. Subsequent seismic surveys confirmed the structure (subsurface high), and drilling yielded hydrocarbon gas in the area.

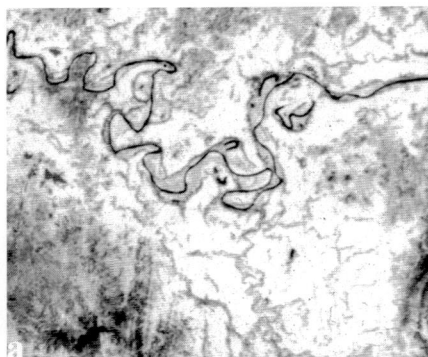 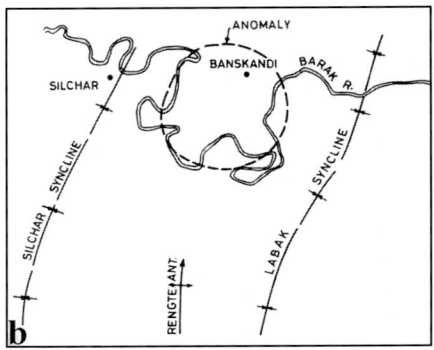

Fig. 16.74. a Drainage pattern marking the Banskandi anomaly, about 20 km in diameter (printed black-and-white from Landsat MSS standard FCC); **b** interpretation map showing the anomaly in relation to other field-mapped geological features in the area. Based on the Landsat anomaly, further investigations were carried out; the structure is yielding hydrocarbon gas. (**a,b** Courtesy of R. P. Agarwal)

2. Geobotanical anomaly, Patrick Draw, Wyoming. Stratigraphic traps located at depths of a few kilometres from the surface are relatively difficult to decipher. The seepage of hydrocarbons from the pool vertically upwards may produce surface manifestations, which could be picked up by multispectral remote sensing data (e.g. Vizy 1974; Deutsch and Estes 1980). These seepages of hydrocarbons may produce rock alterations and geobotanical anomalies. A geobotanical anomaly has been observed at the Patrick Draw site, Wyoming, by Lang et al. (1985) under the NASA–Geosat joint investigation, on which the following description is based.

The Patrick Draw is an oil-producing site from a stratigraphic trap. The area has a semi-arid climate, low relief, and vegetation consisting of primarily indigenous sage and grass. Oil is produced from a sandstone lens, sandwiched between shales, which constitute the stratigraphic trap.

Figure 16.75a is a false-colour image of the Patrick Draw test site generated from the aerial TM simulator scanner. A principal component transformation has been applied to the data (PC1 = red, PC2 = green, and PC3 = blue). A number of spectral units can be distinguished on the false colour image, which can be linked to the type of residual soils developed in the area. A unique feature, circular in outline and lemon–green in appearance on the image, was located on the false-colour image. This area was considered as anomalous, because of its size, shape, location and spectral characteristic. Examination of aerial photographs of the area (Fig. 16.75b) also could not provide any explanation for this unique feature. Ground observations (Fig. 16.75c,d) indicated that this area is marked by stunted sage bushes, which are smaller, less dense and less vigorous than the sage in the adjoining areas. It has been shown that the anomalous zone corresponds to the area vertically above the gas pool at depth in the Patrick Draw reservoir.

Hydrocarbon Exploration

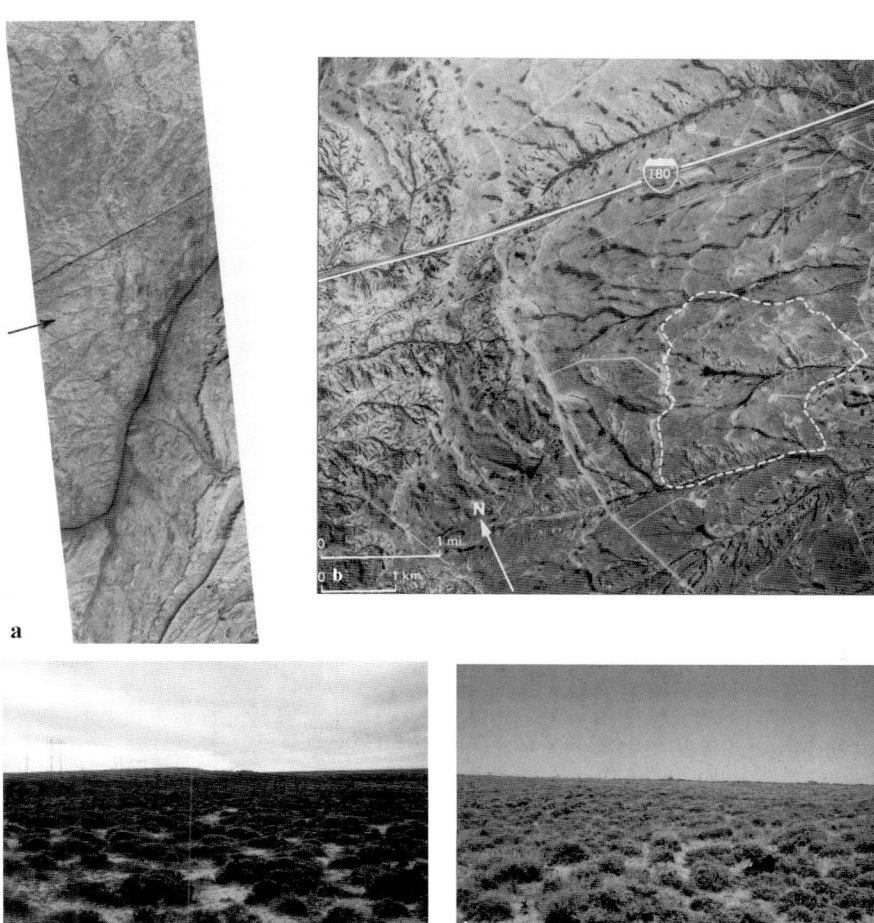

Fig. 16.75. a False-colour image of the Patrick Draw test site based on principal component transformation of aerial TM simulator data; note the peculiar lemon–green near-circular patch on the colour image, which is interpreted to be due to a geobotanical anomaly. **b** Aerial photograph of the area; the area of anomalous spectral property in **a** is also outlined. **c, d** Ground-based photographs of stunted and healthy sage bushes at Patrick Draw. (**a–d** Lang et al. 1985) (see colour plate XI)

It is inferred that seeping hydrocarbon gases from the underlying reservoir have led to vegetation stress, and this geobotanical anomaly has been picked up on the PC image. Field geochemical soil surveys indicated enhanced concentrations of hydrocarbon in and around the affected area, and also the presence of anomalously alkaline soils. Although further research is needed to prove a cause-and-effect relationship between seepage of hydrocarbons and the reported geobotanical anomaly, the results indicate that the syndrome of plant and soil conditions encountered

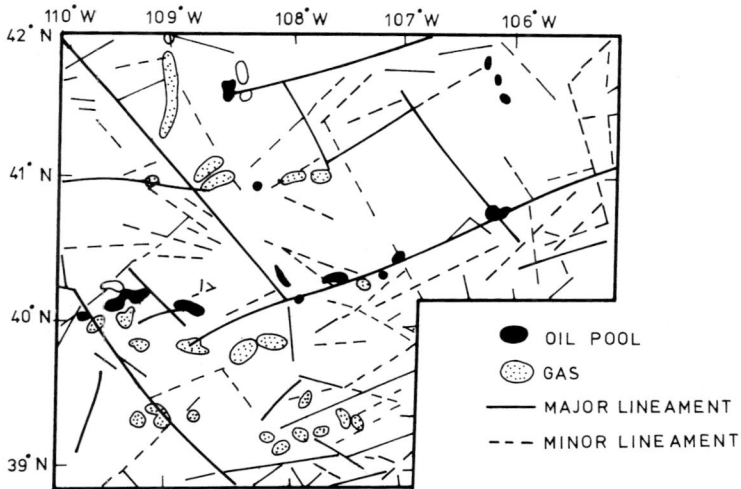

Fig. 16.76. Landsat lineaments and distribution of oil and gas fields in a part of Colorado, USA. (After Saunders et al. in Halbouty 1976)

at the Patrick Draw site could be extended empirically for prospecting for shallow reservoirs of hydrocarbons. It also has numerous implications and reflects the fact that a geobotanical guide may perhaps exist in many forms and needs to be studied more closely in exploration programmes.

3. Lineament control on the distribution of hydrocarbon pools, Colorado. Aerial photographic and satellite image based lineament analysis is a widely utilized technique in petroleum exploration. Lineaments can be considered as fracture zones in the Earth's crust, which may control migration and accumulation of hydrocarbons (e.g. Kozlov et al. 1979). As an example, figure 16.76 shows the relationship between lineaments and the distribution of oil and gas pools in a part of Colorado.

4. Oil slicks. Another important application of remote sensing is in detecting oil slicks, which occur as films on the ocean/sea water surface. Oil slicks originate from both natural and man-induced sources. Natural oil slicks in oceans are of interest for submarine hydrocarbon exploration. Many hydrocarbon reservoirs in offshore basins form slicks, such as those in the Santa Barbara Basin, several basins in Indonesia etc. Persistent or recurrent oil slicks can point towards the presence of undersea oil seeps. Figure 16.77 shows an ERS-SAR image of a suspected natural oil slick. An oil film on the sea surface has a dampening effect on the sea waves, which leads to a lower back-scatter (darker tone); at times, the dampening effect may be so pronounced that it may lead to 'no-show' in the surrounding sea clutter. Man-induced oil slicks in the ocean form due to pollution from oil tankers, drilling rigs, municipal waste etc. Such oil slicks pose environmental hazards to marine fauna and flora and need to be detected and monitored for environ-

Groundwater Investigations

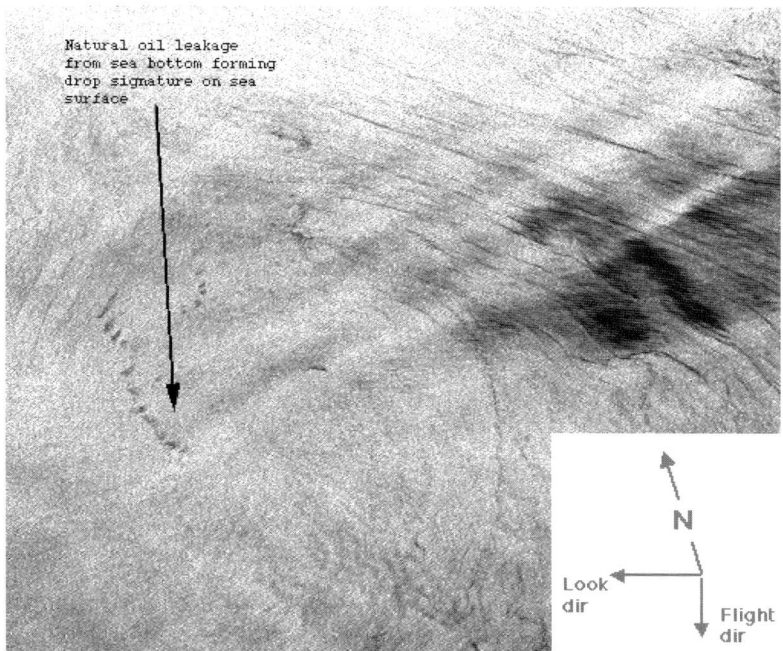

Fig. 16.77. ERS (1994) image showing oil-film signature on the sea surface; this is apparently due to natural oil leakage from the sea bottom (flight direction and look direction shown). (Copyright © 1994, European Space Agency and Tromso Satellite Station)

mental surveillance. As marine oil spills due to man-induced activities are much more common (93%) than the natural seepage (7%), the spectral characteristics of oil spills are discussed in detail in Sect. 16.15.3, dealing with environmental applications.

16.8 Groundwater Investigations

Water is the basic necessity for life. Areas having a good supply of water have always been prosperous. Groundwater constitutes an important source of water supply for various purposes, such as domestic needs, local supplies for industries, agriculture etc. It is generally cool, fresh, hygienic, potable and widely available, and its availability does not generally vary with season as greatly as that of surface water. Groundwater is thus commonly preferred for various applications.

Groundwater constitutes an important component of the hydrological cycle. Water-bearing horizons are called *aquifers*. Good aquifers have good porosity and good permeability. On the other hand, rocks, which do not possess porosity and permeability, are unable to hold and yield water. Therefore, aquifers may be con-

Table 16.5. Important indicators of groundwater on remote sensing data (after Ellyett and Pratt 1975; Singhal and Gupta 1999)

(A) First-order or direct indicators
    (1) Features associated with recharge zones: rivers, canals, lakes, ponds etc.
    (2) Features associated with discharge zones: springs etc.
    (3) Soil moisture
    (4) Vegetation (anomalous)

(B) Second-order or indirect indicators
    (1) Topographic features and general surface gradient
    (2) Landforms
    (3) Depth of weathering and regolith
    (4) Lithology: hard-rock and soft-rock areas
    (5) Geological structure
    (6) Lineaments, joints and fractures
    (7) Faults and shear zones
    (8) Soil types
    (9) Soil moisture
    (10) Vegetation
    (11) Drainage characteristics
    (12) Special geological features, such as karst, alluvial fans, dykes and reefs, unconformities, buried channels, salt encrustations etc., which may have unique bearing on groundwater occurrence and movement

sidered as extensive, subsurface reservoirs of water. The principle sources of groundwater recharge are precipitation and stream flow (influent seepage), and those of discharge include effluent seepage into streams and lakes, springs, evaporation and pumping.

Groundwater aquifers are of two types: unconfined and confined. In *unconfined,* also called *water-table aquifers,* there exists a natural water table, stable under atmospheric conditions; in confined aquifers, the water is contained under pressure greater than the atmospheric pressure, due to overlying and underlying relatively impermeable strata. *Confined aquifers* receive water from a distant area, where the aquifer may be exposed or the overlying confining layer non-existent.

16.8.1 Factors affecting Groundwater Occurrence

One of the most important requirements for groundwater occurrence and flow is that the lithological horizon be porous and permeable, so that it may store and permit easy movement of water. The pores, called voids, are the open spaces in the rock in which water may accumulate, and these voids are of fundamental importance. The porosity could be primary (i.e. developed concurrently with the rock's formation) or secondary (i.e. generated subsequently in the rock by fracturing). The original pores and interstices or intergranular voids in sedimentary rocks constitute primary porosity; on the other hand, joints, fractures, shear zones,

Groundwater Investigations

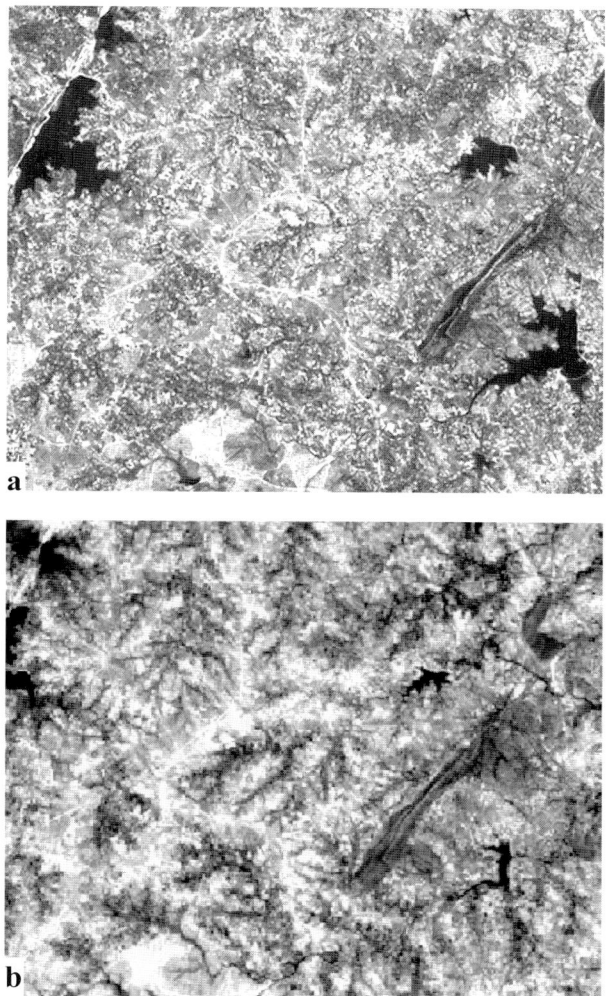

Fig. 16.78. A set of **a** post-monsoon and **b** pre-monsoon IRS-LISS-II red-band images of a part of the Budelkhand granites in Central India. Note that the various landforms (buried pediments, valley fills etc.) and lineaments are better deciphered on the pre-monsoon (summer) image

solution openings etc. constitute secondary porosity. Rocks in which primary porosity is dominant are called *soft rocks*, and those possessing predominantly secondary porosity, *hard rocks*. Unconsolidated sediments also have predominantly primary porosity, but are considered slightly differently when discussing groundwater hydrology. Rock porosity may range from 0 to 50%. Most unconsolidated materials such as gravels and sands, and rocks such as sandstones and conglomerates, possess good primary porosity and are important bearers of groundwater. Secondary porosity may be present in fractured metamorphic and igneous rocks, and in soluble rocks such as limestones. Rocks such as shales, clays, schists etc. may possess varying porosity but in general poor permeability, and serve only as impermeable or confining layers. [For details on groundwater hydrology, refer to Todd (1980); Singhal and Gupta (1999)].

16.8.2 Indicators for Groundwater on Remote Sensing Images

As mentioned elsewhere, remote sensing data provide surface information, whereas groundwater occurs at depth, maybe a few metres or several tens of metres deep. The depth penetration of EM radiation is barely of the order of fractions of a millimetre in the visible region, to barely a few metres in the microwave region (in hyper-arid conditions), at best. Therefore, remote sensing data are unable to provide any direct information on groundwater in most cases. However, the surface morphological–hydrological–geological regime, which primarily governs the subsurface water conditions, can be well studied and mapped on remote sensing data products. Therefore, remote sensing acts as a very useful guide and an efficient tool for regional and local groundwater exploration, particularly as a forerunner in a cost-effective manner (see e.g. Waters et al. 1990; Krishnamurthy et al. 1996; Singhal and Gupta 1999).

In the context of groundwater exploration, the various surface features or indicators can be grouped into two categories: (1) first-order or direct indicators, and (2) second-order or indirect indicators. The first-order indicators are directly related to the groundwater regime (viz. recharge zones, discharge zones, soil moisture and vegetation). The second-order indicators are those hydrogeological parameters which regionally indicate the groundwater regime, e.g. rock/soil types, structures, including rock fractures, landforms, drainage characteristics etc. (see Table 16.5).

16.8.3 Application Examples

Most commonly, the purpose of groundwater investigation is the targeting of water resources for local supply. Remote sensing techniques can provide vital information data, which can be supplemented and verified by other field techniques (geophysical, drilling etc.). In practice, the integrated systematic approach of 'satellite sensor–aerial photographic–geophysical–drilling' has been highly successful for groundwater exploration and widely applied. Basically, it reduces the time factor, risk and expenditure in groundwater development.

16.8.3.1 Image Data Selection

Selection of remote sensing data for groundwater applications has to be done with great care as detection of features of interest is related to spatial and spectral resolution of the sensor as well as seasonal conditions of data acquisition. For example, small-scale image data are good for evaluating the regional setting of landforms, whereas large-scale photographs are required for locating actual borehole sites. Similarly, an understanding of the spectral response of objects is crucial for selecting remote sensing data and interpretation. Further, temporal conditions (rainfall, soil moisture, vegetation etc.) greatly affect the manifestation of features. Figure 16.78 gives an example of the same area (granitic terrain in Central India) imaged by the same sensor (IRS-LISS-II) in two different seasons: post-monsoon and pre-monsoon. The post-monsoon image exhibits a widespread thin vegetation cover, and therefore distribution of various landforms and lineaments is not clear on this image; it is the summer (pre-monsoon) image on which various landforms such as buried pediments, valley fills and lineaments are clearly brought out.

16.8.3.2 Unconsolidated Sediments

Unconsolidated sediments are characterized by the presence of high primary porosity and permeability. Such materials cover extensive areas as deposits of fluvial, aeolian, glacial or marine origin, or as weathered surficial cover. In general, they form good reservoirs of groundwater, the groundwater flow regime being governed by the regional hydrological setting and the morphological–geological evolution of the area.

1. Groundwater seepage patterns in alluvial terrain – the northern Indo-Gangetic plains. For groundwater studies, it is extremely important to map areas of influent (recharge) and effluent (discharge) groundwater seepage. The near–IR, thermal-IR

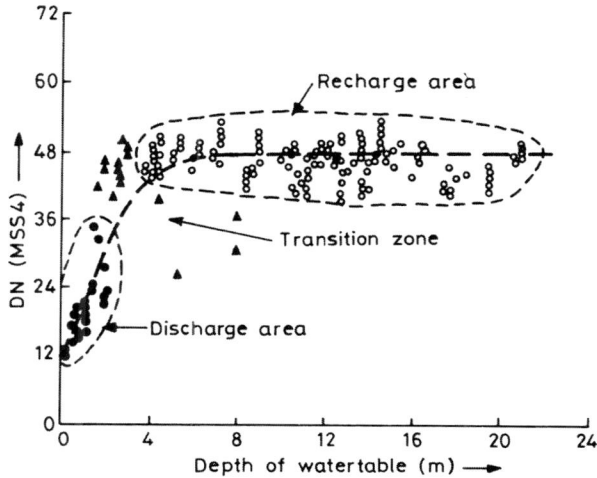

Fig. 16.79. Relationship between NIR reflectance (MSS4 DN values) and depth of water table; the distribution of groundwater recharge and discharge areas is indicated. (Adapted after Bobba et al. 1992)

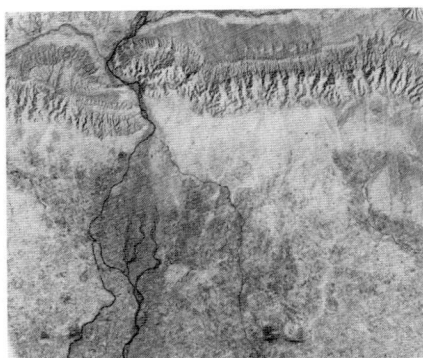 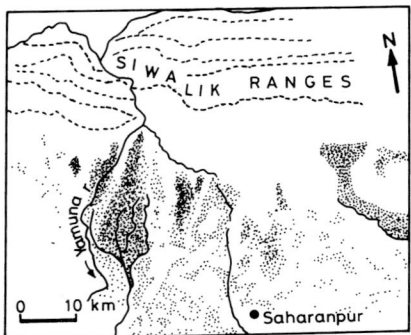

Fig. 16.80. Groundwater seepage pattern in the northern alluvial Gangetic plains. **a** Landsat MSS4 (infrared) image of part of the Gangetic plains and the sub-Himalayas. **b** Interpretation map showing influent and effluent seepage patterns. (Singhal and Gupta 1999) (Reproduced by permission, copyright © 1999, Kluwer Academic Publishers)

and SAR images are highly sensitive to surface moisture and can provide inputs for mapping seepage pattern. Figure 16.79 is a plot of NIR reflectance (Landsat MSS4) against depth of water-table. The groundwater discharge zones have a shallow water-table and lower reflectance than the recharge zones.

The northern alluvial region of the Indo-Gangetic plains provides an interesting example of groundwater seepage pattern and its spatial variation (Fig. 16.80). The vast Indo-Gangetic plains, composed of unconsolidated fluvial sediments, are bordered on the north by the sub-Himalayan (Siwalik) hill ranges. The general topographic slope and groundwater flow is from north to south. The foothill zones of sub-Himalayan ranges and the northern part of the Indo-Gangetic plains possess highly coarse-grained deposits, boulders, gravels and sands, locally called the *bhabhar* zone. The groundwater seepage in this zone is influent, the surface soil moisture is generally very low, and the area has light tones on infrared images (Fig. 16.80). The water infiltrates, moves down the gradient, and emerges as springs about 10–15 km further south. The effluent groundwater seepage and gradual building up of streams are clearly shown on the NIR-band image (Fig. 16.80).

2. *Buried river channel – the 'lost' Saraswati river.* Buried river channels constitute one of the most important targets in a groundwater exploration programme, especially in an area of Quaternary cover. A typical example is furnished by the 'lost' Saraswati river, which is said to have been a mighty river in Vedic or pre-Vedic times, and used to flow in the western part of the present Indo-Gangetic plains, between the Yamuna and Sutluj rivers (Fig. 16.81a). The palaeochannels can be identified in most cases in this area on the basis of higher moisture content in soils, textural characteristics on images and vegetation patterns on former river beds. The old bed of the Saraswati is marked as a 6–8-km-wide zone (Fig. 16.81b). Using Landsat MSS images, Yash Pal et al. (1980) delineated the course

Groundwater Investigations

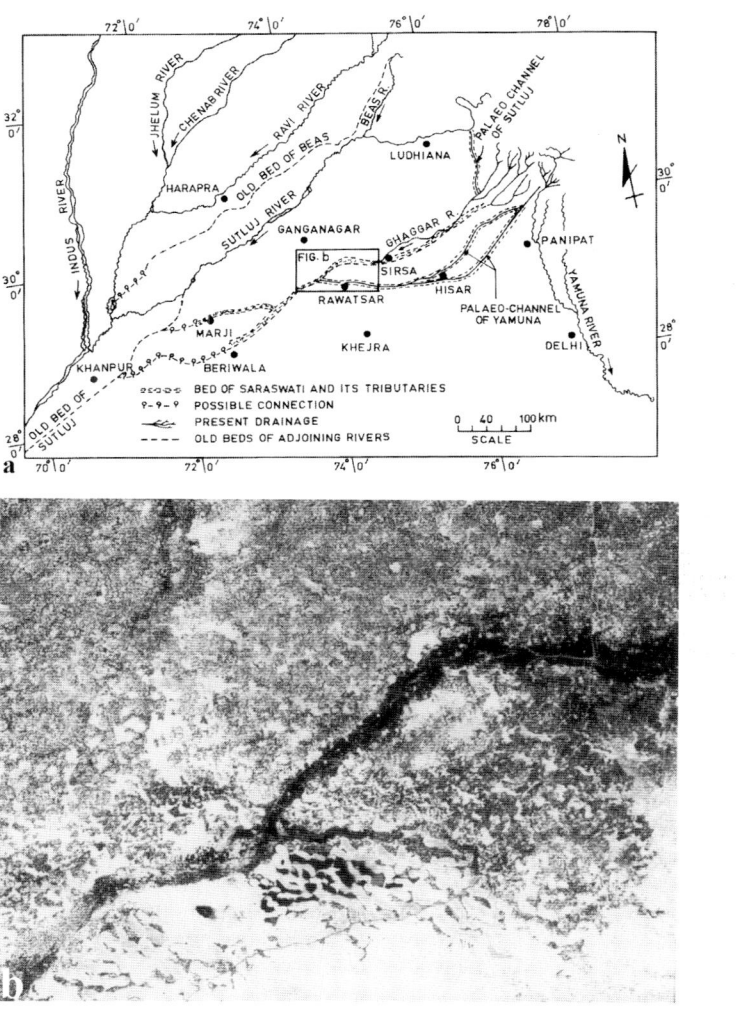

Fig. 16.81a,b. a The northwestern Indian subcontinent with its present river system along with major paleochannels, as deciphered from Landsat imagery. (International boundaries not shown; simplified after Yash Pal et al. 1980). **b** The Landsat MSS2 (red band) image showing the Saraswati river paleochannel, 6-8 km wide, marked by vegetation

of the 'lost' Saraswati river for a distance of about 400 km. They also inferred that sudden westward diversion of the Sutluj river, a former major tributary, led to the drying up of the Saraswati river.

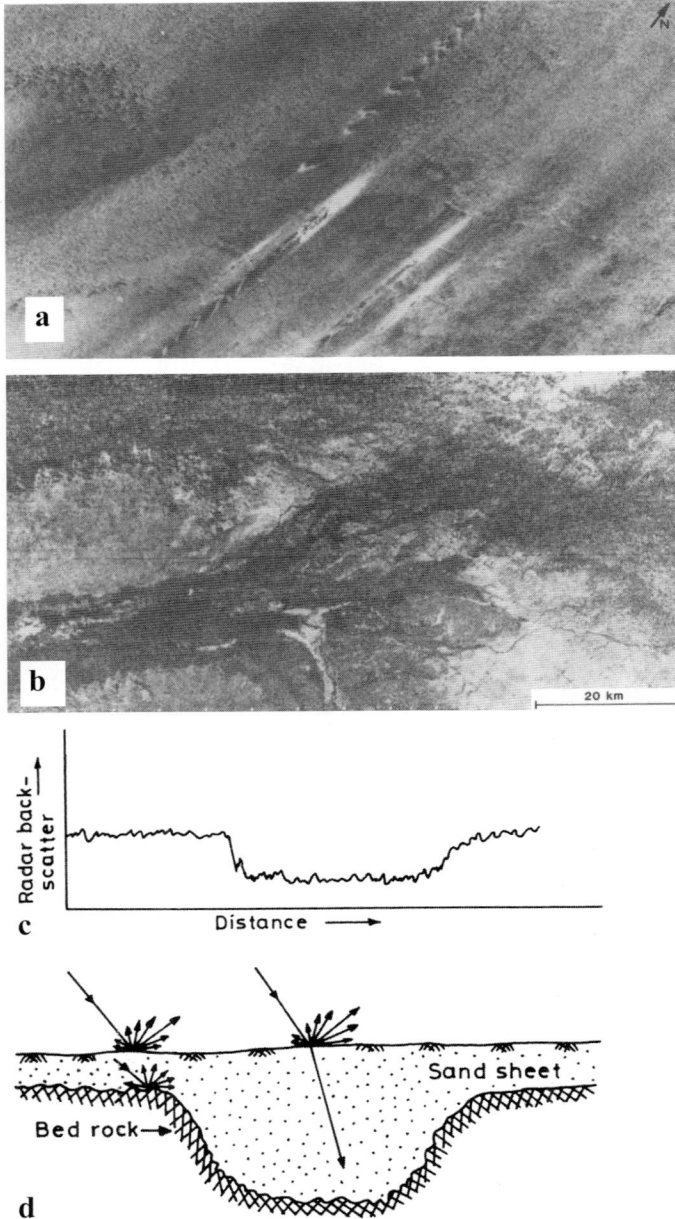

Fig. 16.82. a Landsat image of part of the Libyan desert showing extensive windblown deposits; **b** corresponding SIR-A image showing subsurface channels; **c** relative degree of back-scatter across the area and **d** schematic cross-section (**a,b** Courtesy of J.P. Ford, Jet Propulsion Laboratory, Pasadena)

3. Buried river channels – Eastern Sahara. In arid and hyper-arid regions, buried channels are of added importance, owing to conditions of acute water scarcity. The SIR-A experiment demonstrated, for the first time, the applicability of SLAR data for the delineation of buried channels in hyper-arid regions (McCauley et al. 1982; Elachi et al. 1984). The Eastern Sahara is blanketed by surface deposits of wind-blown sands, which are floored mostly by the Cretaceous Nubia Formation (sandstone and shales) and Precambrian granites. The area has witnessed several climatic cycles during Quaternary times. It is generally inferred by various workers that aridity settled in this area during the early Pleistocene, although many small playas and streams continued to exist even much later – up to Recent times – and that by about 5000 years ago, hyper-aridity had set in the Eastern Sahara Desert, bringing to a halt the fluvial processes in the region (for details see McCauley et al. 1982).

The Landsat MSS images show that the terrain is covered with sand sheets and dunes and is barren and rather featureless (Fig. 16.82a). The corresponding SIR-A images reveal the presence of subsurface buried channels – segments of defunct river systems (Fig. 16.82b). Some of the larger valleys are interpreted to be relicts of a much earlier Tertiary river system, and thus could be palimpsest drainage, related to several previous episodes of fluvial activity.

The buried river channels appear as areas of darker tone on the SAR images and this warrants some further explanation. The thin sand sheets possess no microroughness on the top (in the form of blocky scatterers over large areas), and are

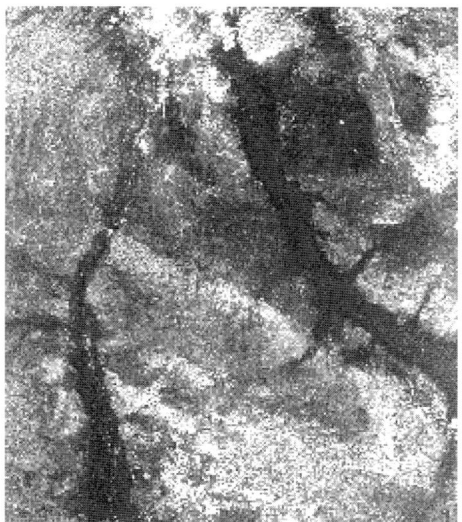

Fig. 16.83. SIR-C image showing palaeochannel of the Wadi Kufra, Libya. Prior to the SIR-C image, only the west branch of the palaeodrainage, known as Wadi Kufra, was recognized. The broader east branch (5 km wide, about 100 km long) was known only after these data from the SIR-C. (Printed black-and-white from FCC) (Courtesy of JPL/ NASA, Pasadena)

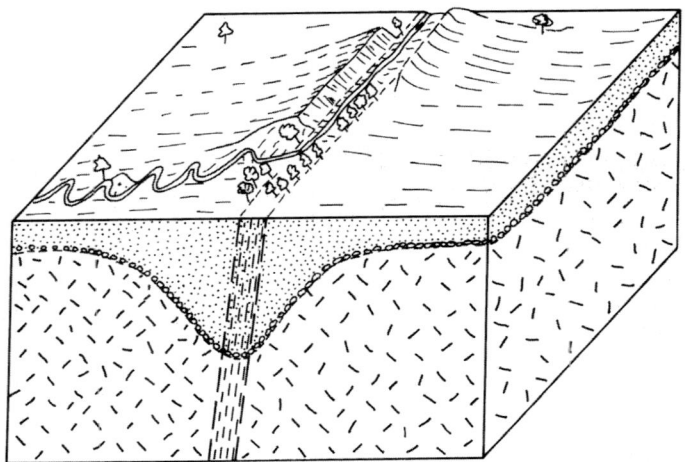

Fig. 16.84. Schematic representation of a lineament. Surface manifestation occurs in terms of alignment of topography, drainage, vegetation etc. The lineament zone possesses a greater depth of weathering. (Singhal and Gupta 1999) (Reproduced by permission, copyright © 1999, Kluwer Academic Publishers)

highly homogeneous throughout. In this situation, the micro-relief of the substrate layer may influence the radar return, if the sand cover is fully penetrated by the incident radar energy. In areas where buried channels are present, the sand cover, which is fine grained and homogeneous, is thicker due to the erosion of the bedrock in comparison to the adjacent areas. This results in weaker return from areas of buried channels (where the bedrock is deeper) and relatively higher radar return from the adjoining areas (where the bedrock is shallower) (Fig. 16.82c,d).

SIR-C image data (Fig. 16.83) furnish another interesting example from an area near Kufra oasis, Libya. The area is now hyper-arid. The valleys and dry 'wadis' or channels are mostly buried under windblown sand. The SIR-C image reveals the system of an old stream valley, now inactive (paleochannel). Prior to the SIR-C mission, only the west branch of the Wadi Kufra was known to exist. The SIR-C data revealed for the first time the existence of a broader (5-km-wide, 100-km-long) east branch of the Wadi Kufra.

16.8.3.3 'Hard-Rock' Terrain

'Hard-rock' terrains are marked by groundwater characteristics quite different from those of 'soft-rock' terrains. As the dominant porosity in such areas is of secondary type, the pattern of groundwater distribution is characterized by non-uniform flow. The most common exploration strategy in such areas is to delineate zones of higher secondary porosity (e.g. fracture systems, faults, shear zones, joints etc.), weathered horizons, vegetation, drainage anomalies and suitable landforms.

Groundwater Investigations

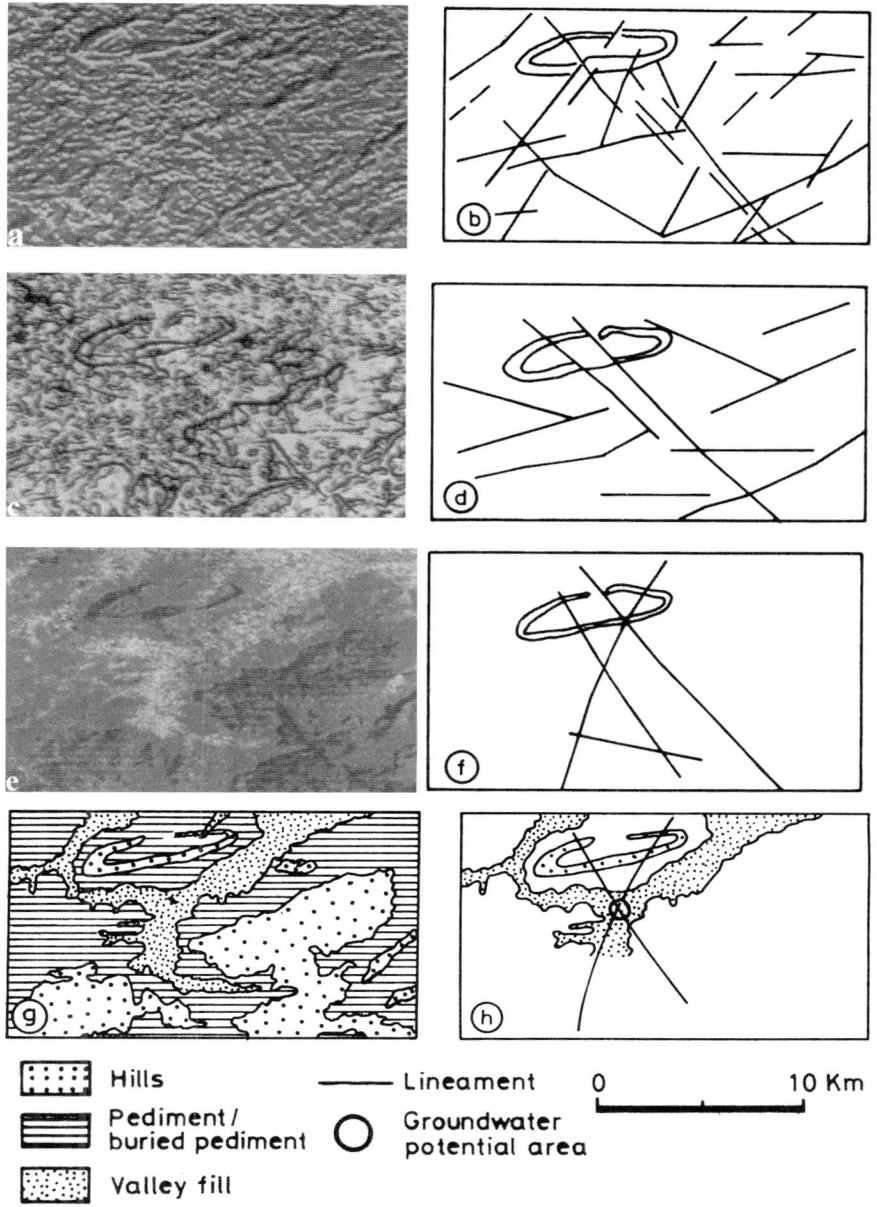

Fig. 16.85. Image processing to enhance lineaments and landforms (Athur valley, India) using MSS data. **a, c** and **e** are Sobel-filtered, Roberts-filtered, and hybrid FCCs respectively (courtesy of A. Perumal); **b, d** and **f** are their respective interpretation maps; **g** is the landform interpretation map; **h** shows the selection of groundwater potential target based on lineament and landform interpretations

1. Fracture traces and lineaments in hard-rock areas. The significance and manifestation of lineaments is discussed elsewhere (see Sect. 16.3.5). Lineaments are surface traces of fractures, faults, shear zones etc., along which topography, drainage, vegetation, soil moisture, springs etc. may become aligned, and they are likely to possess a greater thickness of regolith (Fig. 16.84).

The technique of mapping fracture traces and local lineaments from aerial photographs, which has now been extended to satellite imagery for locating zones of higher permeability in hard-rock terrain, was developed by Parizek (1976; Lattman and Parizek 1964). It has been shown by Parizek (1976) that wells located on fracture traces (lineaments) yield about 10–1000 times more water than the wells in similar rocks and topographical conditions, but located away from fracture traces. Further, wells are found to be more consistent in yield when located on

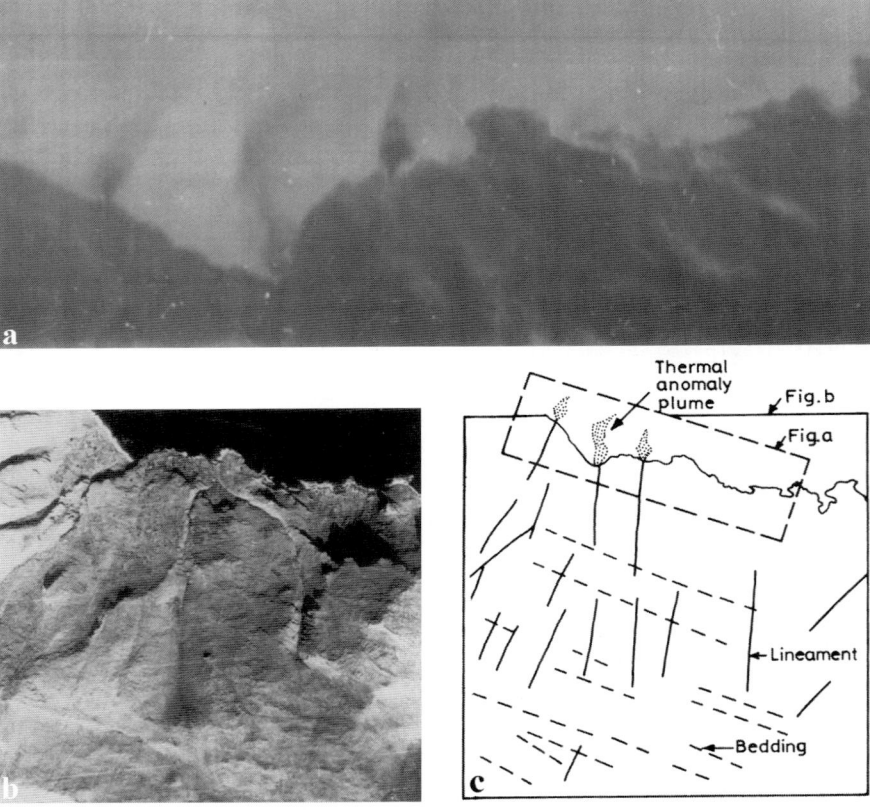

Fig. 16.86a–c. Detection of submarine coastal springs at a test site in Italy using remote sensing data. **a** Thermal-IR aerial scanner (pre-dawn) showing discharge of cooler (darker) groundwater into the warmer (lighter) sea. **b** NIR-band aerial photograph of the same area acquired at noon. **c** Interpretation map. A number of fractures controlling freshwater discharge into the sea can be delineated. (**a–c** Courtesy of J. Bodechtel)

lineaments than under other conditions. Therefore, lineament studies are important for groundwater exploration in hard-rock areas.

In the Athur valley, South India, digital filtering of remote sensing images was carried out to enhance lineaments (Fig. 16.85). These images bring out different lineaments, and the interpretation ought to be done on the basis of commonality and prominence of features.

2. Freshwater springs in coastal areas. Coastal areas are often faced with unique and sometimes severe problems of freshwater supply, especially in hard-rock terrains. In many cases, surface streams may be few or intermittent or with insufficient discharge, and it may be necessary to tap groundwater resources to fulfill the water needs of the area.

In hard-rock terrain, the subsurface water is contained in fractures and zones of secondary porosity. The subsurface water on the land, which is freshwater, moves down the gradient and eventually becomes lost in the sea in the form of submarine springs. Sea water has a relatively higher density owing to higher total dissolved solids. Due to the density contrast, the freshwater rises and spreads over the sea surface, forming a plume, as the mixing process of the freshwater with the seawater goes on concurrently.

Figure 16.86 presents an example. The thermal data provide thermal anomalies (Fig. 16.86a). During the night the land is cooler than the sea, and therefore freshwater, which has a temperature corresponding to that of the land, is cooler than the

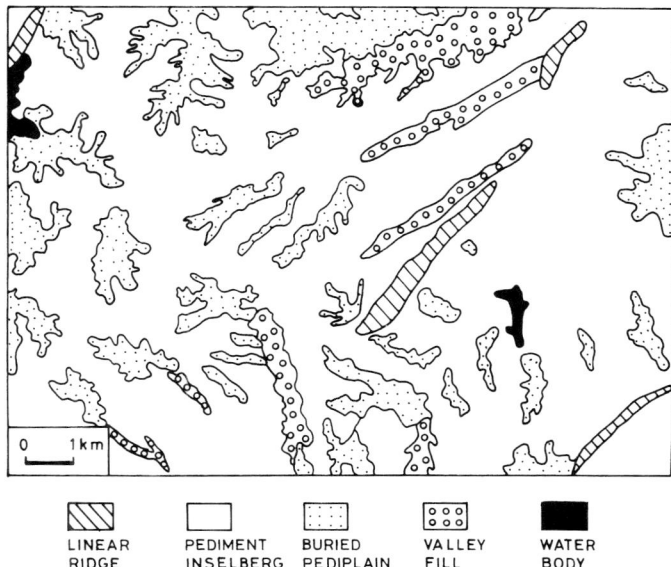

Fig. 16.87. Landform map showing inselbergs/pediments, buried pediments and valley fills in a granitic terrain. The corresponding IRS-LISS-II image is shown in figure 16.78

sea water, the temperature difference being generally of the order of 3–5° C. After eliminating thermal patterns due to surface drainage, the remaining sites of thermal plumes may be attributed only to subsurface discharge. The NIR band image can be interpreted for lineament–geological structure, which can help delineate fractures apparently controlling freshwater discharge into the sea (Fig. 16.86b,c).

3. *Weathered zones and alluvial fills.* For groundwater exploration, identification of suitable geomorphic units is important, as features such as weathered zones, buried pediments and alluvial fills may form potential sites for groundwater targeting in hard-rock terrain. Aerial photographs have conventionally been used for this purpose. Now, with the advent of good-spatial-resolution space sensors, satellite data are also profitably applied for such studies. Figure 16.87 shows an example.

16.9 Engineering Geological Investigations

Remote sensing techniques are now routinely used in engineering geological/ geotechnical investigations. A great value of remote sensing data in such cases lies in their synoptic view, which can be highly useful in predicting likely engineering geological problems and hazards, and suggesting alternative possibilities and solutions (Belcher 1960; Rib 1975). Moreover, repetitive satellite coverage provide vital data on geoenvironmental changes with time. Usually, different stages of engineering investigations require data on different scales and the present-day remote sensing techniques can readily supply inputs on the various scales required. The geological features to be studied depend on the type of engineering project – the commonly required parameters being landform, topography, drainage, lithology, structure, orientation, soil, surface moisture and weathering properties.

16.9.1 River Valley Projects – Dams and Reservoirs

Remote sensing techniques provide a wealth of data, crucial for planning, construction and maintenance of river valley projects. The data are of special utility with respect to the study of the following parameters: (1) topography of various potential sites; (2) shape of the valley (chord to height ratio); (3) size of the catchment and likely river discharge; (4) amount of storage capacity likely to be generated; (5) erosion hazard in the catchment and silt yield; (6) nature of valley slopes and areas of potential landslides; (7) type of bedrock; (8) depth to bedrock or thickness of overburden; (9) structure and orientation of bedding planes/foliation; (10) presence of faults, shear zones, and joints etc.; (11) silting sites in the reservoir; (12) water-tightness of the reservoir and presence of seepage zones, cavities etc.; (13) availability of suitable construction material; (14) sites for

Engineering Geological Investigations 537

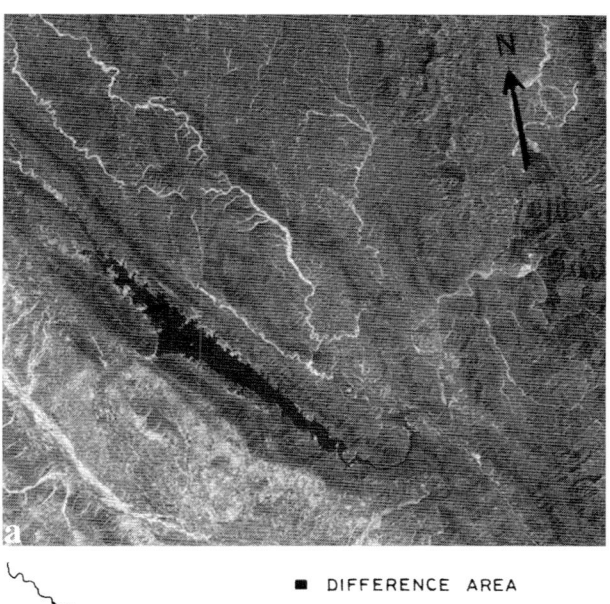

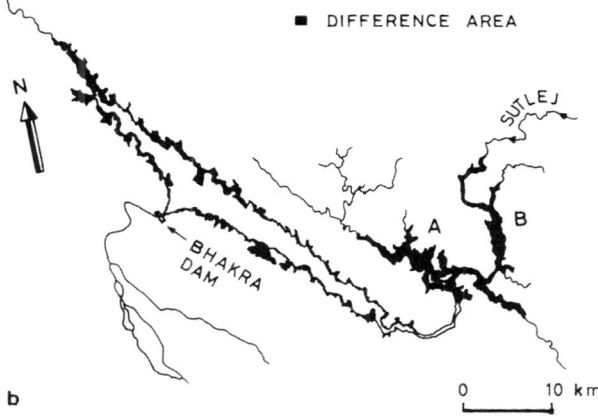

Fig. 16.88. a Landsat MSS2 (red-band) image of the Bhakra dam reservoir in the sub-Himalayas. Note the turbid water and sub-dendritic drainage in broader valley sections, at *A* and *B* (locations shown in **b**). **b** Temporal variations in reservoir lake area; the difference area is marked from a set of post-monsoon and pre-monsoon images. Regions *A* and *B* are identified as silt problematical zones (see text). (**a–b** Gupta and Bodechtel 1982)

ancillary structures; (15) accessibility of the site by road; (16) resources likely to be submerged in the reservoir, such as strategic routes, mineral resources, etc.; and (17) seismic status of the area. Finally, large-scale photographs and images can help in planning the actual location and design of dam structures, such as the alignment of the dam axis and the location of spillway, diversion tunnels, powerhouse, channels etc. Multispectral images can also help identify specific geotechnical problems.

Silting hazards in reservoirs – Bhakra dam reservoir, India. Silting in man-made reservoirs is a well-known hazard in multipurpose river valley projects. It reduces storage capacity of the reservoir and may lead to corrosion of engineering structures. The silt-laden waters exhibit volume scattering at shorter wavelengths (visible), the intensity of scattering and depth of water penetration being greater at shorter wavelengths than at longer ones (Moore 1980). Moreover, the freshly deposited sand/silt commonly has higher albedo in the visible region, due to lack of vegetation compared to the surrounding region. It may often be possible to locate major silt hazard zones in the reservoir on the basis of multiple converging evidence observed on the remote sensing images (Gupta and Bodechtel 1982).

Bhakra dam is a 240-m-high concrete gravity dam, taming the waters of the Satluj river, a tributary of the Indus river. The dam is located in the sub-Himalayan ranges, before the river debouches in the Indo-Gangetic plains. The rocks exposed in the region of the dam foundation and reservoir belong to the sedimentary Siwalik Group of Plio-Pleistocene age. Lithologically, they are poorly sorted clastics, deformed into longitudinal folds and truncated by thrusts, as also observed on the Landsat images (Fig. 16.88a). The dam foundations have been placed on blocky competent sandstones. The lake formed by the dam, known as Govindsagar, spreads over a vast area underlain by sandstones and shales.

The difference in the reservoir lake area demarcated from a set of post-monsoon and pre-monsoon images is shown in figure 16.88b. The same drop in elevation is spread over larger areas at sites A and B than in the steep-walled main reservoir area. This indicates that the ground gradients at sites *A* and *B* are low, and therefore the rocks are likely to be relatively incompetent, possibly clay-shales, in this sedimentary terrain. These areas are repeatedly subjected to loading/unloading and wet/dry cycles, and therefore amenable to increased weathering and degradation. The post-monsoon (high water level) image indicates the presence of turbid/silted water at these sites, and the pre-monsoon (low-water-level) image shows the presence of substantial silted areas. Such features of low gradients, turbid water and silting are absent in other streamlets, or at other places in the reservoir area, indicating that the regions *A* and *B*, constitute the major silt hazard sites in the reservoir (Gupta and Bodechtel 1982).

Environmental studies – Lake Nasser, Egypt. Large river valley projects often have environmental implications associated with them, which need to be monitored. Figure 16.89 presents an example from the Aswan dam, Egypt. Over the past couple of years, four new lakes have been created from the excess water of Lake Nasser. This has brought water to this part of the Sahara for the first time in about 6000 years. At the same time, as also seen from the image, the terrain comprises mainly longitudinal dunes, which would have low permeability. Therefore, the new lakes are likely to lead to problems of water logging and associated hazards. Repetitive remote sensing data can provide useful input in monitoring of such features and phenomena.

Fig. 16.89. Landsat-7 ETM+ sensor image (dated 4 November 2000) showing the area west of Lake Nasser, Southern Egypt. Over the past two years, four new lakes have been created from Nasser's excess water. (Courtesy of NASA)

16.9.2 Landslides

Landslides lead to downward and outward movement of slope-forming material due to gravity and are particularly important in the case of highways, railroads and dam reservoirs. Landslides are best studied on scales of about 1:15,000–1:30,000, which provide spatial resolution of about 5–10 m on the ground. Stereoscopic aerial photographs have conventionally been used for such studies. The higher-resolution satellite sensor data available these days can also be used for such investigations.

Landslides are marked by a number of characteristics on panchromatic photographs (Rib and Liang 1978), viz. sharp lines of break in the topography, hummocky topography on the down-slope side, abrupt changes in tone and vegetation, and drainage anomalies such as a lack of proper drainage on the slided debris. Figure 16.90 gives an excellent example (IRS-1D PAN image) of a landslide occurring in the Gola river (Nainital, Himalayas). The Gola river, following an E–W course, is blocked by landslide debris originating from the southern slopes of the valley. The various photo-characteristics (sharp lines of break, abrupt change in tone, vegetation, lack of drainage on the debris) are very well exhibited.

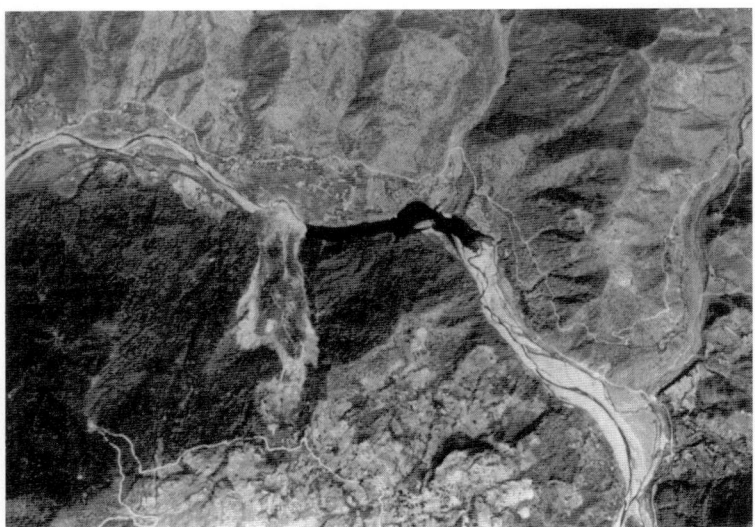

Fig. 16.90. Landslide blocking the Gola river (Kumaon, Himalayas) and forming a lake. Whereas the Gola river is flowing from east to west, the landslide has triggered from the southern bank of the river. (IRS-1D PAN image of 27 March 1999) (Courtesy of A.N. Singh)

Figure 16.91 shows an example of a mudflow at Demizu, Kagoshima, Japan, mapped from the SPOT-2-HRV multispectral sensor.

Debris flow tracks are other features of interest to be considered. Debris flows, which may derive their source material from landslides, move in surges and not in a continuous manner. It is therefore necessary to identify and map debris flow

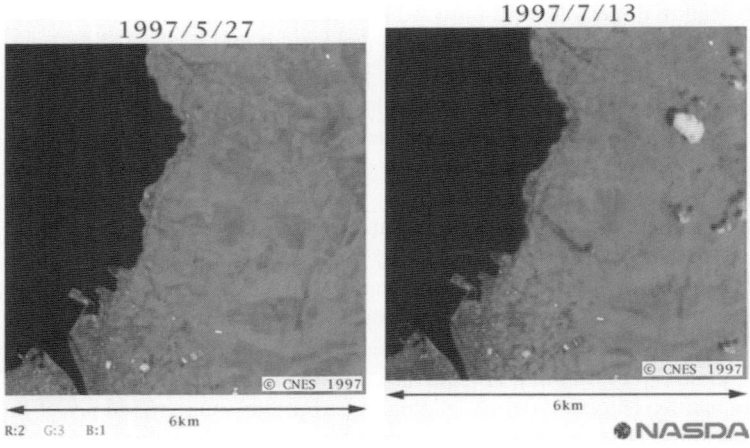

Fig. 16.91. Mudflow at Demizu, Kagoshima, Japan. (SPOT-2-HRV multispectral image data) (Courtesy of NASDA/CNES) (see colour plate XI)

Engineering Geological Investigations 541

Fig. 16.92. A swarm of landslides and debris flow tracks near Uttarkashi, Himalayas (IRS-1D PAN image; bar on lower left corner = 400 m). (Gupta and Saha 2000)

tracks during planning of developmental activities in a mountainous area. (Gupta and Saha 2000). Landslides and the associated debris flow tracks can be easily identified on panchromatic sensor data. Topographic peaks and valleys can be interpreted from differential solar illumination. Light tone (high reflectance) in the panchromatic band is indicative of fresh erosion and absence of vegetation. The light-toned scar edge in higher-elevated areas with a fan apex-downward indicates the source area, and the thin linear light-toned feature is the debris flow track (Fig. 16.92).

While studying remote sensing data for landslides, the most useful strategy is to identify situations and phenomena which lead to slope instability, e.g. (1) presence of weak and unconsolidated rock material, (2) bedding and joint planes dipping towards the valley, (3) presence of fault planes and shear zones etc.; (4) undercutting by streams and steepening of slopes, (5) seepage of water and water saturation in the rock material, and (6) increase in overburden by human activity such as movement of heavy machinery, construction etc. As remote sensing data provide a regional view, areas vulnerable to landslide activity can be delineated for further detailed field investigations (Wieczorek 1984).

For landslide hazard zonation, GIS-based approach is used; this is discussed in Sect. 15.7.

16.9.3 Route Location (Highways and Railroads) and Canal and Pipeline Alignments

Highways, railroads, canals and pipelines are all linear engineering structures, i.e. a single engineering structure is spread over a large distance by length, and a relatively smaller distance by width, so that their investigations have certain aspects in common. Remote sensing data are a powerful aid in planning, exploration, construction and maintenance of such structures.

During the initial planning and feasibility stage, once the terminal points are given by strategic/political/economic considerations, the various possible alternative routes (for highways, railroads, canals or pipelines), can best be selected on photographs, rather than on the ground (Fookes et al. 1985). Aerial photographic mosaic and space images are best used for this purpose. Subsequently, detailed investigations on larger-scale remote sensing data products – preferably with 3-D capability – can throw valuable light on the following relevant aspects: (1) terrain morphology and gradients involved; (2) types of soils and rocks; (3) geological structure; (4) slope stability; (5) volume of earthworks involved; (6) surface drainage characters and water channel crossings etc.; (7) groundwater conditions; (8) availability of suitable construction material; (9) amount and type of clearing required; and (10) property-value compensation. This can give an idea of the total length of the structure and the economics involved.

16.10 Neotectonism, Seismic Hazard and Damage Assessment

Earthquakes cause great misery and extensive damage every year. The technology of earthquake prediction, to enable the sounding of warning alarms beforehand to save people and resources, is still in its infancy. However, the earthquake risk is not the same all over the globe, and therefore seismic risk analysis is carried out in order to design structures (such as atomic power plants, dams, bridges, buildings etc.) in a cost-effective manner. Seismic risk analysis deals with estimating the likelihood of seismic hazard and damage in a particular region. It is based mainly on two types of input data: (1) neotectonism, i.e. spatial and temporal distribution of historical earthquakes, and observation of movements along faults, and (2) local ground conditions, because the degree of damage is linked to the local ground and foundation conditions. Remote sensing can provide valuable inputs to both these aspects. Further, high-resolution remote sensing is becoming a powerful tool in damage assessment. Therefore, the discussion here is divided into three parts: (1) neotectonism, (2) local ground conditions, and (3) damage assessment.

16.10.1 Neotectonism

Earthquakes are caused by rupturing and movement accompanied by release of accumulated strain in parts of the Earth's crust. Most earthquakes are caused by reactivation of existing faults, as they provide the easiest channels of release of strain – the natural lines of least resistance. Remote sensing can help in locating such active and neotectonic fault zones, and this information could be well utilized by earthquake engineers while designing structures.

Neotectonic or active faults are considered to be those along which movements have occurred in the Holocene (past 11,000 years). Seismologists distinguish between neotectonic and active faults, calling neotectonic those which have been active in geologically Recent times, and active those which exhibit present-day activity. However, no such distinction is made in this discussion. Evidence for neotectonic movements may comprise one or more of the following: (1) structural disruption and displacement in rock units of age less than 11,000 years, (2) indirect evidence based on geomorphological, stratigraphic or pedological criteria, and (3) historical record of earthquakes.

1. Structural disruption and displacement in rock units of Holocene age. This forms a direct indication of neotectonic activity. Commonly, aerial remote sensing data coupled with ground data are useful in locating such displacement zones, e.g. in Holocene terraces, alluvium etc. A prerequisite in this case is knowledge of the age of the materials in which the displacement is mapped.

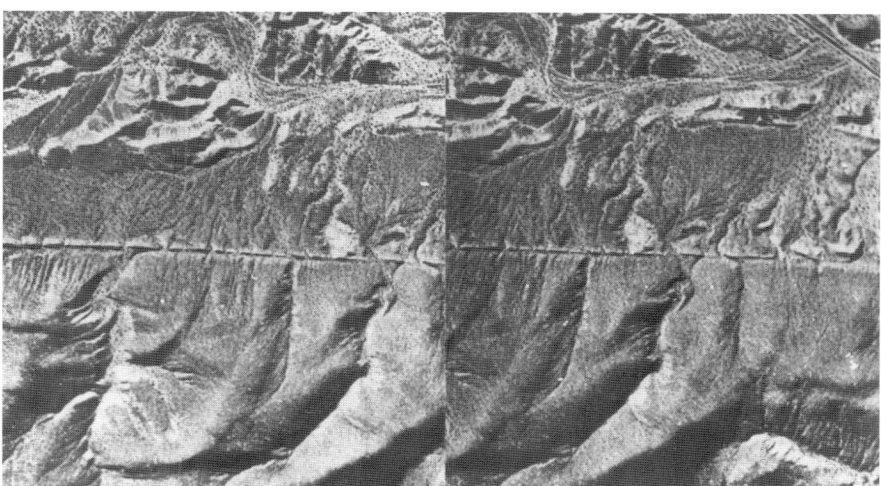

Fig. 16.93. Aerial photographic stereo pair (California, USA) showing numerous off-set streams, drag effects, headless valleys etc. indicating neotectonic faulting; the sense of movement seems to be left-lateral. (USGS GS-GF-139-40)

For example, in the Cottonball Basin, Death Valley, California, Berlin et al. (1980), using 3-cm-wavelength radar images, deciphered two neotectonic faults in evaporite deposits that are less than 2000 years old. The delineation was made possible as the disturbed zone is represented by a somewhat more irregular surface than is found in immediately adjacent areas.

The examples shown in figures 16.94 and 16.96 depicting extensive faults in Holocene sediments could also be grouped under this category.

2. Indirect evidence based on geomorphological features. Mapping of present-day morphological features can provide important, though indirect, clues for delineating neotectonism. Characteristic patterns such as bending and off-setting of streams ridges, sag-ponds, springs, scarps, hanging and headless valleys, river capture etc., and their alignments in certain directions indicate Recent movements. These features may be relatively difficult to decipher in the field, and more readily observed on remote sensing images, due to their advantage of planar synoptic overview. Figure 16.93 shows an aerial photo stereo pair. The off-setting of streams and the alignment of morphological features are very conspicuous. The drag effects imply a left-lateral sense of displacement along the neotectonic fault.

Figure 16.94 shows another interesting example observed on the Landsat images of the Aravalli hills, Rajasthan. The rocks strike NE–SW and are strongly deformed. Although the rocks belong to the Precambrian era, there is evidence of post-Precambrian and neotectonic movements in the area (see e.g. Gupta and Bhaktya 1982; Sen and Sen 1983). The Landsat image shows the presence of an

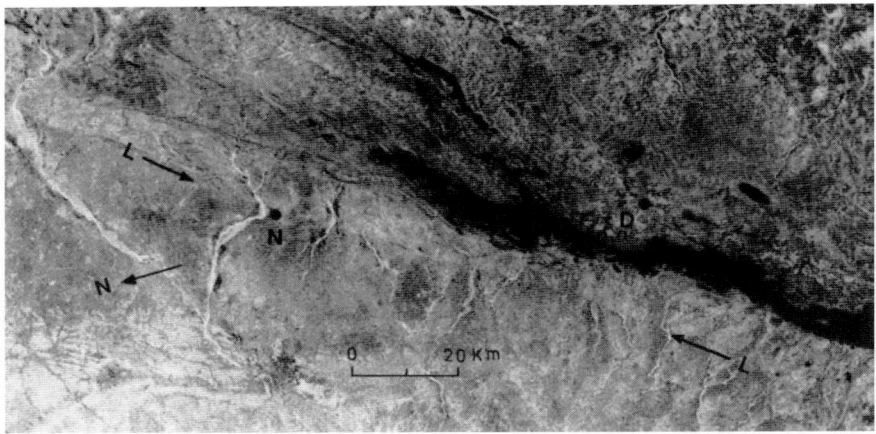

Fig. 16.94. This figure shows a prominent lineament $L-L$ extending for more than 100 km along strike. The lineament is marked by morphological features such as headless valleys, off-set streams and alignment of knick points, indicating it to be a neotectonic fault. It is inferred to have a left-lateral strike-slip component and a vertical component of movement with the eastern block upthrown. [Landsat MSS2 (red-band) image of part of the Aravalli hill ranges, India; N = Nimaj; D = Deogarh]

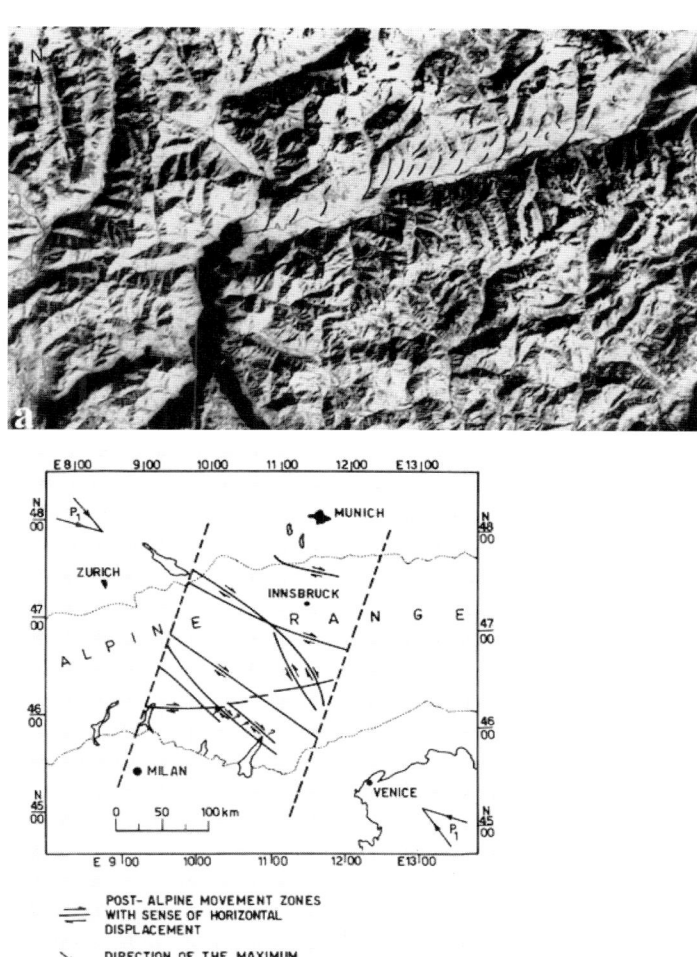

Fig. 16.95. a The Insubric–Tonale Line, Alps. The present-day geomorphological features on either side of the geotectonic boundary are aligned with drag effects indicating a right-lateral sense of displacement. **b** Netectonic lineaments in a section of the eastern Alps interpreted from Landsat images. These lineament features, with their sense of movement are in conformity with the orientation of the present-day stress field (shown as P_1) deduced from fault-plane solutions and in-situ stress measurements (**a,b** Gupta 1977a)

extensive lineament (*L-L* in Fig. 16.94) in the Recent sediments, on the western flank of the Aravalli range. The lineament is marked by numerous headless valleys, off-setting of streams and abrupt changes in gradients of streams (i.e. alignment of knick points), indicating a neotectonic fault. The aerial photographic inte-

Fig. 16.96. The Kunlun fault, one of gigantic strike-slip faults that bound Tibet on the north. In the image, two splays of the fault, both running E–W, are distinctly shown. The northern fault brings sedimentary rocks of the mountains against alluvial fans on the south. The southern fault cuts through the alluvium. Off-sets of young streams (as measured on the image) that cross the fault show 15–75 m of left-lateral displacement. (ASTER image, printed black-and-white from colour composite) (Courtesy of NASA/GSFC/MITI/ERSDAC/ JAROS, and US/Japan ASTER Science Team)

rpretation of Sen and Sen (1983) is in conformity with the above Landsat image interpretation. This fault extends for about 300 km in strike length, parallel to the Aravalli range, and can be called the western Aravalli fault. It is inferred that the fault has a strike-slip displacement with a left-lateral sense of movement, and a vertical component of movement with the eastern block relatively upthrown.

The Insubric–Tonale Line (Fig. 16.95a) is a major tectonic feature in the Alps and runs for a distance of more than 100 km, in a straight E–W direction, disregarding all geological–structural boundaries. On the Landsat image, the Insubric–Tonale Line appears as a well-defined zone, marked by drag effects, indicating a right-lateral sense of displacement. Gansser (1968) wrote of the temptation to speculate a right-lateral displacement along this line, and Laubscher (1971) postulated 300 km of post-Oligocene slip, also with a right-lateral displacement.

A special advantage of satellite data is their synoptic overview, which may facilitate the synthesis of regional features, such as the various neotectonic structural zones. Figure 16.95b shows neotectonic lineament zones deciphered on the basis of Landsat image interpretation in a section of the eastern Alps. The sense of movement along these lineament zones, as interpreted from the Landsat data, is in conformity with the orientation of the present-day stress field as deduced from in-situ stress measurements and fault-plane solution studies in Central Europe (Gupta 1977a).

Figure 16.96 shows the Kunlun fault, which binds Tibet on the north. This is a gigantic strike-slip fault running for a strike length of 1500 km. The geological data show that the Indian plate is moving northwards, which is believed to have resulted in left-lateral motion along the Kunlun fault, uniformly for the last 40,000 years at a rate of 1.1 cm/year, giving a cumulative offset of more than 400 m. On

Fig. 16.97. Aseismic creep along the Hayward fault, California. Based on SAR interferogram from images acquired in June 1992 and September 1997, aseismic creep of 2–3 cm with right-lateral sense of movement has been inferred. (Black-and-white printed from colour image). (After Buergmann et al. 2000) (Source: photojournal.jpl.nasa.gov)

the image, the present-day activity of the fault is clearly manifested in terms of displacement of young streams.

Aseismic creep exhibited by the Hayward fault, California, is another interesting example of neotectonic movement. Figure 16.97 is an interferogram generated from the pair of C-band ERS-SAR data sets acquired in June 1992 and September 1997. A gradual displacement of 2–3 cm, with a right-lateral sense of movement, occurred during the 63-month interval between the acquisition of the two SAR images. The fault movement is aseismic because the movement occurred without being accompanied by an earthquake.

Another example of a neotectonic fault could be Fig. 16.28, depicting Recent vertical movements.

3. Historical record of earthquakes. The data record on past (historical) earthquakes is another type of evidence of seismicity. It can be carefully interpreted in conjunction with data on the structural–tectonic setting in order to derive useful information (see e.g. Allen 1975). The technique of lineament analysis was discus-

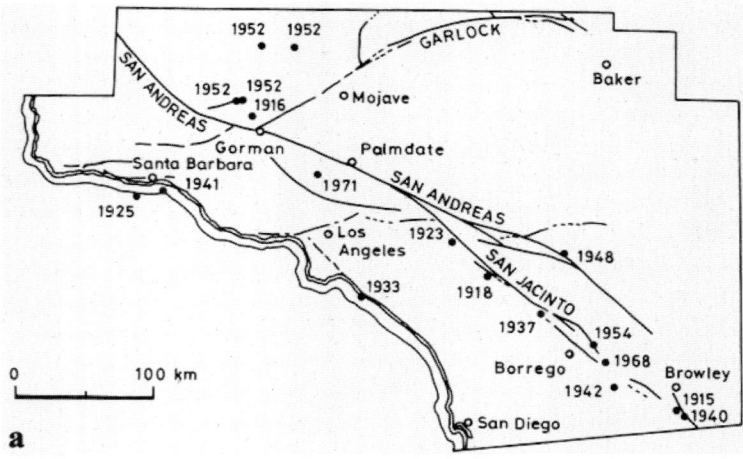

Fig. 16.98. a Relationship of earthquake (magnitude > 6.0, 1912–1974) and Quaternary faulting, southern California (modified after Allen 1975). **b** Perspective view of the San Andreas fault generated from the SRTM (February 2000). The view looks south-east; the fault is the distinctively linear feature to the right of the mountains. (Courtesy of NASA/JPL/NIMA) (Source: photojournal.jpl.nasa.gov)

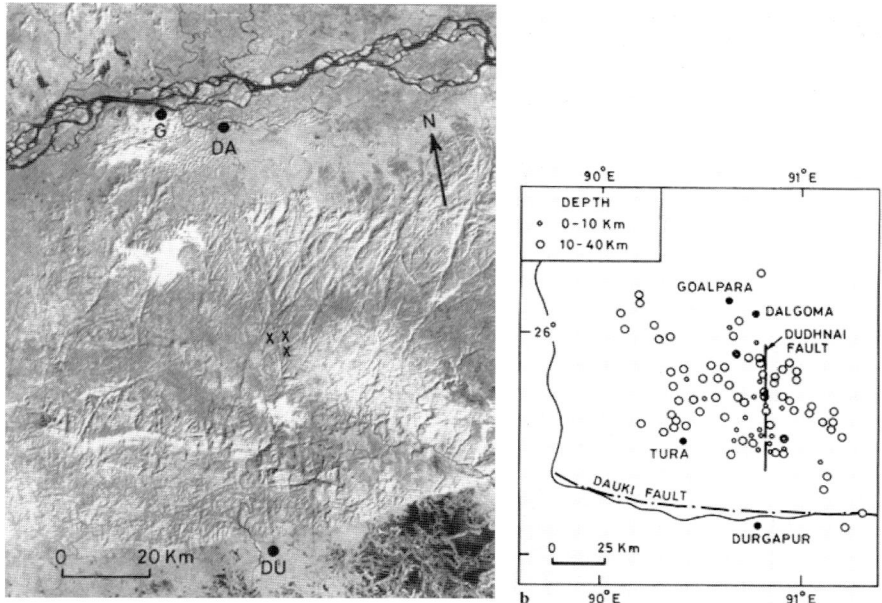

Fig. 16.99. a Landsat MSS image (infrared) of a part of Shillong plateau (India). Note the prominent lineament running between Dalgoma (*DA*) and Durgapur (*DU*), for a distance of more than 60 km. *Cross marks* indicate rocks of carbonatite type reported in the region. In the north is the Brahmputra river. **b** Micro-seismicity map showing alignment of MEQs along the lineament (mapped as Dudhnai fault). (**b** after Kayal 1987)

sed earlier (Sect. 16.3.5). The neotectonic potential of lineaments can be assessed by co-relating historical earthquake data with lineaments. Figure 16.98a shows the distribution of earthquakes (magnitude > 6.0) in the region of the San Andreas fault, California. Figure 16.98b shows an SRTM-data-based perspective view of the San Andreas fault.

Micro-earthquake (MEQ) data can also be utilized in a similar manner for evaluating the neotectonic potential of lineaments. Figure 16.99a is a Landsat MSS image showing the presence of an important lineament in Shillong plateau (India). The segregation of micro-earthquake epicentres along the lineament testifies to the neotectonic activity along this lineament (Fig. 16.99b). Aeromagnetic data also indicate a major tectonic feature coinciding with the lineament (Rama Rao 1999). All this evidence suggests that the lineament has a high neotectonic significance.

4. Delineating extension of known faults. Many faults may splay out and extend laterally, unknown and unmapped. Precise delineation of such features, possible on remote sensing images due to the synoptic views, can be a veritable input in seismic risk-analysis studies.

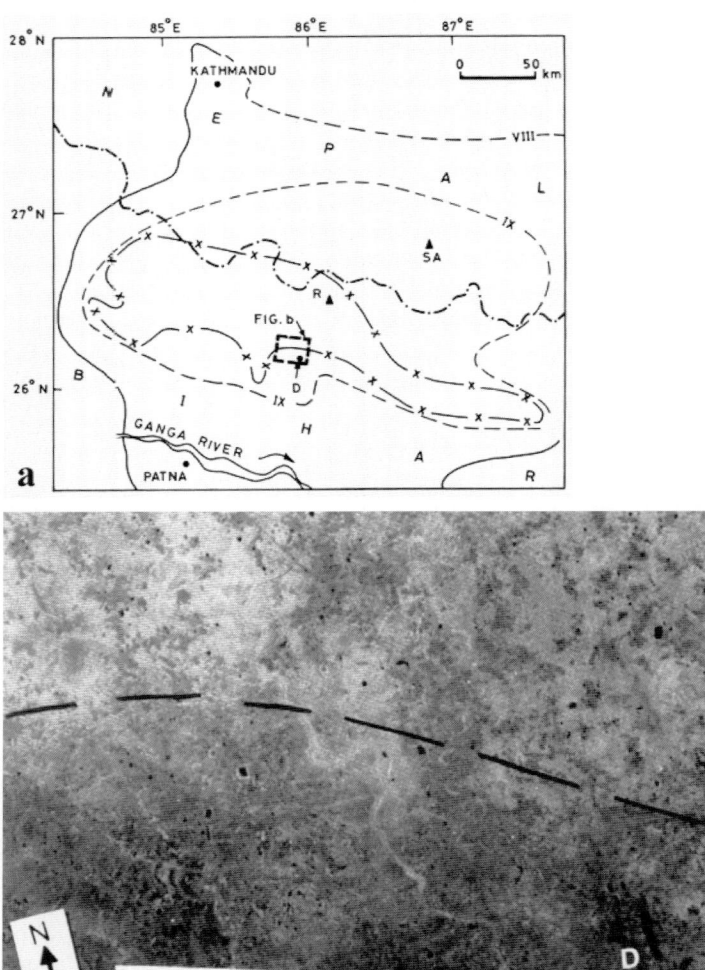

Fig. 16.100. a Disaster map of the north Bihar earthquake, 1934. Isoseismals on Mercalli scale are redrawn after GSI (1939). Much damage occurred due to soil liquefaction in the slump belt. Epicentral estimates of the earthquake after GSI (Roy 1939) (R) and Seeber and Armbruster (1981) (SA) are indicated. Note that the slump belt is located quite a distance from the recent estimates of the earthquake epicentre. **b** Landsat TM image of part of the above area. The dark zone on the Landsat image is a wet clayey zone, north of which lie fine sands, a lithology more susceptible to liquefaction. Note that the boundary passing north of Darbhanga (*D*), seen on the image, matches closely with the southern limit of the slump belt in **a**. (**a,b** Gupta et al. 1998)

16.10.2 Local Ground Conditions

Damage resulting from an earthquake varies spatially. Close to the epicentre, the point directly above the initiation of rupture, disaster is far more severe, and farther away, it generally decreases due to reduced intensity of vibration or motion. Post-earthquake surveys rely on field observations of damage to different types of buildings and structures. Within the same zone of vibration or shock intensity, the damage may vary locally, being a function of both the type of structure and ground conditions. Some of the ground materials forming foundations are more susceptible to damage than others. Remote sensing can aid in delineating different types of foundation materials, such as soil types etc., which may have a different proneness to earthquake damage.

Liquefaction during the north Bihar earthquake (1934). In the north Bihar (India) earthquake of 1934, extensive damage occurred in the northern plains of Bihar (Fig. 16.100a). Based on the initial analysis, it was postulated that the epicentre was located near Madhubani (in Bihar), where the intensity of disaster was also at a maximum. However, subsequent detailed seismological analyses (Chen and Molnar 1977; Seeber and Armbruster 1981) have shown that the 1934 earthquake epicentre was probably located in Nepal, about 100–200 km away. The widespread and severe damage in Bihar was partly a result of liquefaction of soil in the alluvial plains, and the striking feature is that the slump belt – the zone of liquefaction – is located far from the epicentral estimates (Chander 1989).

Liquefaction is a peculiar problem in soils and occurs due to vibrations in saturated, loose alluvial material. It is more severe in fine sands and silts (Prakash 1981) than in other materials. Figure 16.100a shows the zone of soil liquefaction mapped soon after the earthquake of 1934 (GSI 1939). Figure 16.100b is the Landsat TM image (25 May 1986) of part of the area. On the image, a gradational boundary can be marked separating alluvial (fine) sands on the north from a wet clayey zone (dark tone, abundant backswamps etc.) on the south. The areas of fine sands are highly susceptible to soil liquefaction during vibrations, whereas clayey zones are not. It is found that this boundary observed on the Landsat image has a close correspondence with the limit of the liquefaction zone of the 1934 earthquake, at least in the Darbhanga area (Gupta et al. 1998).

Liquefaction during the Bhuj (Kutch) earthquake (2001). A severe earthquake struck western parts of India on 26 January 2001. It caused extensive damage in the area around Bhuj (Kutch), where the epicentre was located. The earthquake was also accompanied by large-scale discharge of water from subsurface to surface, due to soil liquefaction. (It even led to early reports of re-emergence of the mythical Saraswati or Indus river in the area!) The remote sensing data show that it was a temporary phenomenon (Mohanty et al. 2001). IRS-WiFS sensor data are well suited for this study owing to the larger swath and higher revisit cycle. Figure 16.101a,b,c,d give a time series. Figure 16.101a is a pre-earthquake image. Figure

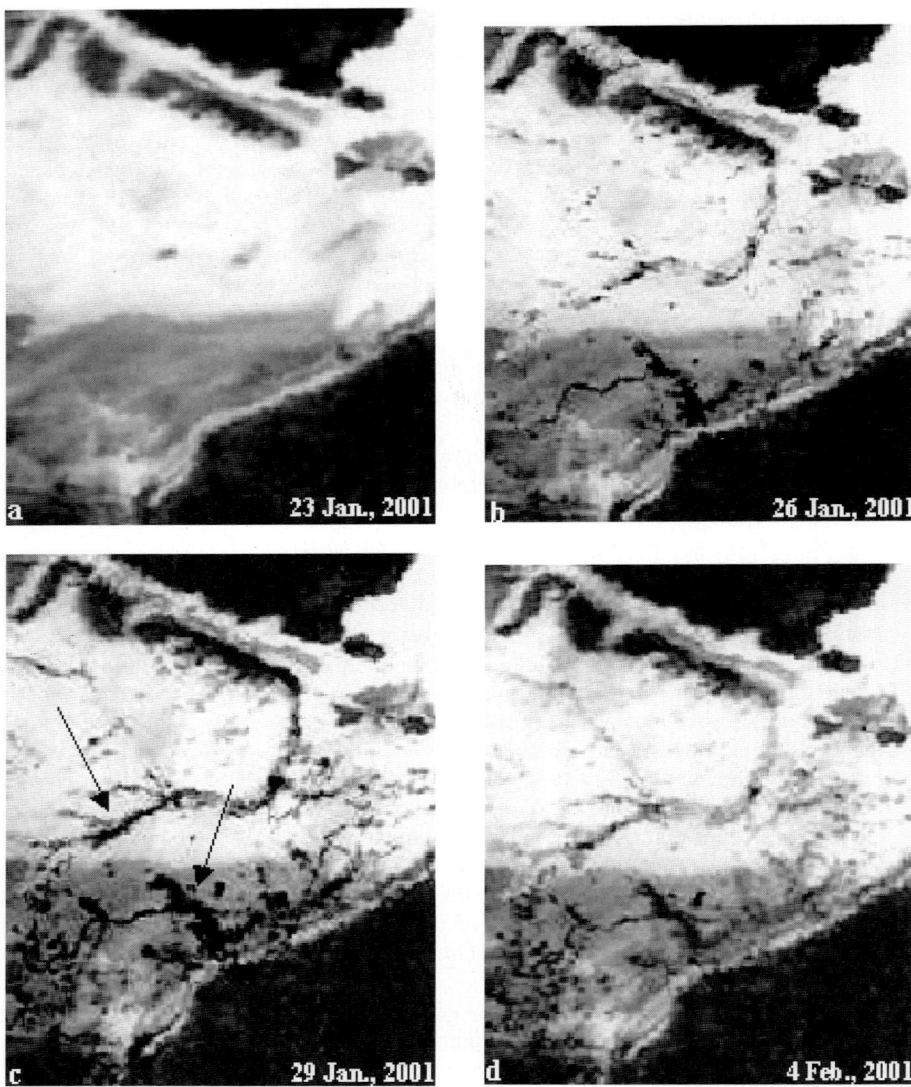

Fig. 16.101. Soil liquefaction during the Bhuj earthquake (26 January 2001); the images are from IRS-WiFS, NIR-band. **a** Image of 23 January 2001, before the earthquake. **b** Image of 26 January 2001, about 100 minutes after the earthquake, shows some water surges on the surface. **c** Image of 29 January 2001 shows the maximum spread of water (*arrows*) among all remote sensing coverages available. **d** Image of 4 Febuary 2001, showing that most of the water channels have dried up. (**a–d** Mohanty et al. 2001)

Fig. 16.102. Damage assessment during the Turkey earthquake (17 August 1999). The figure shows post-earthquake (30 August) and pre-earthquake (8 August) images of parts of the city of Adapazari, one of the worst hit during the earthquake. Buildings that were destroyed by the earthquake can be detected by comparing the two images. (IRS-1D PAN and LISS-III merged image data sets) (Copyright © 1999 Antrix, SI, euromap) (see colour plate XII)

16.101b,c,d were acquired sequentially after the earthquake, and show the emergence of some water on the surface and its gradual drying up.

Thus, the behaviour of ground materials in response to earthquake vibrations is controlled by soil composition and characteristics, and these factors also influence the response of features on remote sensing images. The remote sensing data, therefore, have good potential for providing information on ground materials for earthquake studies.

Fig. 16.103. Damage during the Bhuj earthquake (26 January 2001). This Ikonos PAN (1-m resolution) image acquired on 2 Febuary 2001 shows extensive damage to individual buildings as a result of the earthquake. The image can detect those buildings that have collapsed or have altered rooflines. (Courtesy of Spaceimaging.com)

16.10.3 Disaster Assessment

In a natural disaster spread across a region, management authorities need to have information on sites of heaviest disaster as quickly as possible, so that rescue efforts/emergency relief operations can be optimally planned and provided to the public. Satellite remote sensing with high-resolution data is poised to play a pivotal role in this direction.

On 17 August 1999, parts of the Mediterranean were shaken by a severe earthquake (7.8 magnitude). It was one of the strongest to occur in the last century in the region. The epicentre was located near Izmit, about 80 km east of Istanbul. The city of Adapazari, located only about 40 km from the epicentre, was one of the worst hit. Figure 16.102 shows the utility of pre- and post-earthquake image data sets (IRS PAN and LISS-III, merged) in locating sites of destruction.

Similarly, western parts of India were severely affected by the Bhuj earthquake on 26 January 2001. The high-resolution (1-m resolution) Ikonos image (Fig. 16.103) shows extensive damage to individual buildings and altered rooflines. This type of imagery could be of great assistance to authorities in timely planning of rescue measures, relief operations and damage assessment.

16.11 Volcanic and Geothermal Energy Applications

Areas of volcanic and geothermal energy are characterized by higher ground temperatures, which can be detected on thermal-IR bands from aerial and spaceborne sensors. In usual practice, the thermal-IR data are collected at pre-dawn hours in order to eliminate the direct effect of heating due to solar illumination, and minimize that of topography. However, daytime thermal-IR data can be well utilized for observing volcanic and geothermal energy areas (Watson 1975). The effect of solar heating can be considered to be uniform across a region of flat topography. In the forenoon (09.00–10.00h) and late afternoon ($\approx$16.00h), when thermal crossing pertaining to solar heating occurs, there is no differential effect due to either solar heating or ground physical properties (see Fig. 9.3b); at these hours geothermal areas show significantly higher radiant temperatures than non-geothermal areas. Therefore, thermal-IR remote sensing surveys can be carried out at $\approx$ 09.30h and 16.00h to map volcanic and geothermal energy areas.

16.11.1 Volcano Mapping and Monitoring

Volcanic eruptions are natural hazards that destroy human property and lives and also affect the Earth's environment by emitting large quantities of carbon dioxide and sulpher dioxide into the atmosphere (Fig. 16.104). Monitoring of volcanoes is important in order to understand their activity and behaviour and also possibly predict eruptions and related hazards. Satellite remote sensing offers a means of regularly monitoring the world's sub-aerial volcanoes, generating data on even

Fig. 16.104. Mount Etna in eruption. The image shows prominent white smoke plume rising from the volcano, which had been spewing molten lava, ash and rock for about two weeks prior to the date of image acquisition (31 July 2001). (Black-and-white printed from Ikonos multispectral colour image) (Courtesy of Spaceimaging.com)

inaccessible or dangerous areas. In the Central Andes, for example, using Landsat TM multispectral data, Francis and De Silva (1989) mapped a number of features characteristic of active volcanoes, such as the well-preserved summit crater, pristine lava flow texture and morphology, flank lava flows with low albedo, evidence of post-glacial activity, and higher radiant temperatures (from SWIR bands). This led them to identify the presence of more than 60 major potentially active volcanoes in the region, whereas only 16 had previously been catalogued.

A convenient criterion for regarding a volcano as 'active' or 'potentially active' is that it should exhibit evidence of having erupted during the last 10,000 years. In the absence of isotope data, morphological criteria have to be used. A volcano may be taken as potentially active if it possesses such features as an on-summit crater with pristine morphology or flank lava with pristine morphology (Francis and De Silva 1989). Surface expression of hot magmatic features associated with volcanism is usually of relatively small spatial extent. This implies that the use of thermal-IR imagery with a high spatial resolution would be most appropriate to monitor the surface activity. Our ability to monitor volcanic activity using satellite remote sensing is thus constrained by the sensor resolution – in terms of spatial, spectral and temporal aspects.

Volcanic and Geothermal Energy Applications

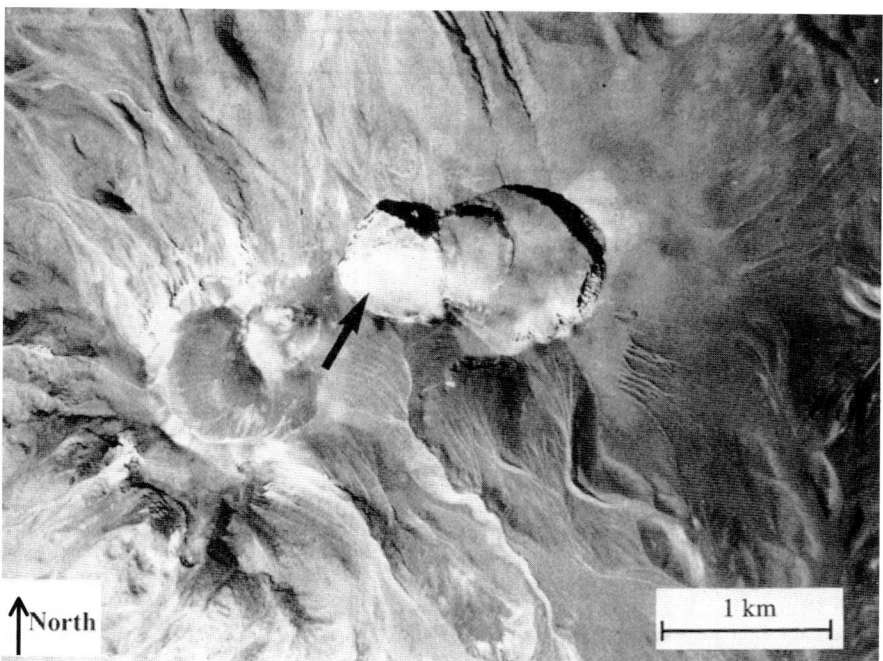

Fig. 16.105. Aerial photograph of summit complex of active volcano at Lascar, Chile (photograph acquired in April 1964). (Oppenheimer et al. 1993)

Lascar volcano, Chile. The Lascar volcano, Chile, has been one of the most active volcanoes recently, and has been a site of many investigations. Figure 16.105 shows an aerial photograph of an active crater at Lascar. Denniss et al. (1998) utilized the high-spatial-resolution data from OPS-JERS-1, in conjunction with high temporal resolution data from NOAA-AVHRR, to document the April 1993 eruption of the Lascar volcano. The pre-eruption image (Fig. 16.106a) shows two major hot spots radiant in both OPS-5 and OPS-8, accompanied by six smaller hot spots. The post-eruption (one day after the eruption) image shows essentially a single radiant source in the middle of an active crater (Fig. 16.106b). The various products of eruption, viz. hot volcanic lava, ash, pumice, plume etc. can be mapped on remote sensing data. Figure 16.107 (FCC of OPS-831 in RGB) shows very clearly the extent of pyroclastic flows during the April 1993 eruption, allowing them to be mapped. Figure 16.107a corresponds to pre-eruption where older lava flows can be seen. Figure 16.107b, acquired one day after the eruption shows the new pyroclastic flows emplaced in the N, NW and SE flanks of the crater.

Temperature estimation. The volcanic vent is found to have a temperature of generally around 1000°C, and emissivity would be in the range 0.6–0.8 (Rothery et al. 1988; Oppenheimer et al. 1991). Features with such high temperatures emit radiation also in the SWIR region (1.0–3.0 µm), as indictated by Planck's law. There-

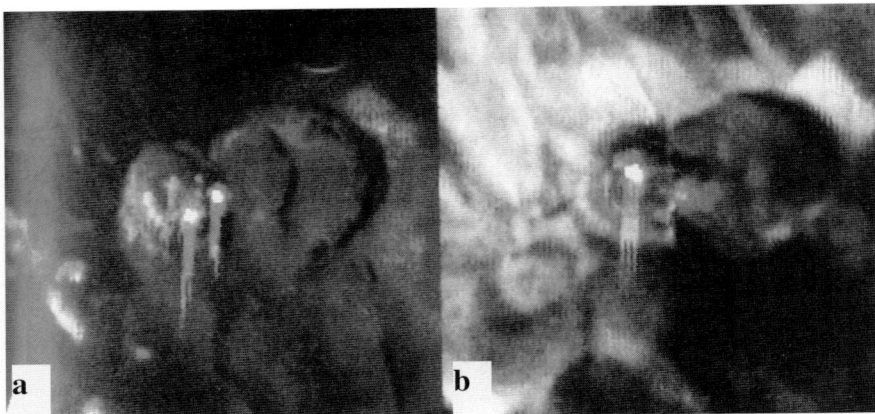

Fig. 16.106. JERS-1-OPS sensor composite of channels 851 (RGB) of Luscar summit craters (images approx. 3 km^2). **a** Pre-eruption; **b** 1-day after the eruption. Thermal hot spots are depicted as yellow if they are radiant in both OPS-8 and OPS-5, or red if radiant only in OPS-8. Severe noise structures are visible in these images, especially vertical and horizontal stripes and along-track blurring. (Denniss et al. 1998) (see colour plate XIII)

fore, although, the SWIR region is generally regarded as suitable for studying reflectance properties of vegetation, soils and rocks, it can also be used for studying high-temperature features.

Although the vent may have a temperature of around 1000°C, it need not occupy the whole of the pixel. For this reason, the temperature integrated over the entire pixel, would be less than the vent temperature, unless the vent covers the pixel fully. The pixel-integrated temperatures (for Landsat TM resolution) are

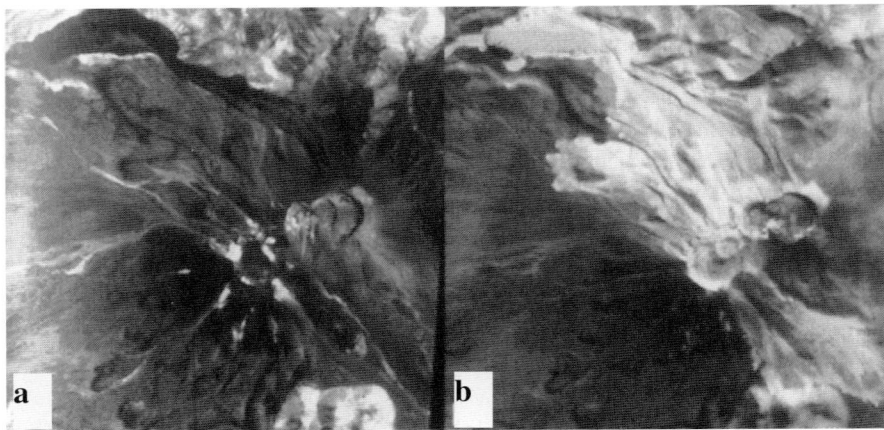

Fig. 16.107 JERS-1-OPS sensor composite of channels 8301 (RGB) of the Luscar volcano, Chile (images approx. 10 km square). **a** Pre-eruption; **b** 1-day after the eruption, showing new pyroclastic flows. (Denniss et al. 1998) (see colour plate XIII)

Volcanic and Geothermal Energy Applications

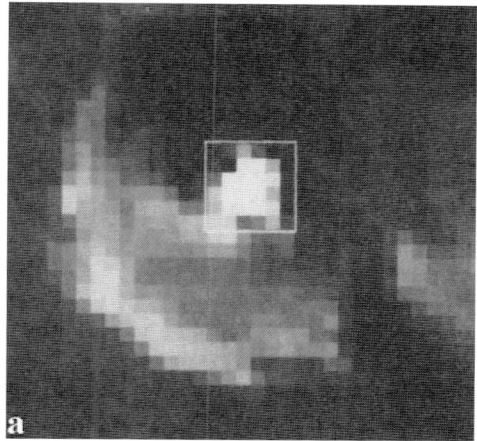

Fig. 16.108. a Anomaly on Landsat TM7 image (16 March 1985) within a crater of the Luscar volcano, Chile; **b** corresponding TM7-based pixel-integrated temperatures. (**a,b** Francis and Rothery 1987)

found to be around 200–400°C.

The spectral regions of different Landsat TM bands together with their sensitivity limits are marked in figure 9.16. The thermal band Landsat TM6 gets saturated at 68°C; therefore, it is not suitable for studying high-temperature objects. TM7 (2.08–2.35 µm) and TM5 (1.55–1.75 µm) together have the capability to measure temperatures integrated over the pixel in the range 160°–420°C and have been used by several workers for such studies.

The procedure of temperature estimation was discussed earlier (Sect. 9.4). It involves the following main steps: (1) determination of emitted radiation for each pixel, including subtraction of radiation from other sources, such as solar reflected radiation, atmospheric scattering etc; (2) conversion of corrected DN values into emitted radiance; (3) conversion of emitted spectral radiance into radiant temperature (pixel-integrated temperature) values. Figure 16.108a shows a Landsat TM7 image exhibiting a thermal anomaly within a crater on Lascar, with pixel-integrated temperatures shown in figure 16.108b.

Sub-pixel temperature estimation. An active lava flow will consist of hot, incandescent, molten material in cracks or open channels, surrounded by a chilled crust. Therefore, thermally, the source pixel will be made up of two distinct surface components: (1) a hot molten component (occupying fraction 'p' of the pixel), and (2) a cool crust component which will occupy the remaining (1 − p) part of the pixel (see Fig. 9.17). Using the dual-band method (Matson and Dozier 1981), the temperature and size of these two sub-pixel heat sources can be calculated. (see Sect. 9.4). Rothery et al. (1988), Glaze et al. (1989) and Oppenheimer (1991) adapted this technique to estimate sub-pixel temperatures at several volcanoes using the SWIR bands of Landsat TM.

Thus, remote sensing using SWIR bands can generate data for understanding the cooling of lava flows – an information that would hardly be available at erupting volcanoes by any other technique.

Plume observations. Plume columns as high as tens of kilometres often accompany large volcanic explosions. Due to such heights, adiabatic cooling takes place. Therefore, the plumes are marked by a negative thermal anomaly and a higher reflectance in the VNIR. Jayaweera et al. (1976) were some of the first to observe a direct eruption of a volcano on remote sensing (NOAA) data and measure the height of the plume from such data as shadow length and sun elevation. The dimensions of a plume at a particular instant can be calculated using simple trigonometric relations and data on plume height. Such data are useful in damage assessment and in planning rehabilitation measures (Holasek and Rose 1991).

At higher latitudes, as there is limited coverage by Landsat.and various other Earth resources satellites, the AVHRR data are usefully employed for real-time monitoring of volcanoes (see e.g. http://www.avo.alaska.edu/).

Differential interferometric SAR is also being applied to volcano monitoring and has been discussed in Section 14.9.

16.11.2 Geothermal Energy

The satellite-acquired thermal-IR data have shown limited utility in the mapping of geothermal areas, mainly due to constraints of ground spatial resolution. Although geothermal areas, as such, have a large areal extent (several tens of km^2), the top surface is by and large relatively cool. Rock material is a poor conductor of heat. The geothermal flux or heat is transported from depth to the surface along narrow fissures and faults, mainly by convection. It is the detection of these narrow warmer zones on the surface that can lead to the geothermal areas, and for this adequate ground spatial resolution in the thermal-IR band is required.

Krafla area, Iceland. An interesting example of application of Landsat TM data in mapping volcanic and geothermal areas is available from the Krafla area, Iceland (courtesy of K. Arnason). In this region the tectono-volcanic activity is manifested through volcanic systems, i.e. a central volcano and an associate fissure swarm passing through it. The activity takes place episodically rather than continuously, with a period of 100–150 years, and each episode of activity lasts about 5–20 years. The most recent activity started in 1975 in the Krafla neovolcanic system and continued for almost a decade. The activity occurred on the 80-km-long N–S-striking fissure swarm, which witnessed an E–W rifting of about 5–7 m. It was accompanied by earthquakes, vertical ground movements, changes in geothermal activity and volcanic eruptions. Although the main magmatic activity remained subsurface, several (nine) phases of short lava eruptions occurred during this time interval. The magma was highly fluid and spread laterally to solidify in a 36-km^2 lava field, no more than 7 m thick on average. A very large eruption took place during 4–18 September 1984, barely about 3 weeks before a Landsat TM pass acquired a set of data on 3 October 1984.

Volcanic and Geothermal Energy Applications 561

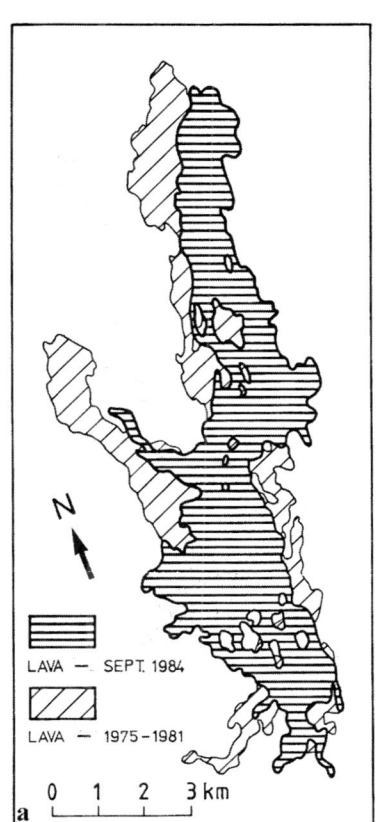 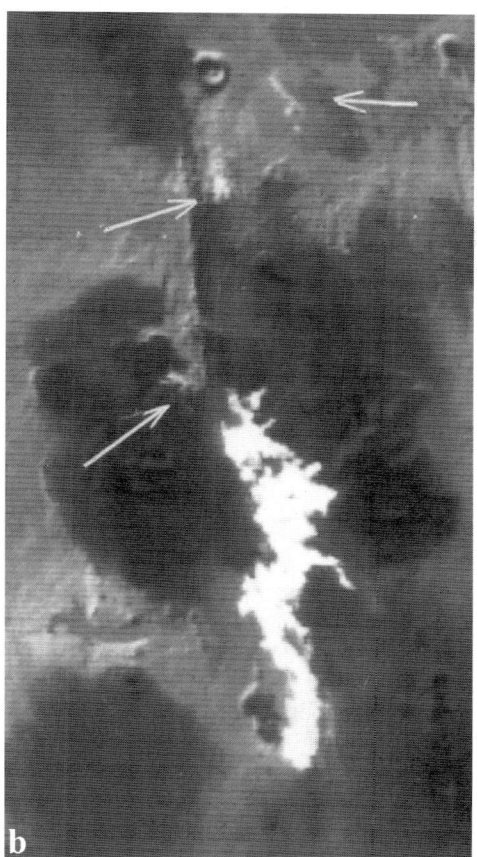

Fig. 16.109. a Map of the areal extent of the lava field in the Krafla area from older (1975–81) and the later (September 1984) eruption. **b** The Landsat TM6 image (3 October 1984) of the lava field; the hot lava from the September 84 eruption is very conspicuous (bright signature), whereas the older lava flows have cooled down and are not discernible. A number of geothermal features are also observed. (**a,b** Courtesy of K. Arnason)

Figure 16.109a shows the extent of the lava fields from older (1975 to 1981) and the later (September 1984) volcanic eruptions. The TM6 image (Fig. 16.109b) shows the hot lava of the September 1984 eruption (bright signature). The older parts of the lava field, which are 3 years older, have already cooled down and are not discernible on the image. Further, some other thermal anomalies are also observed on the Landsat TM6 of the area, where detectability is largely influenced by ground size, and topographic and environmental conditions.

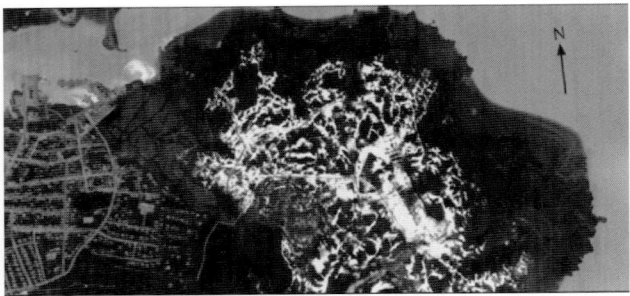 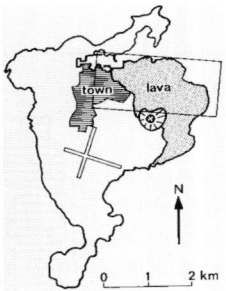

Fig. 16.110. a Thermal-IR (8−14-•m) aerial scanner image (survey 1985) of Heimaey, Iceland, showing the temperature distribution on the lava surface 12 years after the eruption which occurred in 1973. In the west, a village is visible. **b** A map of Heimaey, Iceland with the lava field of 1973; the inserted frame marks the coverage in **a**. (**a,b** Bjornsson and Arnason 1988)

Heimaey Island, Iceland. Figure 16.110a is an 8−14-•m thermal-IR aerial scanner image of Heimaey, an island 10 km off the south coast of Iceland. The image shows part of the local village and a lava flow originating from a volcano (located just outside the image) in an eruption in 1973 (Fig. 16.110b). The aerial scanner image was acquired in 1985. The eruption became famous because of the great efforts to save the village from the flowing magma, as about 6 million tons of sea water was pumped on the part of the flowing magma threatening the village and its fishing harbour. These activities accelerated the solidification and cooling of the lava and are believed to have saved a considerable part of the village. Now the village is served by a central heating system which is operated by pumping water through the porous cracked lava mass, where it evaporates, and the vapour is then collected in heat exchangers at shallow depth.

The image shows strikingly the surface temperature distribution on the lava surface in 1985, 12 years after the eruption. Although the natural pattern of the temperature distribution has been somewhat disturbed by man-made features (roads, pipelines etc.), some important characteristics are obvious. The surface of the lava shows distinct linear or curvilinear structures of high thermal emission. Between these warm structures, the lava surface is cooler. As massive rock is a very poor thermal conductor, the heat reaches the surface mainly by convection through open cracks and fissures. This example shows clearly how important high spatial resolution is in surveying geothermal resources. Although geothermal areas usually extend over tens of square kilometres, the spatial extent of hot sites on the surface, i.e. exposure, is limited, and mostly bound by certain tectonic structures. Mapping of these structures, along which heat is effectively transported to the surface, is important, and this is possible only through sensors of adequate spatial resolution in the TIR.

A 200−500 m broad zone of the lava along the coast appears to have already cooled down completely. This is due to cold seawater penetrating the rock through cracks resulting from extension and shrinking of the cooling lava mass. Also, the eastern-most part of the lava does not show any increased thermal emission. This is due to the cooling activities in 1973 mentioned above. This cold area correlates well with the area which was sprayed with enormous quantities of seawater during the eruption.

16.12 Coal Fires

Coal fires are a widespread problem in coal mining areas the world over, e.g. in Australia, China, India, South Africa, USA, Venezuela and various other countries. These coal fires exist as coal seam fires, underground mine fires, coal refuse fires and coal stack fires. The main cause of such fires is spontaneous combustion of coal occurring whenever it is exposed to oxygen in the air, which may pass through cracks, fractures, vents etc. to reach the coal. The fires burn out a precious energy resource, hinder mining operations and pose a danger to man and machinery, besides leading to environmental pollution and problems of land subsidence. There is, therefore, a need to monitor the distribution and advance of fires in coal fields. Various field methods, such as the delineation of smoke-emitting fractures, thermal logging in bore holes etc., have been adopted with varying degrees of success at different places. However, the study of coal fires is a difficult problem as fire areas are often inaccessible; therefore, remote sensing techniques could provide very useful inputs.

The first documented study of coal fires using thermal-IR remote sensing is that of Slavecki (1964) in Pennsylvania (USA). Ellyett and Fleming (1974) reported a thermal-IR aerial survey for investigating coal fires in the Burning Mountain, Australia. Since then, a number of remote sensing studies have been carried out

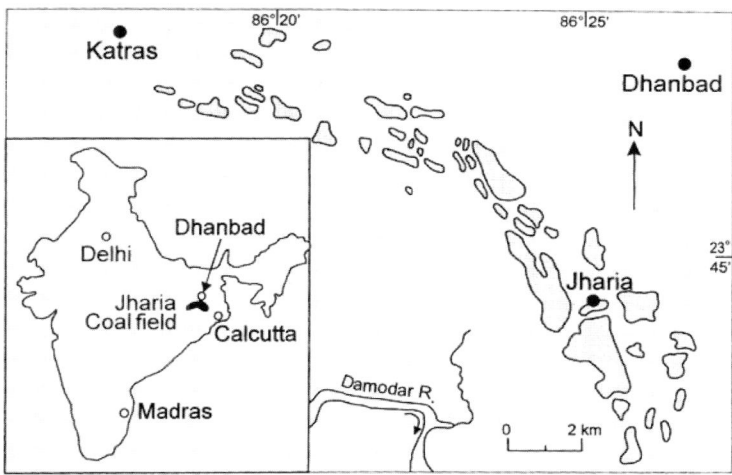

Fig. 16.111. Location map and fire distribution map of the Jharia coal field. (Based on map published by Bharat Coking Coal Limited)

world wide (see e.g. Miller and Watson 1980; Huang et al. 1991; Bhattacharya and Reddy 1994; Mansor et al. 1994; Genderen et al. 1996).

For such problems, information is often sought on a number of aspects, such as: occurrence, distribution and areal extent of fires, whether the fire is surface or subsurface or both, and surface temperature of the ground/fire. Remote sensing can provide useful inputs on all the above aspects.

Fires in the Jharia Coal Field, India

The Jharia coal field (JCF), India, is a fairly large coal field of high-quality coking coal. It covers area of about 450 km^2, where about 70 major fires are reported to be actively burning (Fig. 16.111) (Prasad et al. 1984; Sinha, 1986). Coal fires, both surface and subsurface (Figs. 16.112a,b), are distributed across the entire Jharia coal field. Detailed investigations have been carried out to evaluate the utility of remote sensing technology for the study of coal fires in the JCF (Prakash et al. 1995a,b, 1997; Saraf et al. 1995; Gupta and Prakash 1998; Prakash and Gupta 1998, 1999).

Subsurface fires. The surface temperature of the ground above subsurface coal fires is usually quite low, due to the low thermal conductivity of rocks such as sandstone, shale, coal etc. Landsat TM6 data is best suited for sensing this order of temperature, as the TM6 detector gets saturated only at 68°C (Fig. 9.16), which is

Fig. 16.112a,b. Field photographs showing **a** surface fire and **b** subsurface fire in the JCF. (**a** Prakash and Gupta 1999; **b** courtesy of S. Sengupta)

Fig. 16.113. Processed Landsat TM data of JCF (IHS-processed I = TM4, H = TM6, S = constant, such that red corresponds to highest DNs). The sickle-shaped field above is the Jharia coal field. The relief shown in the background is from TM4. Black linears and patches are coal bands, quarries and dumps. Blue is the background (threshold) surface temperature, anomalous pixels have green, yellow and red colours with increasing temperature. On the lower left is an inset of the psuedocolour TM6 band. The Damodar river appears in the south. (Prakash et al. 1995a) (see colour plate XIV)

much higher a temperature than observed on the ground surface above underground fires. Using density slicing on the Landsat TM6 band, pixels related to

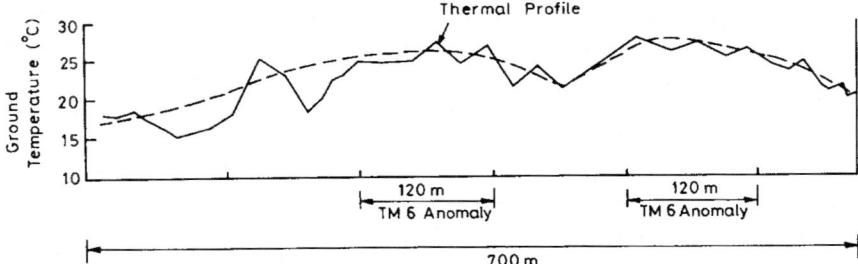

Fig. 16.114. A typical profile of surface temperatures (field measurements) above subsurface fires. The background temperatures are < 24°C. The anomalous ground temperatures reach upto 28°C. (Gupta and Prakash 1998)

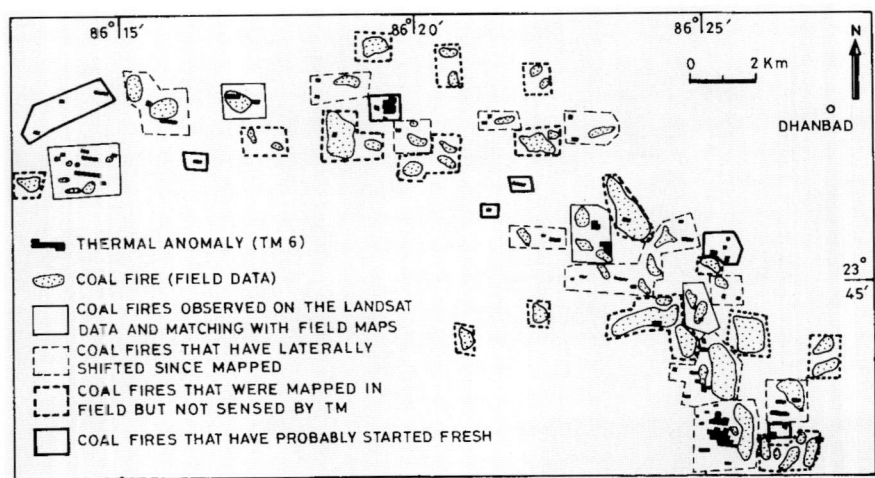

Fig. 16.115. Comparison of thermal anomalies (Landsat TM6) with ground truth on subsurface fires in the Jharia coal field (field data mainly after BCCL and CMPDIL). (Redrawn after Saraf et al. 1995)

higher ground temperatures (subsurface fires) were discriminated from non-fire areas. Estimation of ground surface temperature was carried out from TM6 data, as per the procedure described in Sect. 9.4. The pixels exhibiting thermal anomalies were found to have a temperature range of 25.6–31.6°C. Figure 16.113 shows an IHS-processed image of the JCF. On the lower-left corner is the inset of a subscene in psuedocolour. It is obvious that the thermal anomalies related to various land surface features can be much better located on the IHS image than on the psuedocolour. At selected sites, field temperature measurement profiles were determined, confirming the location of thermally anomalous areas (Fig. 16.114).

Further, figure 16.115 presents a comparison of Landsat TM6 anomalies with ground-reported subsurface fires. It is observed that at many places there is a good correspondence between the two data sets, i.e. thermal anomalies exist at sites of reported subsurface fires. At some places, however, subsurface fire has seemingly migrated since it was mapped (about 15–20 years time interval between time of field data and the Landsat pass). Further, there are areas where a new fire has perhaps started, and also areas where fires could not be detected on the TM6 data, which may be due to either lower sensitivity of the TM6 sensor, or a fire going deeper or getting extinguished in the meantime (a rare occurrence!).

Estimating the depth of subsurface fires. Subsurface fires in the JCF occur at varying depth, ranging from just a few metres up to tens of metres. Estimating the depth of a fire is important not only for combating fire but also for various applications, e.g. for hazard assessment, rehabilitation plans etc. Depth modelling of buried hot features (such as subsurface fire) from remote sensing data is still in its infancy, as it is quite a difficult problem requiring repetitive TIR data and estimates of realistic values of various physical parameters (Prakash et al. 1995b).

Coal Fires

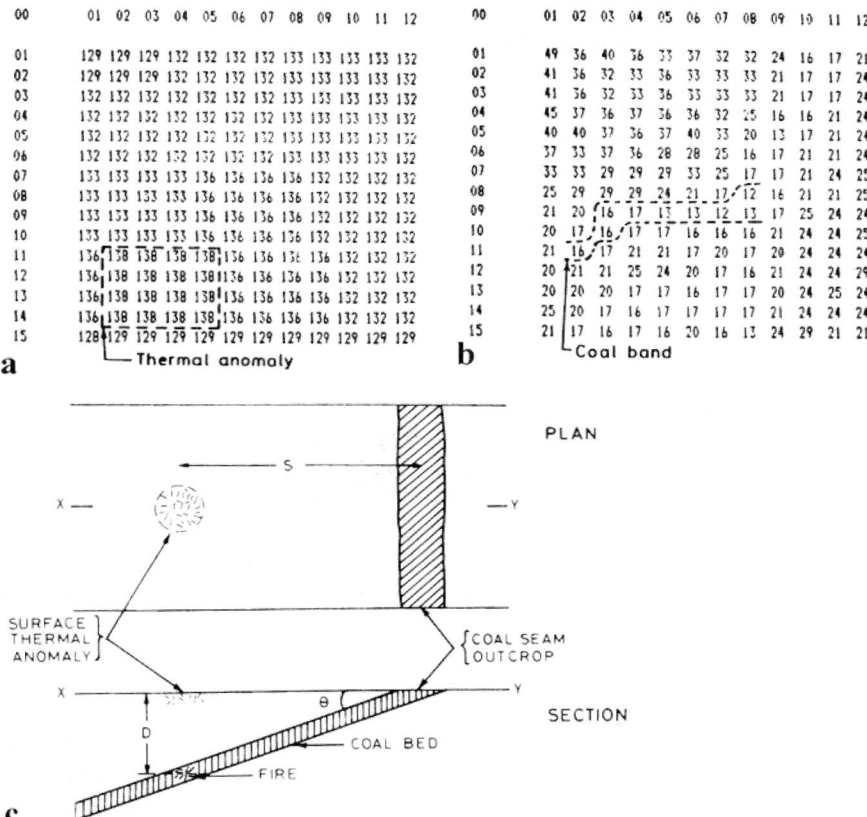

Fig. 16.116a–c. a TM4 and **b** TM6 digital data outputs of the same area showing location of coal seam and thermal anomaly respectively; **c** principle of computing depth of fire from location on thermal anomaly and outcrop. (Saraf et al. 1995)

In simple cases, however, a geometric method can be employed for depth estimation, collectively using information on anomalous thermal pixels and geological–structural setting. The position of a subsurface fire can be determined from thermal anomalies, and VNIR images can provide information on the position of the outcrop (Fig. 16.116a,b). With the field information on the dip of the strata, the depth of a subsurface fire can be computed using simple planar geometry (Fig. 16.116c). The above method has been employed for a few sites and the results have been found to be in reasonable agreement with the field data in the JCF (Saraf et al. 1995). However, the method may have limitations in areas of multiple coal seams, particularly if information on the specific coal seam with the fire is lacking.

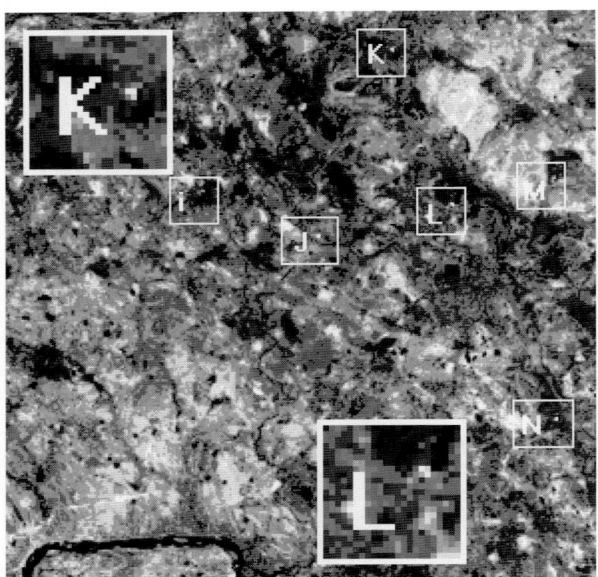

Fig. 16.117. FCC of TM753 (RGB). Windows *I, J, K, L, M* and *N* depict areas of surface fires; yellow pixels correspond to areas of highest temperatures being radiant in both TM5 and TM7; red areas (radiant only in TM7) are relatively lower temperatures (see enlarged windows *K* and *L*). (Prakash and Gupta 1999) (see colour plate XIV)

Surface fires. Surface fires in coal fields are features of high local surface temperature but generally small areal extent. As SWIR bands (TM5 and TM7) have the capability to measure temperatures in the 160–420°C range (Fig 9.16), their data has been used for studying surface fires (cf. Sect. 9.4). The higher DN values in TM7 and TM5 enable identification of surface fires on FCC TM753 (coded in RGB) (Fig. 16.117). Pixels of highest temperature are radiant in both TM7 and TM5, and so appear yellow. Pixels radiant in only TM7 appear red. In many places, a sort of 'zoning effect' is seen where red pixels (of relatively lower temperature) enclose or border the yellow (highest-temperature) pixels.

The procedure of temperature estimation from TM5 and TM7 was described in Sect. 9.4. First, it is necessary to correct the DN values for components of solar reflected radiation and atmospheric scattering etc. The Landsat TM5 and TM7 DN values (corrrected) can be converted into spectral radiance, and then into radiant temperature values. In the JCF, the pixel-integrated temperatures (based on TM7 and TM5) have been found to range between 217°C and 410°C.

Coal Fires 569

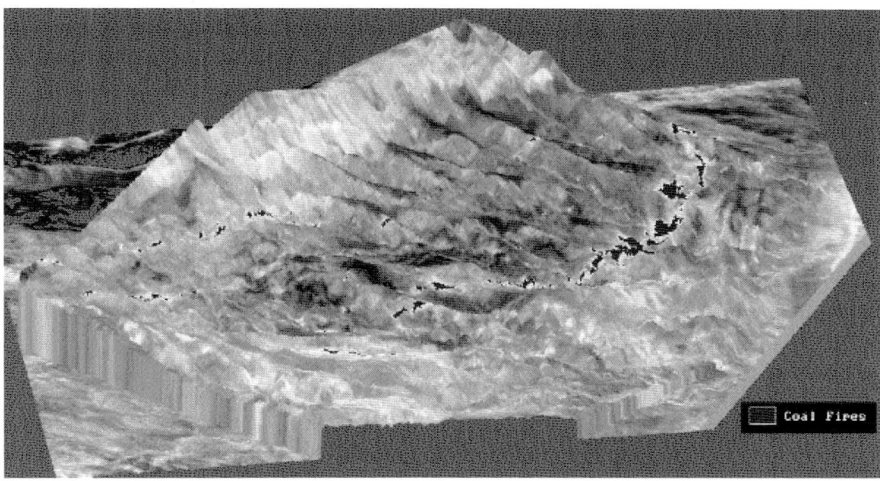

Fig. 16.118. Coal fires in the Xinjiang coal field, China. Airborne thermal-IR image data co-registered and draped over the DEM. (Zhang 1998) (see colour plate XV)

Further, in many cases, fires do not occupy the whole of the pixel, i.e. only a part of the pixel is filled with surface fire. The pixel-integrated temperatures are therefore less than the actual surface temperatures of fires. In cases where the pixels show anomalies in both TM5 and TM7, and no saturation in TM7, it is possible to compute sub-pixel area and temperature using the dual-band method developed by Matson and Dozier (1981) for AVHRR data. The sub-pixel temperatures are found to be in the range of 342–731°C and sub-pixel areas in the range between 0.2 of a pixel (= 180 m^2) and 0.003 of a pixel (= 27 m^2) (Prakash and Gupta 1999).

The above demonstrates the utility of TM data for delineation and mapping of areas affected by surface as well as subsurface fires in coal fields.

Coal Fires in Xinjiang, China

One of the largest deposits of coal in the world occurs in north China, stretching over a region of about 5000 km E–W along strike and 750 km N–S. Coal fires occur in almost all the fields – in scattered or localized or clustered forms. Several workers (e.g. Huang et al. 1991; Genderen and Haiyan 1997; Zhang 1998) have reported coal-fire studies in China using remote sensing.

Coal fires in Xinjiang were investigated by Zhang (1998). Figure 16.118 shows the airborne thermal-IR image draped over the DEM. The coal fires detected from the thermal-IR scanner appear red, and are distributed generally along the NE–SW strike of the coal seam.

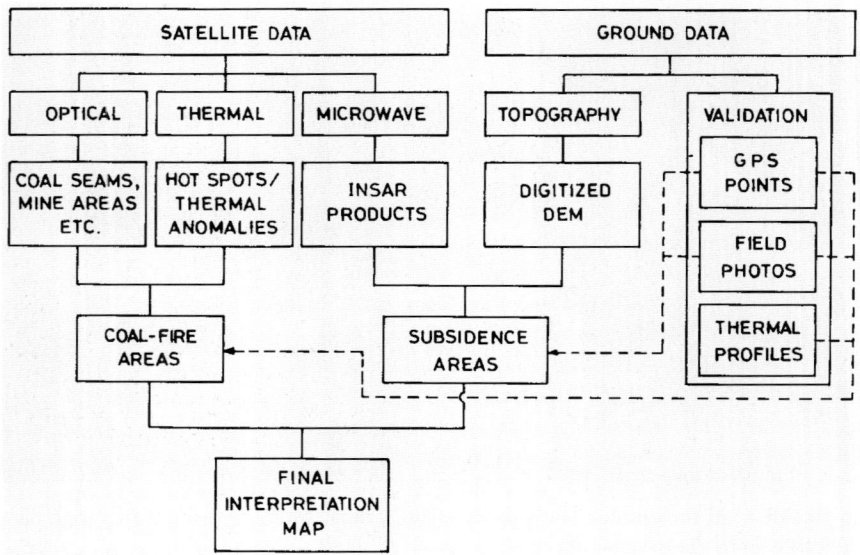

Fig. 16.119. Flow chart of the methodology for data fusion for coal-fire studies (Prakash et al. 2001)

Coal Fires in Ningxia, China

The Ruqigou coal field, Ningxia, is another large coal field in NW China, where numerous coal fires occur. Areas of subsidence due to mining and fires are also widespread. Prakash et al. (2001) used an interesting methodology (Fig. 16.119) for studying coal fires in this area. They used multiple data sets including Landsat TM (day time as well as night-time) data, ERS-1-SAR data and high-resolution (5-m resolution) DEM. The aim was to identify fire areas (from thermal- IR data) and subsidence areas (from InSAR) and examine their correlation.

Figure 16.120a shows an image where TM4 gives the background; the coal fires derived from TM6 are depicted in shades of yellow–orange–red (with increasing temperature). The fires occur broadly in a zone trending NE–SW parallel to the general strike of the beds. Figure 16.120b presents the results of the data fusion study. The colour coding scheme is: R = thermal anomalies from TM6; G = range change showing subsidence (which implies that subsidence areas appear bright green); B = coherence image (such that areas of decreased coherence, e.g. due to dumping etc., appear deep blue). To add spatial information, the FCC was merged with TM4. A yellow colour would imply pixels having thermal anomaly as well as subsidence; however, this is not seen. This implies that at least, during this period in this area, areas of subsidence are not related to coal fires.

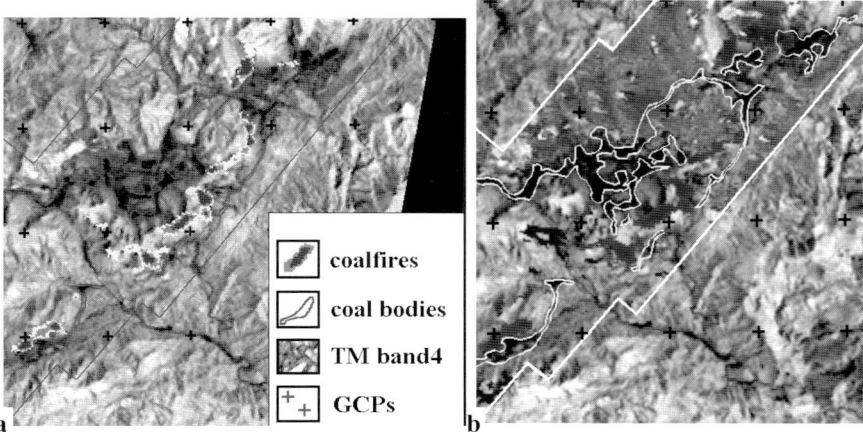

Fig. 16.120. a Processed TM image of a part of Ningxia coal field. Grey background is from TM4. Yellow, orange and red colours depict successively higher temperatures associated with coal fires. The green outlines are the major coal seams, coal dumps and mining areas. **b** Final fusion product with information from VNIR, thermal-IR and microwave regions. TM4 provides the background (gray image). Thermal anomalies from TM6 are in shades of red (coal fires). Bright green corresponds to subsidence from InSAR. Yellow outlines mark the major coal seams, coal dumps and mining areas. (**a,b** Prakash et al. 2001) (see colour plate XV)

16.13 Environmental Applications

Environmental geoscience is a highly interdisciplinary field with offshoots extending into almost all scientific disciplines. The application potential of remote sensing techniques for environmental surveillance stems from their unique advantages: a multispectral approach, synoptic overview and repetitive coverage, i.e. the possibility of examining objects in different EM spectral ranges, in the perspective of the regional setting, and repetitively at certain time intervals. Broadly, the various geo-environmental problems can be related to changes or degradation of land, water, air or vegetation resources. Several aspects for study are possible, for example the following.

1. Land-use changes associated with open-cast strip mining.
2. After-effects of underground mining, such as subsidence etc.
3. Evolution of dumping grounds.
4. Spread and dispersion of smoke plumes from industries and power plants.
5. Discharge of thermal plumes from power plants and industries in rivers and lakes.

6. Deforestation and erosion in river catchments and sediment load studies.
7. General warming of the environment in industrial areas.
8. Discharge from nuclear power plants and associated environmental hazards.
9. Degradation in quality of vegetation, due to atmospheric pollution in industrial areas and metropolitan cities.

In addition to the above, investigations into many other specific problems are possible. Changes in spectral characteristics in the geo-environmental setting in space lead to possibilities for detecting corresponding changes in the environmental setting, through repetitive remote sensing observations. Some examples of applications are given below.

16.13.1 Vegetation

Spectral characteristics of vegetation were discussed in Sects. 3.8 and 16.6.7. Vegetation stress leads to changes in the spectral characteristics of vegetation. Healthy vegetation normally reflects strongly in the near-infrared region, whereas dying or diseased vegetation has a decreased reflectance in the near-infrared region. Further, for stressed vegetation, the red-band absorption feature is shorter and shallower (Singhroy and Kruse 1991).

A widely used parameter for estimating the vigour and density of vegetation is the Normalized Difference Vegetation Index (NDVI). It is expressed as

$$NDVI = (R_{NIR} - R_{RED})/(R_{NIR} + R_{RED})$$

This can also be written as

$$NDVI = (DN_{NIR} - DN_{RED})/(DN_{NIR} + DN_{RED})$$

where DN_{NIR} and DN_{RED} are reflectance values in the NIR and red bands respectively (after atmospheric correction). NDVI values range between -1 and $+1$, a -1 value indicating a water body and $+1$ indicating dense green forest.

The NDVI values are used for a variety of applications – in forestry, agriculture, crop estimation, drought management and change detection. Figure 16.121 is an example of the IHS-processed NDVI difference (change detection) image of a mining area (Jharia coal field). The Landsat TM data sets of 1990 and 1994 were spectrally mutually normalized and used to generate NDVI images, from which a difference image was made (Prakash and Gupta 1998). In figure 16.121 the NDVI difference image is used as hue and TM4 as intensity. The processed image shows vegetation changes in colours (green, yellow, pink); in the background are shown topographic relief, structural features and drainage from TM4.

Environmental Applications

Fig. 16.121. IHS-processed NDVI difference image for change detection in parts of Jharia coal field (India). The Landsat TM 1990 and 1994 data sets were first spectrally normalized and then used for generating NDVI images. The NDVI difference image is used as hue, TM4 as intensity, saturation is constant. Green, yellow and pink indicate areas of vegetation change. Topographic relief, structural features, drainage and transport network appear in the background. (Prakash and Gupta 1998) (see colour plate XVI)

16.13.2 Land Use

Assests obtained from the land surface can be grouped under land resources. Land use deals with the use of the land surface, such as the land area involved in forests, agricultural crops, grazing ground, water reservoirs, waste disposal sites, habitation etc. The degradation of land may result from indiscriminate land use, mining, subsidence and dumping of waste material. Repetitive coverage from remote sensing systems on satellites or aerial platforms is extremely useful for land use monitoring.

16.13.2.1 Strip Mining

Open-cast strip mining is a common method of mining when the deposits are flat-dipping, overburden is thin, and the surface topography does not have much relief. This method involves removal of overburden (stripping) by open excavations on the surface, in order to reach and extract the mineral resource at depth. The mining is usually carried out in benches, and as the earlier sections are mined,

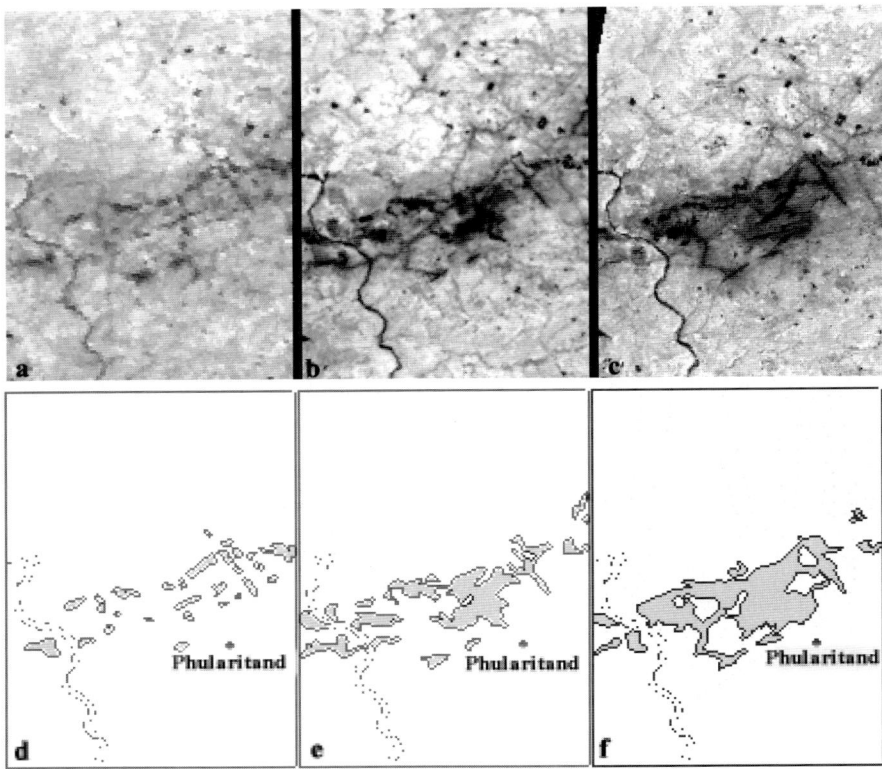

Fig. 16.122a–f. Sequential multisensor NIR-band images showing increase in open-cast mining, Jharia coal field, India. **a** MSS image of 1975; **b** IRS-LISS-II image of 1990; **c** TM image of 1994; **d, e,** and **f** respectively are their interpretation maps. (Prakash and Gupta 1998)

fresh areas are successively stripped. This is accompanied by several environmental problems, such as the following.

1. Degradation of land use due to dumping of mine spoil.
2. Erosion of bare or thinly vegetated spoil dump slopes.
3. Local changes in morphology, landscape and drainage.
4. Discharge of highly mineralized waters from mine sumps, and contamination of surface and subsurface waters.

The Jharia coal field (JCF), which contains India's largest coking-coal reserves, has a long and diverse mining history of both underground and open-cast mining. The land-use pattern is continuously changing due to rehandling of overburden dumps, abandoned mines, mining induced subsidence and surface and subsurface coal fires. Prakash and Gupta (1998) studied land-use change using remote sens-

Environmental Applications 575

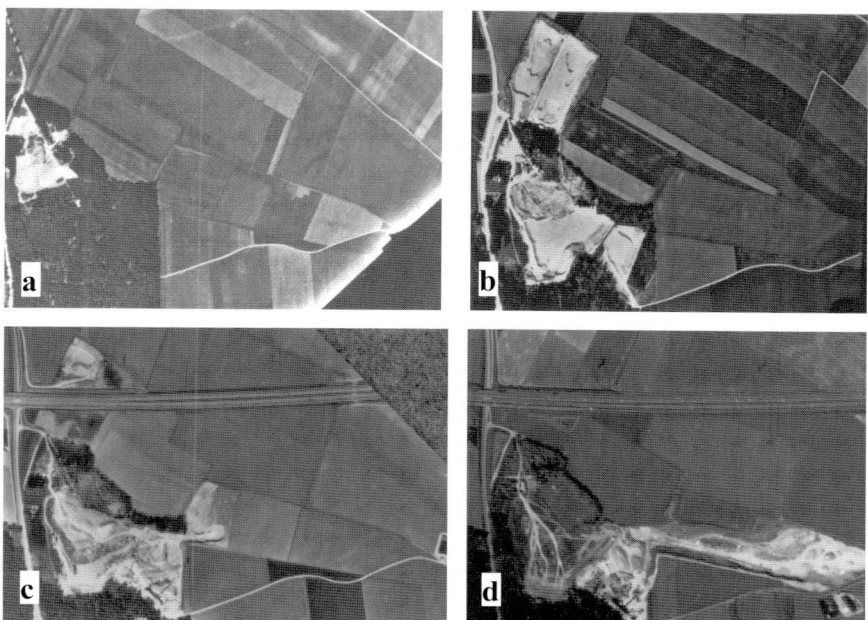

Fig.16.123a-d. Multitemporal aerial photographs showing evolution of a dumping ground at Putzbrunn, SE of Munich, Germany; scale 1:30,000. **a** Photograph, 26 March 1946 (released: Alliierte Komm). **b** Photograph, 10 May 1969 (released 437-69 Reg. präs v. Darmstadt). **c** Photograph, 1 May, 1976 (released: G/7-88566 Reg. V. Obb). **d** Photograph, 28 July 1985 (released: G/z-89540 Reg. V. Obb)

ing –GIS techniques in this area. Figure 16.122 shows the sequential evolution of an open-cast mining area near Phularitand. The extension of mining activity is most noticeable.

16.13.2.2 Dumping Grounds

The evolution of a dumping ground at Putzbrunn, SE of Munich, over a period of about 40 years (1936–1985), is shown in a series of photographs (Fig. 16.123). The dumping area has increased at the expense of forest and agricultural land. Remote sensing can provide information on dimensions, shape, morphology, gradient and surface drainage in the area, which can be used for planning and management.

16.13.3 Soil Erosion

Soil erosion leads to removal of the top fertile humus-bearing soil cover and this problem is quite acute in places. For planning and management, soil scientists

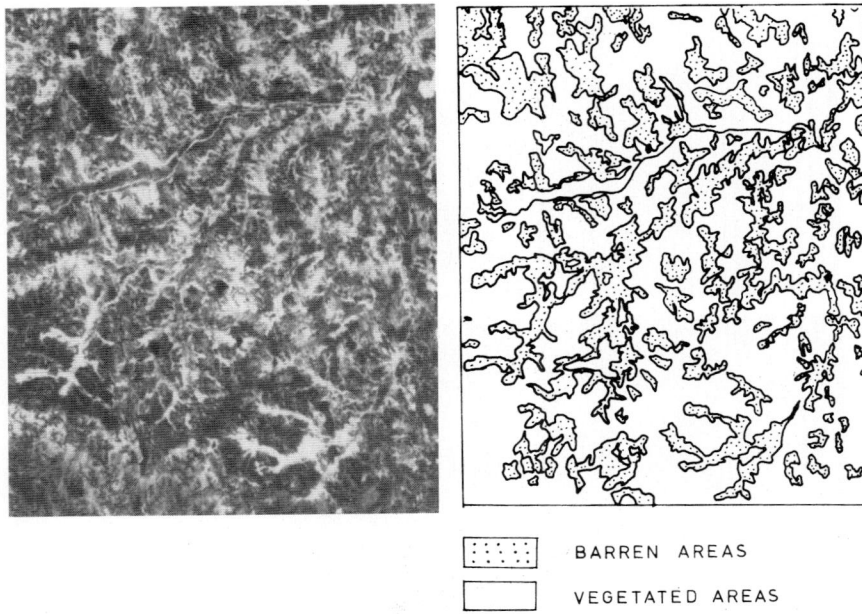

Fig. 16.124a,b. Soil erosion problem in a part of Rajasthan (India). **a** Landsat MSS5 (red-band) image; the dark areas are vegetated; the light-toned areas in dendritic pattern are barren of vegetation and correspond to areas of soil erosion by gullying. **b** Interpretation map

require information on a number of parameters such as: topography, slopes, type of soil, land cover and drainage. Remote sensing can provide data on a number of aspects.

Figure 16.124 shows a typical landscape with a soil erosion problem. The area (about 50 km east of Udaipur in Rajasthan) is covered with a thick blanket of loess. The drainage on the loess is typically dendritic, due to the homogeneous character of the soil. Fine dissection of the loess cover is manifested as barren gullies of varying dimensions, indicating widespread soil erosion.

Coastal areas may be subjected to intensive erosion. Figure 16.125 depicts an example of severe erosion along a 12-km-long stretch, south of Mumbai, on the western coast of India. This figure shows a comparison of the Survey of India topographic map (digitized; surveyed during 1967–68) and the IRS-LISS-III FCC of 4 March 1998. Field data are in conformity with the image interpretation.

16.13.4 Oil Spills

Pollution of marine waters due to oil spills is a common phenomenon worldwide. The various sources of aquatic oil pollution are the following: (a) marine accidents involving collision of oil tankers etc., (b) disposal of oil bearing wastes

Environmental Applications 577

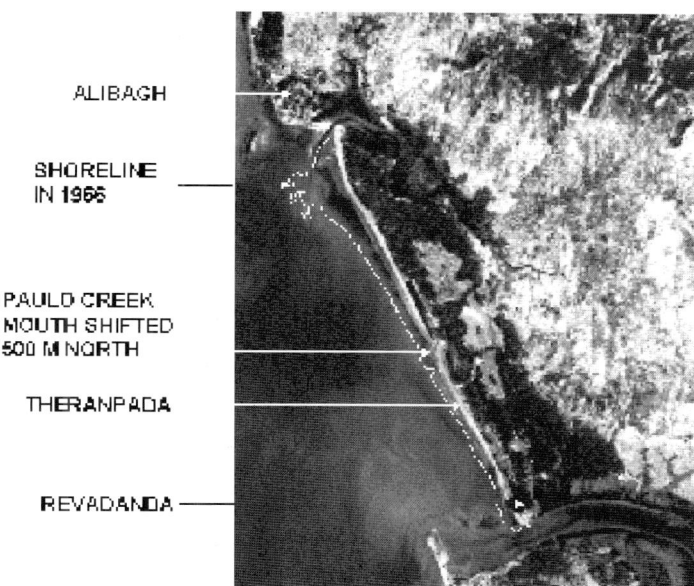

Fig. 16.125. Severe coastal erosion leading to change in shoreline along the 12-km-long coastal stretch between Alibagh and Revadanda, south of Mumbai. Map digitized from Survey of India topographic sheet (surveyed during 1967–68) is compared with IRS-LISS-III image (dated 4 March 1998) co-registered. (Courtesy of A.S. Rajawat and S.R. Nayak)

into the sea, (c) offshore oil drilling operations, and (d) natural submarine seepage. Oil spills lead to a number of environmental effects, such as toxic effects on marine fauna and flora, tainting of seafood, and possible damage to the food chain, as a long-term biological effect. Thus, detection of oil spills is important, for environmental management, as early detection of oil spills can enable timely protection measures.

Types of oil spills. A typical oil spill in water takes the shape of a plume, spreading in the direction of wind and water currents (Fig. 16.126). Concentration, thickness and colour of the oil spill may vary across the plume, and different terms are used for these variations. *Mousse* is the thickest type of oil spill; it forms thick streaks and bands and comprises of a brown emulsion of oil, water and air. *Slick* is a brown- or black-coloured oil layer, relatively thick but less thick than mousse. *Sheen/rainbow* are the terms used to describe a thin, silvery oil layer, commonly exhibiting irridescent multicolour bands, with no brown/black colour.

16.13.4.1 Spectral Characteristics of Oil Spills

1. UV images. UV imaging is a highly sensitive remote sensing method for detecting oil on a water surface and is capable of detecting very thin oil films. The inci-

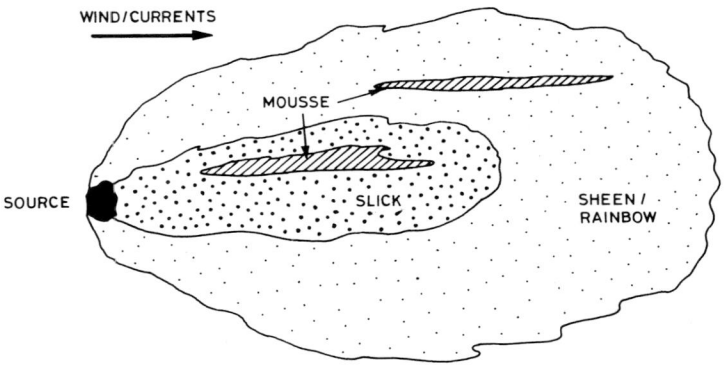

Fig. 16.126. A typical oil spill – shape and terminology

dent UV radiation from the sun (or an artificial source) stimulates fluorescence on the oil surface, but not on the water surface. Therefore, at wavelengths in the range of 0.30 to 0.45 μm (UV–blue region), the oil surface exhibits a brighter signature than the water (Fig. 16.127), due to fluorescence (Table 16.6).

Two types of UV systems are in use – passive and active. The *passive UV* systems record energy stimulated by the sun and therefore can be operated only in the daytime. UV photography (Vizy 1974) and imaging sensors (Maurer and Edgerton 1976) have been used for such applications. The *active UV* systems (such as the

Table 16.6. Remote sensing of oil spills (after Sabins 1997)

Spectral region	Oil signature	Oil property detected	Imaging requirements	False signature
UV (0.3 to 0.4 μm)	Bright	Fluorescence stimulated by sun/laser	Daytime – passive; all-time – active; clear atmosphere	Foam/ seaweed
Visible and reflected IR (0.4 to 3.0 μm)	Mousse – bright Slick – dark Sheen – bright	Reflection and absorption of sunlight	Daytime; clear atmosphere	Wind slicks, discoloured water
Thermal-IR (8 to 14 μm)	Mousse – bright Slick – dark	Radiant temperature controlled by emissivity	Day and night; good weather conditions	Cool currents
Radar (3 to 30 cm)	Dark	Dampening of capillary waves	All-time, all-weather	Current patterns

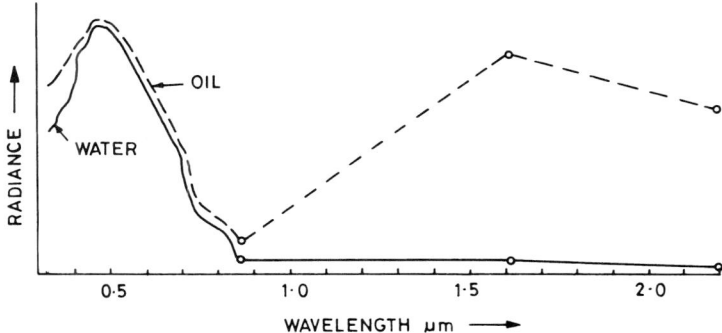

Fig. 16.127. Generalized spectral radiance curves for oil slick and water. (Compiled from data in UV-VIS-NIR region after Vizy 1974; in SWIR region after Stringer et al. 1992)

Airborne Laser Fluorosensor, ALF) record energy stimulated by laser and can acquire images both day and night. Active sensors record fluorescence as a spectrum rather than an image, which can be compared with a reference library of spectra for identifying unknown oil spills (Quinn et al. 1994).

As UV energy is strongly scattered by the atmosphere, a very clear atmosphere and a lower-altitude (< 1000 m) survey are required. On UV images floating patches of foam and seaweed also have bright UV signatures but these can be distinguished from oil on visible-band image data.

2. Visible, near-IR and SWIR images. In the blue spectral band, a thin film of oil on the water surface has a higher reflectance than water (Fig. 16.127). This is responsible for the generally brighter signatures of oil films than water in the visible band. It is not be possible to distinguish between oil film and water on near-IR images. In the SWIR region, mousse and slick have higher reflectance than water (Fig. 16.127), which permits distinction between oil and water.

During the Gulf War in 1991, Iraqi forces released about 5 million barrels of crude oil from a terminal off the coast of Kuwait, into the Arabian Gulf. It formed probably the largest oil spill in history and covered an area of about 1200 km^2 in the Arabian Gulf. Near-real-time monitoring of this oil spill was required to estimate the extent and movement of the slick, which threatened huge desalination plants in Saudi Arabia and Bahrain. Landsat TM images (VIS-NIR-SWIR-TIR) provided valuable data for these studies (Legg 1991).

3. Thermal-IR images. The presence of an oil film on a water surface leads to a lower emissivity. Although the kinetic temperature for oil film/clear water is the same, the difference in emissivity produces a difference in radiant temperature of the order of 1.5°C (Sabins 1997). As most thermal IR detectors have a temperature sensitivity of the order of 0.1°C, this difference in radiant temperature between oil and water can readily be measured. On a thermal-IR image, an oil slick appears

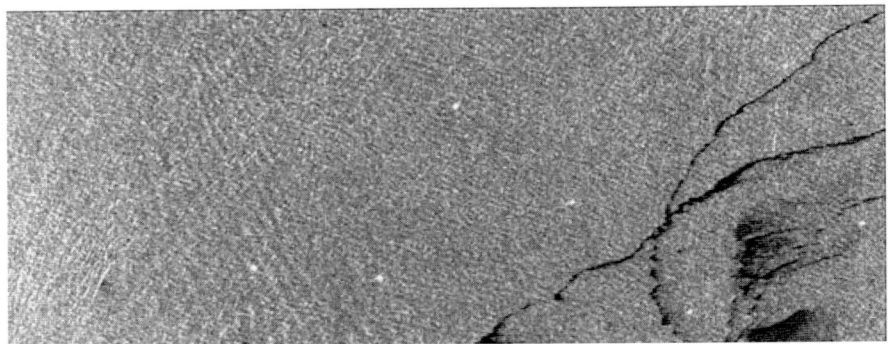

Fig. 16.128. SIR-C image showing oil slick in the 'Bombay High' offshore drilling field, about 150 km west of Mumbai, India. Oil slicks appear dark and drilling platforms appear as bright white spots. Also note the internal waves (left centre) and ocean swell (blue areas adjacent to internal waves). (R = L-band VV, G = average of L-band VV and C-band VV, B = C-band VV; Evans et al. 1997)

cooler (darker) than the surrounding water (brighter signature). However, thicker slicks (mousse) may behave as a blackbody; they may reradiate the absorbed solar energy and thus may show warmer streaks.

The thermal-IR region can provide useful data day and night, although rain and fog may impose constraints on data acquisition. Further, on a TIR image, oil slicks may be confused with cool water currents; this ambiguity can be resolved by using UV and SWIR images.

4. Radar images. Due to the surface tension of an oil film, small-scale surface waves in the sea are dampened, reducing/eliminating the roughness of the sea- water surface. This leads to specular reflection of SAR waves, resulting in low backscatter (dark signature), surrounded by the sea clutter (stronger backscatter, brighter signature) from rough clean sea water. Space-borne radar sensors such as Seasat and ERS -1/-2 owing to their low look angles, are particularly sensitive to differences in water surface roughness. Figure 16.128 is an SIR-C radar image showing oil slick in the offshore drilling field (Bombay High) about 150 km west of Bombay.

On an SAR image, dark streaks may also be caused by smooth water channels (e.g. related to internal waves and shallow bathymetric features), and not only by oil. This ambiguity can be resolved by comparing radar images with image data in other wavelength bands.

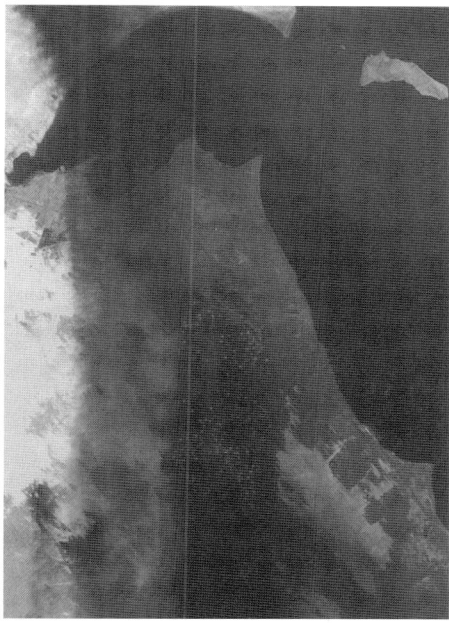

Fig. 16.129. Landsat TM image showing widespread dark smoke from oil well fires in Kuwait (image dated 1 July 1991, printed black-and-white from colour FCC). (Courtesy of EOSAT Inc.)

16.13.5 Smoke from Oil Well Fires

Oil well fires are quite common in oil fields, due to blow-out. In addition, fires may also be set to exhaust an excess of gases present in the reservoir. In the last decade, there were also cases of oil wells catching fire due to the activities of war, as in the case of the Iraq–Kuwait war. All such oil well fires lead to the formation of thick dark smoke which may spread out from the source.

As observation frequency is of paramount importance in monitoring the dynamics of a smoke plume, meteorological satellites (e.g. NOAA-AVHRR) have proven to be very useful for this purpose, with an adequate coverage cycle of about 12 hours. In order to extract the signature of oil fire smoke from AVHRR images, pre-processing and textural analysis are necessary (Khazenie et al. 1993).

On satellite images, in the visible spectrum oil smoke appears black, due to the smoke's high absorptive character and high optical depth. It is quite prominently observed over land-masses (Fig. 16.129). Oil smoke is nearly invisible over water bodies, due to the fact that oil smoke and sea surface both appear dark in the visible wavelengths.

Oil well fire smoke has climatic–environmental implications as it hinders solar insulation, affects the thermal budget of the Earth and leaves toxic effects of gaseous emissions. Limaye et al. (1991), using Meteosat data, reported that smoke from Kuwaiti oil fires could be traced to about 2000 km east of Kuwait. In the case of large fires it is necessary to have accurate estimates on the spread of the smoke plume in order to be able to take civilian/environmental precautions.

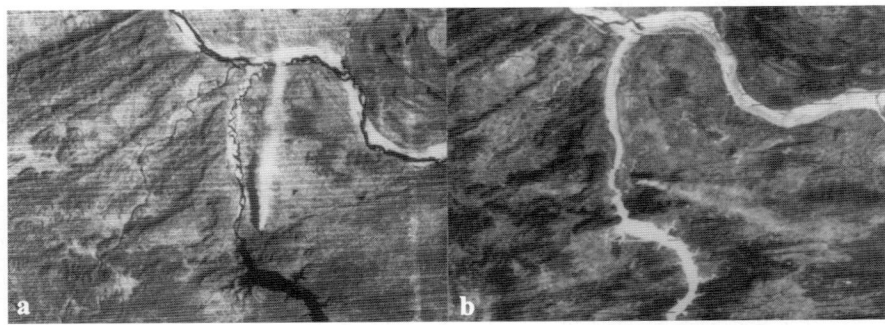

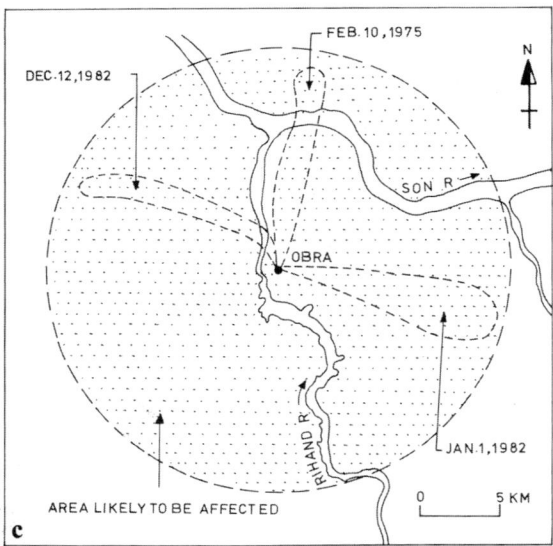

Fig 16.130a–c. Smoke plume emanating from a power plant (Obra, India) and its dispersion in different directions. **a** Smoke plume dispersing in northern direction; Landsat MSS image (red band) dated 25 February 1975. **b** Smoke plume dispersing in eastern direction; Landsat MSS image (red band) dated 1 January 1982. **c** Area around the power plant likely to be affected by dispersion of the smoke plumes. (**a–c** Philip et al. 1988)

Environmental Applications 583

16.13.6 Atmospheric Pollution

Air pollution is caused by the discharge of industrial waste and exhaust into the atmosphere. It affects visibility, and causes acid rain and respiratory problems. Finally, when the particulates from the atmosphere settle on the ground, the deposition of dust on the canopy affects photosynthesis and retards the growth of plants. The dispersion of pollution in the atmosphere depends on the intensity of the industrial discharge, type of discharge, and atmospheric conditions; that some plumes may be carried for very long distances was, in fact, first revealed only by remote sensing data. Conventionally, atmospheric pollution has been investigated by sensors on kites, balloons and aircraft. Now, satellite remote sensing has been found to be very effective in mapping dispersion patterns and regional spread of atmospheric pollutants.

Dispersion of smoke plumes from the coal-fired Obra thermal power plant is shown on Landsat MSS images (Fig. 16.130). The plume is picked up on the shorter-wavelength bands, which are scattered by the suspended particulates (fly ash) contained in the smoke plume discharge. The trend, plume size and its dispersion vary with the prevailing wind direction. The plumes in figure 16.130a,b are differently oriented owing to the prevailing wind directions. Figure 16.130c indicates the area likely to be affected with pollution due to the dispersion of smoke emanating from the coal-fired thermal power plant.

Finally, it can be summarized that the advancements in remote sensing technology have added tremendously to man's ability to monitor the environment.

16.14 Future

Remote sensors have come to stay and have a very bright future. Several modifications and improvements to the existing ones have been proposed, and still others are to come.

What are the functions that a remote sensor might perform in the future? Let us try to imagine. Figure 16.131 shows a cartoon sketch. The EM radiation serves as the most efficient link between the sensor and the sensed, and therefore the remote sensor would have wide 'eyes', open day and night. The potential of 'smell' is demonstrated by numerous animals, and low-altitude air sampling for chemical dispersion patterns has been tested; the remote sensor would have a long 'nose', possibly on the aerial platform. The audible frequencies attenuate quickly through the air column and 'ears' may have only limited use. There would be other 'sense organs', e.g. for magnetic force (magnetometer), weight (gradiometer), radioactivity etc. The remote sensor would certainly have a very powerful computer to isolate signal from noise and recognize meaningful patterns (artificial intelligence!), the entire system being possibly robot controlled.

Fig. 16.131. Future remote sensor – a cartoon sketch. (Courtesy of S.K. Das)

This should help to evaluate the finite natural resources available to humanity for their optimal utilization, to maintain the natural environment, and to preserve the precarious ecological balance on the Earth.

Appendices

Appendix 4.1: What is Colour?

In simple words, colour is the visual effect produced by EM radiation incident on the retina of the human eye. The average human eye is sensitive to radiation from approximately 0.4 to 0.7 μm, which is called the visible region; in exceptional individuals, visibility may extend to slightly shorter and longer wavelengths. The effect of incident radiation, depending upon the collective wavelengths, leads to colour vision.

The visual process can be classified into two types: achromatic and chromatic. The achromatic process is one in which only the cumulative brightness or intensity variation is portrayed. This is a one-dimensional variation. Black-and-white pictures are typical examples. On the other hand, in chromatic processes, the relative intensities of different wavelengths are of interest (and not merely the cumulative intensity of radiation). Colour vision falls in this category.

Colour is described in terms of three parameters: intensity, hue and saturation. *Intensity* is the brightness or scene luminance, i.e. luminous intensity. *Hue* is the dominant wavelength present. In a particular visual process, several wavelengths may be present at the same time, and the most dominant of these is the hue. *Saturation* relates to the proportion of the dominant wavelength vis-à-vis other wavelengths present, i.e. it describes the percentage of the dominant wavelength as contained in a mixture with white light. It gives the relative purity of hue. A pure hue, containing no white component, is a colour with a single wavelength; it is said to be saturated. A colour field thus has a three-dimensional variation – in terms of hue, saturation and intensity.

The human eye can distinguish about only 20–30 gray tones (achromatic vision), whereas it can distinguish more than a million colours (chromatic vision). Due to this high-order difference, scientists prefer colour pictures to facilitate better discrimination and identification of features of interest.

An interesting aspect is that colours are amenable to mixing, i.e. addition/subtraction. Varying visual effects (i.e. colours) can be produced by mixing various colours in different proportions. For example, white light is a visual sensation produced by mixing different wavelengths, e.g. the well-known VIBGYOR. In addition, white light can also be produced by combining blue + green + red, blue + yellow, red + cyan, or green + magenta. It is not possible to definitively

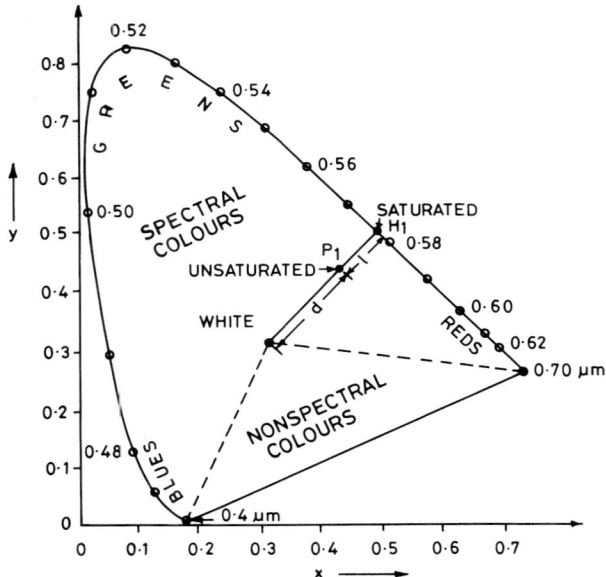

Fig. A.4.1.1. The CIE chromaticity diagram

identify actual colour input components in a particular instance, since the same visual effect can be generated by several alternative combinations. For example, yellow can be produced by green plus red, or by white minus blue.

Colour perception is highly subjective and therefore a standard colour classification scheme, acceptable for all purposes, is difficult to evolve. Several schemes and models of colour description have been formulated (see e.g. Harris et al. 1999). We discuss here three colour models which are more basic in understanding the nature of colour and its applications in digital image processing:

1. CIE colour system
2. RGB colour model
3. IHS colour model.

1. CIE colour system. In 1931, the Commission Internationale del' Echlairage (International Commission on Illumination) adopted a systematic method for colour designation, called the CIE method. The CIE started by specifying a set of X, Y, Z artificial primaries. These primaries can be obtained by mathematical transformations of the real spectrum colours, but as such do not represent the spectrum colours. The relative amounts of X, Y, Z primaries required to match the spectral colour of a particular wavelength are referred to as tristimulus coefficients (x', y', z') at that wavelength. The relative amounts (tristimulus coefficients) can be converted into fractional amounts x, y, z, called the trichromatic coefficients, such that $x + y + z = 1$. Thus, any colour can be represented in terms of X, Y, Z prima-

Appendix 4.1: What is Colour?

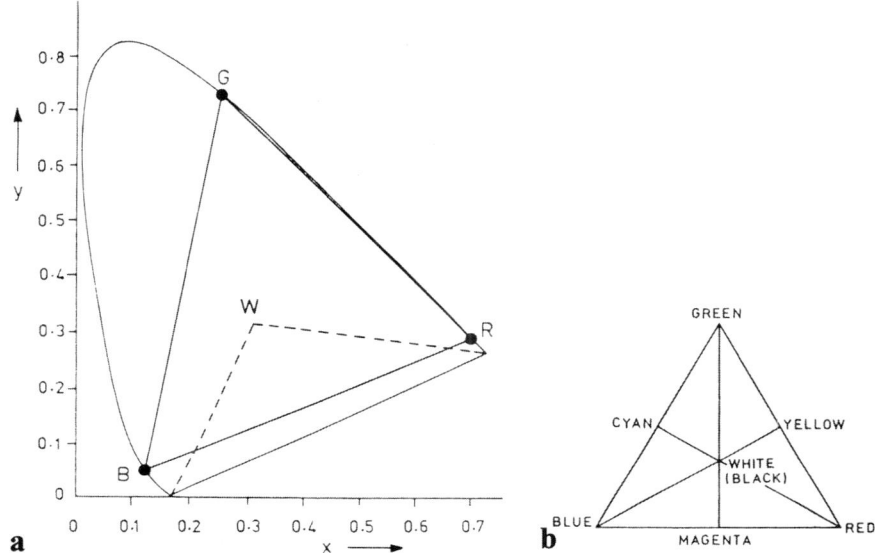

Fig. A.4.1.2. a The colour field generated by B, G, R primaries. **b** Schematic positions of colours in BGR (or RGB) ternary colour diagram

ries by specifying any two of the x, y, z coefficients. This facilitates a two-dimensional representation of the colour field. The graphic representation in terms of chromaticity coordinates x and y is known as the chromaticity diagram (Fig. A.4.1.1).

In the CIE chromaticity diagram spectral colours appear on the boundary of the curve, giving the hue between 0.38 to 0.7 µm. White appears in the central region. The line joining blue/violet with red is called the purple line. The triangular colour field given by white–blue–red–white contains the non-spectral colours.

For any colour, the CIE chromaticity diagram can be used to determine hue and saturation. For a colour P_1 (in Fig. A.4.1.1), the line joining white to P_1 is extended to intersect the curve at H_1. H_1 is the corresponding hue. Saturation is given by the distance away from the white point, i.e. proximity to the curve $\{= d/(1 + d)\}$ in figure A.4.1.1. Saturated colours appear on the boundary of the curve. Unsaturated colours appear on the lines connecting points of pure hue (boundary) with the white point (W).

As mentioned earlier, colours are amenable to mixing. The chromaticity diagram can be used to define the range of colours that would be generated by mixing any set of colours. Even unsaturated colours can be used as end members for mixing purposes.

Blue (B), green (G) and red (R) are called the primary additive colours, since B, G, R, when added together in equal proportion, produce white light (Fig. A.4.1.2a). This colour combination permits generation of a fairly large gamut of

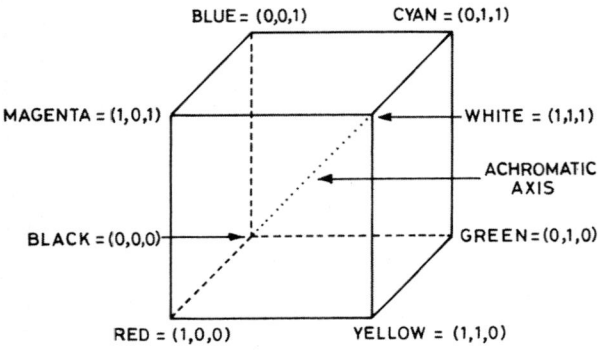

Fig. A.4.1.3. RGB colour model as a cube utilizing a three-dimensional Cartesian coordinate system

colours, although the entire colour field may not be generated by combining any three primary colours.

The B, G, R colour coding is used in superimposing multispectral image data sets (e.g. making FCCs). Figure A.4.1.2b gives the schematic positions of colours in such a ternary colour field. The following points may be noted (mixing in equal proportions):

$B + G + R \rightarrow W$ (B, G, R are called the primary additive colours)

Further:

$(W - B) \rightarrow Y$ ⎫ Yellow (Y), magenta (M) and cyan (C) are called the pri-
$(W - G) \rightarrow M$ ⎬ mary subtractive colours as they are produced by subtract-
$(W - R) \rightarrow C$ ⎭ ing the three primary additive colours, one by one, from white, respectively.

$(B + Y) \rightarrow W$ ⎫ B and Y, G and M, R and C are called mutually complimen-
$(G - M) \rightarrow W$ ⎬ tary colours.
$(R + C) \rightarrow W$ ⎭

It is easy to follow that:
$(B + G) \rightarrow C$
$(G + R) \rightarrow Y$
$(R + B) \rightarrow M$

Further, when the three primary subtractive colours are added together, the result is black:
$Y + M + C \rightarrow (W - B) + (W - G) + (W - R) \rightarrow W - W \rightarrow$ black

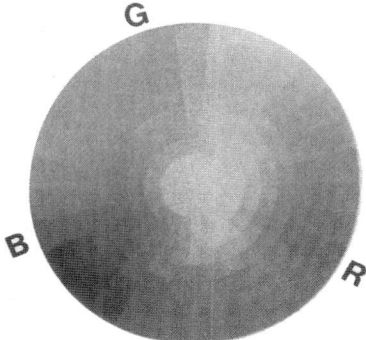

Fig. A.4.1.4. The concept of the Munsell colour circuit or wheel. Hue is represented as the polar angle and saturation on the radial axis (see colour plate I)

2. RGB colour model. The RGB colour model can be visualized as a three-dimensional Cartesian co-ordinate system, the three axes being R, G and B starting from black (Fig. A.4.1.3). The diagonal of the cube from black (0,0,0) to white (1,1,1) is achromatic, the various gray shades lying on this axis.

The three primary subtractive colours (cyan, magenta, yellow) appear on the diagonally opposite corners of the three primary additive corresponding colours.

The RGB model is used in computer hardware and software systems for colour displays.

3. IHS colour model. This is based on the Munsell colour system or wheel (Fig. A.4.1.4). It uses three parameters – intensity (I), hue (H) and saturation (S). The colour space is conceived as a cylinder where hue is represented by the polar angle, saturation by the radius and intensity by the vertical distance on the cylinder axis (Fig. A.4.1.5a). However, as the number of colours that can be perceived decreases with intensity and the contributions of hue and saturation become insignificant at an intensity of zero, a cone could be a better representation of the IHS colour model (Fig. A.4.1.5b).

Transformations between RGB and IHS colour models have been described by many workers (e.g. Buchanan and Pendgrass 1980; Haydn et al. 1982; Gillespie et al. 1986; Edwards and Davis 1994; Harris et al. 1999). If blue is chosen as the reference point for the IHS co-ordinate system, the following equations relate RGB to IHS values in a cylindrical co-ordinate system (Edwards and Davis 1994):

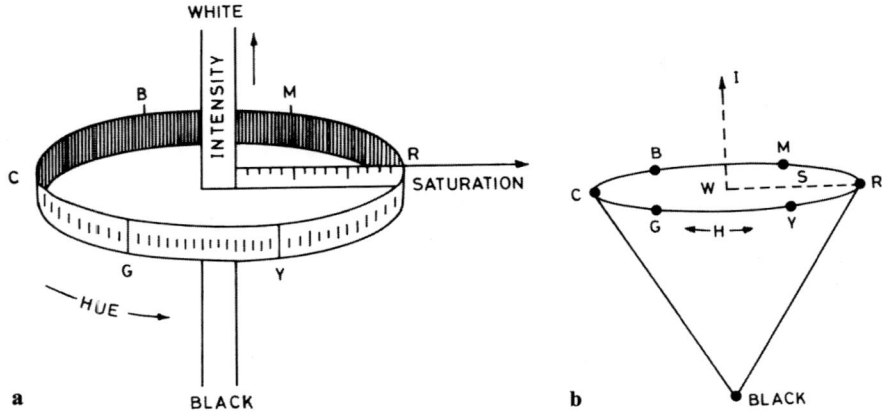

Fig. A.4.1.5a,b. IHS colour model as **a** a cylinder and **b** a cone. Intensity is represented on the vertical axis, hue as the polar angle and saturation on the radial axis

$$I = (DN_R + DN_G + DN_B) \tag{A.1}$$

$$H = \tan^{-1} \frac{(DN_G - DN_R)\sqrt{3}}{(2DN_B - DN_G - DN_R)} \tag{A.2}$$

$$S = \left[\left(DN_B - \frac{1}{3}\right)^2 + \left(DN_G - \frac{1}{3}\right)^2 + \left(DN_R - \frac{1}{3}\right)^2 \right]^{1/2} \tag{A.3}$$

Conversely, the following equations relate a pixel's IHS values to RGB DNs:

$$R = \frac{1}{3} - \frac{S\cos(H)}{\sqrt{6}} - \frac{S\sin(H)}{\sqrt{2}} \tag{A.4}$$

$$G = \frac{1}{3} - \frac{S\cos(H)}{\sqrt{6}} + \frac{S\sin(H)}{\sqrt{2}} \tag{A.5}$$

$$B = \frac{1}{3} + \frac{S\sqrt{6}\cos(H)}{3} \tag{A.6}$$

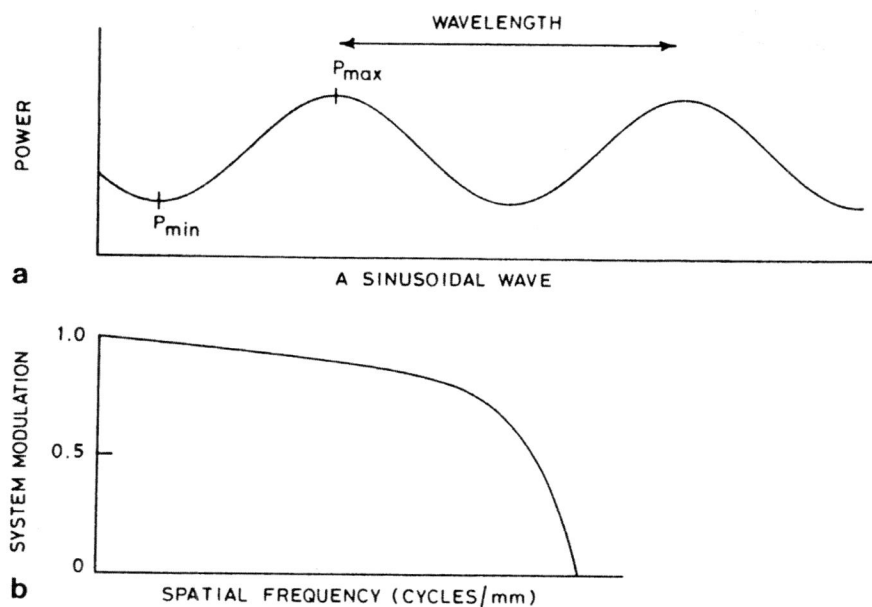

Fig. A.4.2.1a,b. Concept of the modulation transfer function. **a** A sinusoidal wave. **b** Spatial frequency (cycles/mm) vs. system modulation

Appendix 4.2: Modulation Transfer Function (MTF)

The Modulation Transfer Function (MTF) is an important concept in the evaluation of performance of remote sensors. It can be applied to any system, such as a scanner system or photography, or to its part, such as a lens, film etc. MTF is a measure of the faithfulness with which a sensor portrays an object in image form.

Basically, an object scene can be considered to consist of several unit areas, and the spatial variation across the scene may be imagined as temporal variation, if successive pixels are considered one after another. This gives a sine wave in which the frequency is *de facto* spatial and not temporal. The frequency is given here as cycles/mm. If the variation in ground features occurs faster and closely, the frequency is higher, and if the variation occurs slowly, the frequency of the sine wave is lower. The modulation M of a wave of a certain wavelength and frequency is, by definition (Fig. A.4.2.1)

$$M = \frac{P_{max} - P_{min}}{P_{max} + P_{min}} \tag{A.7}$$

How faithfully this modulation is perceived and recorded by the sensor is very important for remote sensing. The ratio of the image modulation (M_i) to object modulation (M_o) is called the MTF.

Therefore,

$$\text{MTF} = \frac{M_i}{M_o} \qquad (A.8)$$

The MTF is frequency dependent; it will have different values for different frequencies. For lower frequencies, i.e. if the variation in ground features is not rapid, the MTF will be nearly unity, meaning that object modulation is being matched image modulation. For higher frequencies, i.e. spatially closer variation, the MTF will have lower values, implying relatively poor portrayal of variations in object scene. Thus, a frequency-dependent evaluation of the MTF for a sensor system or its part is carried out to evaluate its performance (for details, see e.g. Slater 1980).

Appendix 10.1 Image Processing Software

	Name		Web-site
	Company[a]	Software	
1.	ArcView	Image Analysis	www.esri.com
2.	ATLANTIS Scientific Inc.	Earth View	www.atlsci.com
3.	Clark Labs	Idrisi	www.clarklabs.org
4.	ERDAS	Erdas Imagine	www.erdas.com
5.	Earth Resources Mapping	ER Mapper	www.ermapper.com
6.	ITC	Ilwis	www.itc.nl
7.	Micro Images	TNTmips	www.microimages.com
8.	PCI – Geomatics	Easi/Pace	www.pcigeomatics.com
9.	PCI – Geomatics	Geomatica	www.pcigeomatics.com
10.	Research Systems Inc.	Envi	www.rsinc.com
11.	Z/I Imaging	Image Analyst	www.ziimaging.com

[a] in alphabetical order

References

Abbott E (1990) Proceedings of the second Thermal Infrared Multispectral Scanner (TIMS) workshop: Jet Propulsion Laboratory Publ 90-56, Pasadena, CA

Abbott, E (1991) Proceedings of the third Thermal Infrared Multispectral Scanner (TIMS) workshop: Jet Propulsion Laboratory Publ 91-29, Pasadena, CA

Abrams MJ, Brown D (1985) Silver Bell, Arizona, porphyry copper test site. The joint NASA/Geosat test case study, section 4, Am Assoc Petrol Geol, Tulsa, Oklahoma

Abrams MJ, Ashley RP, Rowan LC, Goetz AFH, Kahle AB (1977) Mapping of hydrothermal alteration in the Cuprite Mining District, Nevada, using aircraft scanner images for the spectral region 0.46 to 2.36 µm. Geology 5:713-718

Abrams MJ, Brown D, Lepley L, Sadowski R (1983) Remote sensing for porphyry copper deposits in south Arizona. Econ Geol 78:591-604

Abrams MJ, Kahle AB, Palluconi FD, Schieldge Jp (1984) Geologic mapping using thermal images. Remote Sens Environ 16:13-33

Abrams MJ, Conel JE, Lang HR (1985) The joint NASA/Geosat test case study. Am Assoc Petrol Geol, Tulsa, Oklahoma

Adams JB, Smith MO (1986) Spectral mixture modeling: a new analysis of rock and soil types at the Viking Lander I site. J Geophys Res 91(B8): 8098-8112

Addington JD (1975) A hybrid maximum likeihood classifier using the parallelepiped and Bayesian techniques. Technical Papers, 50th Annu Meet, Am Soc Photogramm, pp 772-784

Agar RA, Villanueva R (1997) Satellite, airborne and ground spectral data applied to mineral exploration in Peru. Proc 12th Int Conf App Geol Remote Sens, Vol I, Env Res Inst Michigan, Ann Arbor, Mich, pp 13-20

Agarwal RP, Misra VN (1994) Application of remote sensing in petroleum exploration – case studies from Northeastern region of India. Ind J Petrol Geol 3(2):45-68

Allen CR (1975) Geological criteria for evaluating seismicity. Geol Soc Am Bull 86: 1041-1057

Allum JAE (1966) Photogeology and regional mapping. Pergamon, Oxford

Almer A, Raggam J, Strobl D (1996) High precision geocoding of remote sensing data of high relief terrain. In: MF Buchroithner (ed) Proc Int Symp High Mountain Remote Sens Cartography, held at Schladming, Austria, 26-28 Sept 1990, Dresden Univ Tech, Dresden, pp 56-65

Alparone L, Cappellini V, Mortelli L, Aiazzi B, Baronti S, Carla R (1998) A pyramid-based approach to multisensor image data fusion with preservation of spectral signatures. Future Trends in Remote Sensing, Gudmandsen (ed), pp 419-426

Anbalagan R (1992) Landslide hazard evaluation and zonation mapping in mountainous terrain. Eng Geol 32:269-277

Arnason K (1988) Geowissenschaftliche Ferner kundung mit Satelittendaten in Island – Möglichkeiten und Grenzen. Doctoralthesis, Ludwig-Maximilians University, Munich

Aronoff S (1989) geographic Information Systems: A management perception. WDL Publ, Ottawa, 294p

Arora MK, Mathur S (2001) Multi-source image classification using neural network in a rugged terrain. Geo Carto Int 16(3):37-44

Avery TE, Berlin GL (1985) Interpretation of aerial photographs, 4th edn Burgress, Minneapolis, Minn

Baker VR (1986) Fluvial landforms. In: Short NM, Blair RW Jr (eds) Geomorphology from space. NASA SP-486,US Govt Printing Office, Washington DC, pp 255-316

Barringer AR (1976) Airborne geophysical and miscellaneous systems. In: Lintz J Jr Simonett DS (eds) Remote sensing of environment. Addison-Wesley, Reading, pp 291-321

Barzegar F (1983) Earth resources remote sensing platforms. Photogramm Eng Remote Sens 49:1669

Batchelor GB (1974) Practical approach to pattern classification. Plenum, London

Batson RM, Edwards K, Eliason EM (1976) Synthetic stereographic and Landsat pictures. Photogramm Eng Remote Sens 42:1279-1284

Baugh WM, Kruse FA (1994) Quantitative geochemical mapping of ammonium minerals using field and airborne spectrometers, Cedar Mountains, Esmeralda County, Nevada. Proc 10th Thematic Conf Geol Remote Sens, Vol II, Env Res Inst Michigan, Ann Arbor, Mich, pp 304-315

Becker F, Li ZL (1995) Surface temperature and emissivity at various scales – definition, measurement, and related problems. Remote Sens Rev 12:225-253

Belcher DJ (1960) Photointerpretation in engineering. In: Colwell RN (ed) Manual of photographic interpretation. Am Soc Photogramm, Falls Church, VA, pp 403-456

Bell R, Singhroy VH, Evans CS, Harrington SE (1989) Geologic lithologic mapping in NW Ontario: remote sensing approaches and caveats. Proc 7th Thematic Conf Remote Sens Explor Geol, Vol II, Env Res Inst Michigan, Ann Arbor, Mich, pp 819-831

Berger Z (1992) Geologic stereo mapping of geologic structures with SPOT satellite data. Am Assoc Pet Geol Bull 76(1):101-120

Berger Z (1994) Satellite Hydrocarbon Exploration and Integration Techniques. Springer Verlag, New York, 319p

Berk A, Bernstein LS, Robertson DC (1989) MODTRAN: A moderate resolution model for LOWTRAN7. Tech Rep GL-TR-89-0122, Geophysics Laboratory, Bedford, Mass

Berlin GL, Schaber GG, Horstman KC (1980) Possible fault detection in Cottonball Basin, California: an application of radar remote sensing. Remote Sens Environ 10:33-42

Bernard R, Taconet O, Vidal-Madjar D, Thony JL, Vauclin M, Chapoton A, Wattrelot F, Lebrun A (1984) Comparison of three in-situ surface soil moisture measurements and application to C-band scatterometer calibration. IEEE Trans Geosc Remote Sens GE-22(4):388-394

Bernstein R (1976) Digital image processing of earth observation sensor data. IBM J Res Develop 20:40-57

Beynon JDE, Lamb DR (eds) (1980) Charge-coupled devices and their applications. McGraw Hill, London

Bezdek JC, Ehrlich R, Full W (1984) FCM: The fuzzy c-means clustering algorithm. Comp Geosci 10:191-203

Bharktya DK, Gupta RP (1981) Regional tectonics and sulphide ore localisation in Delhi-Aravalli belt, Rajasthan, India – use of Landsat imagery. Adv Space Res vol 1, Pergamon, London, pp 299-302

Bharktya DK, Gupta RP (1983) Lineament structures in the Precambrians of Rajasthan as deciphered from Landsat images. Recent researches in geology, vol 10, structure and tectonics of Precambrian rocks. Hindustan Publ, New Delhi, pp 186-197

Bhattacharya A, Reddy S (1994) Underground and surface coal mine fire detection in India's Jharia Coal Field using airborne thermal infrared data. Asian-Pacific Remote Sens J 7(1):59-73

Billings WP (1950) Vegetation and plant growth as affected by chemically altered rocks in the Western Great Basin, Ecology 30:62-74

Biswas SK (1974) Landscape of Kutch: a morpho-tectonic analysis, Indian J Earth Sci 1:177-198

Björnsson S, Arnason K (1988) Strengths and shortcomings in ATM technology as applied to volcanic and geothermal areas in Iceland. Proc 4th Int Conf, Spectral Signatures of Objects in Remote Sensing, Aussois, france, ESA-SP287:189-191

Blom RG, Crippen RE, Elachi C (1984) Detection of subsurface features in SEASAT radar images of Means Valley, Mojave Desert, California, Geology, 12: 346-349

Blom RG, Schenck LR, Alley RE (1987) What are the best radar wavelengths, incidence angles, and polarization for discricrimination among lava flows and sedimentary rocks? a statistical approach. IEEE Trans Geosci Remote Sens, GE-25(2):208-212

Bloom Al (1978) Geomorphology. Prentice Hall, Englewood Cliffs, NJ, p 510

Bloom AL (1986) Coastal landforms. In: Short NM, Blair RW Jr (eds) Geomorphology from space. NASA SP-486 US Govt Printing Office, Washington DC, pp353-406

Bobba AG, Bukata RP, Jerome JH (1992) Digitally processed satellite data as a tool in detecting potential groundwater flow systems. J Hydrol 131:25-62

Bodechtel J, Zilger J (1996) MOMS – history, concepts, goals. Proc MOMS-02 Symp, Cologne, Germany, 5-7 July 1995. Paris: Euro Assoc Remote Sens Lab (EARSeL): 12-25

Bodechtel J, Kley M, Münzer U (1985) Tectonic analysis of typical fold structures in the Zagros Mountains, Iran, by the application of quantitative photogrammetric methods on Metric Camera data. Proc DFVLR-ESA Workshop Oberpfaffenhofen, ESA SP-209: 193-197

Bodechtel J, Haydn R, Zilger J, Meissner D, Seige P, Winkenbach H (1985) MOMS-01:missions and results. In: Schnapf A (ed) Monitoring earth's oceans, land and atmosphere from space. Am Inst Aeronautics Astronautics, New York, pp 524-535

Boerner WM et al. (1998) Polarimetry in radar remote sensing: basic and applied concepts. Principles and Applications of Imaging Radar. Manual of Remote Sensing, 3rd edn, Vol 2, Wiley, New York, pp 271-357

Bonham-Carter GF (1994) Geographic information systems for geoscientists. Pergamon, Oxford

Bonham-Carter GF, Agterberg FP, Wright DF (1988) Integration of geological datasets for gold exploration in Nova Scotia. Photogramm Eng Remote Sens 54:1585-1592

Bonn FJ (1977) Ground truth measurements for thermal infrared remote sensing. Photogramm Engg Remote Sens 43: 1001-1007

Bowers TL, Rowan LC (1996) Remote mineralogic and lithologic mapping of the Ice river alkaline complex, British Columbia, Canada, using AVIRIS data. Photogramm Eng Remote Sens 62(12):1379-1385

Brady M (1982) Computational approaches to image understanding. Association Computer Manufacturers' (ACM) Computing Surveys 14:3-71

Brainard J, Lovett A, Parfitt J (1996) Assessing hazardous waste transport risks using a GIS. Int J Geog Inform Sys 10:831-849

Braithwaite JGN (1966) Dispersive multispectral scanning: a feasibility study, final report. Inst Sci Tech Contract No 14-08-001-10053, Univ Mich, Ann Arbor

Braithwaite JGN, Lowe DS (1966) A spectrum matching technique for enhancing image contrast. Appl Opt 5:893-906

Briole P, Massonnet D, Delacourt C (1997) Post-Eruptive deformation associated with the 1986-87 and 1989 lave flows of Etna, detected by radar interferometry. Geophys Res Lett 24:37-40

Bristow Q (1979) Gamma ray spectrometric methods in uranium exploration airborne instrumentation. In: Hood PJ (ed) Geophysics and geochemistry in the search for metallic areas. Geol Surv Can, Econ Geol Report 31:135-146

Brooks RR (1972) Geobotany and biogechemistry in mineral exploration. Harper and Row, New York, 290 pp

Bruzewicz AJ (1994) Remote sensing and GIS for emergency management. Proc 1st Federal Geographic Tech Conf. Wasington DC, GIS World Inc 1:161-164

Buchanan MD (1979) Effective utilization of colour in multidimensional data presentation. Proc Soc Photo Opt Instrument Eng 199:9-19

Buchanan MD, Pendgrass R (1980) Digital image processing: can intensity hue and saturation replace red, green and blue? Electro-Optical Systems Design 12(3):29-36

Buchroithner MF, Granica K (1997) Applications of imaging radar in hydro-geological disaster management – a review. Remote Sens Rev 16:1-134

Buergmann R et al. (2000) Earthquake potential along the northern Hayward fault, California. Science, 18 Aug., 2000, 289(5482):1178-1182

Burns KL. Brown GH (1978) The human perception of geological lineaments and other discrete features in remote sensing imagery: signal strength, noise levels and quality, Remote Sens Environ 7:163-167

Burns KL. Shepherd J, Berman M (1976) Reproducibility of geological lineaments and other discrete features interpreted from imagery: measurement by a coefficient of association. Remote Sens Environ 5:267-301

Burrough PA (1986) Principles of geographical information systems for land resources assessment. Oxford, Clarendon Press

Byrne GF, Davis JR (1980) Thermal inertia, thermal admittance and the effect of layers. Remote Sens Environ 9:295-300

Byrne GF, Crapper PF, Mayo KK (1980) Monitoring land-cover change by principal component analysis of multitemporal Landsat data. Remote Sens Environ 10:175-189

Campbell AN, Hollister VF, Dutta RV, Hart PE (1982) Recognition of a hidden mineral deposit by an artificial intelligence program. Science 217(4563):927-928

Campbell JB (1996) Introduction to Remote Sensing. 2nd ed, Guildford, New York, pp 399-409

Campbell NA (1996) The decorrelation stretch transform. Int J Remote Sens, 17:1939-1949

Carlsaw HS, Jaegar JC (1959) Conduction of heat in solids, 2nd edn, Oxford Univ Press New York

Carper WJ, Lillesand TM, Kiefer RW (1990) The use of Intensity Hue-Saturation Transformations for Merging SPOT Panchromatic and Multispectral Image Data. Photogramm Eng Remote Sens 56(4):459-467

Castleman KR (1977) Digital image processing. Prentice-Hall, Englewood Cliffs, NJ

Catlow DR, Parsall RJ, Wyutt BK (1984) The integrated use of digital cartographic data and remotely sensed imagery. Proc integrated approaches in remote sensing, Guildford, UK ESA-SP-214, pp 41-66

Chander R (1989) Southern limits of major earthquake ruptures along the Himalaya between longitudes 75° and 90° E. Tectonophysics 170: 115-123

Chang SH, Collins W (1983) Confirmation of the airborne biogeophysical mineral exploration technique using laboratory methods. Econ Geol 78:723-736

Chattopadhyay N, Hashimi S (1984) The Sung Valley alkaline-ultramafic-carbonatite Complex, East Kasi and Jaintia Hills Districts, Meghalaya. Rec Geol Surv Ind 113 (IV): 24-33

Chavez PS Jr (1988) An improved dark object subtraction technique for atmospheric scattering correction of multispectral data. Remote Sens Environ 24: 459-479

Chavez PS Jr (1996) Image-based atmospheric corrections revisited and improved. Photogramm Eng Remote Sens 62:1025-1036

Chavez PS Jr, Kwarteng AY (1989) Extracting spectral contrast in Landsat Thematic Mapper image data using selective principal component analysis. Photogramm Eng Remote Sens 55:339-348

Chen WP, Molnar P (1977) Seismic moments of major earthquakes and average rates of slip in central Asia. J Geophys Res 82:2945-2968

Chhikara RS (1984) Effect of mixed pixels on crop proportion estimation. Remote Sens Environ 14:207-218

Chiu HY, Collins W (1978) A spectroradiometer for airborne remote sensing. Photogramm Eng Remote Sens 44: 507-517

Christensen PR (1986) A study of filter selection for the thematic mapper thermal infrared enhancement. Commercial applications and scientific research requirements for thermal infrared observations of terrestrial surfaces, NASA-EOSAT Joint Report, pp 105-114

Civco DL (1989) Topographic normalization of Landsat Thematic Mapper digital imagery. Photogram Eng Remote Sens 55(9):1303-1309

Clark RN (1999) Spectroscopy of rocks and minerals, and principles of spectroscopy. In: Rencz AN (ed) Remote Sensing for the Earth Sciences, Manual of Remote Sensing, 3rd edn, vol 3, Am Soc Photogramm Remote Sens, John Wiley, pp 3-58

Clark RN, Roush TL (1984) Reflectance spectroscopy: quantitative analysis techniques for remote sensing applications. J Geophys Res 89(B7): 6329-6340

Clark RN, Swayze GA (1995) Automated spectral analysis: mapping minerals, amorphous materials, environmental materials, vegetation, water, ice and snow, and other materials: the USGS tricorder algorithm (abstract), Lunar and Planetary Science XXVI, pp 255-256

Clark RN, Swayze GA (1996) Evolution in Imaging Spectroscopy Analysis and Srus or Signal to Noni: An examination of knows flow we have come. Summaries, 6th Annual JPL Airborne Earth Science Workshop, March 4-8, 1996 http://speclab.cr.usgs.gov

Clark RN, King TVV, Klejwa M, Swayze G, Vergo N (1990a) High spectral resolution reflectance spectroscopy of minerals. J Geophys Res 95:12653-12680

Clark RN, Gallagher AJ, Swayze GA (1990b) Material absorption band depth mapping of imaging spectrometer data using complete band shape least-squares bit with library reference spectra. Proc 2nd Airborne Visible/Infrared Imaging Spectrometer (AVIRIS) Workshop, JPL Publ 90-54, Jet propulsion Laboratory, California Inst Tech, Pasadena, CA, pp 176-186

Colby JD (1991) Topographic normalization in rugged terrain. Photogramm Eng Remote Sens 57(5):531-537

Coleman JM, Roberts HH, Huh OK (1986) Deltaic landforms. In: Short NM, Blair RW Jr (eds) Geomorphology from space NASA-SP-486, US Govt Printing Office, Washington, DC, pp317-352

Collins W, Chang SH, Kuo JT (1981) Detection of hidden mineral deposits by airborne spectral analysis of forest canopies. NASA Contract NSG-5222, Final Rep, p 61

Collins W, Chang SH, Raines G, Channey F, Ashley R (1983) Airborne biogeochemical mapping of hidden mineral deposits, Econ Geol 78:737-749

Colvocoresses AP (1979) Multispectral linear array as an alternative to Landsat-D. Photogramm Eng Remote Sens 45:67-69

Colwell RN (ed) (1960) Manual of photographic interpretation. Am Soc Photogramm, Falls Church, VA

Colwell RN (1976) The visible portion of the spectrum. In: Lintz J Jr, Simonett DS (eds) Remote sensing of environment. Addison-Wesley, Reading, pp 134-154

Colwell RN (ed) (1983) Manual of remote sensing, vols I, II, 2nd edn. Am Soc Photogramm, Falls Church, VA

Condit CD, Chavez PS (1979) Basic concepts of computerised digital image processing for geologists. US Geol Surv Bull No. 1462, US Govt Printing Office, Washington DC, 16 pp

Conel JE, Alley RE (1985) Lisbon Valley, Utah, uranium test case report. The joint NASA/Geosat test case study, section 8, Am Assoc Petrol Geol, Tulsa, Oklahoma

Congalton RG (1991) A review of assessing the accuracy of classifications of remotely sensed data. Remote Sens Environ 37:35-46

Conradsen K, Harpoth O (1984) Use of Landsat multispectral scanner data for detection and reconnaissance mapping of iron oxide staining in mineral exploration, Central East Greenland. Econ Geol 79: 1229-1244

Conradsen K, Nilsson G (1984) Application of integrated Landsat, geochemical and geophysical data in mineral exploration. Int Symp Remote Sens Environ, 3rd Thematic Conf Remote Sens Explor Geol Ann Arbor, MI, pp 499-511

Couloigner I, Ranchin T, Valtonen VP, Wald L (1998) Benefit of the future SPOT 5 and of data fusion to urban roads mapping. Int J Remote Sens 19(8):1519-1532

Coulson S (1993) SAR interferometry with ERS-1. ESA-publ, Earth Observation Quarterly, No 40, April 1993, pp 20-23

Cox D, Singer DA (eds) (1986) Mineral Deposit Models. USGS Bull 1693, U S Geol Surv, Washington D C

Cracknell AP (1998) Review Article: Synergy in remote sensing-what's in a pixel? Int J Remote Sens 19:2025-2074

Craib KB (1972) Synthetic aperture SLAR systems and their application for regional resources analysis Conf Earth resources observation and information analysis system in remote sensing of earth resources. Space Inst, Univ Tennessee, Tullahoma, pp 152-178

Crane RB (1971) Preprocessing techniques to reduce atmospheric and sensor variability in multispectral scanner data. Proc 7th Int Symp on Remote Sens of Environ, vol II. Ann Arbor, MI, pp 1345-1355

Crippen RE (1987) The regression intersection method of adjusting image data for band ratioing. Int J Remote Sens 9:767-776

Crippen RE (1989) Selection of Landsat TM band and band-ratio combination to maximize lithological information in color composite displays. Proc 7th Thematic Conf Remote Sens Expl Geol, pp 917-921

Croft FC, Faust NL, Holcomb DW (1993) Merging of Radar and VIS/IR Imagery. 9th Thematic Conf Geol Remote Sens, Pasadena, CA, February 8-11, pp 379-381

Crosta AP and Moore JM (1989) Enhancement of Landsat Thematic mapper imagery for residual soil mapping in SW Minas Gerais State, Brazil: a prospecting case history in greenstone belt terrain. Proc 17th Thematic Conf Remote Sens Explor Geol, pp 1171-1187

Crowley JK, Hook SJ (1996) Mapping playa evaporite minerals and associated sediemnts in Death Valley, California, with multispectral theramal images. J Geophys Res 99B:643-660

Cudahy TJ, Connor PM, Hausknecht P, Hook SJ, Huntington JF, Kahle AB, Phillips RN, Whitbourn LB (1994) Airborne CO_2 laser spectrometer and TIMS TIR data for mineral

mapping in Australia. Proc 7th Australian Remote Sens Conf, Melbourne, Victoria, Australia, March 1-4, pp 918-924
Curran PJ (1985) Principles of remote sensing. Longman, London
Daily MI (1983) Hue-saturation-intensity split-spectrum processing of Seasat radar imagery. Photogramm Eng Remote Sens 49:349-355
Dancak C (1979) Temperature caliberation of test infrared scanner. Photogramm Remote Sens 45:749-751
Davis JC (1986) Statistics and data analysis in geology. 3rd edn, Wiley, New York, 646 pp
Davis LS (1975) A survey of edge detection techniques. Computer Graphics Image Processing 4:248-270
de Azevedo LHA (1971) Radar in the Amazon project Radam. Proc 7th Int Symp Remote Sensing of Environ. Ann Arbor, MI, pp 2303-2306
De Loor GP (1981) The observation of tidal parameters, currents and bathymetry with SLAR imagery of the sea. IEEEJ Oceanic Eng 6:124-129
Dellwig LF (1969) An evaluation of multifrequency radar imagery of the Pisgah crater area, California. Mod Geol 1:65-73
Dellwig LF, Moore RK (1966) The geological value of simultaneously produced like-and cross-polarized radar imagery. J Geophy Res 71:3597-3601
Denniss AM, Harris AJL, Rothery DA, Francis PW, Carlton RW (1998) Satellite observation of the April 1993 eruption of Lascar volcano. Int J Remote Sens 19(5):801-821
Desio A (1974) Karakorum Mountains, Mesozoic Cenozoic orogenic belts. Geol Soc Spec Publ 4, London, pp 255-266
Deutsch M, Estes JE (1980) Landsat detection of oil from natural seeps. Photogramm Eng Remote Sens 46:1313-1322
Dickerhof C, et al. (1999) Mineral Identification and Lithological Mapping on the Island of Naxos (Greece) using DIAS 7915 Hyperspectral Data. EARSeL Advances in Remote Sens 1(1):255-273
Dixon TH (1995) SAR interferometry and surface change detection. Report of the workshop held in Boulder, Colorado, February 3-4, 1994
Doimer J, Strahler AH (1983) Ground investigations in support of remote sensing. In: Colwell RN (ed) Manual of remote sensing, 2nd edn: ch 23, p969-989. Am Soc Photogramm Remote Sens, Falls Church, VA
Doyle FJ (1985) The Large Format Camera on shuttle mission 41-G. Photogramm Eng Remote Sens 51:200
Dozier J, Frew J (1981) Atmospheric corrections to satellite radiometric data over rugged terrain. Remote Sens Environ 11:191-205
Drury SA (1993) Image interpretation in Geology, 2nd edn. London:Allen and Unwin
Duval JS (1983) Composite color images of aerial gamma-ray spectrometric data. Geophysics 48:722-735
Eastman Kodak Company (1981) Applied infrared photography. Eastman Kodak publ, Rochester, New York
Eastman Kodak Company (1990) Handbook of Kodak photographic filters. Eastman Kodak publ, Rochester, New York
Eastman Kodak Company (1992) Kodak data for aerial photography. Eastman Kodak publ, 6th edn, Rochester, New York
Edwards K, Davis PA (1994) The use of intensity-hue-saturation transformation for producing color shaded-relief images. Photogramm Eng Remote Sens 60:1369-1374
Ehlers M (1991) Multisensor Image Fusion Techniques in Remote Sensing. ISPRS J Photogramm Remote Sens 4671(1):19-30
Elachi C (1980) Spaceborne imaging radar: geologic and oceanographic applications. Science 209(4461):1073-1082

Elachi C (1983) Microwave and infrared satellite remote sensors. In: Colwell RN (ed) Manual of remote sensing. 2nd edn, vol 1, Am Soc Photogramm, Falls Church, VA

Elachi C, Ulaby FT (1990) Radar polarimetry for geosicence applications. Artech House, Norwood, MA, 364 pp

Elachi C, Brown WE, Cinino JB, Dixon T et al. (1982a) Shuttle imaging radar experiment. Science 218:996-1003

Elachi C, Roth LE, Schaber GG (1984) Spaceborne radar subsurface imaging in hyperarid regions. IEEE Trans GE-22: 382-387

Ellyett CD, Fleming A W (1974) Thermal Infrared Imagery of the Burning Mountain Coal Fire. Remote Sens Environ 3(1):79-86

Ellyett CD, Pratt DA (1975) A review of the potential applications of remote sensing techniques to hydrogeological studies in Australia. Aust Water Res Council Tech Pap No. 13, 147 pp

Elvidge CD (1982) Affect of vegetation on airborne thematic maper imagery of the Kalamazoo propyry copper deposit, Arizona. Int Symp Remote Sens Environ, 2nd Thematic Conf Remote Sens Explor Geol, Fort Worth, Texas, pp 661-667

EOSAT (1994) Merge process enhances value of data sets. EOSAT Notes, Vol.9, pp 5

Eppes TA, Rouse JW Jr (1974) Viewing-angle effects in radar images. Photogramm Eng 40:169-173

Evans DL, Farr TG, Ford JP, Thompson TW, Werner CL (1986) Multipolarization radar images for geologic mapping and vegetation discrimination. IEEE Trans Geosci Remote Sens 24:246-257

Evans DL, Farr TG, Zebker HA, Mouginis-Mark PJ (1992) Radar interferometry studies of the earth's topography. EOS, Transactions, Am Geophys Union 73, 533:557-558

Evans DL, Plant JJ, Stofan ER (1997) Overview of the Spaceborne Imaging Radar-C/X-band Synthetic Aperture Radar (SIR-C/X-SAR) missions. Remote Sens Environ 59: 135-140

Everett JR, Morisawa M, Short NM (1986) Tectonic landform. In: Short NM, Blair RW Jr (eds) Geomorphology from space, NASA SP-486, US Govt Printing Office, Washington D C, pp 27-184

Fabbri AG (1984) Image processing of geological data. Van Nostrand Reinhold, New York, 244 pp

Farag AA (1992) Edge-based image segmentation. Remote Sens Rev 6:95-122

Farmer VC (ed) (1974) The infrared spectra of minerals. Mineralogical Soc Publ, London

Farr TG (1983) Use of radar image texture in geologic mapping, Proc Symp Spaceborne Imaging Radar. Jet Propul Lab Pasadena, CA, pp 73-75

Farrand WH, Harsanyi JC (1997) Mapping the distribution of mine tailings in the Coeur d' Alene River valley, Idaho, through the use of a Constrained Energy Minimization Technique. Remote Sens Environ 59:64-76

Farrand WH, Seelos A (1996) Using mineral maps generated from imaging spectrometer data to map faults: an example from summitville, Colorado. Proc 11th Thematic Conf Geol Remote Sens, Vol II, Env Res Inst Michigan, Ann Arbor, Mich, pp 222-230

Fielding EJ, Blom RG, Goldstein R.M (1997) Detection and monitoring of rapid subsidence over Lost Hills and Belridge oil fields by SAR interferometry. Proc 12th Int Conf Workshops Appl Geol Remote Sens, Denver, Colorado, vol I, p 84

Fischer WA (1975) History of remote sensing. In:Reeves RG(ed) manual of remote sensing. Am Soc Photogramm, Falls Church, VA, pp27-50

Fisher PF, Pathirana S (1990) The evaluation of fuzzy membership of land cover classes in the suburban zone. Remote Sens Environ 34:121-132

Foody GM (1992) A fuzzy sets approach to representation of vegetation continua from remotely sensed data: An example from Lowland heath. Photogramm Eng Remote Sens 58:221-225

Foody GM (1995) Land cover classification by an artificial neural network with ancillary information. Int J Geog Inform Sys 9:527-542

Foody GM, Arora MK (1996) Incorporating Mixed Pixels in the Training, Allocation and Testing Stages of Supervised Classifications. Pattern Recog Lett 17:1389-1398

Foody GM, Arora MK (1997) An Evaluation of Some Factors Affecting the Accuracy of Classification by an Artificial Neural Network. Int J Remote Sens 18(4):799-810

Fookes PG, Sweeney M, Manby CND, Martin RP (1985) Geological and geotechnical engineering aspects of low-cost roads in mountainous terrain. Eng Geol 21:1-152

Ford JP (1982) Resolution versus speckle relative to geologic interpretability of spaceborne radar images – a survey of user preferences. IEEE Trans Geosci and Remote Sens GE-20 (4): 434-444

Ford JP (1998) Radar Geology. In: Henderson FM, Lewis AJ (eds) Principles and Applications of Imaging Radar. Manual of Remote Sensing, 3rd edn, Vol 2, Wiley, New York, pp 511- 565

Ford JP, Cimino JB, Elachi C (1983) Space shuttle Columbia views the world with imaging radar: the SIR-A experiment. Jet Prpul Lab Publ No 82-95, Pasadena, CA, 179pp

Ford JP, Cimino JB, Holt B, Ruzek MR (1986) Shuttle imaging radar views the Earth from Challenger: the SIR-B experiment. Jet Propul Lab Publ No 86-10, Pasadena, CA, 135pp

Franceschetti G, Lanari R (1999) Synthetic Aperture Radar Processing. CRC Press, Baco Raton, Florida, p 307

Francis PW, De Silva SL (1989) Application of theLandsat Thematic Mapper to the identification of potentially active volcanoes in the Central Andes. Remote Sens Environ 28:245-255

Francis PW, Rothery DA (1987) Using the Landsat Thematic Mapper to detect and monitor active volcanoes: an example from Lascar volcanoe, northern Chile. Geology 15:614-617

Franklin SE (1994) Discrimination of subalpine forest species and canopy density using digital CASI, SPOT PLA and Landsat TM data. Photogramm Eng Remote Sens 60:1233-1241

Fraser RS, Curran RJ (1976) Effects of the atmosphere on remote sensing. In: Lintz J Jr, Simonett DS (eds) Remote sensing of environment. Addison-Wesely, Reading, pp 34-84

Fraser SJ (1991) Discrimination and identification of ferric oxides using satellite Thematic Mapper data: A Newman case study. Int J Remote Sens 12(3):635-641

Fraser SJ, Green AA (1987) A software defoliant for geological analysis of band ratios. Int J Remote Sens 8:525-532

Gabriel AK, Goldstein RM (1988) Crossed orbit interferometry. Theory and experimental results from SIR-B. Int J Remote Sens 9:857-872

Gabriel AK, Goldstein RM, Zebker HA (1989) Mapping small elevation changes over large areas. Differential radar interferometry. J Geophys Res 94(B7):9183-9191

Gangopadhyay PK (1967) Structural framework of Alwar region with special reference to the occurrence of some rock types. Proc Symp Upper Mantle Project, Nat Geophys Res Inst, Hyderabad, pp 420-429

Gansser A (1968) The Insubric Line – a major geotectonic problem. Schweiz Mineral Petrogr Mitt 48:123-143

Gates DM (1970) Physical and physiological properties of plants. In: Remote sensing with special reference to agriculture and forestry. Nat Acad Sci, Washington, DC, pp 224-252

Gausmann HW, Escobar DE, Everitt JH, Richardson AJ, Rodriguez RR (1978) Distinguishing succulent plants from crop and woody plants. Photogramm Eng Remote Sens 44:487-491

Gausmann HW, Escobar DE, Knipling EB (1977) Relation of *Peperomia obtusifolia's* anomalous leaf reflectance to its leaf anatomy. Photogramm Eng Remote Sens 43:1183-1185

Genderen JLvan, Haiyan G (1997) Environmental monitoring of spontaneous combustion in the North China Colafields. ITC Enschede, 244 pp

Genderen JLvan, Cassells CJS, Zhang XM (1996) The synergistic use of remote sensed data for the detection of underground coal fires. Int Archieves Photogramm Remote Sens, vol xxxi, part 7, Viena, 9-19, July 1996

Gens R (1998) Quality Assessment of SAR Interferometric Data. ITC Publication No 61, pp 141

Gens R, Genderen JLvan (1996) SAR interferometry – issues, techniques, applications. Int J Remote Sens 17:1803-1835

Ghiglia DC, Pritt MD (1998) Two-dimensional phase unwrapping: theory, algorithms and software. Wiley Interscience, 493p

Gillespie AR (1980) Digital techniques of image enhancement. In: Seiegal BS, Gillespie AR (eds) Remote sensing in geology, Wiley, New York, pp 139-226

Gillespie AR (1992) Enhancements of multispectral thermal infrared images: de-correlation contrast stretching, Remote Sens Environ 42:147-156

Gillespie AR, Kahle AB (1977) Construction and interpretation of a digital thermal inertia image. Photogramm Eng Remote Sens 43:983-1000

Gillespie AR, Kahle AB, Palluconi FD (1984) Mapping alluvial fans in Death Valley, California using multichannel thermal infrared images. Geophys Res Lett 11:1153-1156

Gillespie AR, Kahle AB Walker RE (1986) Color enhancement of highly correlated images: I-decorrelation and HIS contrast stretches, Remote Sens Environ 20:209-235

Gillespie AR, Kahle AB, Walker RE (1987) Colour enhancement of highly correlated images:II-Channel ratio and chromaticity transformation techniques, Remote Sens Environ 22:343-365

Glaze LS, Francis PW, Self S, Rothery DA (1989) The 16 September 1986 eruption of Lascar volcano, north Chile: satellite investigations Bull. Volcanology 51P:146-160

Glikson AY, Creasey JW (1995) Application of Landsat-5 TM imagery to mapping of the Giles Complex and associated granulites, Tomkinson Ranges, western Musgaves Block, central Australia. J Aust Geol Geophy 16:173-193

Goetz AFH (1980) Stereosat: a global digital stereo imaging mission. Int archieves of photogrammetry, pt B9 XIV Congress, Hamburg, pp 563-570

Goetz AFH, Rowan LC (1981) Geologic remote sensing. Science 211:781-791

Goetz AFH, Srivastava V (1985) Mineralogic mapping in the Cuprite mining district, Nevada. Proc Airborne Imaging Spectrometer Data Analysis Workshop. JPL Publ 85-41, Jet Propulsion Laboratory, Pasadena, CA, pp 22-31

Goetz AFH, Rowan LC, Kingston MJ (1982) Mineral identification from orbit: initial result from the Shuttle Multispectral Infrared Radiometer. Science 218:1020-1024

Goetz AFH, Rock BN, Rowan LC (1983) Remote sensing for exploration: an overview. Econ Geol 79:573-590

Gold DP (1980) Structural geology. In: Siegal BS, Gillespie AR (eds) Remote sensing in geology. Wiley, New York, pp 419-483

Goldstein RM, Barnett TP, Zebker HA (1989) Remote sensing of ocean currents. Science 246:1282-1285

Goldstein RM, Engelhardt H, Kamb B, Frolich RM (1993) Satellite radar interferometry for monitoring ice sheet motion. Application to an Antarctic ice stream. Science 262:1525-1530

Gong P (1996) Integrated analysis of spatial data for multiple sources: using evidential reasoning and artificial neural network techniques for geological mapping. Photogramm Eng Remote Sens 62:513-523

Gonzales RC, Woods RE (1992) Digital Image Processing. Addison-Wesley, Reading

Goosens MA (1991) Integration of remote sensing data and ground data as an aid to exploration for granite related mineralization, Salamance province, W-Spain. Proc 8th Int Conf Geol Remote Sens, Vol I, Environ Res Inst Mich, Ann Arbor, Mich, pp 393-406

Graham LC (1974) Synthetic interferometer radar for topographic mapping. Proc IEEE 62:763-768

Green AA, Berman M, Switzer P, Graig MD (1988) A transformation for ordering multispectral data in terms of image quality with implications for noise removal. IEEE Trans Geosci Remote Sens 26:65-74

Green RO (1992) Determination of the in-flight spectral and radiometric characteristics of the Airborne Visible/Infrared Imaging Spectrometer (AVIRIS). In: Toselli F, Bodechtel J (eds) Imaging Spectrometry: Fundamentals and Prospective Applications. Kluwer, Dordrecht:103-123

Guilbert JM, Park CF Jr (1986) The geology of ore deposits. Freeman, New York, 985pp

Guild PW (1972) Metallogeny and the new global tectonics. Proc 24th Int Geol Cong Sect 4, mineral deposits, pp 17-24

Gupta RP (1977a) Delineation of active faulting and some tectonic interpretations in Munich-Milan section of eastern Alps-use of Landsat imagery. Tectonophysics 38:297-315

Gupta RP (1977b) Neue geologische Strukuren in Himalaja entdekt, Umschau 77:329-330

Gupta RP (1991) Remote sensing geology. 1st edn, Springer-Verlag, Heidelberg, pp 356

Gupta RP, Bharktya DK (1982) Post-Precambrian tectonism in the Delhi-Aravalli belt, Precambrian Indian Shield-evidences from Landsat images. Tectonophysics 85:T9-T19

Gupta RP, Bodechtel J (1982) Geotechnical applications of Landsat image analysis of Bhakra dam reservoir, India. Remote Sens Environ 12:3-13

Gupta RP, Joshi BC (1990) Landslide hazard zoning using the GIS approach – a case study from the Ramganga Catchment, Himalayas. Eng Geol 28:119-131

Gupta RP, Prakash A (1998) Reflectance aureoles associated with thermal anomalies due to subsurface mine fires in the Jharia caolfield, India. Int J Remote Sens (under publication)

Gupta RP, Saha AK (2000) Mapping debris flows in the Himalayas. GIS@Development, IV(12):26-27 http://www.gis-development.net

Gupta RP, Sen AK (1988) Imprints of the Ninety-East Ridge in the Shillong Plateau, Indian Shield. Tectonophysics 154:335-341

Gupta RP, Saraf AK, Chander R (1998) Discrimination of areas susceptible to earthquake induced liquefaction from Landsat data. Int J Remote Sens 19(4):569-572

Gupta RP, Saha AK, Arora MK, Kumar A (1999) Landslide hazard zonation in a part of the Bhagirathi Valley, Garhwal Himalayas, using integrated remote sensing – GIS. Himalayan Geol 20(2):71-85

Halbouty MT (1976) Application of Landsat imagery to petroleum and mineral exploration. Am Assoc petrol Geol 60:745-793

Halbouty MT (1980) Geologic significance of Landsat data of 15 giant oil and gas fields. AAPG Bulletin 64(1):8-36

Hall DK, Martinec J (1985) Remote Sensing of Ice and Snow. Chapman and Hall, London, 189p

Halsema D Van, Kooij MWA Van der, Groenewoud W, Huising J, Ambrosius BAC Klees R (1995) SAR interferometric in Nederlands. Remote Sens Nieuwesbrief, Juni 1995, pp 31-34

Haralick RM (1979) Statistical and structural approaches to texture. Proc IEEE 67:786-804

Haralick RM, Fu K (1983) Pattern recognition and classification. In: Colwell RN (ed) Manual of remote sensing. Am Soc Photogramm Remote Sens, Falls Church, VA, pp 793-805

Harding AE, Forrest MD (1989) Analysis of multiple geological data sets from English Lake District. IEEE Trans Geosci Remote Sens 27:732-739

Harris PM, Ventura SJ (1995) The integration of geographic data with remotely sensed imagery to improve classification in urban area. Photogramm Eng Remote Sens 61:993-998

Harris JR, Murray R, Hirose T (1990) IHS transform for the integration of radar imagery and other remotely sensed data. Photogramm Eng Remote Sens 56:1631-1641

Harris JR, David WV, Andrew NR (1999) Integration and visualization of geoscience data. Remote Sensing for the Earth Sciences, Manual of Remote Sensing, 3rd ed, Vol 3, Am Soc Photogramm Remote Sens pp 307-354

Haydn R (1985) A concept for the processing and display of Thematic Mapper data. Proc Symp Landsat-4 Science Characterization Early Results, NASA publ 2355 Greenbelt, MD. pp 217-237

Haydn R, Dalke GW, Henkel J, Bare JE (1982) Application of the IHS colour transform to the processing of multisensor data and image enhancement. Proc Int Symp Remote Sens of Arid and Semi-Arid Lands, Cairo, pp 599-616

Henderson FM, Lewis AJ (eds) (1998) Principles and Applications of Imaging Radar. Manual of Remote Sensing, 3rd edn, Vol 2, Wiley, New York

Hepner GF, Logan T, Ritter N, Bryant N (1990) Artificial neural network classification using a minimal training set: comparison to conventional supervised classification. Photogramm Eng Remote Sens 56:469-473

Heron AM (1922) Geology of the western Jaipur. Rec Geol Surv Ind vol LIV, pp345-397

Heron AM (1953) The geology of central Rajputana. Mem Geol Surv Ind 79:1-389

Heywood I, Cornelius S, Carver S (1998) An Introduction to Geographical Information Systems. Addison Wesley Longman Limited, 279pp

Hill J (1991) A quantitative approach to remote sensing: sensor calibration and comparison. In Belward AS and Valenzuela CR (eds) (1991):97-110

Hiller K (1984) MOMS-01 experimental missions on space shuttle flights STS-7 June'83, STS-II Feb.'84-data catalogue. DFVLR, Oberpfaffenhofen

Hintz A (1999) Large image format aerial cameras: film based systems and digital perspectives. In: Nieuwenhuis, Vaughan and Molenaar (eds), Operational Remote Sensing for Sustainable Development, Balkema, Rotterdam, pp 231-236

Hixson M, Scholz D, Fuchs N, Akiyama T (1980) Evaluation of several schemes for classification of remotely sensed data. Photogramm Eng Remote Sens 46:1547-1553

Hobbs WH (1904) Lineaments of the Atlantic border region. Geol Soc Am Bull 15:483-506

Hodgson R, Cady B, Pairman D (1981) A solid-state airborne sensing system for remote sensing. Photogramm Eng Remote Sens 47:177-182

Hofmann O, Nave P, Ebner H (1984) DPS-a digital photogrammetric system for producing digital elevation models and orthophotos by means of linear array scanner imagery. Photogramm Eng Remote Sen 50:1135-1142

Holasek RE, Rose WI (1991) Anatomy of Augustine volcano eruptions as recorded by multispectral image processing of digital AVHRR weather Satellite. Bull Volcanology 53:420-435

Holben BN, Justice C (1980) An examination of spectral band ratioing to reduce the topographic effect on remotely sensed data. NASA Technical Memorandum 80640, Goddard Space Flight Center, Greenbelt, MD, p28

Hook SJ, Kahle AB (1996) The micro Fourier transform interferometer (µFTIR): a new field spectrometer for acquisition of infrared data of natural surfaces. Remote Sens Environ 56:172-181

Hook SJ, Gabell AR, Green AA, Kealy PS (1992) A comparison of techniques for extracting for emissivity information from thermal infrared data for geologic studies. Remote Sens Environ 42:123-135

Hook SJ, Abbott EA, Grove C, Kahle AB, Palluconi F (1999) Use of multispectral thermal infrared data in geological studies. In: Rencz AN (ed) Remote Sensing for the Earth Sciences, Manual of Remote Sensing, 3rd edn, vol 3, Am Soc Photogramm Remote Sens, John Wiley, pp 59-110

Hoover G, Kahle AB (1987) A thermal emission spectrometer for field use. Photogramm EngRemote Sens 53:627-632,
http://southport.jpl.nasa.gov/scienceapps/dixon/index.html

Hord RM (1982) Digital image processing of remotely sensed data. Academic Press, New York, 256pp

Horler DNH, Barber J, Barringer AR (1980) Effects of heavy metals on the absorbance and reflectance spectra of plants. Int J Remote Sens 1:121-136

Horler DNH, Dockray M, Barber J, Barringer AR (1983) Red edge measurements for remote sensing plant chlorophyll content. Proc Symp Remote Sens Mineral Expl Commun on Space Research, Ottawa

Howard AD (1967) Drainage analysis in geological interpretation: a summation. Am Assoc Petrol Geol Bull 51:2246-2259

Hsu S (1978) Texture-tone analysis for automated landuse mapping. Photogramm Eng Remote Sens 44:1393-1404

Huang Yongfang, Huang Hai, Chen Wei, Yinxi LI (1991) Remote Sensing Approaches for Zunderground Coal Fire Detection. Presented at the Beijing International Conference on Reducing of Geological Hazards, pp 634-641

Huguenin RL, Jones JL (1986) Intelligent information extraction from reflectance spectra: absorption band positions. J Geophy Res 91:9585-9598

Hunt GR (1977) Spectral signatures of particulate minerals in the visible and near-infrared. Geophysics 42:501-513

Hunt GR (1979) Near-Infrared (1.3-2.4µm) spectra of alteration minerals potential for use in remote sensing. Geophysics 44:1974-1986

Hunt GR (1980) Electromagnetic radiation: the communication link in remote sensing. In: Siegal BS. Gillepie AR (eds) Remote Sensing in geology. Wiley, New York, pp 5-45

Huntington JF, Raiche AP (1978) A multi-attribute method for comparing geological lineament interpretations. Remote Sens Environ 7: 145-161

Hutchinson CF (1982) Techniques for combining Landsat and ancillary data for digital classification improvement. Photogramm Eng Remote Sens 48:123-130

Hutsinpiller A (1988) Discrimination of hydrothermal alteration mineral assemblages at Virginia City, Nevada, using the airborne imaging spectrometer. Remote Sens Environ 24:53-66

Ichoku C, Karnieli A (1996) A review of mixture modeling techniques for sub-pixel land cover estimation. Remote Sens Rev 13:161-186

Itten KI, Meyer P (1993) Geometric and radiometric correction of TM data of mountainous forested areas. IEEE Trans Geosc Remote Sens 31:764-770

Jakobsson SP (1979) Petrology of Recent basalts of the eastern volcanic zone, Iceland. Acta Naturalia Islandica, No 26, Icelandic Museum of Natural History, Reykjavik

Janza FJ (1975) Interaction mechanisms. In: Reeves RG (ed) Manual of remote sensing. 1st edn, Am Soc Photogramm, Falls Church, VA, pp 75-179

Jayaweera K, Seifert R, Wendler G (1976) Satellite observations of the eruption of Tolbackhik Volcano. Trans Am Geophys Union 57: 196-200

Jensen JR (1986) Introductory digital image processing. Prentice Hall, Englewood Cliffs, New Jersey, 379pp

Jiang D, Wang P, Meng F (1994) Application of Landsat TM data into exploration for porphyry copper deposits in forested area. In: Proc 10th Thematic Conf Geol Remote Sens, Vol II, Environ Res Inst Michigan, Ann Arbor, Mich, pp 611-618

Johnson PE, Smith MO, Taylor-George S, Adams JB (1983) A semiempricial method for analysis of the reflectance spectra of binary mineral mixtures. J Geophys Res 88 (B4): 3557-3561

Jones RC (1968) How images are detected. Sci Am 219:111-117

Jonsson S, Adam N, Bjornsson H (1998) Effects of geothermal activity observed by satellite radar interferometry. Geophysic Res Lett 25(7):1059-1062

Joria PE, Jorgenson JC (1996) Comparison of three methods for mapping Tundra with Landsat digital data. Photogramm Eng Remote Sens 62:163-169

Joseph G (1996) Imaging sensors for remote sensing. Remote Sens Rev 13:257-342

Joughin et al. (1998) Interferometric estimation of three-dimensional ice-flow using ascending and descending passes. IEEE Trans Geosci Remote Sens 36(1):25-35

JPL (Jet Propulsion Laboratory) (1986) Shuttle imaging radar-C science plan. NASA-Jet Propulsion Laboratory, Pasadena, Publ 86-29

Justice C, Wharton SW, Holben BN (1981) Application of digital terrain data to quantify and reduce the topographic effect on Landsat data. Int J Remote Sens 2:213-230

Kahle AB (1977) A simple thermal model of the Earth's surface for geologic mapping by remote sensing. J Geophys Res 82: 1673-1690

Kahle AB (1980) Surface thermal properties. In: Siegal BS, Gillespie AR (eds) Remote sensing in geology. Wiley, New York, pp 257-273

Kahle AB (1983) The new airborne thermal infared multispectral scanner (TIMS). Proc Int Geosci Remote Sens Symp (IGARSS), San Francisco, Sect FA-4, pp 7.1-7.3

Kahle AB, Goetz AFH (1983) Mineralogic information from a new airborne thermal infrared multispectral scanner. Science 222:24-27

Kahle AB, Rowan LC (1980) Evaluation of multispectral middle infrared aircraft images for lithologic mapping in the East Tintic Mountains, Utah. Geology 8:234-239

Kahle AB, Gillespie AR, Goetz AFH (1976) Thermal inertia imaging: a new geologic mapping tool. Geophys Res Lett 3:26-28

Kahle AB, Madura DP, Soha JM (1980) Middle infrared multispectral aircraft scanner data: analysis for geological applications. Appl Optics 19:2279-2290

Kahle AB, Schieldge JP, Alley RE (1984a) Sensitivity of thermal inertia calculations to variations in environmental factors. Remote Sens Environ 16:211-232

Kahle AB, Shumate MS, Nash DB (1984b) Active airborne infrared laser system for identification of surface rock and minerals, Geophys Res Lett 11:1149-1152

Kahle AB, Christensen P, Crawford M, Cuddapah P, Malila W, Palluconi F, Podwysocki M, Salisbury J, Vincent R (1986) Geology panel report. Commercial Applications and Scientific Research Requirements for TIR Observations of Terrestrial Surfaces, EOSAT-NASA Thermal IR Working Group, Aug 1986, pp 17-34

Kahle AB, Pallucani FD, Hook SJ, Realmuto VJ, Bothwell G (1991) The advanced spaceborne thermal emission and reflectance radiometer (ASTER). Int J Imaging Systems Tech 3:144-156

Kaneko T (1978) Colour composite pictures fom principal axis components of multispectral scanner data IBM J Res Dev 22:386-392

Kaplon ED (ed) (1996) Understanding GPS: Principles and Applications. Artech House Publ, Boston

Karr C Jr (ed) (1975) Infrared and Raman Spectroscopy of lunar and terrestrial minerals. Academic Press, New York

Kasischke ES, Schuchman AR, Lyzenga RD, Meadows AG (1983) Detection of bottom features on Seasat synthetic aperture radar imagery. Photogramm Eng Remote Sens 49:1341-1353

Kats YAG, Ryabukhin AG, Trofimov DM (1976) Space methods in geology. Moscow State University, Moscow, 248pp (in Russian)

Kaufmann HJ (1985) Rechnergestützte methodische Untersuchungen Multispekturaler Satellitenbiddaten (Landsat/HCMM) for geologische Fragestellungen am Beispiel Anti-Atlas, Marokko. PhD Dissertation, Univ Munich

Kaufmann HJ (1988) Mineral exploration along the Aqaba-Levant structure by use of TM-data- concepts, processing and results. Int J Remote Sens 9(10-11):1639-1658

Kayal JR (1987) Microseismicity and source mechanism study: Shillong Plateau, northeast India. Bull Seism Soc Am 77(1): 184-194

Key JR, Maslanik JA, Barry RG (1989) Cloud classification from satellite data using a fuzzy set algorithm: A polar example. Int J Remote Sens 10:1823-1842

Khazenie N, Richardson KA (1993) Detection of oil fire smoke over water in the Persion Gulf region. Photogramm Eng Remote Sens 59:1271-1276

King DJ (1995) Airborne multispectral digital camera and video sensors: a critical review of system designs and applications. Canadian J Remote Sens, Special Issue on Aerial Optical Remote Sensing 21 (3):245-273 web-site: www.carleton.ca/~dking/papers.html

Konecny G (1984) The photogrammetric camera experiment on Spacelab 1. Bildmessung und Luftbildwesen 52:195-200

Konecny G, Lohmann P, Engel H, Kruck E (1987) Evaluation of SPOT imagery on analytical photogrammetric instruments. Photogramm Eng Remote Sens 53:1223-1230

Koopmans BN (1983a) Spaceborne imaging radars: present and future. ITCJ (1983-3): 223-231

Kowalik WS, Lyon RJP, Switzer P (1983) The effects of additive radiance terms on ratios of Landsat data. Photogramm Eng Remote Sens 49:659-669

Kozlov VV, Romashov AA, Volchegurskiv L, Vorobeyer VT (1979) Use of satellite photographs to study deep crustal structures of petroliferous regions. Pt-III Lineaments of the Aral-Caspian Region. Int Geol Rev 11:1337-1344

Kreiter VM (1968) Geological prospecting and exploration. Mir, Moscow, 361pp

Krishnamurthy J, Venkatesa Kumar N, Jayaraman V, Manivel M (1996) An approach to demarcate groundwater potential zones through remote sensing and a geographic information system. Int J Remote Sens 17:1867-1884

Krishnamurthy P (1985) Petrology of the carbonatites and associated rocks of Sung Valley, Jaintia Hills District, Meghalaya, India. J Geol Soc Ind 26:361-379

Krüger G, Erzinger J, Kaufmann H (1998) Laboratory and airborne reflectance spectroscopic analyses of lignite overburden dumps. J Geochem Explor 64:47-65

Kruse FA (1997) Characterization of active hot-springs environments using multispectral and hyperspectral remote sensing. Proc 12th Int Conf App Geol Remote Sens, Vol I, Env Res Inst Michigan, Ann Arbor, Mich, pp 214-221

Kruse FA (1999) Visible Infrared sensors case studies. In: Rencz AN (ed) Remote Sensing for the Earth Sciences, Manual of Remote Sensing, 3rd ed, vol 3, Am Soc Photogramm Remote Sens, Wiley, pp 567-611

Kruse FA, Boardman JW (1997) Characterizing and mapping of kimberlites and related diatremes in Utah, Colorado, and Wyoming, USA, using the Airborne Visible/Infrared Imaging Spectrometer (AVIRIS). Proc 12th Int Conf App Geol Remote Sens, Vol I, Env Res Inst Michigan, Ann Arbor, Mich, pp 21-28

Kruse FA, Lefkoff AB (1993) Knowledge Based Geologic Mapping with Imaging Spectrometers. Remote Sens Rev 8:3-28

Kruse FA, Calvin WM, Seznec O (1988) Automated extraction of absorption features from airborne visible/infrared imaging spectrometer (AVIRIS) and Geophysical Environmental Research imaging spectrometer (GERIS) data. In: Proc AVIRIS Performance Evaluation Workshop, JPL Publ 88-38, Jet Propulsion Laboratory, California Inst Tech, Pasadena, CA, pp 62-75

Kruse FA, Kierein-Young KS, Boardman JW (1990) Mineral mapping at cuprite, Nevada with a 63-channel Imaging Spectrometer. Photogramm Eng Remote Sens 56(1):83-92

Kruse FA, Lefkoff AB, Boardman JW, Heidebrecht KB, Shapiro AT, Barloon PJ, Goetz AFH (1993) The Spectral Image Processing System (SIPS)-interactive visualization and analysis of imaging spectrometer data. Remote Sens Environ 44:145-163

Labovitz ML, Masuoka EJ, Bell R, Nelson RF, Latsen EA, Hooker LK, Troensegaard KW (1985) Experimental evidence for spring and autumn windows for the detection for geobotanical anomalies through the remote sensing of overlying vegetation. Int J Remote Sens 6:195-216

Lang HR (1999) Stratigraphy. In: Rencz AN (ed) Remote Sensing for the Earth Sciences, Manual of Remote Sensing, 3rd ed, vol 3, Am Soc Photogramm Remote Sens, Wiley, pp 357-374

Lang HR, Alderman WH, Sabins FF (1985) Patrick Draw, Wyoming, petroleum test case report. The joint NASA/Geosat test case project, Sect 11, Am Assoc Petrol Geol Tulsa, Oklahoma

Lang HR, Adams SL, Conel JE, McGuffie BA, Paylor Ed, Walker RE (1987) Multispectral remote sensing as stratigraphic and structural tool, Wind River Basin and Big Horn River Basin areas, Wyoming. Am Assoc Petrol Geol Bull 71(4):389-402

Lathram EH, Gryc G (1973) Metallogenic significance of Alaskan geostructures seen from space. Proc 8th Int Symp of Environ Ann Arbor, MI, pp 1209-1211

Lattman LH, Parizek RR (1964) Relationship between fracture traces and the occurrence of groundwater in carbonate rocks. J Hydrol 2:73-91

Laubscher HP (1971) The large-scale kinematics of the western Alps and the western Appennines and its palinspastic implications. AM J Sci 271:193-226

Leatherdale JD (1978) The practical contribution of space imagery to topographic mapping. Symp ISP Commission IV, Ottawa, Canada

Leberl FW (1998) Radargrammetry. In: Henderson FM, Lewis AJ (eds) Principles and Applications of Imaging Radar. Manual of Remote Sensing, 3rd ed, Vol 2, Wiley, New York, pp 183-269

Leberl FW, Olson D (1982) Raster scanning for operational digitizing of graphical data. Photogramm Eng Remote Sens 48:615-627

Leckie DG (1982) An error analysis of thermal infrared line scan data for quantitative studies. Photogramm Eng Remote Sens 48:945-954

Leckie DG (1990) Synergism of Synthetic Aperture Radar and Visible/Infrared Data for Forest Type Discrimination. Photogramm Eng Remote Sens 56(9):1237-1246

Leckie DG (1998) Forestry applications using imaging radar. Principles and Applications of Imaging Radar. Manual of Remote Sensing, 3rd ed, Vol 2, Wiley, New York, pp 435-509

Legg CA (1991) The Arabian Gulf oil slick, January and February 1991.Int J Remote Sens 12:1795-1796
Lei Q, Henkel J, Frei M, Mehl. H, Lörchner G, Bodechtel J (1996) Radiometric noise correction of panchromatic high resolution data of MOMS-02. Proc MOMS-02 Symp, Cologne, Germany, 5-7 July 1995. Paris: Euro Assoc Remote Sens Lab (EARSeL):303-313
Leick A (1995) GPS Satellite Surveying. 2nd ed, Wiley, NewYork
Leuder DR (1959) Aerial photographic interpretation. McGraw Hill, New York
Lewis AJ (ed) (1976) Geoscience applications of imaging radar systems. Remote Sensing of the electromagnetic spectrum, vol 3, No 3
Lewis AJ, Henderson FM (1998) Radar fundamentals: the geoscience perspective. In: Henderson FM, Lewis AJ (eds) Principles and Applications of Imaging Radar. Manual of Remote Sensing, 3rd ed, Vol 2, Wiley, New York, pp131-181
Li M, Daels L, Antrop M (1996) Lambertian and Minnaert relation simulation for topographic normalization. Proc 11th Thematic Conf Workshops App Geol Remote Sens, Las Vegas, Nevada, 27-29 February 1996. Ann Arbor Mich:Environ Res Inst Michigan 2:133-141
Li ZR, Mcdonnell MJ (1988) Atmospheric Correction of Thermal Infrared Images. Int J Remote Sens 9(1):107-121
Light DL (1981) Satellite photogrammetry. In: Slama CC (ed) Manual of photogrammetry. 4th edn. Am Soc Photogramm, Falls Church, VA, pp 883-977
Lillesand TM, Kiefer RW (1987) Remote sensing and image interpretation, 2nd edn, Wiley, New York, 721 pp
Lillesand TM, Keifer RW (2000) Remote Sensing and Image Interpretation. 4th edn, Wiley, New York, 724 pp
Limaye SS, Suomi VE, Velden C, Tripoli G (1991) Satellite observation of smoke from oil fires in Kuwait. Science 252:1536-1539
Lockwood JP, Lipman PW (1987) Holocene eruptive history of Mauna Loa volcano. In: Decker RW, Write TL, Stauffer PH (eds) Volcanism in Hawaii. Vol 1, USGS Prof Pap 1350, U S Geol Surv, Washington D C, pp 509-535
Longley PA, Goodchild MF, Maguire DJ, Rhind DW (eds) (1999) Geographical Information Systems. Wiley, NewYork, 1-2
Lou Y, Kim Y, Zyl J van (1996) The NASA/JPL airborne synthetic aperture radar system. Summaries of the 6th Annual AIRSAR Earth Science Workshop, JPL Publ 96-4, Vol 2, Jet Propulsion Laboratory, California Inst Tech, Pasadena, CA, pp 51-56
Loughlin WP (1991) Principal Component Analysis for Alteration Mapping. Photogramm Eng Remote Sens 57(9):1163-1169
Lowe DS (1969) Line scan devices and why we use them. Proc 5th Int Symp Remote Sens Environ, Ann Arbor, MI, pp 77-101
Lowe DS (1976) Non-photographic optical sensors. In: Lintz J Jr, Simonett DS (eds) Remote sensing environment. Addison-Wesley, Reading, pp 155-193
Lowman PD Jr (1969) Geologic orbital photography: experience from the Gemini program. Photogrammetrica 24: 77-106
Lyon RJP (1962) Minerals in the infrared a critical bibliography. Stanford Res Inst Publ, Palo Alto CA, 76 pp
Lyon RJP (1965) Analysis of rocks by spectral infrared emission (18 to 25 microns). Econ Geol 60:715-736
Lyon RJP, Patterson JW (1966) Infared spectral signatures – a field geological tool. Proc 4th Int Symp Remote Sens Environ, Ann Arbor, MI, pp 215-220
Lyon RJP, Elvidge C, Lyon JG (1982) Practical requirements for operational use of geobotany and biogechemistry in mineral exploration. Proc Int Symp Remote Sens Environ, 2nd Thematic Conf Remote Sens Expl Geol, Fort Worth, Texas, pp 85-91

MacDonald HC (1969) Geologic evaluation of radar imagery from Darien Province, Panama. Mod Geol 1:1-63

MacDonald HC (1969) The influence of radar look direction on the detection of selected geologic features. Proc 6th Int Symp Remote Sens Environ, Ann Arbor, MI, pp 637-650

MacDonald HC (1980) Techniques and applications of imaging radars. In: Siegal BS, Gillespie AR (eds) Remote sensing in geology. Wiley, New York, pp 297-336

MacDonald HC, Waite WP (1973) Imaging radars provide terrain texture and roughness parameters in semi-arid environments, Mod Geol 4: 145-158

Macias LF (1995) Remote sensing of mafic–ultramafic rocks: examples from Australian Precambrian terranes. J Aust Geol Geophys 16:163-171

Madsen SN, Zebker HA (1998) Imaging radar interferometry. In: Henderson FM, Lewis AJ (eds) Principles and Applications of Imaging Radar, Manual of Remote Sensing, 3rd ed, Wiley, New York, 2:359-380

Maguire DJ, Goodchild, MF, Rhind DW (eds) (1991) Geographic Information Systems – Principles and Applications. Harlow, Essex:Longman Scientific and Technical

Mansor SB, Cracknell AP, Shilin BV, Gornyi VI (1994) Monitoring of Underground Coal Fires Using thermal Infrared Data. Int J Remote Sens 15(8):1675-1685

Markham BL, Barker JL (1983) Spectral characterization of the Landsat-4 MSS sensors. Photogramm Eng Remote Sens 49(6): 811-833

Markham BL, Barker JL (1986) Landsat MSS and TM post calibration dynamic ranges, exoatmospheric reflectances and at- satellite temperatures, EOSAT Landsat Technical Notes 1, August 1986, Earth Observation Satellite Co, (Lanham, Maryland), pp 3-8

Marsh SE, Schieldge JP, Kahle AB (1982) An instrument for measuring thermal inertia in the field. Photogramm Eng Remote Sens 48:605-607

Maselli F, Conese G, Petkov L, Resti R (1992) Inclusion of prior probabilities derived from a nonparametric process into the maximum likelihood classifier. Photogramm Eng Remote Sens 58:201-207

Massonnet D, Feigl K (1998) Radar interferometry and its applications to changes in the earth's surface. Rev Geophysics 36(4):441-500

Massonnet D, Rossi M, Carmona C, Adranga F, Peltzer G, Feigl K, Rabaute T (1993) The displacement field of the Landers earthquake mapped by radar interferometry. Nature 364:138-142

Massonnet D, Feigl D, Rossi M, Adranga F (1994) Radar interferometric mapping of deformation in the year after the Landers earthquake. Nature 369:227-230

Massonnet D, Briole P, Arnaud A (1995) Deflation of Mount Etna monitored by spaceborne radar interferometry. Nature 375:567-570

Masuoka EJ, Labovitz ML, Bell R, Nelson RF, Broderick PW, Ludwig RW (1982) The application of remote sensing in geobotanical exploration for metal sulfides. Proc Int Symp Remote Sens Environ, 2nd Thematic Conf Remote Sens Explor Geol, Fort Worth, Texas, 669pp

Mather PM (1987) Computer processing of remotely sensed images, an introduction. Wiley, Chicester, 352 pp

Mather PM (1999) Computer processing of remotely sensed images, an introduction. 2nd ed, Wiley, Chicester, 292 pp

Matson M, Dozier J (1981) Identification of subresolution high temperature sources using a thermal infrared sensor. Photogramm Eng Remote Sens 47(9):1311-1318

McCauley JF, Schaber GC, Breed CS, Grolier MJ, Haynes CV, Issawi B, Elachi C, Blom R (1982) Subsurface valleys and geoarchaeology of the eastern Sahara revealed by shuttle radar. Science 218:1004-1019

McDonald RA (1995) Opening the cold war sky to the public-declassifying satellite reconnaissance imagery. Photogramm Eng Remote Sens 61:385-390
Mckinstry HE (1948) Mining geology. Prentice Hall, Englewood Cliffs, NJ, 680 pp
Mees CEK, James TH (1966) The theory of the photographic processes. 3rd edn. Macmillan, New York
Mekel JFM (1970) ITC textbook of photo-interpretation. Chap VIII.1 The use of aerial photographs in geology. ITC, Enschede
Mekel JFM (1978) ITC textbook of photo-interpretation. Chap VIII. The use of aerial photographs and other images in geological mapping. ITC, Enschede
Melen R, Buss D (ed) (1977) Charge-coupled devices: technology and applications. IEEE press (selected reprint series), New York, 415 pp
Meyer P (1994) A parametric approach for the geocoding of Airborne Visible/Infrared Imaging Spectrometer (AVIRIS) data in rugged terrain. Remote Sens Environ 49:118-130
Miller SH, Watson K (1977) Evaluation of algorithms for geological thermal inertia mapping. Proc 11th Symp Remote Sens Environ, Ann Arbor, MI, pt2, pp 1147-1160
Miller SH, Watson K (1980) Thermal Infrared Aircraft Scanner Data of the Area of Underground Coal Fires, Sherodean, Wyoming, July 1975 and October 1978. Open Fire Report, p 5
Miller VC, Miller CF (1961) Photogeology. McGraw-Hill, New York
Milton NM, Collins W, Chang SH, Schmidt RG (1983) Remote detection of metal anomalies on Pilot Mountain, Randolph County, North Carolina. Econ Geol 78:605-617
Miranda FP, McCafferty AE, Taranik JV (1994) Reconnaissance geologic mapping of a portion of the rain-forest-covered Guiana Shield, northwestern Brazil, using SIR-B and digital aeromagnetic data. Geophysics 59:733-743
Moffitt FH, Mikhail EM (1980) Photogrammetry, 3rd edn. Harper & Row, New York
Mohanty KK, Maiti K, Nayak S (2001) Monitoring water surges. GIS@Development Vol (3):32-33, http://www.gis-development.net
Moik JG (1980) Digital processing of remotely sensed images. NASA SP-431, US Govt Printing Office, Washington, DC
Moore GK, Waltz FA (1983) Objective procedures for lineament enhancement and extraction. Photogramm Eng Remote Sens 49: 641-647
Morain SA (1976) Use of Radar for Vegetation analysis. Remote Sensing of the Electromagnetic Spectrum 3, CL 4:61-78
Mouat DA (1982) The response of vegetation to geochemical conditions. Proc Int Symp Remote Sens Environ, 2nd Thematic Conference Remote Sens Explor Geol, Fort Worth, Texas, pp 75-84
Murakami M, Fujiwara S, Nemoto M, Saito T (1995) Application of the interferometric JERS-1 SAR for detection of crustal deformations in the Izu Peninsula, Japan, Eos Trans, AGU, 76 suppl, F63
Mustard JF, JM Sunshine (1999) Spectral analysis for earth science: Investigations using remote sensing data. In: Rencz AN (ed) Remote Sensing for the Earth Sciences, Manual of Remote Sensing, 3rd ed, Vol 3, Am Soc Photogramm Remote Sens, Wiley, New York, pp 251-306
Nagarajan R, Mukherjee A, Roy A, Khire MV (1998) Temporal remote sensing data and GIS application in landslide hazard zonation of part of Western Ghat, India. Int J Remote Sens 19(4):573-585
Naraghi M, Stromberg W, Daily M (1983) Geometric rectification of radar imagery using digital elevation models. Photogramm Eng Remote Sens 49:195-199
Nelson R (1985) Reducing Landsat MSS scene variability. Photogramm Eng Remote Sens 51:583-593

Nishidai T (1993) Early results from 'Fuyo-1' Japan's Earth Resources Satellite (JERS-1). Int J Remote Sens 14:1825-1833

O'Leary DW, Friedman JD, Pohn HA (1976) Lineament, linear and lineation: some proposed new standards for old terms. Geol Soc Am Bull 87:1463-1469

O'Leary DW, Johnson GR, England AW (1983) Fracutre detection by airborne microwave radiometery in parts of the Mississippi embayments, Missouri and Tennessee. Remote Sens Environ 13: 509-523

Offield TW, Abbott EA, Gillespie AR, Loguercio SO (1977) Structure mapping on enhanced Landsat images of southern Brazil: tectonic control of mineralization and speculation on metallogency, Geophysics 42:482-500

Oppenheimer C (1991) Lava flow cooling estimated from Landsat Thematic Mapper infrared data: the Lonquimay eruption, Chile, 1989. JGeophys Res 96:21,865-21,878

Oppenheimer C, Francis PW, Rothery DA, Carlton RW, Glaze LS (1993) Infrared image analysis of volcanic thermal features: Volcan Lascar, Chile, 1984-1992. J Geophys Res 98:4269-4286

Ortega GE (1986) Intrduction to the geology and metallogeny of the Almaden area, Castro-Iberian zone, Spain. Proc 2nd European Workshop on Remote Sensing in Mineral Exploration EEC, Brussels

Ouadrari Hassan, Vermote Eric F (1999) Operational atmospheric correction of Landsat TM Data. Remote Sens Environ 70: 4-15

Pachauri AK, Pant M (1992) Landslide hazard mapping based on geological attributes. Eng Geol 32:81-100

Palluconi FD, Meeks GR (1985) Thermal Infrared Multispectral Scanner (TIMS): An Investigator's Guide to TIMS Data, JPL Publ 85-32, Jet Propulsion Laboratory, California Inst Tech, Pasadena, CA

Pandey SN (1987) Principles and applications of photogeology. Eastern Wiley, New Delhi, 366 pp

Parasnis DS (1980) Principles of applied geophysics, 3rd edn, Chapman and Hall, London, 275pp

Parizek RR (1976) Lineaments and groundwater. In: McMurthy GT, Petersen GW (eds) Interdisciplinary application and interpretations of EREP data within the Susquehanna River basin. Pennsylvania State Univ, pp 4-59 to 4-86

Parlow E (ed) (1996a) Progress in Environmental Remote Sensing Research Applications. Proc 15th EARSeL Symp, Basle Switzerland, 4-6 September 1996. Rotterdam: A.A. Balkema

Parvis M (1950) Drainage pattern significance in airphoto identification of soils and bedrocks. Photogramm Eng 16:387-409

Peak WH, Oliver TC (1971) The response of terrestrial surfaces at microwave frequencies. Ohio State Univ Electrosic Lab 2440-7 Tech Rep AFAL-TR70-301, p 255

Peddle DR (1993) An empirical comparison of evidential reasoning, linear discriminant analysis, and maximum likelihood algorithms for land cover classification. Can J Remote Sens 19:31-44

Peli T, Malah D (1982) A study of edge detection algorithms. Computer Graphics Image Processing 20:1-21

Peters WC (1978) Exploration mining and geology. Wiley, New York, 644pp

Peterson DL, Aber JD, Matson PA, Card DH, Swanberg N, Wessman C, Spanner M (1988) Remote sensing of forest canopy and leaf biochemcial contents. Remote Sens Environ 24:85-108

Philip G, Gupta RP, Manickavasgam RM (1988) Geoenvironmental studies in singrauli coal belt (MP and UP). Geosciences J IX:205-214

Philip G, Gupta RP, Bhattacharya A (1989) Channel migration studies in the middle Ganga basin, India, using remote sensing data. Int J Remote Sens 10:1141-1149

Pichel W, Bristor CL, Brower R (1973) Artificial stereo: a technique for combining multichannel satellite image data. Bull Am Meteorol Soc 54:688-690

Pitas I (1993) Digital Image Processing Algorithms. Prentice-Hall, Englewood Cliffs, NJ

Podwysocki MH, Segal DB, Abrams MJ (1983) Use of multispectral scanner images for assessment of hydrothermal alteration in the Marysvale, Utah mining area. Econ Geol 78: 675-687

Pohl C, van Genderen JL (1998) Multisensor image fusion in Remote sensing: concepts, methods and application. Int J Remote Sens 19:823-854

Pohn HA, Offield TW, Watson K (1974) Thermal inertia mapping from satellite – discrimination of geologic units in Oman. J Res US Geol Surv 2:147-158

Prakash A, Gupta RP (1998) Land-use mapping and change detection in coal mining area - a case study in the Jharia Coalfield, India. Int J Remote Sens 19(3):391-410

Prakash A, Gupta RP (1999) Surface fires in the Jharia coalfield, India- their distribution and estimation of area and temperature from TM data. Int J Remote Sens 20(10): 1935-1946

Prakash A, Saraf AK, Gupta RP, Dutta M, Sundaram RM (1995a) Surface thermal anomalies associated with underground fires in Jharia coal mines, India. Int J Remote Sens 16(12):2105-2109

Prakash A, Sastry RGS, Gupta RP, Saraf AK (1995b) Estimating the depth of buried hot features from thermal IR remote sensing data: a conceptual approach. Int J Remote Sens 16(13):2503-2510

Prakash A, Gupta RP, Saraf AK (1997) A landsat TM based comparative study of surface and subsurface fires in the Jharia coalfield, India. Int J Remote Sens 18(11):2463-2469

Prakash A, Fielding EJ, Gens R, van Genderen JL, Evans DL (2001) Data fusion for investigating land subsidence and coal fire hazards in a coal mining area. Int J Remote Sens 22:921-932

Prakash S (1981) Soil dynamics. McGraw-Hill, New York, pp 274-339

Prasad SN, Rao MNA, Mookherjee A (1984) Mine planning- A step towards modernisation, Coal Mining in India, CMPDIL publication, 12th World Mining Congress, New Delhi

Prata AJ (1995) Thermal remote sensing of land surfaces temperature from satellites- current status and future prospects. Remote Sens Rev 12:175-224

Pratt WK (1978) Digital image processing. Wiley, New York

Price JC (1977) Thermal inertia mapping: a new view of the Earth. J Geophys Res 81:2582-2590

Price JC (1983) Estimating Surface Temperatures from Satellite Thermal Infrared Data-A Simple Formulation for the Atmospheric Effect. Remote Sens Environ 13:353-361

Price JC (1988) An update on visible and near infrared calibration of satellite instruments. Remote sens Environ 24:419-422

Price JC (1998) An update on visible and near infrared calibration of satellite instruments. Remote Sens Environ 24:419-422

Price NJ, Cosgrove J (1990) Analysis of Geological Structures. Cambridge University Press, 502pp

Proy C, Tanre D, Deschamps PY (1989) Evaluation of topographic effects in remotely sensed data. Remote Sens Environ 30:21-32

Puglisi G, Coltelli M (1998) SAR interferometry applications on active volcanoes: state of the art and perspectives for volcano monitoring. Workshop Synthetic Aperture Radar, 25-26 February 1998, Florence, Italy

Quinn MF et al. (1994) Measurement and analysis procedures for remote identification of oil spills using a laser fluorosensor. Int J Remote Sens 15:2637-2658

Rahman H, Dedieu G (1994) SMAC: a simplified method for the atmospheric measurements in the solar spectrum. Int J Remote Sens 15(1):123-143

Rama Rao JV (1999) Geological and structural inference of Shillong massif from aeromagnetic data. J Geophysics, Assoc Explor Geophysicists, vol xx (1), pp 21-24

Ranchin T, Wald L (1998) Fusion of airborne and spaceborne images in visible range. Operational Remote Sensing for Sustainable Development, Noeuwenhuis, Vaughan & Molennar (eds), pp 255-260

Ranson KJ Sun G (1994) Northern forest classification using temporal multifrequency and multipolarization SAR images. Remote Sens Environ 47(2):142-153

Rao DP, Subramanian SK (1997) Geomorphic analysis of IRS-1C PAN stereo image. Interface, (NRSA, Hyderabad), Vol 8, No 2 (April-June), pp 4-5

Rao YSN, Rahman AA, Rao DP (1974) On the structure of the Siwalik range between the rivers Yamuna and Ganga. Himalayan Geol, vol 4, pp 137-150

Rast M (1992) ESA's Activities in the field of imaging spectroscopy. In: Toselli F, J Bodechtel (eds) Imaging Spectroscopy: Fundamentals and Prospective Applications. Kluwer, Dordrecht, pp 167-191

Rast M, Hook SJ, Elvidge CD, Alley RE (1991) An evaluation of techniques for the extraction of mineral absorption features from high spectral resolution remote sensing data. Photogramm Eng Remote Sens 57:1303-1309

Ray RG (1965) Aerial photographs in geologic interpretation and mapping. USGS Prof Paper 373

Realmuto VJ (1990) Separating the effects of stemperature and emissivity:emissivity spectrum normalization in Proceedings of the 2nd TIMS Workshop, JPL Publ. 90-55, Jet Propulsion Laboratory, California Inst Tech, Pasadena, CA

Rebillard P, Evans P (1983) Analysis of coregistered Landsat, Seasat and SIR-A images of varied terrain types. Geophys Res Lett 10(4):277-280

Reeves RG (1968) Introduction to electromagnetic remote sensing with emphasis on applications to geology and hydrology. Am Geol Inst, Washington, DC

Reeves RG (ed) (1975) Manual of remote sensing. Ist edn, Am Soc Photogramm, Falls Church, VA

Reinhackel G, Kruger G (1998) Combined use of laboratory and airborne spectrometry from the reflective to Thermal wavelength range for a quantitative analysis of lignite overburden dumps. 27th R S E Int Symp, R S E Tromsoe, Norway, pp 507-512

Rencz AN, Bowie C, Ward B (1996) Application of thermal imagery from Landsat data to identify kimberlites, Lac de Gras area, District of Mackenzie, N.W.T. In: LeChaimant AN, Richardson DG, DiLabio RNW, Richardson KA (eds). Searching for diamonds in Canada, Geol Surv Canada, Open File 3228:255-257

Rib HT (1975) Engineering: regional inventories, corridor surveys and site investigations. In: Reeves RG (ed)Manual of remote sensing. Am Soc Photogramm, Falls Church, VA, pt 2, pp 1881-1945

Rib HT, Liang TA (1978) Recognition and identification. In: Schuster RL, Krizek RV (eds) Landslides analysis and control. Trans Res Board Nat Res Council USA Spec Rep 176:34-80

Ricci M (1982) Dip determination in photogeology. Photogramm Eng Remote Sens 48:407-414

Rice DP, Malila WA (1983) Investigation of radiometric properties of Landsat-4 MSS. Proc Symp Landsat-4 Characterization, Early Results, Goddard Space Flight Center, Greenbelt, Maryland

Richards JA, Jia X (1999) Remote Sensing Digital Image Analysis. 3rd ed, Springer Verlag, Heidelberg, 363pp

Richter R (1996) A spatially adaptive fast atmospheric correction algorithm. Int J Remote Sens 17:1201-1214

Robbins J, Seigel HO (1982) The luminex method – a new active remote sensing method for exploration for mineral deposits. Proc Int Symp Remote Sens Environ, 2nd Thematic Conf Remote Sens Explor Geol, Fort Worth, Texas, pp 203-204

Robinove CJ (1982) Computation of physical values from Landsat digital data. Photogram Eng Remote Sens 48:781-784

Rocca F, Prati C, Feretti (1997) An overview of ERS-SAR interferometry, 3rd ERS Symposium, Space at the Service of Our Environment, Florence, 17-21 March, 1997, ESA-SP-414, Vol 1, pp xxvii-xxxvi http://florence97.erssymposium.org

Rock BN, Hoshizaki T, Miller Jr (1988) Comparison of in-situ and airborne spectral measurements of the blue-shift associated with forest decline. Remote Sens Environ 24:109-127

Rogers AE, Ingalls RP (1969) Venus: Mapping the surface Reflectivity by Radar Interferometry. Science 165: 797-799

Rosen PA, Hensley S, Joughin IR, Li F, Madsen SN, Rodriguez E, Goldstein RM (1999) Synthetic aperture radar interferometry. Proc IEEE XX(Y):1-110

Rosenfeld A, Kak AC (1982) Digital picture processing, 2nd edn, Academic Press, Orlando, FLA

Rothery DA, Francis PW, Wood CA (1988) Volcano monitoring using short wavelength IR data from satellites. J Geophys Res 93(B7):7993-8008

Rowan LC, Bowers TL (1995) Analysis of linear features mapped in Landsat thematic mapper and side-looking radar images of the Reno, Nevada-California 1° X 2° quadrangle: implications of mineral resource studies. Photogramm Eng Remote Sens 61:749-759

Rowan LC, Wetlaufer PH, Goetz AFH, Billingsley FC, Stewart JH (1974) Discrimination of rock types and detection of hydrothemally alerted areas in south-central Nevada by sue of computer-enhanced ERTS images, USGS prof Pap 883, p35

Rowan LC, Goetz AFH, Ashley RP (1977) discrimination of hydrothermally altered and unaltered rocks in visible and near-infrared multispectral images. Geophysics 42:522-535

Rowan LC, Goetz AFH, Crowley JK, Kingston MJ (1983) Identification of hydrothermal mineralization in Baja California. Mexico, from orbit using the shuttle multispectral infrared radiometer. Proc Int Geosci Remote Sens Symp (IGARRS) vol 1, pp3.1-3.9

Rowan LC, Watson K, Crowley JK, Anton-Pancheco C, Gumiel P, Kingston MJ, Miller SH, Bowers TL (1993) Mapping lithologies in the Iron Hill, Colorado, carbonatite alkalic igneous rock complex using thermal infrared multispectral scanner and airborne visible-infrared imaging spectrometer data. Proc 9th Thematic Conf Geol Remote Sens, Vol I, Env Res Inst Michigan, Ann Arbor, Mich, pp 195-197

Roy Chowdhary MK, Das Gupta SP (1965) Ore localization in Khetri copper belt. Econ Geol 60:69-88

Roy SC (1939) Seismometric study. Mem Geol surv India 73:49-75

Ruiz-Armenta JR, Prol-Ledesma RM (1998) Techniques for enhancing the spectral reponse of hydrothermal alteration minerals in Thematic Mapper images of Central Mexico. Int J Remote Sens 19(10):1981-2000

Russ JC (1995) The Image Processing Handbook. 2nd ed., CRC Press, Boca raton, FL

Ryerson RA, Morain SA, Budge AM, (eds) (1997) The Manual of Remote Sensing, 3rd edn, Earth Observing Platforms and Sensors (CD-ROM). Am Soc Photogramm Engg Remote Sens

Sabine C (1999) Remote sensing strategies for mineral exploration. Remote Sensing for the Earth Sciences, Manual of Remote Sensing, 3rd edn, Am Soc Photogramm Remote Sens, John Wiley, Vol 3, pp 375-447

Sabine C, Realmuto VJ, Taranik JV (1994) Quantitative estimation of granitoid composition from thermal infrared multispectral scanner (TIMS) data, Desolation Wilderness, northern Sierra Nevada, California. J Geophys Res 99(B3):4261-4271

Sabins FF Jr (1969) Thermal infrared imagery and its application to structural mapping in southern California. Geol Soc Am Bull 80:397-404

Sabins FF Jr (1983) Geologic interpretation of space shuttle radar images of Indonesia. Am Assoc Petrol Geol Bull 67:2076-2099

Sabins FF Jr (1987) Remote sensing principles and interpretation. 2nd edn, Freeman, San Francisco, 449 pp

Sabins FF Jr(1997) Remote Sensing–Principles and Interpretation. 3rd ed, Freeman & Co, NY

Sabins FF Jr, Blom R, Elachi C (1980) Seasat radar image of San Andreas fault, California. Am Assoc Petrol Geol Bull 64:619-628

Saha AK, Gupta RP, Arora MK (2002) GIS-based landslide hazard zonation in the Bhagirathi (Ganga) Valley, Himalayas. Int J Remote Sens 23(2):357-369

Salisbury JW, Hunt GR (1974) Remote sensing of rock type in the visible and near infrared. Proc 9th Int Symp Remote Sens Environ, Ann Arbor, MI, vol III, pp 1953-1958

Saraf AK, Prakash A, Sengupta S, Gupta RP (1995) Landsat-TM data for estimating ground temperature and depth of subsurface coal fire in the Jharia coalfield, India. Int J Remote Sens 16(12):2111-2124

Sawchuka AA (1978) Artificial stereo. App Optics 17:3869-3873

Scarpace FL, Madding RP, Green T III (1975) Scanning thermal plumes. Photogramm Eng 41:1223-1231

Schaber GG, Berlin GL, Brown WE Jr (1976) Variations in surface roughness within Death Valley, California, geologic evaluation on 25-cm wavelength radar images. Geol Soc Am Bull 87:29-41

Schalkoff RJ (1992) Pattern Recognition: Statistical, Structural and Neural Approaches. Wiley, New York

Schanda E (1986) Physical fundamentals of remote sensing. Springer, Berlin Heidelberg New York Tokyo, 187 pp

Schetselaar EM (1998) Fusion by the IHS transform: should we use cylindrical or spherical coordinates? Int J Remote Sens 19:759-765

Schmugge T (1980) Techniques and applications of microwave radiometry. In: Siegel BS, Gillespie AR (eds) Remote sensing in geology. Wiley, New York, pp 337-361

Schott JR (1989) Image processing of thermal infrared images. Photogramm Eng Remote Sens 55(9):1311-1321

Schott JR, Volchok WJ (1985) Thematic Mapper thermal infrared calibration. Photogramm Eng Remote Sens 51:1351-1357

Schowengerdt RA (1997) Remote Sensing: Models and Methods for Image processing. 2nd edn, San Diego: Academic Press

Schultejann PA (1985) Structural trends in Borregeo Valley, California: interpretations from SIR-A and SEASAT SAR. Photogramm Eng Remote Sens 51:1615-1624

Seeber L, Armbruster JG, Quitmeyer RC (1981) Seismicity and continental subduction in the Himalayan arc. Inter Union Commission on Geodynamics, Working Group 6:215-242

Segal DB (1982) Theoretical basis for differentiation of ferric-iron bearing minerals, using Landsat MSS data. Proc Symp Remote Sens Environ, 2nd Thematic Conf Remote Sens Explor Geol, Fort Worth, Texas, pp 949-951
Segal DB (1983) Use of Landsat multispectral scanner data for the definition of limonitic exposures in heavily vegetated areas. Econ Geol 78:711-722
Sellers PJ (1985) Canopy reflectance – photosynthesis and transpiration. NASA Contract Rep 177822, Greenbelt, MD, NASA Goddard Space Flight Center.
Sellers WD (1965) Physical climatology. Univ Chicago Press, Chicago, 272 pp
Sen D, Sen S (1983) Post-Neogene tectonism along the Aravalli range, Rajasthan, India. Tectonophysics 93:75-98
Settle JJ, Drake NA (1993) Linear mixing and the estimation of ground cover proportions. Int J Remote Sens 14:1159-1177
Settle M (1984) The Joint NASA/Geosat test case project. Executve summary, pt 1, Am Assoc Petrol Geol, AAPG Book store, Tulsa, Oklahoma, p 30
Sharma RP (1977) The role of ERTS-1 multispectral imagery in the elucidation of tectonic framework and economic potentials of Kumaun and Simla Himalaya. Himalayan Geol 7: 77-99
Sharma RS (1988) Patterns of metamorphism in the Precambrian rocks of the Aravalli mountain belt. Mem Geol Soc Ind 7:33-75
Shaw GB (1979) Local and regional edge detectors: some comparisons. Computer Graphics Image Processing 9:135-149
Sheffield C (1985) Selecting band combinations from multispectral data. Photogramm Eng Remote Sens 51:681-687
Short NM (1986) Volcanic landforms. In: Short NM, Blair RW Jr (eds) Geomorphology from space. NASA SP-486 US Govt Printing Office, Washington, DC, pp 185-254
Short NM, Blair RW Jr (eds) (1986) Geomorphology from space NASA SP-486 US Govt Printing Office, Washington, DC
Short NM, Stuart LM Jr (1982) The Heat Capacity Mapping Mission (HCMM) anthology. NASA SP-465, US Govt Printing Office, Washington, DC, p 264
Short NM, Lowman PD Jr. Freden SC, Finch WA Jr (1976) Mission to the Earth: Landsat views the world. NASA SP-360 US Govt Printing Office, Washington, DC, p 459
Siegal BS (1977) Significance of operator variation and the angle of illumination in lineament analysis on synoptic images. Mod Geol 6:75-85
Siegal BS, Abrams MJ (1976) Geologic mapping using Landsat data. Photogramm Eng Remote Sens 42:325-337
Siegal BS, Goetz AFH (1977) Effect of vegetation on rock and soil type discrimination. Photogramm Eng Remote Sens 43:191-196
Silva LF (1978) Radiation and instrumentation in remote sensing. In: Swain PH, Davis SM (eds) Remote sensing: the quantitative approach. McGraw Hill, New York, pp 21-135
Singhal BBS, Gupta RP (1999) Applied Hydrogeology of Fractured Rocks. Kluwer, Dordrecht, 393p
Singhroy VH, Kruse FA (1991) Detection of metal stress in boreal forest species using the 0.67 µm chlorophyll absorption band. Proc 8th Thematic Conf Geol Remote Sens, Vol I, Env Res Inst Michigan, Ann Arbor, Mich, pp 361-372
Sinha PR (1986) Mine fires in Indian coalfields. Energy 11(11/12):1147-1154
Slama CC (ed) (1981) Manual of photogrammetry. 4th edn, Am Soc Photogramm, Fall Church, VA, 1056 pp
Slater PN (1975) Photographic remote sensing. In: Reeves RG (ed) Manual of remote sensing. Am Soc Photogramm, Falls Church, VA, pp 235-323
Slater PN (1979) A re-examination of the Landsat MSS. Photogramm Eng Remote Sens 45: 1479-1485

Slater PN (1980) Remote sensing – optics and optical systems. Addison Wesley, Reading, 575 pp
Slater PN (1983) Photographic systems for remote sensing. In: Colwell RN (ed) Manual of remote sensing. 2nd edn, Am Soc Photogramm, Falls Church, VA, pp 231-291
Slater PN (1985) Survey of multispectral imaging systems for earth observations. Remote Sens Environ 17:85-102
Slavecki RJ (1964) Detection and location of subsurface coal fire. Proc 3rd Symp, Remote Sens Environ, October 14-16, 1964, pp 537-547, University of Michigan, Ann Arbor, MI
Small D, Pasquali P, Füglistaler S (1966) A comparison of phase to height conversion methods for SAR interferometry. Proc IGARSS '96, Lincoln, Nebraska, pp 342-344
Smirnov V (1976) Geology of mineral deposits. Mir, Moscow
Smith HTU (1943) Aerial photographs and their application. Appleton-Century, New York, 372 pp
Smith JA, Lin TL, Ranson KJ (1980) The lambertian assumption and Landsat Data. Photogramm Eng Remote Sens 46(9):1183-1189
Smith JT Jr (1968) Filters for aerial photography. In: Smith JT Jr, Anson A (ed) Manual of colour aerial photography. Am Soc Photogramm, Falls Church, VA, pp 189-195
Smith JT Jr, Anson A (eds) (1968) Manual of colour aerial photography. Am Soc Photogramm, Falls Church, VA
Smith MO, Johnson PE, Adams JB (1985) Quantitative determination of mineral types and abundances from reflectance spectra using principal component analysis. J Geophys Res 90 (Suppl): C797-C804
Solaas GA (1994) ERS-1 interferometric baseline algorithm verification. ESA Technical Report ES-TN-DPE-OM-GS02
Stanton RL (1972) Ore petrology. McGraw Hill, New York, 713 pp
Star J, Estes J (1990) Geographic Information Systems: An Introduction. Prentice Hall, Englewood Cliffs, New Jersey
Stofan ER, Evans DL, Schmullius C, Holt B, Plaut J, Zyl J van, Wall SD, Way J (1995) Overview of results of Spaceborne Imaging Radar-C, X-band Synthetic Aperture Radar (SIR-C/X-SAR). IEEE Trans Geosci Remote Sens 33:817-828
Stohr CJ, West TR (1985) Terrain and look angle effects upon multispectral scanner response. Photogramm Eng Remote Sens 51:229-235
Stow RJ, Wright P (1997) Mining subsidence land survey by SAR interferometry. 3rd ERS Symposium, ESA-SP 414, Vol 1, pp 525-530
Strahler AH (1980) The use of prior probabilities in maximum likelihood classification of remotely sensed data. Remote Sens Environ 10:135-163
Strahler AH, Logan TL, Bryant NA (1978) Improving forest cover classification accuracy from Landsat by incorporating topographic information. Proc 12th Symp Remote Sens Environ, Ann Arbor, MI, vol II, pp 927-942
Strahler AH, Estes JE, Maynard PF, Mertz FC, Stow DA (1980) Incorporating collateral data in Landsat classification and modelling procedures. vol II, 14th Symp, Remote Sens Environ, Ann Arbor, Michigan, pp 1009-1026
Strauss GK, Roger G, Lecolle M, Lopera E (1981) Geochemical and geological study of the volcano-sedimentary sulfide orebody of La Zarza-Huelva, Spain. Econ Geol 76:1975-2000
Stringer WJ et al. (1992) Detection of petroleum spilled from the MV Exxon Valdez. Int J Remote Sens 13:799-824
Suits GH (1983) The nature of electromagnetic radiation. In: Colwell RN (ed) Manual of remote sensing. Am Soc Photogramm, Falls Church, VA, pp 37-60

Sundaram RM (1998) Integrated GIS Studies for Delineation of Earthquake-Induced Hazard Zones in Parts of Garhwal Himalaya. Ph.D thesis, Department of Earth Sciences, University of Roorkee, 135pp

Swain PH (1978) Fundmentals of pattern recognition in remote sensing. In: Swain PH, Davis SM (eds) Remote sensing: the quantitative approach. Mcgraw Hill, New York, pp136-187

Tanre D, Deroo C, Duhaut P, Herman M, Morcrette JJ, Perbos J, Deschamps P Y (1990) Description of a computer code to simulate the satellite signal in the solar spectrum: the 5S code. Int J Remote Sens 11(4):659-668

Taranik DL, Kruse FA, Goetz AFH, Atkinson WW (1991) Remote sensing of ferric iron minerals as guides for gold exploration. Proc 8th Thematic Conf Geol Remote Sens, Vol I, Env Res Inst Michigan, Ann Arbor, Mich, pp 197-205

Taranik JV (1978) Principles of computer processing of Landsat data for geologic applications. USGS Open File Rep 78-177, Sioux Falls, South Dakota

Tator BA (1960) Photo interpretation in geology. In: Colwell RN (ed) Manual of photographic interpretation. Am Soc Photogramm, Falls Church, VA, pp 169-342

Tauch R, Kähler M (1988) Improving the quality of satellite images maps by various processing techniques. Int Archives of Photogrammetry & RS, Proc. XVI ISPRS Congress, Tokyo, Japan, pp IV238-IV247

Teillet PM, Fedosejevs G (1995) On the dark target approach to atmospheric correction of remotely-sensed data. Canadian J Remote Sens 21:374-387

Teillet PM, Guindon B, Goodenough DG (1982) On the slope-aspect correction of multispectral data. Canadian J Remote Sens 8:84-106

Teng WL et al. (1997) Fundamentalsof photographic interpretation. In: Philipson WR (editor-in-chief) Manual of Photographic Interpretation, 2nd ed, Am Soc Photogramm Remote Sens, Bethesda, Md, pp 49-113

Thiruvengadachari S, Kalpana AR (eds) (1986) Indian Remote Sensing Satellite – data users handbook. Department of Space Govt of India, NRSA data centre, Hyderabad, 110 pp

Thomas IL, Howorth R, Eggers A, Fowler ADW (1981) Textural enhancement of a circular geological feature. Photogramm Eng Remote Sens 47:89-91

Thome KJ, Gellman DI, Parada RJ, Biggar SF, Slater PN, Moran MS (1993) Absolute radiometric calibration of Thematic Mapper. SPIE Proc 600:2-8

Thompson LL (1979) Remote sensing using solid state array technology. Photogramm Eng Remote Sens 45:47-55

Thornbury WD (1978) Principles of geomorphology. 2nd edn, Wiley, New York

Todd DK (1980) Groundwater hydrology. 2nd edn, Wiley, New York

Tom Ch, Miller LD (1984) An automated land-use mapping comparison of the Bayesian maximum likelihood and linear discriminant analysis algorithms. Photogramm Engg Remote Sens 50:193-207

Townshend JRG (ed) (1981) Terrain analysis and remote sensing. George Allen & Unwin, London, pp 38-54

Tracy RA, Noll RE (1979) User-oriented data processing considerations in linear array applications. Photogramm Eng Remote Sens 45:57-61

Trevett JW (1986) Imaging radar for resources surveys. Chapman and Hall, London, 313 p

Turner RE, Malila WA, Nalepka RF, Thomson FJ (1974) Influence of the atmosphere on remotely sensed data. Proc Soc Photo-Optical Instrument Eng. Vol 51, Scanners and Imagery Systems for Earth Observations, San Diego, CA, pp 101-114

Ulaby FT, Goetz AFH (1987) Remote sensing techniques. Encyclopedia of physical science and technology, vol 12. Academic Press, New York, pp 164-196

Ulaby FT, Moore RK, Fung AK (1981) Microwave remote sensing-active and passive, vol I, microwave remote sensing fundamentals and radiometry. Addison-Wesley, Reading

Ulaby FT, Moore RK, Fung AK (1982) Microwave remote-sensing active and passive, vol II, radar remote sensing and surface scattering and emission theory. Addison – Wesley, Reading

Ulaby FT, Brisco B, Dobson MC (1983) Improved spatial mapping of rainfall events with spaceborne SAR imagery. IEEE Trans Geosci Remote Sens GE–21:118-121

Ulaby FT, Moore RK, Fung AK (1986) Microwave remote senisng: active and passive, vol III, from theory to applications. Artech House, Delham, Mass

van der Meer F (1994) Mapping the degree of serpentinization within ultramafic rock bodies using imaging spectrometer data. Int J Remote Sens 15(18):3851-3857

van der Meer F (1999) Imaging spectrometry for geologic remote sensing. Geologie en Mijnbouw 77:137-151

van der Meer F, Bakker W (1997) CCSM: Cross Correlogram Spectral Matching. Int J Remote Sens 18:1197-1201

van der Meer F, Bakker W (1998) Validated surface mineralogy from high-spectral resolution remote sensing: a review and a novel approach applied to gold exploration using AVIRIS data. Terra Nova 10:112-119

van der Meer F, Yang H, Lang H (2001) Imaging spectrometry and geological applications. In: van der Meer F & de Jong S (ed.) Imaging spectrometry: Basic Principles and Prospective Applications. Kluwer, Dordrecht

van Westen CJ (1994) GIS in landslide hazard zonation: a review, with examples from the Andes of Columbia. In: Price M, Heywood I (eds) Mountain Environments and Geographic Information System. Taylor & Francis, Basingstoke, UK, pp 135-165

Vane G, Goetz AFH (1988) Terrestrial imaging spectroscopy. Remote Sens Environ 24:1-29

Vane G, Goetz AFH, Wellman JB (1983) Airborne imaging spectrometer: a new tool for remote sensing. Proc IEEE Int Gremote Sens Symp (IGARSS) FA-4:6.1-6.5

Vane G, Green RO, Chrien TG, Enmark HT, Hansen EG, Porter (1993) The Airborne Visible/Infrared Imaging Spectrometer (AVIRIS). Remote Sens Environ 44:127-143

Varnes (1984) Landslide hazard zonation: a review of principles and practice. UNESCO, Paris, pp 1-63

Vermote EF, Tanre D, Denze JL, Herman M, Morcrette JJ (1997) Second simulation of the satellite signal in the solar spectrum, 6S: an overview. IEEE Trans Geosc Remote Sens 35:675-686

Verstappen HT (1983) Applied geomorphology. Elsevier, Amsterdam 437 pp

Vickers RS, Lyon RJP (1967) Infrared sensing from spacecraft – a geological interpretation. Proc Thermophysics spec conf, Am Inst Aeronautics Astronautics, Pap 67-284, p10

Vincent RK, Thomson FJ (1972) Rock type discrimination from ratioed infrared scanner images of Pisgah Crater, California. Science 175:986-989

Vincent RK, Thomson FJ, Watson K (1972) Recognition of exposed quartz sand and sandstone by two-channel infrared imagery. J Geophys Res 77:2473-2477

Vizy KN (1974) Detecting and monitoring oil slicks with aerial photos. Phtogramm Eng 40:697-708

Vlcek J (1982) A field method for determination of emissivity with imaging radiometers. Photogramm Eng Remote Sens 48:609-614

Vlcek J, King D (1985) Development and use of a 4-camera video system for resource surveys. Proc 19th Int Symp Remote Sens Environ (Environ Res Inst Michigan), University of Michigan, Ann Arbor, MI, pp 483-489

Volk P (1985) Untersuchungen an geologischen, fernerkundlichen und geophysikalischen Daten und deren interpretationsoptimierten Darstellungen für die Exploration in der SE-Iberischen Kiesprovinz. PhD dissertation, Univ Munich, 154 pp

Volk P, Haydn R, Bodechtel J (1986) Integration of remote sensing and other geodata for ore exploration – a SW Iberian case study. Proc Int Symp Remote Sens Environ, 5th Thematic Conf, Remote Sens Explor Geol Reno, Nevada

von Bandat HF (1983) Aerogeology. Gulf Publ, Houston, Texas, pp 70-85

Walker AS (1986) Eolian landforms. In: Short NM, Blair RW Jr (eds) Geomorphology from space. NASA SP-486, US Govt Printing Office, Washington, DC, pp 447-520

Wang F (1990) Fuzzy supervised classification of remote sensing images. IEEE Trans Geosci Remote Sens 28:194-201

Wang JR, O'Neill PE, Jackson TJ, Engman ET (1983) Multifreqency measurements of the effect of soil moisutre, soil texture and surface roughness. IEEE Trans, GE-21, No 1

Warwick D, Hartopp PG, Viljoen RP (1979) Application of the thermal infrared line scanning technique to engineering geological mapping in South Africa. Q J Eng Geol 12:159-179

Waters P, Greenbaum D, Smart PL, Osmaston H (1990) Applications of remote sensing to groundwater hydrology. Remote Sens Rev 4(2):223-264

Watson K (1971) A computer program of thermal modelling for interpretation of infrared images. US Geol Surv Rep PB Washington, DC

Watson K (1973) Periodic heating of a layer over a semi-infinite half solid. J Geophys Res 78:5904-5910

Watson K (1975) Geologic applications of thermal infrared images. Proc IEEE 63(1):128-137

Watson K (1982a) Regional thermal inertia mapping from an experimental satellite. Geophysics 47:1681-1687

Watson K (1982b) Topographic slope correction for analysis of thermal infrared images. Nat Tech Infor Serv E82-10214

Watson K, Hummer-Miller S (1981) A simple algorithm to estimate the effective regional atmospheric parameters for thermal inertia mapping. Remote Sens Environ 11:455-462

Watson K, Hummer-Miller S, Offield T (1982) Geologic thermal inertia mapping using HCMM satellite data. Int Geosci Remote Sens Symp (IGARSS) Munich 1:2.1-2.6

Watson K, Kruse F, Hummer-Miller S (1990) Thermal infrared exploration in the Carlin Trend, northern Nevada. Geophysics 55:70-79

Watson K, Rowan LC, Bowers TL, Anton-Pacheco C, Gumiel P, Miller SH (1996) Lithologic analysis from ultispectral thermal infrared data of the alkalic rock complex at Iron Hill, Colorado. Geophysics 61:706-721

Welch R (1980) Measurements from linear array camera images. Photogramm Eng Remote Sens 46:315-318

Welch R (1983) Impact of geometry on height measurements from MLA digital image data. Photogramm Eng Remote Sens 49:1437-1441

Welch R, Ehlers M (1987) Merging multiresolution SPOT HRV and Landsat TM data. Photogramm Eng Remote Sens 53(3):301-303

Welch R, Marko W (1981) Cartographic potential of spacecraft line array camara system: Stereosat. Photogramm Eng Remote Sens 47:1173-1185

Wharton Se, Irons JR, Huegel F (1981) LAPR: an experiemntal pushbroom scanner. Photogramm Eng Remote Sens 47:631-639

Whitney GG, Abrams MJ, Goetz AFH (1983) Mineral discrimination using a portable ratio-determining radiometer, Econ Geol 78:688-698

Wieczorek GF (1984) Preparing a detailed landslide-inventory map for hazard evaluation and reduction. Bull Assoc Eng Geol 21(3):337-342

Williams RS Jr (1986) Glaciers and glacial landforms. In: short NM, Blair RW Jr (eds) Geomorphology from space. NASA SP-486, US Govt Printing Office, Washington, DC, pp 521-596

Williams RS Jr, Southworth CS (1984) Remote sensing makes important gains. Geotimes 8:13-15

Windeler DS, Lyon RJP (1991) Discriminating dolomitization of marble in the ludwig scarn near Yerington, Nevada, using high-resolution airborne infrared imagery. Photgramm Eng Remote Sens 57:1171-1178

Wise DU (1982) Linesmanship and practice of linear geo-art. Geol Soc Am Bull 93:886-888

Woldai T (1983) Lop-Nur (China) studied from Landsat and SIR-A imagery. J ITC (1983-3):253-257

Wolf PR (1983) Elements of photogrammetry. 2nd edn, McGraw-Hill, New York

Woodcock CE, Strahler AH (1987) The factor of scale in Remote sensing. Remote Sens Environ 21:311-332

Woodham RJ (1989) Determining intrinsic surface reflectance in rugged terrain and changing illumination. Proc Int Geosci Remote Sens symp (IGARSS89), Vancouver, British Columbia, Canada, 10-14 July 1989, New York: IEEE Press, vol 1, 1-5

Wyszecki G, Stiles WS (1967) Color science. Wiley, New York

Yang H, Zhang J, Van Der Meer F, Kroonenberg SB (1999) Spectral characteristics of wheat associated with hydrocarbon microseepages. Int J Remote Sens 20(4):807-813

Yang H, Meer FVD, Zhang J, Kroonenberg SB (2000) Direct detection of onshore hydrocarbon microseepages by remote sensing techniques. Remote Sens Rev 18:1-18

Yash Pal, Sahai B, Sood RK, Agrawal DP (1980) Remote sensing of the 'lost' Saraswati river. Proc Ind Acad Sci (Earth Planet Sci) 89(3):317-331

Zebker HA, Goldstein RM (1986) Topographic mapping from interferometry synthetic aperture radar observations. J Geophys Res 91(B5):4993-4999

Zebker HA, van Zyl JJ, Held DN (1987) Imaging Polarimetry From Wave Synthesis. J Geophys Res 92 (B1):683-701

Zebker HA, Werner CL, Rosen PA, Hensley S, (1994) Mapping the World's Topography with Radar Interferometry. Proc IEEE 82(12): 1774-1786

Zernitz ER (1932) Drainage patterns and their significance. J Geol 40:498-521

Zhang XM (1998) Coal Fires in North China- detection, monitoring and prediction using remote sensing data. ITC Publ No 58, 133p

Zhou Z, Civco DL, Silander JA (1998) A wavelet transform method to merge Landsat TM and SPOT panchromatic data. Int J Remote Sensing 19:743-757

Zobrist AL, Blackwell RJ, Stromberg WD (1979) Integration of Landsat, Seasat and other geo-data soruces. Proc 13th Int Symp Remote Sens Environ, Ann Arbor, MI, pp 271-279

Zobrist AL, Bryant NA, McLeod RG (1983) Technology for large digital mosaics of Landsat data. Photogramm Eng Remote Sens 49:1325-1335

ILLUSTRATIONS – LOCATION INDEX

1. **Algeria**

   Fig. 16.19 Page 448
   Fig. 16.37 Page 473
   Fig. 16.47 Page 487

2. **Alps**

   Fig. 8.5 Page 171
   Fig. 16.39 Page 476
   Fig. 16.95 Page 545

3. **Andes**

   Fig. 16.50 Page 491
   Fig. 16.42 Page 480
   Fig. 16.105 Page 557
   Fig. 16.106 Page 558
   Fig. 16.107 Page 558
   Fig. 16.108 Page 559

4. **Atlantic Ocean**

   Fig. 13.23 Page 365

5. **Australia**

   Fig. 16.15 Page 444

6. **Austria-Germany**

   Fig. 16.39 Page 476

7. **Bangladesh**

   Fig. 16.11 Page 442

8. **Botswana**

   Fig. 15.12 Page 409

9. **Brazil**

   Fig. 13.20 Page 362
   Fig. 16.43 Page 483

10. **Chad**

    Fig. 16.41 Page 479

11. **China**

    Fig. 16.27 Page 457
    Fig. 16.36 Page 472
    Fig. 16.96 Page 546
    Fig. 16.118 Page 569
    Fig. 16.120 Page 571

12. **Canada**

    Fig. 13.14b Page 354
    Fig. 16.16 Page 445

13. **Egypt**

    Fig. 16.89 Page 539

14. Equador

Fig. 13.1 Page 338

15. France

Fig. 8.5 Page 171

16. Germany

Fig. 4.4 Page 57
Fig. 9.10 Page 198
Fig. 11.27 Page 313
Fig. 13.12 Page 352
Fig. 16.12 Page 442
Fig. 16.123 Page 575

17. Greenland

Fig. 16.63 Page 508

18. Guinea

Fig. 16.10 Page 441

19. Hawaii Island

Fig. 10.35 Page 263
Fig. 16.49 Page 490

20. Iceland

Fig. 13.4 Page 341
Fig. 14.16 Page 390
Fig. 16.17 Page 446
Fig. 16.29 Page 461
Fig. 16.57 Page 499
Fig. 16.109 Page 561
Fig. 16.110 Page 562

21. India

Fig. 1.2 Page 5
Fig. 6.4 Page 127
Fig. 6.13 Page 135
Fig. 7.5 Page 152
Fig. 8.2 Page 164
Fig. 8.6 Page 174
Fig. 8.7 Page 175
Fig. 8.8 Page 176
Fig. 8.9 Page 177
Fig. 8.10 Page 178
Fig. 9.14 Page 204
Fig. 10.1 Page 218
Fig. 10.7 Page 228
Fig. 10.8b Page 230
Fig. 10.14 Page 239
Fig. 10.19 Page 245
Fig. 10.20 Page 246
Fig. 10.21 Page 247
Fig. 10.34 Page 261
Fig. 13.2 Page 339
Fig. 13.7 Page 346
Fig. 14.8 Page 379
Fig. 15.15 Page 414
Fig. 15.19 Page 421
Fig. 16.5 Page 435
Fig. 16.7 Page 438
Fig. 16.9 Page 440
Fig. 16.23 Page 452
Fig. 16.24 Page 453
Fig. 16.26 Page 456
Fig. 16.30 Page 462
Fig. 16.31 Page 464
Fig. 16.34 Page 469
Fig. 16.35 Page 470
Fig. 16.38 Page 475
Fig. 16.48 Page 488
Fig. 16.59 Page 504
Fig. 16.62 Page 507
Fig. 16.74 Page 520
Fig. 16.78 Page 525
Fig. 16.80 Page 528
Fig. 16.81 Page 529
Fig. 16.85 Page 533
Fig. 16.88 Page 537
Fig. 16.90 Page 540
Fig. 16.92 Page 541
Fig. 16.94 Page 544
Fig. 16.99 Page 549
Fig. 16.100 Page 550

Illustrations – Location Index

Fig. 16.101 Page 552
Fig. 16.103 Page 554
Fig. 16.113 Page 565
Fig. 16.117 Page 568
Fig. 16.121 Page 573
Fig. 16.122 Page 574
Fig. 16.124 Page 576
Fig. 16.125 Page 577
Fig. 16.128 Page 580
Fig. 16.130 Page 582

22. **India- Pakistan**

Fig. 16.28 Page 459

23. **Indonesia**

Fig. 16.3 Page 433

24. **Iran**

Fig. 16.25 Page 455

25. **Italy**

Fig. 5.26 Page 117
Fig. 13.21 Page 363
Fig. 13.22 Page 364
Fig. 14.7 Page 379
Fig. 16.86 Page 534
Fig. 16.104 Page 556

26. **Japan**

Fig. 16.91 Page 540

27. **Korea**

Fig. 10.41 Page 273

28. **Kuwait**

Fig. 16.129 Page 581

29. **Libya**

Fig. 16.82 Page 530
Fig. 16.83 Page 531

30. **Mauritania**

Fig. 1.3 Page 6
Fig. 16.21 Page 451

31. **Mexico**

Fig. 10.30 Page 257

32. **Morocco**

Fig. 9.9 Page 197
Fig. 10.13 Page 238
Fig. 10.40 Page 272
Fig. 16.22 Page 452

33. **North Sea**

Fig. 16.77 Page 523

34. **Sahara**

Fig. 7.1 Page 147
Fig. 16.82 Page 530

35. **Saudi Arabia – Yemen**

Fig. 10.9 Page 233
Fig. 10.27 Page 254
Fig. 10.29 Page 256
Fig. 16.13 Page 443
Fig. 16.14 Page 443

36. **Somalia**

Fig. 8.4 Page 166
Fig. 16.4 Page 434

37. **South Africa**

   Fig. 9.12   Page 202
   Fig. 16.58 Page 503

38. **Spain**

   Fig. 15.16 Page 415
   Fig. 15.17 Page 417
   Fig. 16.40 Page 478

39. **Switzerland**

   Fig. 14.14 Page 388

40. **Thailand**

   Fig. 13.16 Page 357

41. **Turkey**

   Fig. 14.13 Page 387
   Fig. 16.102 Page 553

42. **U.S.A.**

   Fig. 9.15   Page 205
   Fig. 11.26 Page 312
   Fig. 11.28 Page 315
   Fig. 13.3   Page 340
   Fig. 13.5   Page 343
   Fig. 13.8   Page 349
   Fig. 13.18 Page 360
   Fig. 14.12 Page 385
   Fig. 14.15 Page 389
   Fig. 14.17 Page 391
   Fig. 16.46 Page 486
   Fig. 16.54 Page 495
   Fig. 16.68 Page 512
   Fig. 16.69 Page 513
   Fig. 16.72 Page 516
   Fig. 16.75 Page 521
   Fig. 16.93 Page 543
   Fig. 16.97 Page 547
   Fig. 16.98 Page 548

43. **Venezuela**

   Fig 7.3 Page 149

Colour-Plate I

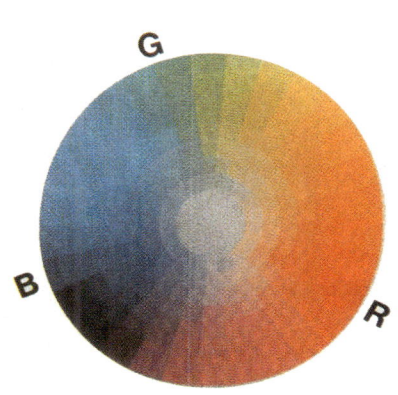
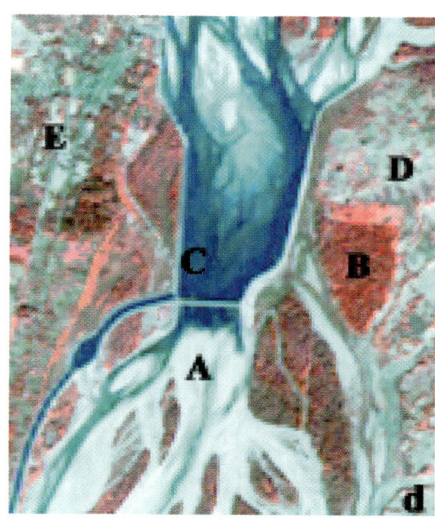

Fig. A.4.1.4. The concept of the Munsell colour circuit or wheel. Hue is represented as the polar angle and saturation on the radial axis

Fig. 8.10d. FCC generated from green, red and near IR spectral band images coded in blue, green and red respectively, (for details see figures 8.10 and 8.11)

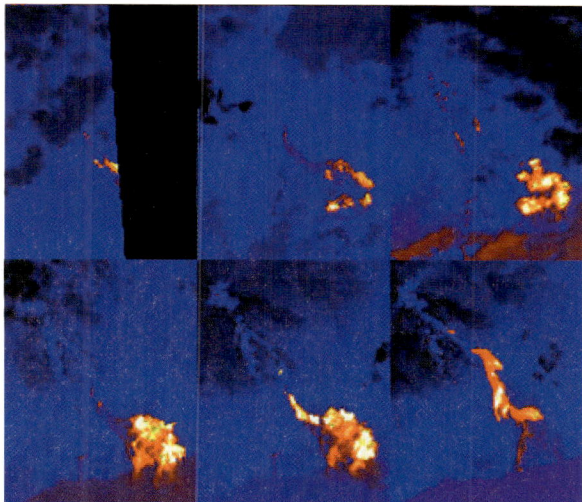

Fig. 10.35. Pseudocolour coding. A sequence of night-time thermal images (ASTER band 14) shows lava flows entering the sea, Hawaii Island. Colour coding from black (coldest) through blue, red, yellow, white (hottest). The first five images show a time sequence of a single eruptive phase; the last image shows flows from a later eruptive phase. (Courtesyof-NASA/GSFC/MITI/ERSDAC/ JAROS and US/Japan ASTER Science Team)

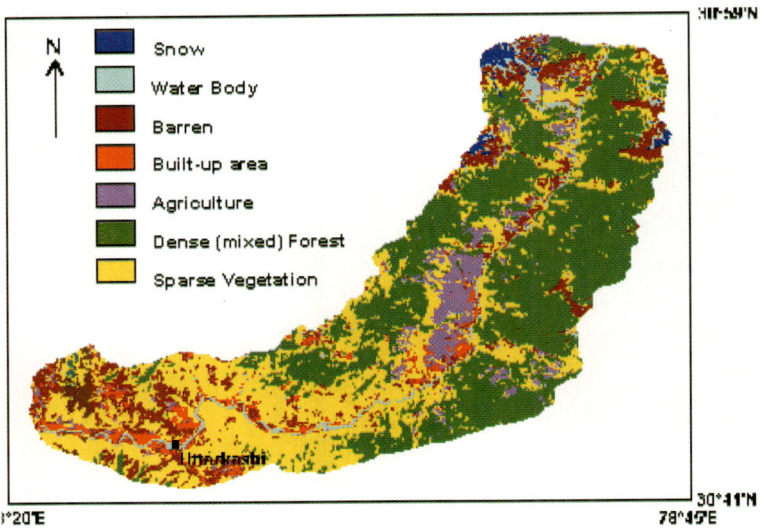

Fig. 10.44. Land-use/land-cover classification of remote sensing data (IRS-LISS-II) (using MLC) of a part of the Ganges valley, Himalayas. (Gupta et al. 1999)

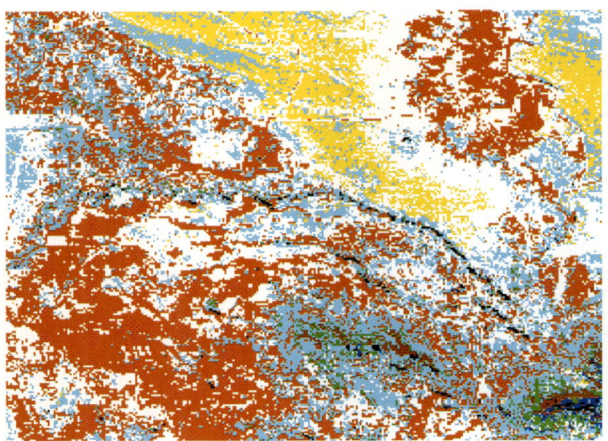

Fig. 11.28. Mineral alteration map based on Probe1 (HyMAP) hyperspectral data, Bluff test site, Utah; the red and brown colours represent the bleached zones related to hydrocarbon seepage. (van der Meer et al. 2001) (Reproduced by permission, copyright © 2001, Kluwer Academic Publishers)

Colour-Plate III 629

Fig. 13.16. Radar image showing landscape in the Hoei Range of north-central Thailand; the plateau terrain has been dissected by fluvial erosion; note the fine morphological details brought out by combining multiple SAR images. (SIR-C image; colour coding: R = L-band HH, G = L-band HV, B = C-band HV). (Courtesy of NASA/JPL)

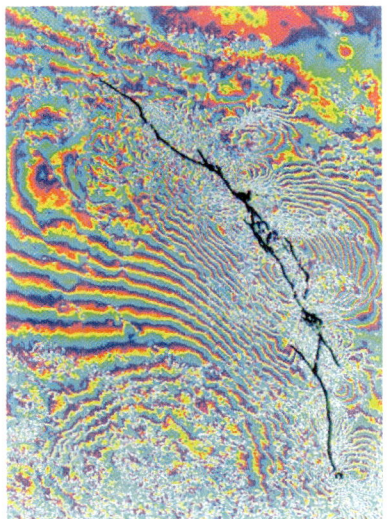

Fig. 14.12. Co-seismic interferogram of the Landers earthquake (28 June 1992) generated from ERS SAR images. The *dotted black lines* denote known faults; *solid white lines* show the surface ruptures mapped in the field. One cycle of colour represents one fringe equivalent to 2.8-cm vertical change. (Massonnet et al. 1993)

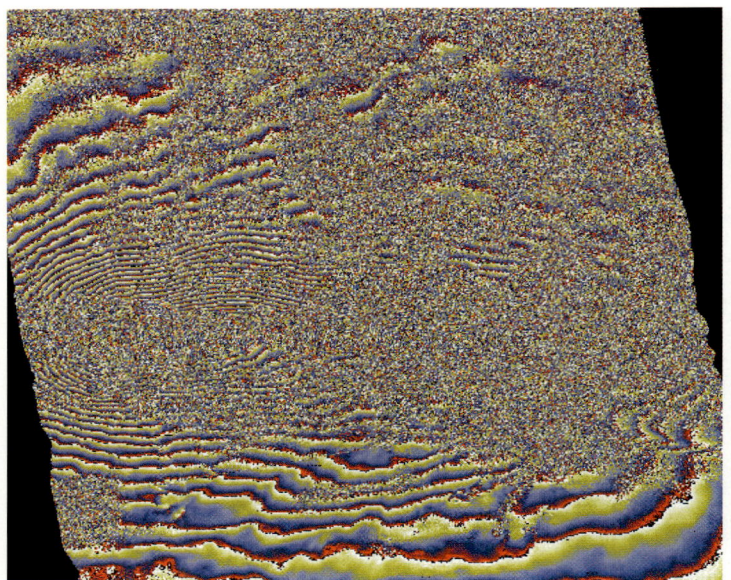

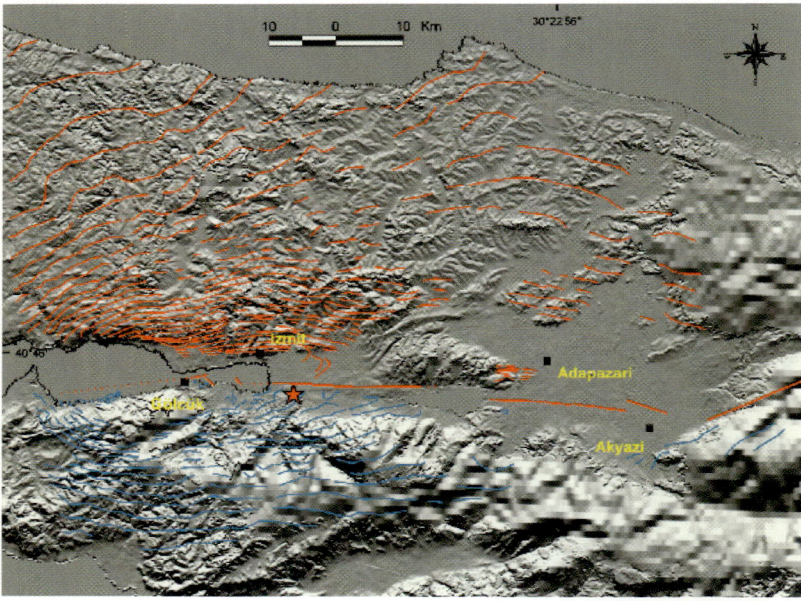

Fig. 14.13. a. Interferogram for the Izmit earthquake, Turkey, 17 August 1999 (generated from 13 August 1999 and 17 September 1999 ERS-2 image pair, with normal baseline of 65 m). The event occurred along the North Anatolian Fault with right-lateral displacement; **b** shows the fringe traces as superimposed over a shaded relief map for a better appreciation of the situation. (**a, b** courtesy of S. Stramondo)

Colour-Plate V   631

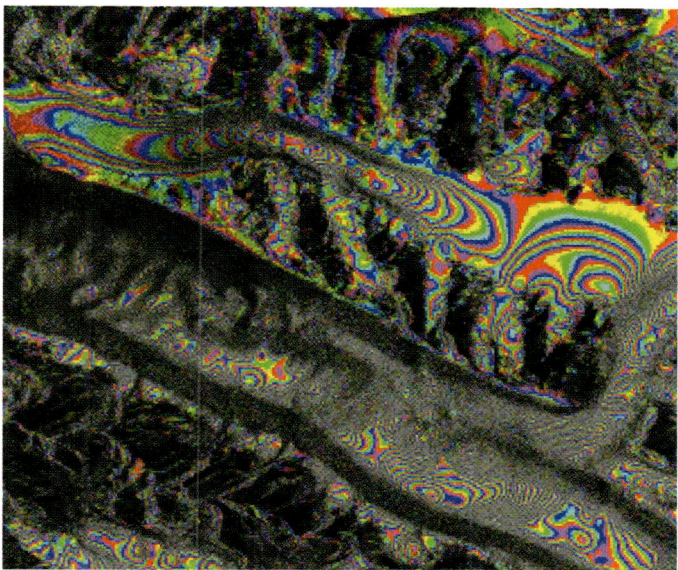

Fig. 14.15. Interferogram of Bagley Ice Field, Alaska; phase fringes show both topography and motion influences. (D.R. Fatland, http://www.asf.alaska.edu/step/insar/absracts/ fatland.html)

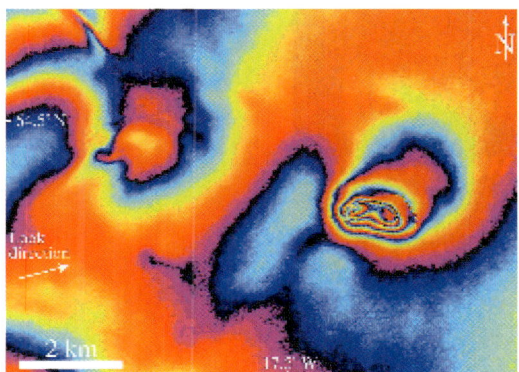

Fig. 14.16. Differential interferogram generated from ERS-1/-2 data (March 27–28, 1996, ascending orbit) over cauldrons in the Vatnajokull ice cap. Fringes indicate uplift in ice surface during an interval of one day. (Data copyright © ESA, processing by DLR, Jonsson et al. 1998)

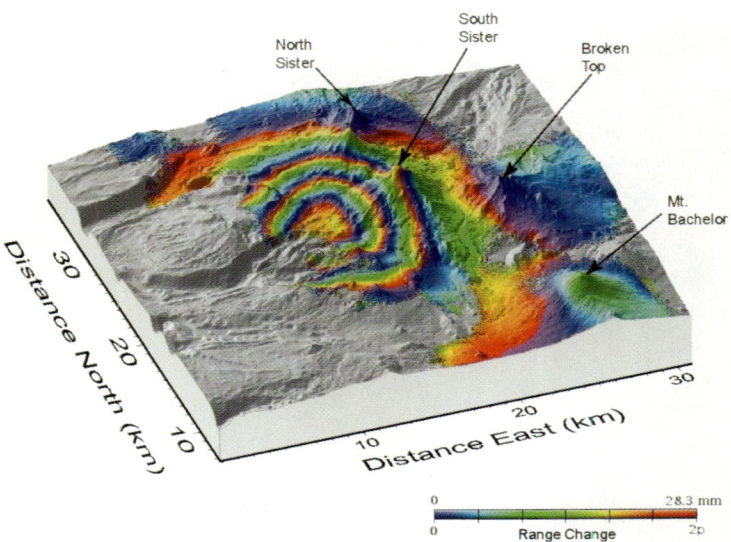

Fig. 14.17. Differential InSAR from satellite SAR data (passes in 1996, 2000) of the Three Sisters volcano region, Oregon. A broad uplift of the ground surface over an area of about 15–20-km diameter, with maximum uplift of about 10 cm at its centre, is detected. (Courtesy of USGS; interferogram by C. Wicks; http://vulcan.wr.usgs.gov/volcanoes/sisters/)

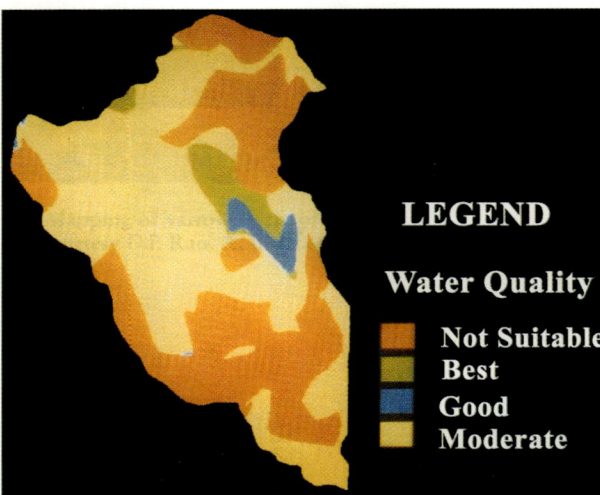

Fig. 15.19. Classification of groundwater. The groundwaters have been classified with respect to suitability for drinking purpose per WHO norms, using data on chemical quality of groundwater (TDS, Cl, Na + K, and Ca + Mg), and applying Boolean logic. (Courtesy of N.K. Srivastava)

Colour-Plates VII

Fig. 15.17. Colour synthetic stereo pair demonstrating integration of remote sensing, geochemical and geophysical exploration data in the Mayasa Concession, Spain. The colour coding scheme uses IHS, where intensity (I) = TM4, hue (H) = Hg values in ppm (so that low values correspond to the blue end and high values correspond to the red end of the colour wheel), and saturation (S) = constant. The gravimetric data is represented as parallax, so that higher positive gravity anomaly is shown as higher positive relief. The Pb values in ppm are indicated by the white contours (for details see text). (Courtesy of Minas de Almaden y Arrayanes Sa, Madrid)

Fig. 16.49. AIRSAR image showing Kilauea volcano, Hawaii. The main caldera is seen on the right side of the image. Lava flows of different ages exhibit differences in backscatter due to variation in surface roughness. (Multiband image with colour coding: R = P-band HV; G = L-band HV, B = C-band HV) (Courtesy of NASA/JPL)

Fig. 16.50. ASTER FCC image of part of the Andes. The scene is dominated by the Pampa Luxsar lava complex (mostly upper right of the scene). Lava flows are distributed around remnants of large dissected cones. On the middle-left edge are the Olca and Paruma stratovolcanoes appearing in blue due to lack of vegetation. (Courtesy of NASA/GSFC/MITI/ERSDAC/JAROS, and US/Japan ASTER Science Team)

Colour-Plate IX 635

Fig. 16.54. FCC generated from TIMS images after decorrelation stretch (R = TIMS5, G = TIMS3, B = TIMS1). Note the prominent compositional layering; red to red–orange are silica-rich, and blue–green are areas rich in carbonates; also see Fig. 16.53. (Lang et al. 1987)

Fig. 16.68. The CRC of a part of the Silver Bell porphyry copper deposit (see text). The arcuate yellowish-orange band that trends from left to right through the centre of the area is the alteration zone comprising dominantly iron oxide and clays; field information is in agreement with these data. (Abrams and Brown 1985)

Fig. 16.69. ASTER image showing the open-pit Escondida mine, Chile. ASTER-SWIR bands 4, 6 and 8 are coded in RGB respectively. The Al-OH-bearing minerals appear in shades of blue •purple, Mg-OH-bearing minerals appear in shades of green •yellow and Al-Mg-OH-bearing minerals appear in shades of red (for details on colour interpretation, refer to figures 16.70 and 16.71). (Courtesy: NASA/GSFC/MITI/ERSDAC/JAROS, and US/Japan ASTER Science Team)

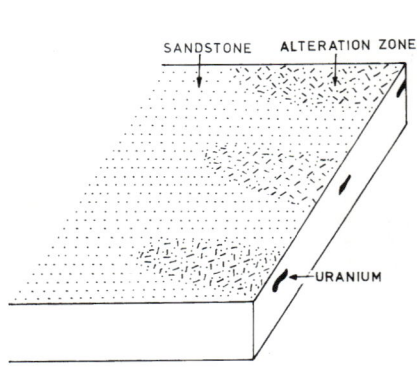

Fig. 16.72. a Schematic showing alteration•bleaching zone in the overlying Wingate sandstone associated with the uranium mineralization at depth. **b** False-colour image of the Lisbon valley, generated from canonically transformed aerial scanner data. Red •olive green colour = bleached parts of overlying Wingate Formation; shades of grey and beige = unbleached Wingate Formation. (**b** Conel and Alley 1985)

Fig. 16.75. a False-colour image of the Patrick Draw test site based on principal compo-nent transformation of aerial TM simulator data; note the peculiar lemon -green near-circu-larpatch on the colour image, which is interpreted to be due to **a** geobotanical anom-aly. **b** Aerial photograph of the area; the area of anomalous spectral property in a is also outlined. (Lang et al. 1985)

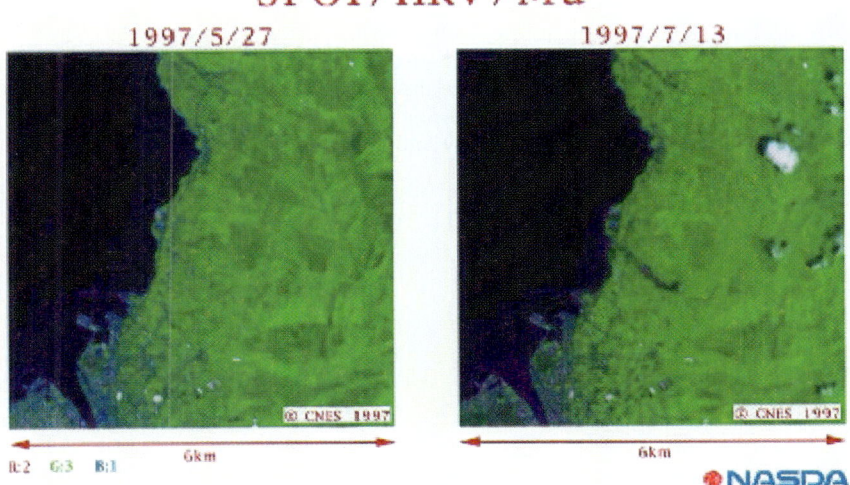

Fig. 16.91. Mudflow at Demizu, Kagoshima, Japan. (SPOT-2-HRV multispectral image data) (Courtesy of NASDA/CNES)

Fig. 16.102. Damage assessment during the Turkey earthquake (17 August 1999). The figure shows post-earthquake (30 August) and pre-earthquake (8 August) images of parts of the city of Adapazari, one of the worst hit during the earthquake. Buildings that were destroyed by the earthquake can be detected by comparing the two images. (IRS-1D PAN and LISS-III merged image data sets) (Copyright © 1999 Antrix, SI, euromap)

Colour-Plate XIII

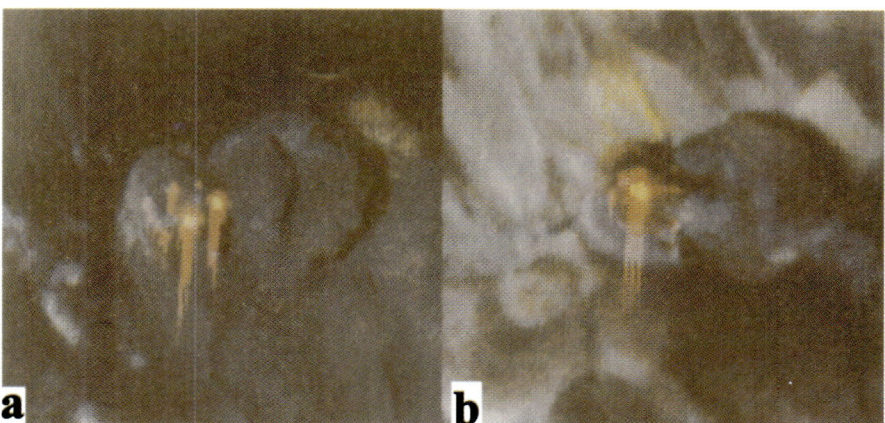

Fig. 16.106. JERS-1-OPS sensor composite of channels 851 (RGB) of Luscar summit craters (images approx. 3km2) **a** Pre- eruption; **b** 1-day after the eruption. Thermal hot spots are depicted as yellow if they are radiant in both OPS-8 and OPS-5, or red if radiant only in OPS-8. Servere noise structures are visible in these images, especially vertical and horizontal stripes and along-track blurring. (Denniss et al. 1998)

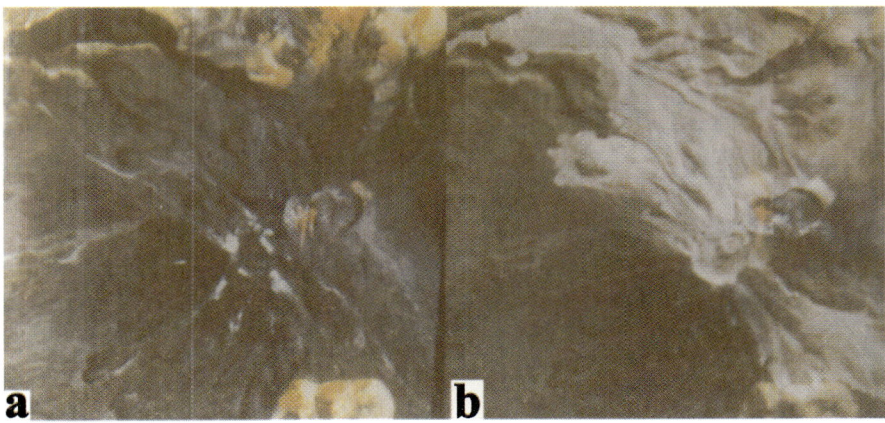

Fig. 16.107. JERS-1-OPS sensor composite of channels 8301 (RGB) of the Luscar volcano, Chile (images approx. 10 km square). **a** Pre-eruption; **b** 1-day after the eruption, showing new pyroclastic flows. (Denniss et al. 1998)

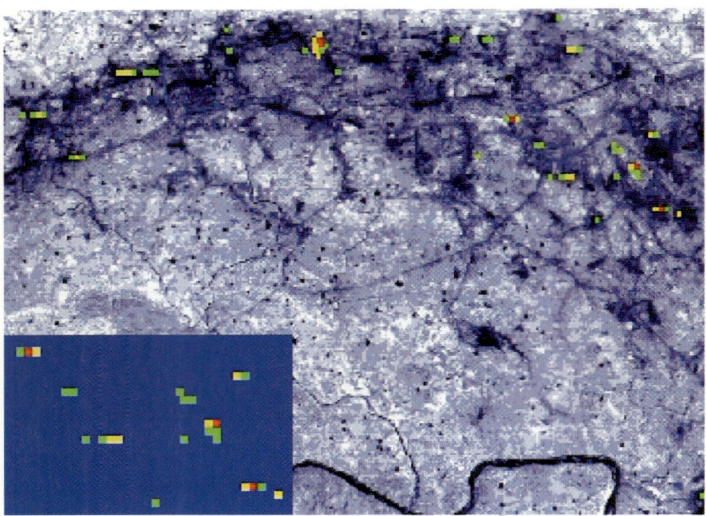

Fig. 16.113. Processed Landsat TM data of JCF (IHS-processed I = TM4, H = TM6, S = constant, such that red corresponds to highest DNs). The sickle-shaped field above is the Jharia coal field. The relief shown in the background is from TM4. Black linears and patches are coal bands, quarries and dumps. Blue is the background (threshold) surface temperature, anomalous pixels have green, yellow and red colours with increasing temperature. On the lower left is an inset of the psuedocolour TM6 band. The Damodar river appears in the south. (Prakash et al. 1995a)

Fig. 16.117. FCC of TM753 (RGB). Windows I, J, K, L, M and N depict areas of surface fires; yellow pixels correspond to areas of highest temperatures being radiant in both TM5 and TM7; red areas (radiant only in TM7) are relatively lower temperatures (see enlarged windows K and L). (Prakash and Gupta 1999)

Colour-Plate XV

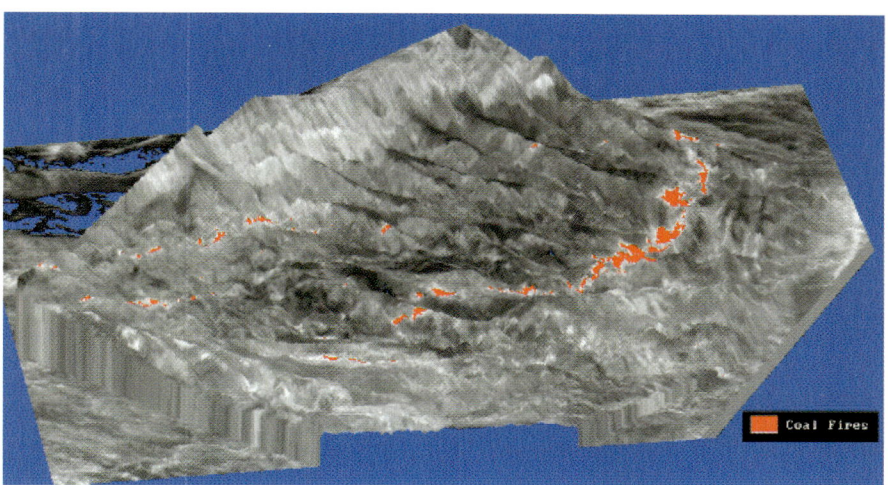

Fig. 16.118. Coal fires in the Xinjiang coal field, China. Airborne thermal-IR image data co-registered and draped over the DEM. (Zhang 1998)

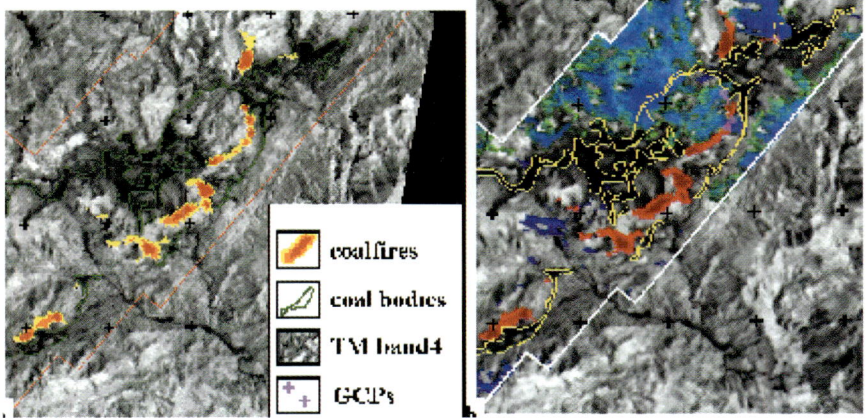

Fig. 16.120. a Processed TM image of a part of Ningxia coal field. Grey background is from TM4. Yellow, orange and red colours depict successively higher temperatures associated with coal fires. The green outlines are the major coal seams, coal dumps and mining areas. **b** Final fusion product with information from VNIR, thermal-IR and microwave regions. TM4 provides the background (gray image). Thermal anomalies from TM6 are in shades of red (coal fires). Bright green corresponds to subsidence from InSAR. Yellow outlines mark

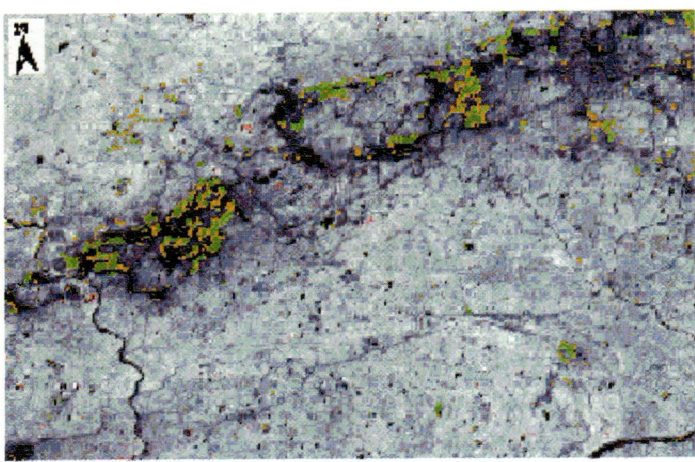

Fig. 16.121. IHS-processed NDVI difference image for change detection in parts of Jharia coal field (India). The Landsat TM 1990 and 1994 data sets were first spectrally normalized and then used for generating NDVI images. The NDVI difference image is used as hue, TM4 as intensity, saturation is constant. Green, yellow and pink indicate areas of vegetation change. Topographic relief, structural features, drainage and transport network appear in the background. (Prakash and Gupta 1998)

Fig. 16.128. SIR-C image showing oil slick in the 'Bombay High' offshore drilling field, about 150 km west of Mumbai, India. Oil slicks appear dark and drilling platforms appear as bright white spots. Also note the internal waves (left centre) and ocean swell (blue areas adjacent to internal waves). (R = L-band VV, G = average of L-band VV and C-band VV, B = C-band VV; Evans et al. 1997)

Subject Index

2

2.5-D 136, 270

3

3-D 136, 270

A

absolute phase 380
- registration 234
absorption band characterization 307, 308
absorption bands 35
- depth 290
- feature 290
- filters 68
- mechanism 32
absorptivity, spectral 22
acidic extrusive rocks 489
across-track interferometry 369
active faults 543
- sensor 24
- system 320
- UV systems 579
addition image 253
additive colour viewer 154
aeolian landforms 442
air pollution 583
airborne laser fluorosensor 579
airborne magnetic 401
aircraft 8
AIRSAR 373
AIS 300
albedo 29, 168, 186,
ALF 579
alluvial fills 536
Almaz-1 334
Al-OH 40

along-track interferometry 371
ALOS 119
alteration haloes 505
- zones 47, 505
anastomotic drainage 437
angular distortion 128
- unconformity 480
anisotropic filters 463
ANN classification 283
annular drainage 435
anomaly, geomorphic 519
-, geobotanical 520
antenna gain 347
anti-vignetting filter 69
apparent thermal inertia 199
applications software 221
aquifers 523
artificial neural network 283
aspect angle 349
- ratio 126, 233
-, topographic 165
association 157
ASTER 113, 211
ATI 199
atmospheric absorption 26
- contribution 224
- effects 24
- emission 28
- interaction 25
- models 226, 305
- pollution 583
- scattering 25, 224
- windows 26
attribute data 398
AVIRIS 300
azimuth angle 325
- direction 322, 323
- resolution 323

B

B/H-ratio 137, 345
balloons 8
band shape, absorption 308
band-pass 299
- filter 69
barbed drainage 437
basalt 489
base image 235
- projection 412
baseline 372
- decorrelation 383
basic intrusives 488
basins 479
beam polarization 329, 348
bedding 448
bicubic interpolation 237
BIL format 220
bilinear interpolation 237
binary mode 220
bistatic radar 321
bit-data 217
bits 220
black-and-white film 59
blackbody 20, 22
- radiation 20
blue-shift 294
boolean conditions 420
Bovery transform 268
box filter 243
braided drainage 437
BRDF 12, 226
brightness contrast ratio 147
- temperature 318
broad-band thermal IR 192
browse products 122
BSQ format 220
buried river 528, 531

C

calibration 76, 96
-, sensor 228
-, spectral 302
cameras 55
camouflage detection film 67
carbonates 42, 293, 510
cartographic camera 58
- manual digitizer 405
CARTOSAT 119
CBERS 112
CBERS 119
CCD area array 92, 298
- linear array 88, 297
- matrix sensor 92
CCRS-SAR 373
central perspective geometry 144
change-detection 94, 253
channel plan-form 439
charge coupled device 88, 89
charge transfer effect 37
chlorosis 517
CIE chromaticity diagram 265, 586
CIR film 67, 177
circular features 477
Civco method 228
classification 274, 416, 423
clays 175
cluster analysis 276, 281
- diagram 251
clutter 338
coal fires 563
co-albedo 186
coastal erosion 577
- landforms 441
coatings 295
collateral ancillary data 393
colour 155
- coding 413
-, complementary 65
- dye 64
- enhancement 262
- films 64
- infrared film 67
- models 264
- negative film 64
- positive film 67
- ratio composites 511
- reversal films 67
- ternary diagram 494, 511, 514
- transformations 269
complementary colours 65
complex dielectric constant 31, 354
- signal 368
- volume-scattering coefficient 355
conduction band effect 37
confined aquifers 524

Subject Index

confusion matrix 285
connectivity functions 422
conservation of energy 28
contiguity 422
contingency table 285
continuum 290
- removal 307
contrast enhancement 240-242
- ratio 148
-, film 62
conventional camera 56
convergence of evidence 157
corner reflector 339, 359, 376
corona photographs 73
cosmetic operations 232
coverages 398
craters 479
CRC 511
crisp classification 282
cross folding 479
- polarization 329
Crosta-technique 256
CRs 359, 376
crystal field effect 38
cultural features 171

D

damage assessment 553
damp terrain 196
dark-object subtraction 224
data bank 394
- base 397
- disk 220
- format 399
- input 405
-, laboratory 49
- management 413
- structures 399
DBMS 398, 413
debris flow tracks 540
decision level fusion 268
decorrelation 383
decorrelation stretch 258
deltaic landforms 441
DEM 381, 383
dendritic drainage 435
density, film 61
density, rock 188
density slicing 242

density slicing colour coding 263
depolarization 348
depression angle 324, 329
depth of absorption 290, 308
- penetration 31, 365
- perception 136
deranged drainage 437
destriping 229-231, 304
detailed exploration stage 501
detectors 76
diagonal edge image 246
dielectric property 319
differential GPS 15
- SAR interferometry 381
diffuse reflection 29
- scattering 338
digital camera 92
- data bank 394
- electronic imaging 92
- elevation model 383
- image 217
- image classification 274
- number 83, 217
- photographic camera 92
- snapshot system 92
digitization 405
digitizer 407
digitizing camera 219
dilational fracture 468
DInSAR 381
directed PCT 509
disaster assessment 555
discontinuities 447
discriminant function 280
distortion 123
-, geometric 342
-, panoramic 232
-, radiometric 224
distributory drainage 435
D-log E curve 61
DN 83, 217
dolerite 489
dolomite 484
domes 479
DPCT 509
drainage 158
- anomaly 437, 519
- density 158
- pattern 158, 435
- texture 158
dunes 443

dwell time 91
dyke 488
dynamic range 94, 148

E

Earth Resources Technology Satellite 2, 97
- Terrain Camera 72
Earth's shape, effect of 131
- curvature, effect of 131
- emission 32
- rotation 131, 232
earthquakes 385, 542-555
edge enhancement 243
edges 243
effluent seepage 528
elastic scattering 25
electromagnetic spectrum 23
electronic bandwidth 78
- energy 36
- processes 36
- transition 37
EM induction 403
- radiation 19
- spectrum 23
emissivity 22, 190
-, spectral 22, 210
energy budget, SOR 162
-, thermal 185
energy levels 26
enhancement, colour 262
enhancement, edge 243
environmental applications 571
Envisat-1 335, 376
EPP 143
equivalent positive plane 143
EROS 120
error matrix 284
ERS series 374
ERTS 2, 97
ESA 302
ETC 57, 72
ETM 105
ETM+ 105
exploration programme 500
-, hydrocarbon 519
exposure, film 60, 61
extrusive igneous rocks 486

F

false colour film 67
far-range 323
fault 456
FCC 177, 180
feasibility aspect 4
feature axes 250
- level fusion 268
- mapping 307
- selection 278
- space 250
Fe-O 37
Fe-O mapping 507-508
fiducial marks 143
field data 9, 358
- scatterometer 358
- spectra 49
field-of-view 56
film contrast 62
- format 55
- processing 60
- resolution 62
- sensitometry 61
- speed 62
films 59
filter weight matrices 243
filtering 242
filters 68
fires, coal 563-570
-, forest 205
-, oil well 581
flat-field correction 306
flattening 378
fluvial landforms 435
- placers 502
fluvio-glacial landforms 445
focal plane array 92
fold 450
foreshortening 342
forest fires 205
forests 175
format, data 399
fourier analysis 242, 248
fourier spectrum 248
FOV 56, 87
FPA-system 92
frame transfer technique 95
frequency domain filtering 242
fringe image 379

Subject Index

full-width at half maximum 80
Fuyo 111
fuzzy classification 282
FWHM 80, 299

G

gamma 62
- radiation 403
gas pools 519
gaussian stretching 241
GCPs 234
geobotanical anomaly 520
- guides 515
geochemical data 404
geocoding 304
geological guides 502
geometric distortion 342
- rectification 344
- transformation 412
geomorphic anomalies 519
geomorphological guides 502
geomorphology 431
geophysical data 400
geotechnical elements 157
geothermal 187, 389
- energy 555, 560
GIS 393, 395
- functions 419
glacial landforms 444
glacier 388
glare 171
gneisses 493
goethite 508
gossan 510
GPS 14
GRACE 403
gradient image 244
grain 360
granites 487
gravity 403
graybody 23
GRD 71
ground swath 87
- measurements 9, 194
- radiance 25
- range 323
- resolution 83
- - cell 83
- - distance 71

- truth 9
groundwater 523
guides, geological 501

H

haloes, alteration 505
handheld cameras 73
hard rock 525, 532
hard targets 338
hardware 221
haze component 225
HCMM 117, 200
HCMR 117
heat balance 187
- capacity 188
- energy 185
- transfer 186
helicopters 8
hematite 508
high resolution spectra 291
high contrast film 62
high-pass filter 242
histogram equalization stretching 242
HRV 108
HRVIR 110
hydrocarbon 315
- exploration 519
hydroxyl ion 40, 506
hyperspectral sensing 287

I

IFOV 79, 83, 87, 91
IFSAR 367
igneous rocks 45, 486
IHS model 265, 589
IKONOS 115
image 83, 93
- classification 274
- filtering 242
- fusion 267
- motion 94, 96, 150
- quality 147
- segmentation 274
- skew 131, 232
- smoothing 248

- transformation 250
imaging radar 320
- sensors 83
- spectrometers 297
- spectrometry 287
- systems 75
- tube 84
incidence angle 12, 325, 349
indexing 151
influent seepage 528
infrared film 59
InSAR 367
in-situ spectra 49
instrument error 125
integration period 90
interference filter 69
interferogram 377, 378
interferometric fringes 379
- SAR 367
interframe distortions 124
interline distortions 124
- transfer technique 95
internal average relative reflectance 306
interpolation 408, 422
interpretoscope 141
interval scale 398
intimate mixtures 295
intraframe distortions 124
intraline distortions 124
intrinsic parameter 10
intrusive igneous rocks 486
intrusives 479
iron oxide 507
irradiance 20
IRS series 105
ISAR 367

J

jarosite 508
JERS 111, 334, 374

K

kaolinite 292
Karhunen-Loeve analysis 255

kernels 243
kinetic temperature 185, 191
Kirchoff's law 22, 190
KOSMOS 73
krigging 237
KVR-1000 73

L

laboratory data 49
- spectra 291
Lambertian reflection 29
- surface 29
land subsidence 386, 563
- use 573
landform 157, 433-445
Landsat 97
landslide hazard zonation 424
landslides 386, 539
laplacian image 244
Large Format Camera 73
large format cameras 55
- particle scattering 26
laser radar 214
lava fields 561
- flow 562
layover 342
leaf cell structure 51
- pigments 51
lens angle 57
- centre 143
- stereoscope 141
LIDAR 215
limestone 484
limonite 175, 507
lineament 460 – 477, 522, 532
- intersection 467
- - density 467, 471, 505
linear array 89
- CCD-array 88
- contrast stretching 241
- mixture 295
- - modeling 283
liquefaction 551
LISS 106-107
lithological guides 502
lithology 481
LMM 283
local incidence angle 325, 350

Subject Index 649

- indicators   518
- operation   223, 422
location data   397
loess   444
logarithmic stretch   241
long wavelength pass filter   68
- wave radiation   186
look angle   167, 324, 329, 349
- direction effects   344
- up tables, see LUTs
low contrast film   62
- frequency image subtraction   246
- pass   242
- sun angle   165
LOWTRAN   208
luminex   180
LUTs   222, 231

M

magma   562
magmatic activity   560
magnetic data   400
MAGSAT   401
map layer   398
mapping camera   58
maps   398
marble   491
master image   235
maximum likelihood classifier   279
meandering drainage   435
measured phase   380
measurement function   420
- space   276
mega-geomorphology   432
MERIS   302
metamorphic rocks   47, 490
meteoritic impacts   479
metric camera   58
Metric Camera   73
Mg-OH   40
microearthquake   549
microlineament   474, 477
microwave radiation   318
- radiometer   318
- range   23, 317
- sensors   317-18
middle-IR   23
Mie scattering   26
mine exploration stage   501

mineral assemblages   493
- exploration   498
- species quantification   312
mineralogic identification   312
minimum distance to means classifier   278
Minnaert correction   227
MIR   24
mirror stereoscope   141
misclassification matrix   285
missing lines   231
mixed baseline   373
mixtures   294
MNF   304
MODTRAN   208
molecular scattering   25
MOMS   3, 110, 166
morphostructural anomalies   519
morphotectonics   434
MOS-1   117
mosaic   152
mosaic, radar   344
mousse   577
MSS   2, 98-100
MTF   71, 80
MTI   114
mudflow   540
multiband camera   58
- products   174
- radiometer   81-82
multichannel radiometers   12
multidisciplinary applications   6
- geodata   393
multiple linear stretch   241
- pass interferometry   372
multi-polarization   349
multispectral camera   58
- digital cameras   96
- imaging systems   75
- products   174

N

nadir line   143
- point   143
NASA   2
natural hazards   555
NDVI   572
NDVI difference image   572

NEΔP_λ 81
NEΔT 81
nearest neighbour method 237
near-IR 23
near-range 323
neighbourhood function 421
neotectonic lineament 546
neotectonism 542, 543
neovolcanic rift zone 460
- system 560
NEP$_\lambda$ 78
network function 422
neutral density filter 69
NIR 24
noise equivalent power difference 80
- - spectral power 78
- - temperature difference 81
nominal scale 398
nonlinear mixture 295
non-photographic sensors 75
nonselective scattering 25
non-specular reflection 29
non-systematic distortions 123
normal baseline 372
normalised difference vegetation index 572
no-show 340

O

object contrast ratio 71
oblique photographs 71, 143
obsidian 489
oil pools 519
- slicks 522
- spills 523, 576
- well fires 581
OM line scanner 86, 297
opacity, film 61
OPS 111
optical range 23, 75
- thickness 25
- mechanical line scanner 86
OrbView 119
ordinal scale 398
overlay function 420
over-scanning 129

P

paleochannels 528
palimpsest drainage 437
panchromatic 59, 171
panoramic camera 58
- distortion 126, 232
PAN-sensor, IRS 107
parallactic angle 136
parallax 144, 271
parallel drainage 435
- polarization 329
parallelopiped classifier 279
passive microwave sensor 318
- sensors 24
path radiance 25, 149, 224
path-and-rows 121
pathlength 24
pattern 157
- recognition 274
PCT 255
penetration 31
penetration depth 31
periodic noise 231
perspective angles 136
- view 272, 423
phase 368
- difference 368
- unwrapping 379
-, absolute 380
-, measured 380
phenological changes 517
photo-conductive detectors 77
photo-conductors 77
photodiodes 77
photoelectric effect 76
photo-gate 89
photogrammetric camera 58
photogrammetry 142
photographic scale 57
photography 2, 53
photo-interpretation, elements of 154
photometer 81
photons 20
photo-scale 95
phyllites 492
piece-wise linear stretch 241
pitch 128
pixel 83, 86, 217
- integrated temperature 209, 559, 568

Subject Index

placer deposits   502
planar discontinuity   447
Planck's Law   22
- radiation equation   206
platforms   8
plume, smoke   582, 583
-, volcanic   560
PMT   76
point operations   223
polarimetry   357
polarization diversity   357
- filter   69
-, beam   329, 348
pollution, atmospheric   583
portable mirror stereoscope   141
positive film   67
post-classification sorting   418
precision products   122
pre-classification stratification   416
preliminary investigation stage   500
preprocessing   302
primary additive colours   64
- colours   64
- subtractive colours   64
principal component analysis   255
- phase   380
- point   143
products, image   122
prospecting stage   500
proximity function   422
pseudocolour   263, 566
pseudo-gossan   510
pseudoscopy   137
pulse rectangle   323, 327
pumice   489
pushbroom imaging spectrometer   298
pushbroom scanner   88

Q

quality, image   147
-, radiometric   148
quantization levels   81
quantum detectors   76
- nature   20
quartzite   490
QuickBird   115

R

radar   320
- equation   346
- mosaic   344
- polarimetry   357
- return   346
- shadow   340
- signal   367
- wavelength   328, 347
radargrammetry   345
Radarsat-1   334, 374
radial drainage   435
radiance   20, 169
radiant energy   20
- flux   20
- temperature   191
radiation   19
radiometer   81, 317
radiometric calibration   94
- distortions   224
- effects   96
- irregularities   341
- quality   148
- resolution   63, 80
radiometry   318
rainbow   577
Raleigh scattering   25
Raleigh-Jean's law   22, 318
range baseline   372
- direction   323
- distance   323
- resolution   323, 326
ranking index   264
RAR   327
raster structure   399
ratio scale   398
ratioing   258
raw data   220
RBV   85, 101
real aperture radar   327
receivers, GPS   15
recharge zones   528
reconnaissance camera   57
- stage   500
rectangular drainage   435
red edge   294
reflectance   29, 169
reflection   28
regional exploration stage   501
registration   233, 412

relative registration 234
relief displacement 132
remote sensing 1
-- platforms 8
-- principle 3
-- programme 6
repeat track interferometry 372
- pass interferometry 372
reseau marks 58
resolution 71, 80, 329
-, radar 323
-, radiometric 63
RESURS-1 112
retrieval function 420
return beam vidicon 85
reversal films 67
RGB 180, 264, 589
rhyolite 489
rift zone 460
ringed targets 505
ring-like fractures 479
river piracy 439
- valley projects 536
rock alteration 505
- attributes 429
rockets, manned 8
roll 128
route location 542
RTI 372
rubber-sheet stretching 235

S

S/N ratio 79, 91, 304
sampling 10
sand dunes 443
sandstone 482
SAR 327
SAR interferometry 367
satellites, unmanned 8
scale 143, 398
scale, photographic 57
scanning 83, 408
- mirror stereoscope 141
- radiometer 82, 319
scan-time shift 127
scatter-diagram 251
scattering 25, 29
-, diffused 338

scatterogram 251
scatterometer 14, 358
schist 492
sea bottom features 366
- echo 340
search function 421
Seasat 331, 374
sedimentary rocks 46, 482
seepage patterns 528
seismic hazard 542
selective absorption 26, 33
- emission 33
- reflection 33
- scattering 25
semi-diffused reflection 29
sensitometry, film 61
sensor calibration 228
- look angle 167
serpentine 293
shaded relief model 227, 271
shading 150
shadow 155
shale 484
shape 155
- of spectral feature 308
sharpening 243
shear fracture 468
sheen 577
shift register 89, 95
shoreline of emergence 441
-- submergence 441
short wavelength blocking filter 68
short-wave IR 23
side-lobe banding 341
signal-to-noise ratio 79
silicates 43
silting, reservoir 538
single lens frame camera 56
- look complex 377
SIR series 331
SIR-C/X-SAR 331, 374
SIS 300
size 155
skew 131, 232
SKYLAB 116
slant range 323, 349
-- display 343
SLAR 3, 320, 321
- imagery 337
- stereoscopy 345

Subject Index 653

slate 493
slave image 235
SLC 377
slick 577
slope, topographic 165
small format camera 55
SMIRR 300
smoke 205
- plume 583
smoothing, image 248
Snell's Law 29
snow 171, 174
soft rocks 525
software 221
soil 159, 171, 175
- erosion 575
- moisture 365
solar heating 185
- illumination 148, 226
- reflection region 161
solid solution 293
SONAR 403
SOR region 161
space photography 57
- shuttle 2
spatial baseline 372
- domain filtering 242
- resolution 80, 83, 95
specific heat 187
speckle 342
spectra, field 49
-, high resolution 291
spectral absorptivity 22
- angle mapping 309
- band-width 80, 299
- calibration 302
- curves 3
- data 11
- detectivity 78
- differences 518
- emissivity 13, 22, 210
- feature fitting 308
- feature, shape of 308
- features 39
- filters 68
- irradiance 20
- libraries 296
- mixing 295
- range 299
- resolution 80
- response curve 3, 33

- signature 33
- slopes 260
- unmixing 309
spectrometry 287
spectroradiometer 11, 12, 82
spectroscopy 34
spectrozonal photography 69
specular reflection 29, 340
spike noise 231
SPIN-2 74
SPOT 108, 120
spot size 83, 323
spread function 422
squint angle 325
- mode 323, 325
SRTM 335, 375
standard FCC 177
- film 59
- photographic products 122
staring arrays 92
steeply dipping beds 449
Stefan-Boltzmann Law 21, 191
stereo-coverages 138
stereoscope 140
stereoscopy 136
stratification 416
stratigraphical guides 502
stratigraphic traps 519
stratigraphy 497
stream function 422
stressed vegetation 293
stretching, contrast 240-242
strip camera 58
- mining 573
striping 151, 229
structural guides 503
- traps 519
structure 445
structures, data 399
subduction zone 468
subhorizontal beds 448
submarine springs 534
subpixel 210
- temperature 209, 559, 569
subsurface fires 564
subtraction image 253
sulphides 37
sun attitude 162
sun angle 165
supervised classification 275
surface fires 568

- roughness   350
- temperature   185
swath width   323, 329
SWIR   24, 39, 45, 206-209
synoptic overview   4
synthetic aperture radar   327
- stereo   271, 415
systematic distortions   123

T

target reflectance   168
tectonic landforms   433
tectono-volcanic activity   560
tectosilicates   510
temperature estimation   206, 557-559
template matching   236
templates   235
temporal baseline   372
- decorrelation   383
- resolution   81
Terra-Server   74
terrestrial platforms   8
textural enhancement   247
texture   155
-, image   274
thematic data   11, 405
- map   11, 274
thermal conductivity   187
- diffusivity   188
- energy   185
- inertia   189, 198
- infrared region   183
- IR   42, 48
- properties   187
three-dimensional view   136
time amplitude signals   322
- saving   4
- variant parameter   10
TIMS   211, 494, 495
TIR region   183
TK-350   73
TM   2, 102-05
tone   155
topographic effects   226
- slope   165
topology   399
training areas   275
- data   277

transfer-gate   89
transformation, image   250
translational energy   36
transmission mechanism   30
transmissivity, atmospheric   25
traps   519
trellis drainage   435
tripack film   64

U

ultrabasic intrusives   488
unconfined aquifers   524
unconformity   480
unconsolidated sediments   527
underground mine fires   563
under-scanning   129
universal indicators   518
unmanned satellites   8
unsupervised classification   276
unwrapped phase   380
UV photography   578
UV systems   579

V

V/H ratio   86, 129
VE   138, 345
vector   399
vegetation   159, 171, 572
- index   51-52
- stress   293, 516
vertical beds   450
- exaggeration   137
- photographs   142
vibrational energy   36
- process   39
vidicon   84
vignetting   96, 150
VIS   24
VLF   403
VNIR   39, 45, 171
volcanic landforms   434
volcano   390-91, 479, 555-60
volume scattering   31, 348

Subject Index

W

wadis   532
water logging   538
- molecules   41
- table aquifers   524
wave characteristics   19
wavelet transform   269
whiskbroom imaging spectrometer   297
- scanner   86
Wien's displacement law   21
WiFS   107
wrapped phase   380

Y

yaw   128

Z

zoned targets   505

Printing (Computer to Plate): Saladruck Berlin
Binding: Stürtz AG, Würzburg

Johnny Rutherford, 1974 winner of the Indianapolis 500 race, poses in Victory Lane with the bomb-shaped Borg-Warner Trophy, which is studded with small portraits of the previous winners.

SPORTS CLASSIC

THE INDIANAPOLIS 500

By JULIAN MAY

Creative Education
Mankato, Minnesota

Photograph and Illustration Credits

ILLUSTRATORS

John Keely, Minneapolis 42

PHOTOGRAPHERS

Indianapolis Motor Speedway 2, 8, 10, 15, 17, 18, 22, 25, 29, 31, 37, 45
UPI 1, 8, 13, 16, 20, 24, 30, 33, 36, 39, 41, 46
STP Corp. 26
Image International 34

Published by Creative Educational Society, Inc., 123 South Broad Street, Mankato, Minnesota 56001. Copyright © 1975 by Creative Educational Society, Inc. International copyrights reserved in all countries. No part of this book may be reproduced in any form without written permission from the publisher. Printed in the United States. Distributed by Childrens Press, 1224 West Van Buren Street, Chicago, Illinois 60607.

Library of Congress Number: 75-8502   ISBN: 0-87191-441-7
Library of Congress Cataloging in Publication Data
   May, Julian.   Indianapolis 500
    (Sports classic)
    SUMMARY: Gives a brief history of the Indianapolis 500 from 1911 to the present day.
    1. Indianapolis Speedway Race—Juvenile literature. 2. Automobile racing—Juvenile literature. (1. Indianapolis Speedway Race. 2. Automobile racing) I. Title.
    GV1033.5.155M39   796.7'2'0977252   75-8502   ISBN 0-87191-441-7

Contents

 7 The Old Brickyard
10 Most Dangerous Sport
14 Funny Car Revolution
19 Call Him A.J.
23 Racer's Edge
27 Unser Back-To-Back
30 Backstage At Indy
34 The Worst Race Ever
40 Johnny Comes Back
44 Indianapolis 500 Winners

The Old Brickyard

The blimp flies overhead. Colored balloons sail in the sky. Fireworks explode. A voice sings *Back Home Again in Indiana*. There is an age-old command: "Gentlemen, start your engines!"

With a roar, a pack of shiny race cars pulls onto the track. They take their allotted positions for the parade lap as more than a quarter of a million fans cheer.

The cars set the pace. A green flag signals: GO! Engines howl and another Indianapolis 500-mile race has begun. It has been called racing's greatest spectacle. It has been called the grimmest sport since the ancient Romans fed Christians to the lions. Every Memorial Day, the largest sports crowd ever to gather in North America heads for the Indianapolis Motor Speedway to watch the most famous automobile race of all.

It's the Indy 500.

The Speedway was built back in 1909, at a time when auto racing in the United States was moving off country roads and onto tracks with grandstand seats. At first, short races were run at the Speedway. The first 500-mile event was held on Memorial Day in 1911.

Forty automobiles competed in that race, thundering around the brick-paved track before an audience of 80,000. The winner was Ray Harroun, driving a 6-cylinder Marmon at the death-defying average speed of 74.59 mph.

Ray Harroun's yellow-and-black Marmon Wasp sported the first rear-view mirror in automotive history. He put it on because he alone did not carry a "riding mechanic" to watch for the approach of other drivers. In winning the first Indy 500 in 1911, Harroun traveled the 200 laps in 6 hours and 42 minutes.

The race was such a success that they repeated it the next year . . . and the next. The traditional Indy 500 has been held every Memorial Day since—except for two years during World War I and four during World War II.

The track became known as "The Old Brickyard" because of its pavement—the finest available in 1911.

Wilbur Shaw was the guiding spirit of the Indianapolis 500 race until his death in an aircraft accident in 1954.

Even today, when the old surface has long been covered with smooth asphalt, a symbolic strip of brick remains at the starting line.

Two famous racing-car designers dominated at Indy during the 1920's. Fred Duesenberg and Harry Miller were just about the only Americans in the race-car business until the Great Depression hit in 1929. Then the speedway changed its rules again and allowed cars with modified stock engines to compete.

The result was a flood of new-style cars. Buicks, Chryslers, Studebakers and Stutzes appeared on the Indy track, many sponsored by "little guys" who could never have afforded the expensive Millers and Duesies.

But the specialists still won the prizes! The best finish by a modified stock was in 1932, when a Studebaker came in third.

Fred Offenhauser acquired the old Miller racing car factory during the 1930's. The famous four-cylinder engines carrying his name were destined to take over at Indianapolis and rule the track for nearly 25 years.

World War II forced the Speedway to close down. It almost never opened again. But Wilbur Shaw, an Indy star of the 30's and one of the greatest drivers of all time, saved both the Speedway and the Indy 500.

Shaw urged a businessman named Tony Hulman to buy the old track and restore it. Wilbur Shaw himself became the Speedway's new president and general manager. Under his guidance, the Indy 500 race resumed in 1946 and has been going strong ever since.

Most Dangerous Sport

The 2½-mile track of the Indianapolis Motor Speedway is a rectangle with rounded corners, not a true oval. The main stretch is on the left, flanked by white-roofed grandstands. Two parallel white roofs in the infield mark the location of "Gasoline Alley," garages where the cars are prepared for the race. The Goodyear Blimp, a perennial feature of the race, cruises at lower left.

Who would defy death just for fun?

Lots of people. All through human history, daredevils have risked their necks for the sport of it. And many other people, too timid to take a chance themselves, have enjoyed second-hand thrills from the sidelines.

Automobile racing has always been a hazardous sport, and the Indy 500 is the most risky race of all. Those who are honest admit that the danger adds to the excitement. Both drivers and spectators have accepted the risk of death as part of the game.

In the very first Indy 500, a riding mechanic was killed when his race car overturned. In the years that followed, far too many drivers, crewmen, and even spectators have died or been seriously injured during the race itself or its preliminaries.

Speedway officials have tried to make the race safer. They have improved the track — but it is still designed for speeds of 75 mph, not 145. They have reduced engine size and fuel allowances in an attempt to slow the cars down—but the designers built better engines, and cars go faster every year.

Automobile racing does not have to be deadly to be fun — the drag racers proved that. But safety is mostly a matter of driver attitude, and Indy drivers themselves must take most of the blame for the hazards of the race. Tracks can be made safe and cars can be made safe. But in the long run it must always be the drivers who either seek risks or avoid them.

Today's race cars look small and fragile compared to the fire-belching monsters of 50 years ago. But the new little cars are really many times safer than the old ones. Their ground-hugging design and

wide tires make them more stable. Wheels may fall off in a crash, but the car's strong body, equipped with a roll-bar, usually protects the driver from serious injury. Flame-resistant suits and crash helmets have also helped to save many lives.

Two Indy 500 winners contrast the old and the new in racing headgear. Mauri Rose (top) won in 1941, 1947 and 1948. Bobby Unser, dressed for the 1974 contest, won in 1968.

Funny Car Revolution

From the late 1930's until 1963, race cars with Offenhauser engines dominated the Indy 500 completely. There were only a few challengers.

The most interesting was the Novi. It was named after Novi, Michigan, hometown of its sponsor, Lew Welch. The Novi was a supercharged V-8 that first appeared at Indy in 1941. It was the most powerful racing engine ever seen. After World War II it set records in the time trials and twice won the post position.

But despite its impressive speed, the Novi never came home a winner. It was not reliable enough to go the 500 miles. Both drivers and fans loved the bellowing monster machines, but the best Novi finish was third in 1948.

A new kind of challenger showed up in 1961 — and set Indianapolis laughing! It was a rear-engine Cooper Climax — small, lightweight, frail when compared to the Indy-class roadsters. Australian Jack Brabham brought the little car in ninth — which was good, but a fluke.

Right?

In 1963, more little "funny cars" from across the ocean invaded Indianapolis. Once again the Indy boys laughed. Why, the British cars were powered by modified Ford stock engines! How could they hope to conquer the powerful Offenhausers and Novis?

The laughter soon stopped. The performance of the little Lotus-Ford cars was far from funny. Their

Rufus Parnell Jones won the 1963 Indianapolis 500 with an average speed of 143.137 mph. His car leaked oil on the last laps when Jim Clark was only a few seconds behind. Rather than risk an accident on the slippery pavement, the prudent Clark slowed down and settled for second place. He said later: "I would rather be second than dead."

The most powerful racers ever seen at Indy were the Novi Specials. Designed by Ed Winfield, the earliest models put out 500 hp. In later years Novis achieved an incredible 837 hp in a compact 167 cubic inches. This Novi, driven by Herb Ardinger, finished fourth in the 1947 Indy 500. In the background of the picture is the Speedway's "pagoda tower," a vanished Indy landmark.

wide-track, ultra-light little shapes hugged the curves. Their efficient, reliable engines pushed them into the top ranks at the time trials — and kept on running when fancier engines failed.

The 1963 Indy 500 was a contest between the good old Indy roadster and the new-style European Grand Prix racing car. Parnelli Jones, in an Offy roadster, won.

Scotland's Jim Clark, in a Lotus-Ford, finished 34 seconds behind him.

There were only two rear-engine Fords in the 1963 race. Next year there were seven. Jim Clark's Lotus qualified for the pole but dropped out of the race when the suspension failed. A. J. Foyt won the 1964 Indy 500 in an Offy-powered roadster.

On the opposite page, Graham Hill's Number 24 can be seen in the lead, having narrowly escaped the big smash. His average speed was 144.317 mph. Although a rookie at Indy, Hill was a champion driver in Europe. Jim Clark finished second in 1966.

It was the sunset of the dinosaurs.

In 1965, only four starters were Offy roadsters and only two were Novis. All the rest were funny cars. The winner of that race was Jim Clark and the runner-up Parnelli Jones.

Both drove Lotus-Fords. The funny car revolution was complete.

Only a single roadster qualified in 1966, only to be wrecked before it had run a single lap. The new breed of spidery little machines had taken over. After a start marred by the most spectacular pileup in Indy history, the 1966 race was won by Graham Hill of England, who was racing for the first time at the Old Brickyard. His car was a Lola-Ford, sponsored by John Mecom, Jr.

Jim Clark of Scotland won the 1965 Indy 500 with an average speed of 150.686 mph. He was the first non-American driver to take the classic since France's Dario Resta won in 1916. The Lotus-Fords were designed by Colin Chapman of England.

Call Him A. J.

Anthony Joseph Foyt, Jr. was born in Texas in 1935. When he was a teen-ager, he began racing motorycycles and hot rods on dirt tracks. Later he raced midgets, finally graduating to big cars in 1957. A rookie at Indianapolis in 1958, he spun out on the 148th lap and was awarded 16th place.

In his subsequent trips to the Brickyard, Foyt was 10th and 25th. Then came 1961. It was a roadster contest that year, and the climax was fought by three popular drivers. Rodger Ward had won the Indy in 1959 and come in second in 1960. Eddie Sachs, known as the "clown prince" of the Brickyard, was a crowd favorite although he had never come close to winning. The third driver was A. J. Foyt.

Foyt led with ten laps to go when he felt his car running out of fuel. He had to go to the pit, allowing Eddie Sachs to forge into the lead.

Back on the track, Foyt tried vainly to catch up to Sachs. Only three laps remained and Sachs seemed a sure winner. But then one of Eddie's tires began to shred. He was forced to stop for a change, giving up the advantage to A. J. once more.

Foyt surged ahead, winning over Eddie Sachs by just eight seconds. His average speed was 138.130 mph.

In the years that followed, A. J. Foyt became known as an expert but hot-tempered driver. He won many other races and came in third in the 1963 Indy 500, which featured the first Offy-Lotus duel.

A. J. was interested in the new rear-engined cars

In 1961, A. J. Foyt won the Indy 500 by a margin of only 8 seconds, a record tight squeak.

but suspicious of them, too. He was afraid the lightweight chassis wasn't safe. A. J. drove an Offy roadster in the 1964 race — and almost immediately, his fears about the funny cars seemed to be confirmed. Rookie driver Dave MacDonald lost control of his rear-engined Thompson-Ford. MacDonald crashed in flames, and Eddie Sachs's car slammed into him. Both men died.

The race was stopped, then started again. A. J. Foyt breezed home the winner, nearly three miles ahead of Rodger Ward's rear-engine Ford.

But the Lotus handwriting was on the wall. The next year, Foyt gave up his roadster in favor of a funny car. He didn't win in 1965 or 1966 at Indy, but 1967 was destined to be Foyt's greatest year. Running second to a turbine-engined car throughout most of the race, A. J. took over the lead when the Turbocar's gearbox broke. Steering through a last-lap crash, Foyt was the only driver to complete the 200 laps. It was his third Indy victory.

A third trip to Victory Lane brings a smile to the face of A. J. Foyt. He was still going strong in 1974, when he qualified for the pole with a speed of 191.632 mph.

Racer's Edge

Once there were three drag-racing brothers who lived in Chicago and dreamed of competing in the Indy 500. In 1946, they spent their savings on an ancient race car, restored it, then didn't have enough money to ship it to the race.

So they put license plates on it and *drove* it to Indianapolis.

The brothers were named Granatelli. Their old car was in the first ten for awhile. Then the driver quit. Young Andy Granatelli leaped into the cockpit, drove through the infield and tried to continue the race.

He was disqualified.

"We'll be back!" Andy shouted.

And so he was. He and his brothers, Bob and Vince, built many more Indy race cars. Andy became president of the STP Corporation, but he never forgot his first love — auto racing. He raced Novis and he raced funny cars, always seeking to be one jump ahead of the others, seeking the racer's edge of victory.

In 1966, the STP team almost made it. They thought their driver, Jim Clark, had won and they headed for Victory Lane. But there had been a mistake in scoring. Graham Hill was awarded the win.

In 1967, the Granatelli team introduced the

A. J. Foyt sits in the Sheraton-Thompson Special that won the 1967 Indy 500. He set a new track record of a peculiar kind, becoming the first one-man finisher in Speedway history. An accident on his last lap caused the other cars to be red-flagged. Foyt's average speed was 151.207, also a record.

Parnelli Jones drove the revolutionary Turbocar in the 1967 race. The aircraft-type turbine engine is mounted beside the driver. Instead of roaring, the turbine gave out a quiet sound that caused the car to be dubbed the "Whooshmobile."

Jim Clark sits in the Lotus-Ford that almost won the 1966 Indy 500. Andy Granatelli (in jacket) stands in back with the crew. The man with the moustache is Colin Chapman of England, brilliant designer of the Lotus racing cars.

funniest car of all. Its engine was an aircraft-type turbine. The Turbocar was leading in the 197th lap when a ball bearing in the gear box failed. Once again, by the narrowest edge, Granatelli lost the Indy 500.

Turbocars were called "unfair" and banned from the race after 1968. For 1969, Granatelli had two "turbocharged" Offies and one Ford, the latter driven by Mario Andretti.

The race developed into a duel between Andretti and A. J. Foyt, who was seeking his fourth Indy win. Then Foyt's turbocharger began to act up and he dropped out of it. Andretti took the lead. Nursing his car to prevent overheating, he ran a prudent race and finally licked the last-minute jinx that had seemed to plague Andy Granatelli.

When the STP team gathered in Victory Lane, Andy gave Mario Andretti a big kiss! It was a good old Italian custom.

Unser Back-To-Back

It isn't easy, being the little brother in a famous racing family. Al Unser's two older brothers, Jerry and Bobby, had both raced at Indy. Jerry was killed in a practice session in 1959. But Bobby came in a winner in 1968, driving a rear-engined turbocharged Offy-Eagle.

Meanwhile, little brother Al was racing at the Old Brickyard, too. He was shorter and quieter than Bobby, but obviously a top-quality driver. In his first race, in 1965, he came in ninth. In 1967, he was second.

Despite the fact that they often raced against each other, Bobby and Al Unser remained very close. In the 1968 race, Al bounced off the wall after losing two wheels. He was unhurt and didn't want Bobby to worry. So he climbed up on the wall and waved as his brother sped by.

The next year, Al broke his leg and was sidelined. He came back in 1970, driving a Johnny Lightning Special owned by Parnelli Jones, and earned pole position.

The race shaped up as a battle among the Unsers, Johnny Rutherford, and Mark Donohue. A. J. Foyt lost his bid when he broke his gearbox. The others

After long years of seeing his cars also-rans, Andy Granatelli rejoices in Victory Lane with driver Mario Andretti in 1969. The car, a turbocharged Ford Hawk, was Andretti's backup vehicle. His original car, a new four-wheel-drive Lotus, was demolished during qualification. Andretti suffered face burns. The Hawk came home with an average speed of 156.867 mph.

never really challenged Al, who led for all but 10 of the 200 laps.

At the end, obeying Parnelli's advice to drive coolly, little brother Al Unser came home to the wave of a checkered flag. Later, in Victory Lane, nobody applauded him harder than his brother Bobby.

In 1971, Al Unser returned with a new Johnny Lightning Ford. He wasn't on the pole this time. The honor belonged to Peter Revson and his McLaren. He wasn't even favored to win. The smart money was on Mark Donohue in another McLaren.

The 1971 race was marked for disaster even before the green flag fell. An inexperienced pace-car driver crashed into the photographers' stand. Then, on lap 16, a spectacular three-car accident wiped out Mel Kenyon, Gordon Johncock, and Mario Andretti. No one was badly hurt.

The favorite, Mark Donohue, fell out of contention when a gear tooth broke. Al Unser took over the lead.

His brother, Bobby, now provided a sort of rerun of the 1968 crash that Al had suffered. Smashing into the wall after spinning to avoid another wrecked car, Bobby climbed out unhurt. He waved from the sidelines so that Al wouldn't worry.

And Al rolled on to win the 1971 Indy 500, setting a new record of 157.735 mph. It was the first back-to-back victory since 1954.

Mark Donohue, 1972 winner, makes a pit stop during the 1974 race. Cloth hoods protect the crew from flash fires as they service the winged race car. Modern tires actually melt a little at high speed. The sticky surface helps glue the car to the track at speeds in excess of 190 mph.

Backstage At Indy

Not all of the excitement at the Indianapolis Speedway occurs on the track. And not all the races are won on the asphalt-covered bricks. No matter how hot the driver, his victory really depends upon the mechanics and the other members of the pit crew.

When a shiny new race car first rolls into Indy's Gasoline Alley garage, the pit crew tests it, improves it, and tinkers with it to get it into the best possible shape. Strangely enough, Indy cars must go fastest not during the race but during the qualification runs before Memorial Day. A car's qualification speed determines whether or not it will be allowed in the

In the first Indy 500 race, the pit crew changes a tire on Bob Burman's Benz. Worn and exploding tires were a great hazard in early racing. One of the great contributions to automotive safety that racing has achieved is the development of tough, reliable tires.

Crews gather around to start the engines at the 1973 Indy 500 race. Car 66 belongs to Mark Donohue; Car 8 is Bobby Unser's; Car 7 is driven by Johnny Rutherford.

race. It also determines the position of the car at the start of the race: the fastest cars are in front, the slowest in the rear.

The field is fixed at 33 cars, so some racers are sure to be disappointed as they try to qualify. A year's hard work and a small fortune may go down the drain when a slower car is "bumped" out of the starting lineup by one that performs better.

But that's the Indy 500!

Rookie drivers must pass a special test before they are allowed to compete. No one is excused — not even pilots who have won other important races. The Speedway track is different from any other in the world, requiring special driving techniques. A driver who does not keep this in mind is a menace not only to himself but to the other competitors as well.

During the actual race, the work of the pit crew is all-important. The men have practiced long and hard so that they will be able to refuel the car, change its tires, and perform other maintenance jobs during the 500-mile ordeal.

When a racer comes in for a pit stop, men swarm over the car like ants on a piece of candy. If the crew is a good one, the pit stop will take less than 20 seconds. In the 1974 race, A. J. Foyt's crew did the job in 16 seconds.

Returning with a rear-engine Turbocar in 1968, Andy Granatelli was frustrated again when the vehicle's fuel pump gave out Driver Joe Leonard had held the lead with only 32.5 miles to go. Bobby Unser whipped past to win. Airplane turbines were effectively banned after the 1968 race.

The Worst Race Ever

At the start of the 1973 race, Salt Walther's car veered to the outside and slammed into the wire fence. Spectators sitting less than 10 feet away were sprayed with burning fuel. Nine of them went to the hospital along with Walther. Next year, the dangerous rows of seats were removed.

In 1964, Johnny Rutherford's second Indy season, he barely escaped being involved in the terrible crash that took the lives of Eddie Sachs and Dave MacDonald. After being trapped briefly, Johnny's number 86 (left) cleared the flames. Johnny was not hurt, but his car was too badly damaged to continue.

As cars sped faster and faster around the ancient Speedway track, some people predicted that Indy was ripe for a real disaster. There were many seats very close to the barrier wall — and the wall itself was not very high. The pit entrance was narrow and confusing. And the new, winged race cars loaded with 80 gallons of fuel were tricky to handle.

Added to all that was the rain.

Because the track is slippery and banked shallowly on the curves, the Indy 500 cannot be run in rain. But it rained all morning on Memorial Day, 1973, and part of the afternoon, too. Some four hours late, the race finally started. Everyone was nervous and impatient. The drivers knew that the race would have to be less than 500 miles long. This meant that they would have to charge hard at the start and keep on pushing. The car that led when the officials finally halted the race would be declared the winner.

The start was very ragged, and gray clouds began

In the early days of the Indy 500, spectators were allowed to stand close to the track. It was exciting, but very dangerous. In the 1927 race, Norman Batten's Miller Special caught fire. Batten could have jumped to safety, but his flaming car might have crashed into the fans at trackside. Batten bravely remained with the car and steered it to a safe stop, suffering painful burns in the process.

to build up over the grandstand again. In the sixth row, young driver David (Salt) Walther lost control when he was hit from behind by another car trying to bully its way through the pack.

Walther's car crashed in a ball of flame and pinwheeled along the track. Most of the other drivers managed to avoid the wreck.

Injured fans and Walther, who was badly burned, were rushed to the hospital. Then the sky opened up, drowning out the race for the rest of the day.

It rained all through Tuesday and into Wednesday. Finally, on the afternoon of the third day, the race started again. Less than 100,000 of the original crowd of 300,000 remained to watch.

This time the start was perfect, and the race went smoothly. At the 42nd lap, a young driver named Swede Savage took the lead, which he clung to for 12 circuits.

Then it was time for a pit stop. Swede's red STP Eagle went out of control as he sought to go into the narrow pit entrance. He hit the wall and the car exploded into a thousand pieces.

An STP crewman, Armando Teran, ran to the flaming wreck. A speeding fire truck hit and killed him. Swede Savage, critically injured, died in the hospital a month after the race.

When the race resumed, rain threatened again. Gordon Johncock — also a member of Andy Granatelli's STP team — took the lead on lap 73 and hung onto it. Rain began to fall on the 130th lap, and not long after the red flag stopped the race. It had been the shortest in Indy history, only 133 laps long, a total of 332.5 miles.

David Earle (Swede) Savage was 26 years old when he raced in the 1973 Indy 500. Here his car is shown as it disintegrated upon hitting the inside wall on the fourth turn.

Johnny Comes Back

After the 1973 disaster, there were some who said the Indy 500 should be abolished. Fortunately, the race continued — but changes were made. The dangerous seats near the track were removed and the protective wall built higher. A new, safer pit was constructed. Rear wings on the cars were reduced in size to lessen turbulence that made cars hard to control. Fuel loads were cut in half. A new starting tower and new starting rules were intended to prevent botched beginnings.

So the race drivers gathered again. There was veteran A. J. Foyt, still seeking that fourth Indy win, sitting on the pole. The Unser boys were on deck, and so were Johncock, Andretti, and Billy Vukovich, son of the late, great Bill.

And there in the fifth row was Salt Walther! He had recovered completely from his terrible burns and was back, rarin' to go.

Far back in the 25th spot was veteran Johnny Rutherford. He had qualified at 190.446, just behind the polesitting Foyt, but a technicality had banished him to the back of the pack. It was a dangerous place to be, for he would have to thread his way through slower cars to get out into contention.

Sunshine shone on the Old Brickyard as the front row of cars set the pace. The green flag fell, this time from the top of a new tower instead of from the track.

And they were off, with STP's Wally Dallenbach in the lead.

A. J. Foyt's Coyote holds a slim lead over Johnny Rutherford in his McLaren.

Driving as hard as he dared, Johnny Rutherford began working his way up on the inside. Meanwhile, Dallenbach dropped out with a broken valve and Mario Andretti's car burned a piston. A. J. Foyt took over the lead, with Bobby Unser in second place.

It took Johnny Rutherford 35 laps to get up with the big boys, where he belonged. He passed Unser and challenged Foyt. On lap 65, Johnny finally passed Foyt.

He kept this position until lap 125, when Foyt overtook him again. Meanwhile, the only serious accident of the race took place when Jerry Karl hit the wall in the number-three turn. He escaped with a hurt foot.

As Rutherford and Foyt diced for first, Salt Walther's engine was starting to go. And he wasn't the only one in trouble. Black smoke began to pour from the orange Coyote of A. J. Foyt. Giving up his lead in the 140th lap, he was black-flagged for inspection.

Foyt's crew couldn't discover the trouble, so he roared out for another lap. But a pump had given way inside his blower. Smoke and oil poured out of his car and he was flagged out of the race. His face an angry red, A. J. Foyt stormed into the pit, changed his clothes, and zoomed out of the Speedway on his motorcycle.

Out on the track, Johnny Rutherford had the lead all to himself. Bobby Unser tried hard, but finished 22.3 seconds behind. Young Billy Vukovich was third.

The 1974 race was the first Indy 500 that Johnny Rutherford had ever been able to finish. After eleven years of frustration, he had finally made it to the top.

Indianapolis 500 Winners

Year	Driver	Car Chassis/Engine	Average (mph)
1911	Ray Harroun	Marmon Wasp	74.59
1912	Joe Dawson	National	78.72
1913	Jules Goux	Peugeot	75.933
1914	Rene Thomas	Delage	82.47
1915	Ralph DePalma	Mercedes	89.84
1916	Dario Resta	Peugeot	84.00*
1917	NO RACE		
1918	NO RACE		
1919	Howdy Wilcox	Peugeot	88.05
1920	Gaston Chevrolet	Frontenac	88.16
1921	Tommy Milton	Frontenac	89.62
1922	Jimmy Murphy	Duesenberg/Miller	94.48
1923	Tommy Milton	Miller	90.95
1924	L. Corum-J. Boyer	Duesenberg	98.23
1925	Pete DePaolo	Duesenberg	101.13
1926	Frank Lockhart	Miller	95.904**
1927	George Souders	Duesenberg	97.545
1928	Louis Meyer	Miller	99.482
1929	Ray Keech	Miller	97.585
1930	Billy Arnold	Summers/Miller	100.448
1931	Louis Schneider	Stevens/Miller	96.629
1932	Fred Frame	Wetteroth/Miller	104.144
1933	Louis Meyer	Miller	104.162
1934	Bill Cummings	Miller	104.863
1935	Kelly Petillo	Wetteroth/Offenhauser	106.240
1936	Louis Meyer	Stevens/Miller	109.069
1937	Wilbur Shaw	Shaw/Offenhauser	113.580
1938	Floyd Roberts	Wetteroth/Miller	117.200
1939	Wilbur Shaw	Maserati	115.035
1940	Wilbur Shaw	Maserati	114.277
1941	F. Davis-M. Rose	Wetteroth/Offenhauser	115.117
1942	NO RACE		
1943	NO RACE		
1944	NO RACE		
1945	NO RACE		
1946	George Robson	Adams/Sparks	114.820
1947	Mauri Rose	Deidt/Offenhauser	116.338
1948	Mauri Rose	Deidt/Offenhauser	119.814
1949	Bill Holland	Deidt/Offenhauser	121.327
1950	Johnny Parsons	Kurtis/Offenhauser	124.002
1951	Lee Wallard	Kurtis/Offenhauser	126.224
1952	Troy Ruttman	Kuzma/Offenhauser	128.922
1953	Bill Vukovich	KK500A/Offenhauser	128.740
1954	Bill Vukovich	KK500A/Offenhauser	130.840
1955	Bob Sweikert	KK500C/Offenhauser	128.209
1956	Pat Flaherty	Watson/Offenhauser	128.490
1957	Sam Hanks	Epperly/Offenhauser	135.601

1958	Jimmy Bryan	Epperly/Offenhauser	133.791
1959	Rodger Ward	Watson/Offenhauser	135.857
1960	Jim Rathmann	Watson/Offenhauser	138.767
1961	A. J. Foyt	Watson/Offenhauser	139.130
1962	Rodger Ward	Watson/Offenhauser	140.293
1963	Parnelli Jones	Watson/Offenhauser	143.137
1964	A. J. Foyt	Watson/Offenhauser	147.350
1965	Jim Clark	Lotus/Ford	150.686
1966	Graham Hill	Lola/Ford	144.317
1967	A. J. Foyt	Coyote/Ford	151.207
1968	Bobby Unser	Eagle/Offenhauser	152.882
1969	Mario Andretti	Brawner/Ford	156.867
1970	Al Unser	Lola/Ford	155.749
1971	Al Unser	Lola/Ford	157.735
1972	Mark Donohue	McLaren/Offenhauser	162.962
1973	Gordon Johncock	Eagle/Offenhauser	159.036****
1974	Johnny Rutherford	McLaren/Offenhauser	158.586

* 1916 race cut to 300 miles because of fuel restriction.
** 1926 race cut to 400 miles because of rain.
*** 1950 race cut to 345 miles because of rain.
****1973 race cut to 332.5 miles because of rain.

WINNER Bill Vukovich

One of the most talented and hard-pressing drivers of the 1950's was Bill (The Mad Russian) Vukovich. Driving a hot new Roadster, he qualified for the pole position in the 1953 race by driving 138.392 mph on a wet track, then ran away with the race itself. Returning in the same car the next year, Vukovich won a second time. In 1955 he was back — but took one chance too many. Attempting to slide past three spinning cars, he went off the track after striking one of the other vehicles. He died instantly.

The start of the 1966 Indy 500 saw a spectacular pileup just as the green flag dropped. The cars circle the track in rows of three, according to their qualifying position, then accelerate at the drop of the flag. This time, one driver did not get going fast enough, causing a chain-reaction smash. Eleven cars were wrecked, including the only Offy roadster in the race. The only injury among the drivers was suffered by A. J. Foyt, who scratched his finger!

SPORTS CLASSICS

WORLD SERIES
U.S. OPEN GOLF CHAMPIONSHIP
WIMBLEDON TENNIS TOURNAMENT
KENTUCKY DERBY
INDIANAPOLIS 500
OLYMPIC GAMES
SUPER BOWL
MASTERS TOURNAMENT OF GOLF
STANLEY CUP
NBA PLAYOFFS

CREATIVE EDUCATION